W9-BOB-011

Art Notebook for

BIOLOGY

Seventh Edition

Neil A. Campbell • Jane B. Reece

PEARSON

Benjamin Cummings

San Francisco Boston New York
Cape Town Hong Kong London Madrid Mexico City
Montreal Munich Paris Singapore Sydney Tokyo Toronto

Editor-in-Chief: Beth Wilbur
Biology Media Producer: Christopher Delgado
Biology Marketing Manager: Jeff Hester
Managing Editor, Production: Erin Gregg
Production Services: TechBooks/GTS
Manufacturing Buyer: Stacy Wong
Printer: Courier Company

ISBN 0-8053-7183-4

2 3 4 5 6 7 8 9 10—CRK—07 06 05

www.aw-bc.com

PEARSON
Benjamin Cummings

Figure 1.4 Basic scheme for energy flow through an ecosystem, page 6

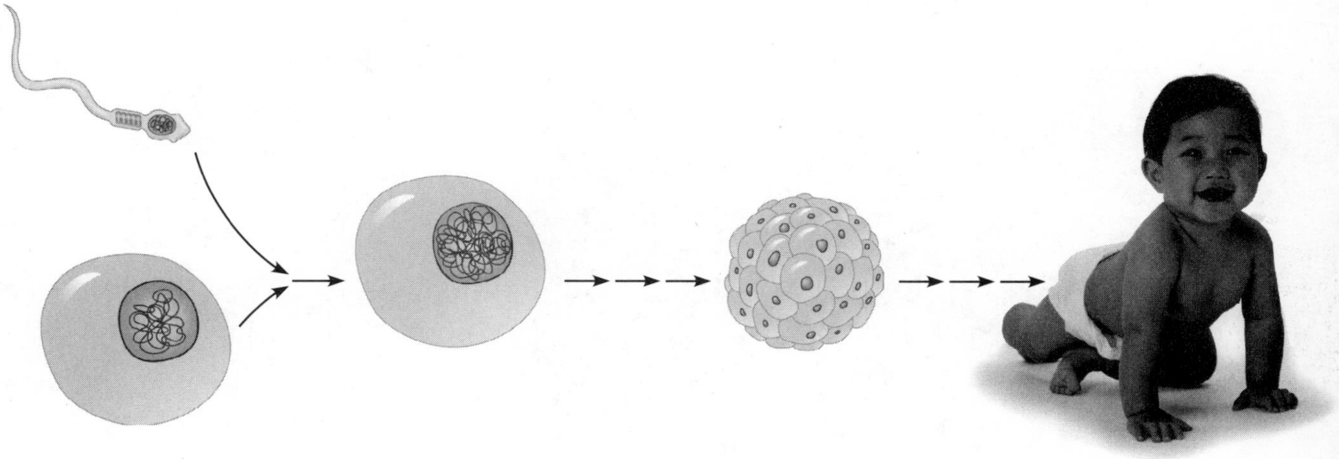

Figure 1.6 Inherited DNA directs development of an organism, page 7

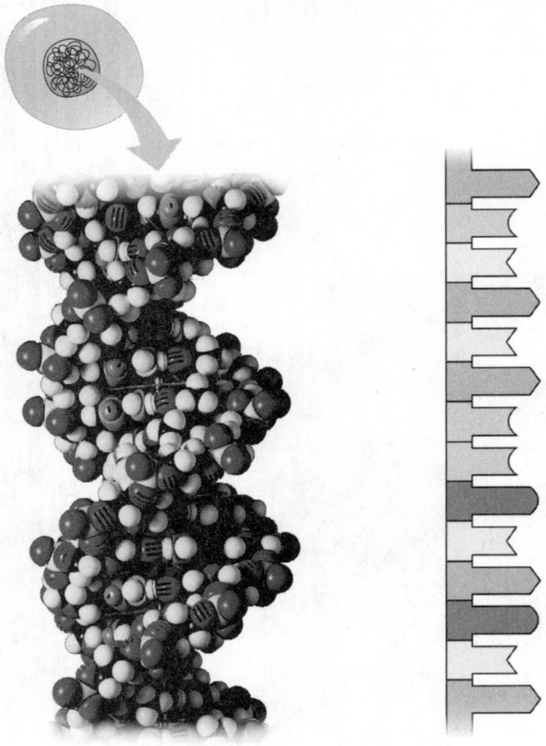

Figure 1.7 DNA: the genetic material, page 7

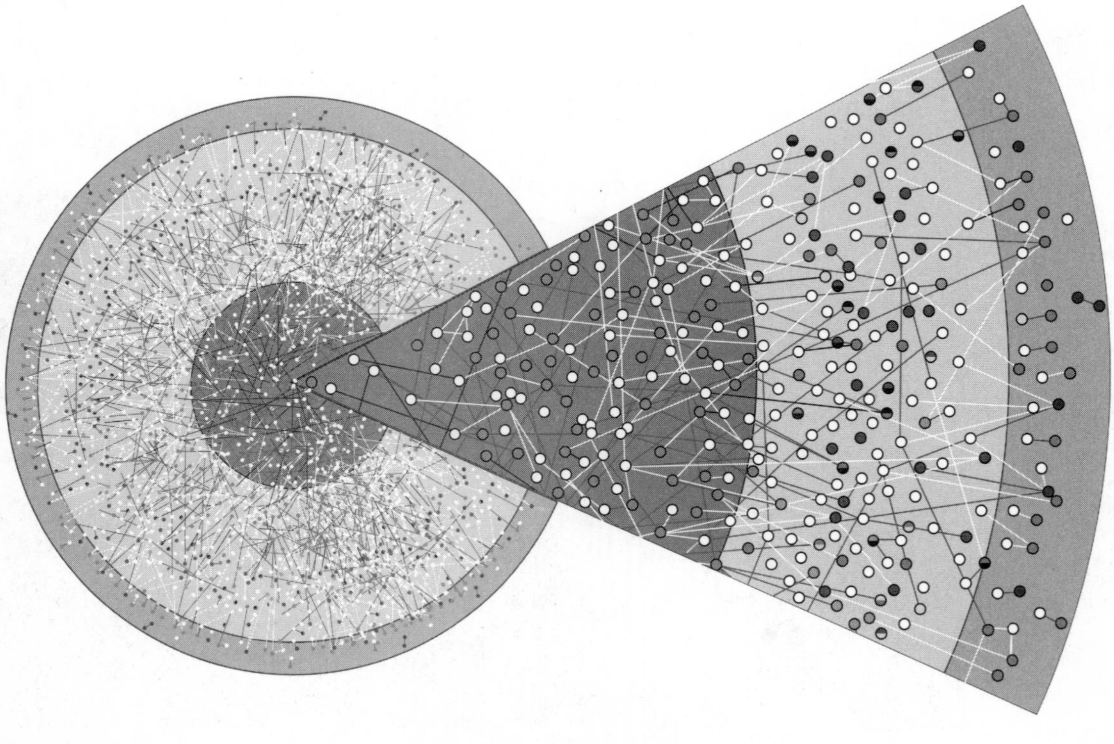

Figure 1.10 A systems map of interactions between proteins in a cell, page 10

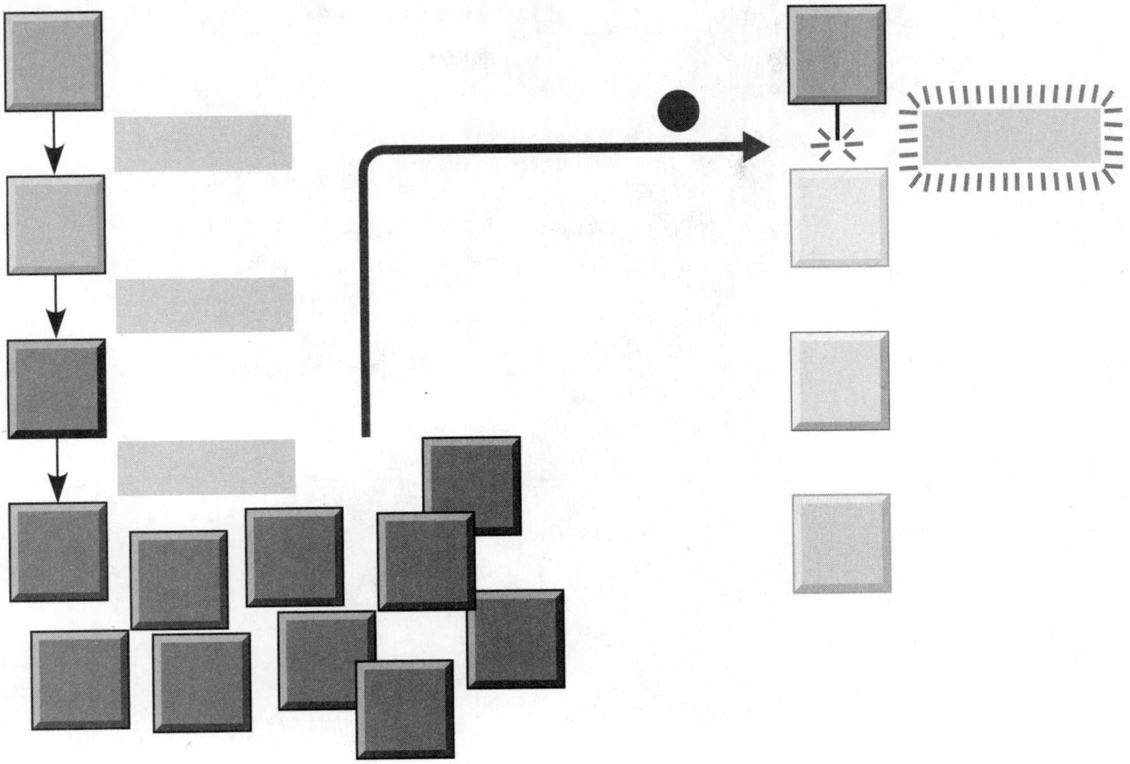

Figure 1.11 Negative feedback, page 11

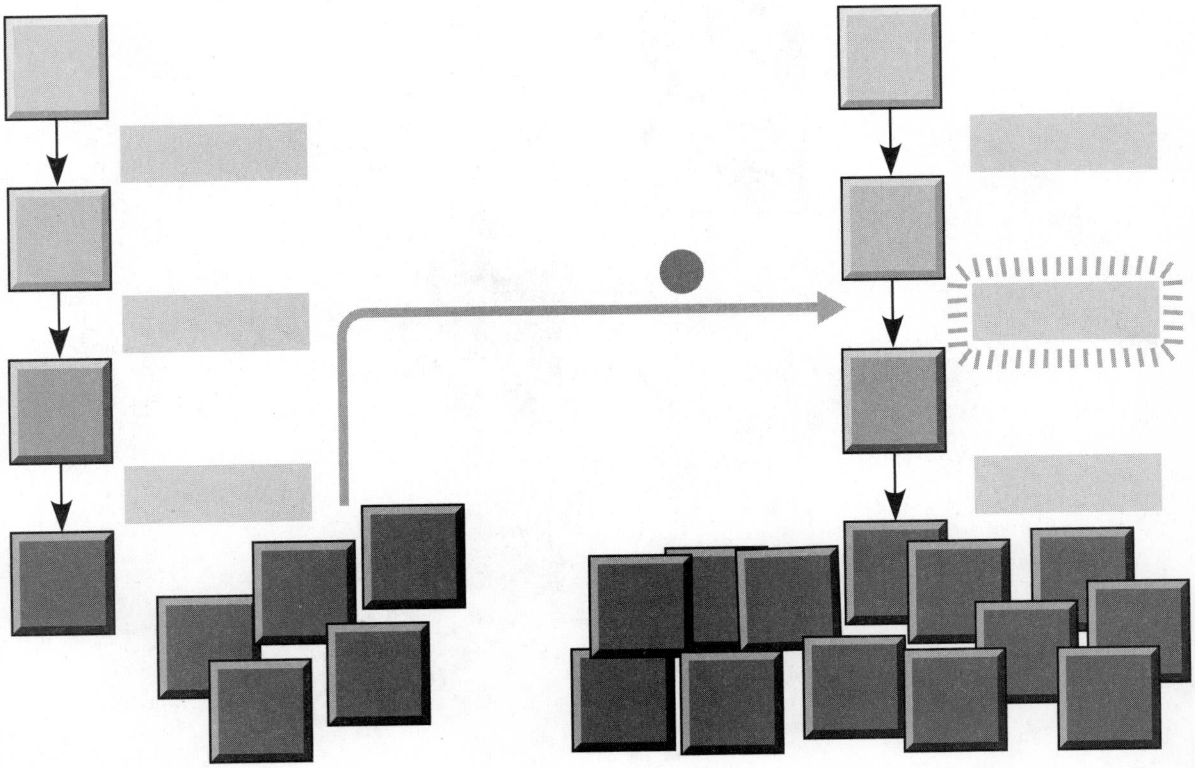

Figure 1.12 Positive feedback, page 12

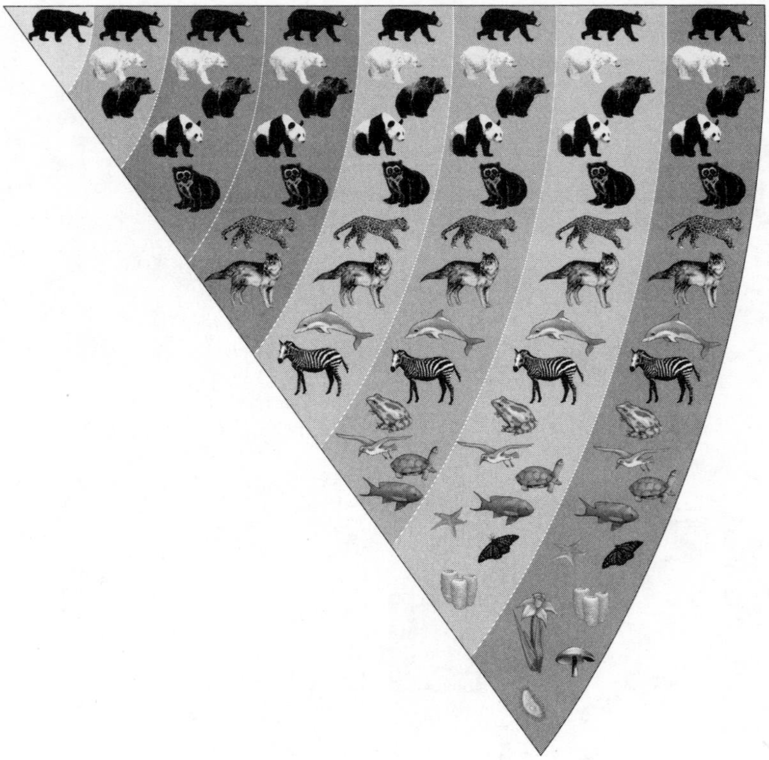

Figure 1.14 Classifying life, page 13

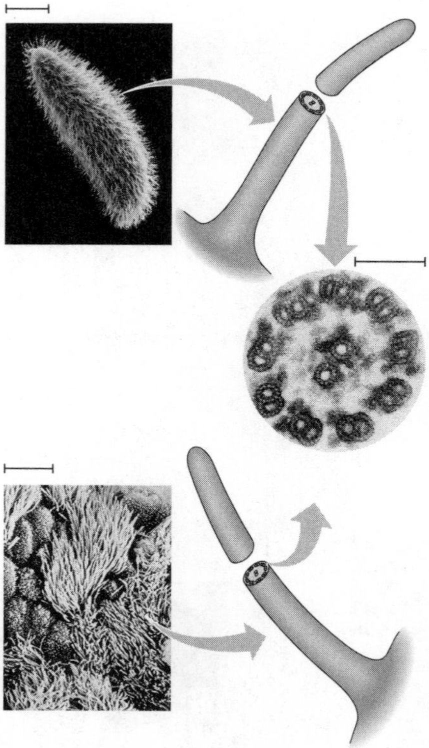

Figure 1.16 An example of unity underlying the diversity of life: the architecture of cilia in eukaryotes, page 15

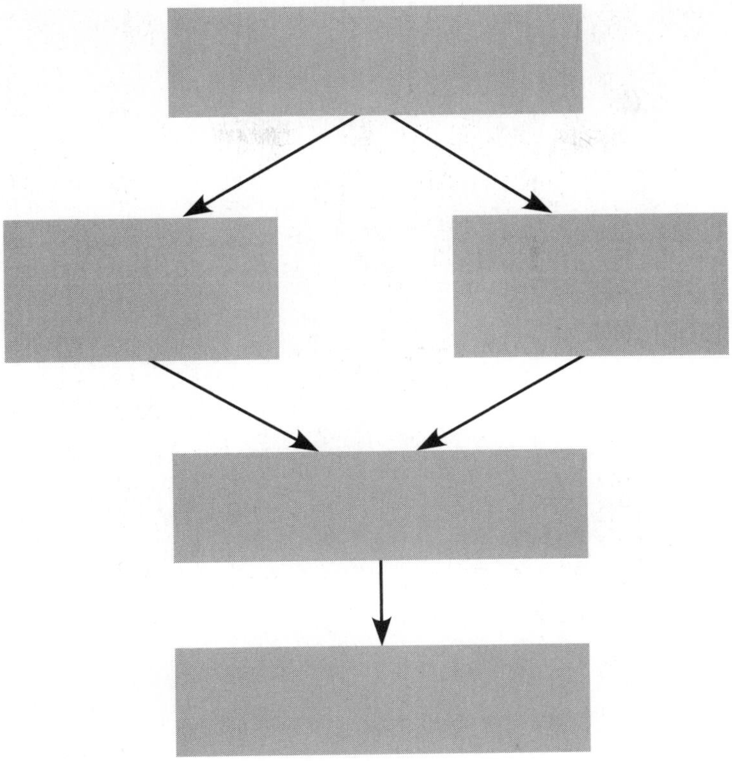

Figure 1.20 Summary of natural selection, page 16

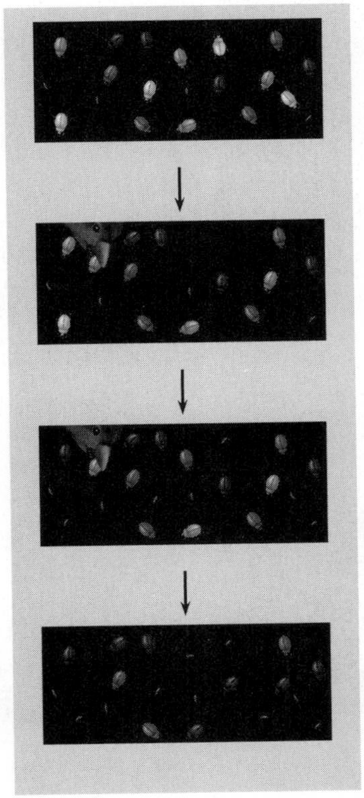

Figure 1.21 Natural selection, page 17

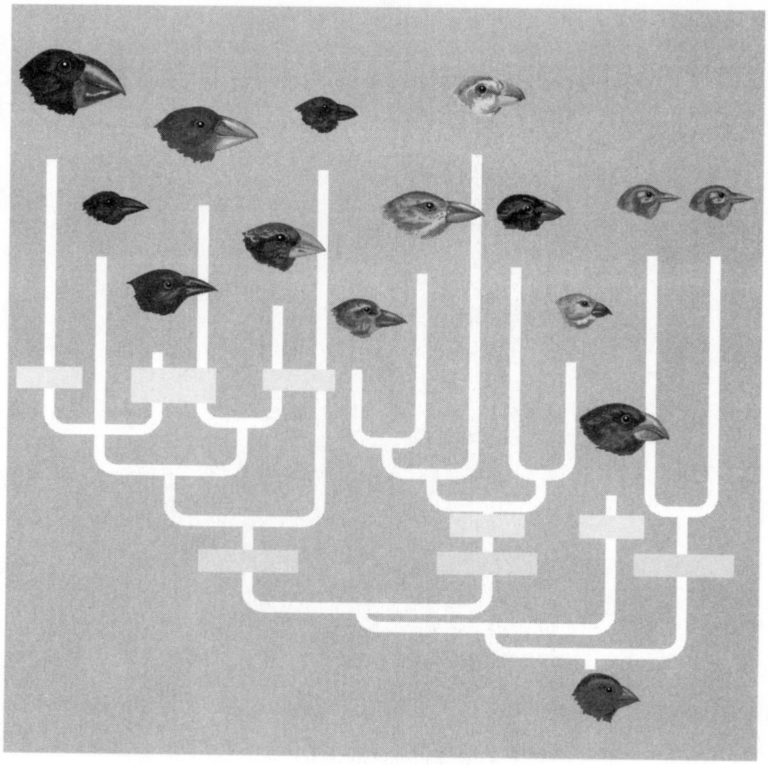

Figure 1.23 Descent with modification: adaptive radiation of finches on the Galápagos Islands, page 18

Figure 1.25 A campground example of hypothesis-based inquiry, page 20

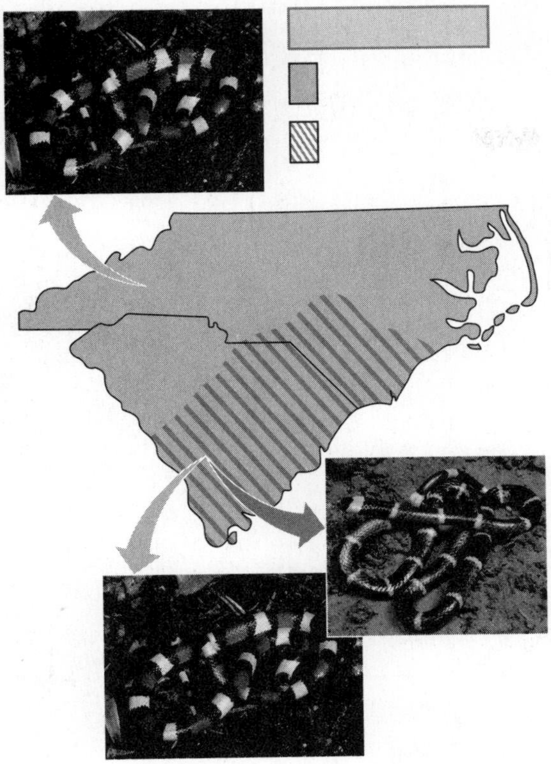

Figure 1.27 Geographic ranges of Carolina coral snakes and king snakes, page 22

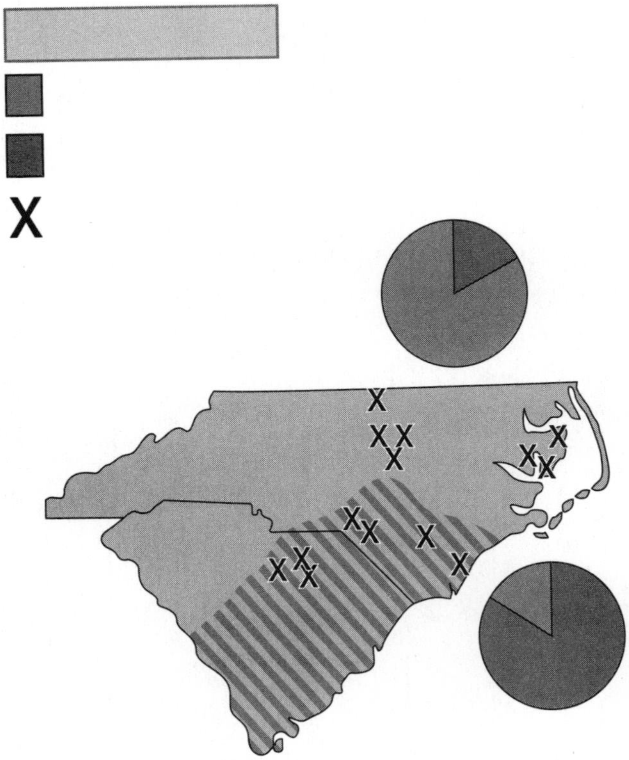

Figure 1.29 Does the presence of poisonous coral snakes affect predation rates on their mimics, king snakes?, page 23

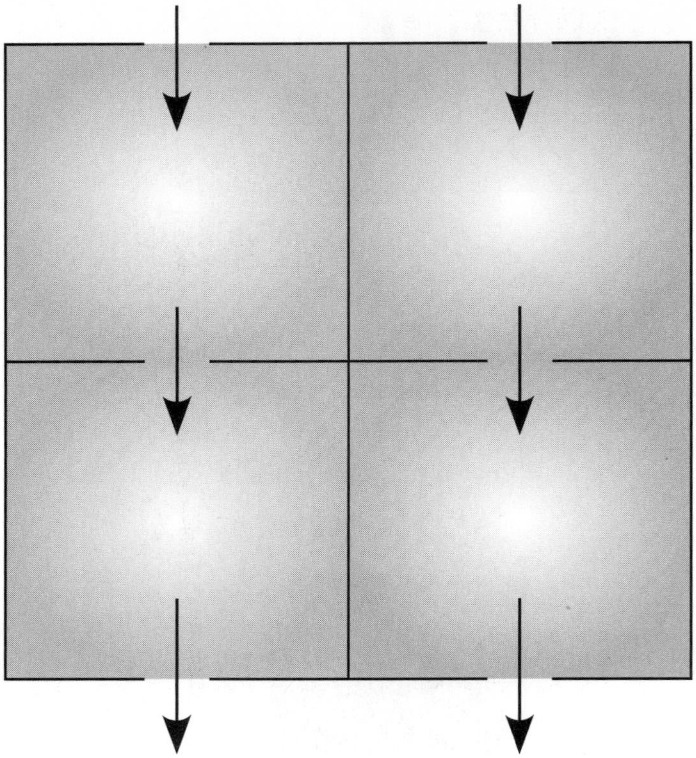

Figure 1.30 Modeling the pattern of blood flow through the four chambers of a human heart, page 25

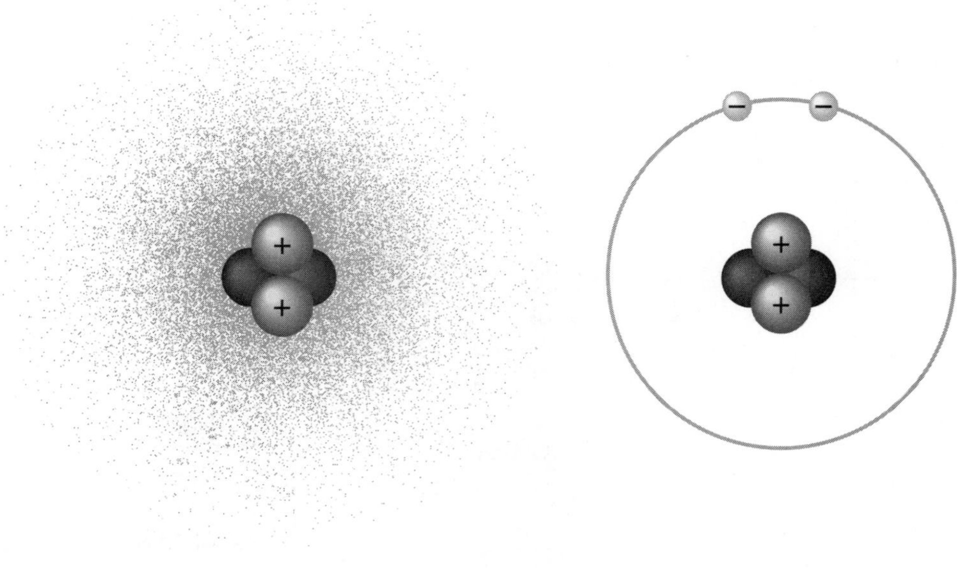

Figure 2.4 Simplified models of a helium (He) atom, page 34

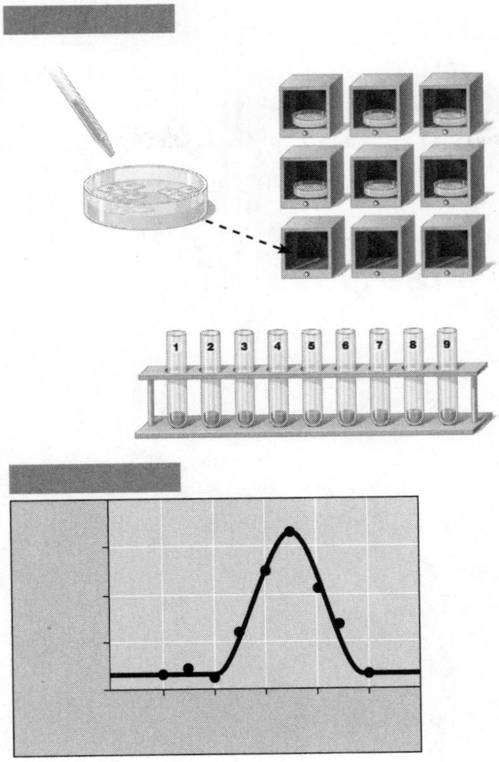

Figure 2.5 Research Method: Radioactive Tracers, page 35

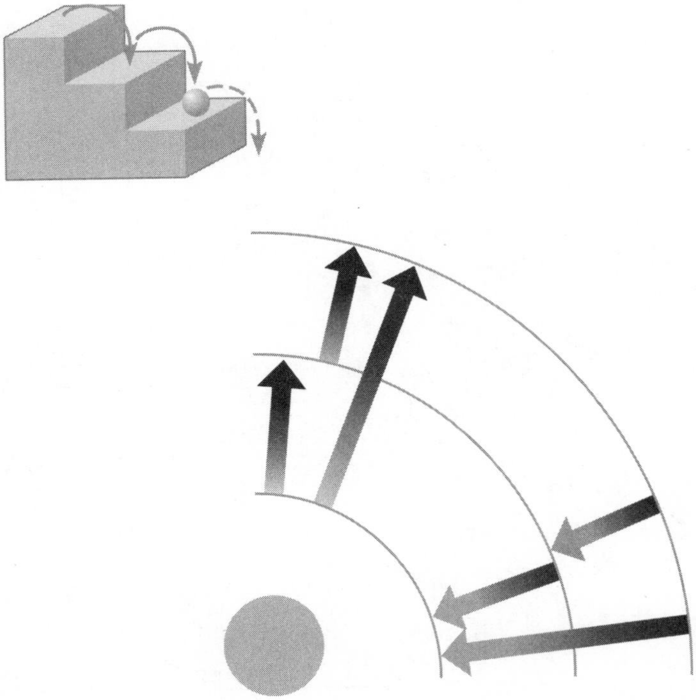

Figure 2.7 Energy levels of an atom's electrons, page 37

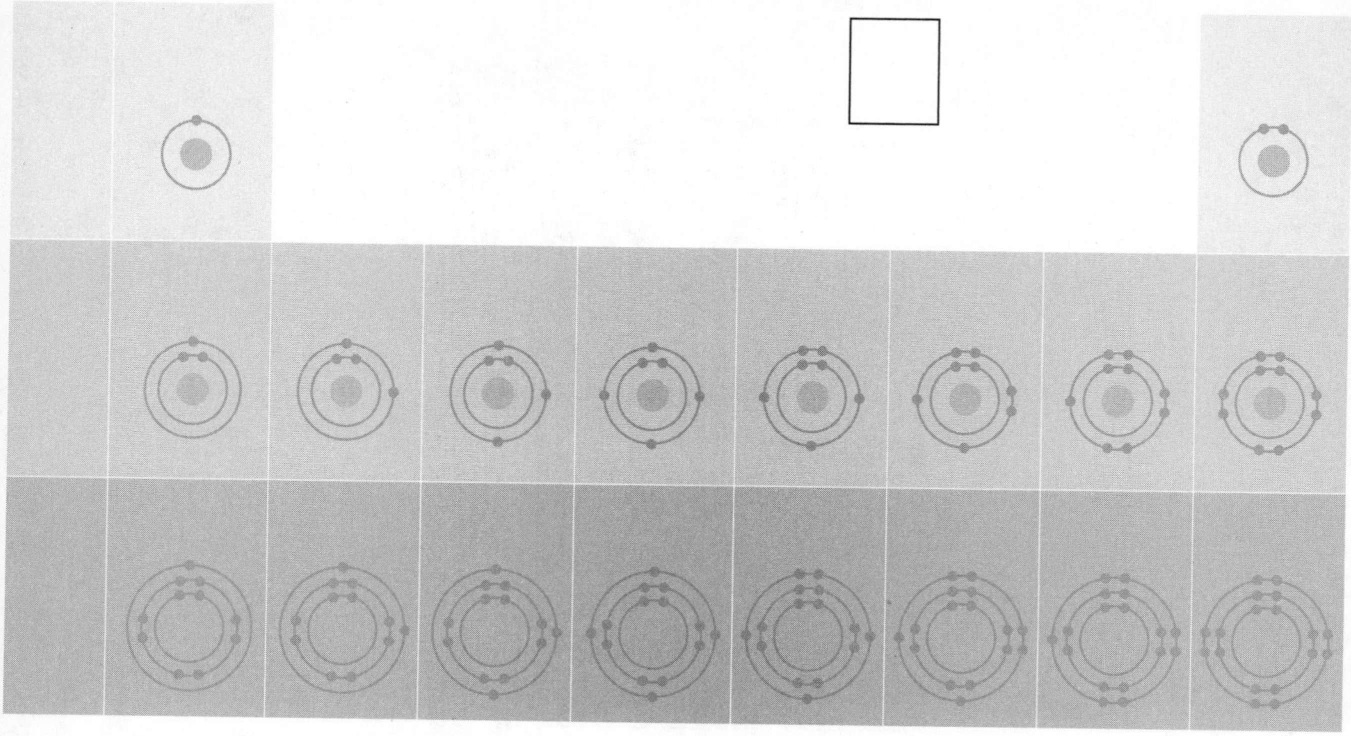

Figure 2.8 Electron-shell diagrams of the first 18 elements in the periodic table, page 37

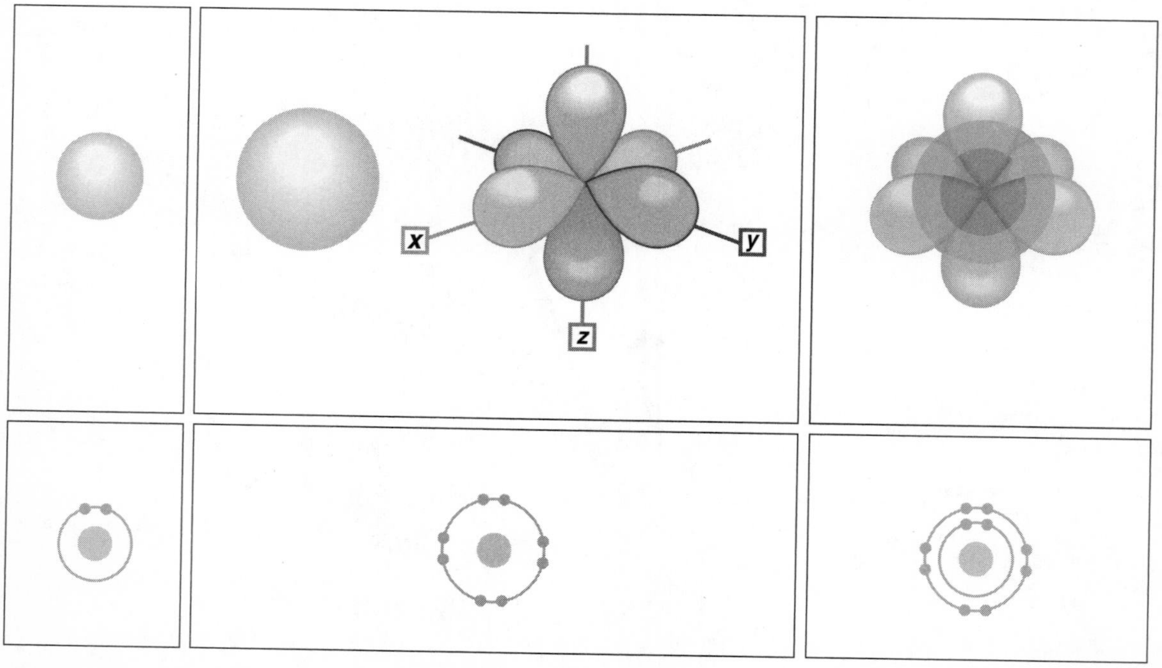

Figure 2.9 Electron orbitals, page 38

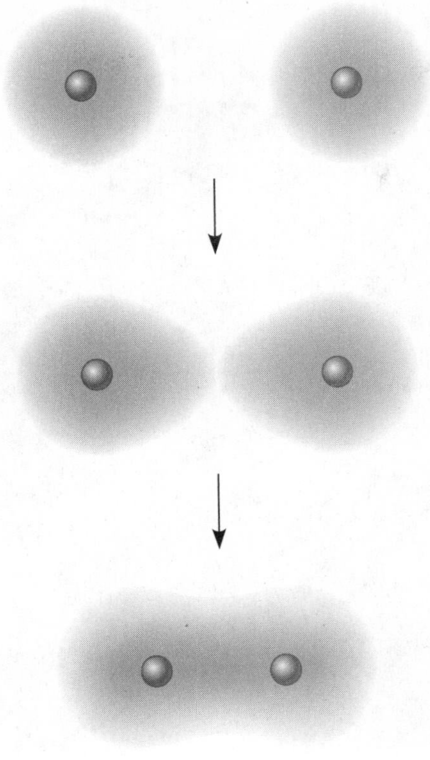

Figure 2.10 Formation of a covalent bond, page 39

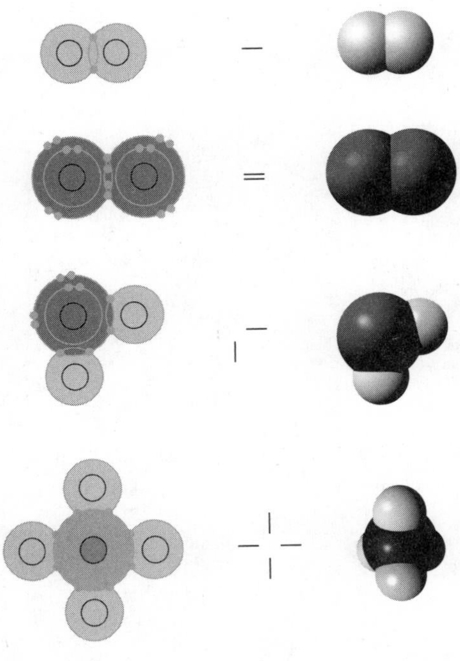

Figure 2.11 Covalent bonding in four molecules, page 40

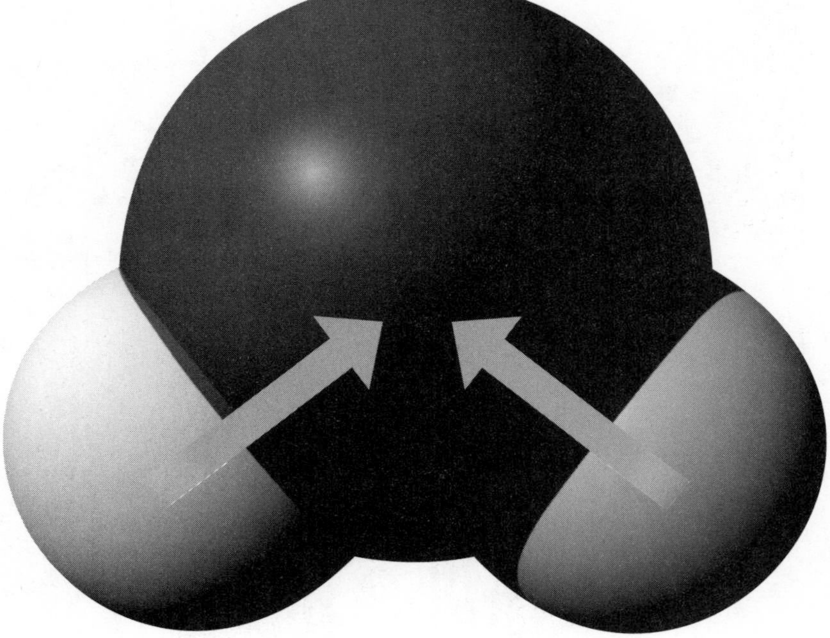

Figure 2.12 Polar covalent bonds in a water molecule, page 41

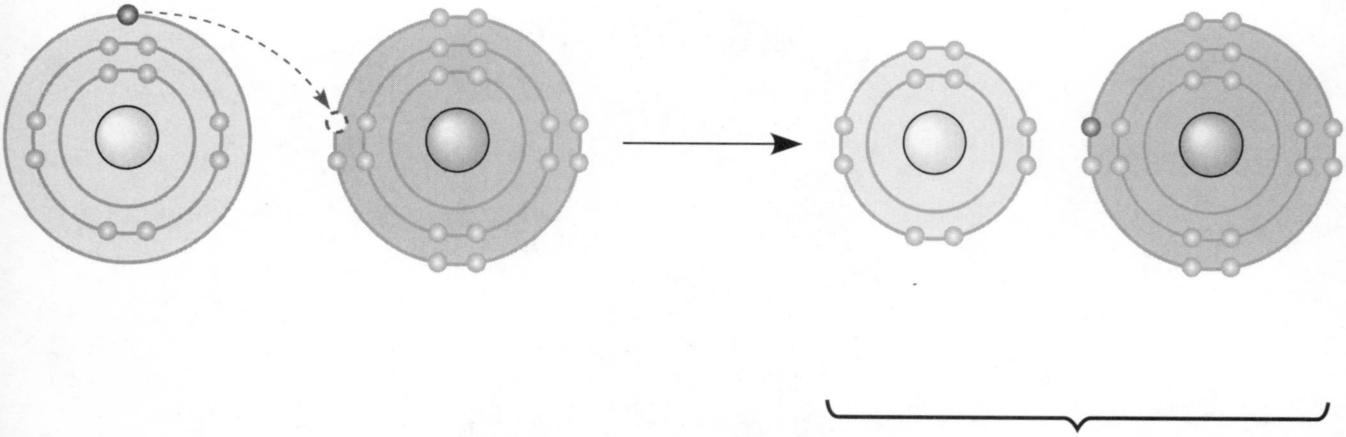

Figure 2.13 Electron transfer and ionic bonding, page 41

Figure 2.14 A sodium chloride crystal, page 41

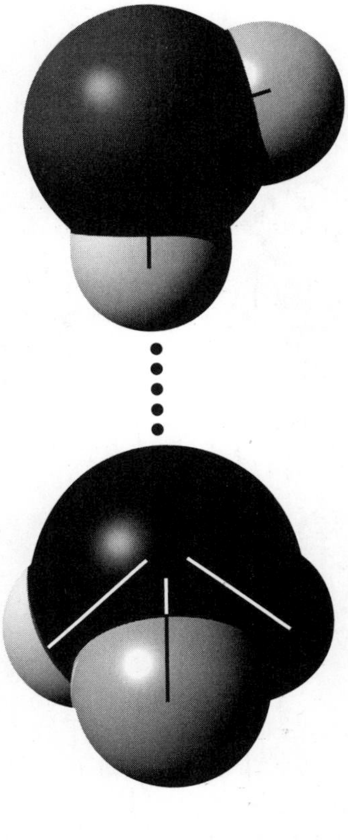

Figure 2.15 A hydrogen bond, page 42

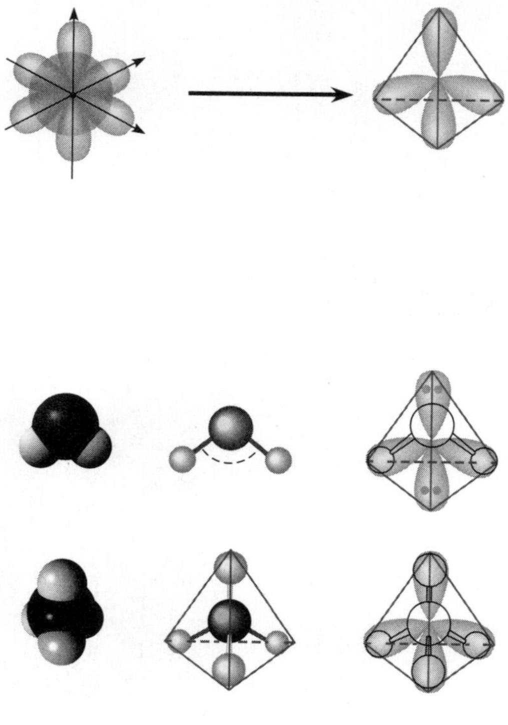

Figure 2.16 Molecular shapes due to hybrid orbitals, page 43

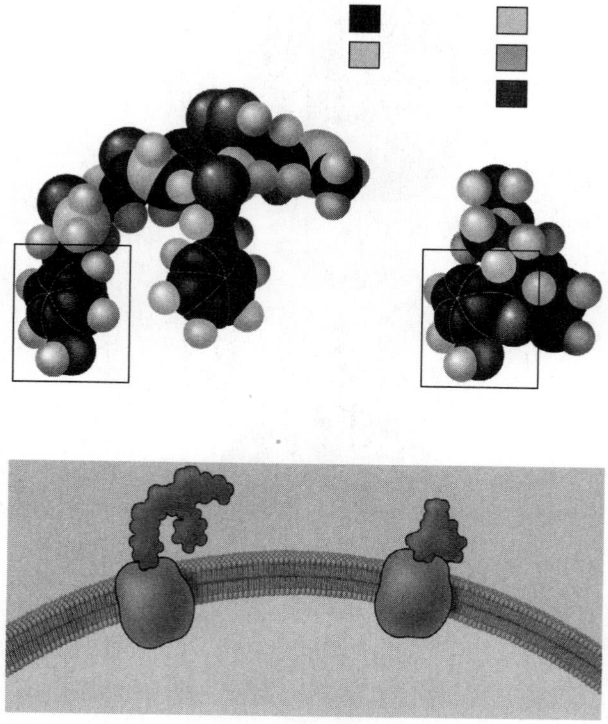

Figure 2.17 A molecular mimic, page 43

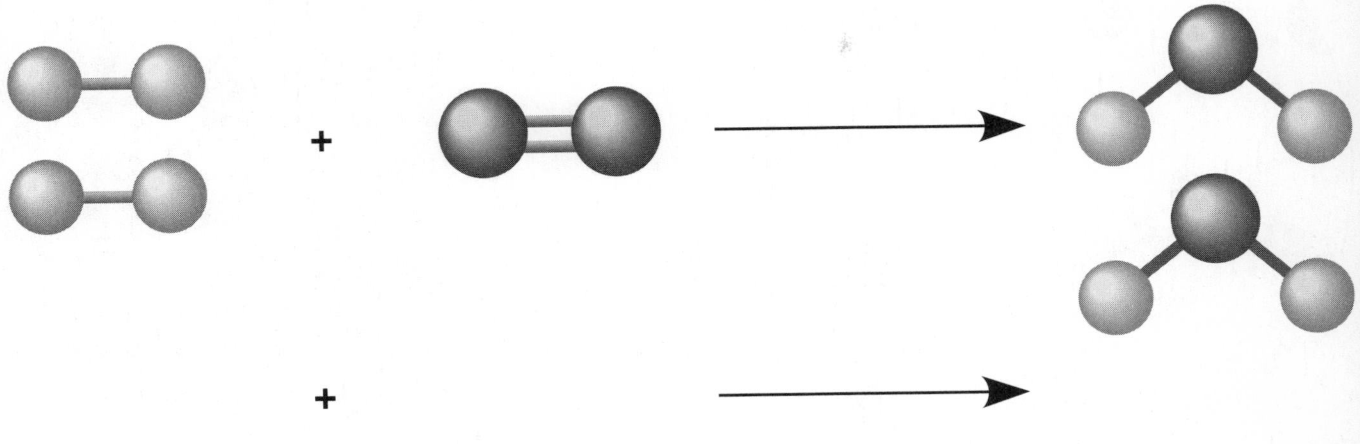

Figure 2.UN1 An example of a chemical reaction, page 44

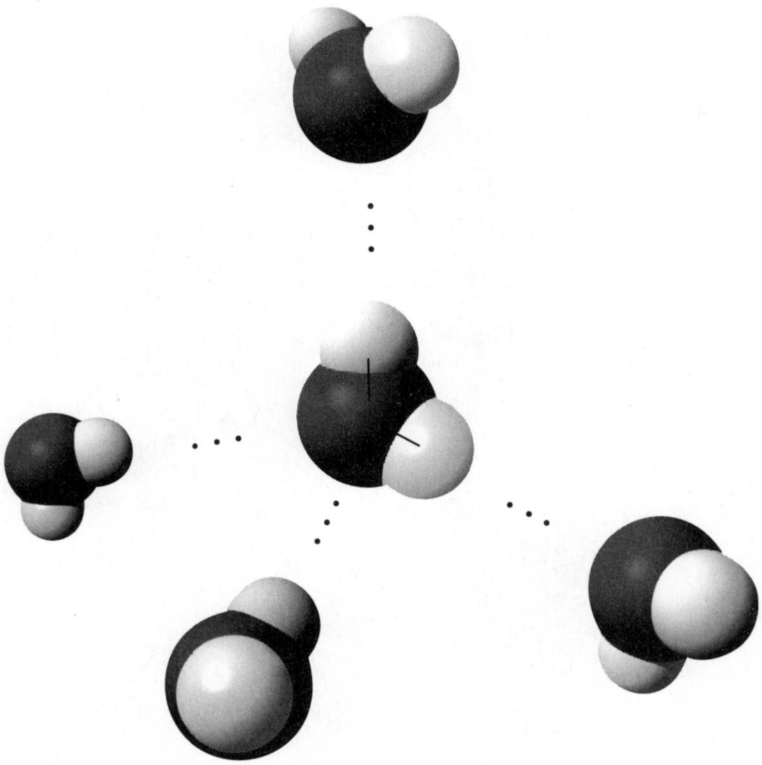

Figure 3.2 Hydrogen bonds between water molecules, page 48

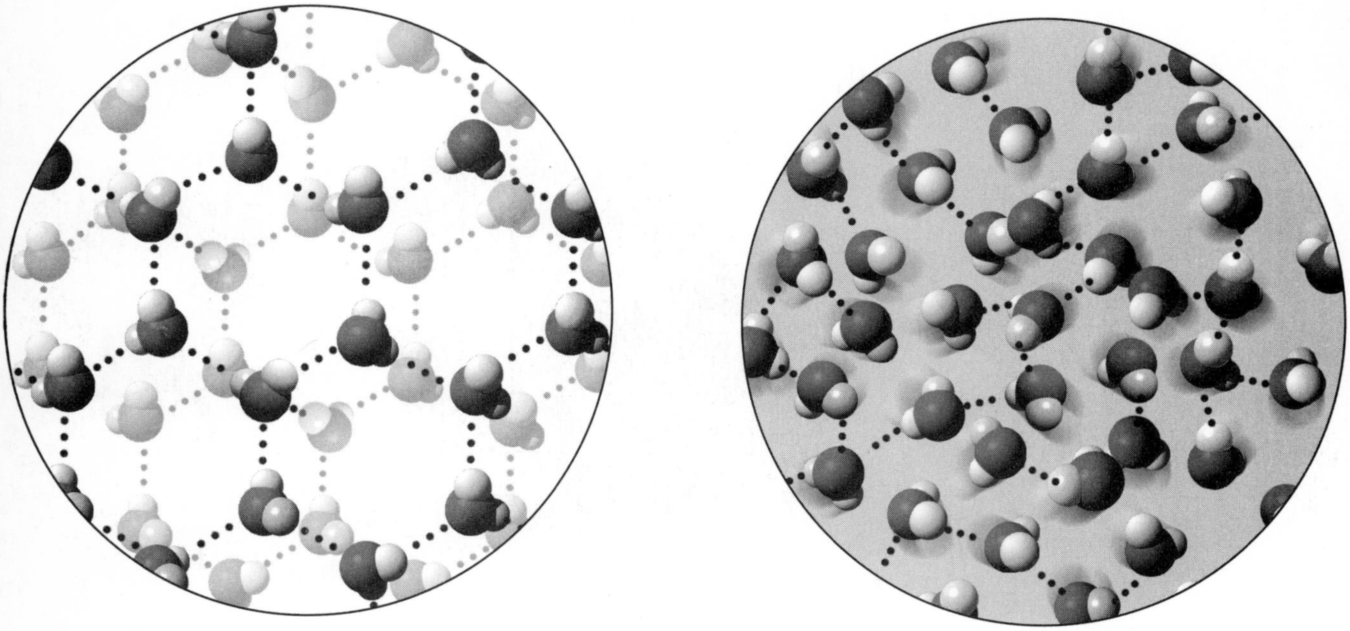

Figure 3.5 Ice: crystalline structure and floating barrier, page 51

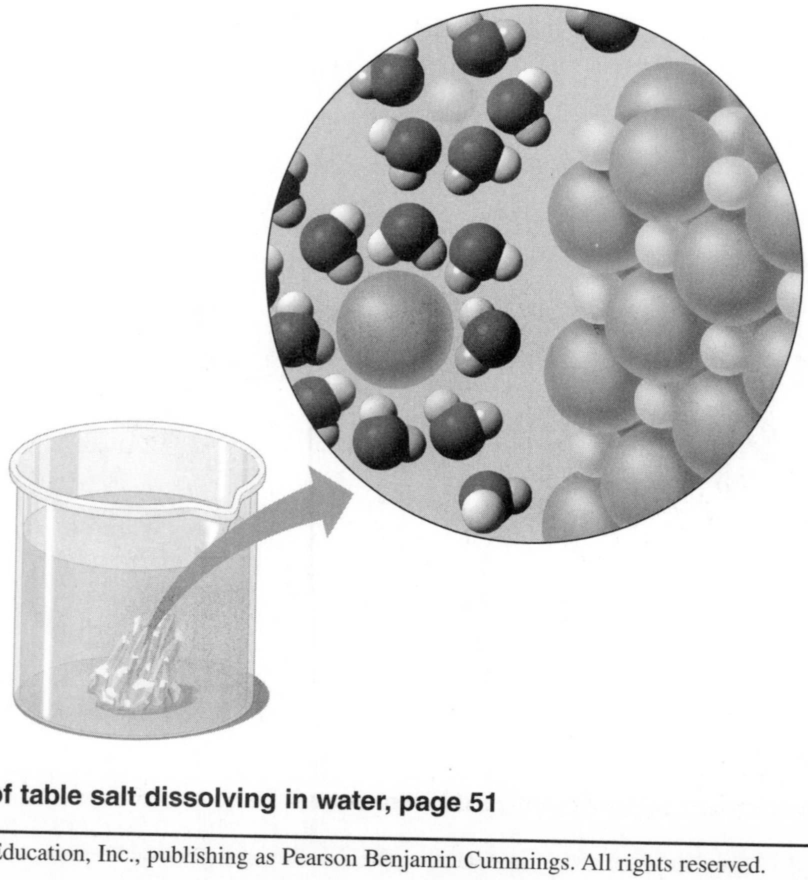

Figure 3.6 A crystal of table salt dissolving in water, page 51

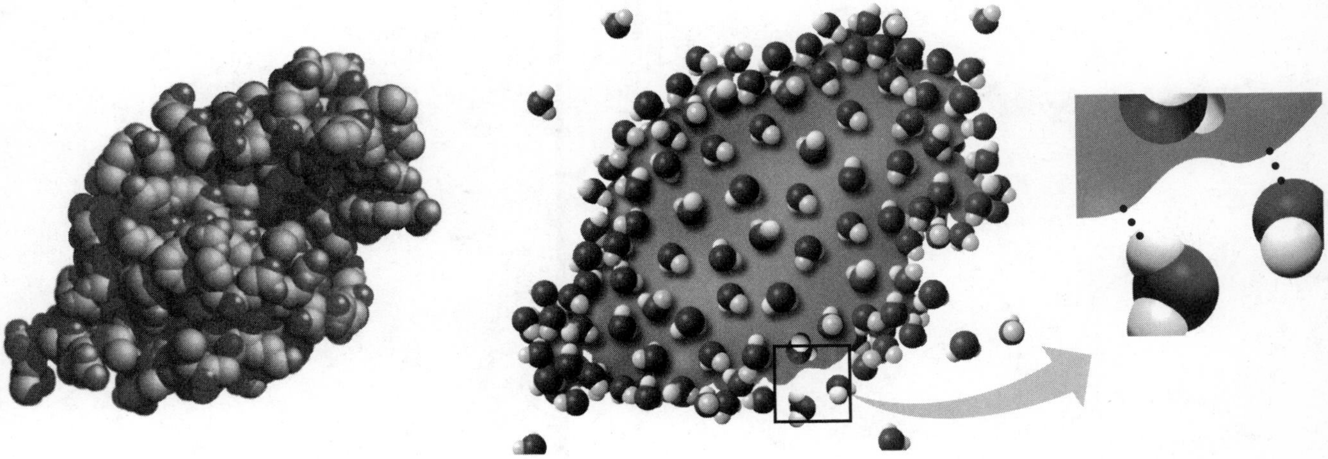

Figure 3.7 A water-soluble protein, page 52

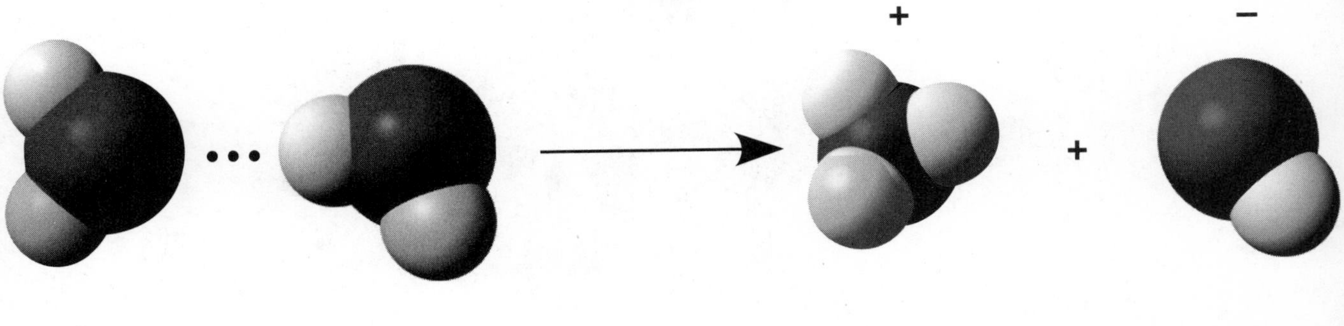

Figure 3.UN1 Reaction producing hydroxide and hydronium ions, page 53

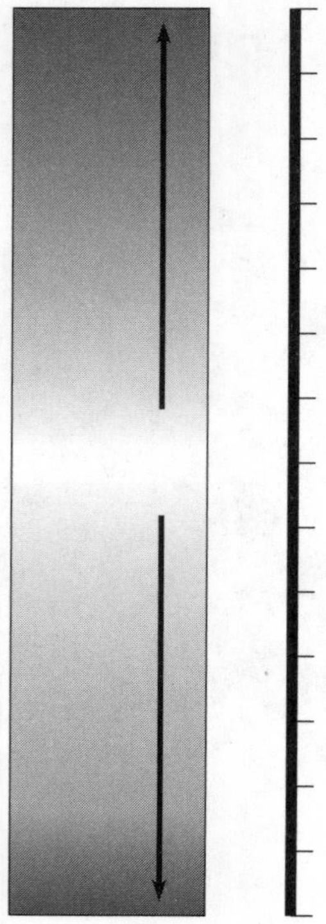

Figure 3.8 The pH scale and pH values of some aqueous solutions, page 54

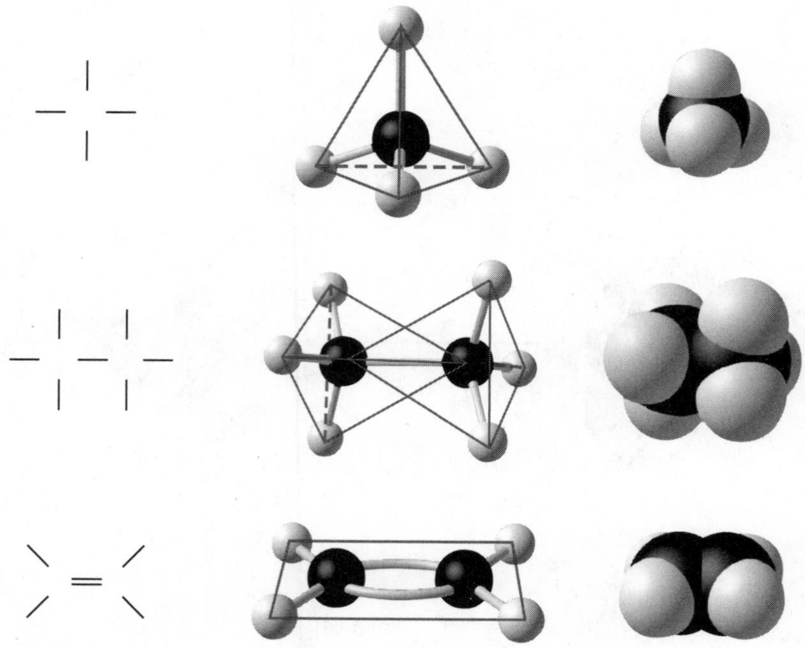

Figure 4.3 The shapes of three simple organic molecules, page 60

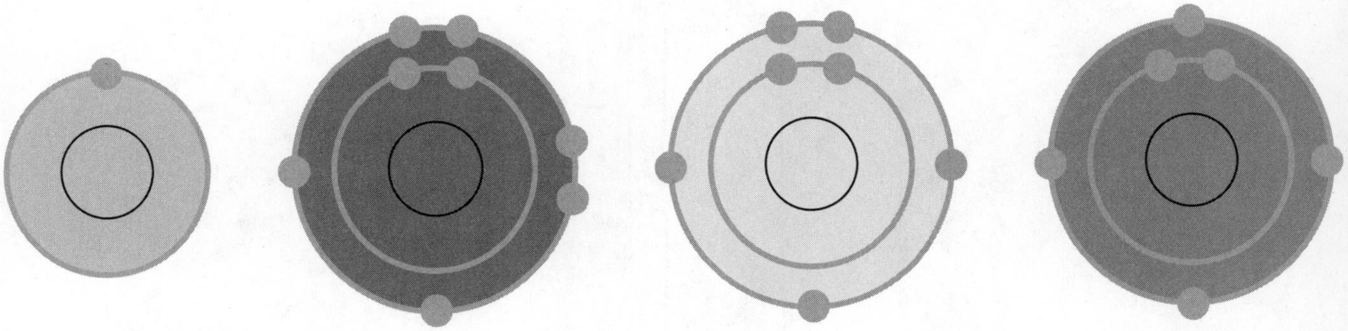

Figure 4.4 Electron-shell diagrams showing valences for the major elements of organic molecules, page 60

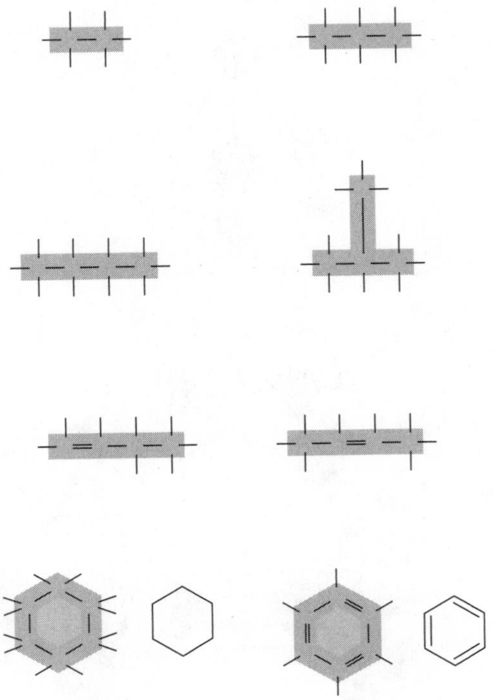

Figure 4.5 Variations in carbon skeletons, page 61

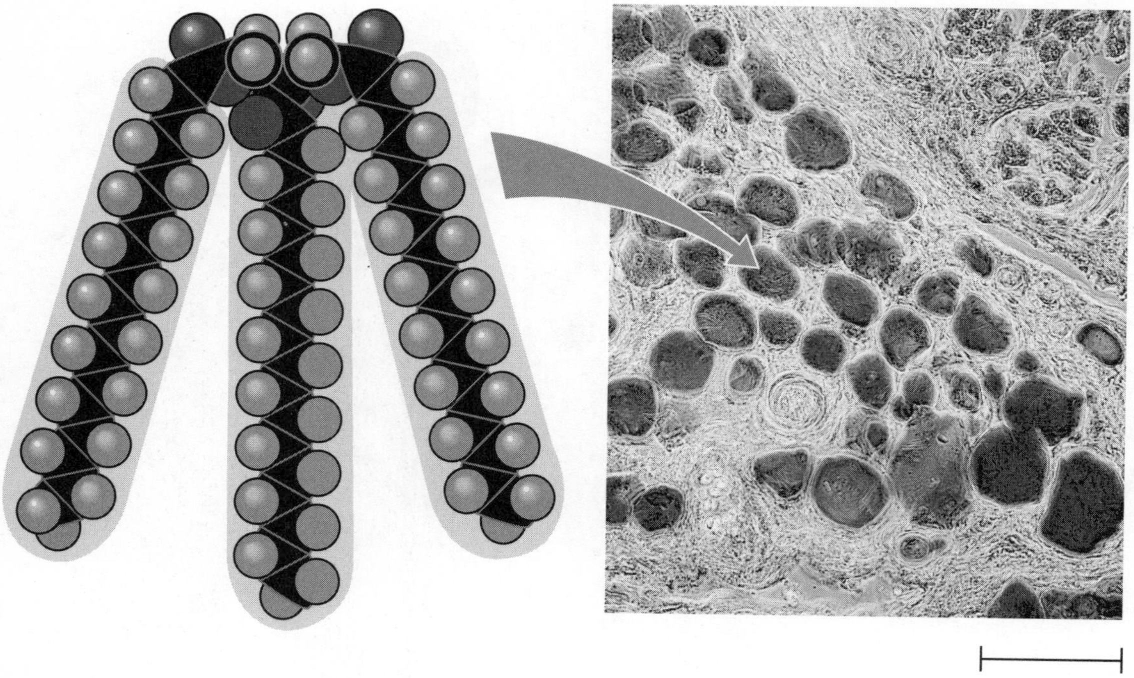

Figure 4.6 The role of hydrocarbons in fats, page 61

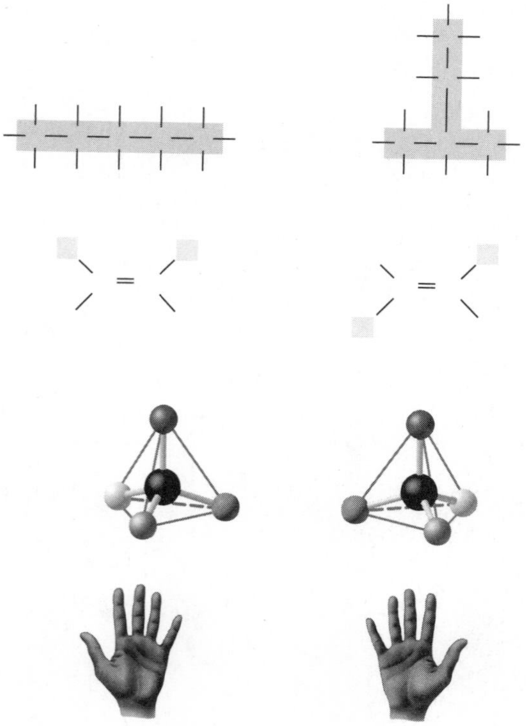

Figure 4.7 Three types of isomers, page 62

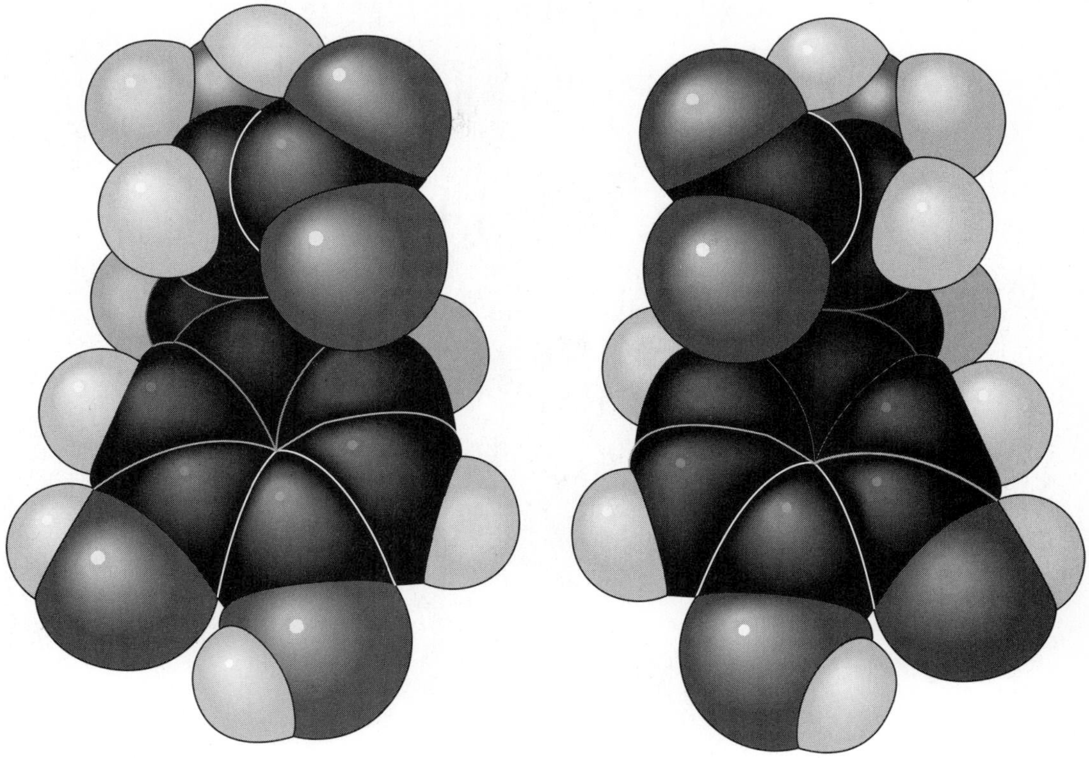

Figure 4.8 The pharmacological importance of enantiomers, page 63

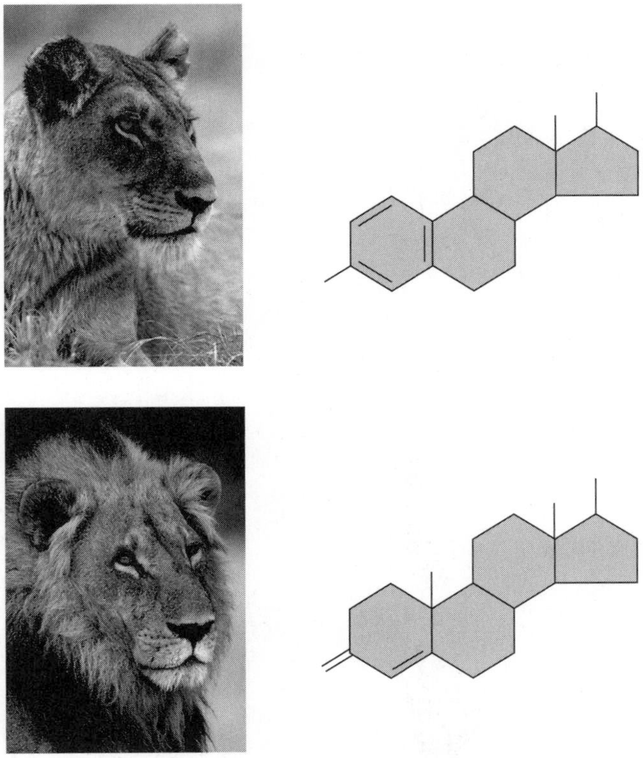

Figure 4.9 A comparison of functional groups of female (estradiol) and male (testosterone) sex hormones, page 63

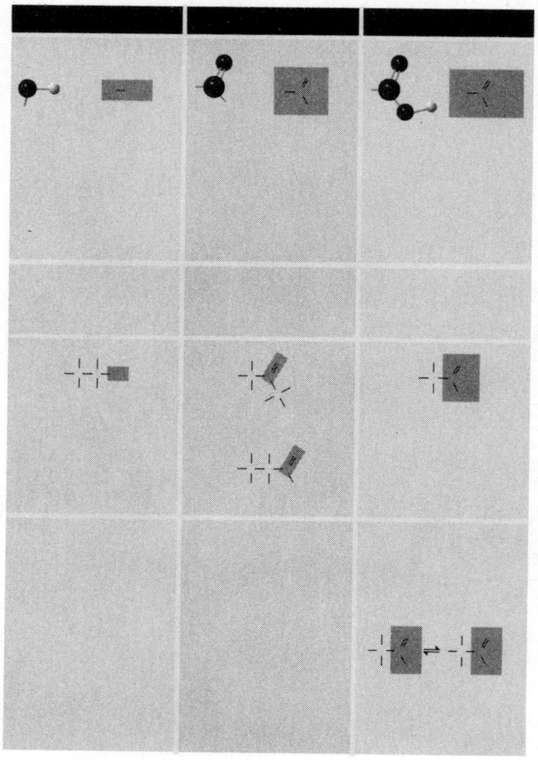

Figure 4.10 Exploring some important functional groups of organic compounds (part 1), page 64

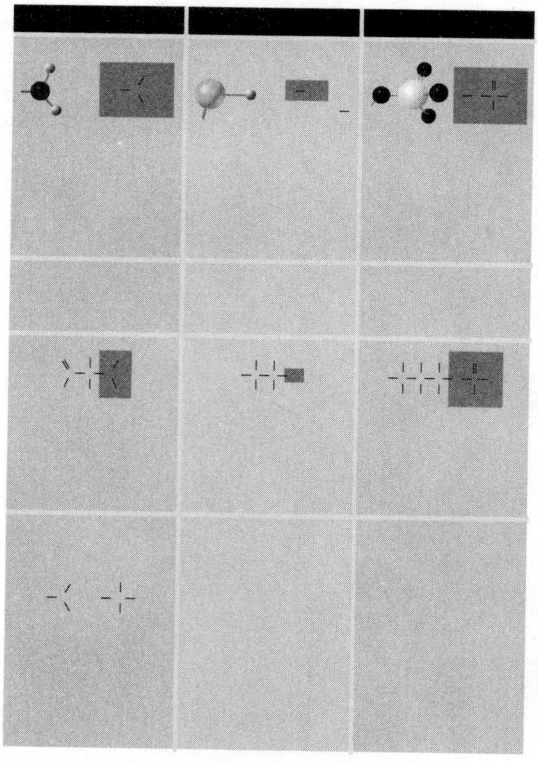

Figure 4.10 Exploring some important functional groups of organic compounds (part 2), page 65

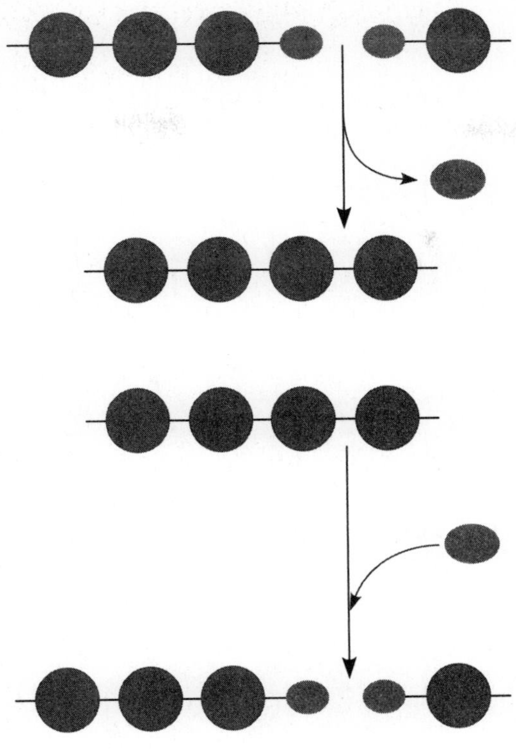

Figure 5.2 The synthesis and breakdown of polymers, page 69

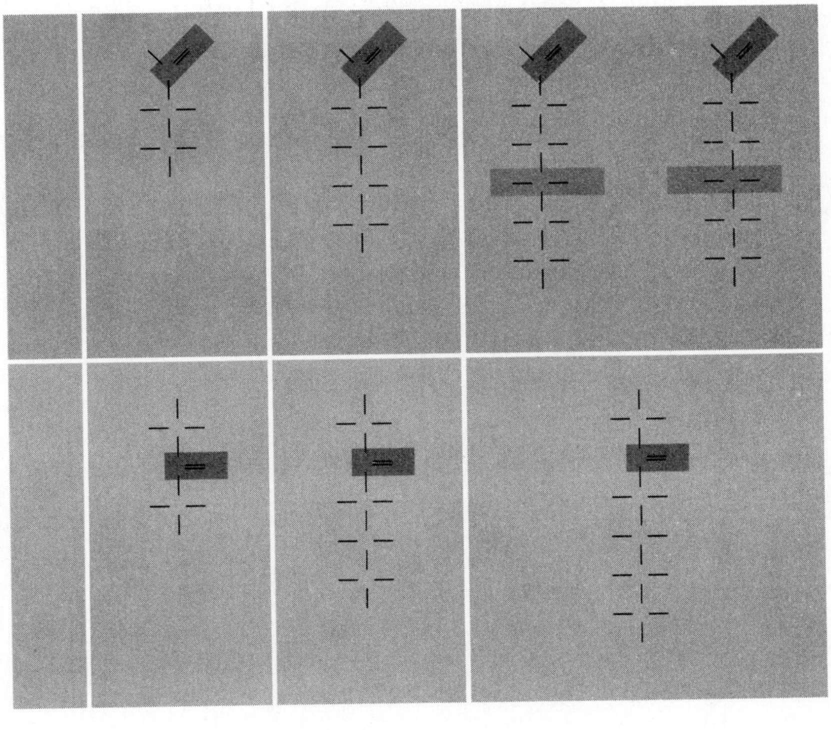

Figure 5.3 The structure and classification of some monosaccharides, page 70

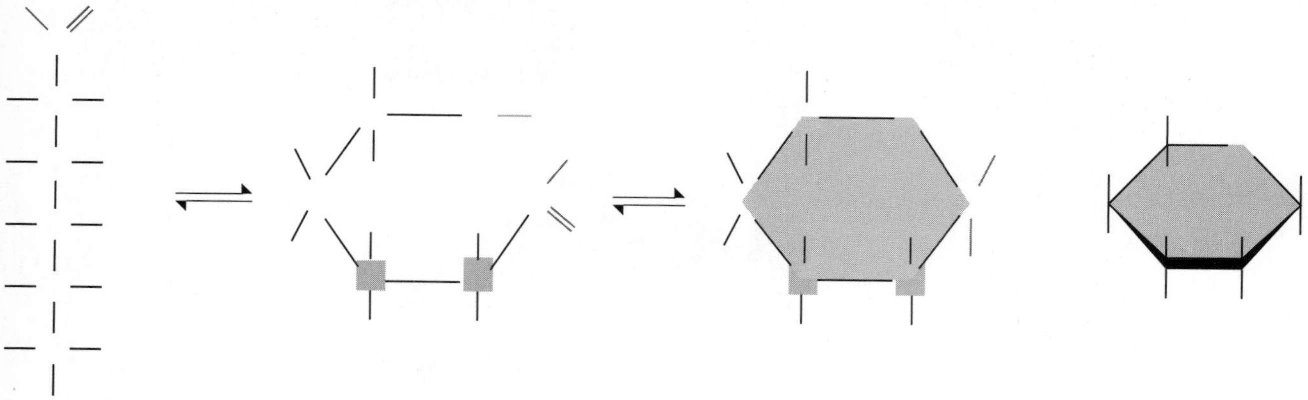

Figure 5.4 Linear and ring forms of glucose, page 71

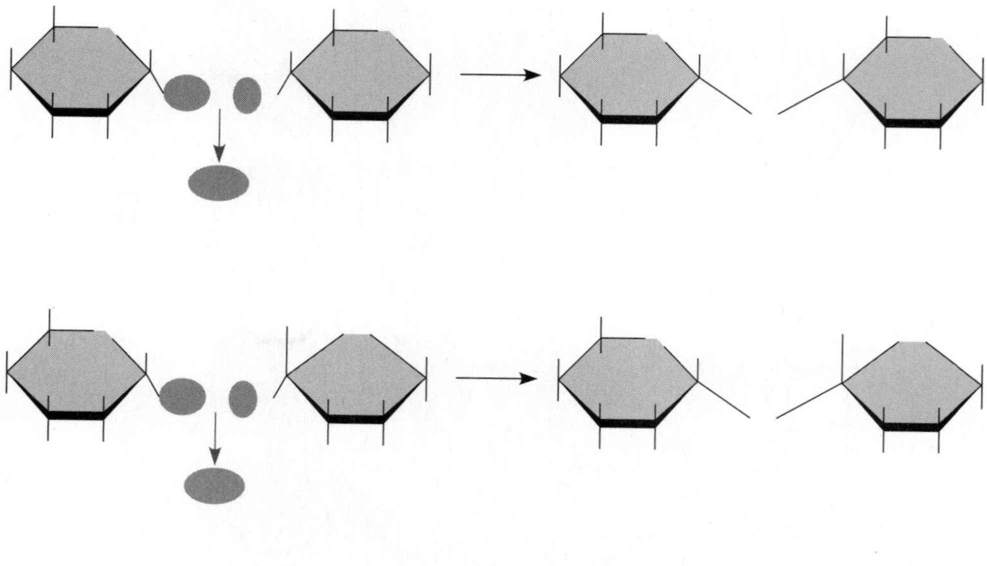

Figure 5.5 Examples of disaccharide synthesis, page 71

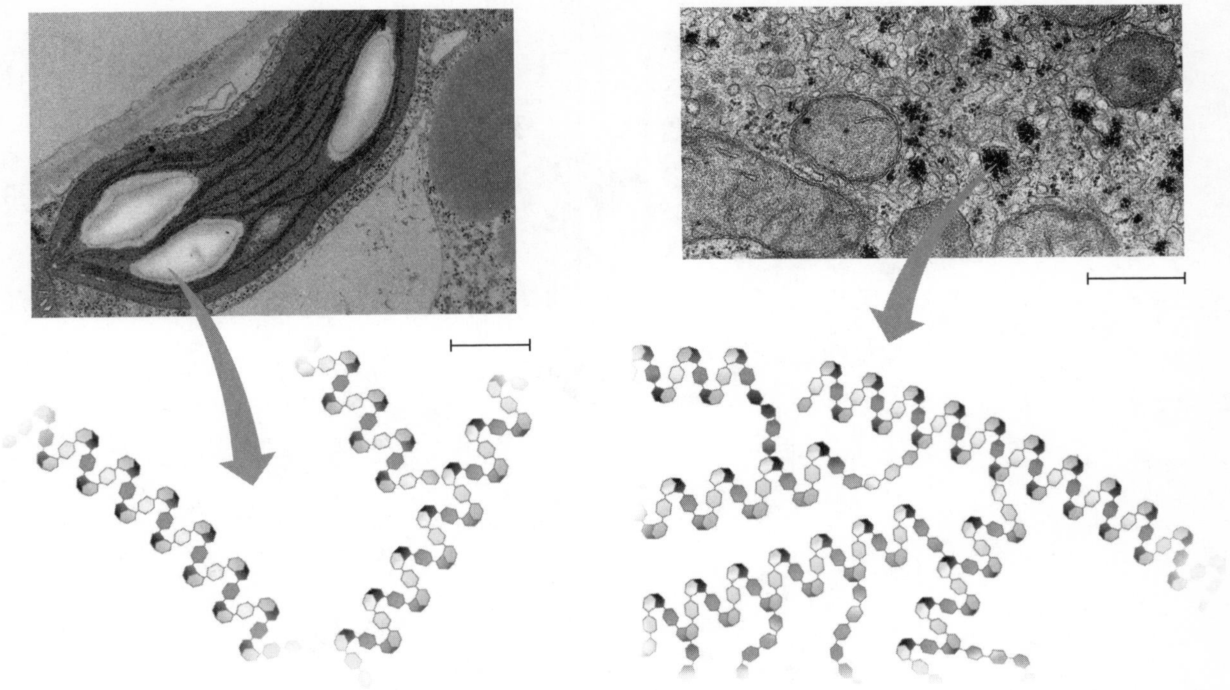

Figure 5.6 Storage polysaccharides of plants and animals, page 72

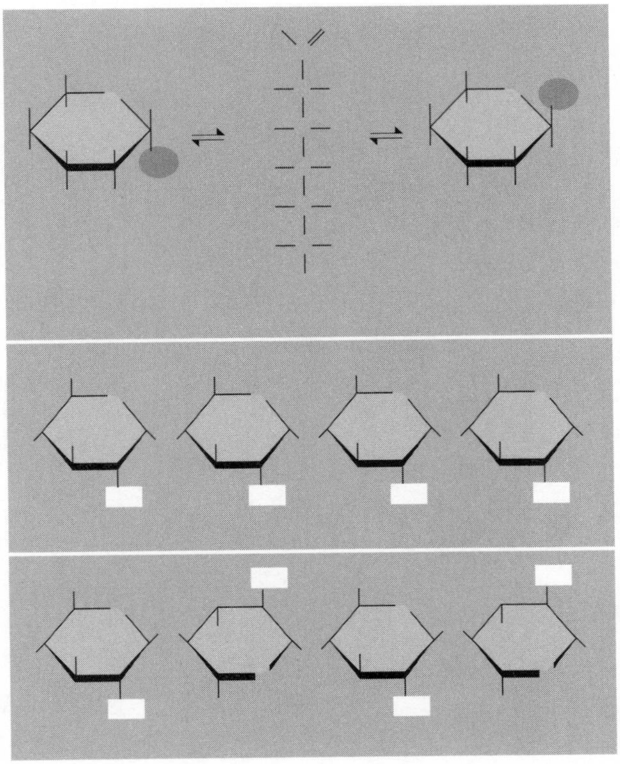

Figure 5.7 Starch and cellulose structures, page 73

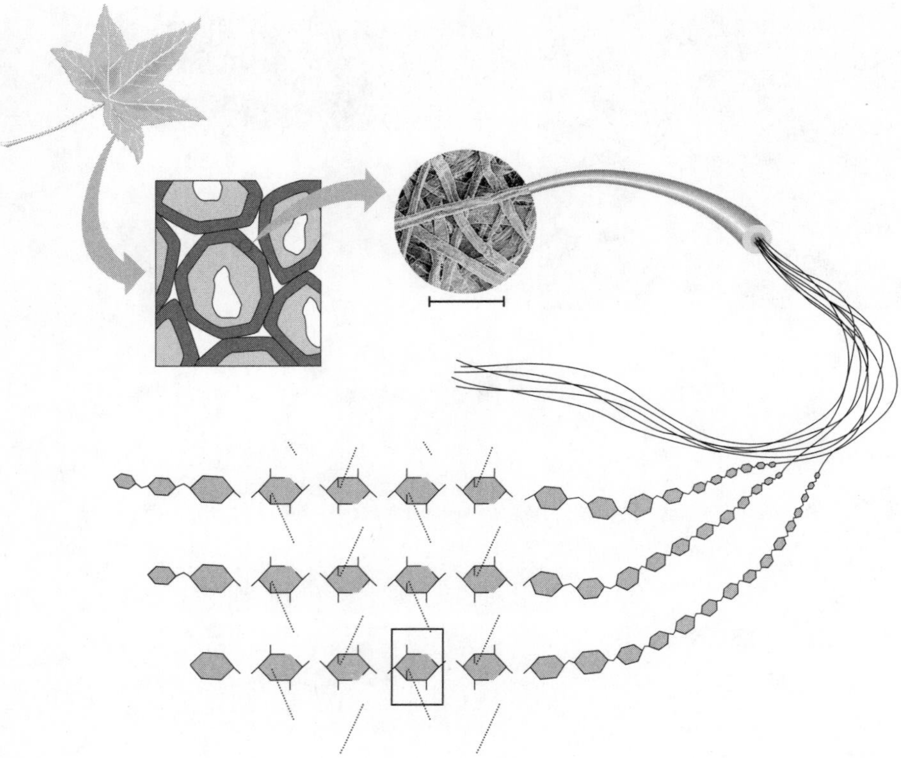

Figure 5.8 The arrangement of cellulose in plant cell walls, page 73

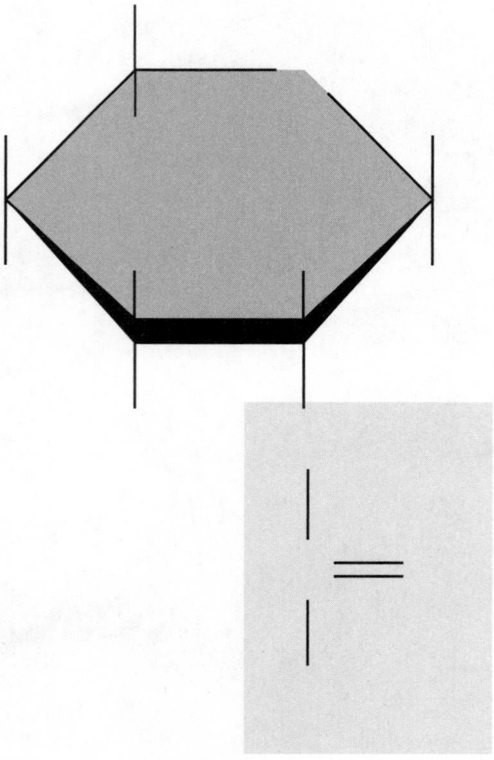

Figure 5.10 Chitin, a structural polysaccharide, page 74

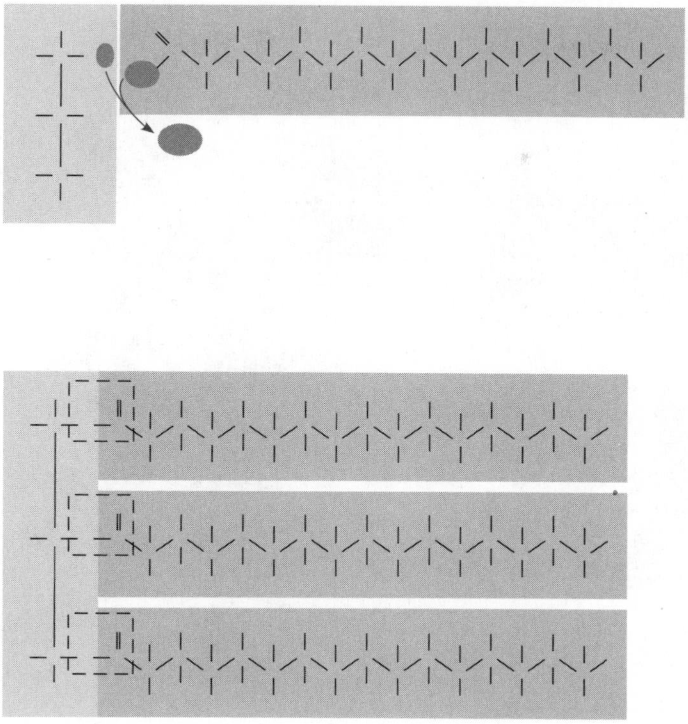

Figure 5.11 The synthesis and structure of a fat, or triacylglycerol, page 75

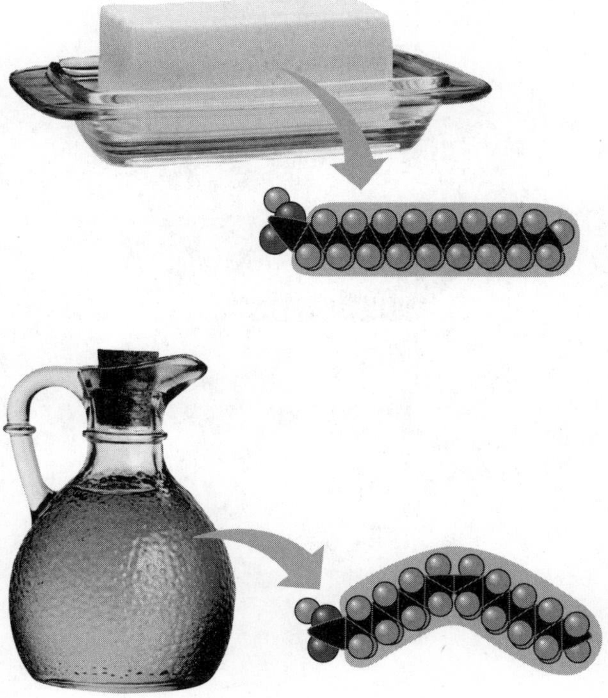

Figure 5.12 Examples of saturated and unsaturated fats and fatty acids, page 75

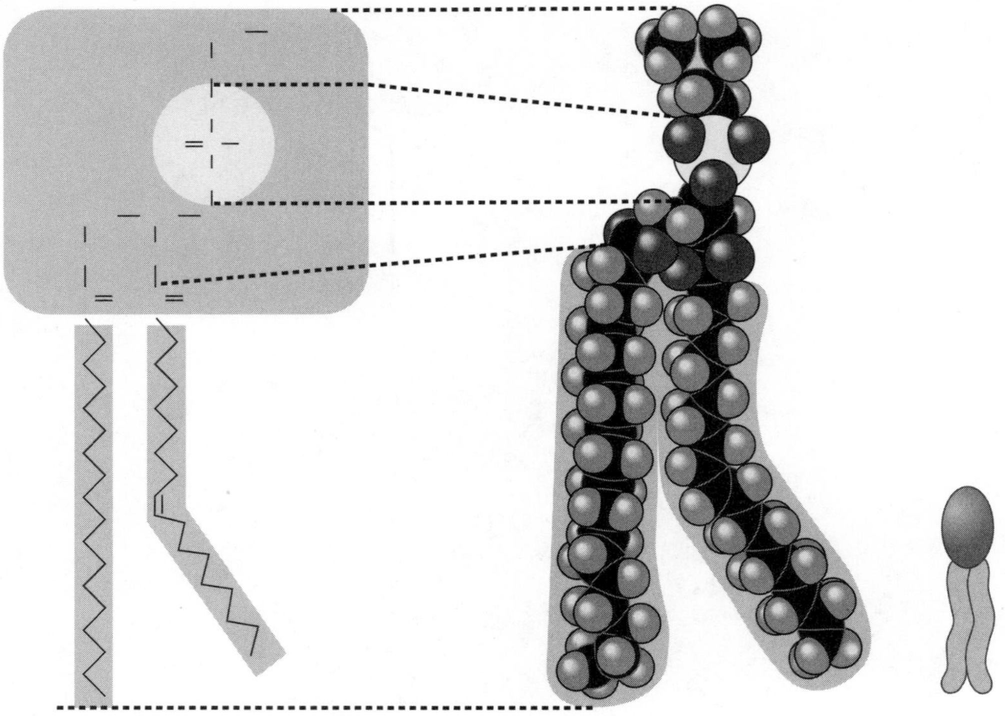

Figure 5.13 The structure of a phospholipid, page 76

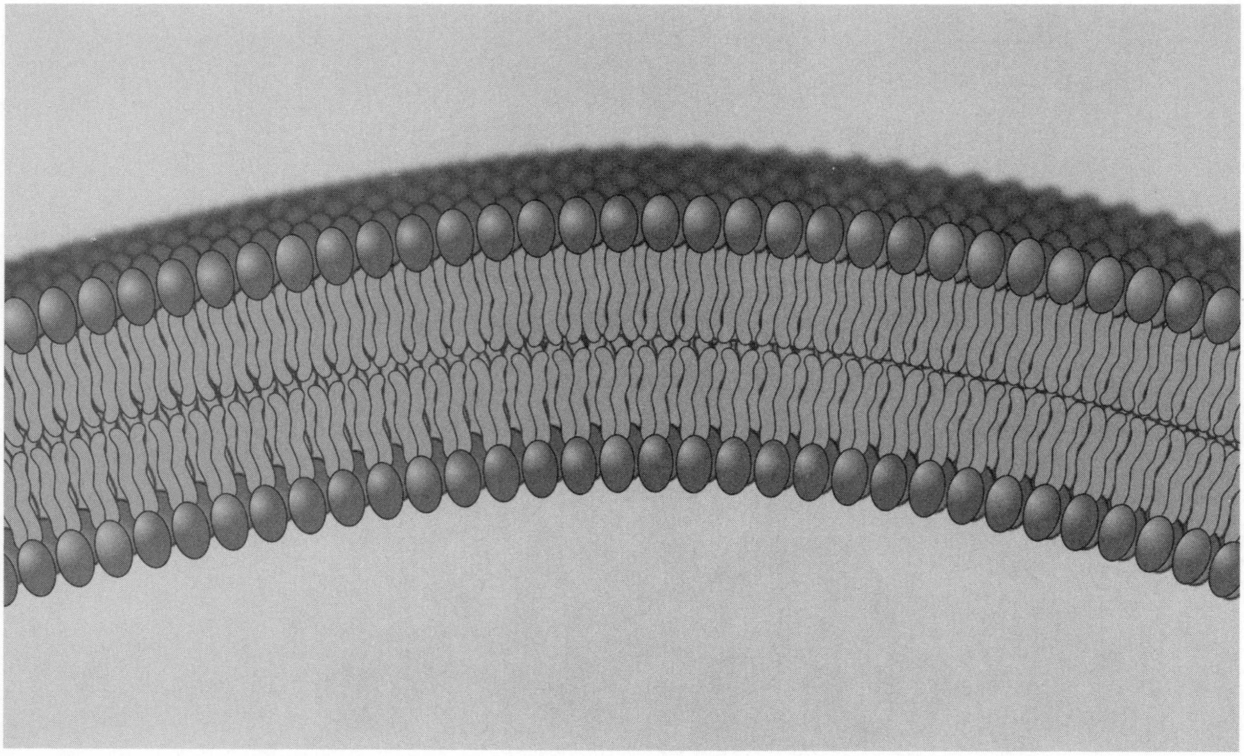

Figure 5.14 Bilayer structure formed by self-assembly of phospholipids in an aqueous environment, page 77

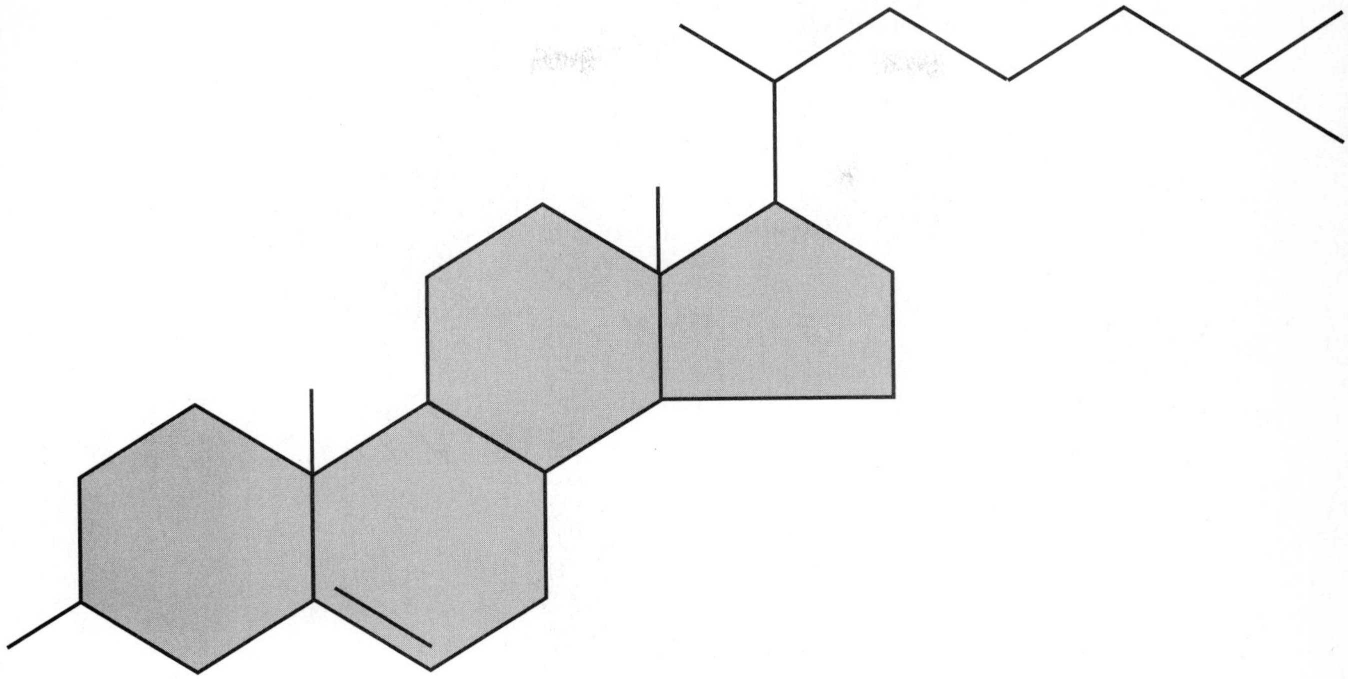

Figure 5.15 Cholesterol, a steroid, page 77

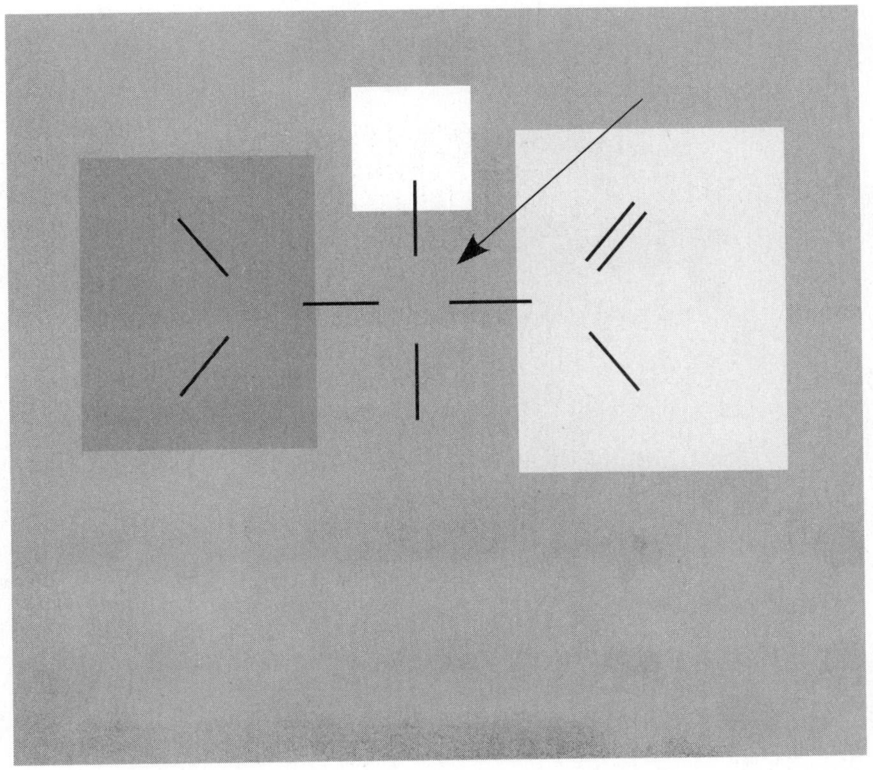

Figure 5.UN1 An amino group and a carboxyl group, page 78

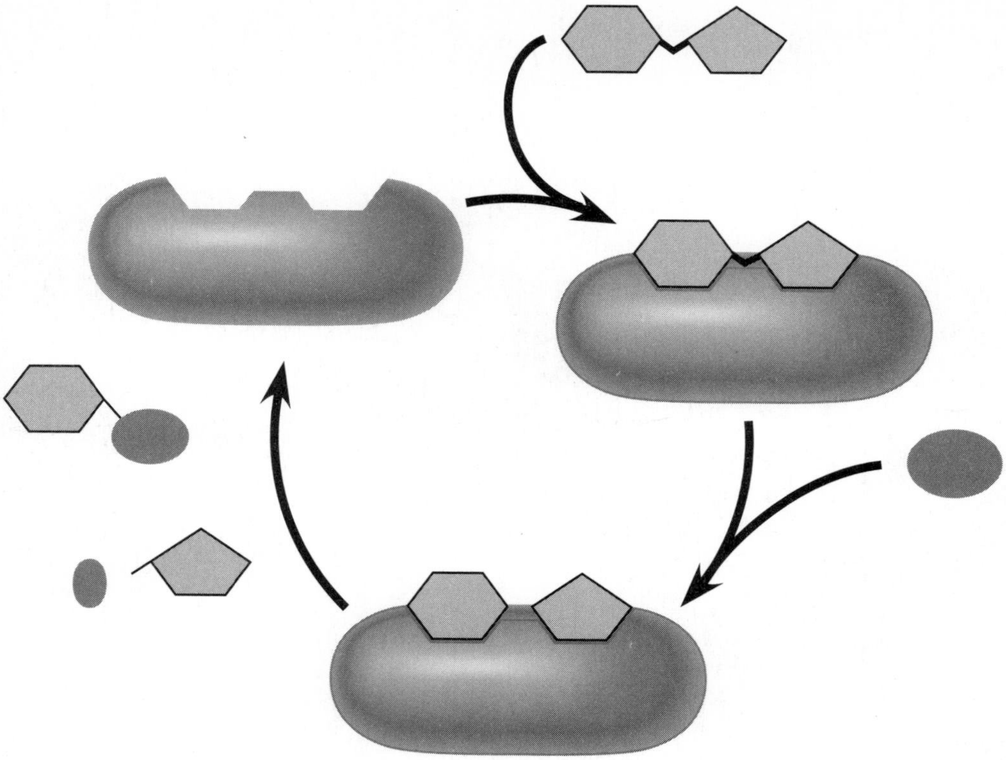

Figure 5.16 The catalytic cycle of an enzyme, page 78

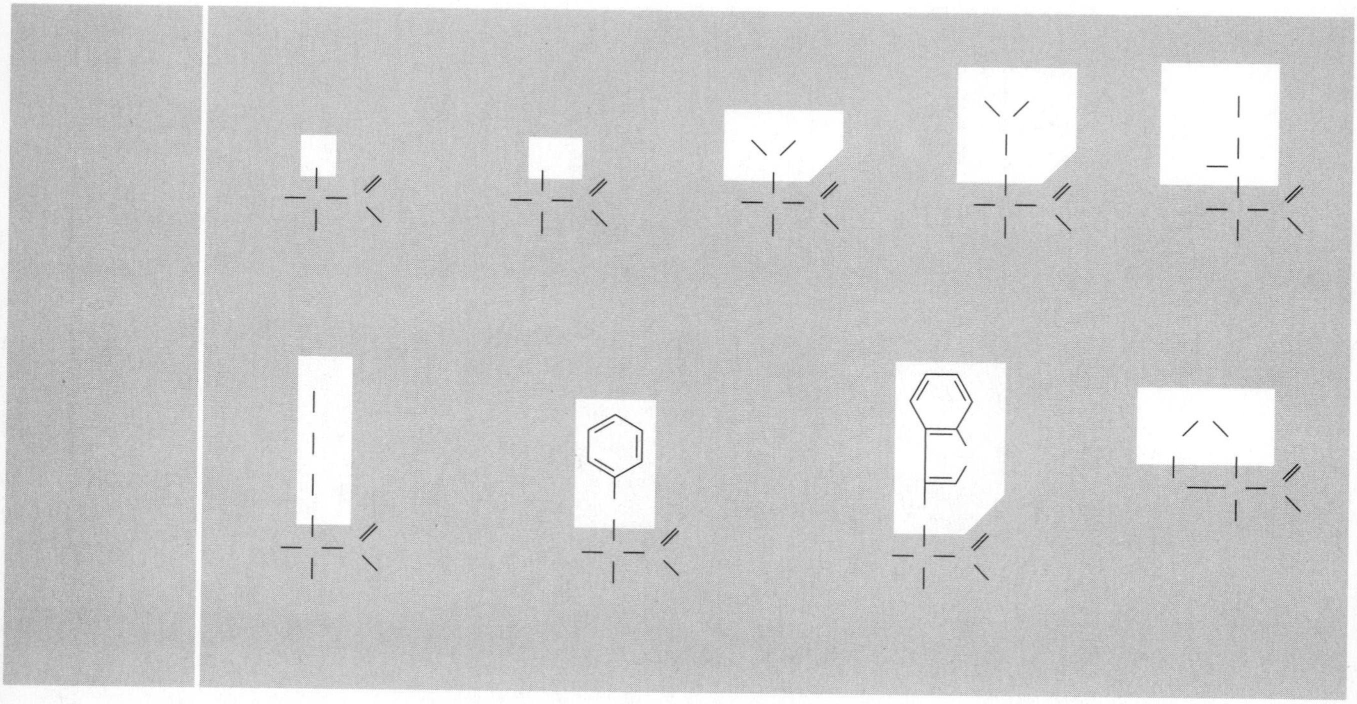

Figure 5.17 The 20 amino acids of proteins: Nonpolar, page 79

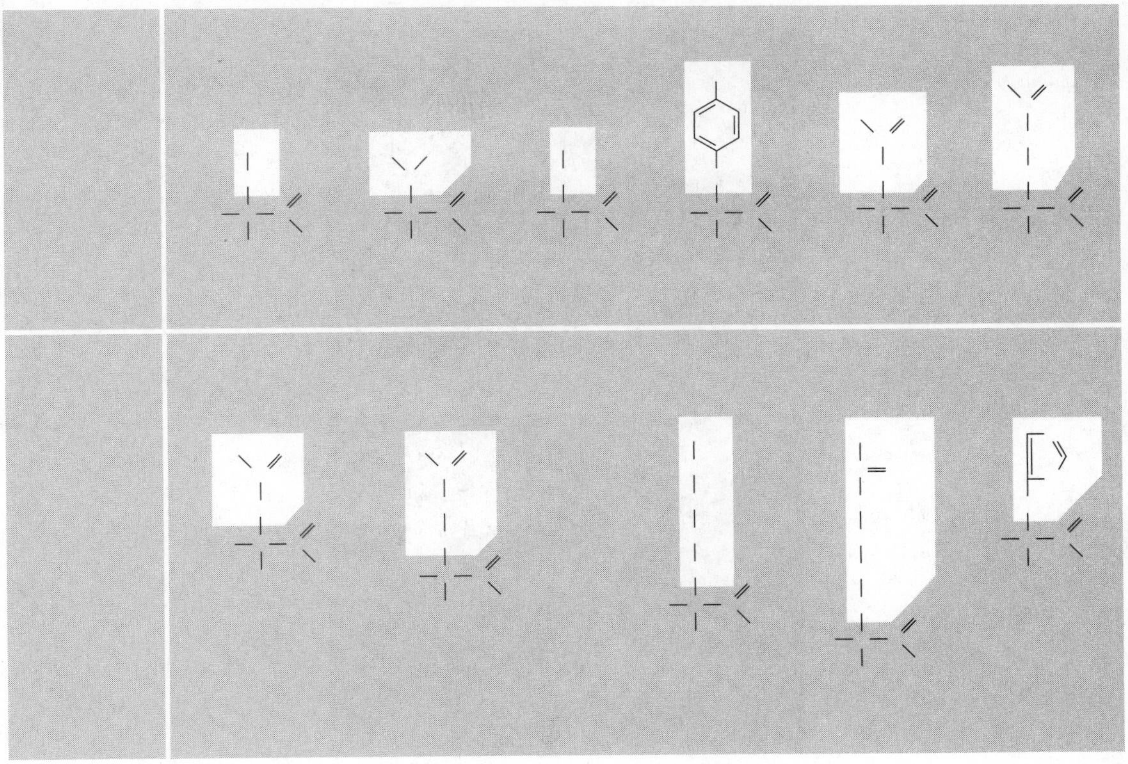

Figure 5.17 The 20 amino acids of proteins: Polar and electrically charged, page 79

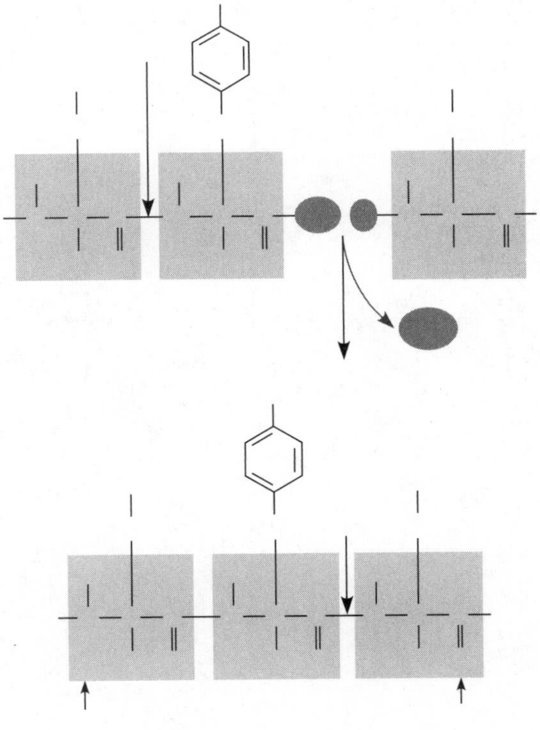

Figure 5.18 Making a polypeptide chain, page 80

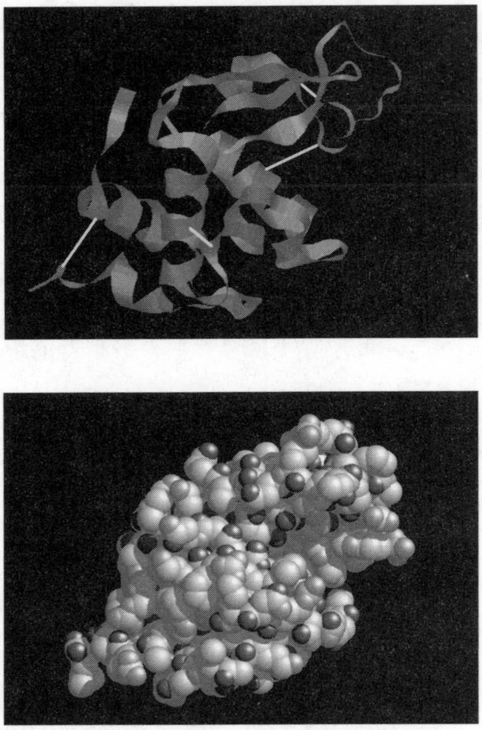

Figure 5.19 Conformation of a protein, the enzyme lysozyme, page 81

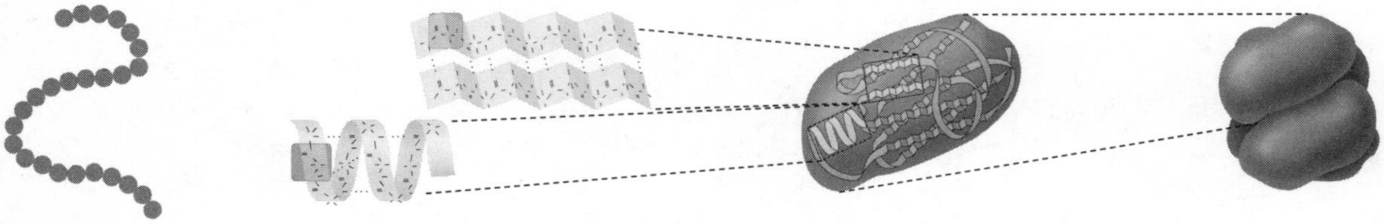

Figure 5.20 Levels of protein structure, page 82–83

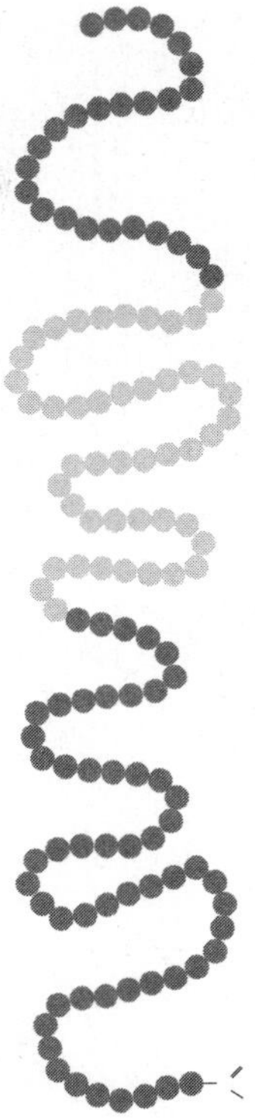

Figure 5.20 Levels of protein structure: primary structure, page 82

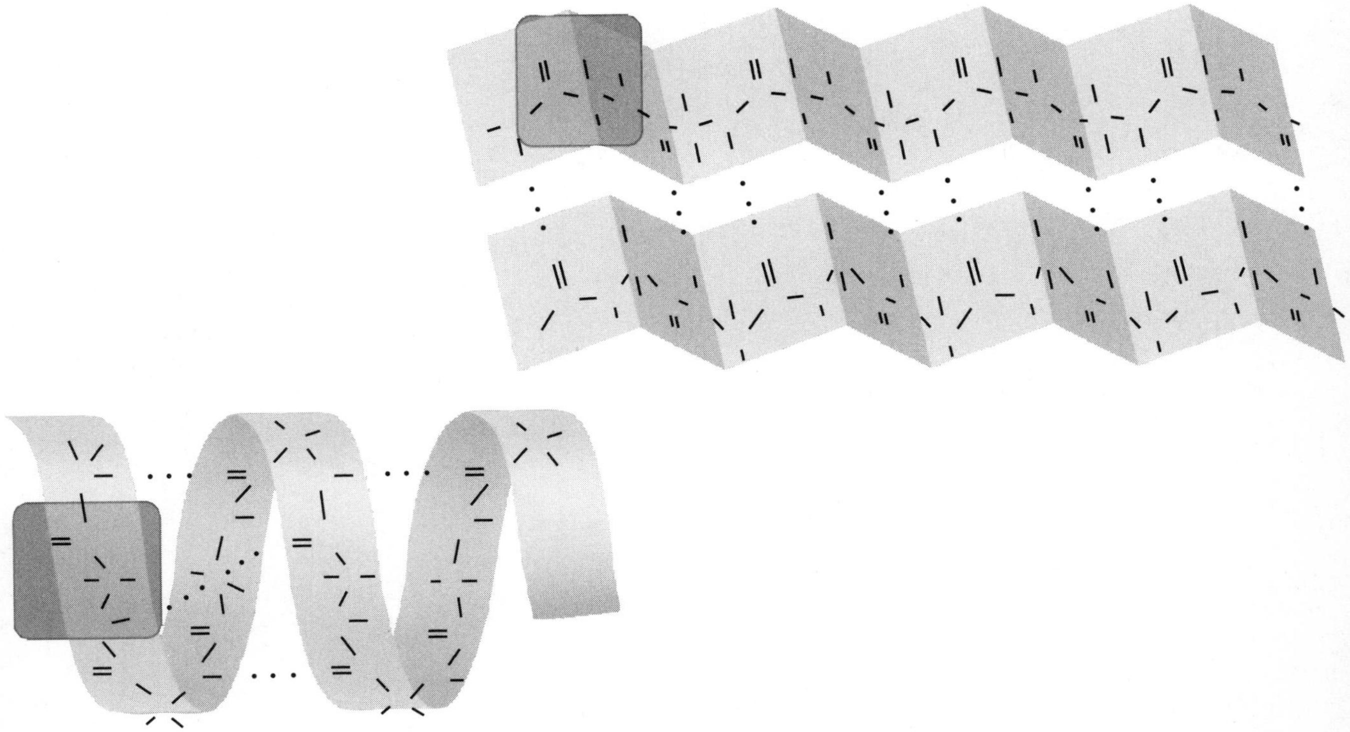

Figure 5.20 Levels of protein structure: secondary structure, page 82

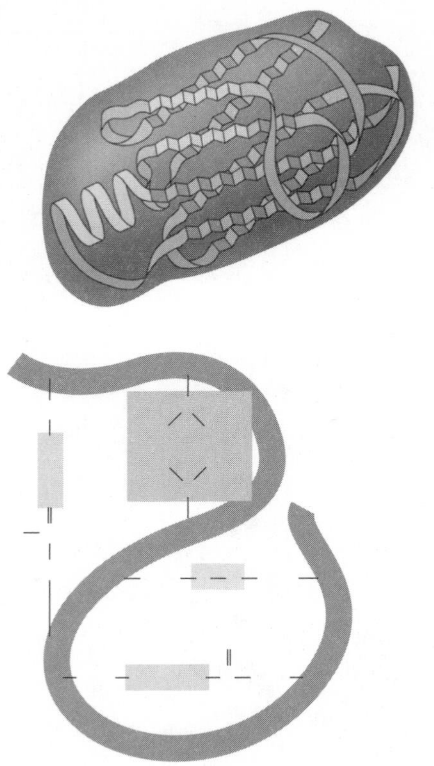

Figure 5.20 Levels of protein structure: tertiary structure, page 83

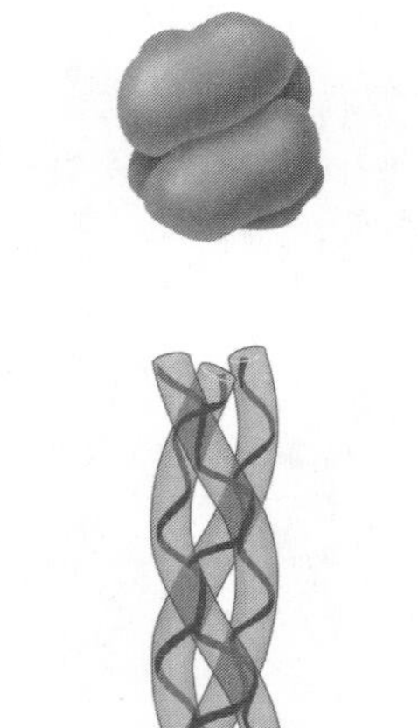

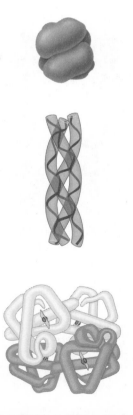

Figure 5.20 Levels of protein structure: quaternary structure, page 83

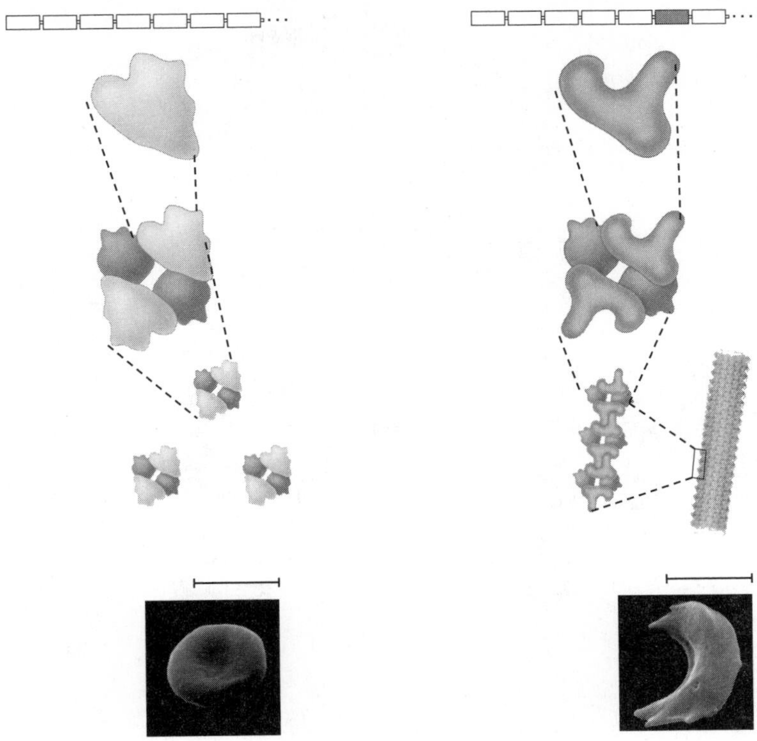

Figure 5.21 A single amino acid substitution in a protein causes sickle-cell disease, page 84

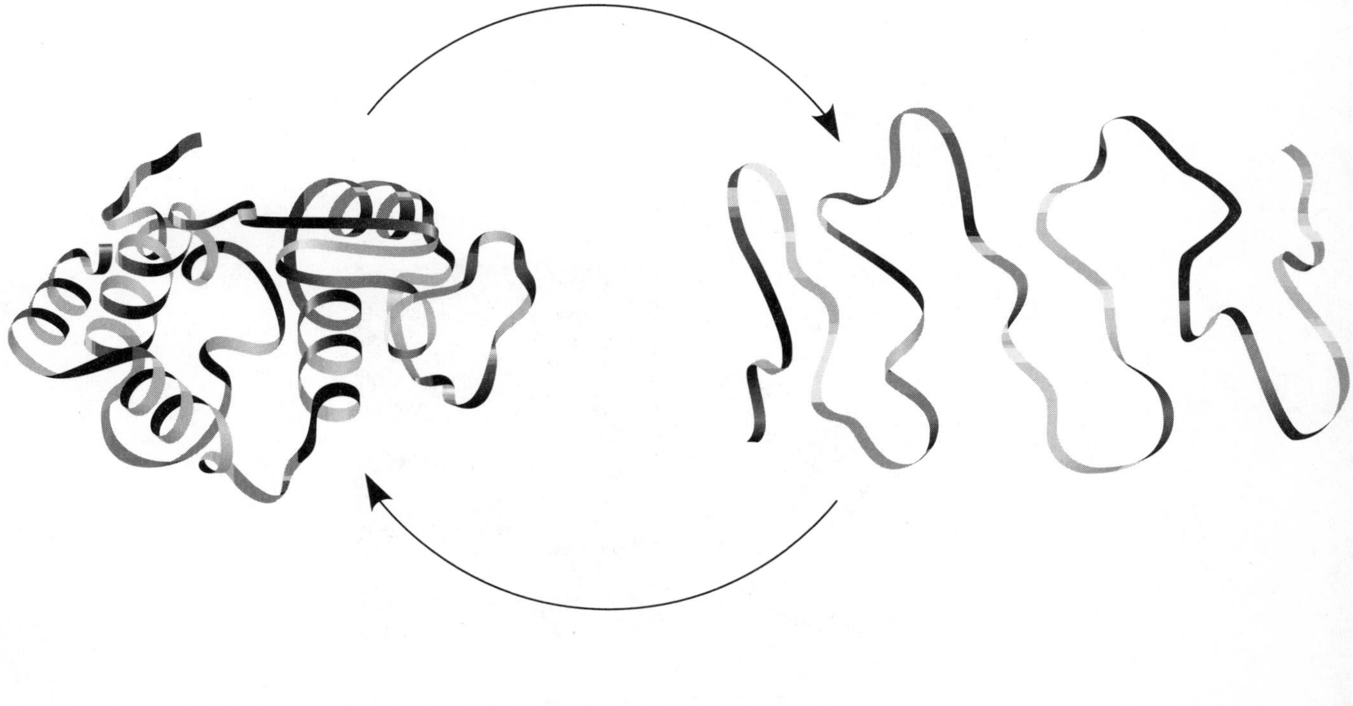

Figure 5.22 Denaturation and renaturation of a protein, page 85

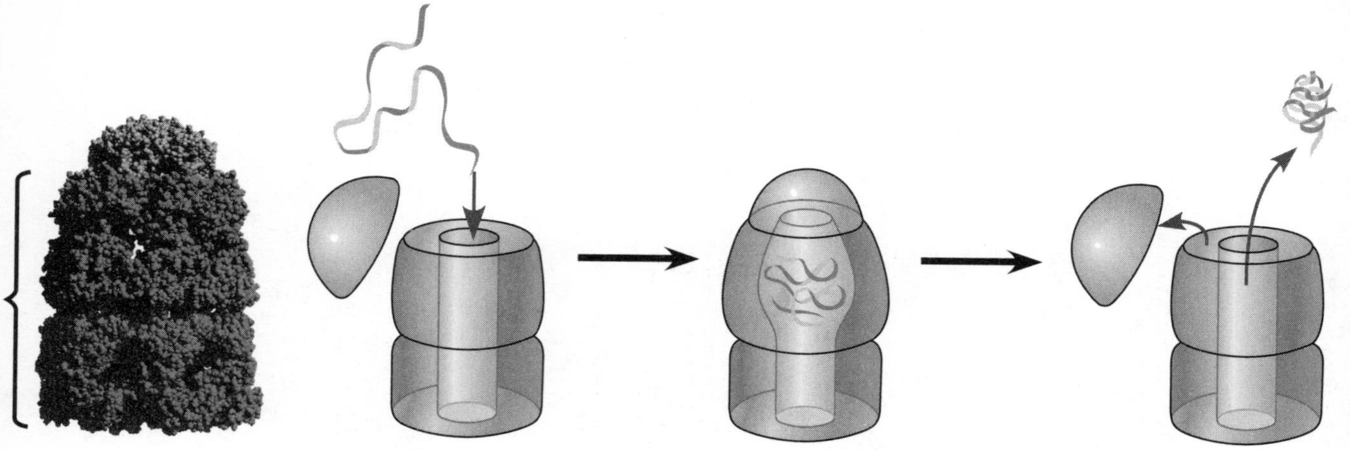

Figure 5.23 A chaperonin in action, page 85

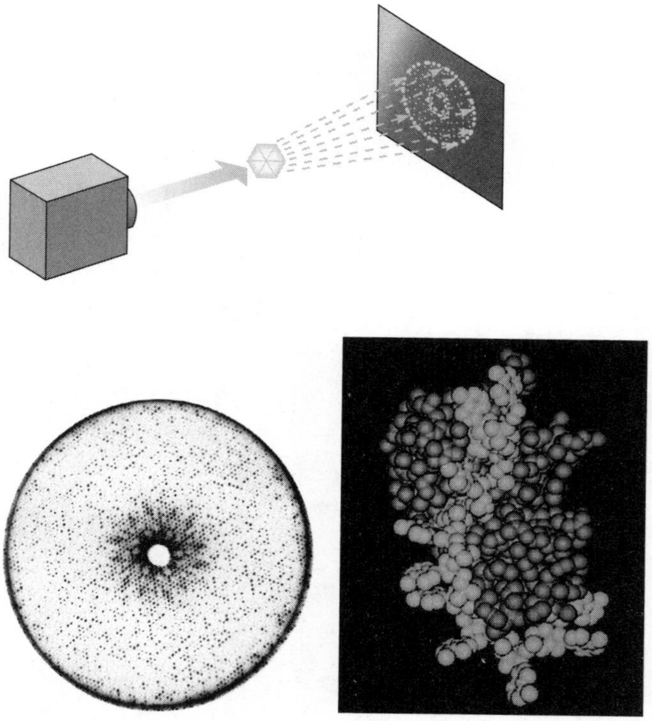

Figure 5.24 X-ray crystallography, page 86

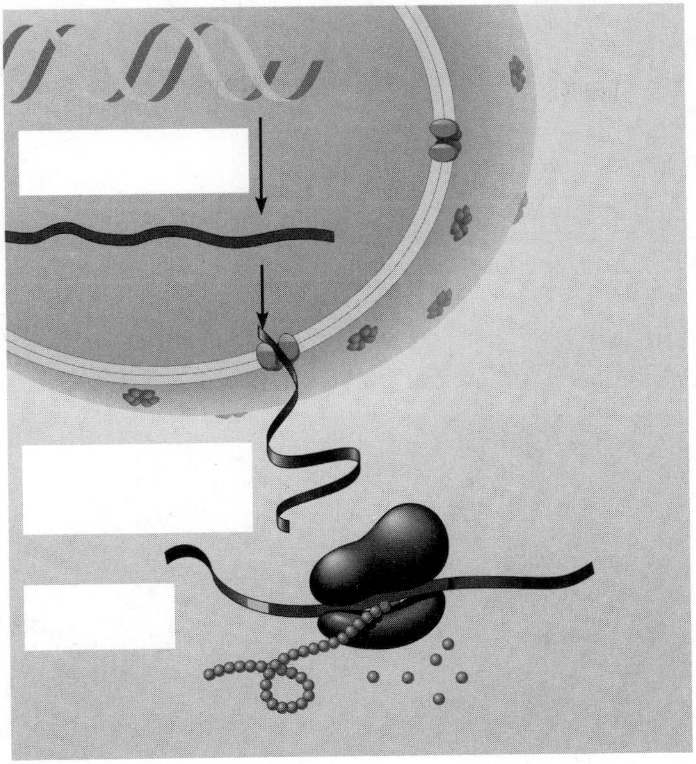

Figure 5.25 DNA → RNA → protein: A diagrammatic overview of information flow in a cell, page 86

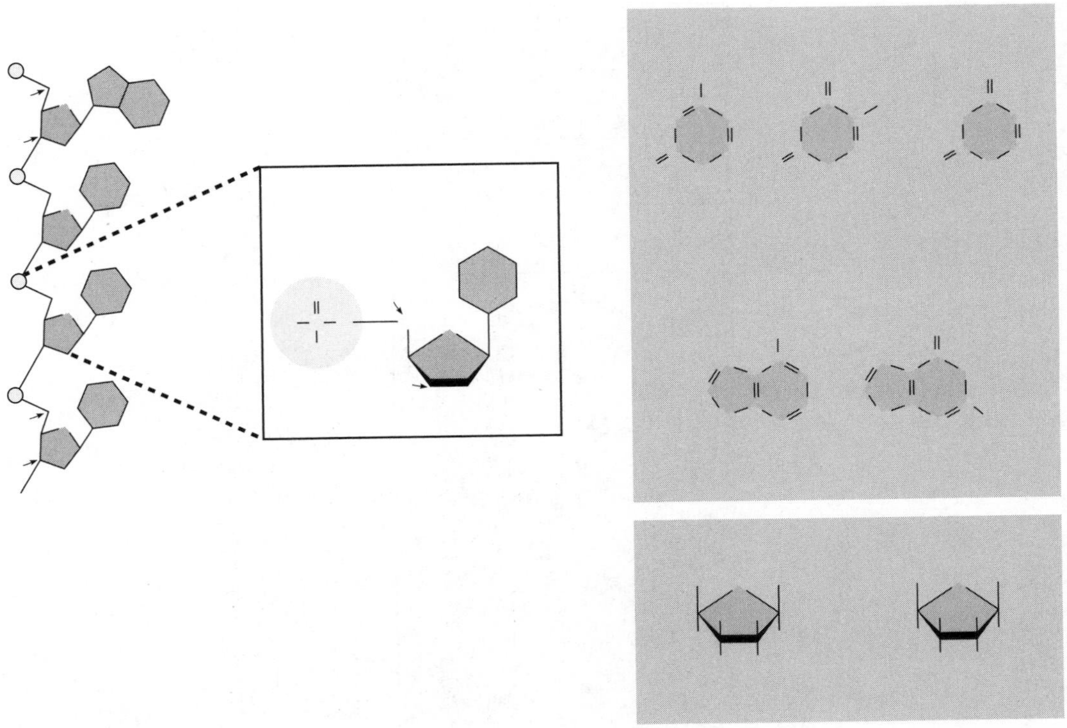

Figure 5.26 The components of nucleic acids, page 87

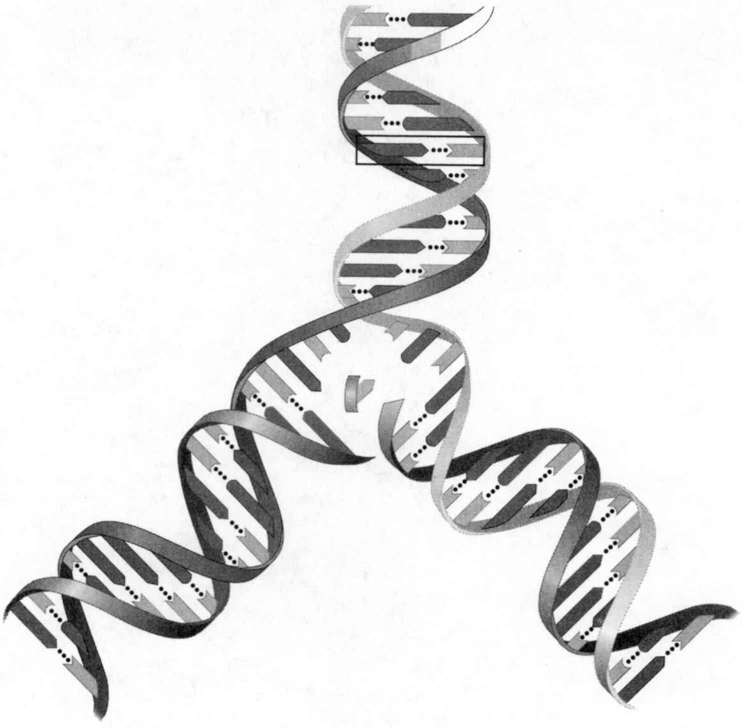

Figure 5.27 The DNA double helix and its replication, page 88

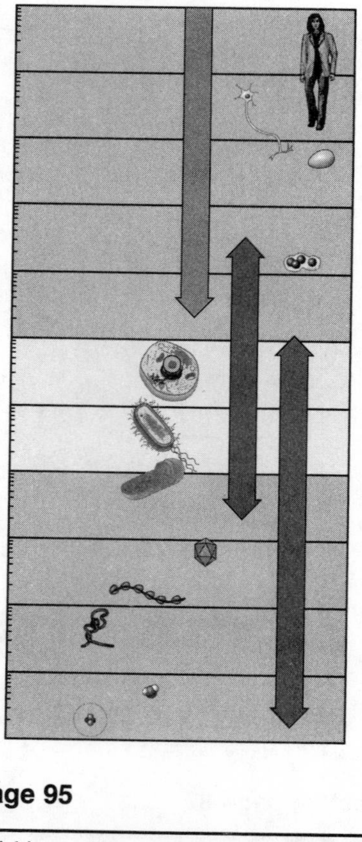

Figure 6.2 The size range of cells, page 95

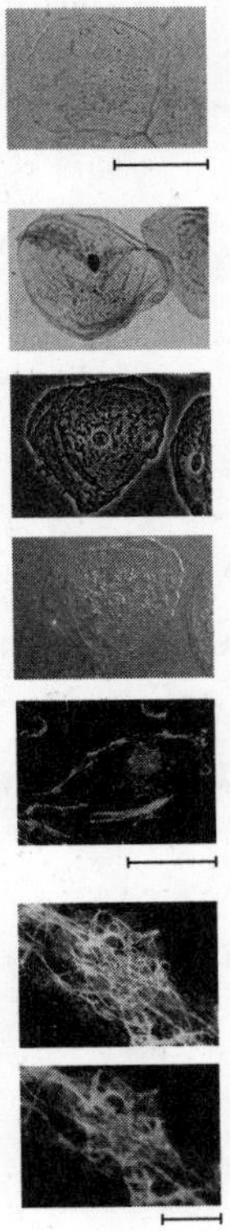

Figure 6.3 Light Microscopy, page 96

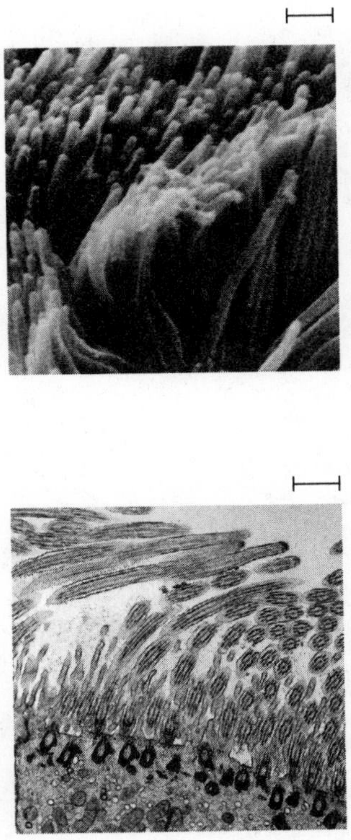

Figure 6.4 Electron Microscopy, page 96

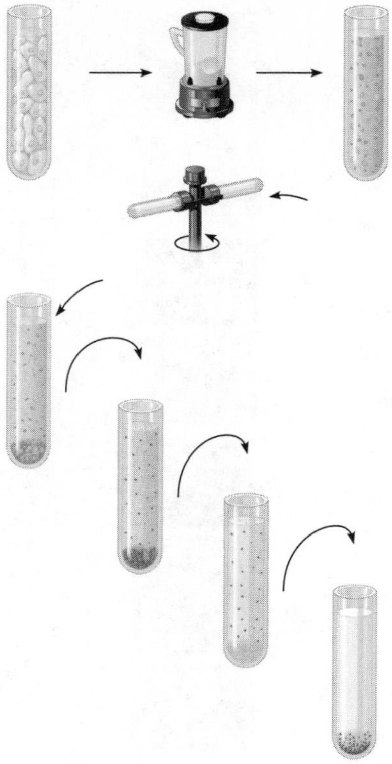

Figure 6.5 Cell Fractionation, page 97

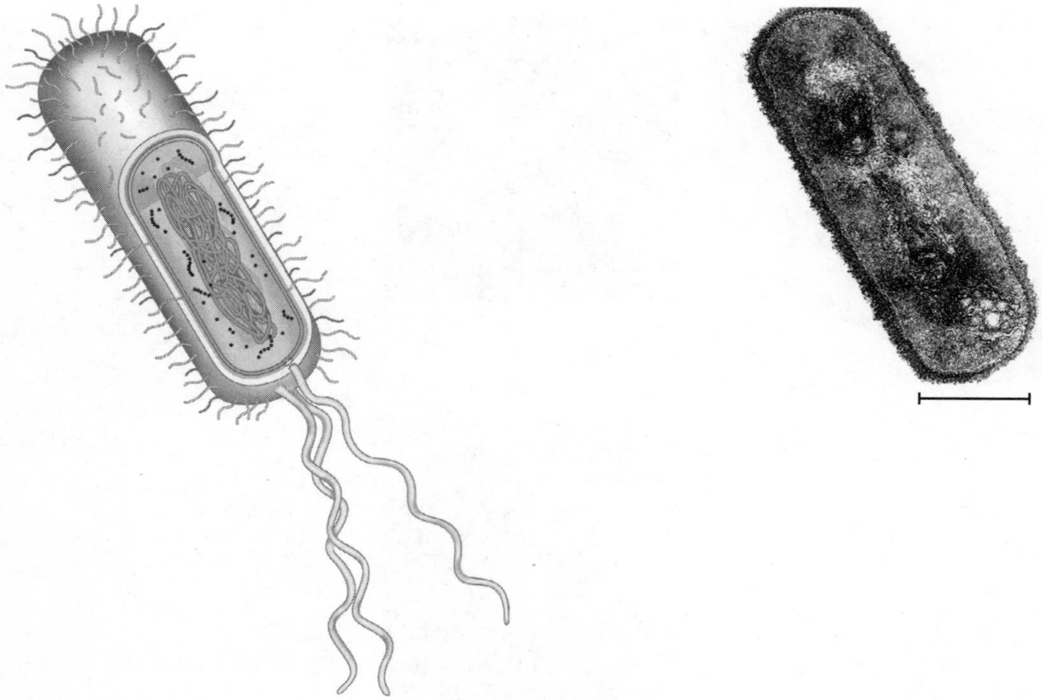

Figure 6.6 A prokaryotic cell, page 98

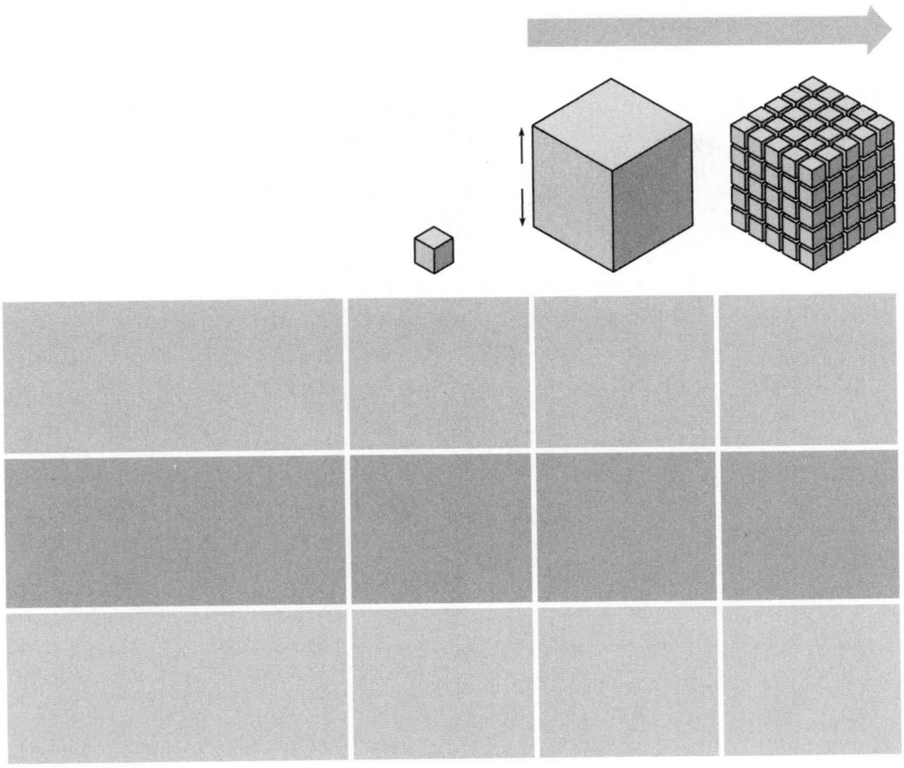

Figure 6.7 Geometric relationships between surface area and volume, page 99

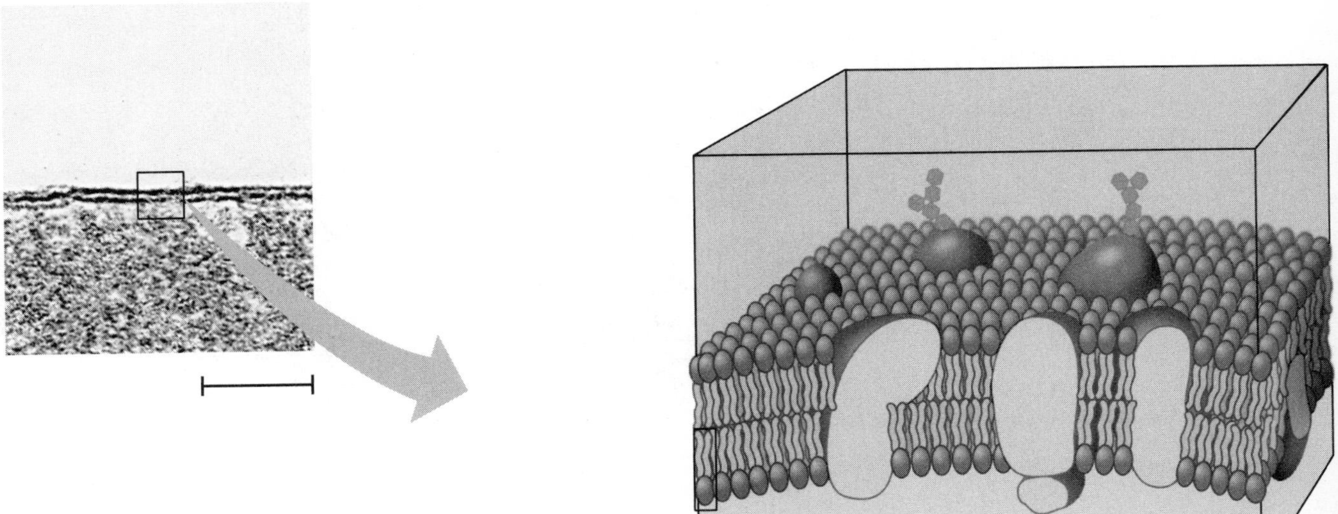

Figure 6.8 The plasma membrane, page 99

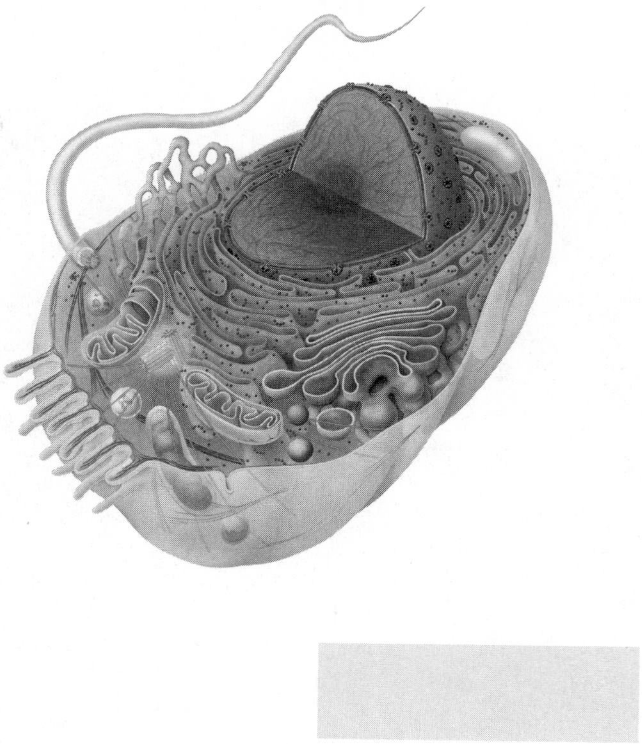

Figure 6.9 Animal and plant cells: the animal cell, page 100

Figure 6.9 Animal and plant cells: the plant cell, page 101

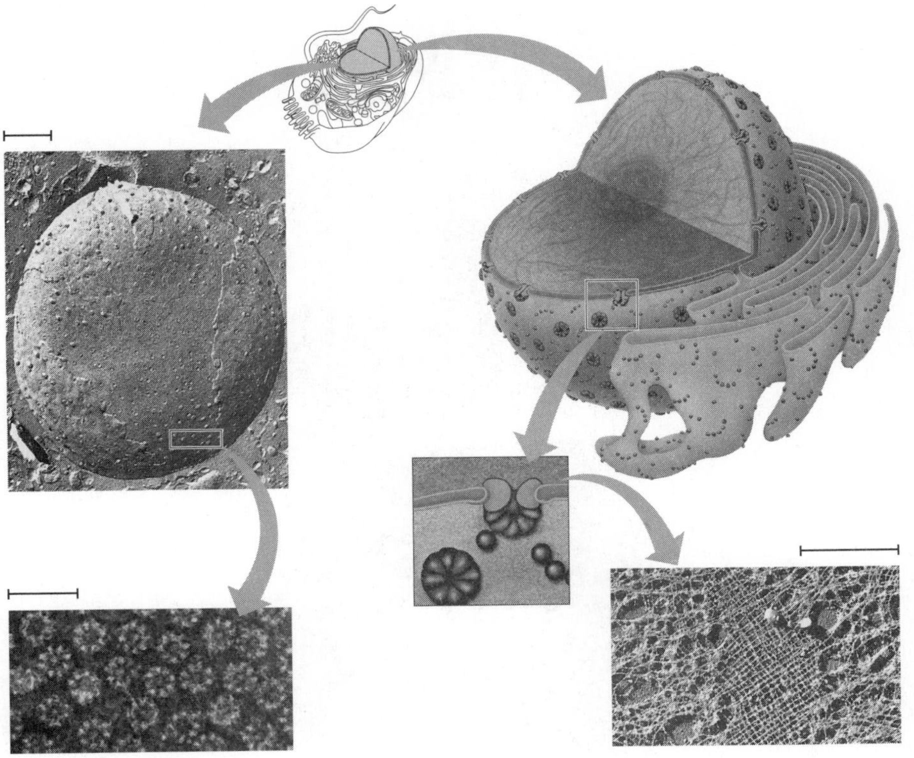

Figure 6.10 The nucleus and its envelope, page 103

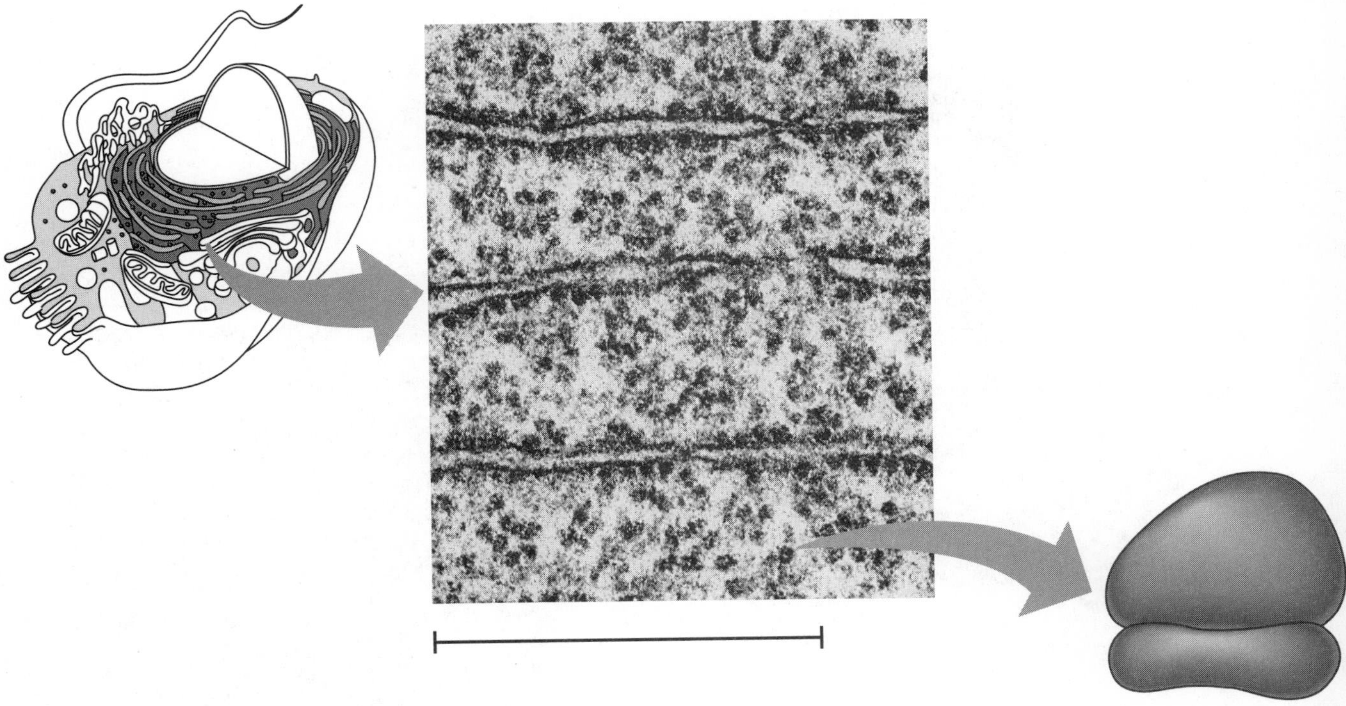

Figure 6.11 Ribosomes, page 103

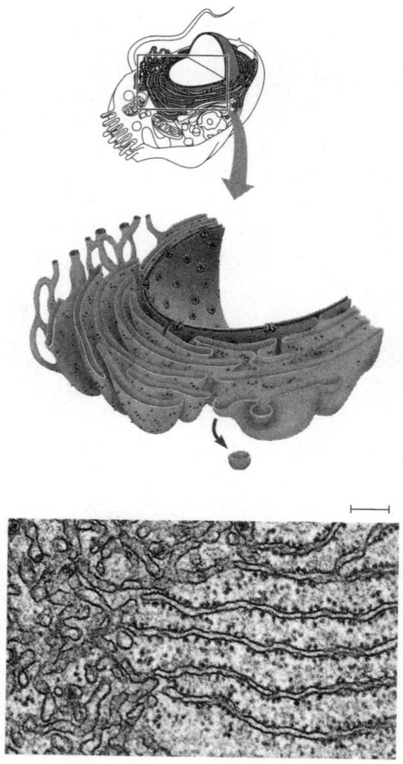

Figure 6.12 Endoplasmic reticulum (ER), page 105

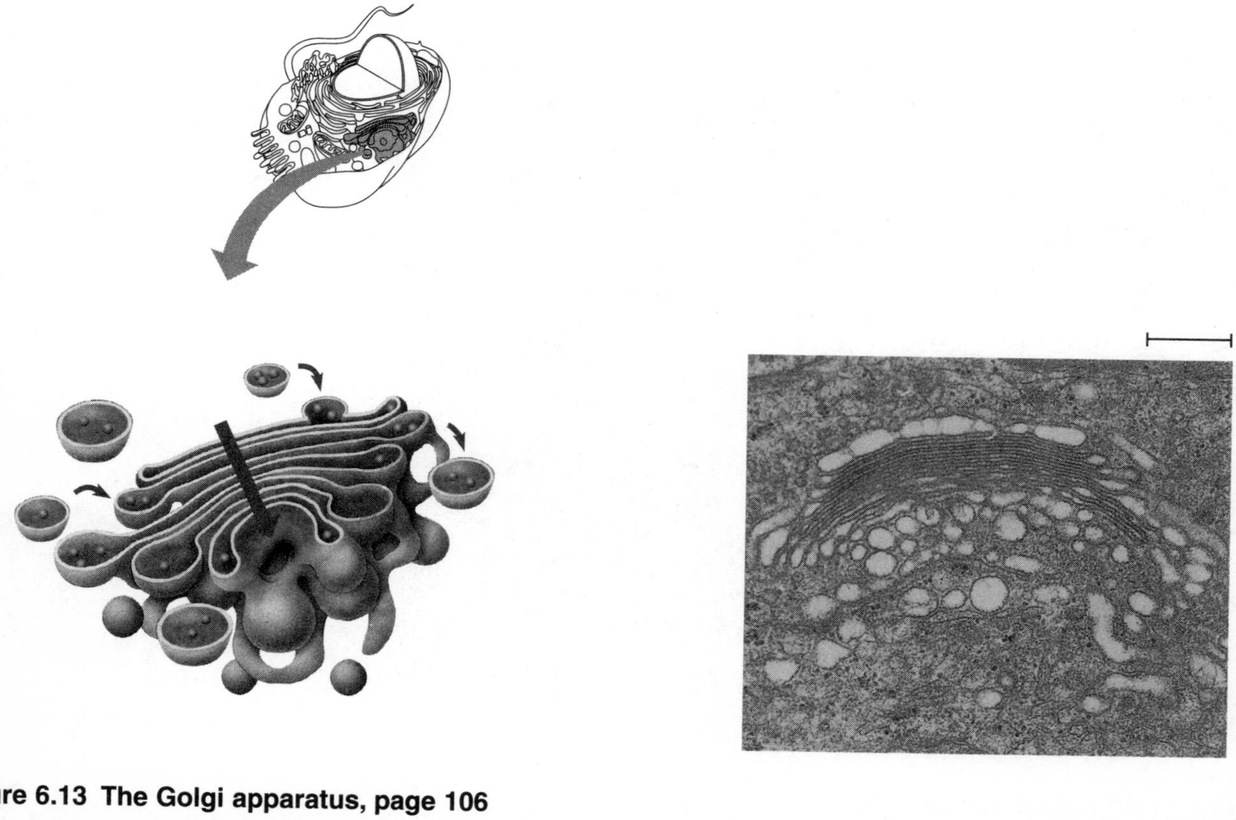

Figure 6.13 The Golgi apparatus, page 106

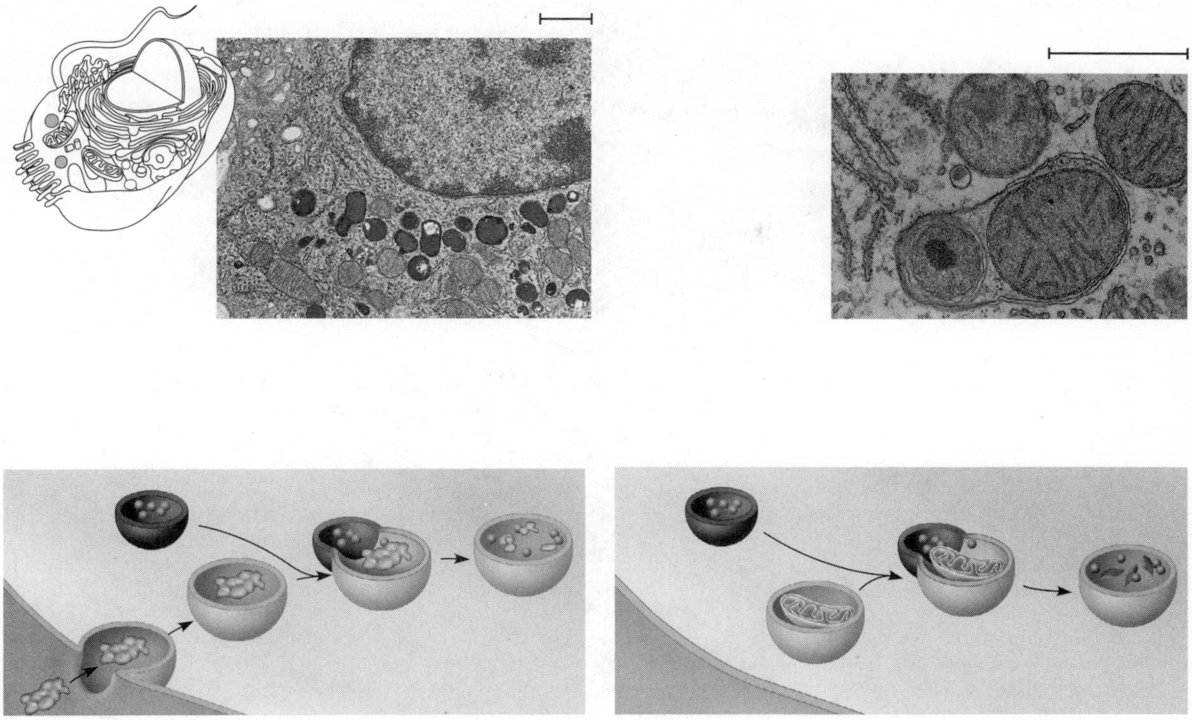

Figure 6.14 Lysosomes, page 107

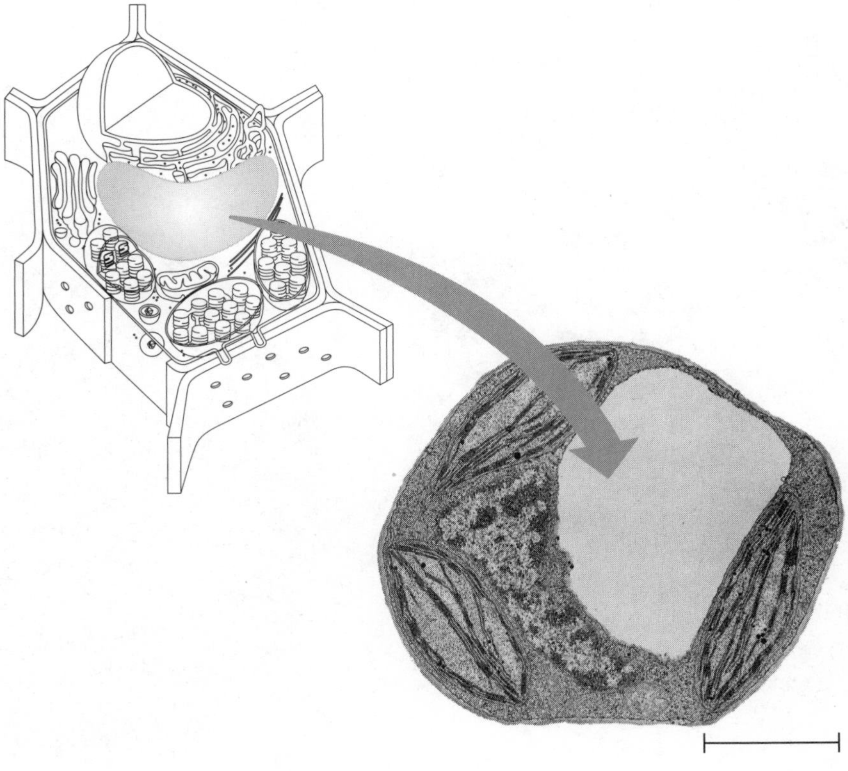

Figure 6.15 The plant cell vacuole, page 108

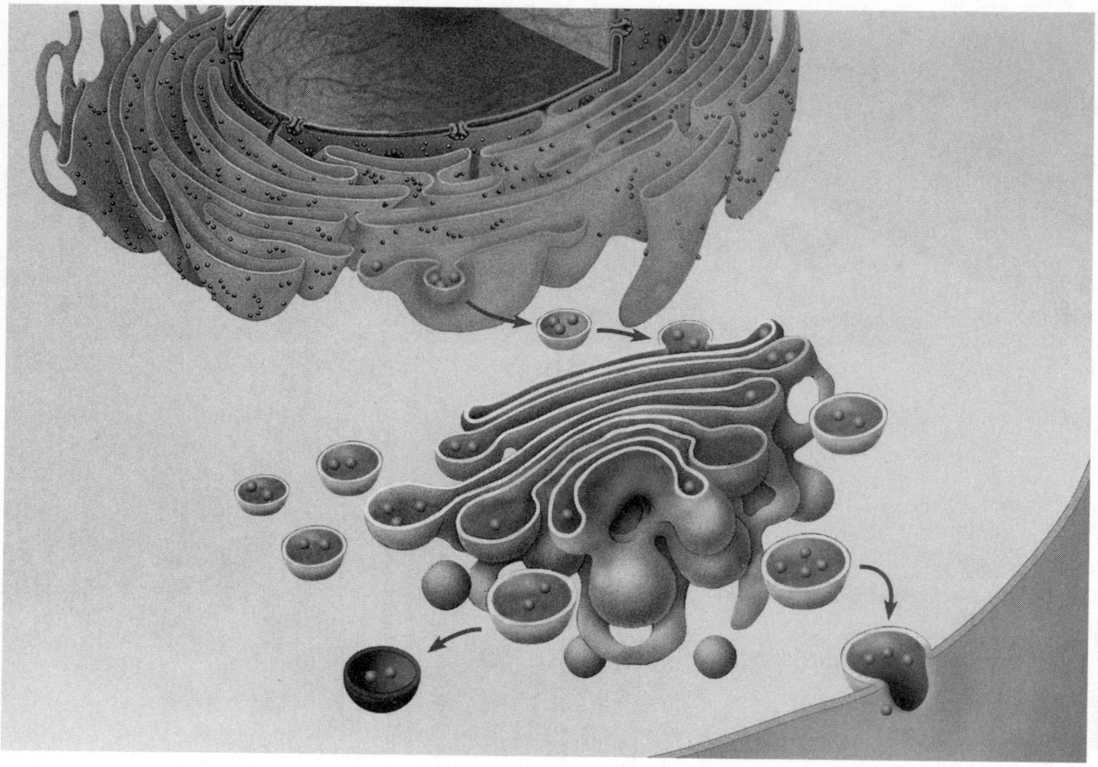

Figure 6.16 Review: relationships among organelles of the endomembrane system, page 109

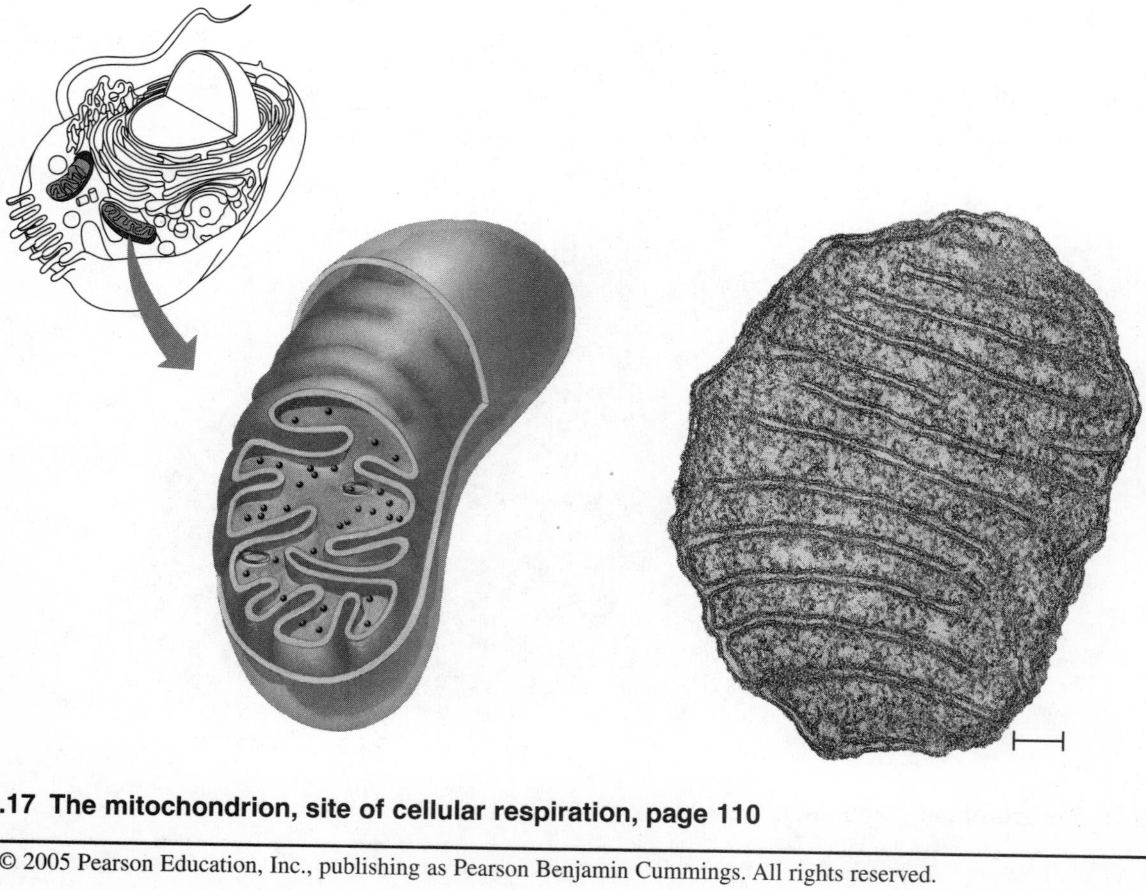

Figure 6.17 The mitochondrion, site of cellular respiration, page 110

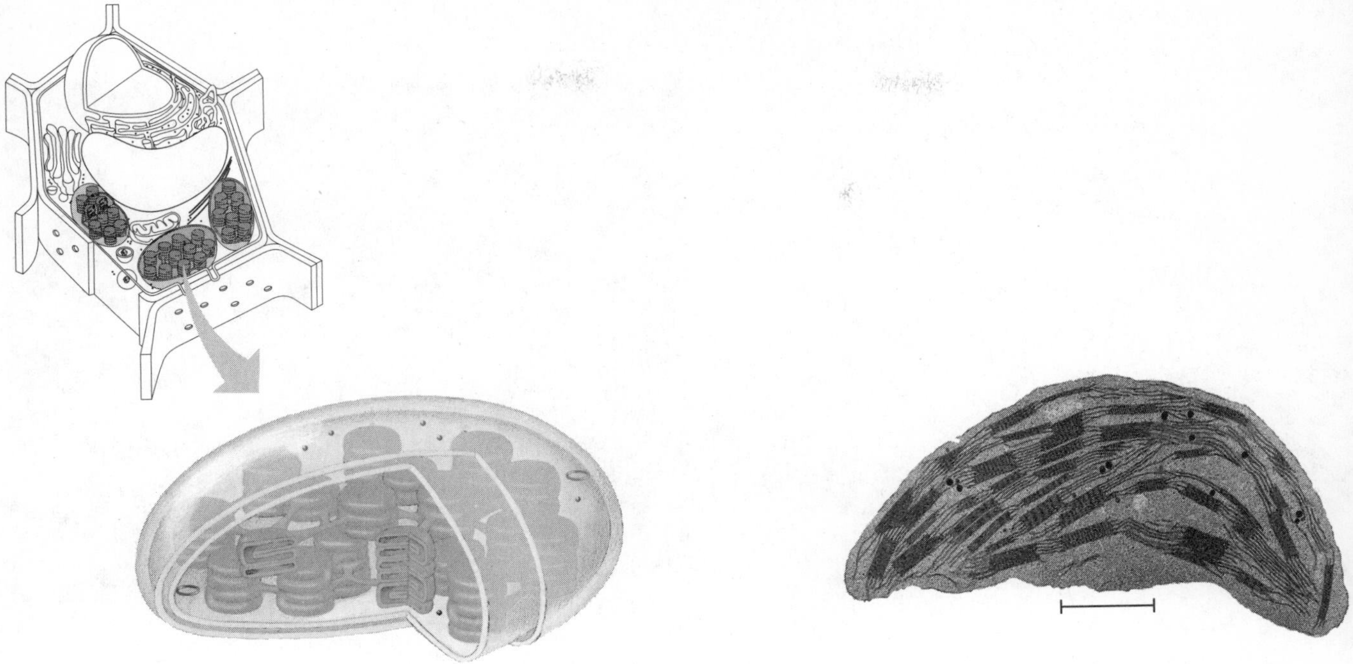

Figure 6.18 The chloroplast, site of photosynthesis, page 111

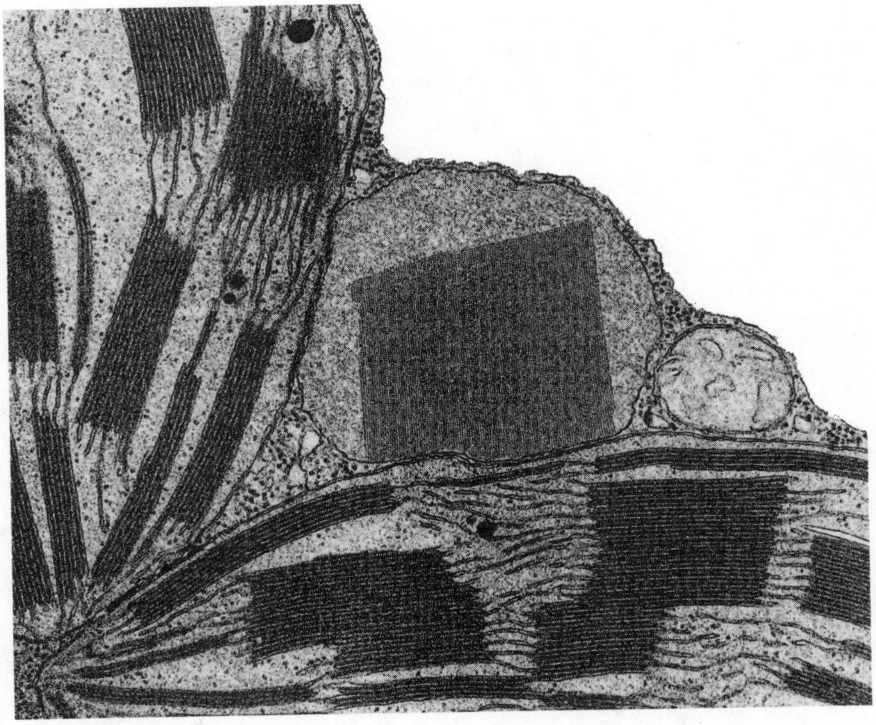

Figure 6.19 Peroxisomes, page 111

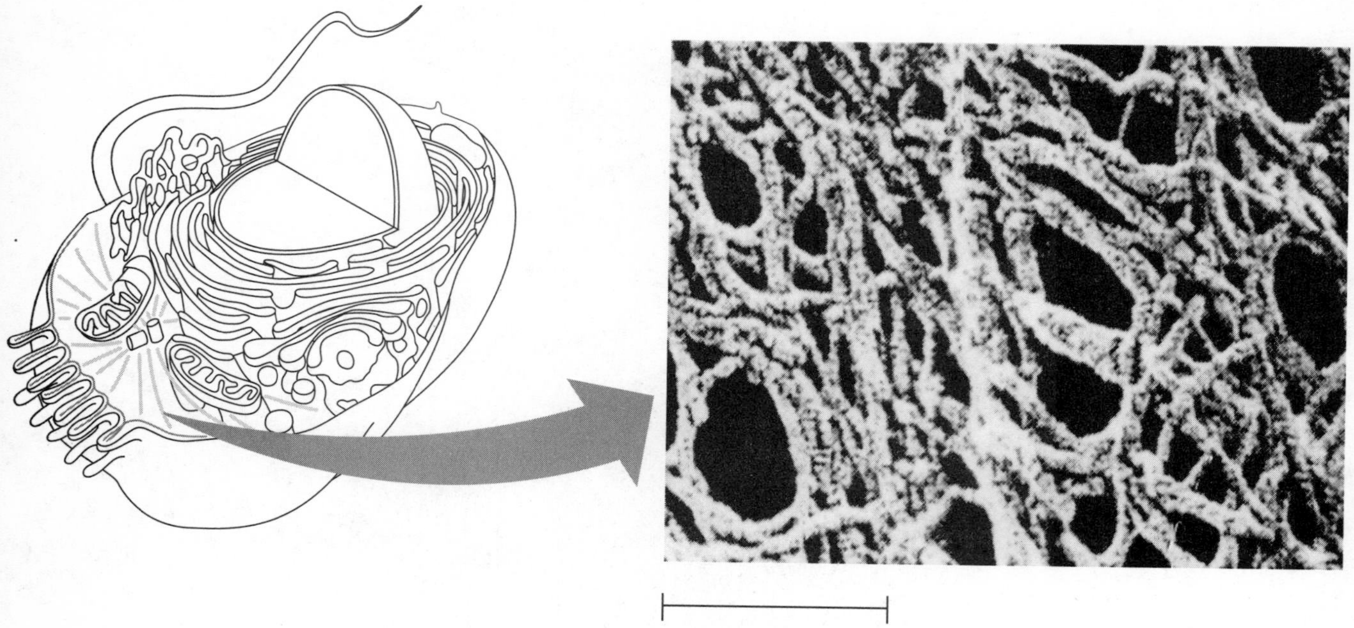

Figure 6.20 The cytoskeleton, page 112

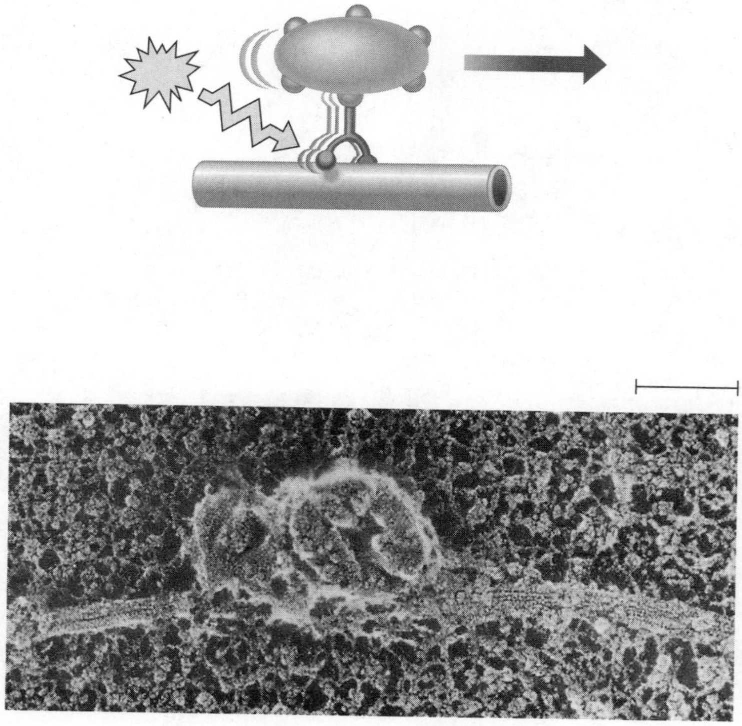

Figure 6.21 Motor proteins and the cytoskeleton, page 112

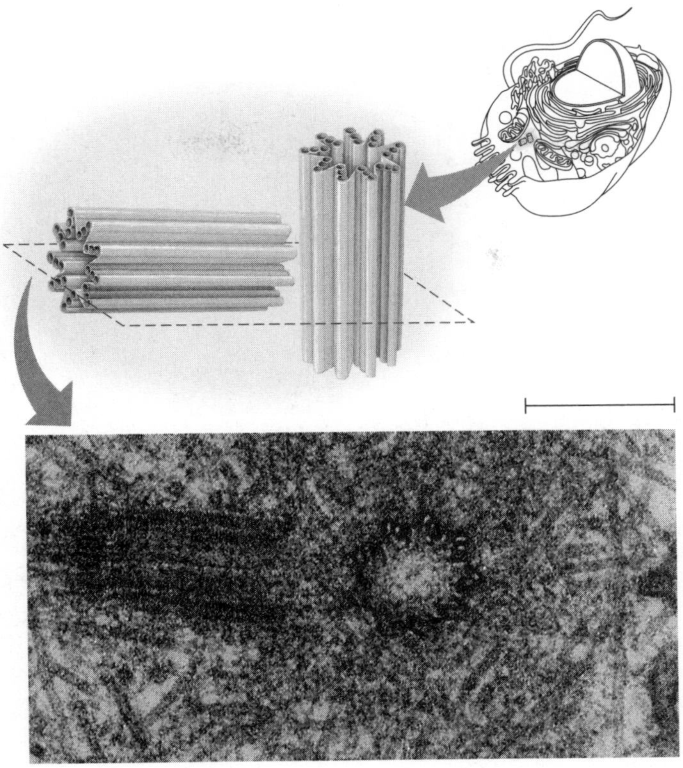

Figure 6.22 Centrosome containing a pair of centrioles, page 114

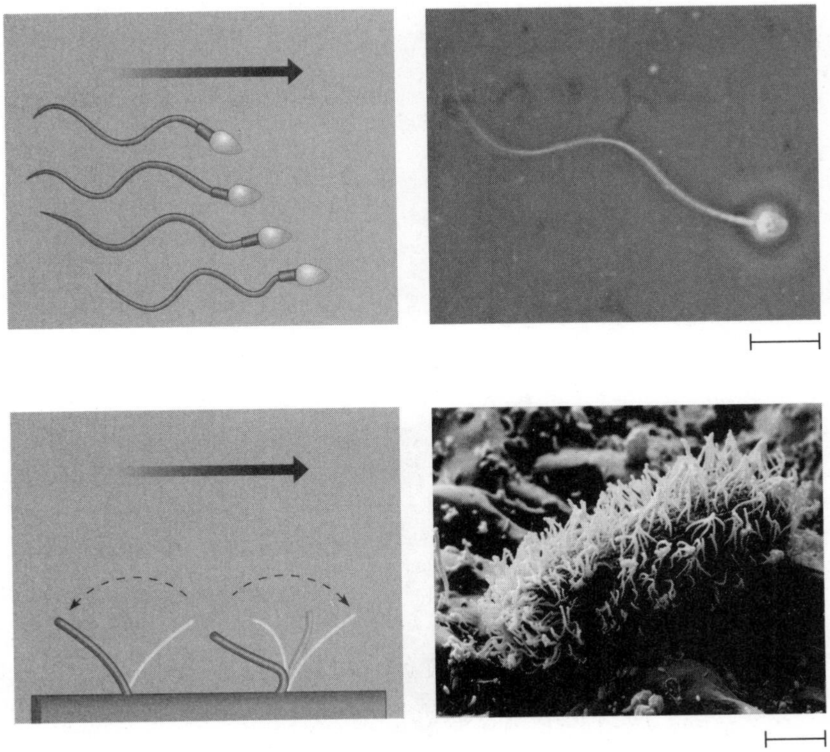

Figure 6.23 A comparison of the beating of flagella and cilia, page 115

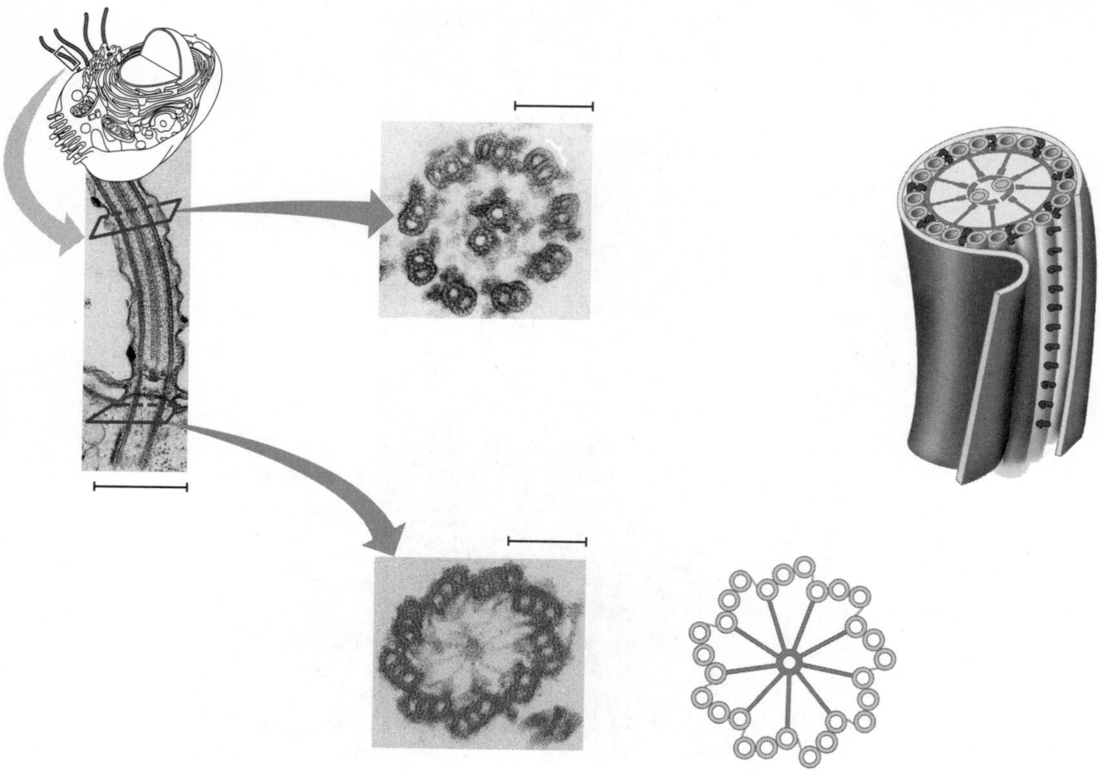

Figure 6.24 Ultrastructure of a eukaryotic flagellum or cilium, page 115

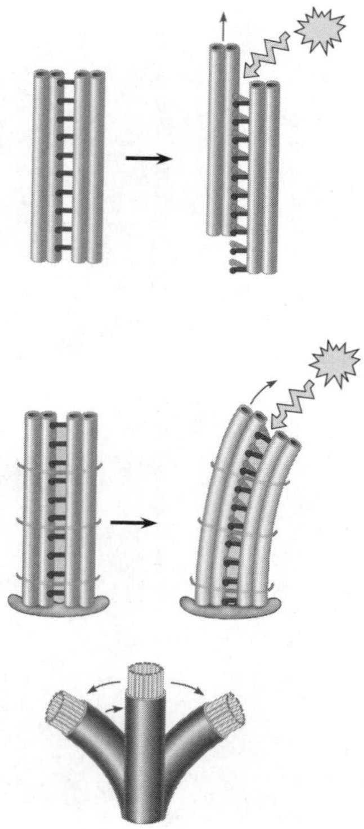

Figure 6.25 How dynein "walking" moves flagella and cilia, page 116

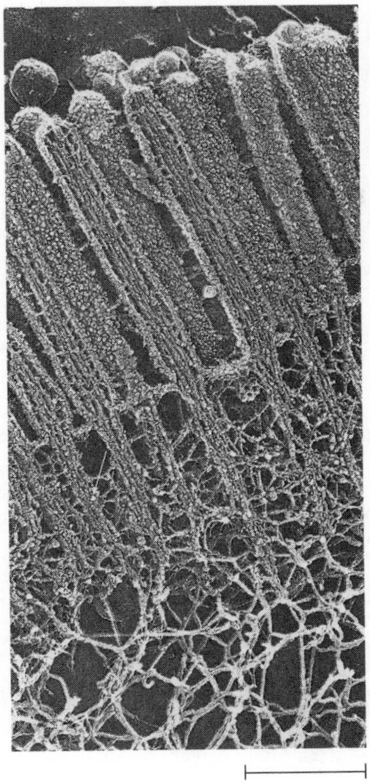

Figure 6.26 A structural role of microfilaments, page 117

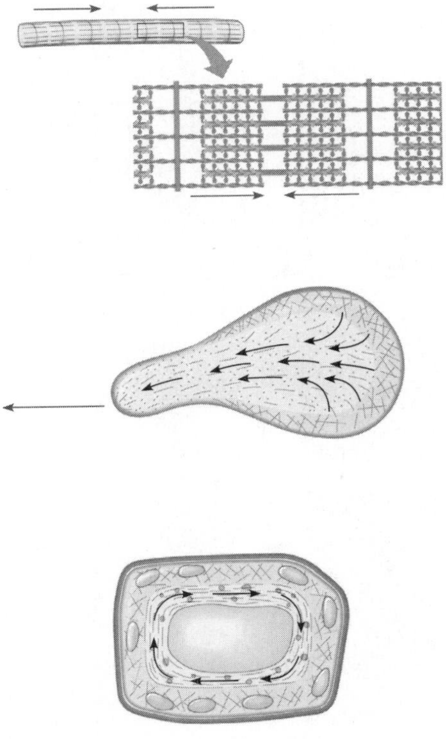

Figure 6.27 Microfilaments and motility, page 117

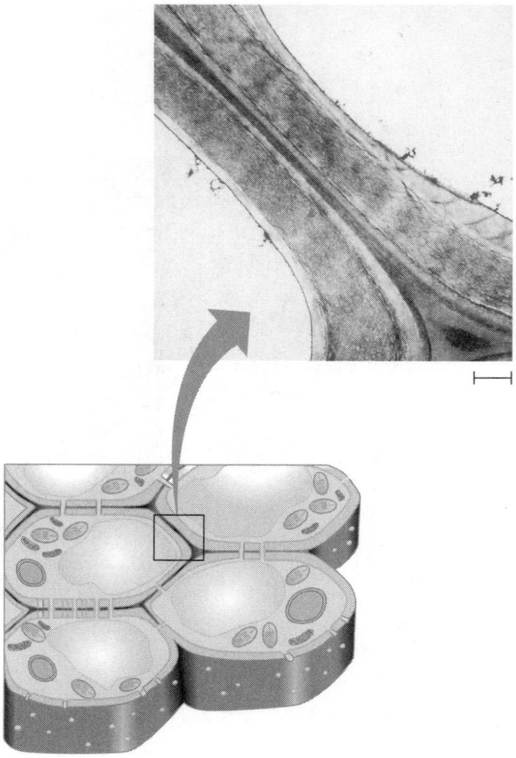

Figure 6.28 Plant cell walls, page 119

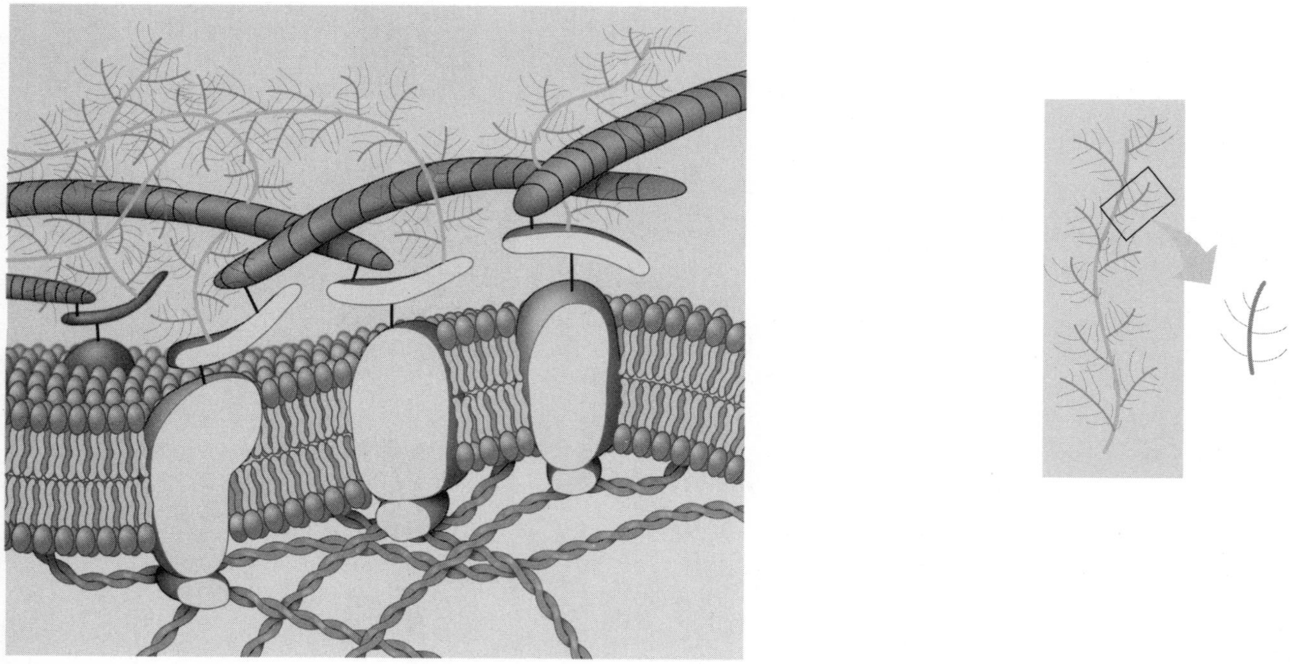

Figure 6.29 Extracellular matrix (ECM) of an animal cell, page 119

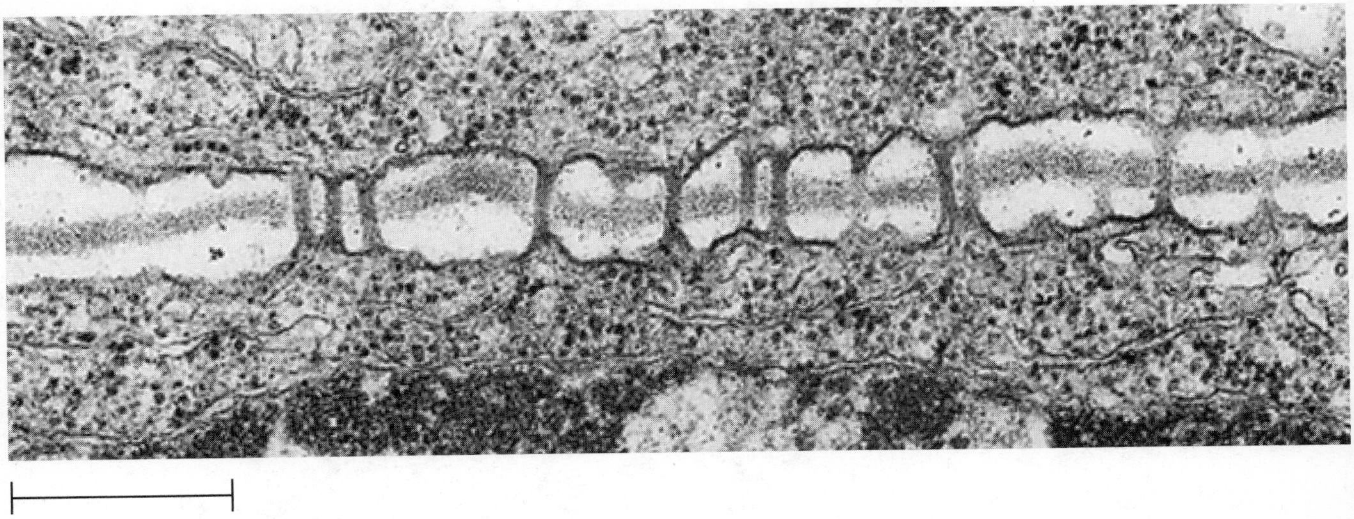

Figure 6.30 Plasmodesmata between plant cells, page 120

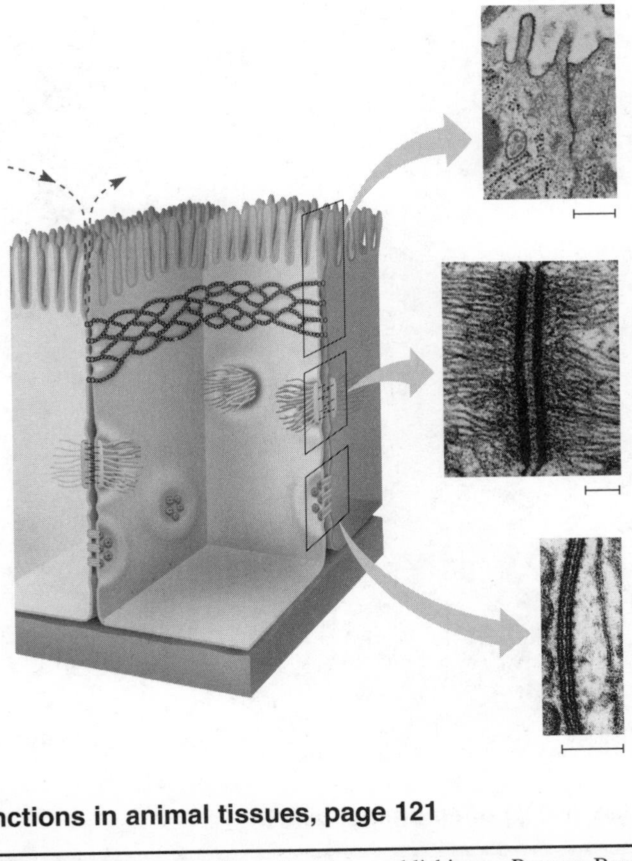

Figure 6.31 Intercellular junctions in animal tissues, page 121

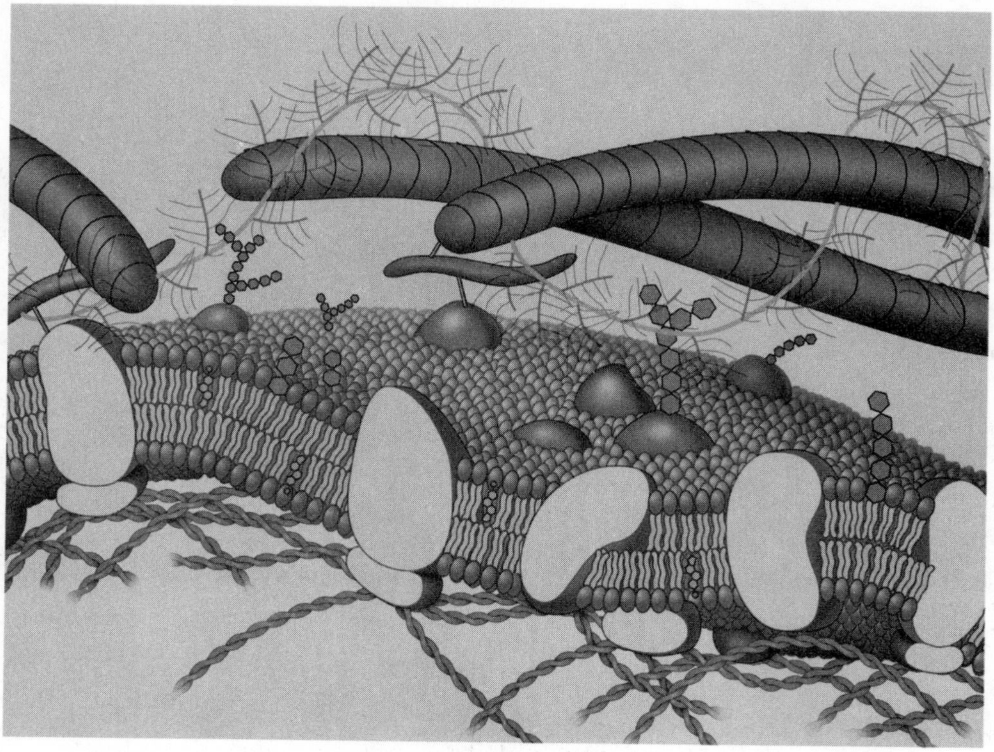

Figure 7.1 The plasma membrane, page 124

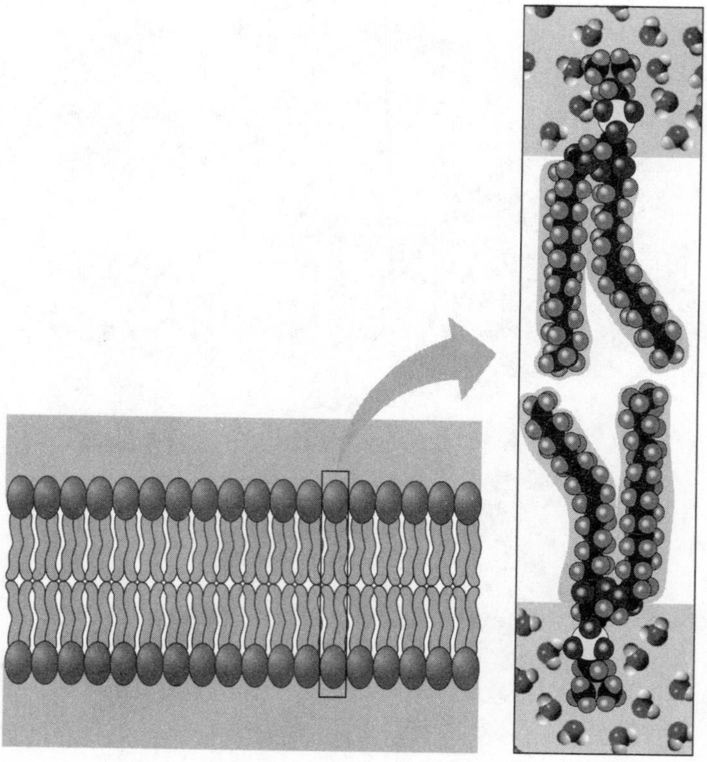

Figure 7.2 Phospholipid bilayer (cross section), page 125

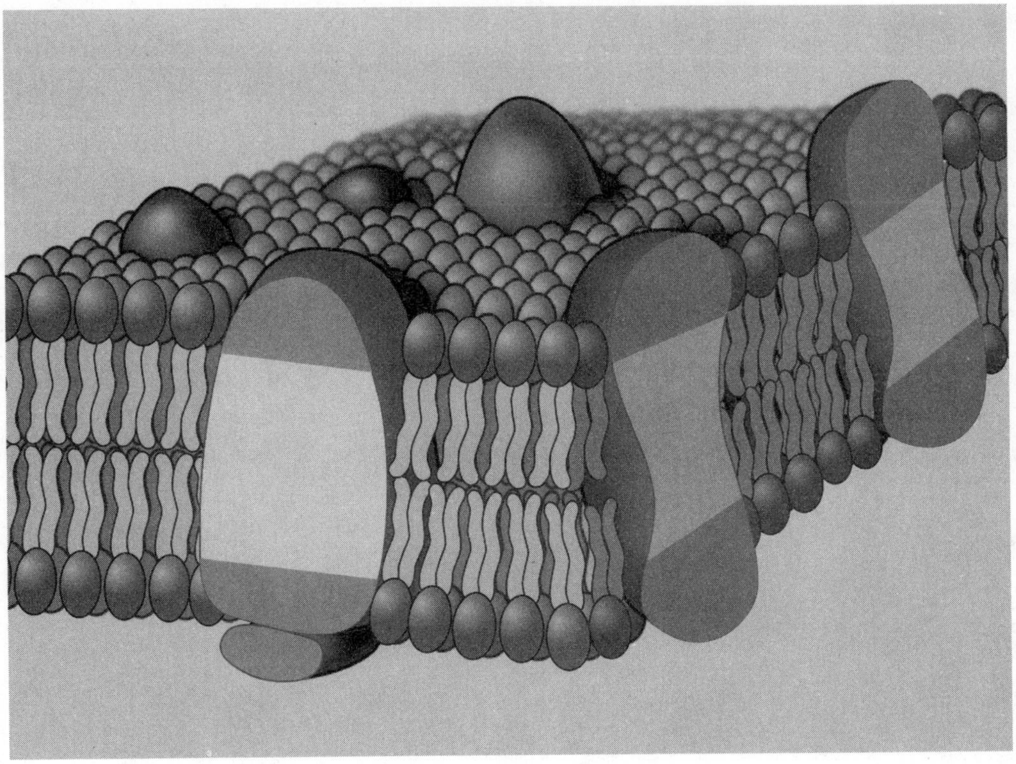

Figure 7.3 The fluid mosaic model for membranes, page 125

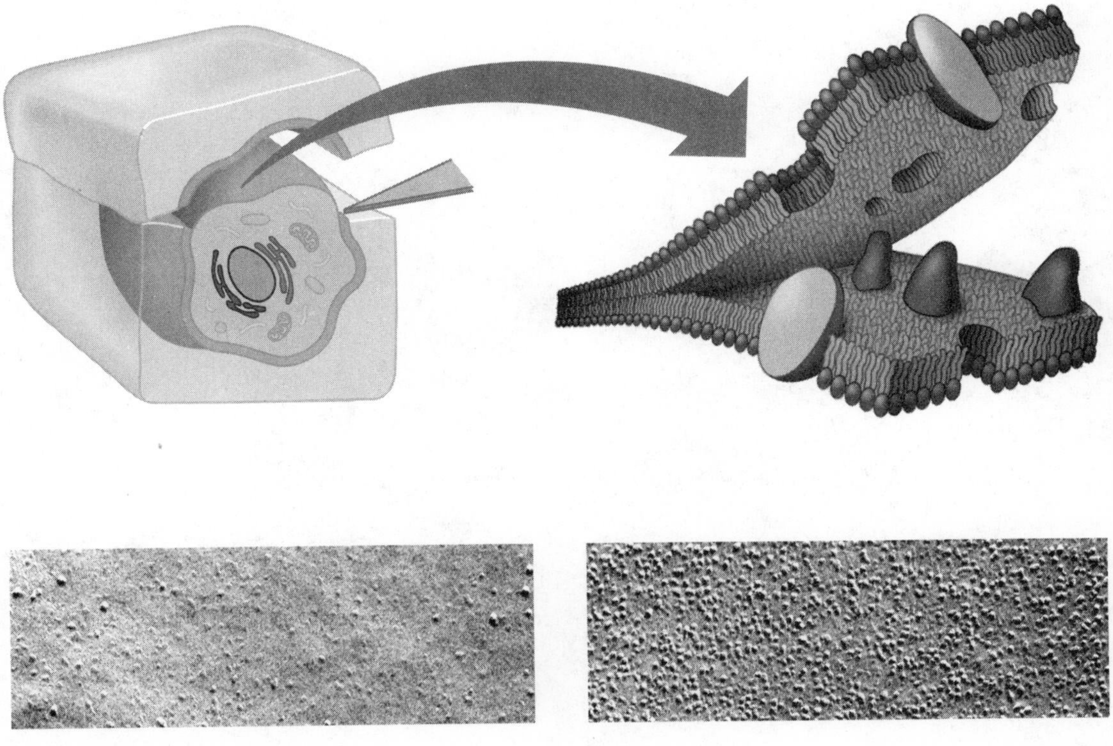

Figure 7.4 Freeze-fracture, page 126

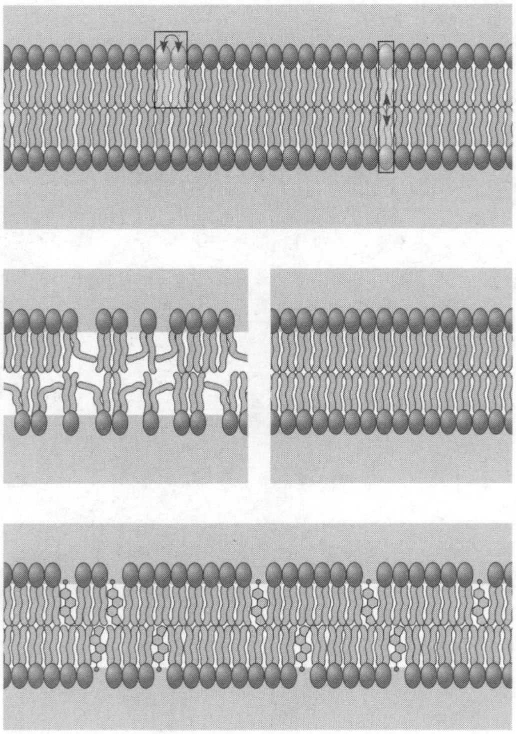

Figure 7.5 The fluidity of membranes, page 126

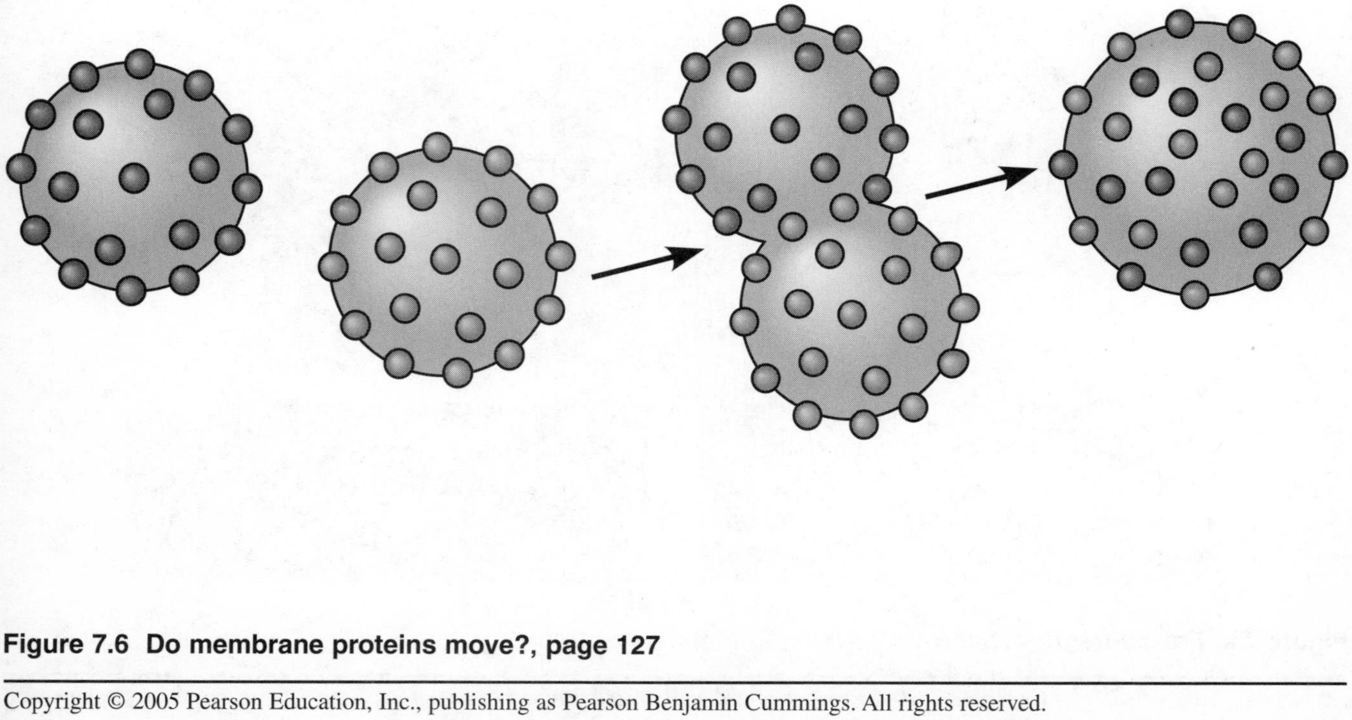

Figure 7.6 Do membrane proteins move?, page 127

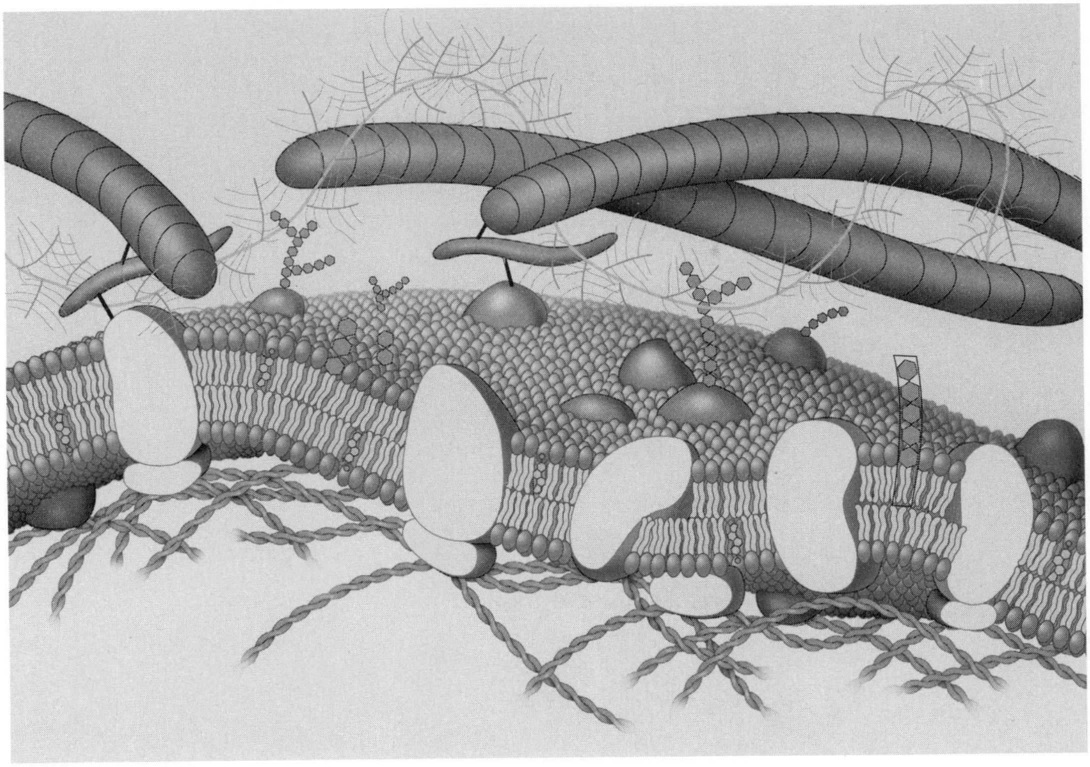

Figure 7.7 The detailed structure of an animal cell's plasma membrane, in cross section, page 127

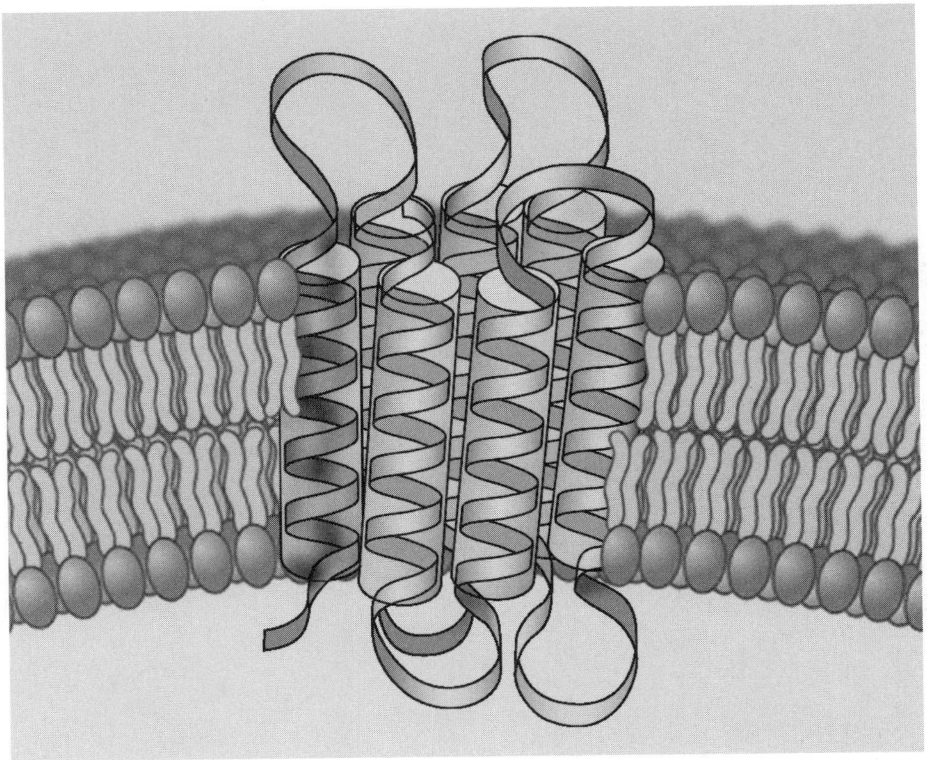

Figure 7.8 The structure of a transmembrane protein, page 128

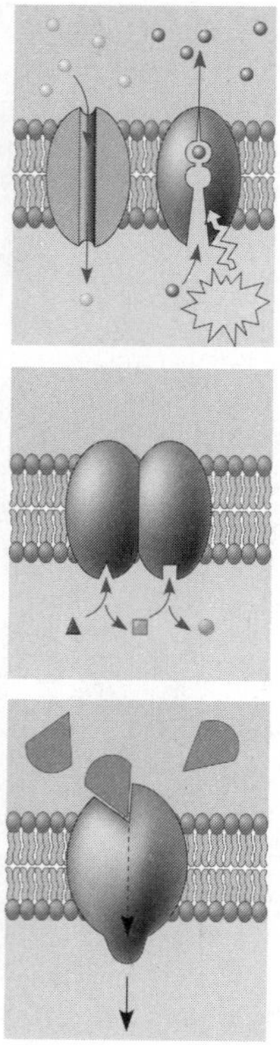

Figure 7.9 Some functions of membrane proteins, part 1, page 128

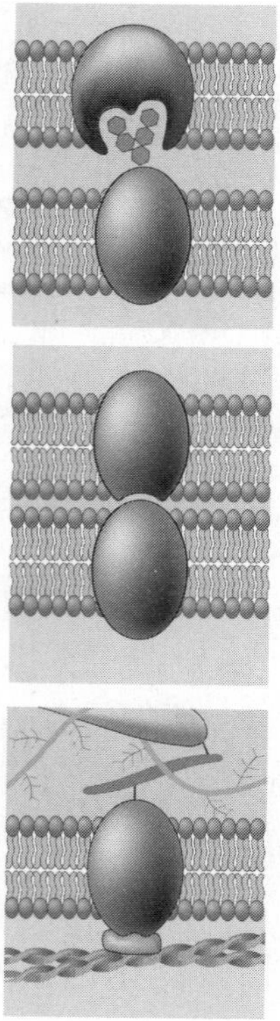

Figure 7.9 Some functions of membrane proteins, part 2, page 128

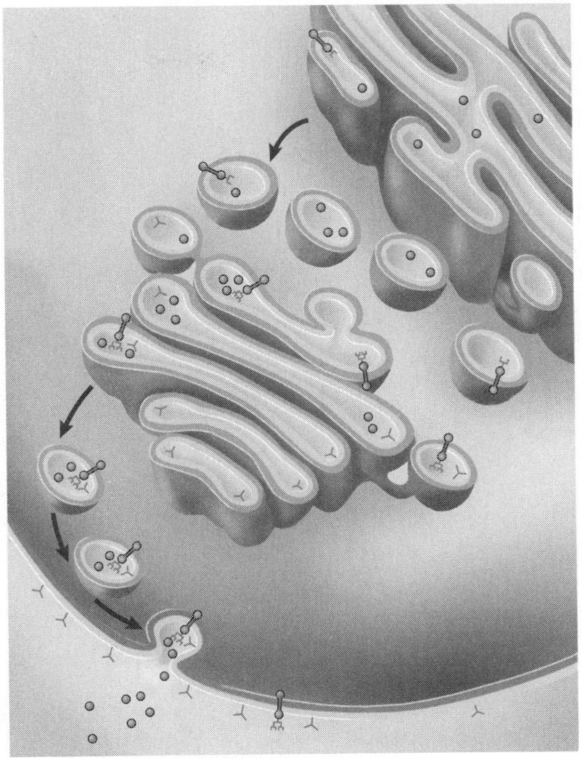

Figure 7.10 Synthesis of membrane components and their orientation on the resulting membrane, page 129

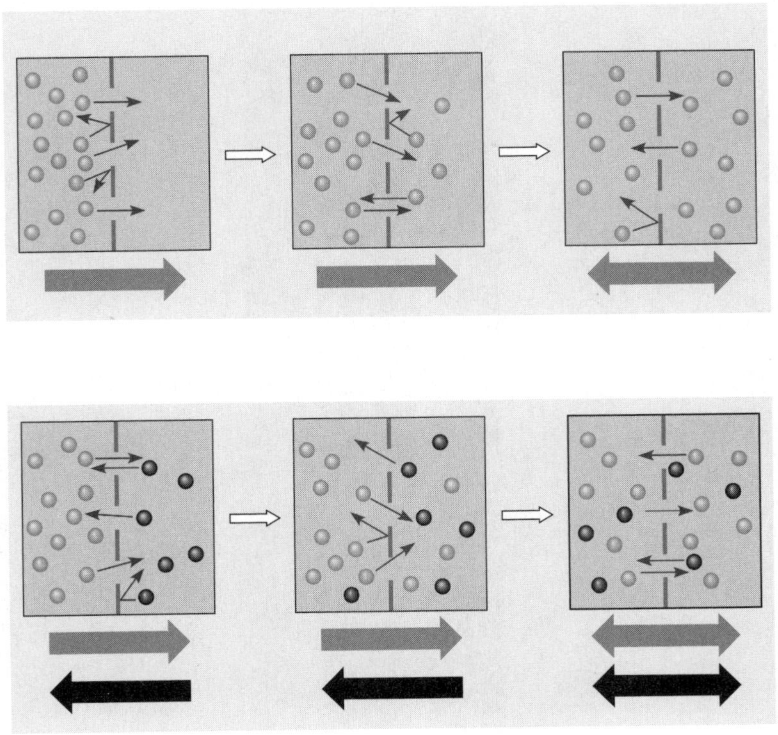

Figure 7.11 The diffusion of solutes across a membrane, page 131

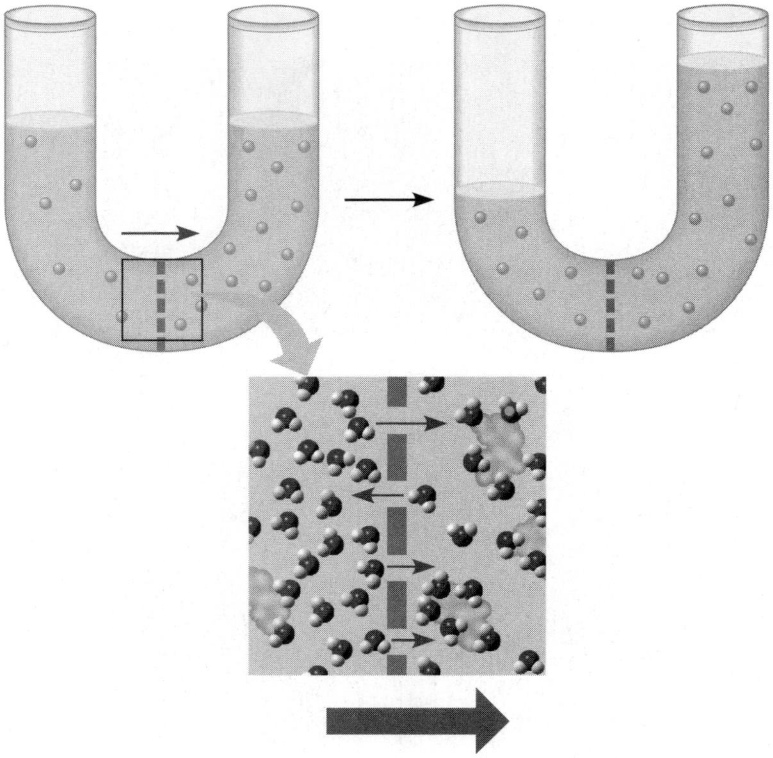

Figure 7.12 Osmosis, page 132

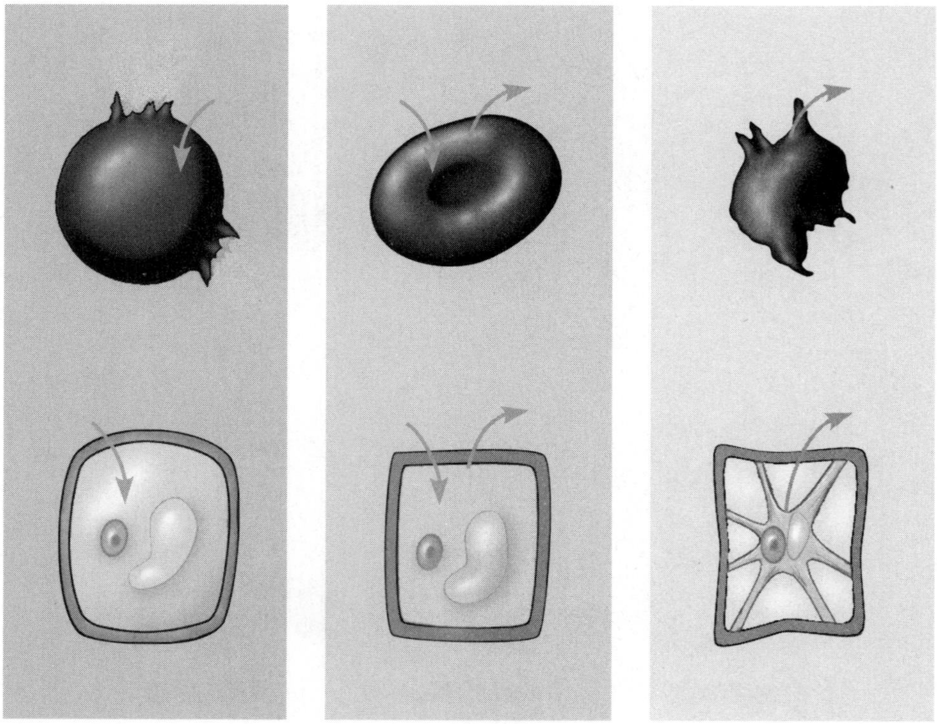

Figure 7.13 The water balance of living cells, page 133

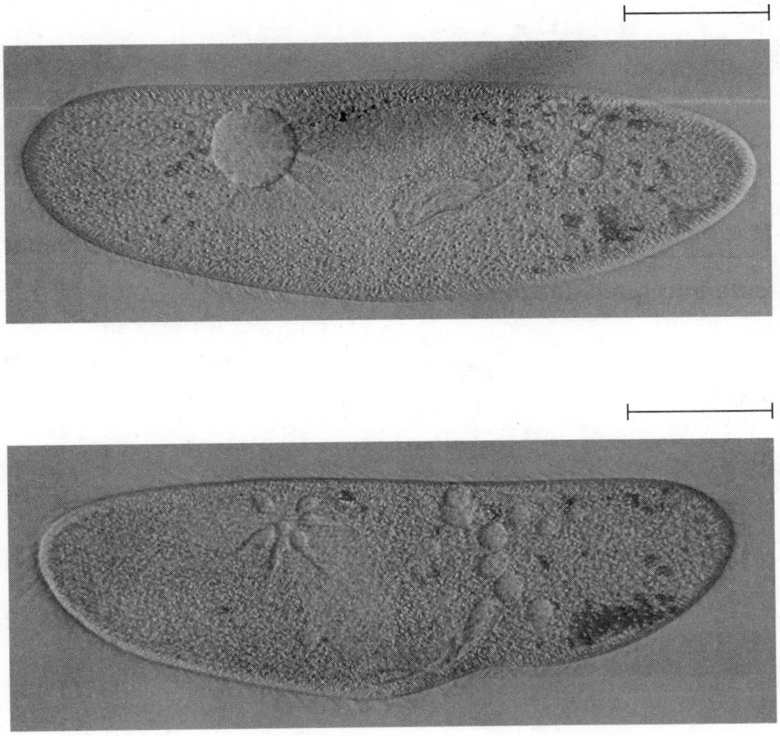

Figure 7.14 The contractile vacuole of *Paramecium:* an evolutionary adaptation for osmoregulation, page 133

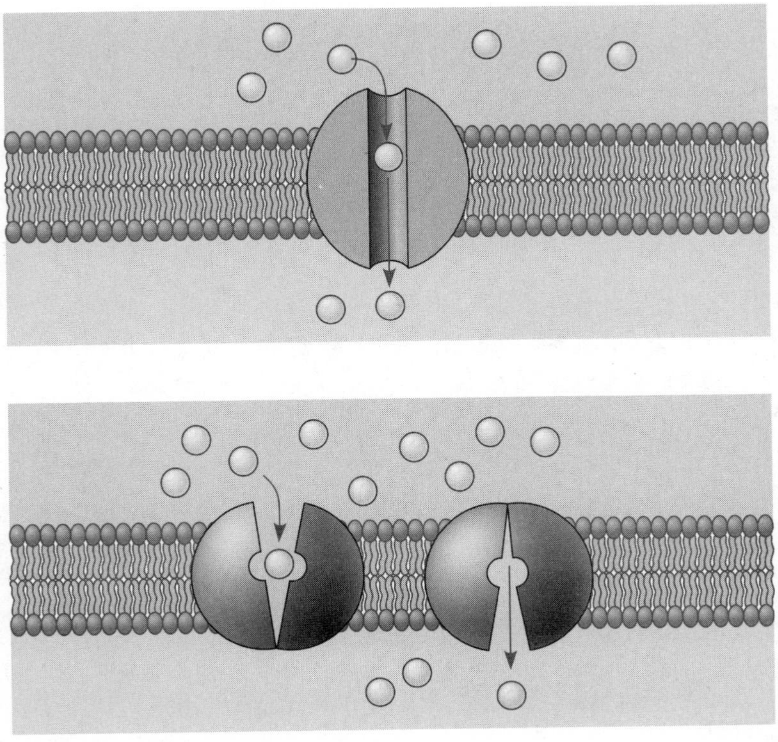

Figure 7.15 Two types of transport proteins that carry out facilitated diffusion, page 134

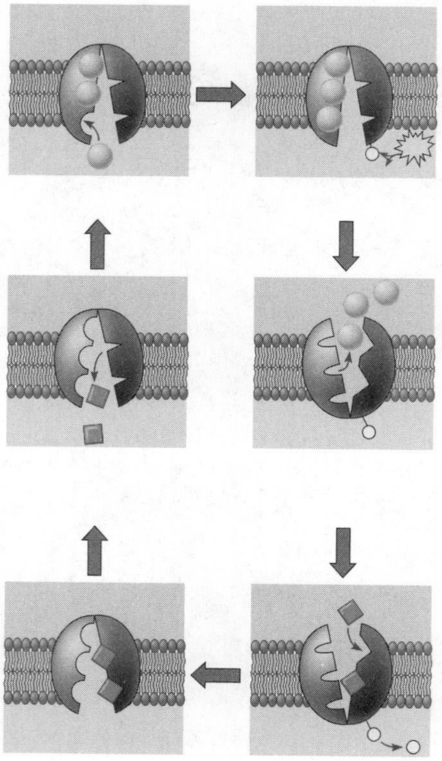

Figure 7.16 The sodium-potassium pump: a specific case of active transport, page 135

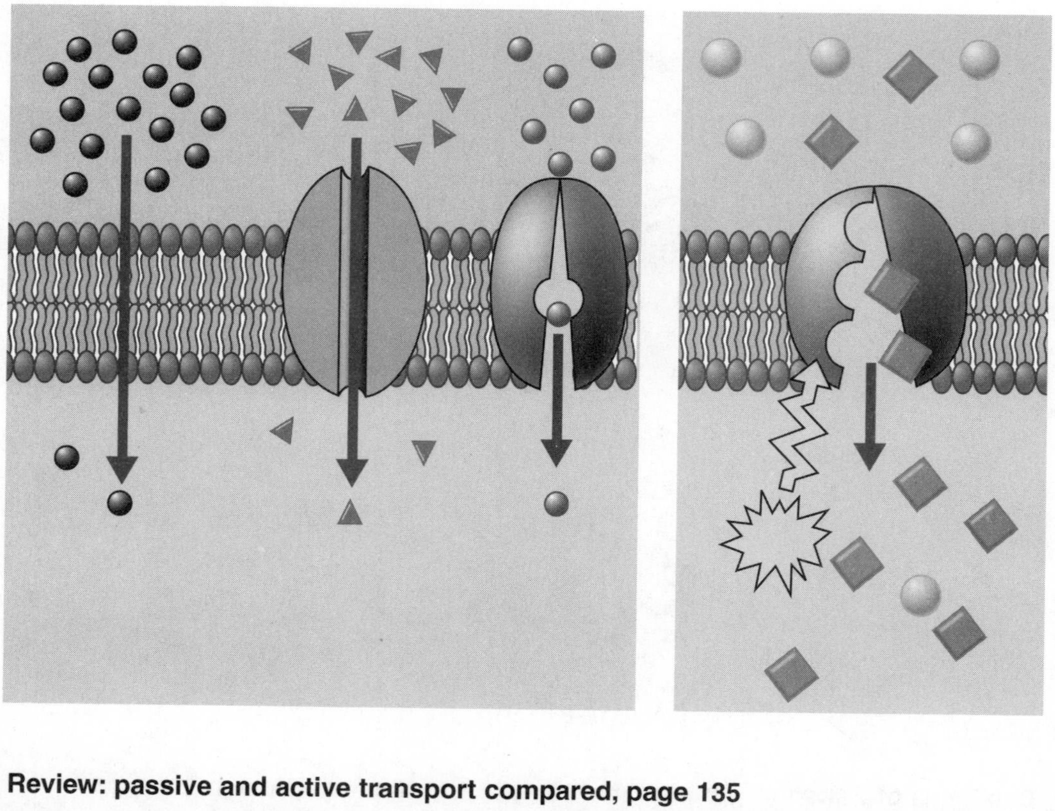

Figure 7.17 Review: passive and active transport compared, page 135

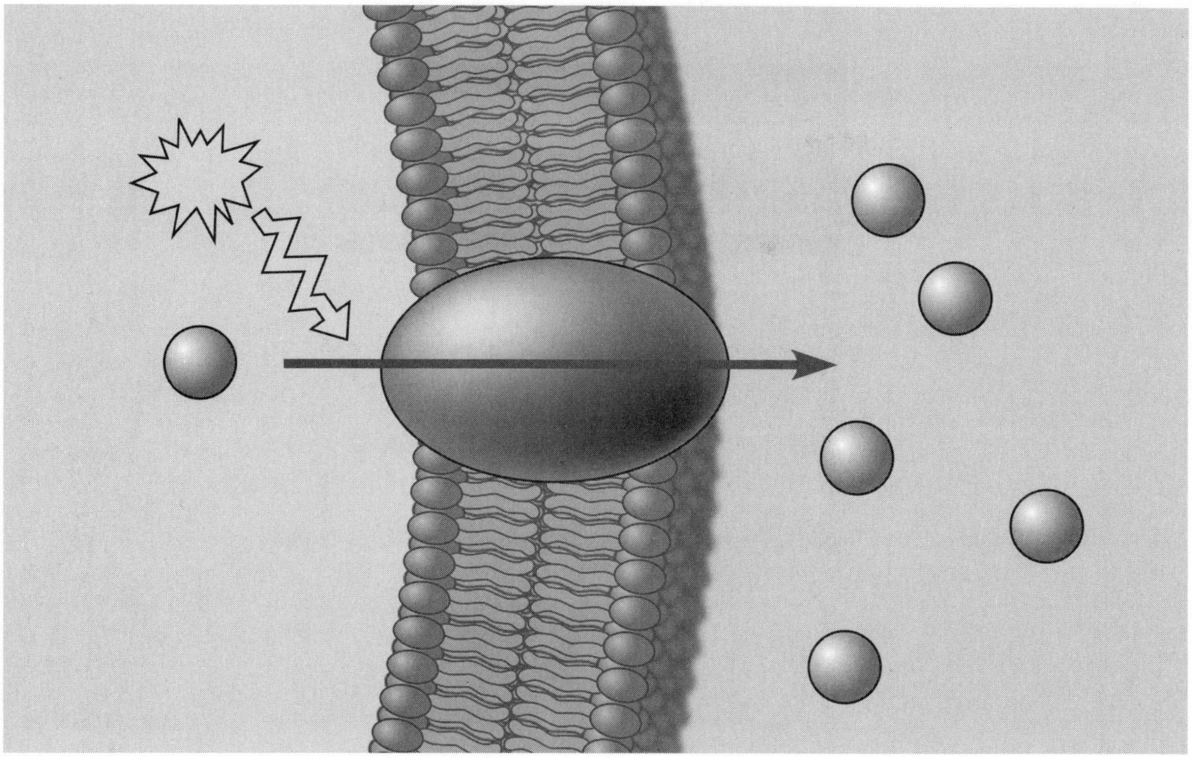

Figure 7.18 An electrogenic pump, page 136

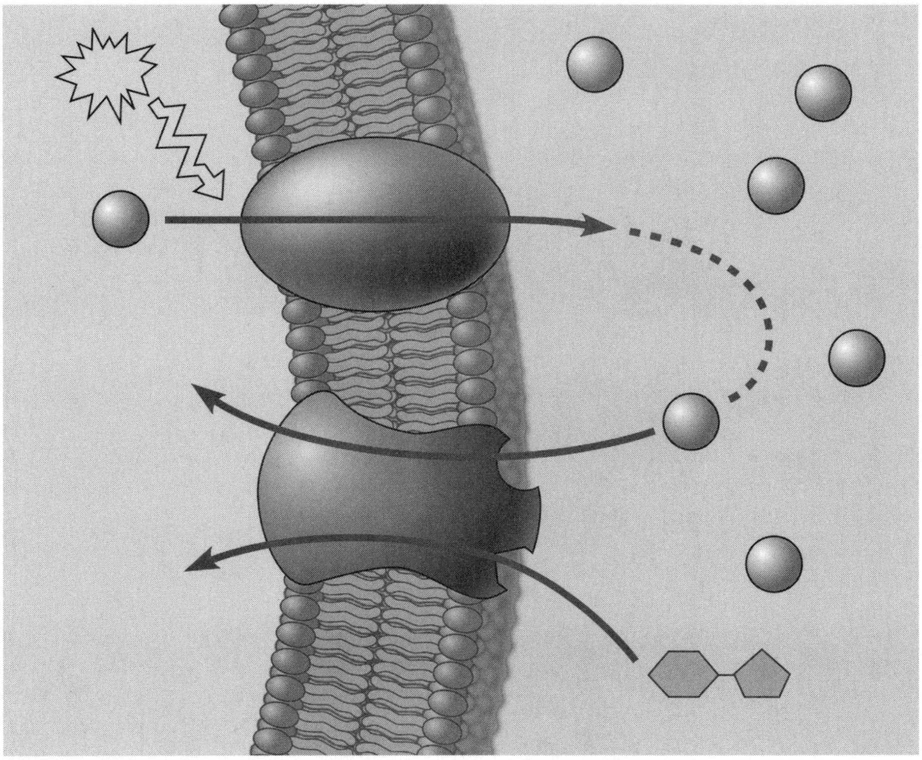

Figure 7.19 Cotransport: active transport driven by a concentration gradient, page 136

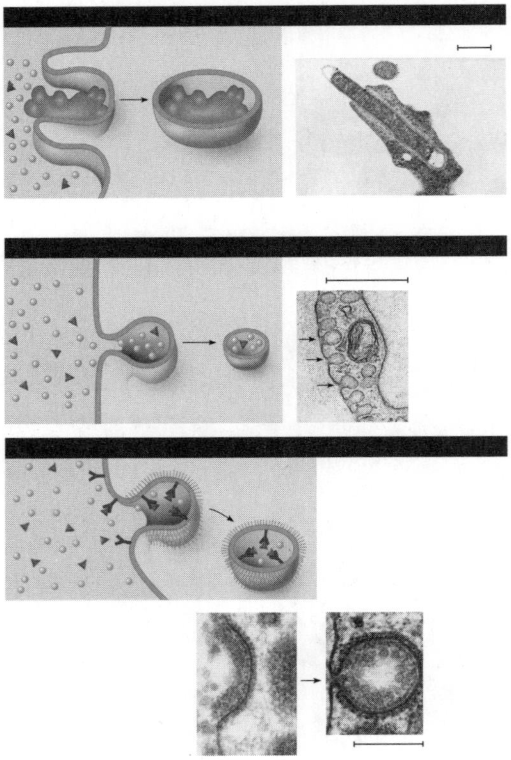

Figure 7.20 Endocytosis in animal cells, page 138

Figure 7.UN1 An artificial cell immersed in a solution, page 140

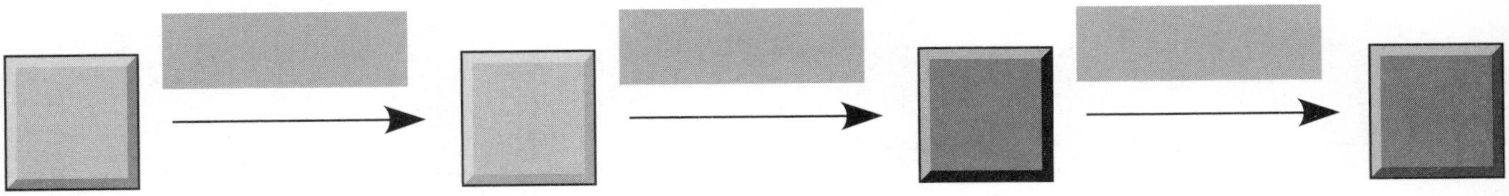

Figure 8.UN1 A metabolic pathway, page 141

Figure 8.3 The two laws of thermodynamics, page 143

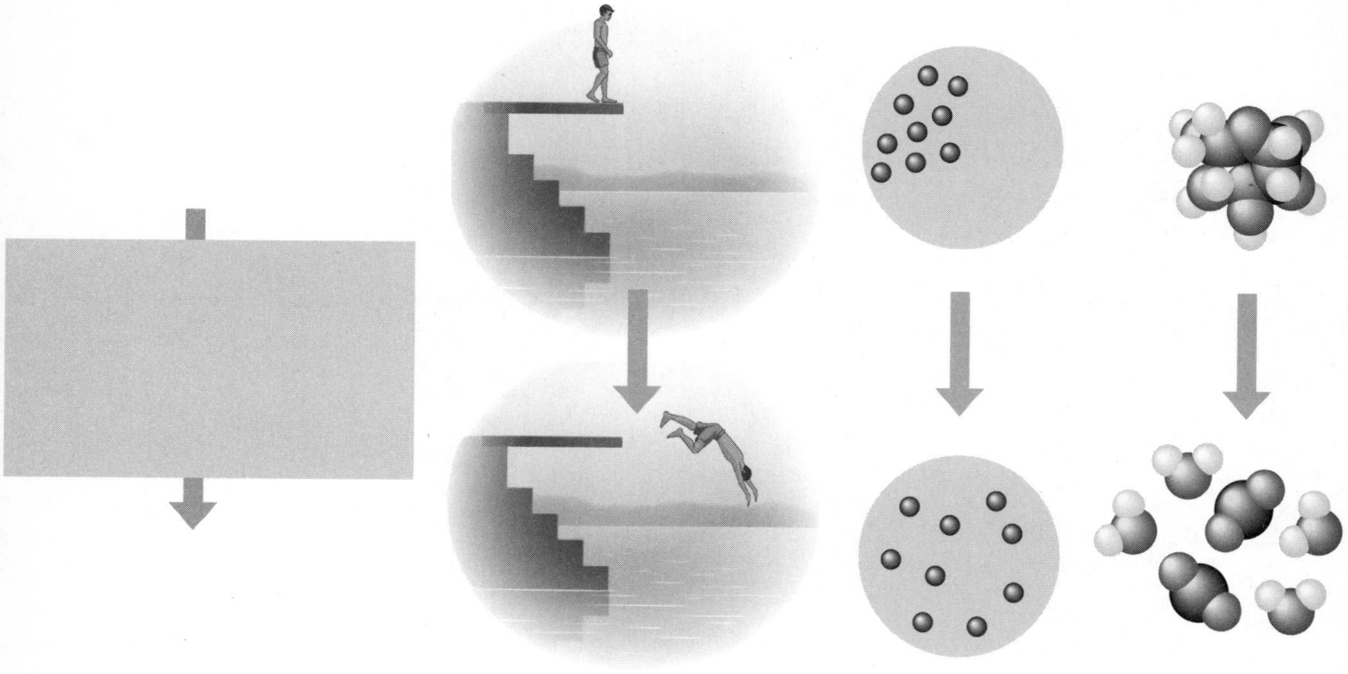

Figure 8.5 The relationship of free energy to stability, work capacity, and spontaneous change, page 146

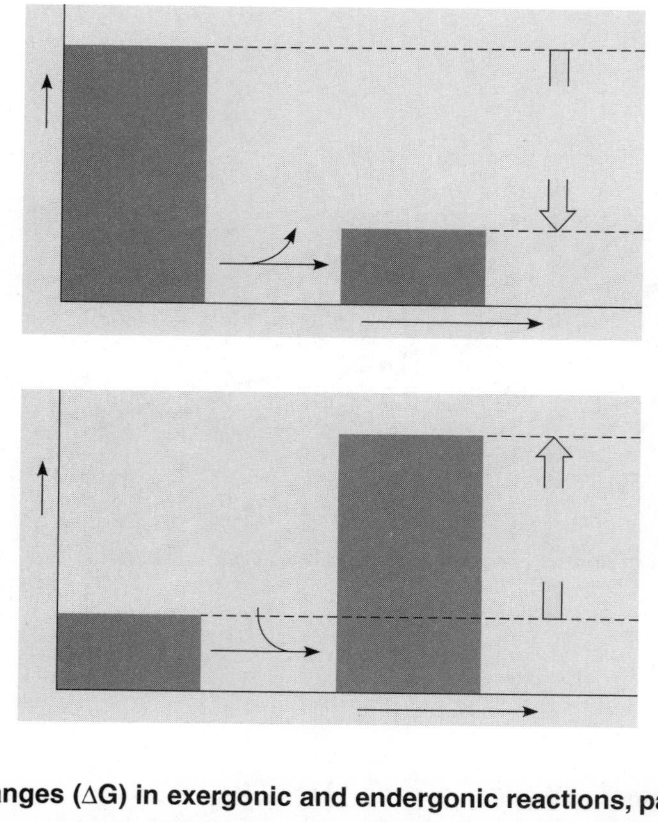

Figure 8.6 Free energy changes (ΔG) in exergonic and endergonic reactions, page 147

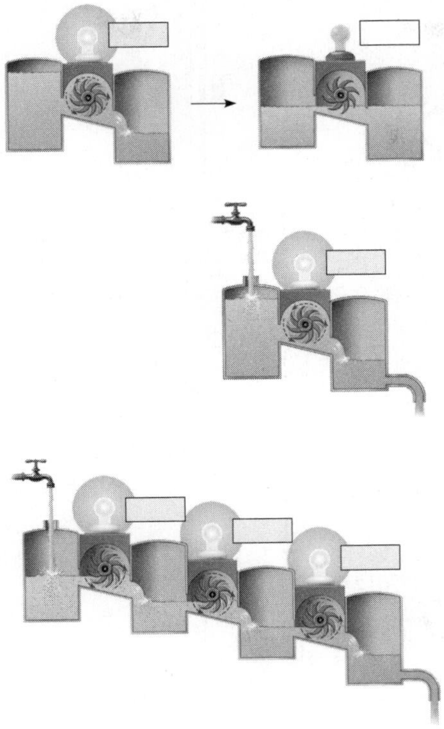

Figure 8.7 Equilibrium and work in closed and open systems, page 147

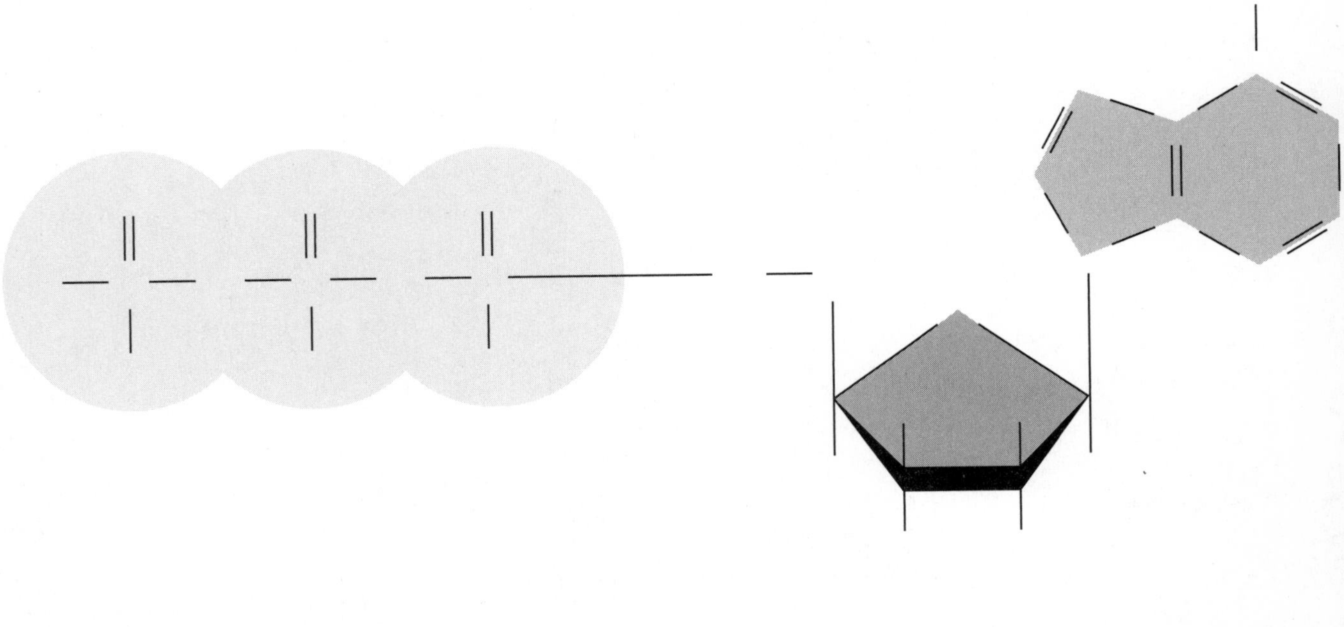

Figure 8.8 The structure of adenosine triphosphate (ATP), page 148

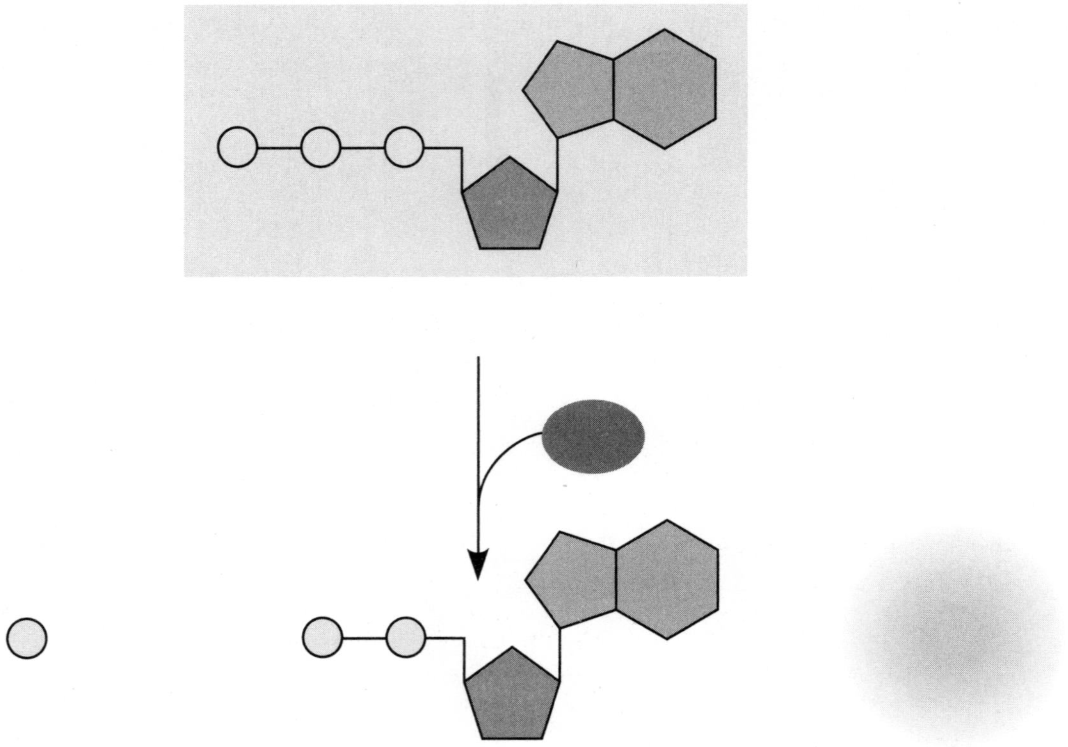

Figure 8.9 The hydrolysis of ATP, page 148

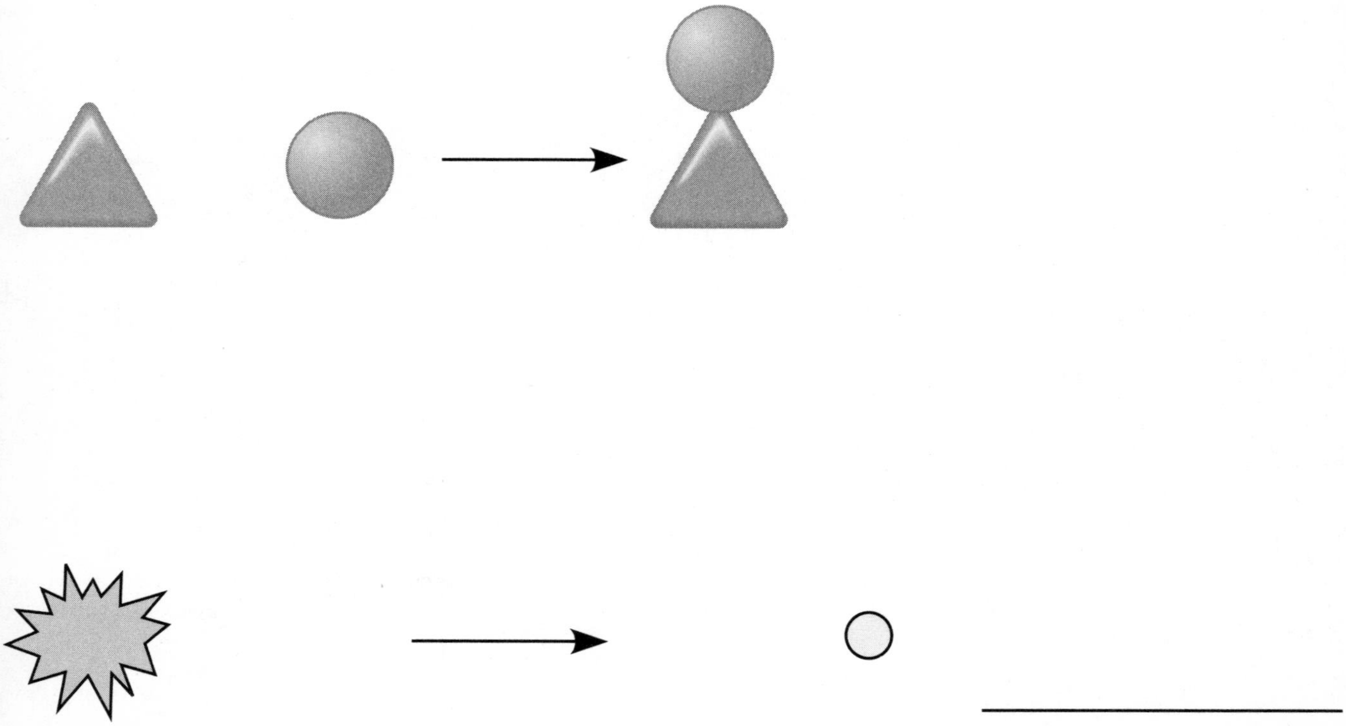

Figure 8.10 Energy coupling using ATP hydrolysis, page 149

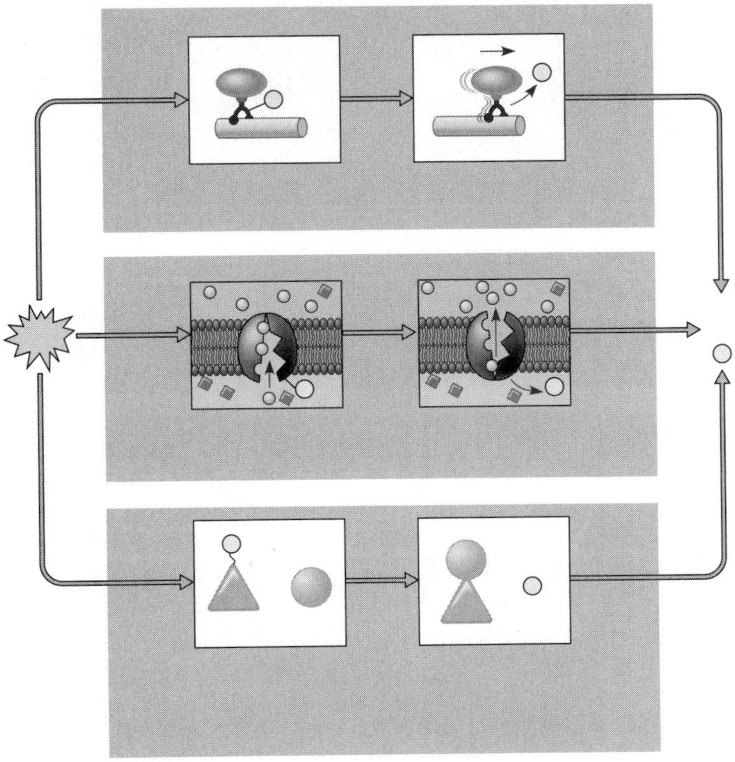

Figure 8.11 How ATP drives cellular work, page 149

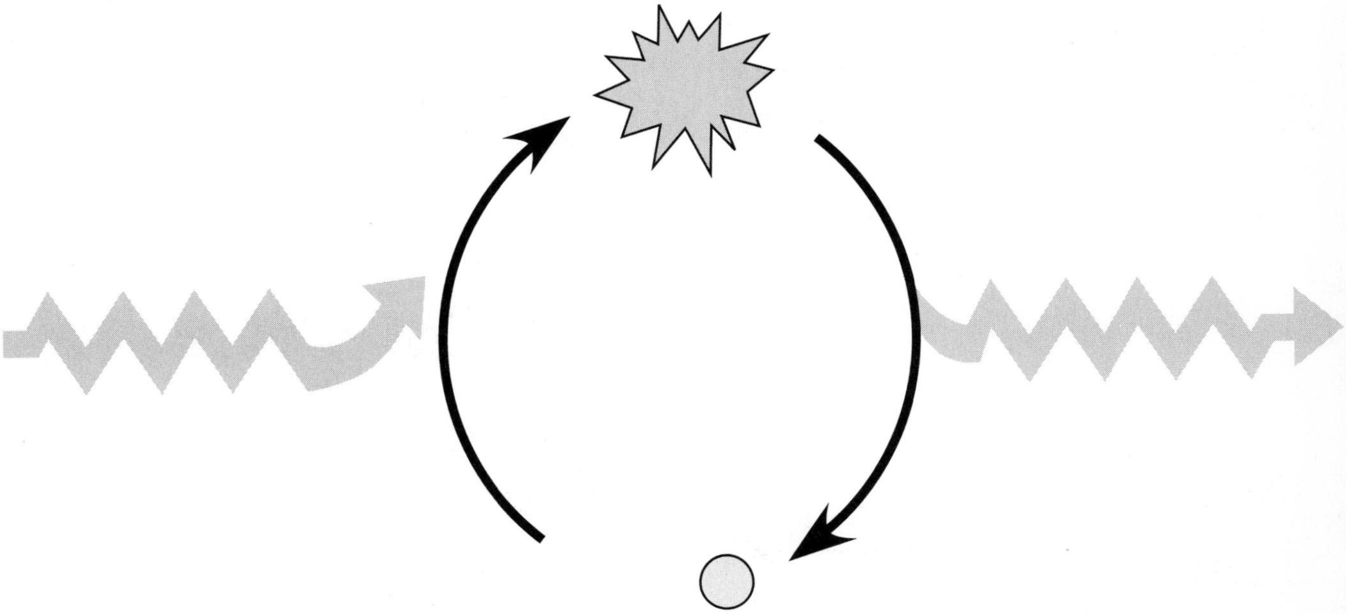

Figure 8.12 The ATP cycle, page 150

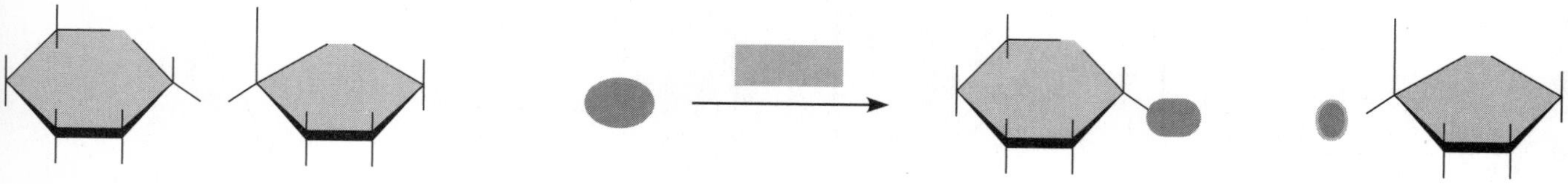

Figure 8.13 Example of an enzyme-catalyzed reaction: hydrolysis of sucrose by sucrase, page 151

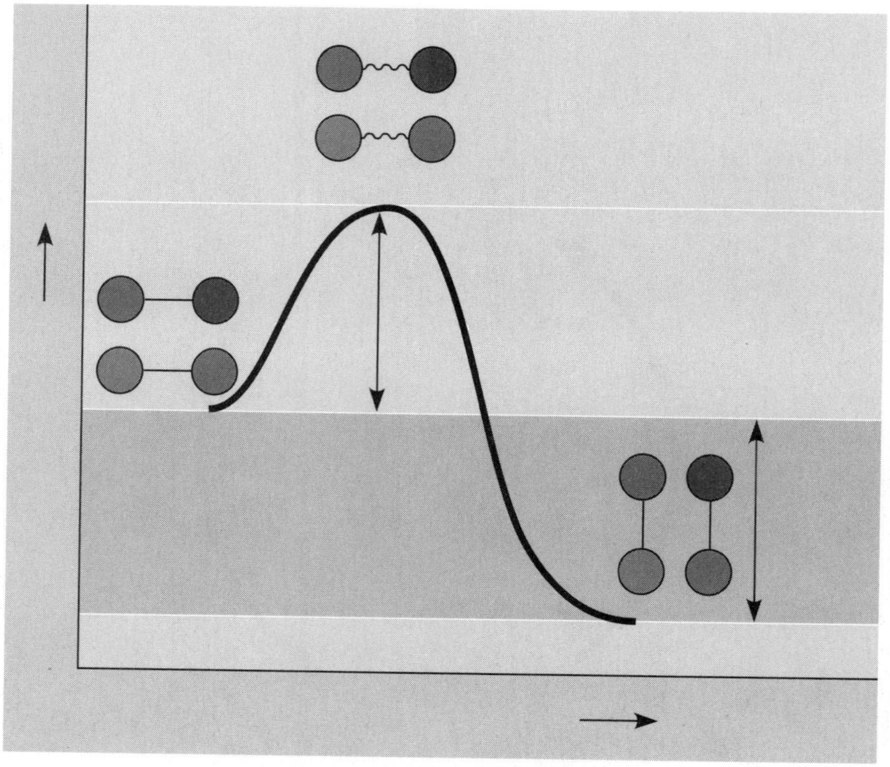

Figure 8.14 Energy profile of an exergonic reaction, page 151

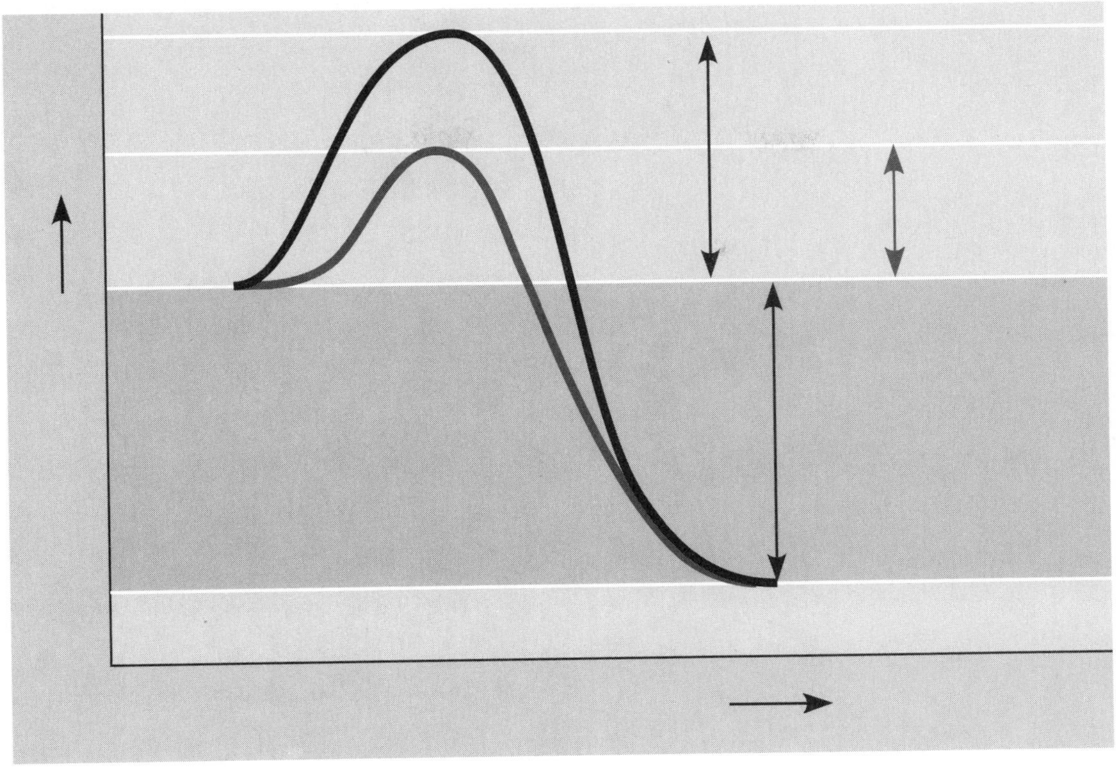

Figure 8.15 The effect of enzymes on reaction rate, page 152

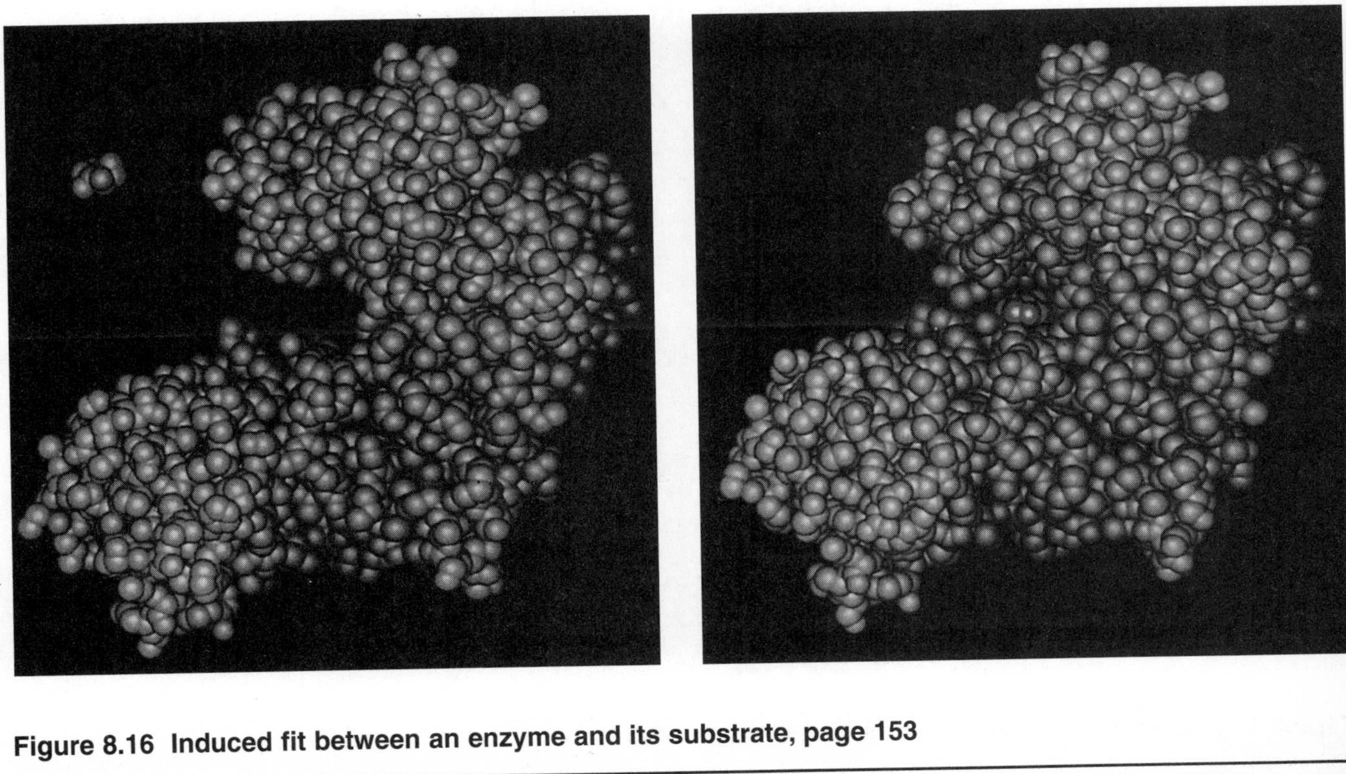

Figure 8.16 Induced fit between an enzyme and its substrate, page 153

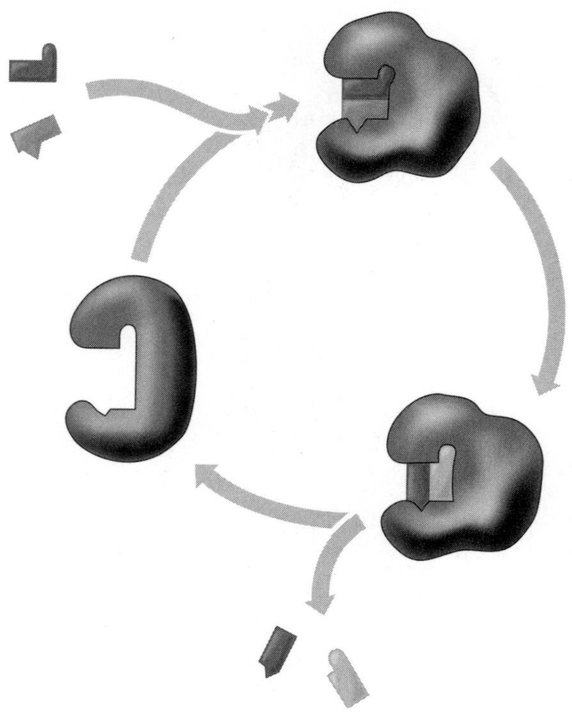

Figure 8.17 The active site and catalytic cycle of an enzyme, page 153

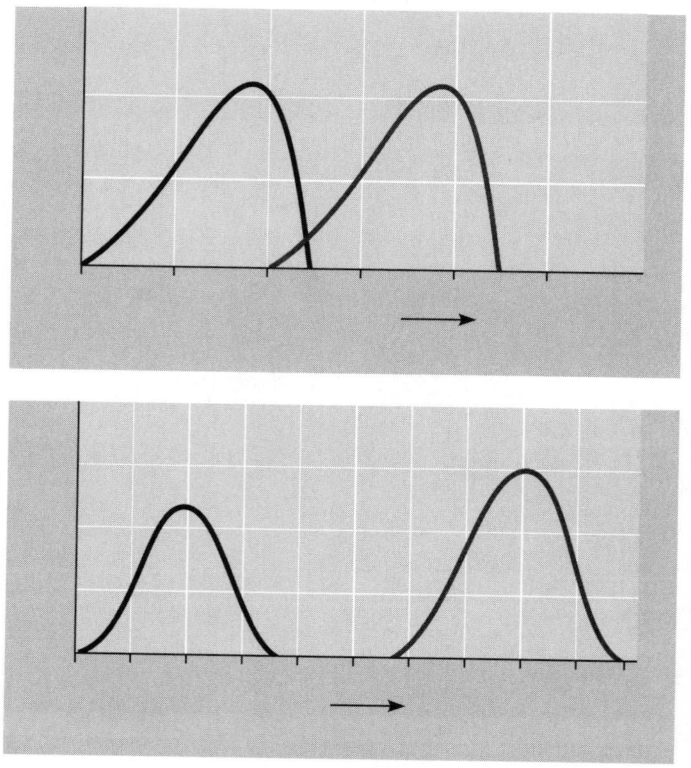

Figure 8.18 Environmental factors affecting enzyme activity, page 154

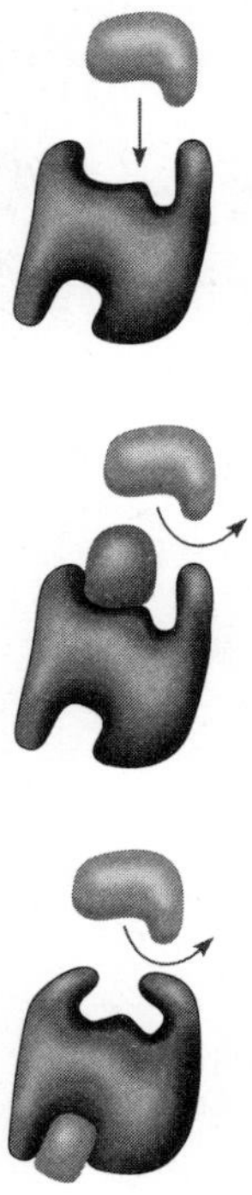

Figure 8.19 Inhibition of enzyme activity, page 155

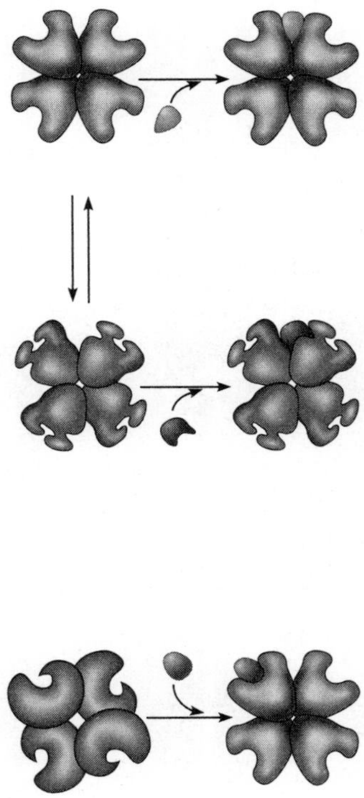

Figure 8.20 Allosteric regulation of enzyme activity, page 156

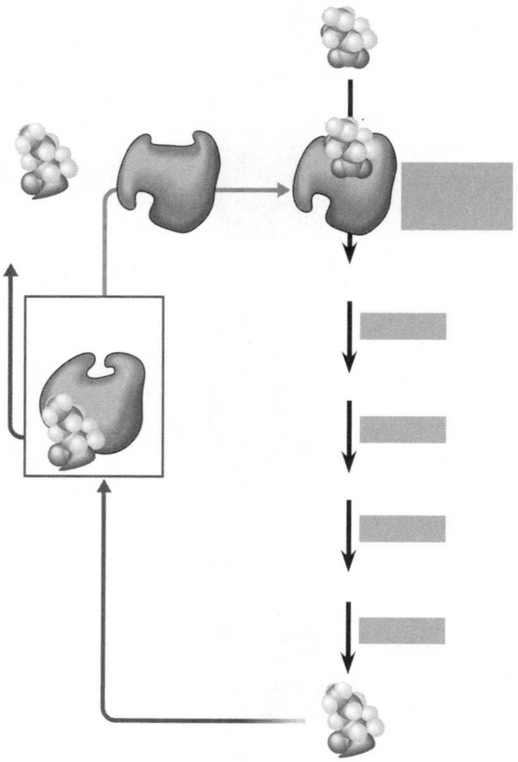

Figure 8.21 Feedback inhibition in isoleucine synthesis, page 157

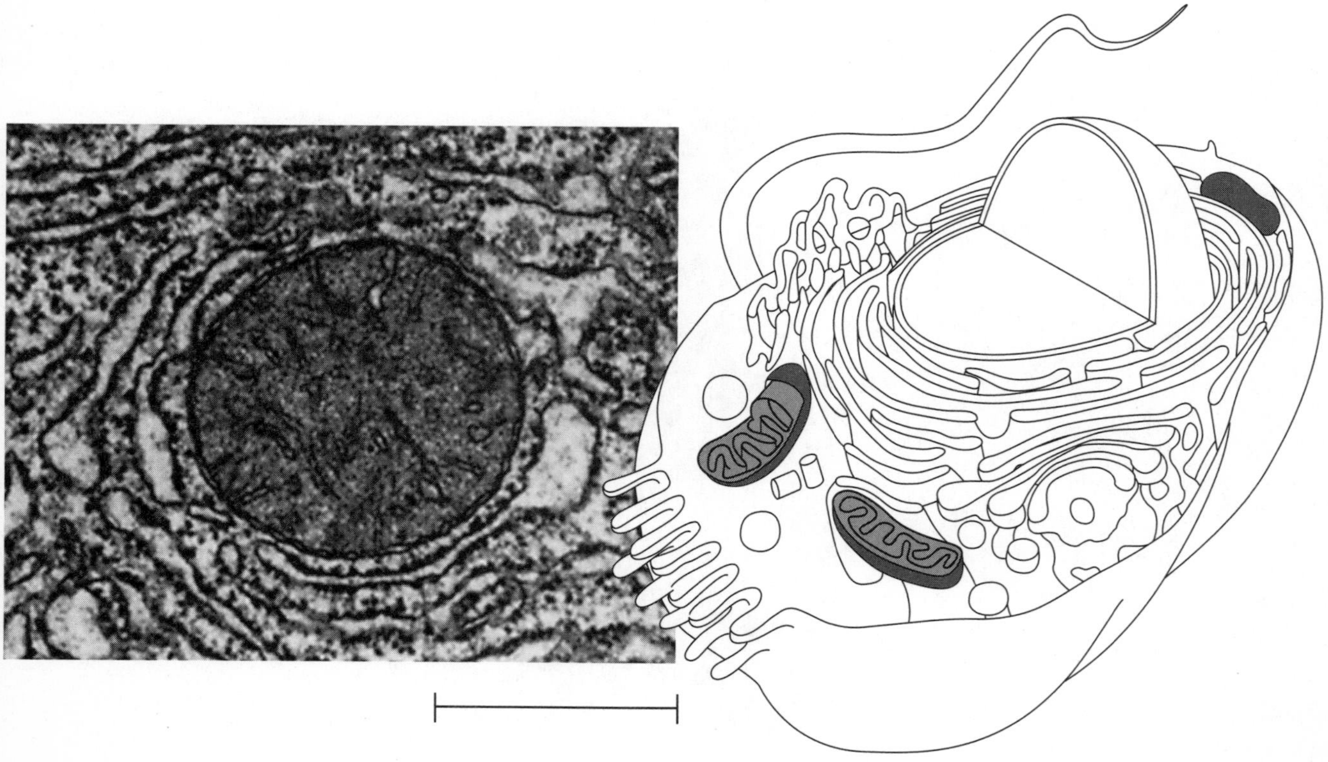

Figure 8.22 Organelles and structural order in metabolism, page 156

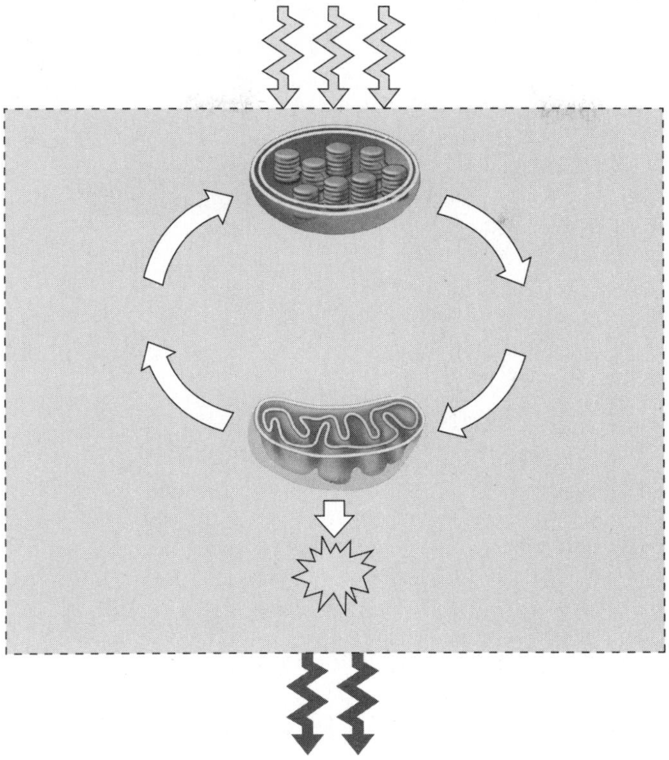

Figure 9.2 Energy flow and chemical recycling in ecosystems, page 160

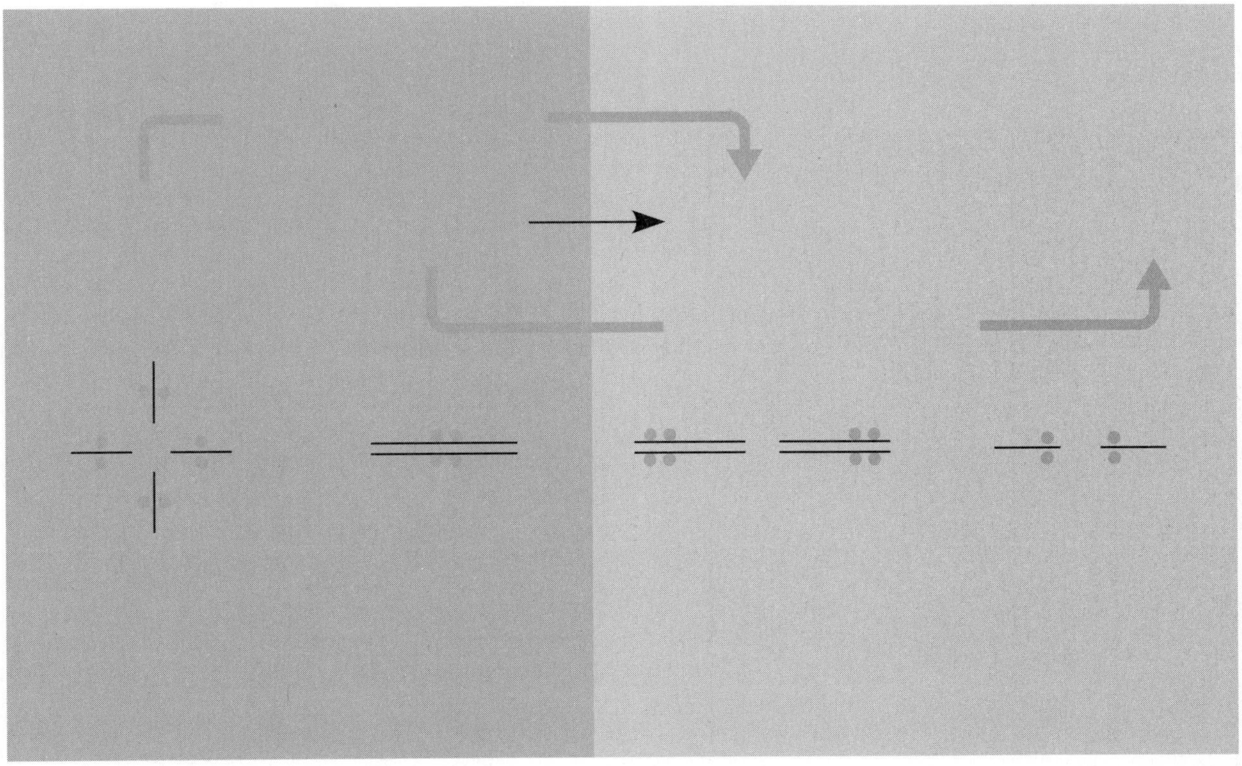

Figure 9.3 Methane combustion as an energy-yielding redox reaction, page 162

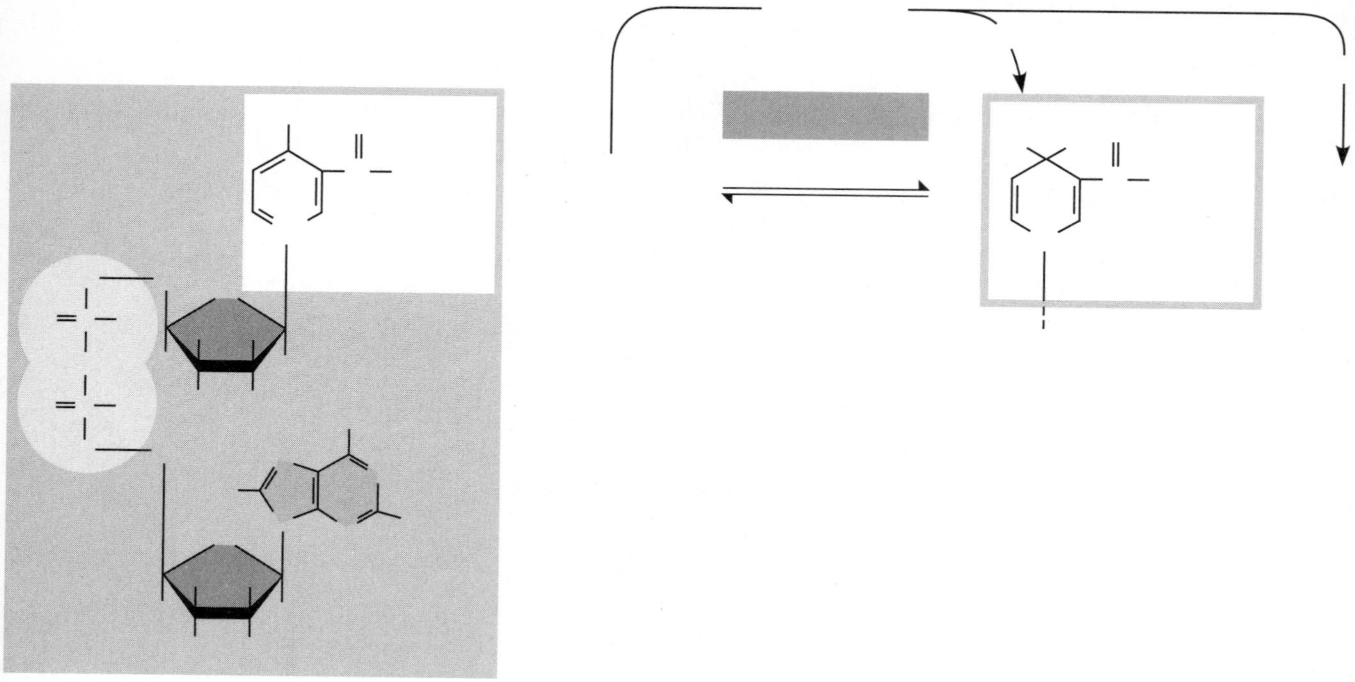

Figure 9.4 NAD⁺ as an electron shuttle, page 163

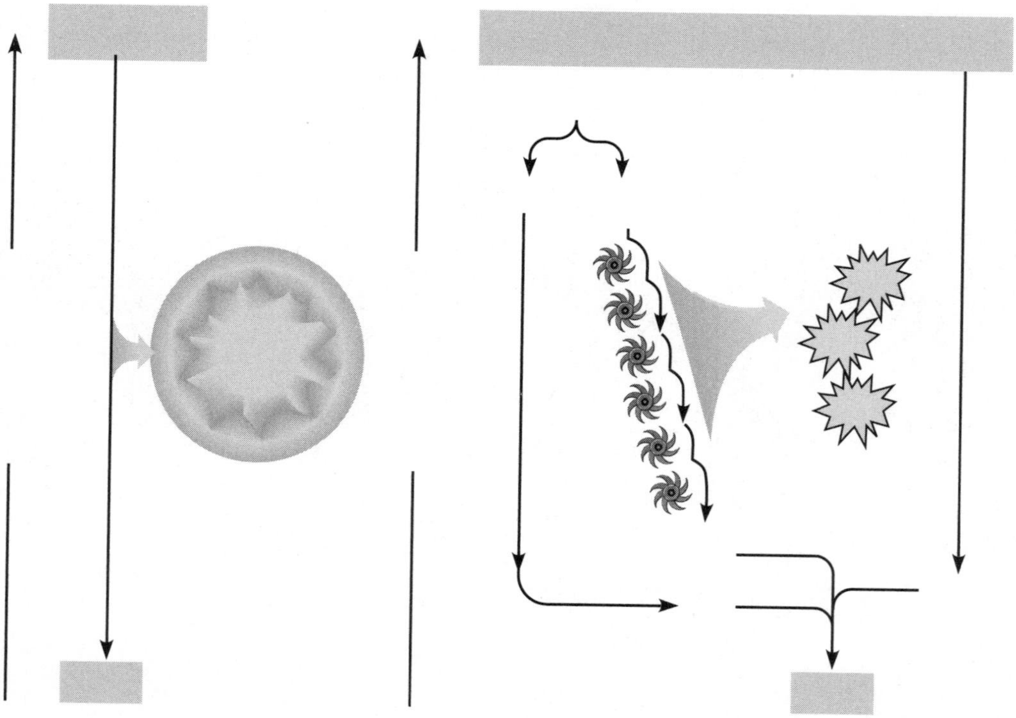

Figure 9.5 An introduction to electron transport chains, page 163

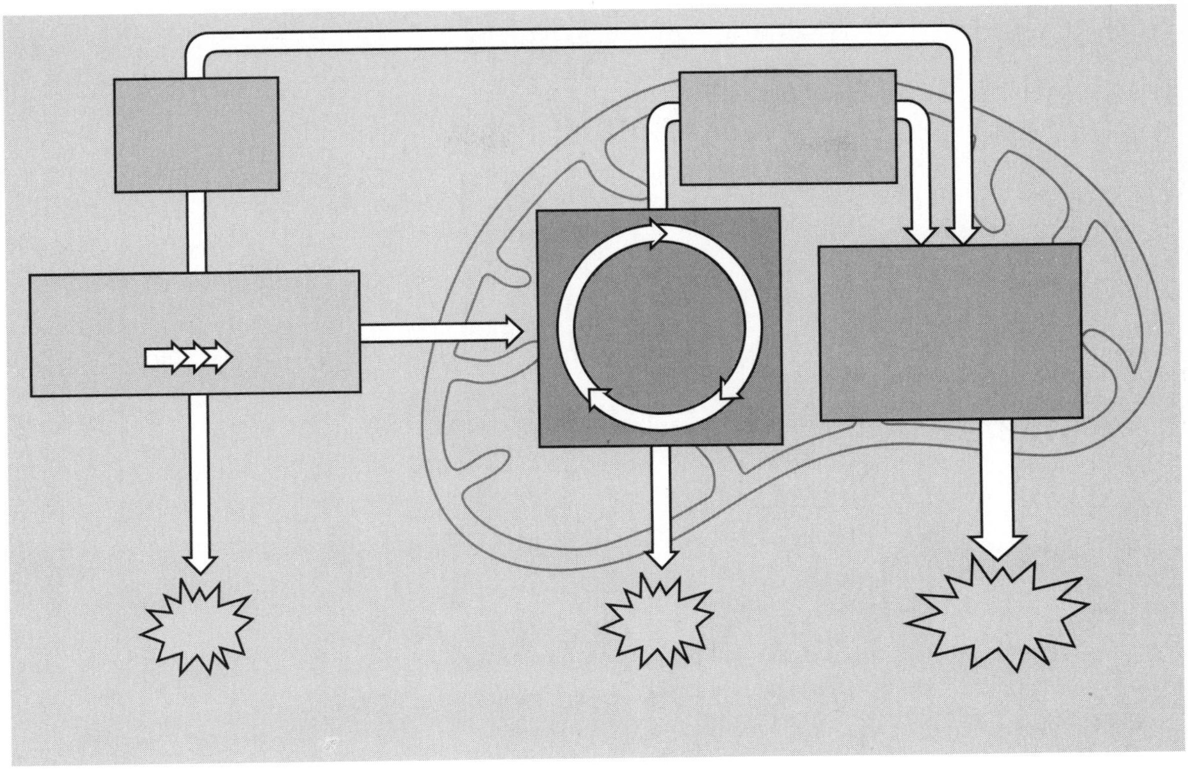

Figure 9.6 An overview of cellular respiration, page 164

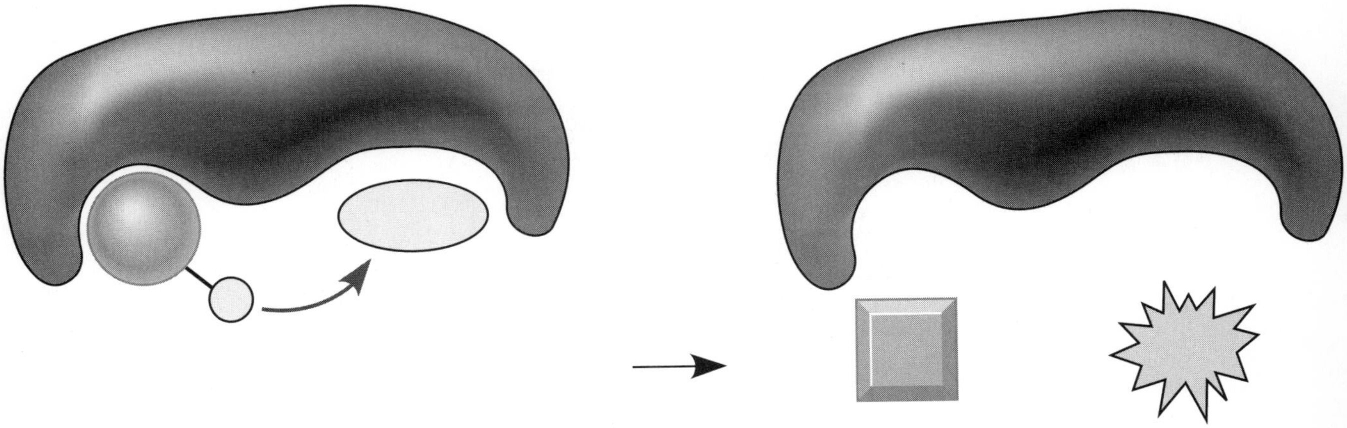

Figure 9.7 Substrate-level phosphorylation, page 165

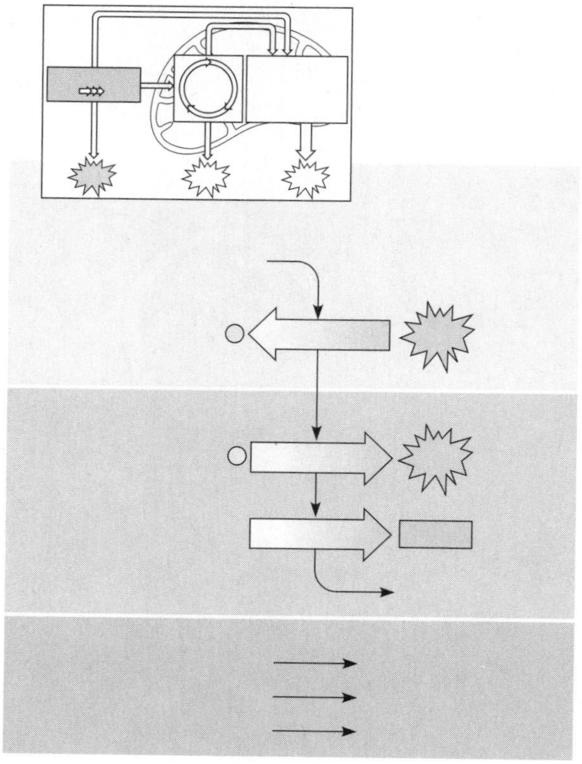

Figure 9.8 The energy input and output of glycolysis, page 165

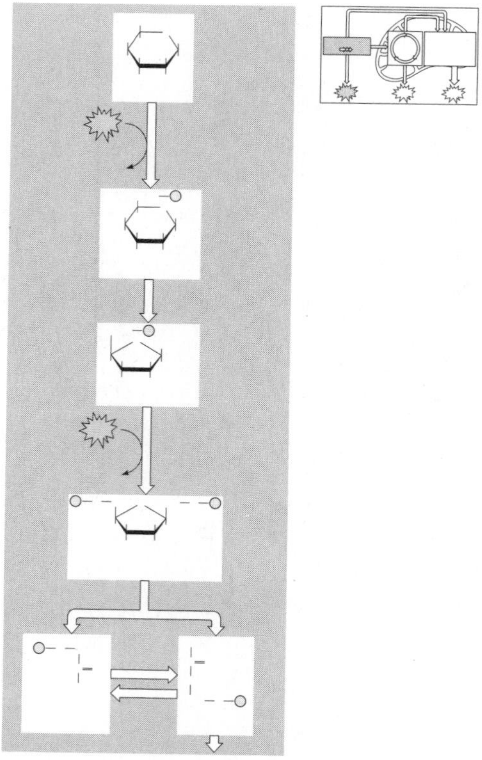

Figure 9.9 A closer look at glycolysis: energy investment phase, page 166

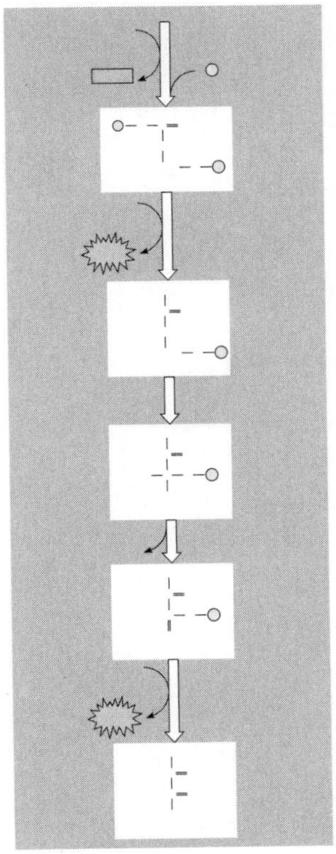

Figure 9.9 A closer look at glycolysis: energy payoff phase, page 167

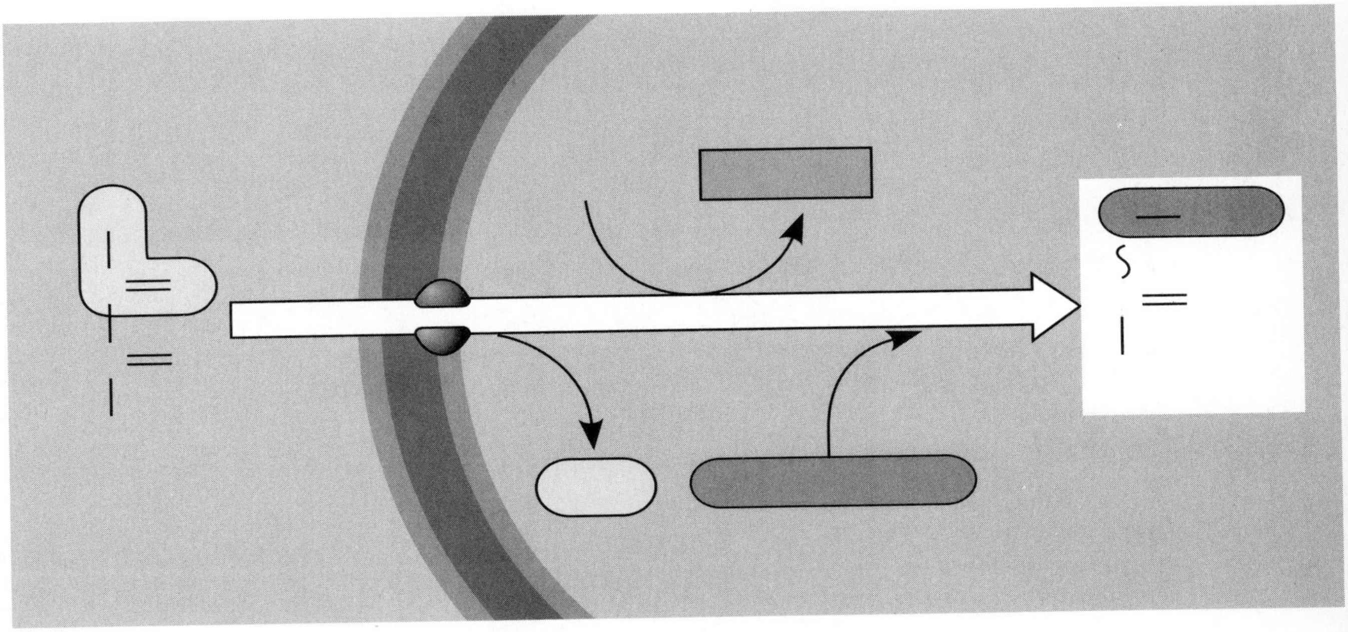

Figure 9.10 Conversion of pyruvate to acetyl CoA, the junction between glycolysis and the citric acid cycle, page 168

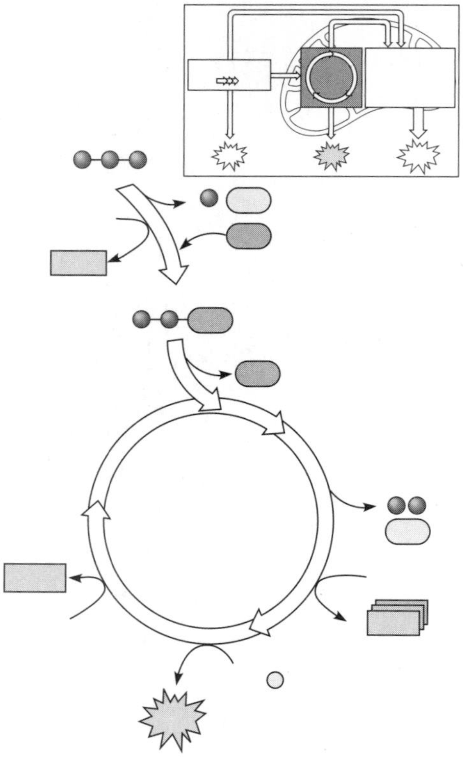

Figure 9.11 An overview of the citric acid cycle, page 168

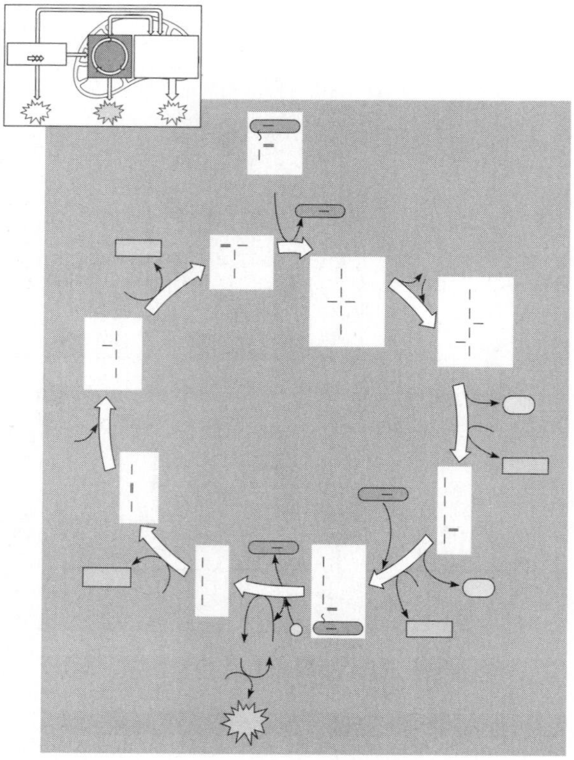

Figure 9.12 A closer look at the citric acid cycle, page 169

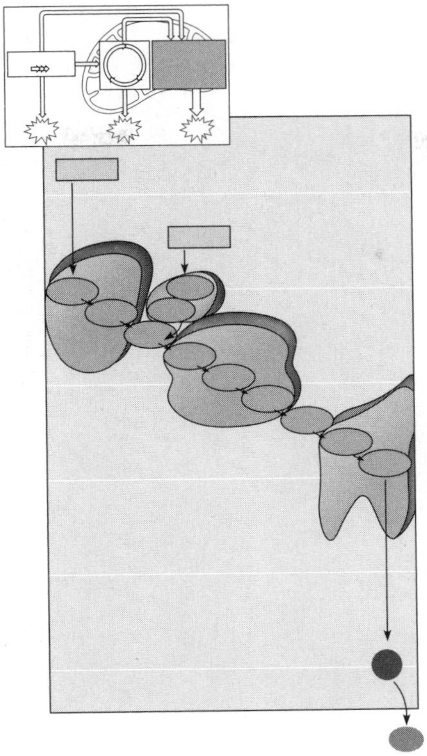

Figure 9.13 Free-energy change during electron transport, page 171

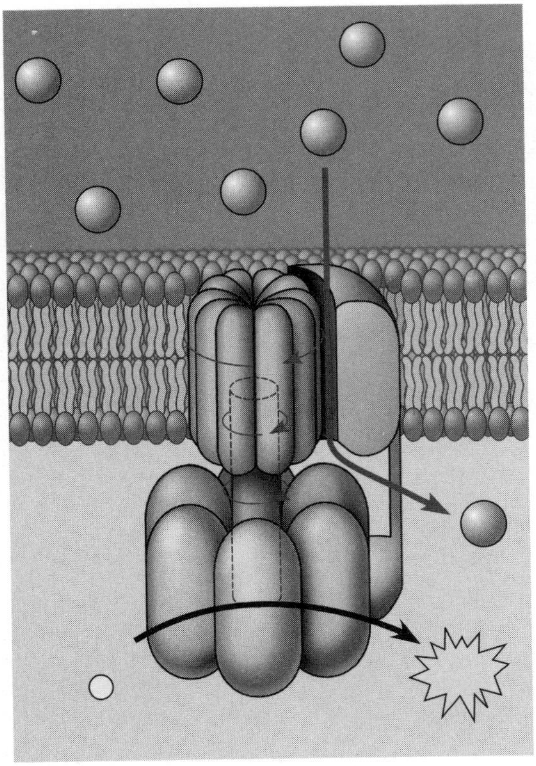

Figure 9.14 ATP synthesis, a molecular mill, page 171

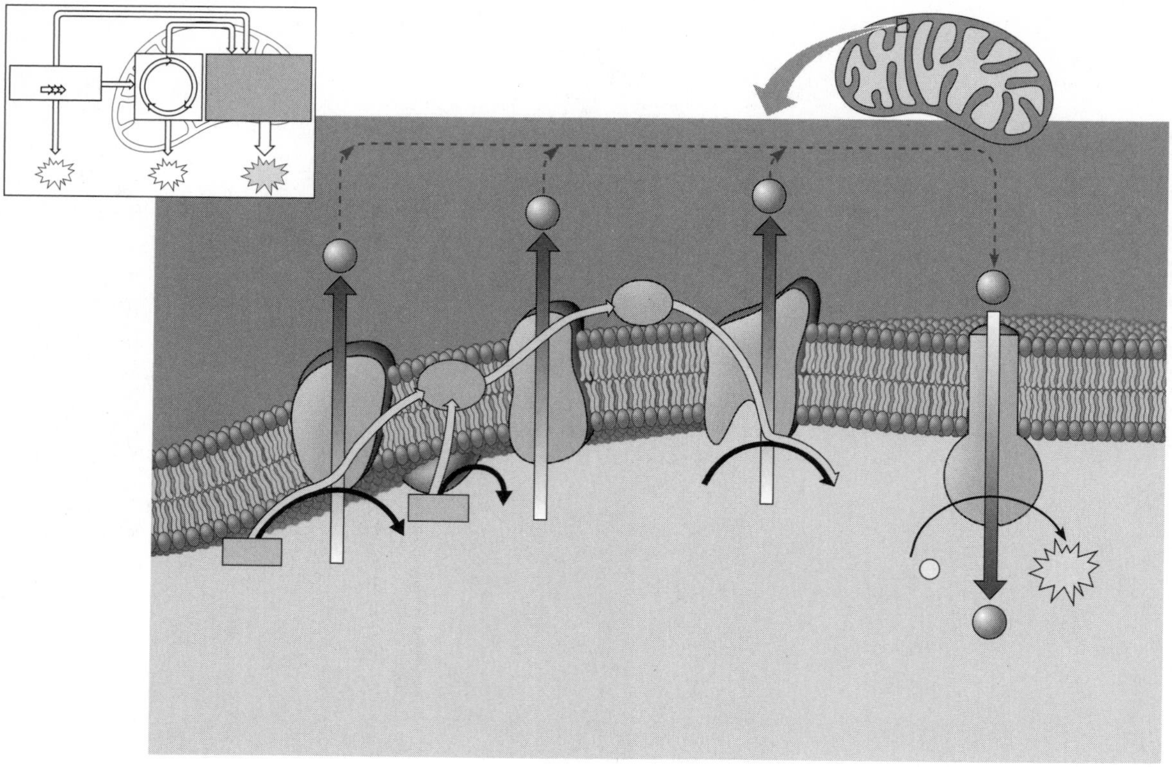

Figure 9.15 Chemiosmosis couples the electron transport chain to ATP synthesis, page 172

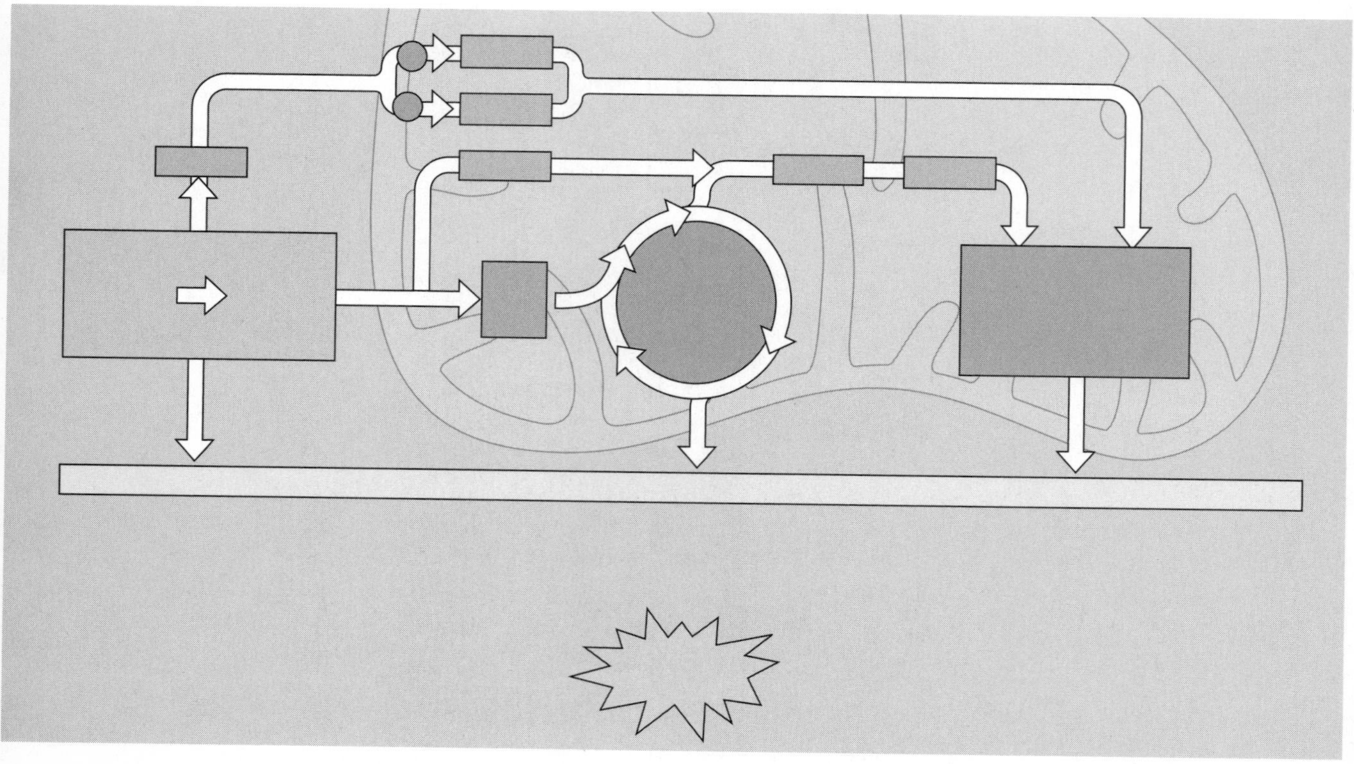

Figure 9.16 ATP yield per molecule of glucose at each stage of cellular respiration, page 173

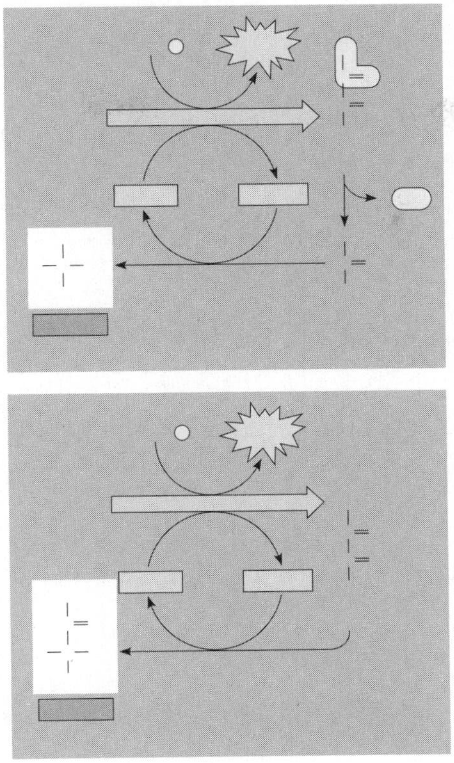

Figure 9.17 Fermentation, page 175

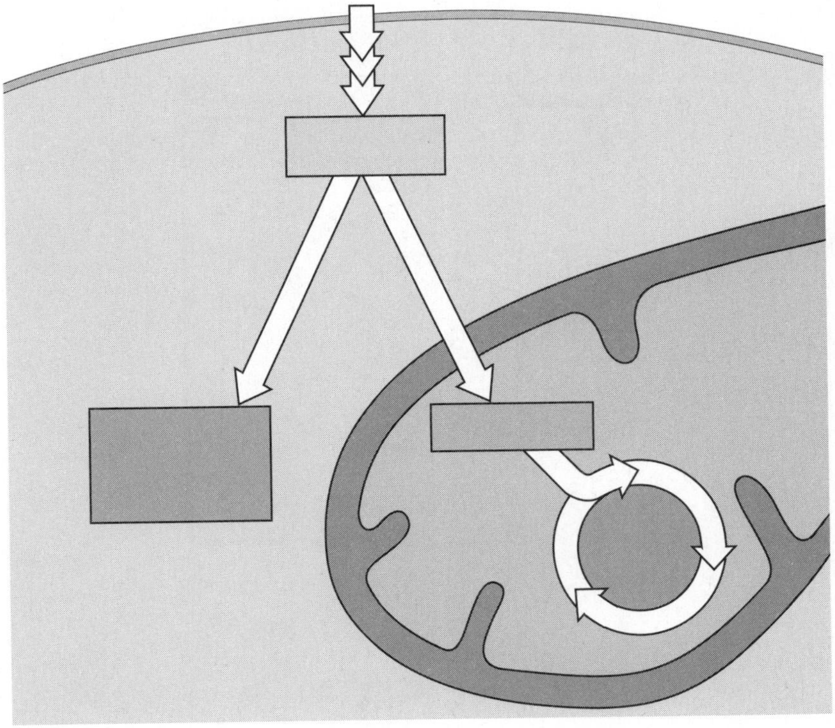

Figure 9.18 Pyruvate as a key juncture in catabolism, page 176

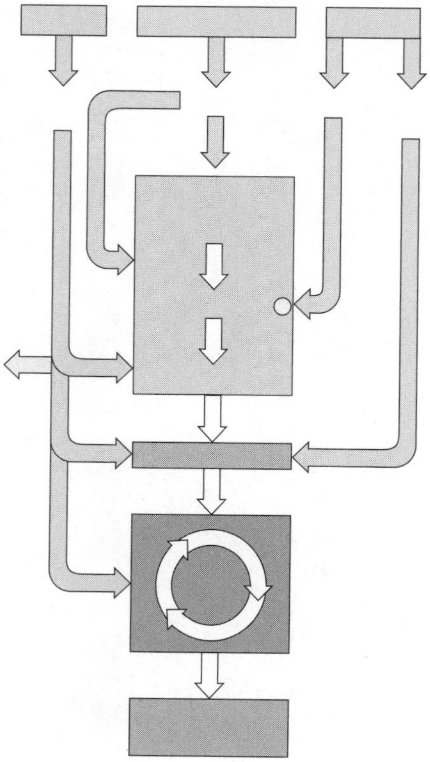

Figure 9.19 The catabolism of various molecules from food, page 177

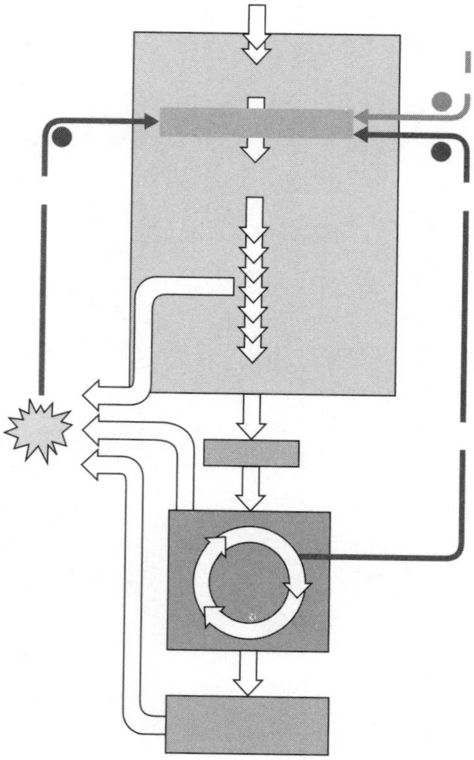

Figure 9.20 The control of cellular respiration, page 178

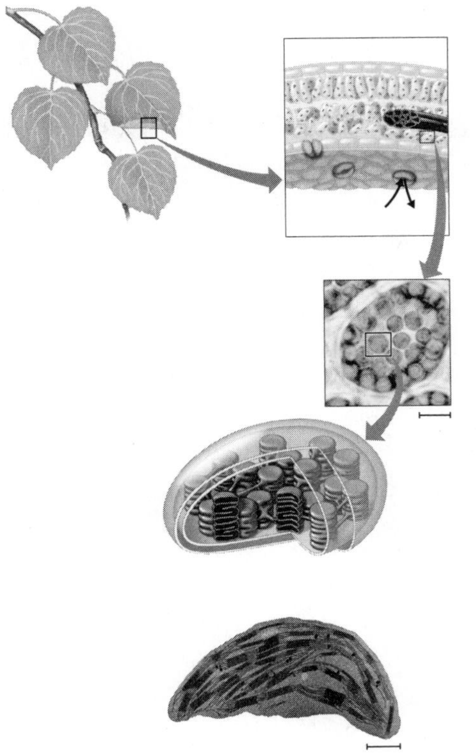

Figure 10.3 Focusing in on the location of photosynthesis in a plant, page 183

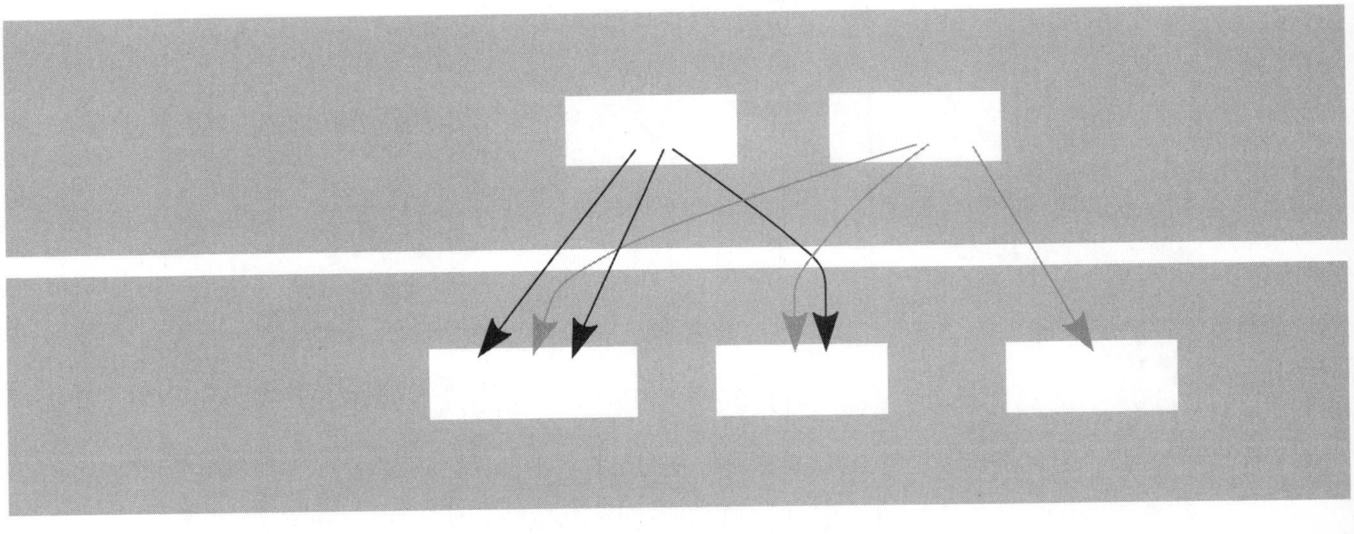

Figure 10.4 Tracking atoms through photosynthesis, page 184

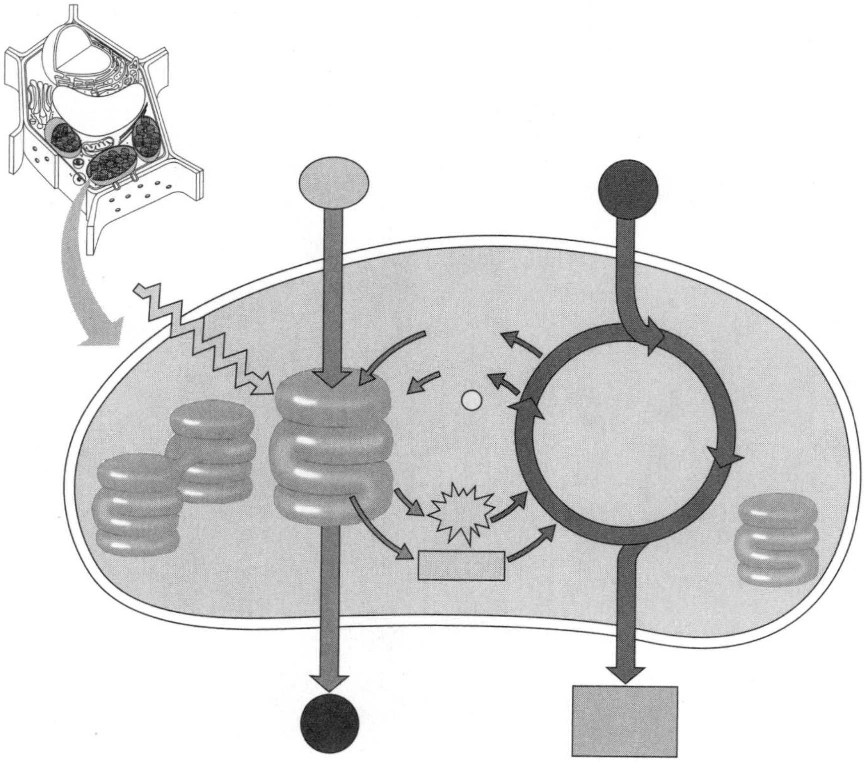

Figure 10.5 An overview of photosynthesis: cooperation of the light reactions and the Calvin cycle, page 185

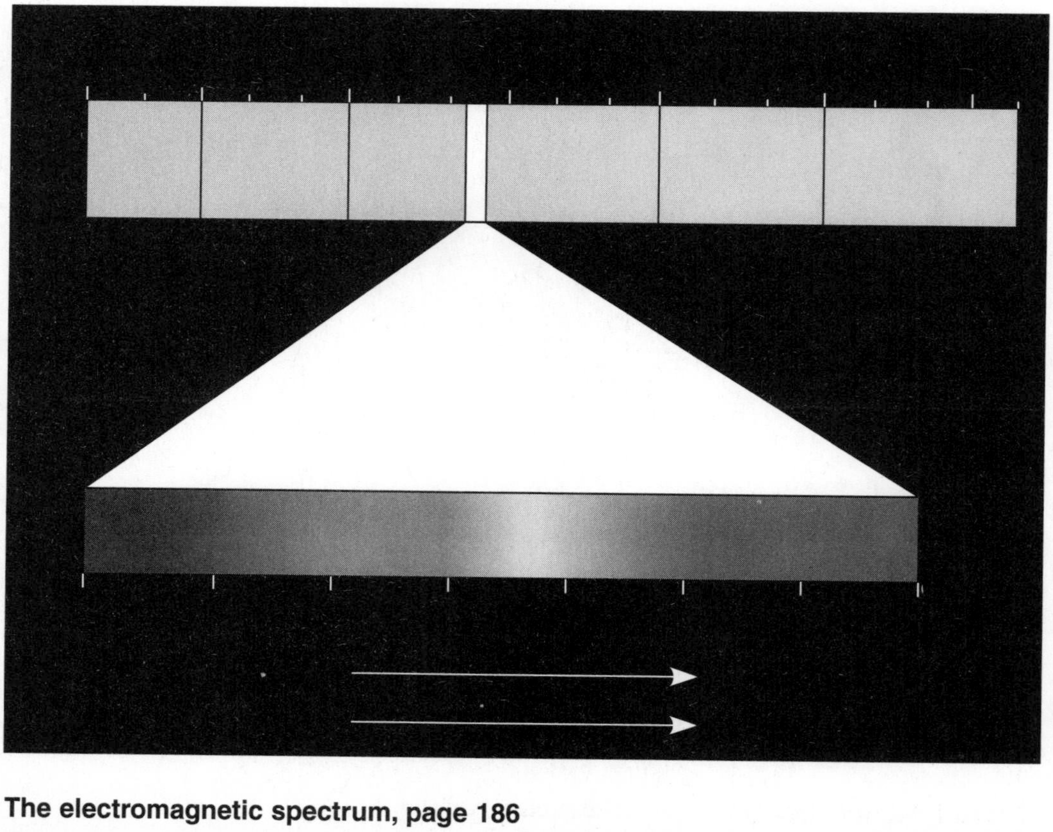

Figure 10.6 The electromagnetic spectrum, page 186

Figure 10.7 Why leaves are green: interaction of light with chloroplasts, page 186

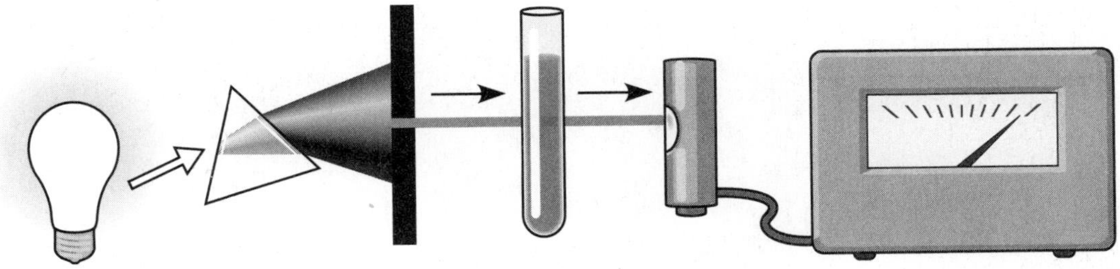

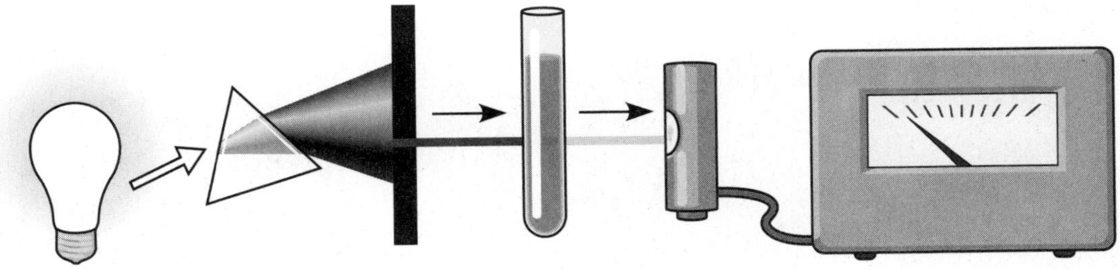

Figure 10.8 Determining an absorption spectrum, page 187

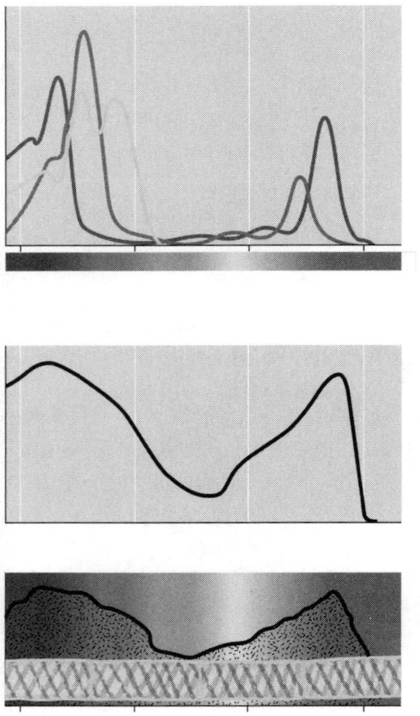

Figure 10.9 Which wavelengths of light are most effective in driving photosynthesis?, page 187

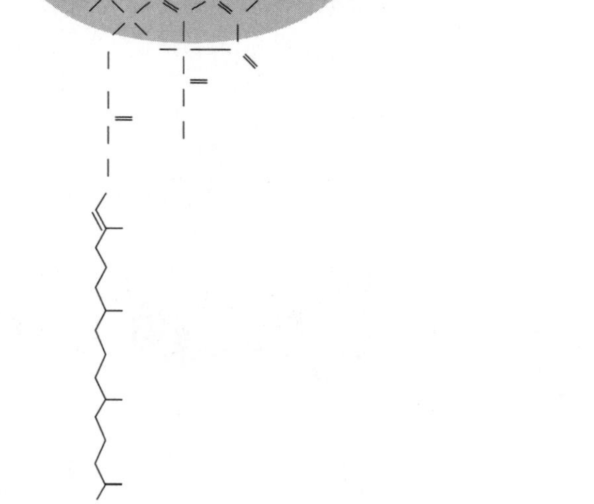

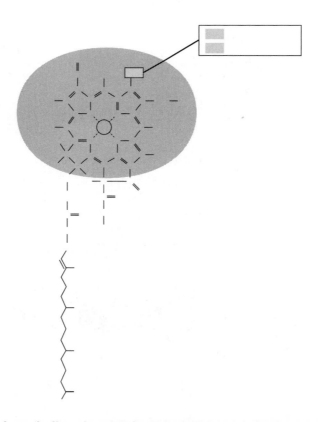

Figure 10.10 Structure of chlorophyll molecules in chloroplasts of plants, page 188

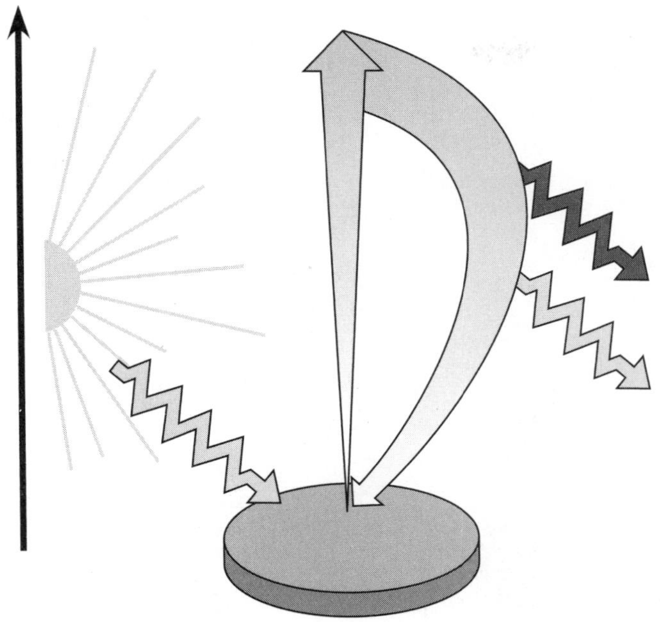

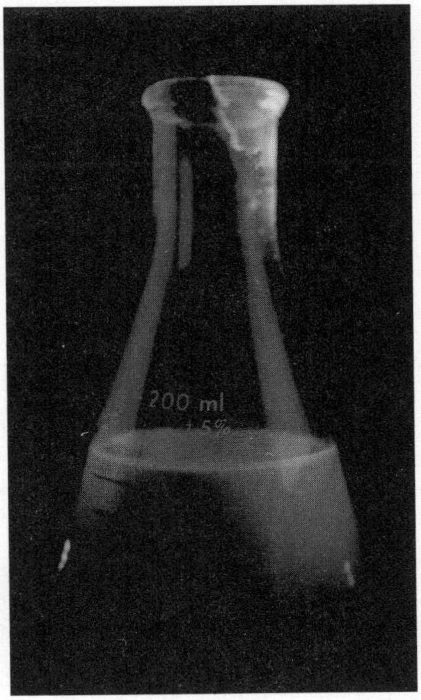

Figure 10.11 Excitation of isolated chlorophyll by light, page 189

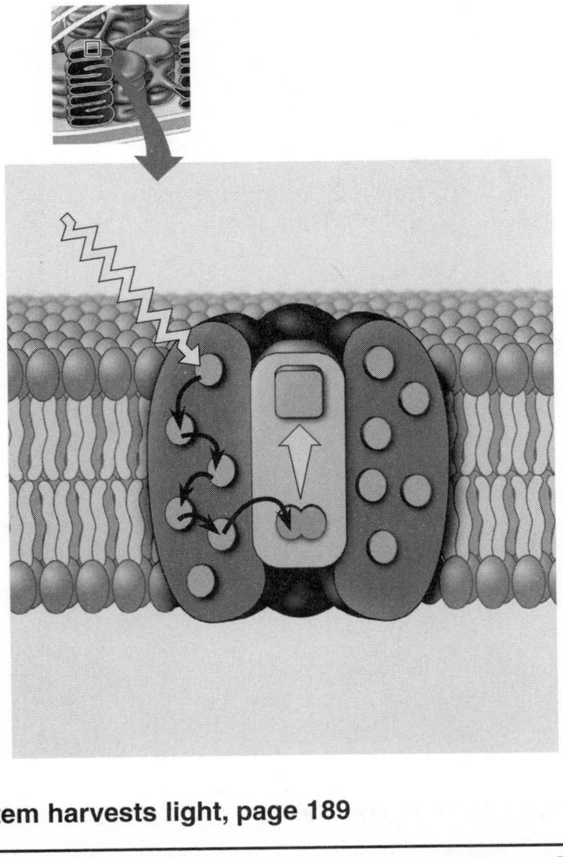

Figure 10.12 How a photosystem harvests light, page 189

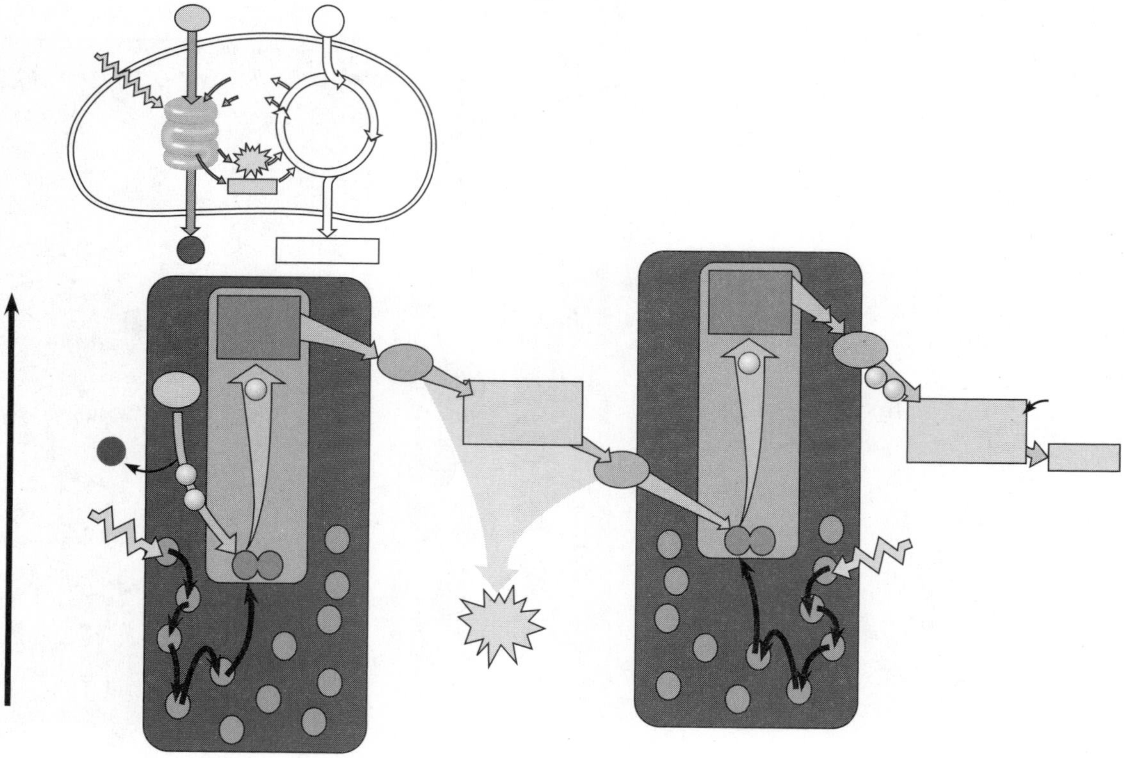

Figure 10.13 How noncyclic electron flow during the light reactions generates ATP and NADPH, page 190

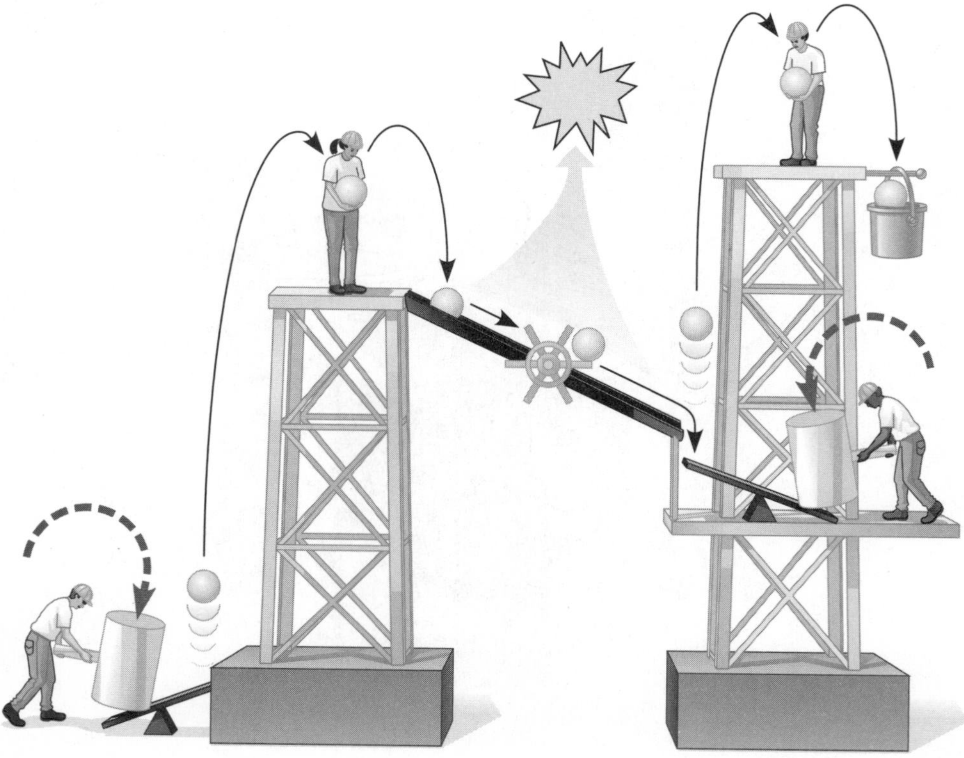

Figure 10.14 A mechanical analogy for the light reactions, page 191

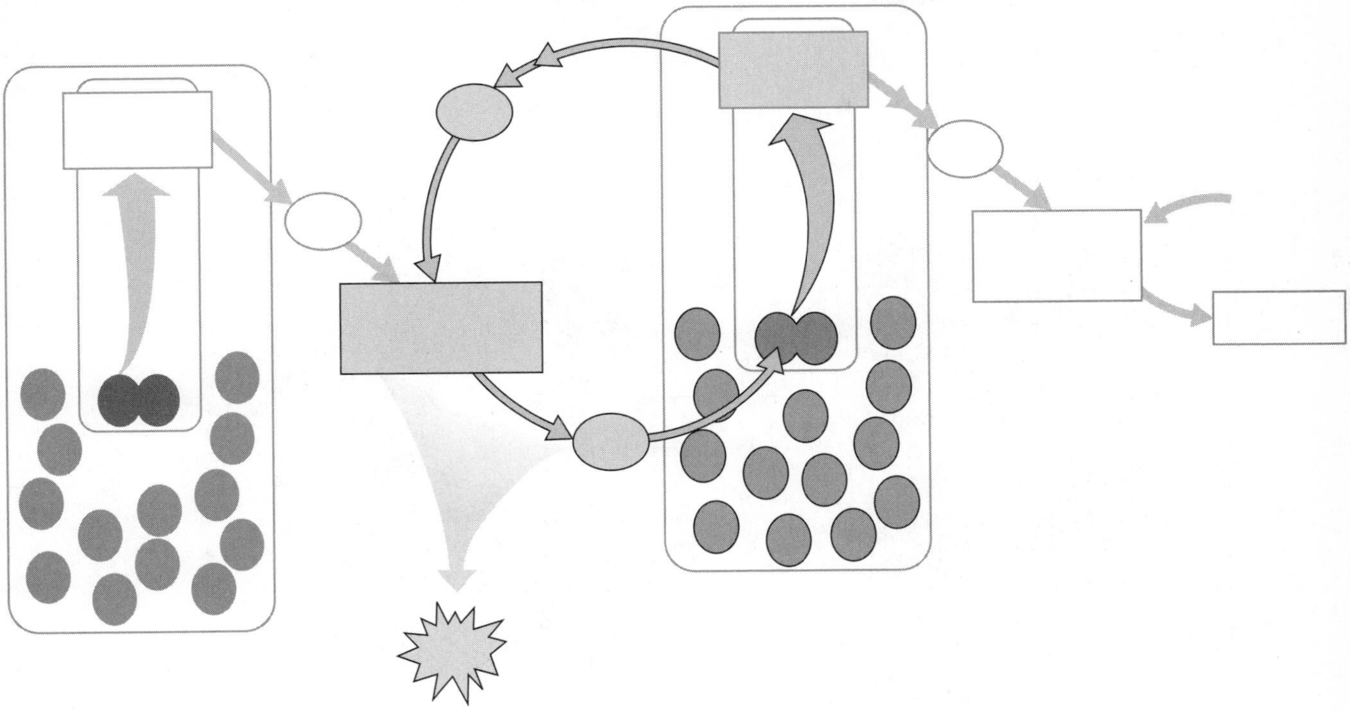

Figure 10.15 Cyclic electron flow, page 191

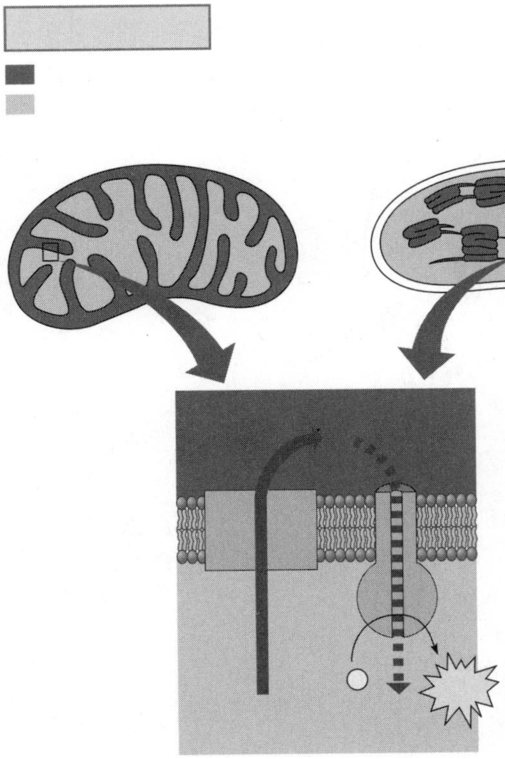

Figure 10.16 Comparison of chemiosmosis in mitochondria and chloroplasts, page 192

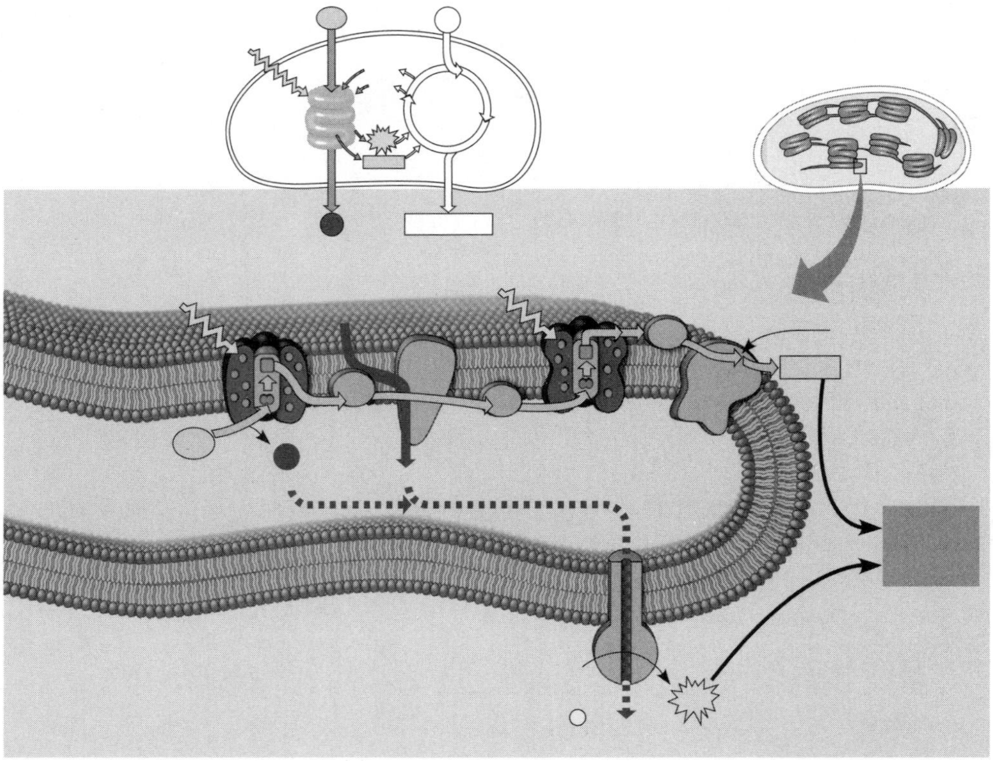

Figure 10.17 The light reactions and chemiosmosis: the organization of the thylakoid membrane, page 193

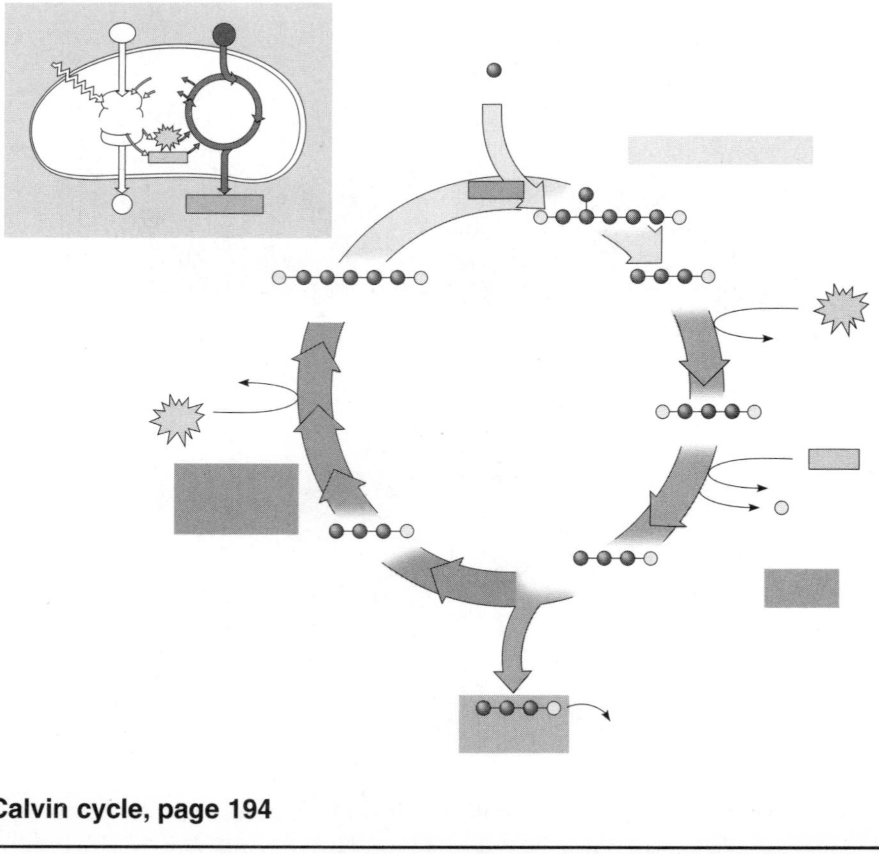

Figure 10.18 The Calvin cycle, page 194

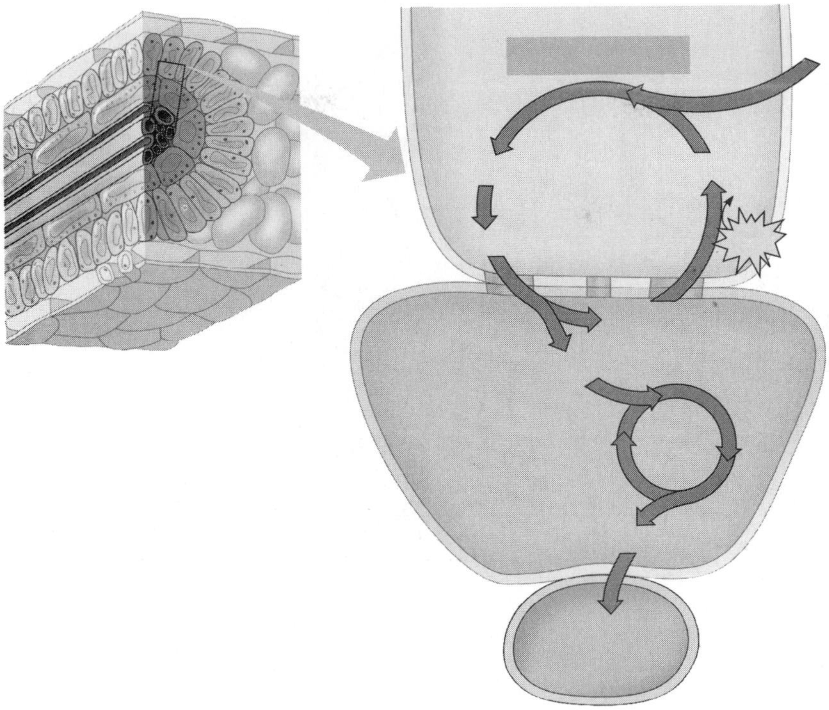

Figure 10.19 C_4 leaf anatomy and the C_4 pathway, page 196

Figure 10.20 C_4 and CAM photosynthesis compared, page 197

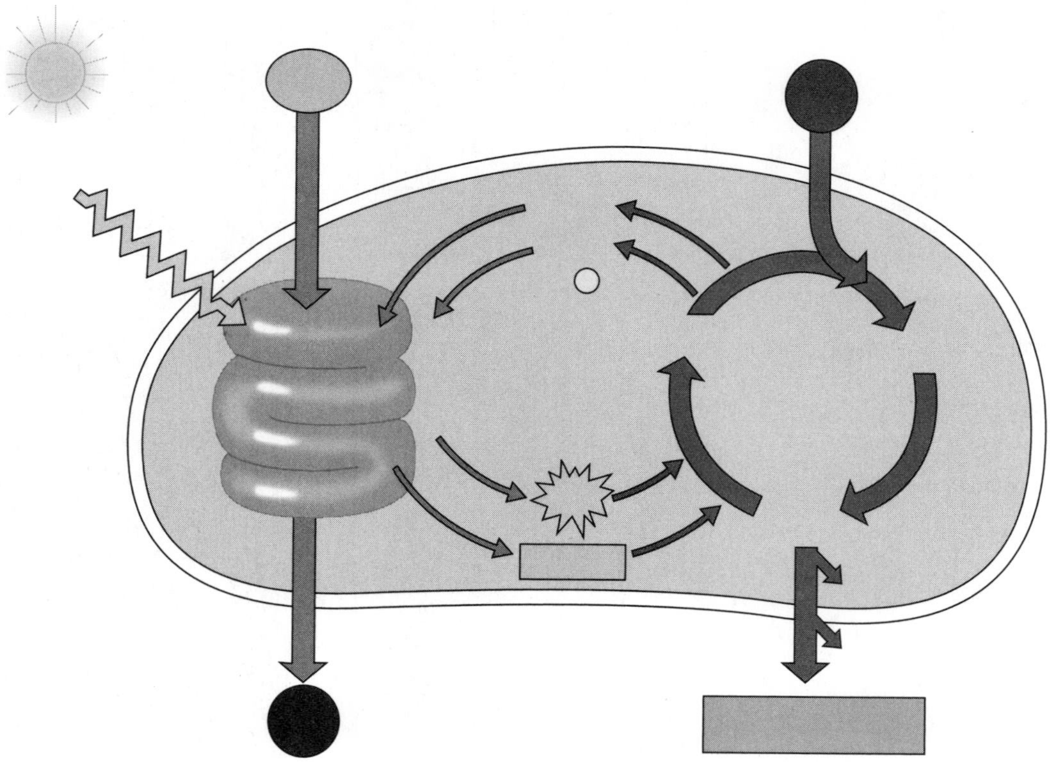

Figure 10.21 A review of photosynthesis, page 198

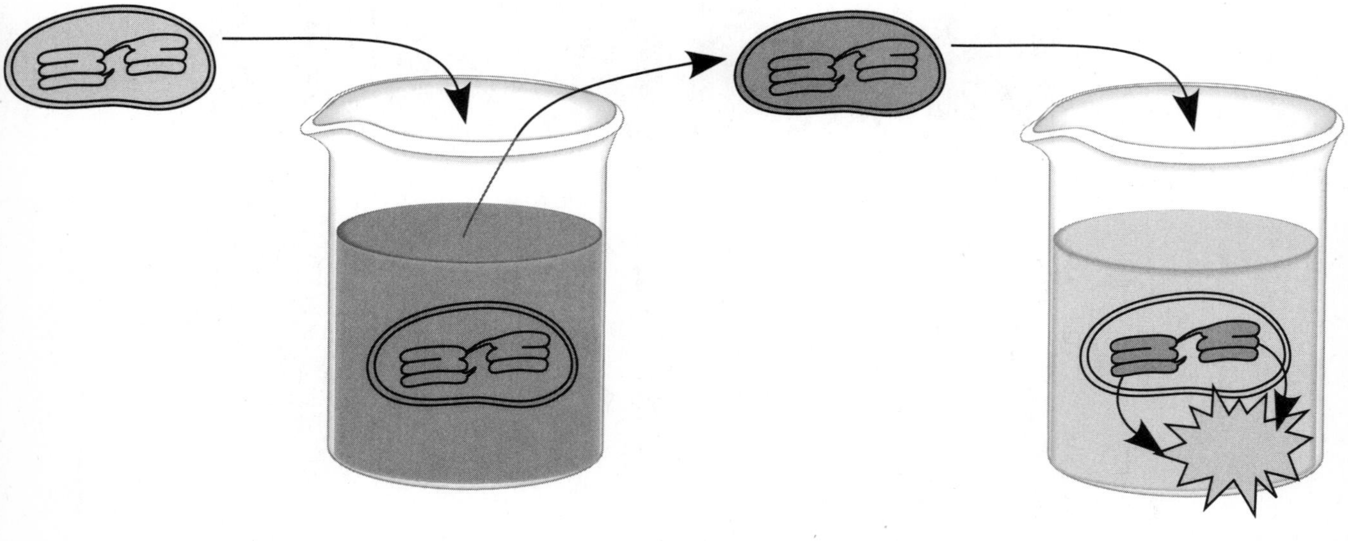

Figure 10.UN1 An experiment with isolated chloroplasts, page 200

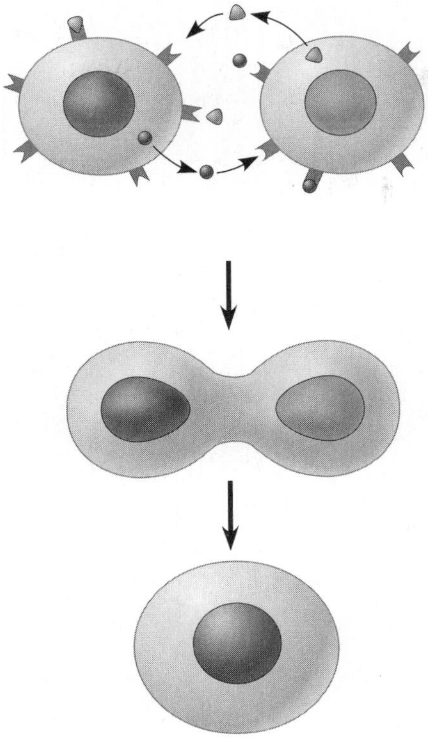

Figure 11.2 Communication between mating yeast cells, page 202

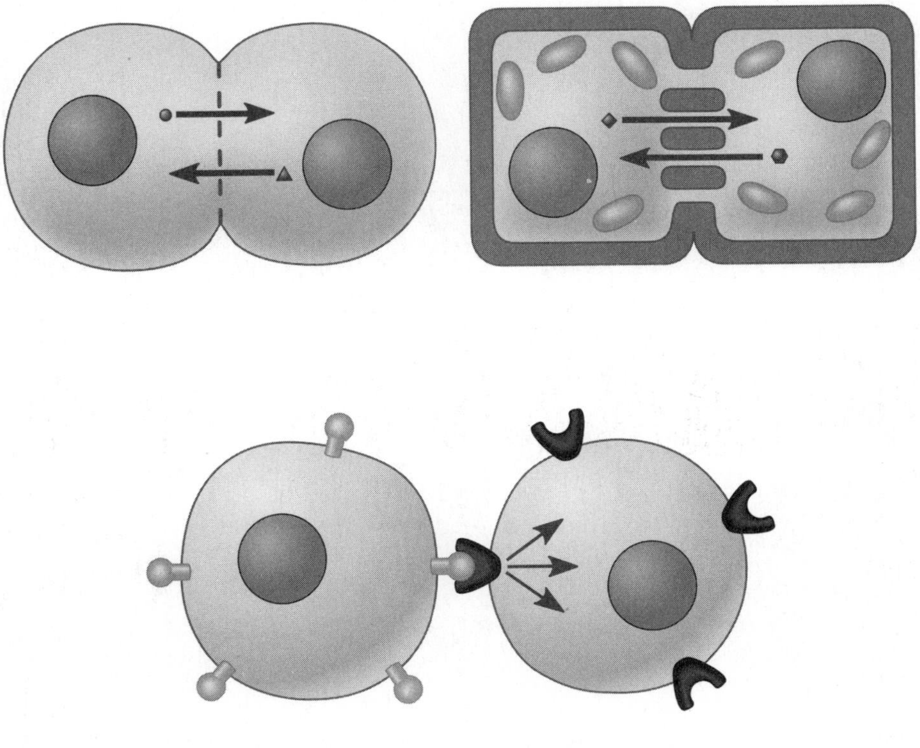

Figure 11.3 Communication by direct contact between cells, page 202

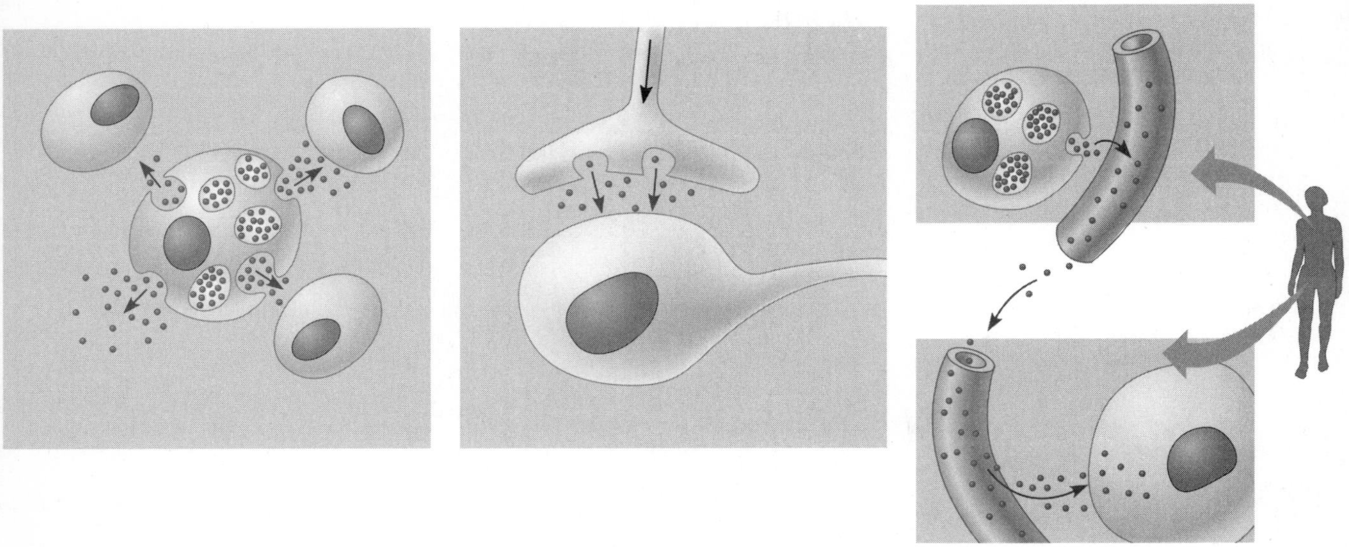

Figure 11.4 Local and long-distance cell communication in animals, page 203

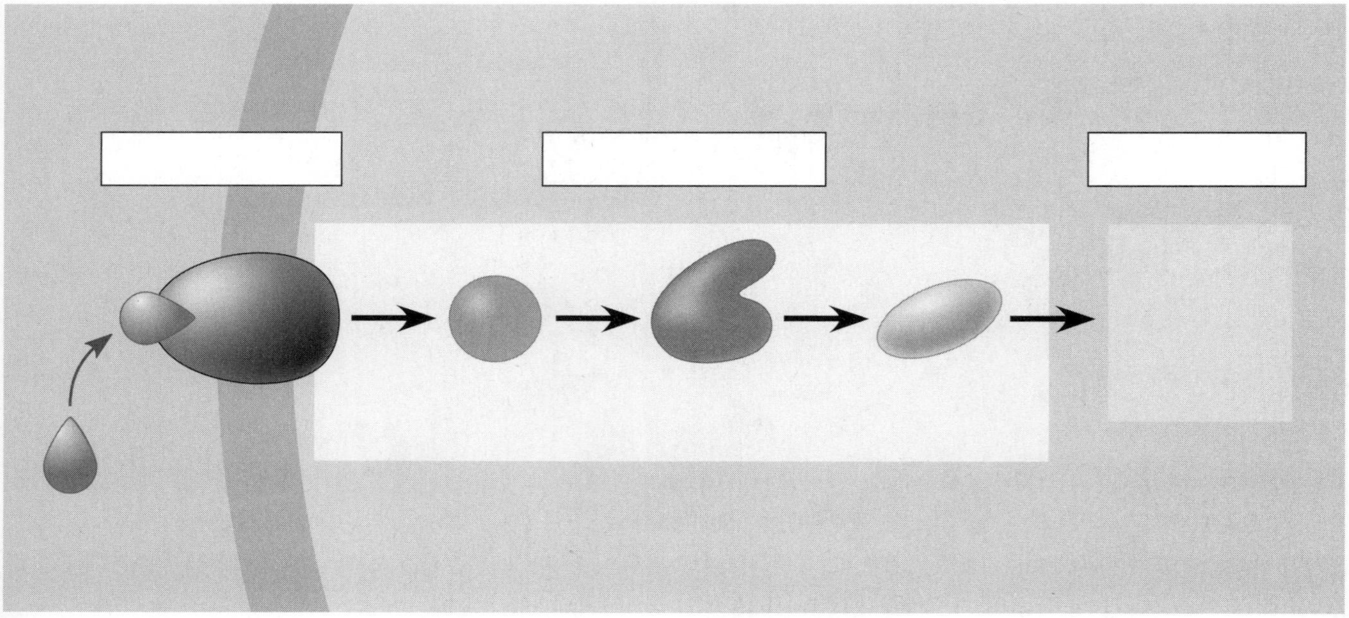

Figure 11.5 Overview of cell signaling, page 204

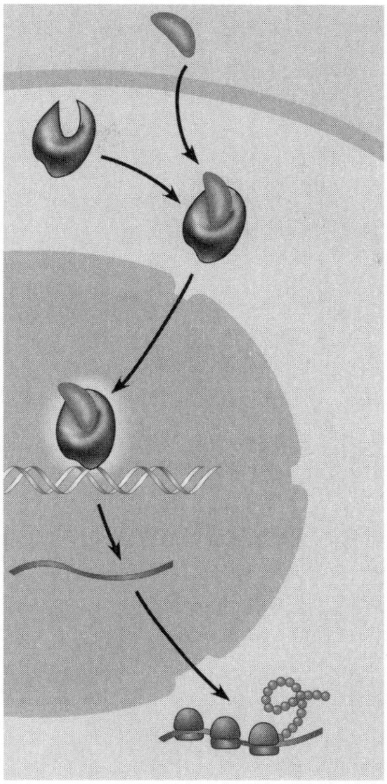

Figure 11.6 Steroid hormone interacting with an intracellular receptor, page 205

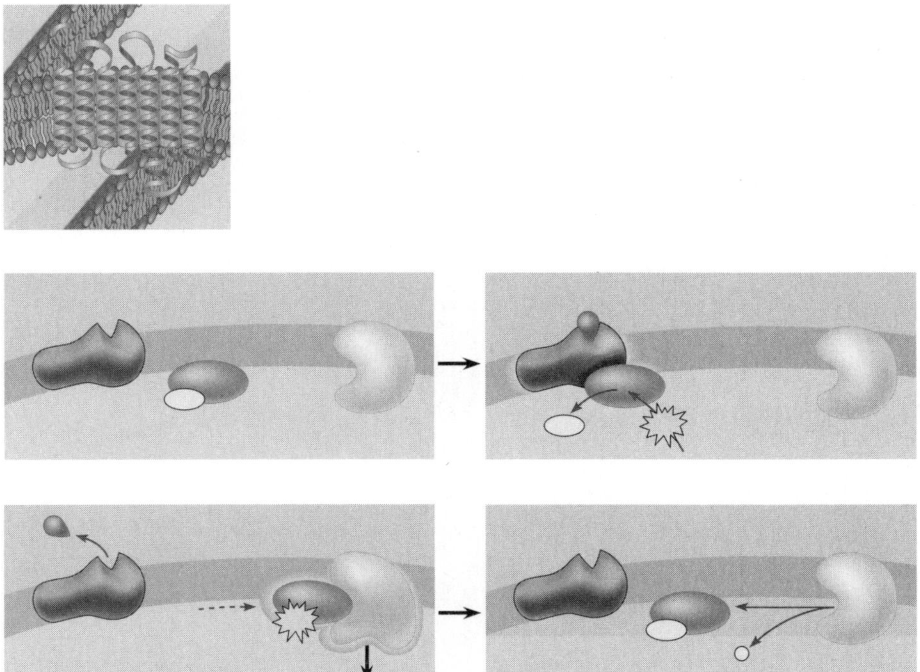

Figure 11.7 Membrane receptors: G-protein-linked receptors, page 206

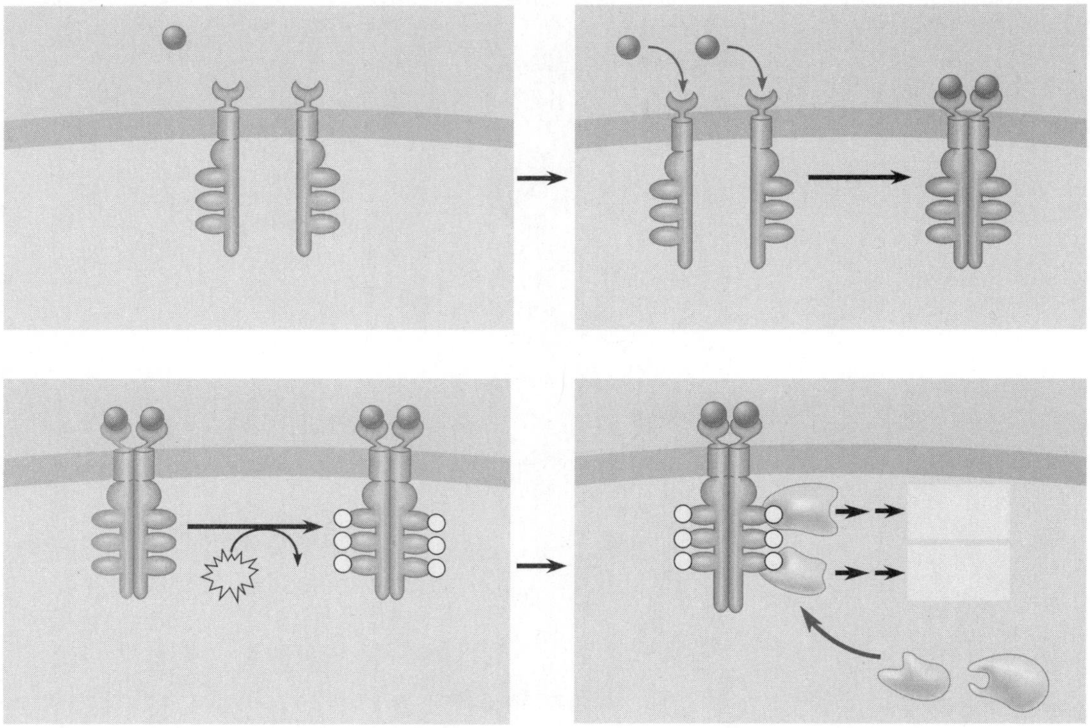

Figure 11.7 Membrane receptors: receptor tyrosine kinases, page 207

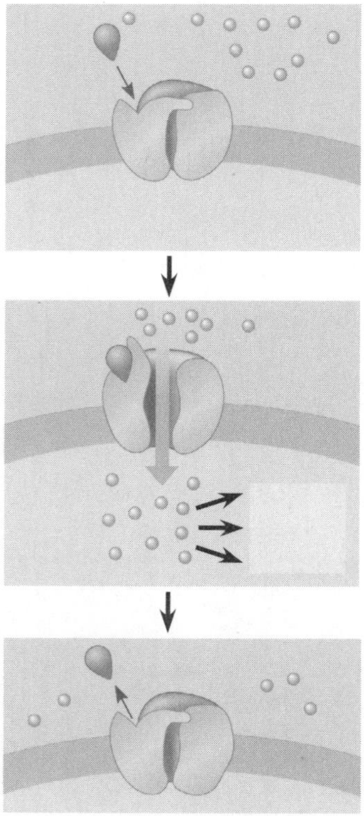

Figure 11.7 Membrane receptors: ion channel receptors, page 208

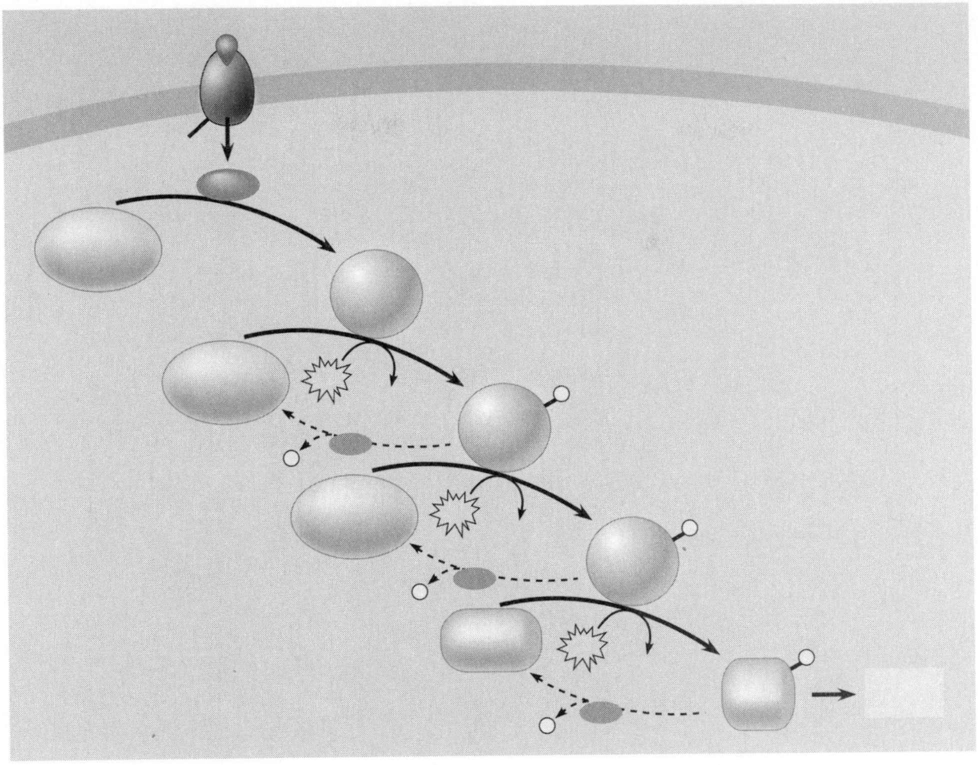

Figure 11.8 A phosphorylation cascade, page 209

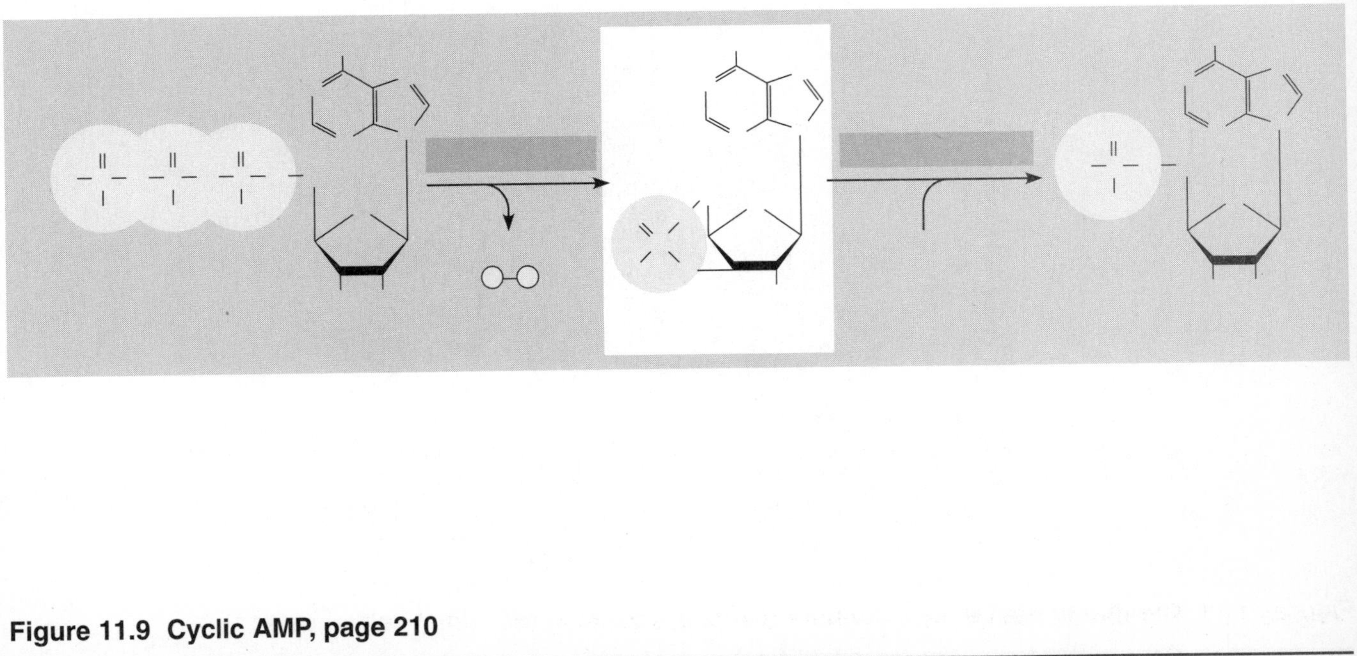

Figure 11.9 Cyclic AMP, page 210

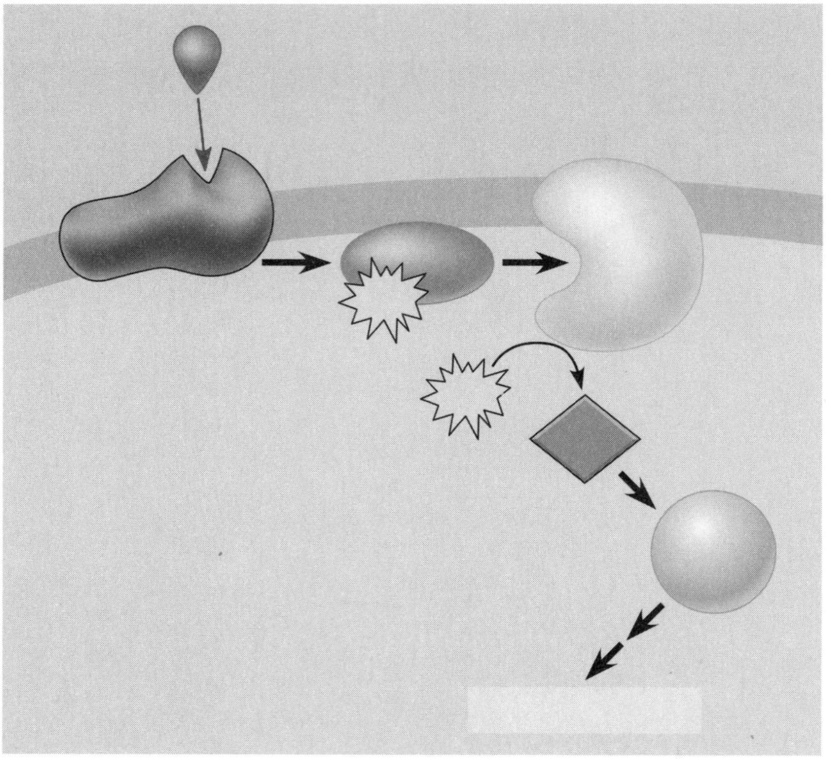

Figure 11.10 cAMP as a second messenger in a G-protein-signaling pathway, page 211

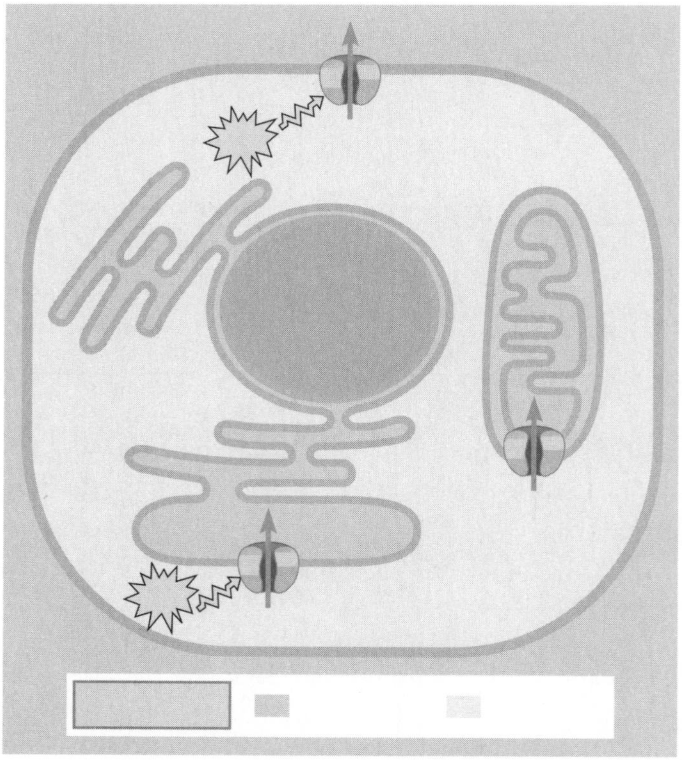

Figure 11.11 The maintenance of calcium ion concentrations in an animal cell, page 211

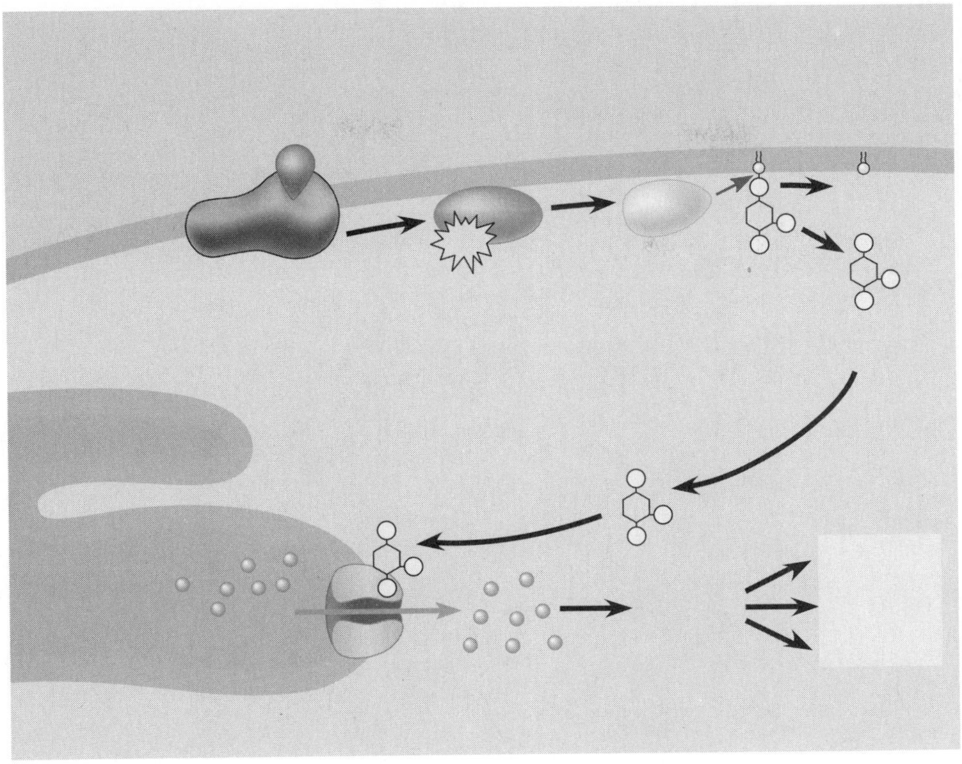

Figure 11.12 Calcium and IP_3 in signaling pathways, page 212

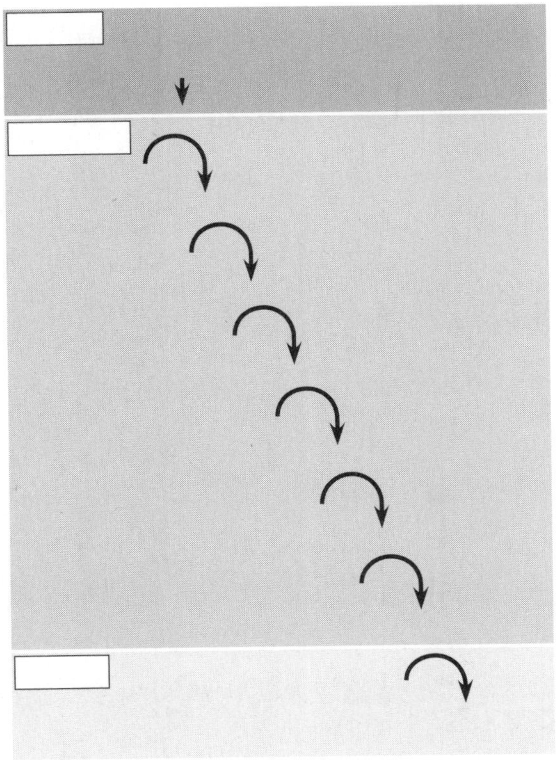

Figure 11.13 Cytoplasmic response to a signal: the stimulation of glycogen breakdown by epinephrine, page 213

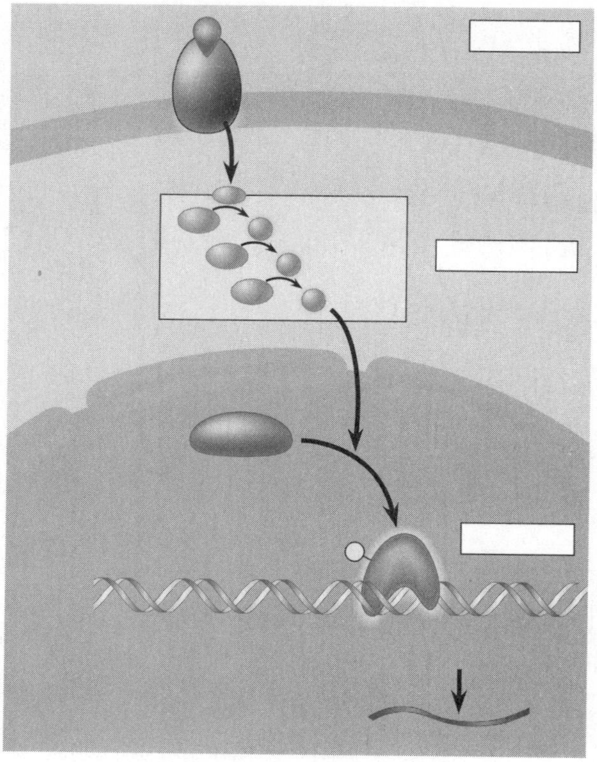

Figure 11.14 Nuclear responses to a signal: the activation of a specific gene by a growth factor, page 213

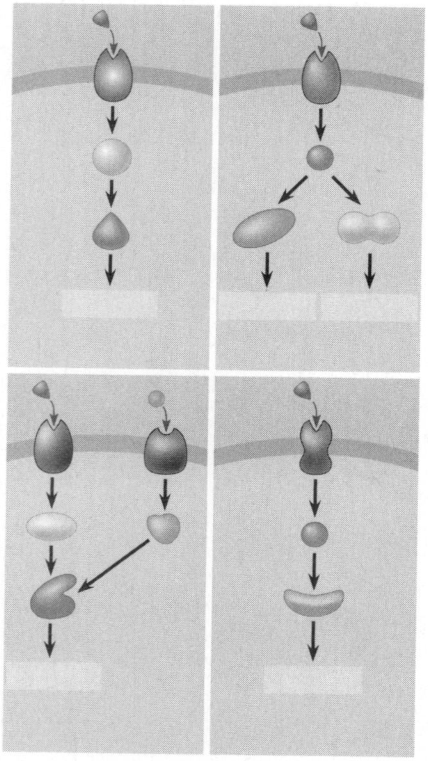

Figure 11.15 The specificity of cell signaling, page 214

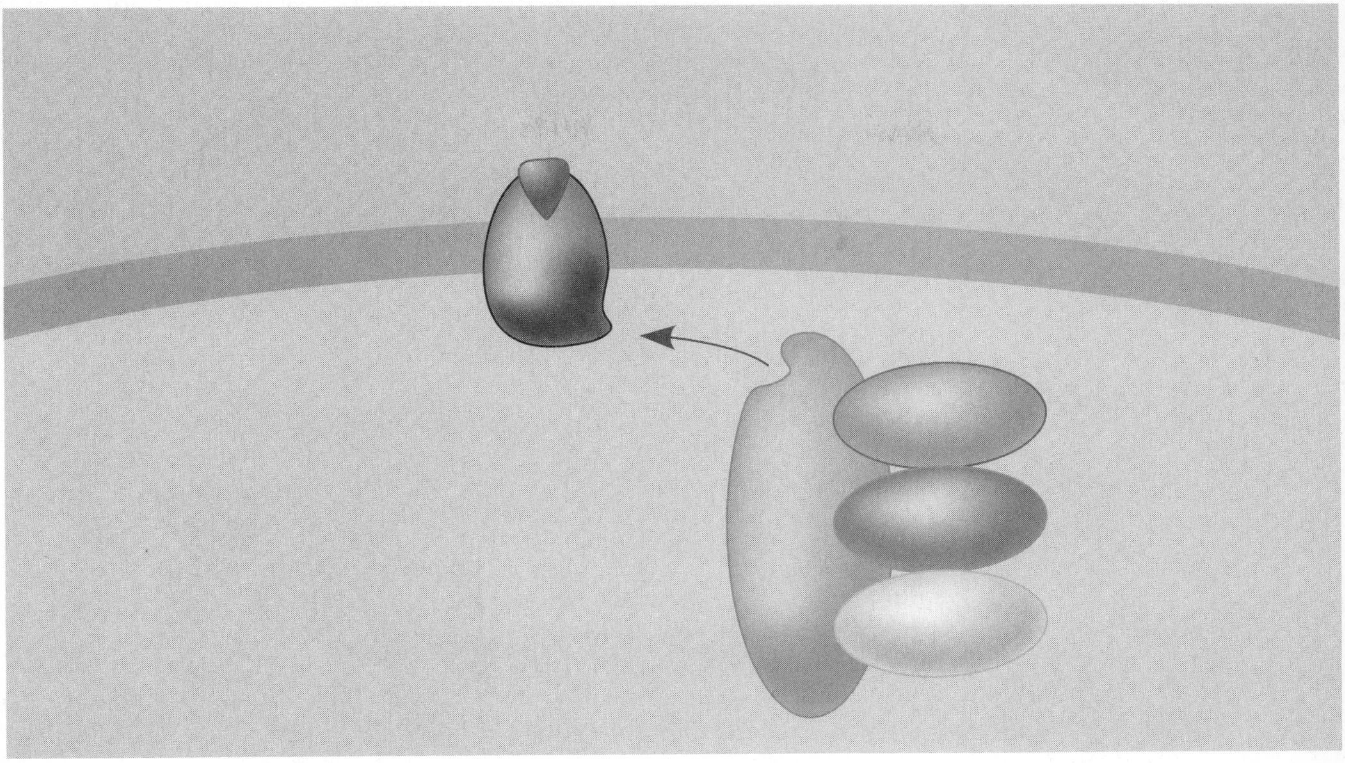

Figure 11.16 A scaffolding protein, page 215

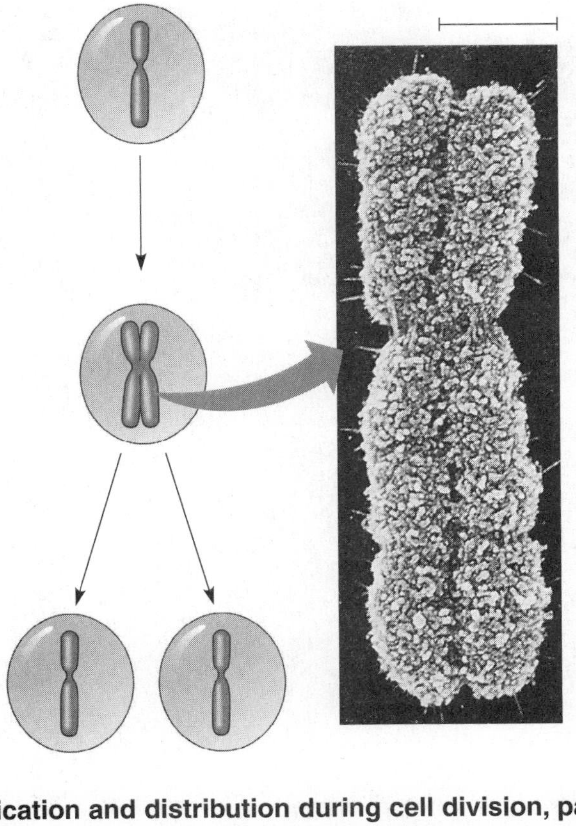

Figure 12.4 Chromosome duplication and distribution during cell division, page 220

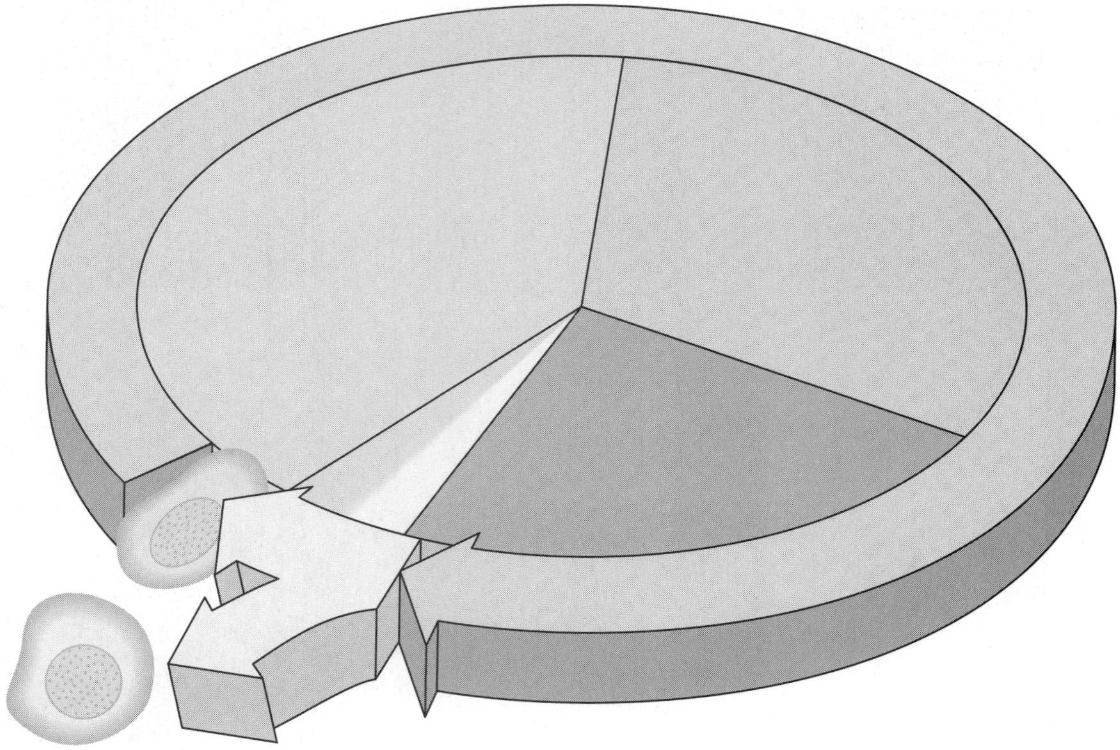

Figure 12.5 The cell cycle, page 221

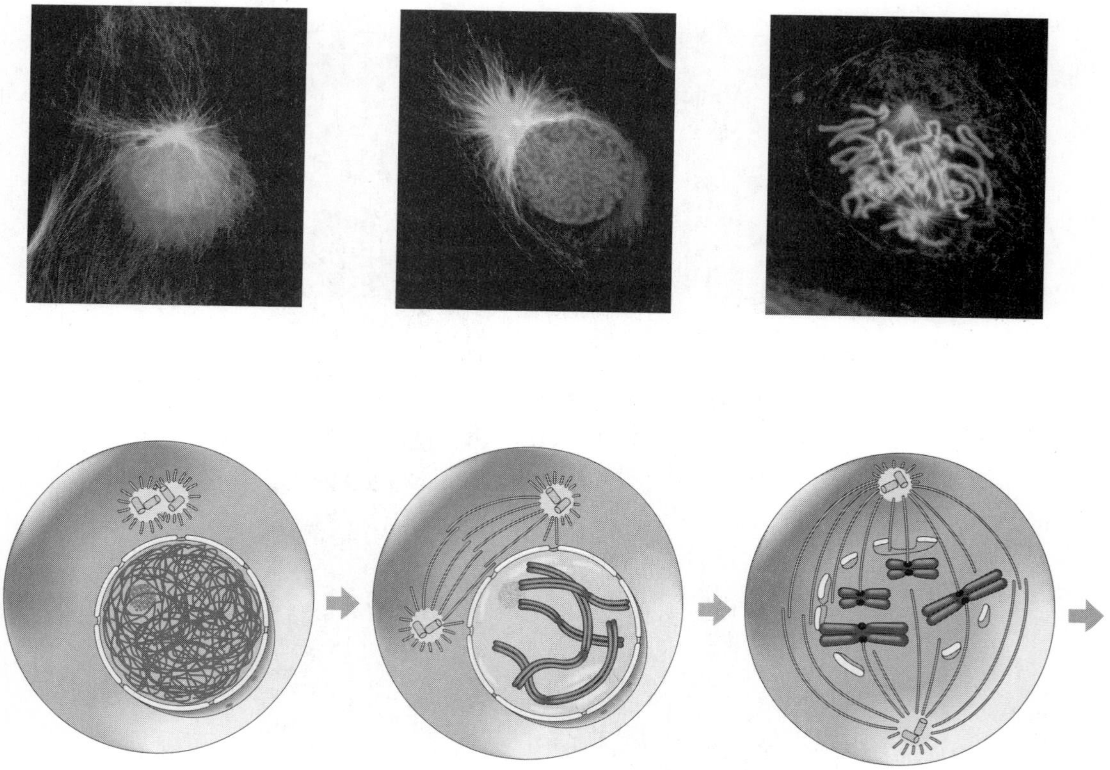

Figure 12.6 The mitotic division of an animal cell: G_2 of interphase; prophase; prometaphase, page 222

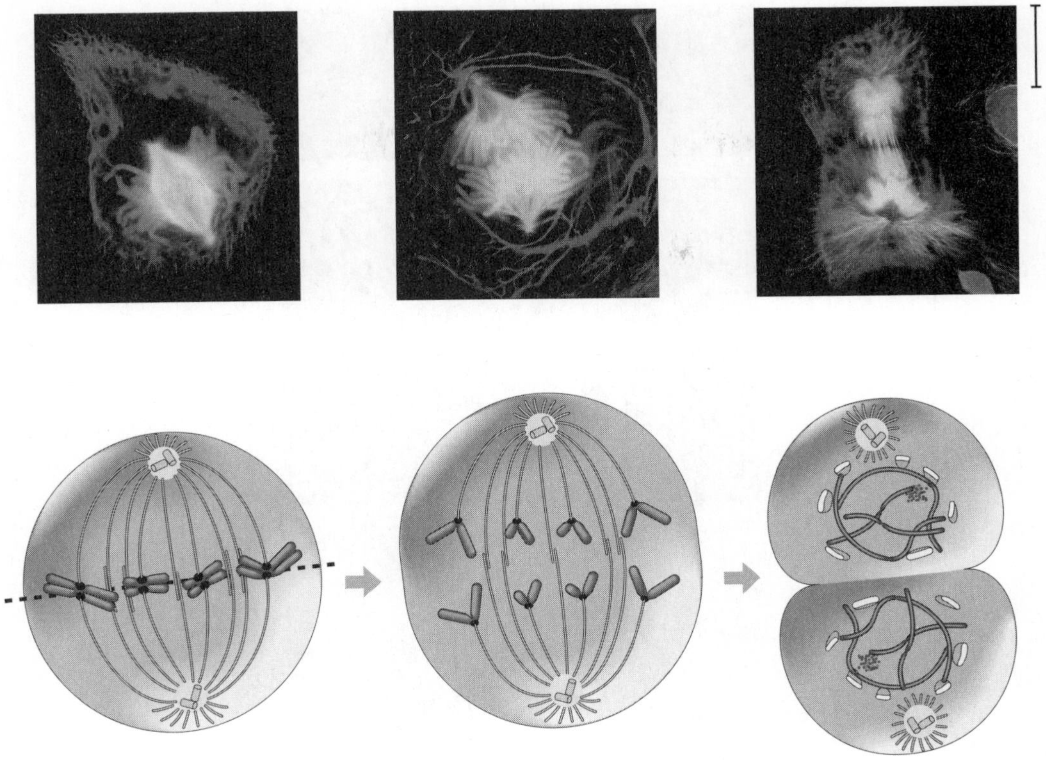

Figure 12.6 The mitotic division of an animal cell: metaphase; anaphase; telophase and cytokinesis, page 223

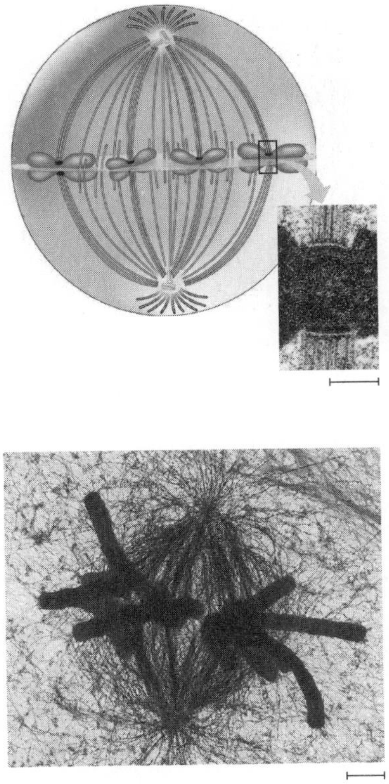

Figure 12.7 The mitotic spindle at metaphase, page 224

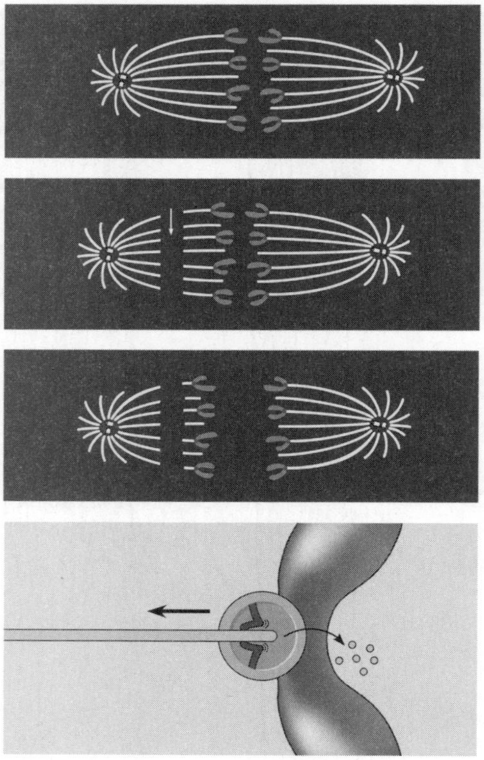

Figure 12.8 During anaphase, do kinetochore microtubules shorten at their spindle pole ends or their kinetochore ends?, page 225

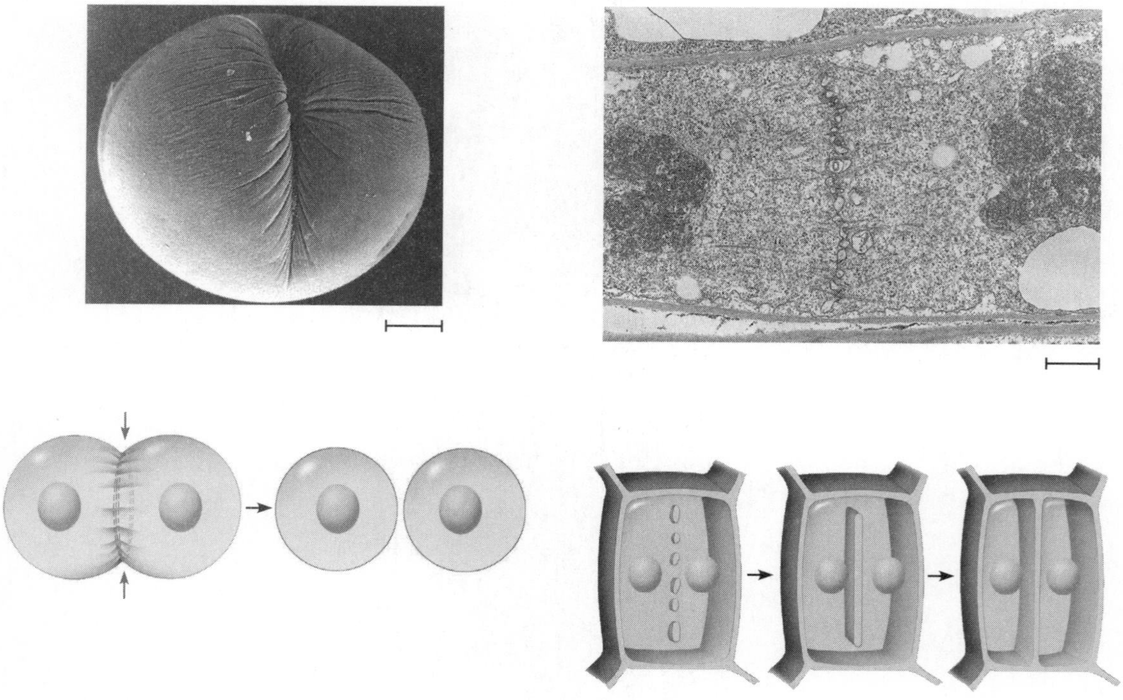

Figure 12.9 Cytokinesis in animal and plant cells, page 225

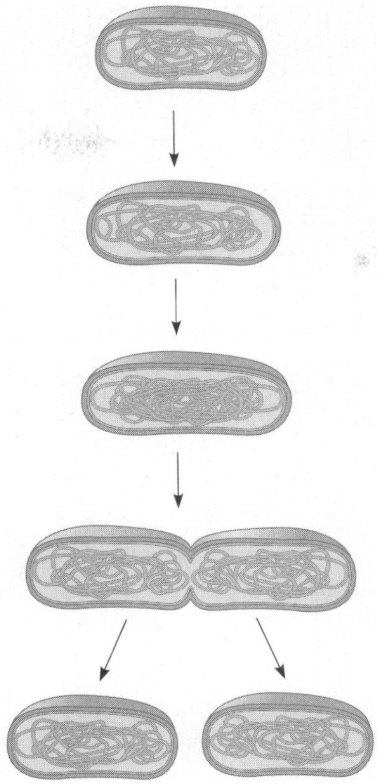

Figure 12.11 Bacterial cell division (binary fission), page 227

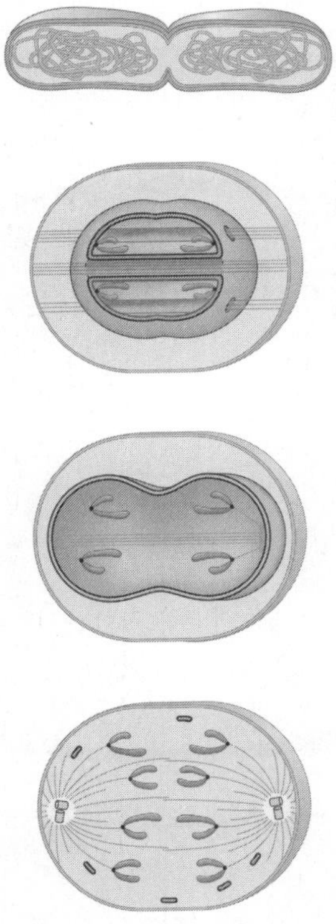

Figure 12.12 A hypothetical sequence for the evolution of mitosis, page 227

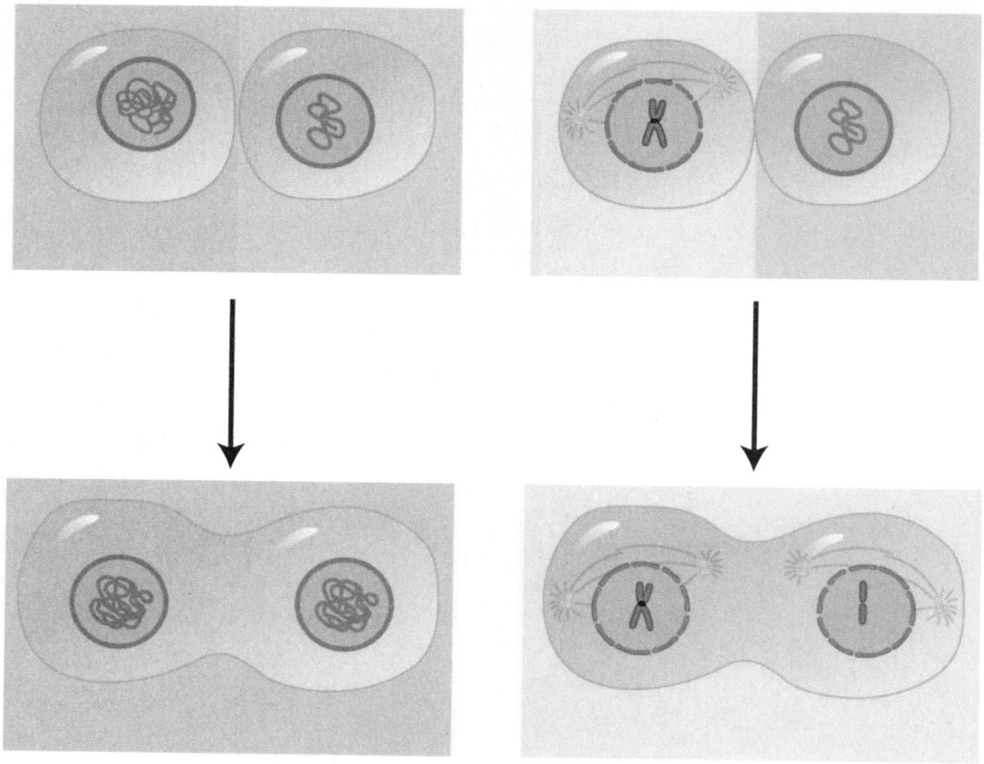

Figure 12.13 Are there molecular signals in the cytoplasm that regulate the cell cycle?, page 228

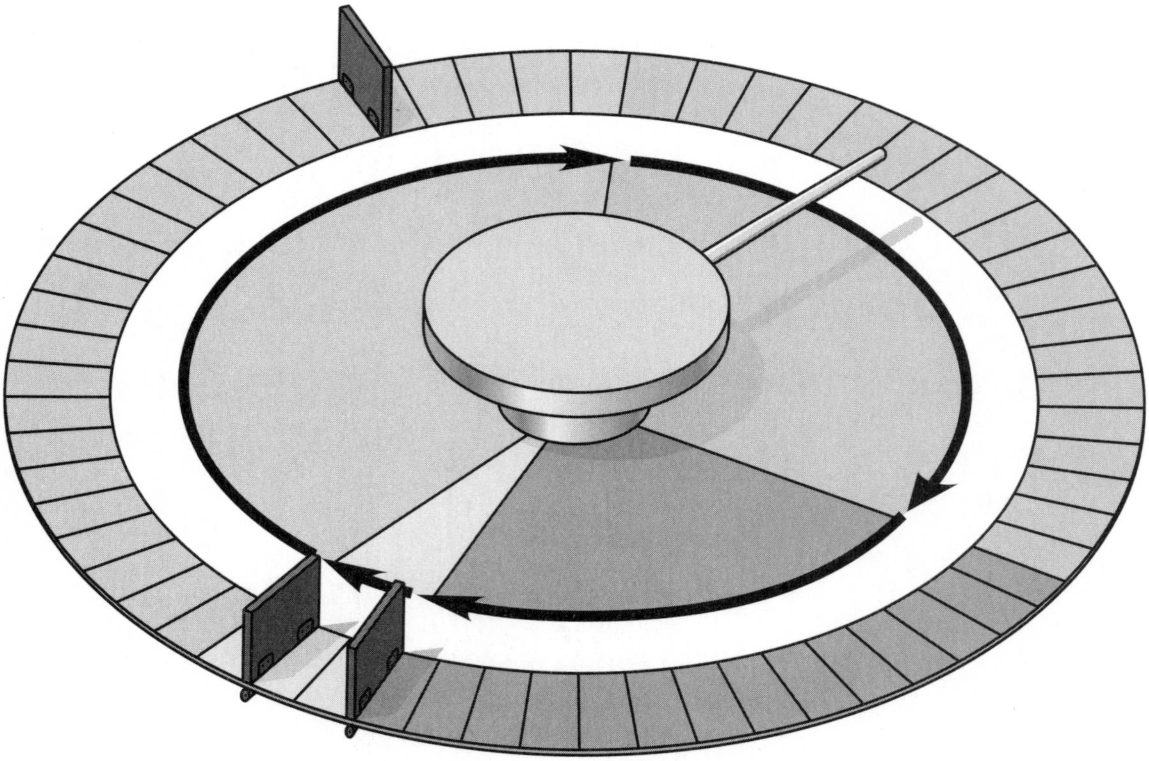

Figure 12.14 Mechanical analogy for the cell cycle control system, page 229

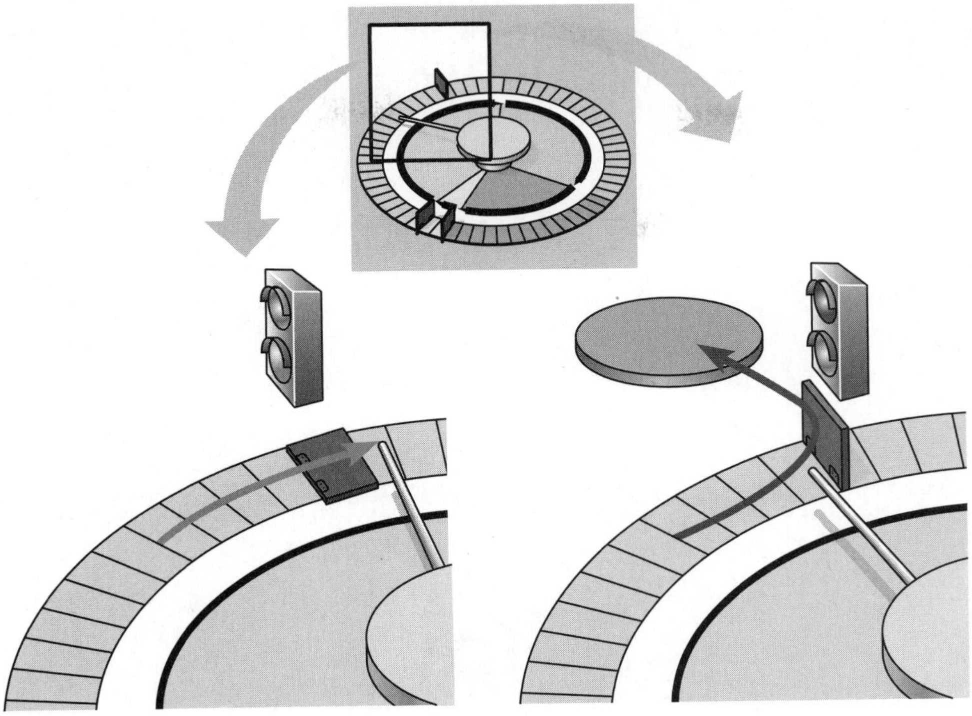

Figure 12.15 The G_1 checkpoint, page 229

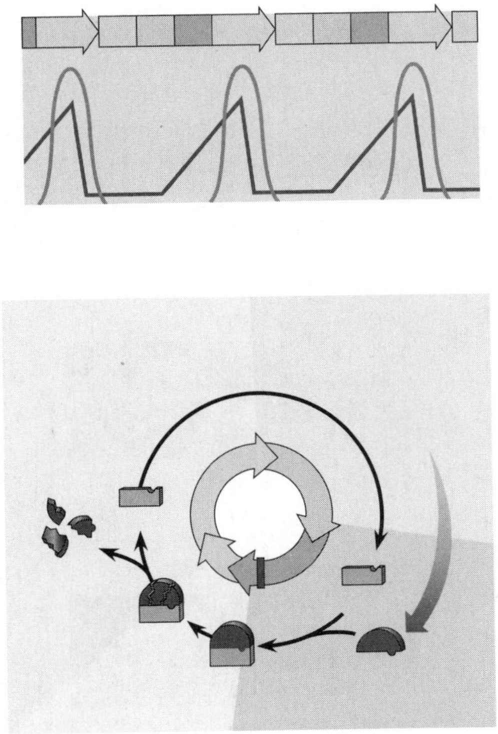

Figure 12.16 Molecular control of the cell cycle at the G_2 checkpoint, page 230

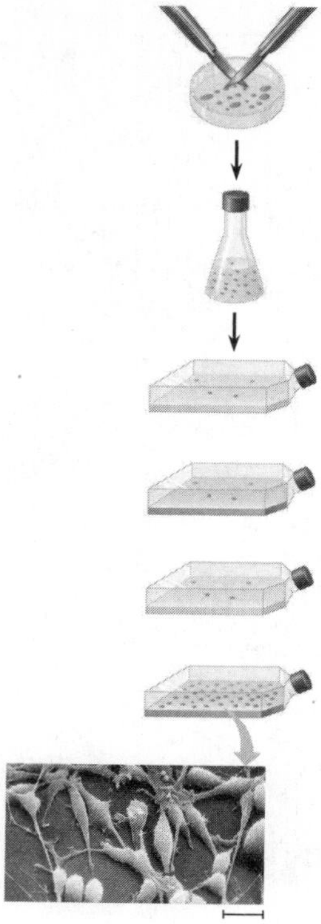

Figure 12.17 Does platelet-derived growth factor (PDGF) stimulate the division of human fibroblast cells in culture?, page 231

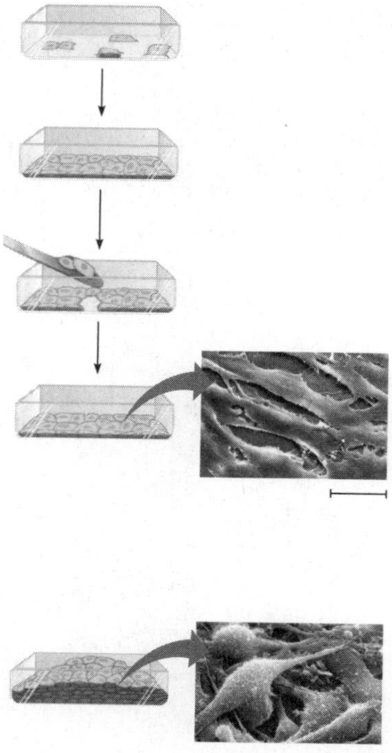

Figure 12.18 Density-dependent inhibition and anchorage dependence of cell division, page 232

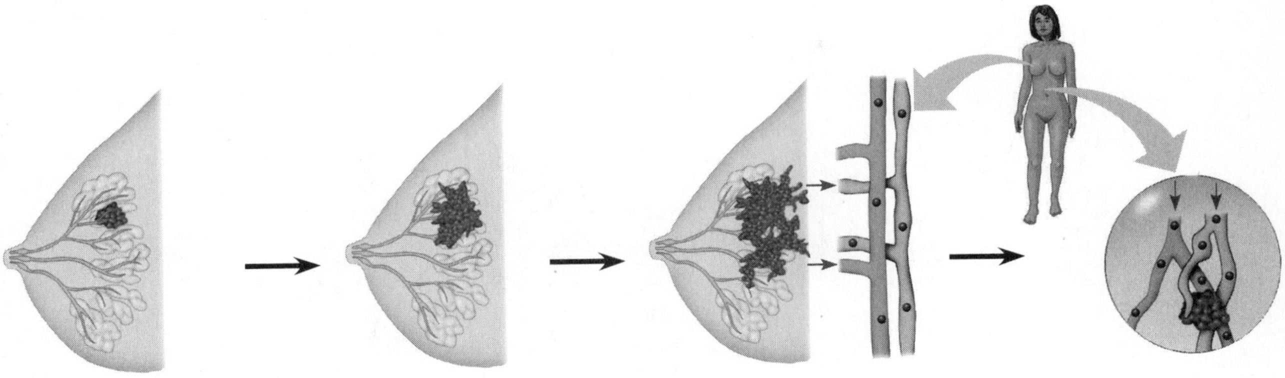

Figure 12.19 The growth and metastasis of a malignant breast tumor, page 233

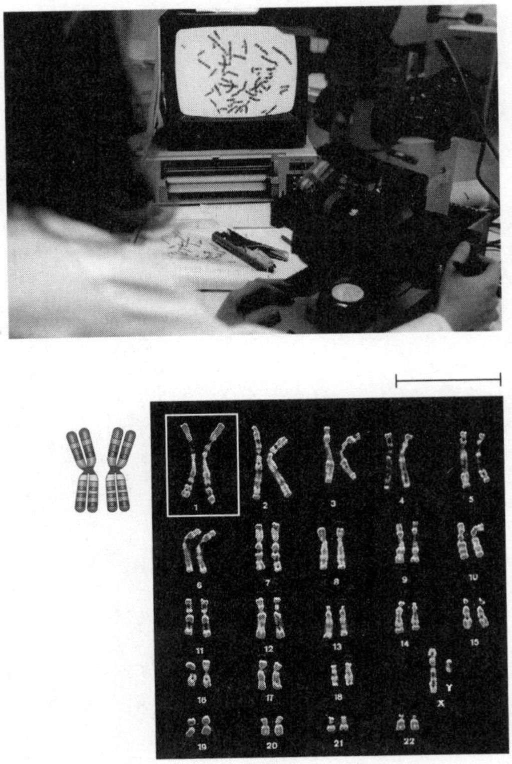

Figure 13.3 Preparing a karyotype, page 240

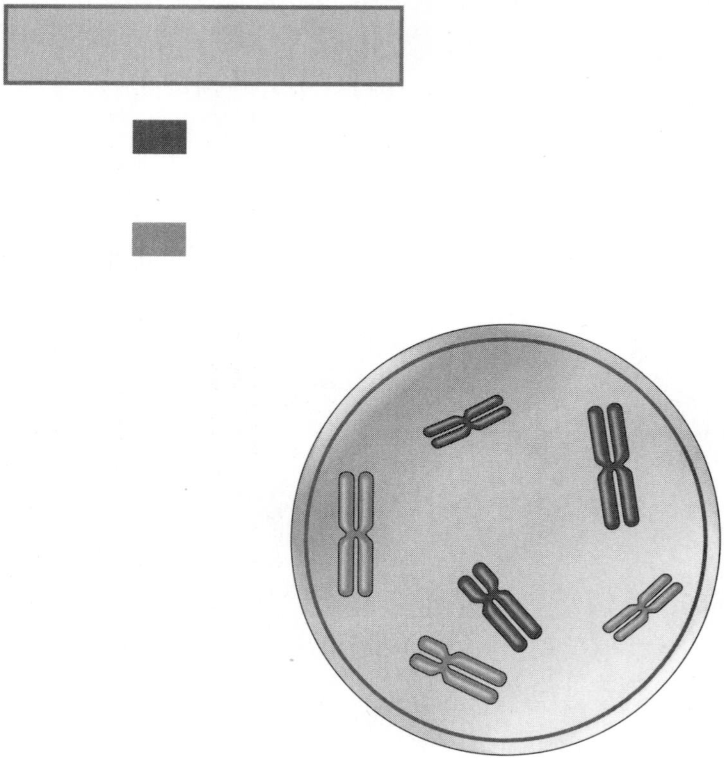

Figure 13.4 Describing chromosomes, page 241

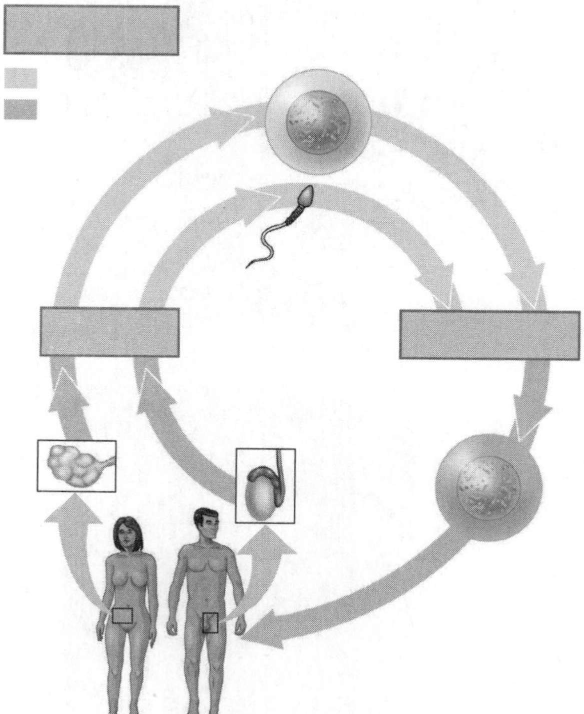

Figure 13.5 The human life cycle, page 241

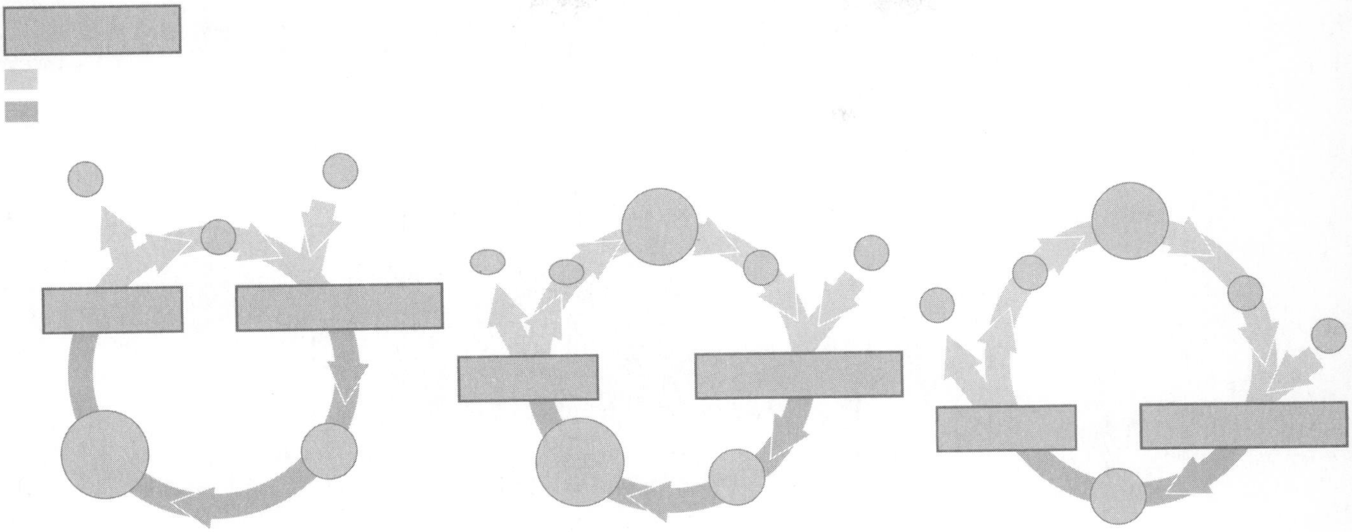

Figure 13.6 Three types of sexual life cycles, page 242

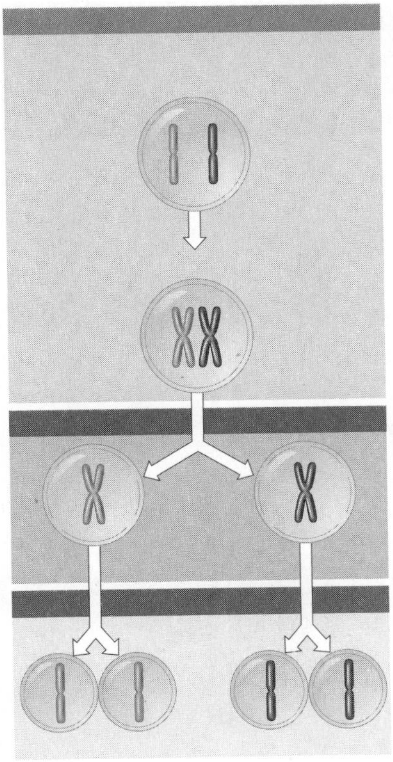

Figure 13.7 Overview of meiosis: how meiosis reduces chromosome number, page 243

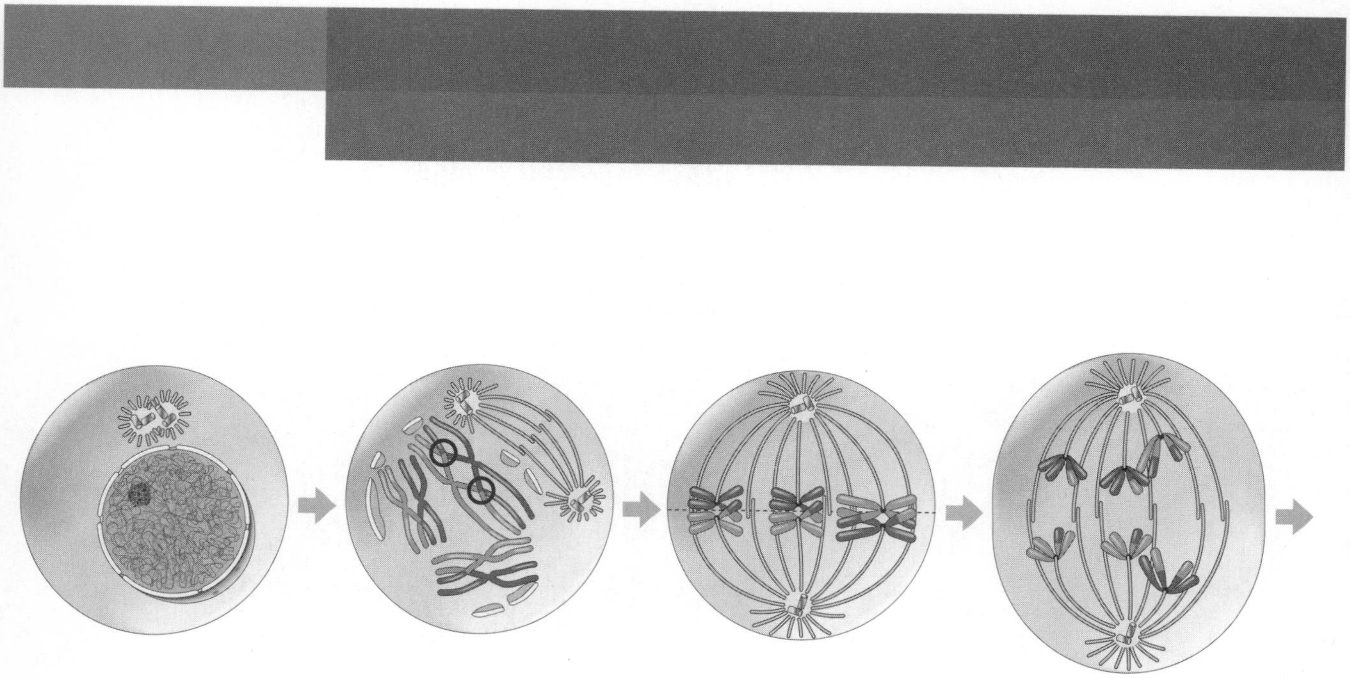

Figure 13.8 The meiotic division of an animal cell (part 1), page 244

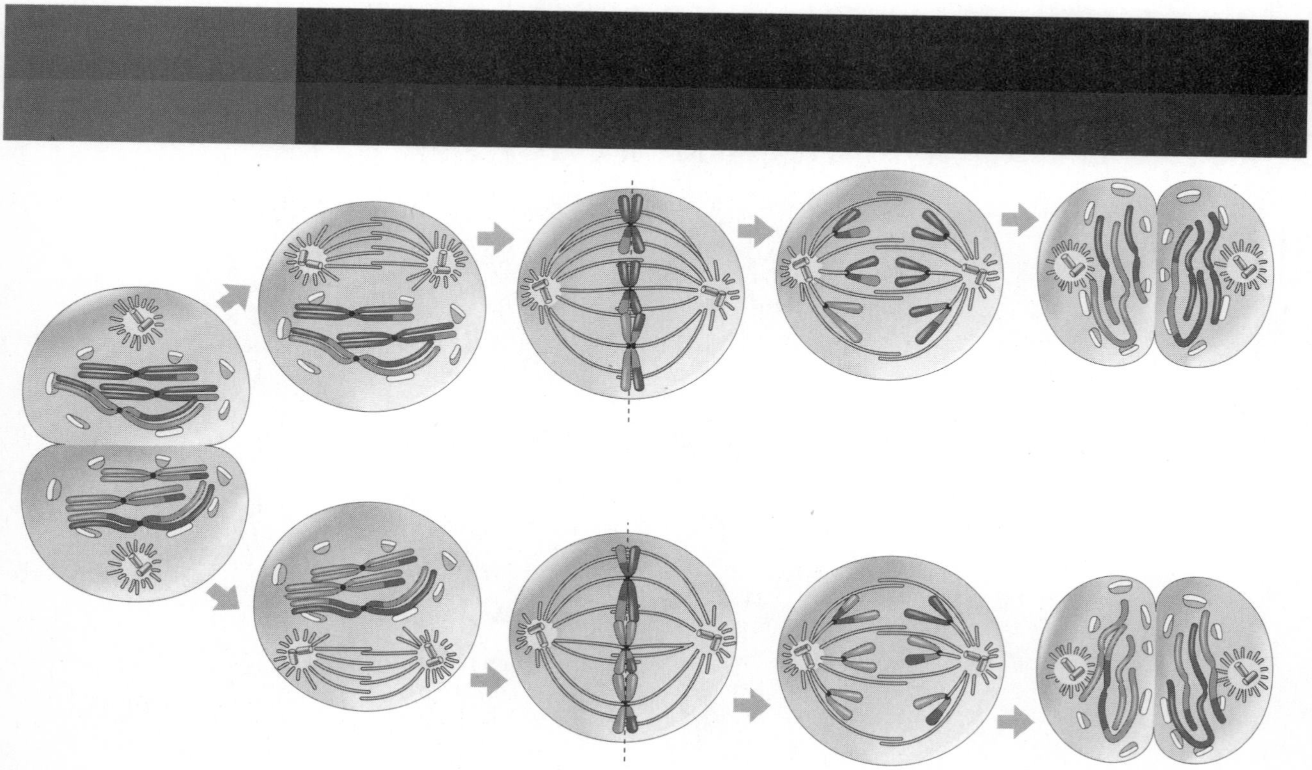

Figure 13.8 The meiotic division of an animal cell (part 2), page 245

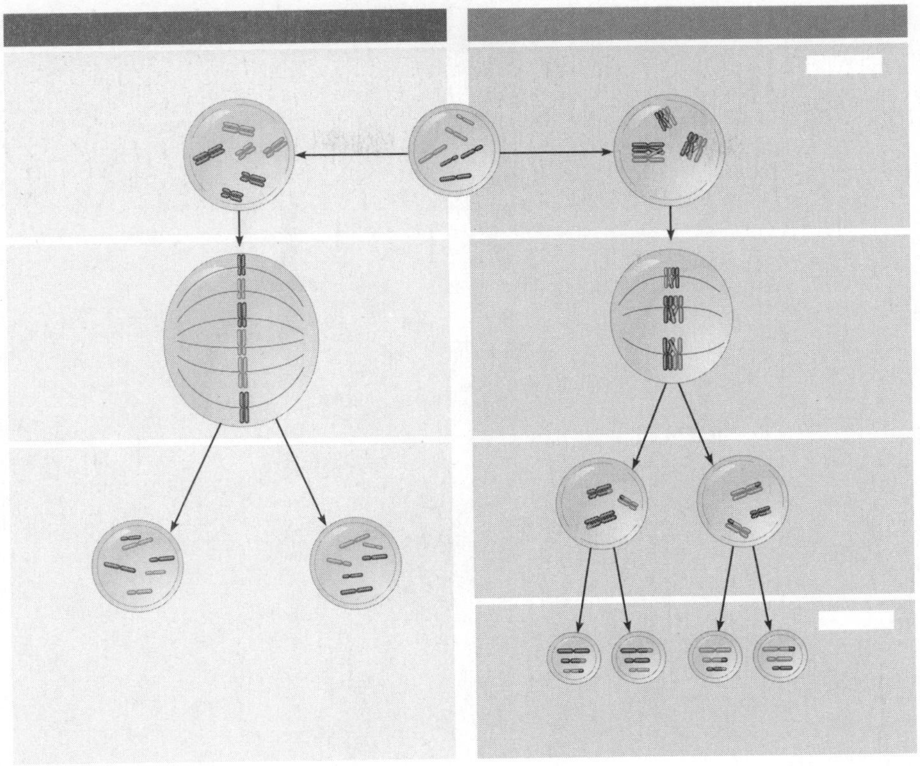

Figure 13.9 A comparison of mitosis and meiosis, page 246

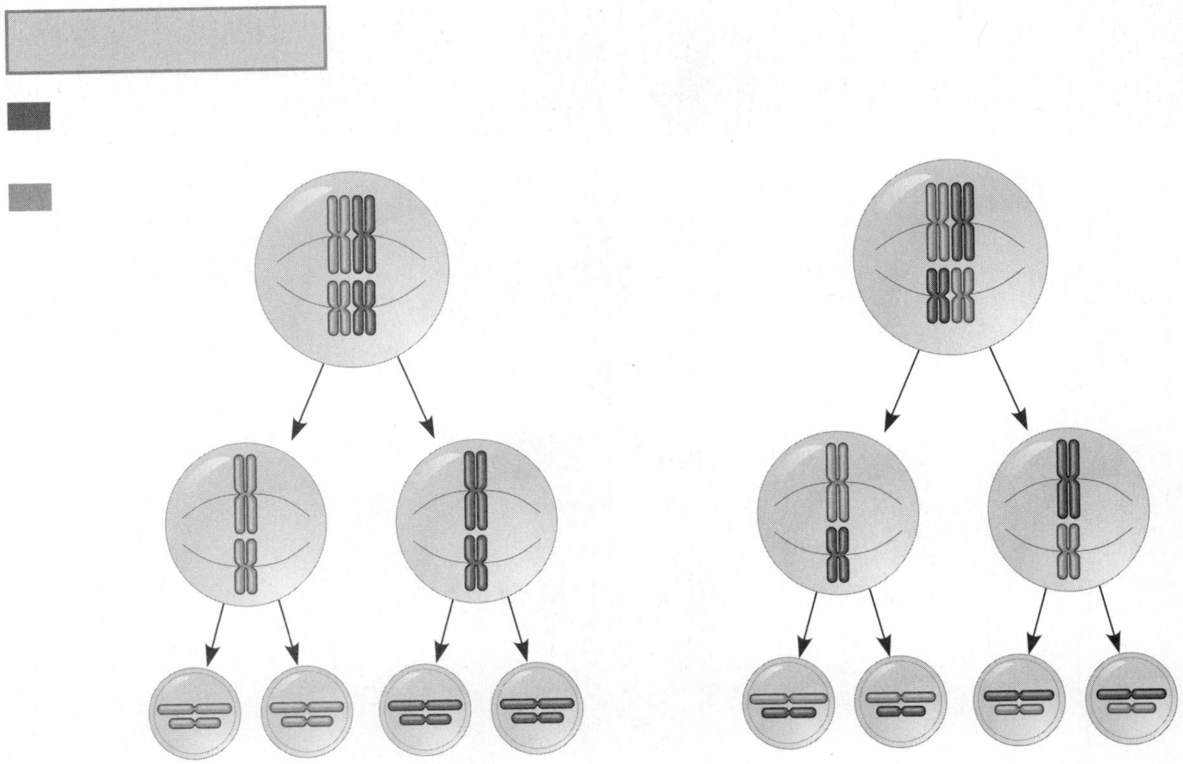

Figure 13.10 The independent assortment of homologous chromosomes in meiosis, page 248

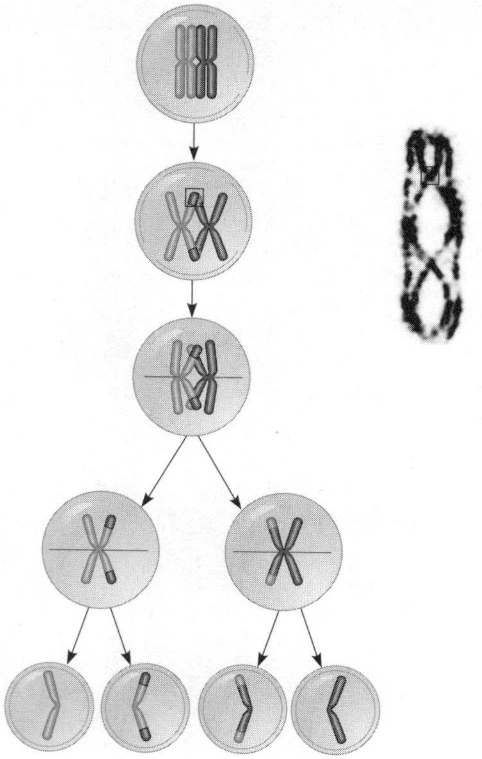

Figure 13.11 The results of crossing over during meiosis, page 249

Figure 14.2 Crossing pea plants, page 252

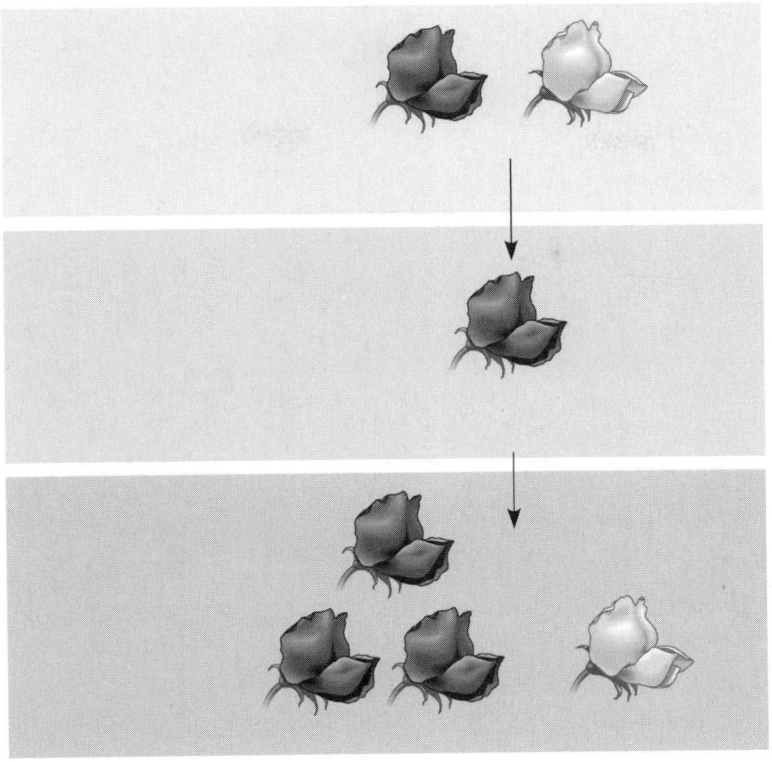

Figure 14.3 When F_1 pea plants with purple flowers are allowed to self-pollinate, what flower color appears in the F_2 generation?, page 253

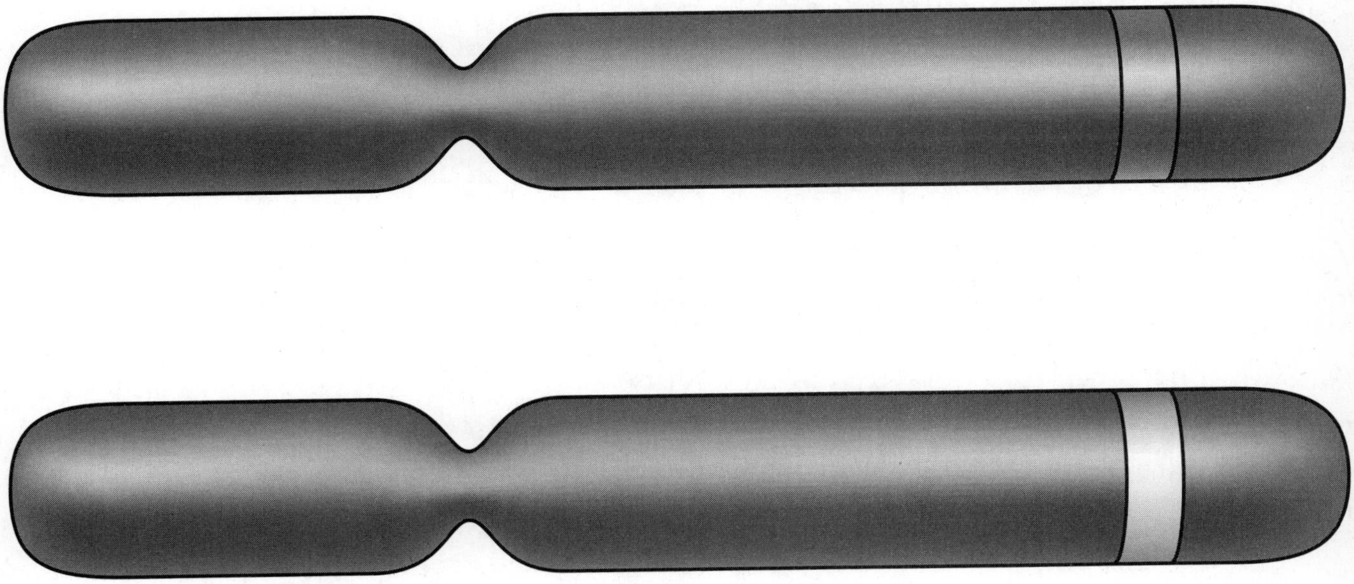

Figure 14.4 Alleles, alternative versions of a gene, page 255

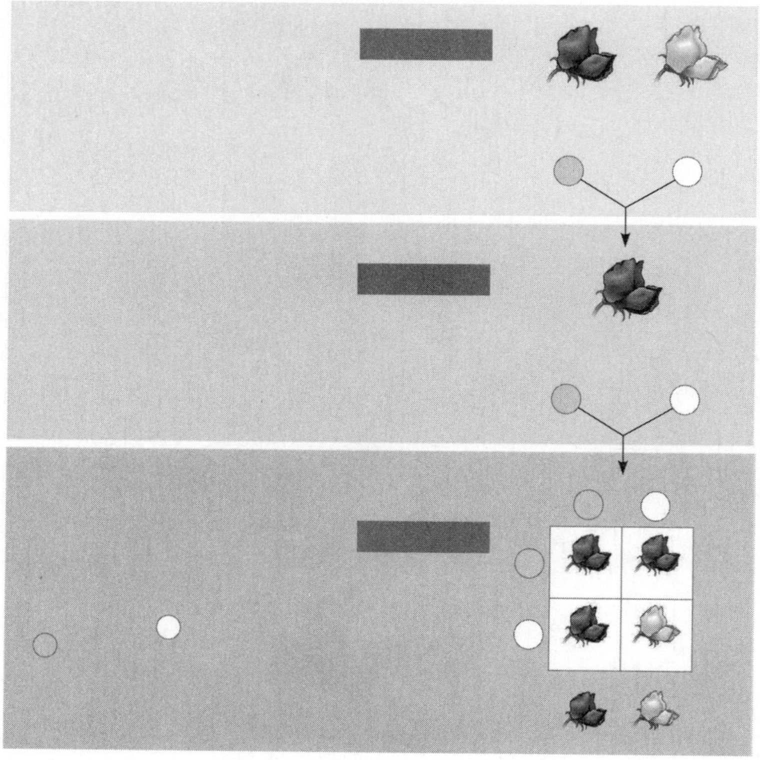

Figure 14.5 Mendel's law of segregation, page 255

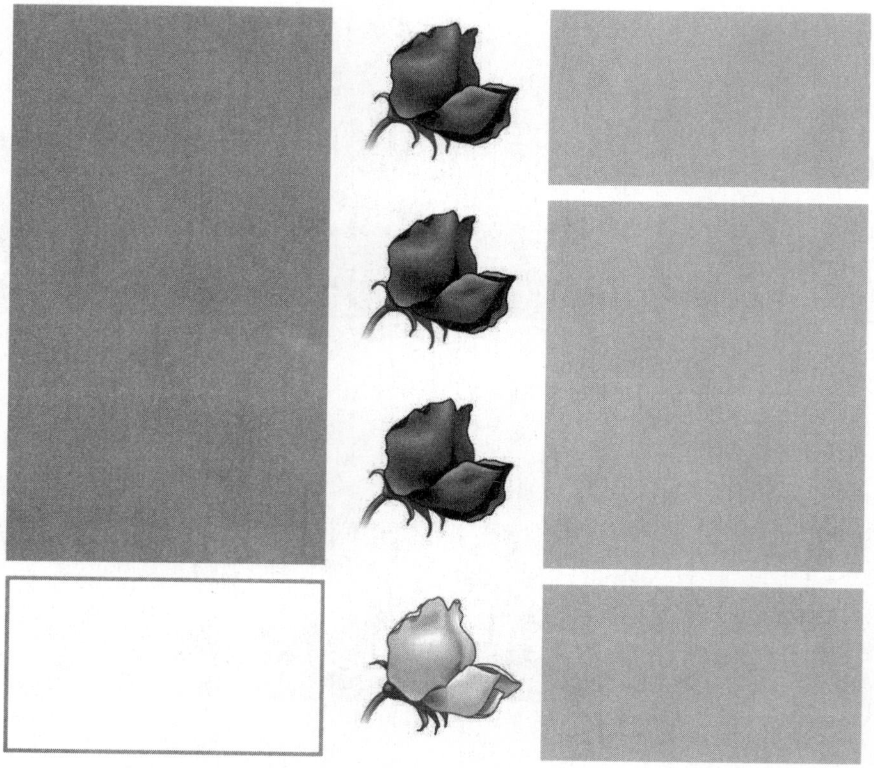

Figure 14.6 Phenotype versus genotype, page 256

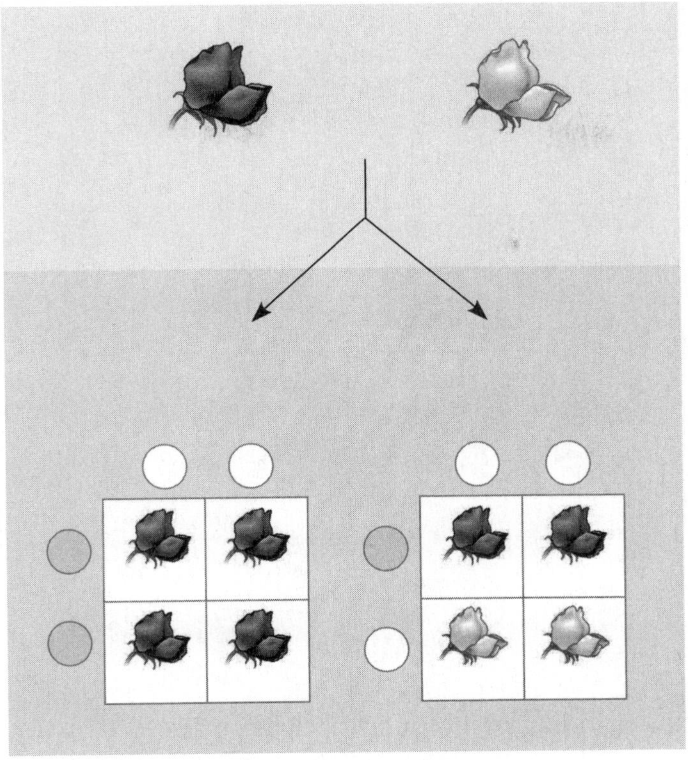

Figure 14.7 The testcross, page 256

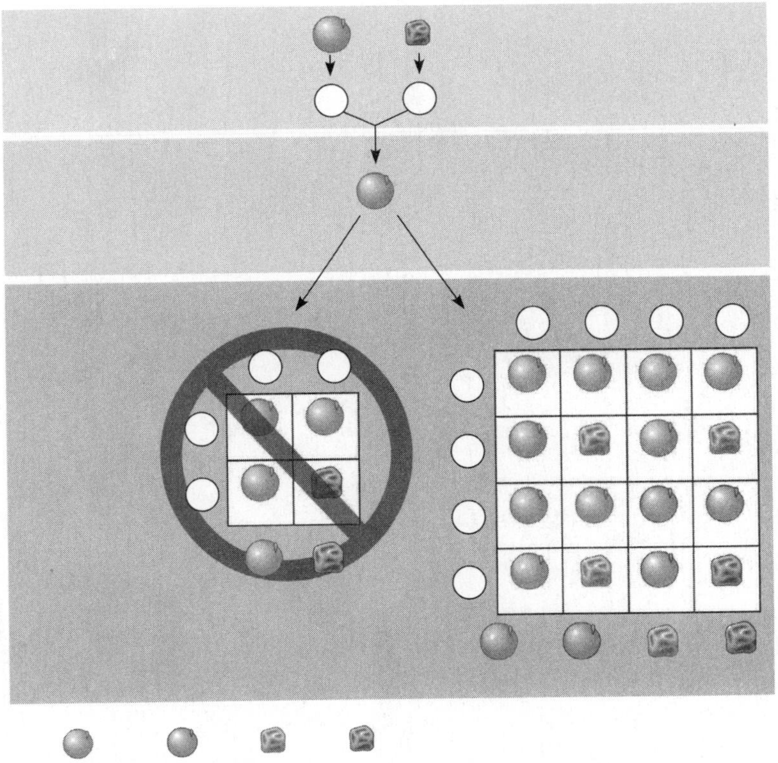

Figure 14.8 Do the alleles for seed color and seed shape sort into gametes dependently (together) or independencly? page 257

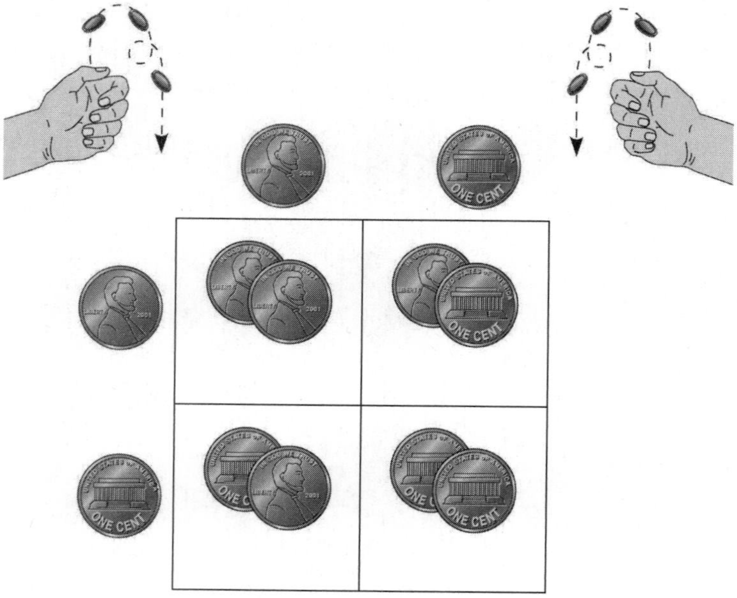

Figure 14.9 Segregation of alleles and fertilization as chance events, page 259

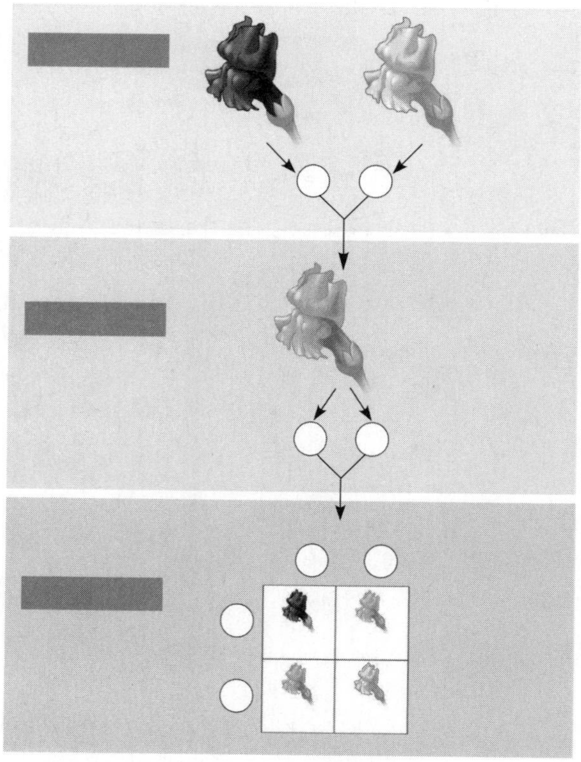

Figure 14.10 Incomplete dominance in snapdragon color, page 261

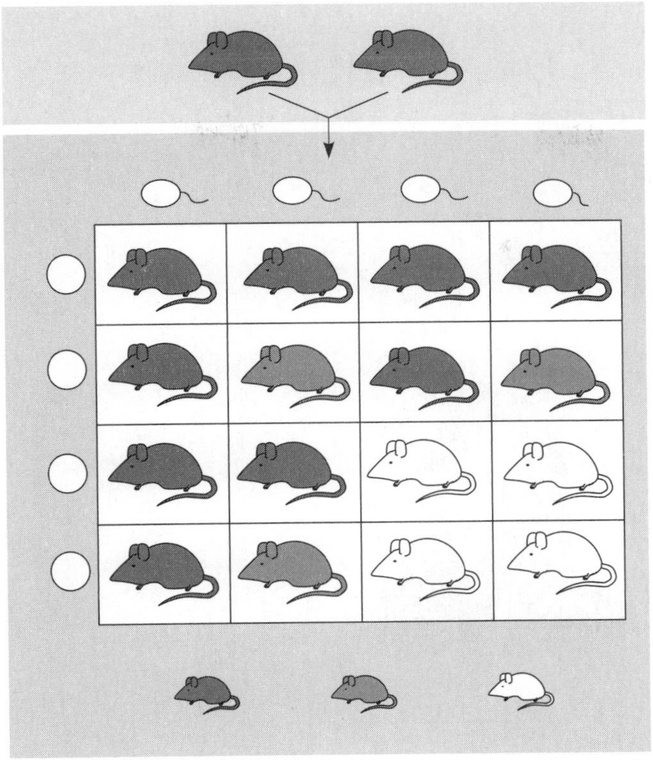

Figure 14.11 An example of epistasis, page 263

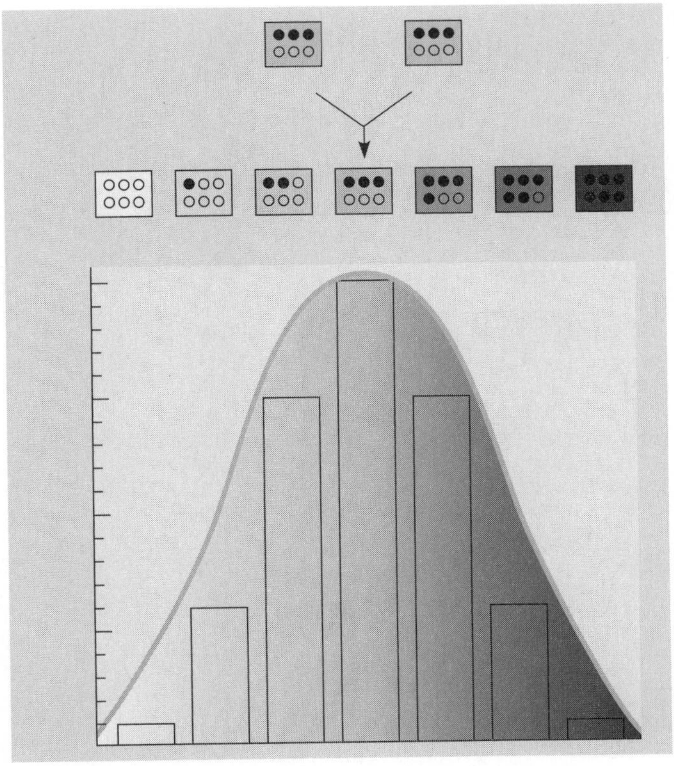

Figure 14.12 A simplified model for polygenic inheritance of skin color, page 263

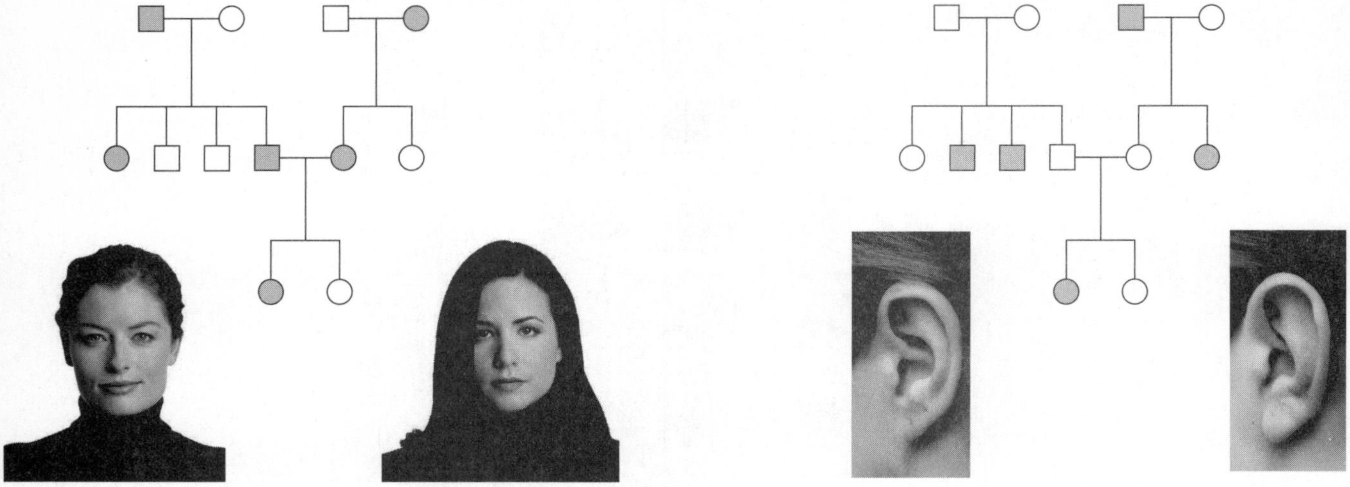

Figure 14.14 Pedigree analysis, page 265

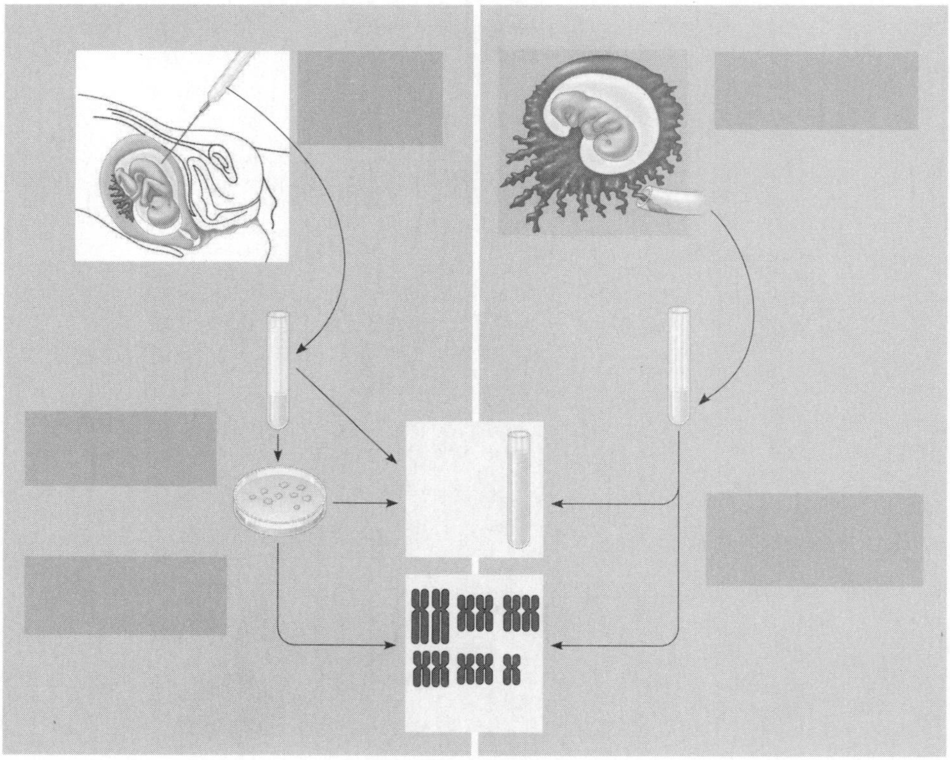

Figure 14.17 Testing a fetus for genetic disorders, page 270

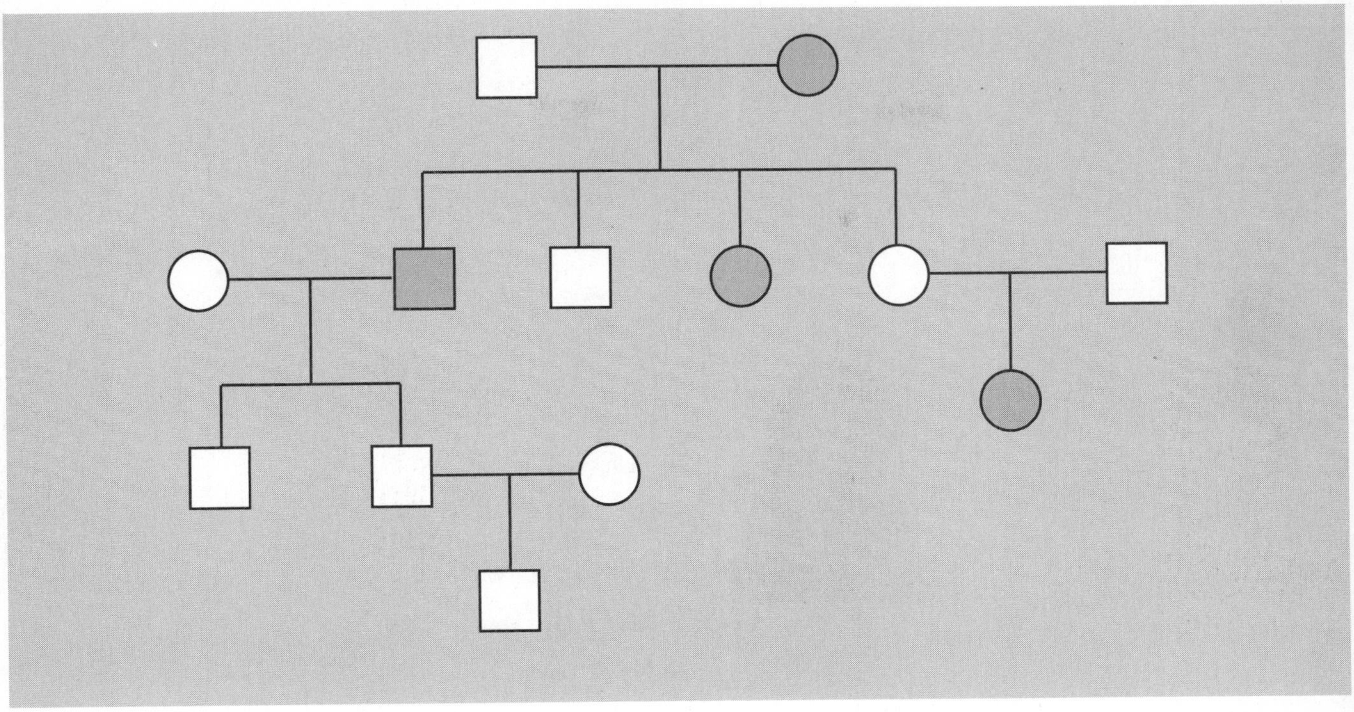

Figure 14.UN1 A pedigree tracing the inheritance of alkaptonuria, page 273

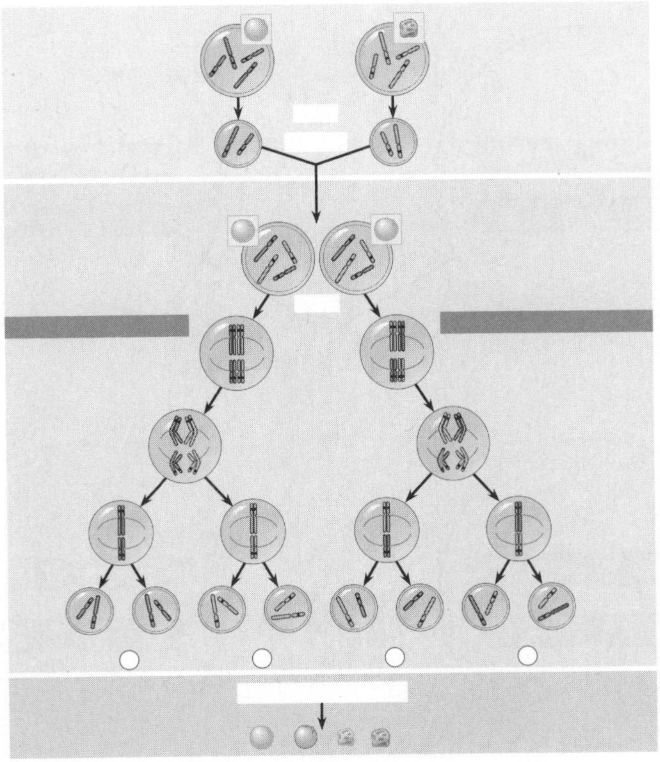

Figure 15.2 The chromosomal basis of Mendel's laws, page 275

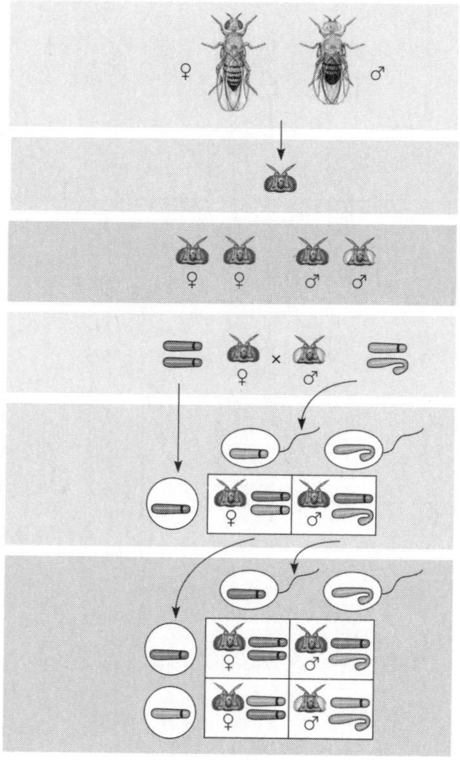

Figure 15.4 In a cross between a wild-type female fruit fly and a mutant white-eyed male, what color eyes will the F_1 and F_2 offspring have?, page 277

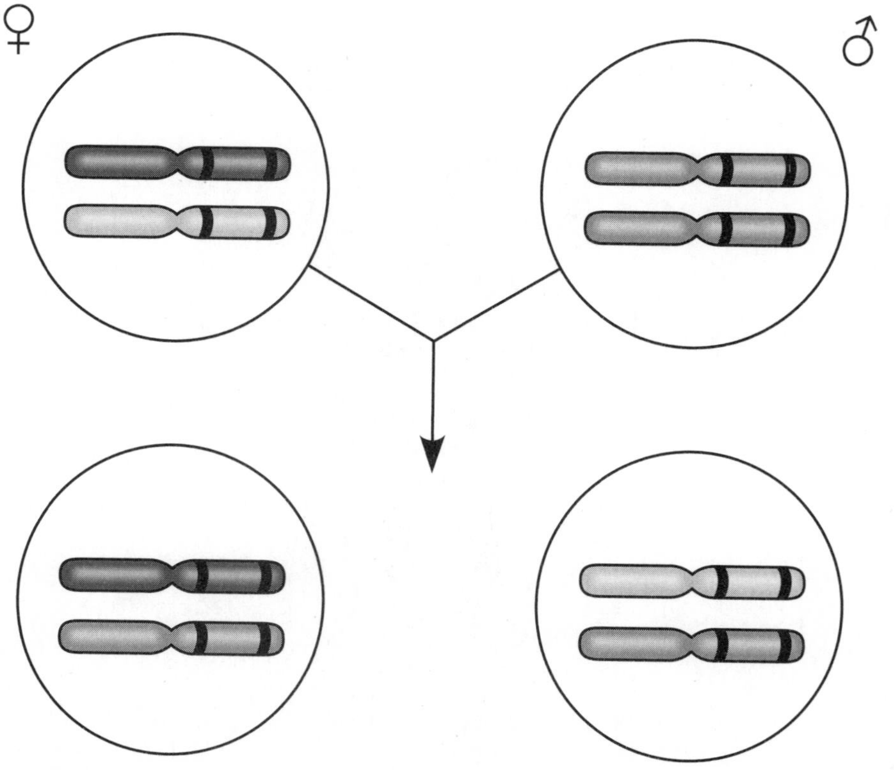

Figure 15.UN1 *Drosophila* testcross, page 278

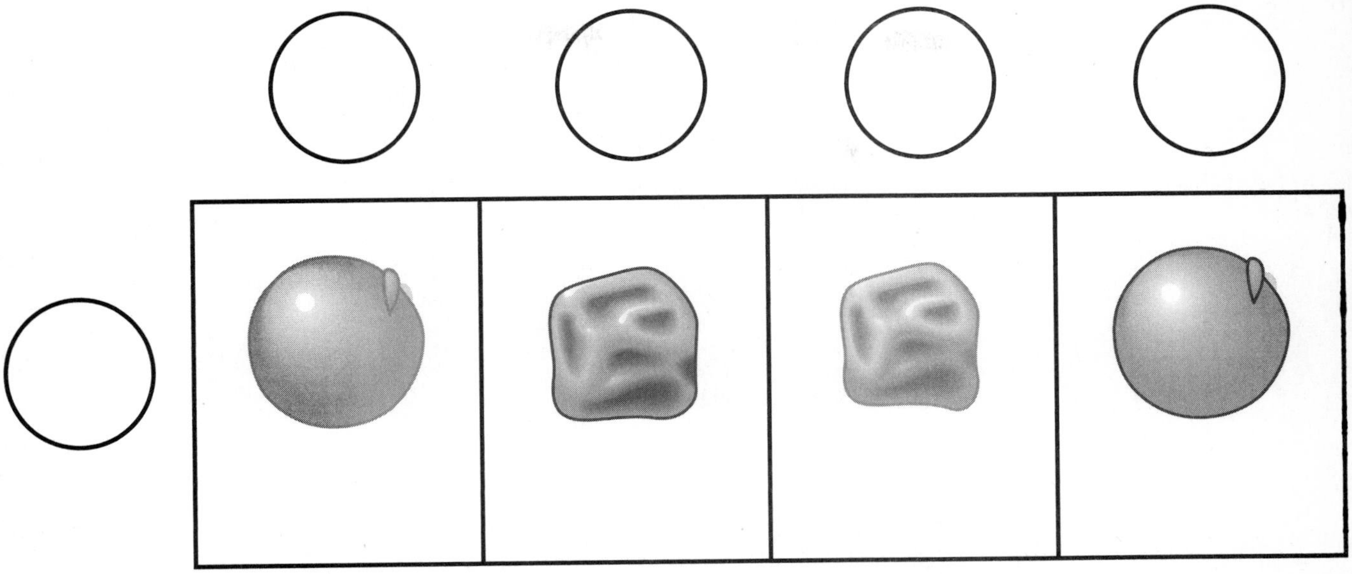

Figure 15.UN2 Punnett square showing recombination of unlinked genes, page 278

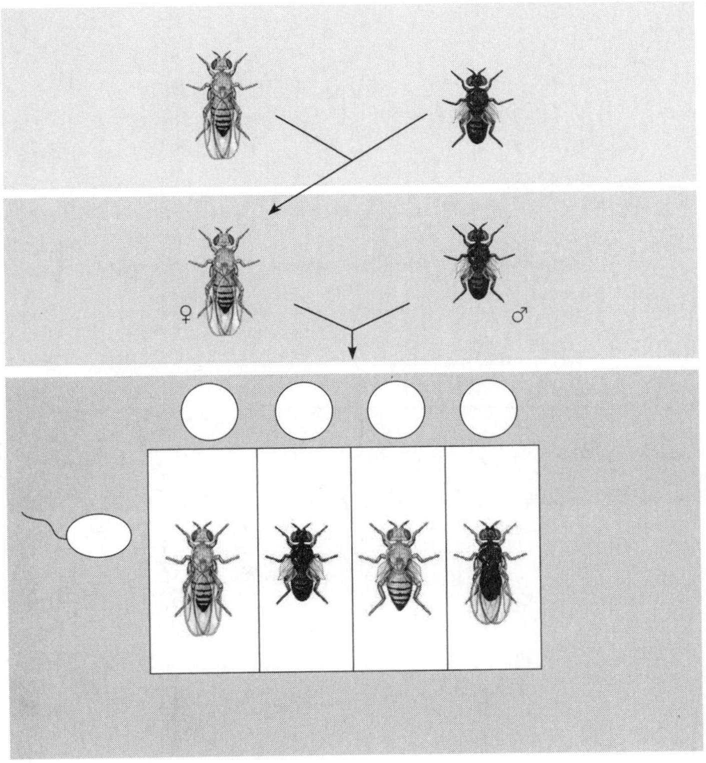

Figure 15.5 Are the genes for body color and wing size in fruit flies located on the same chromosome or different chromosomes?, page 279

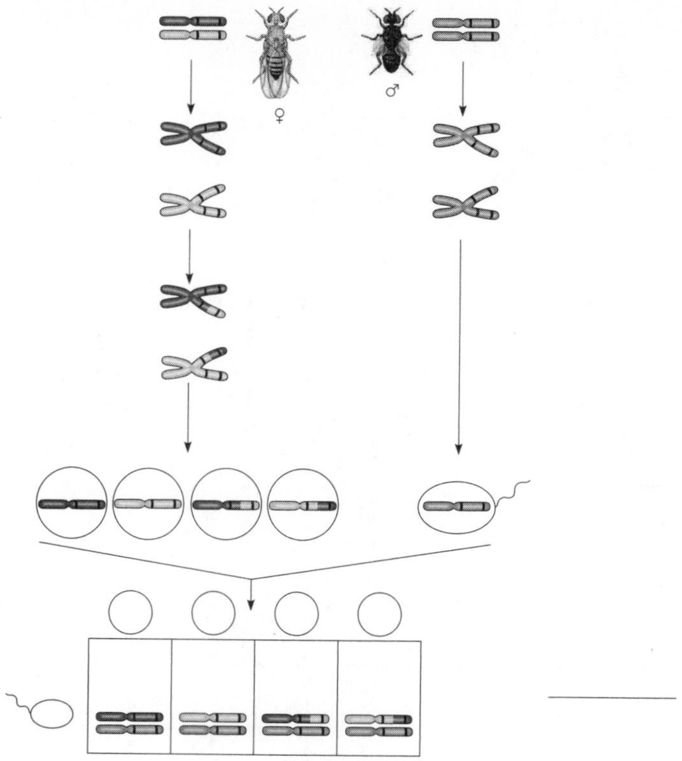

Figure 15.6 Chromosomal basis for recombination of linked genes, page 280

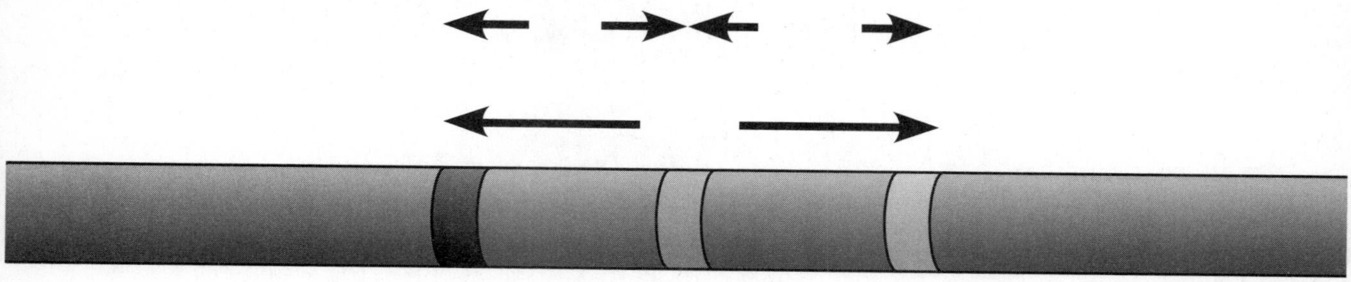

Figure 15.7 Constructing a linkage map, page 281

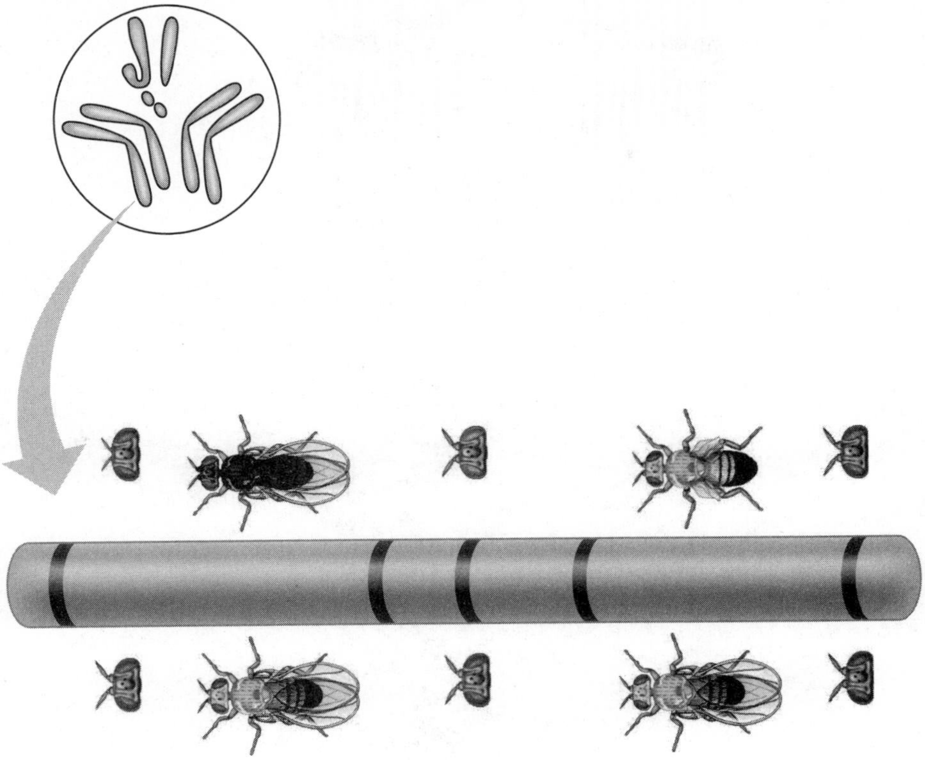

Figure 15.8 A partial genetic (linkage) map of a *Drosophila* chromosome, page 281

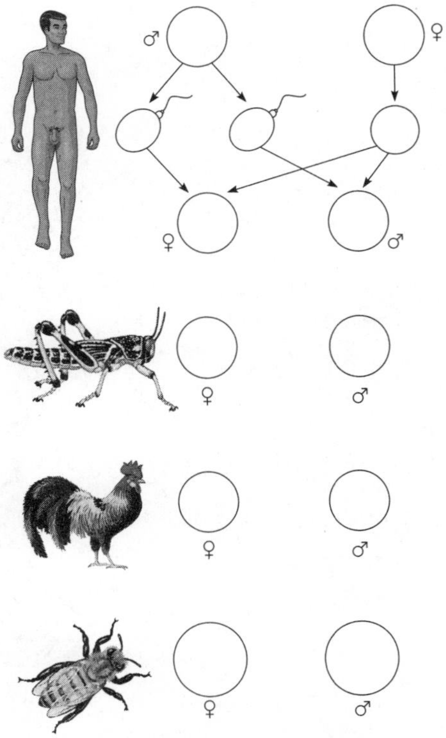

Figure 15.9 Some chromosomal systems of sex determination, page 282

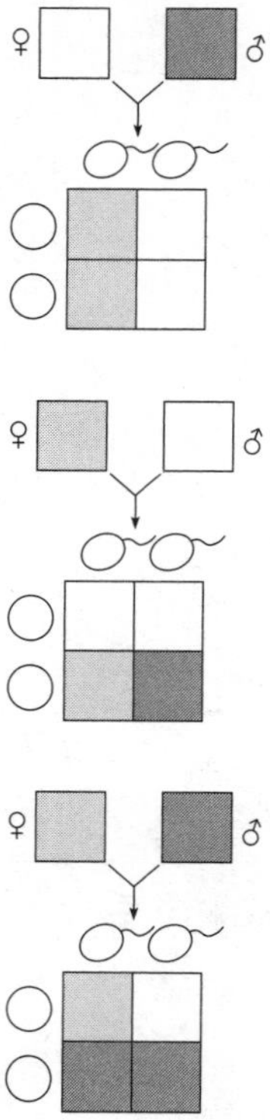

Figure 15.10 The transmission of sex-linked recessive traits, page 283

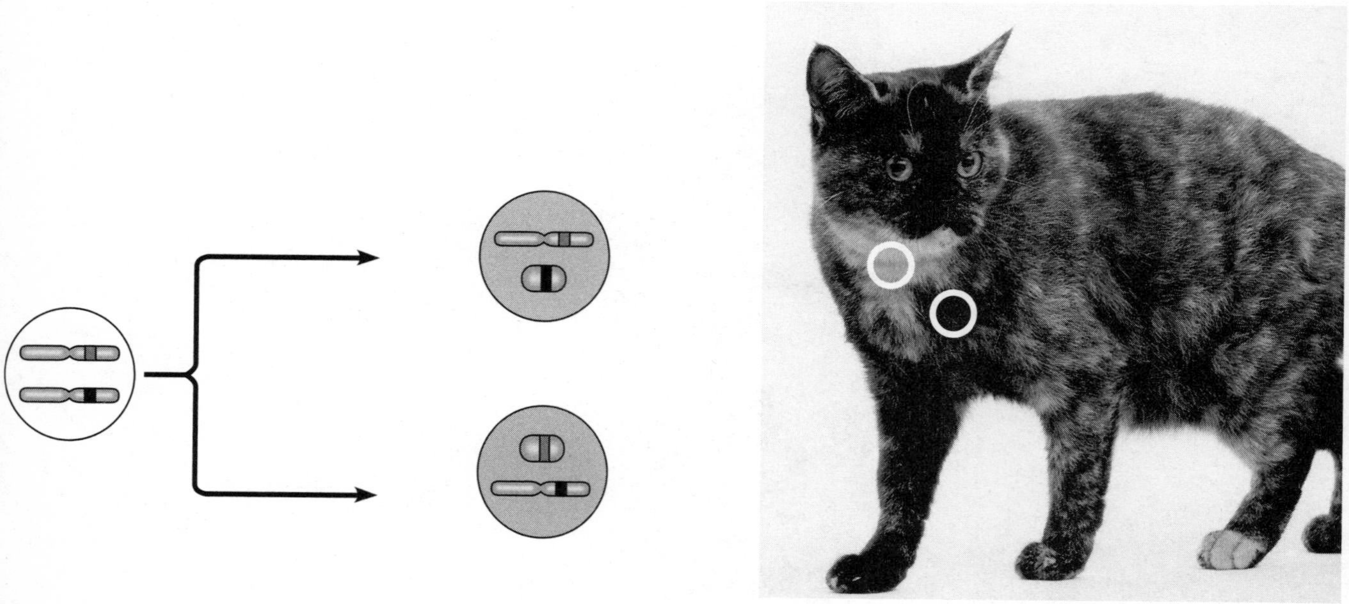

Figure 15.11 X inactivation and the tortoiseshell cat, page 284

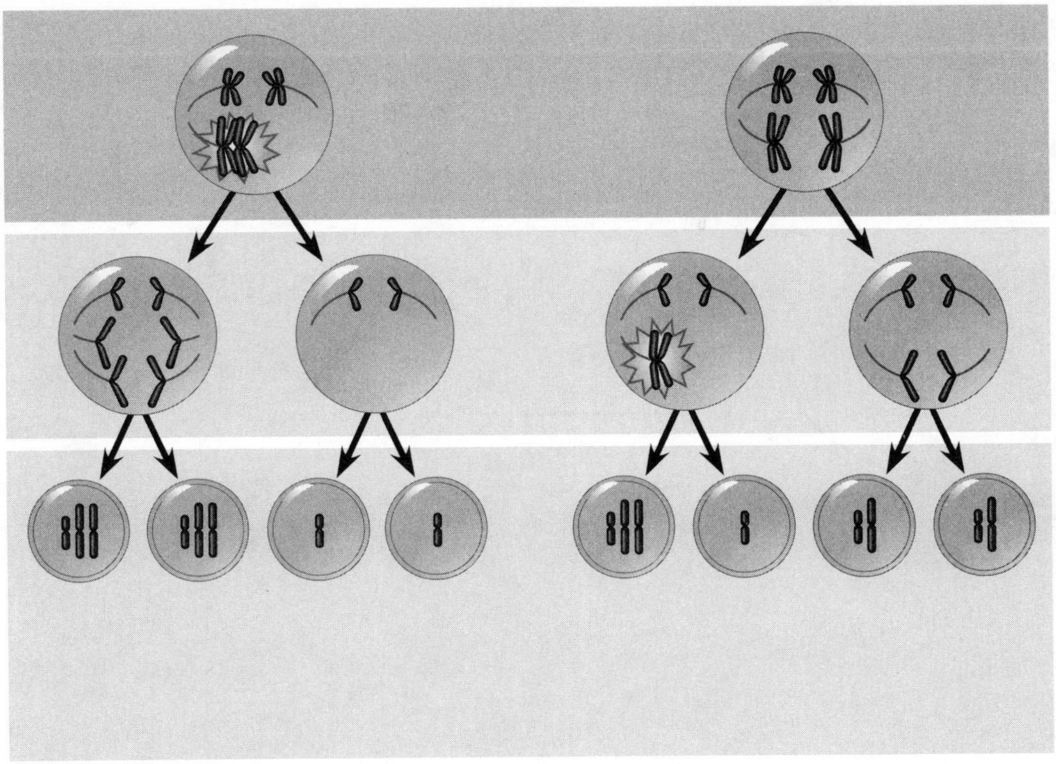

Figure 15.12 Meiotic nondisjunction, page 285

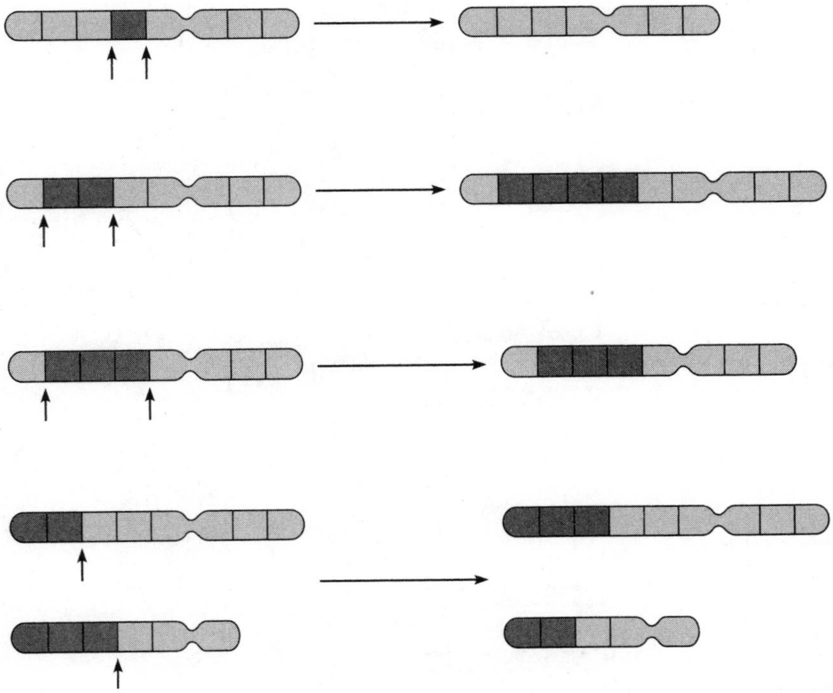

Figure 15.14 Alterations of chromosome structure, page 286

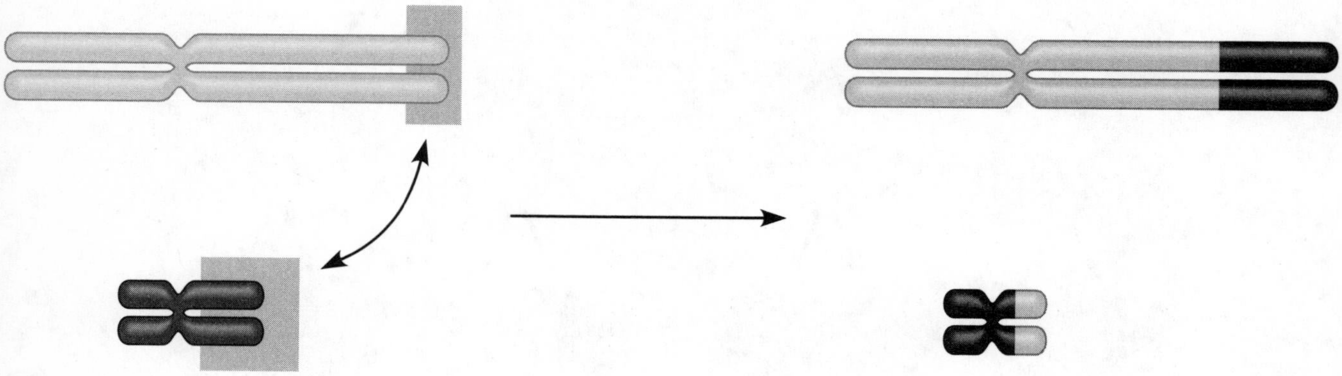

Figure 15.16 Translocation associated with chronic myelogenous leukemia (CML), page 288

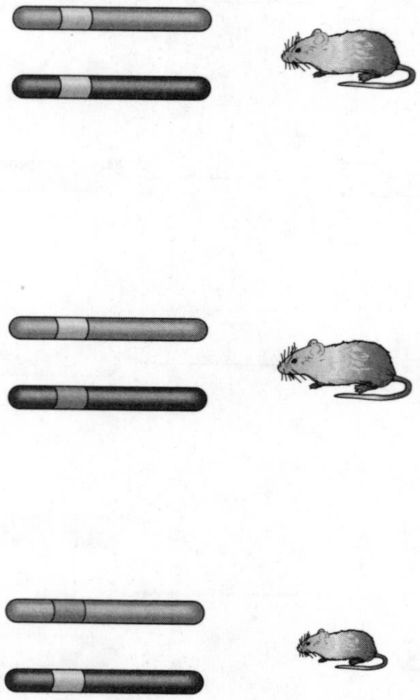

Figure 15.17 Genomic imprinting of the mouse *Igf2* gene, page 289

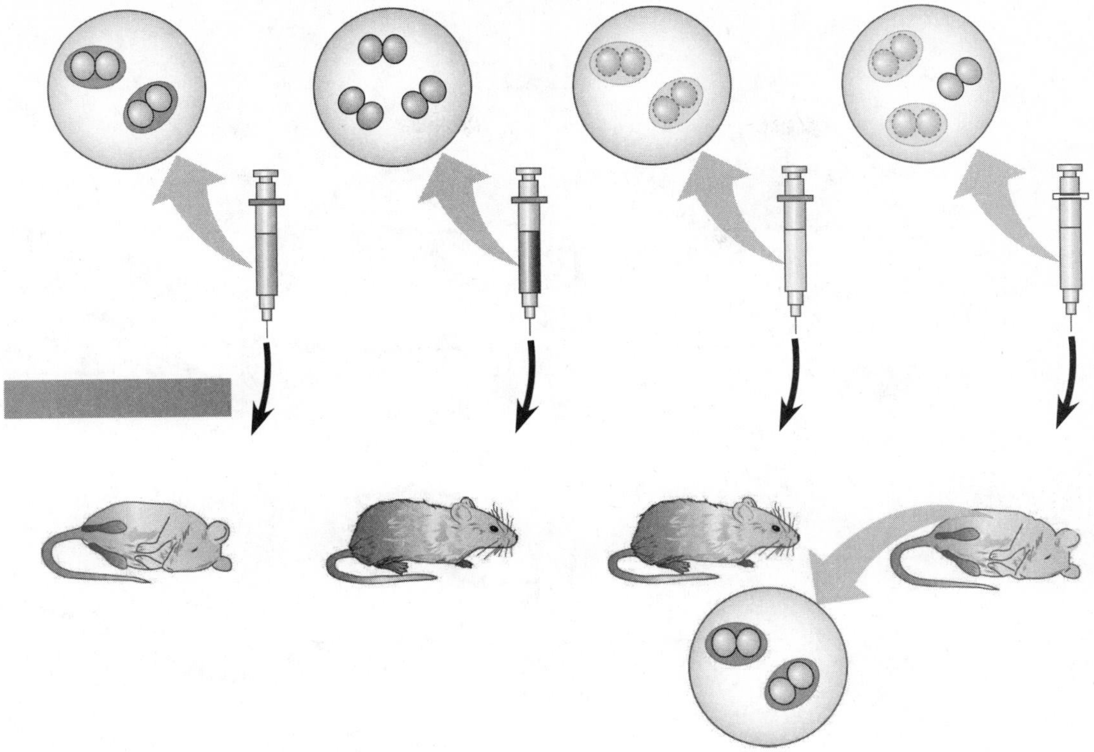

Figure 16.2 Can the genetic trait of pathogenicity be transferred between bacteria?, page 294

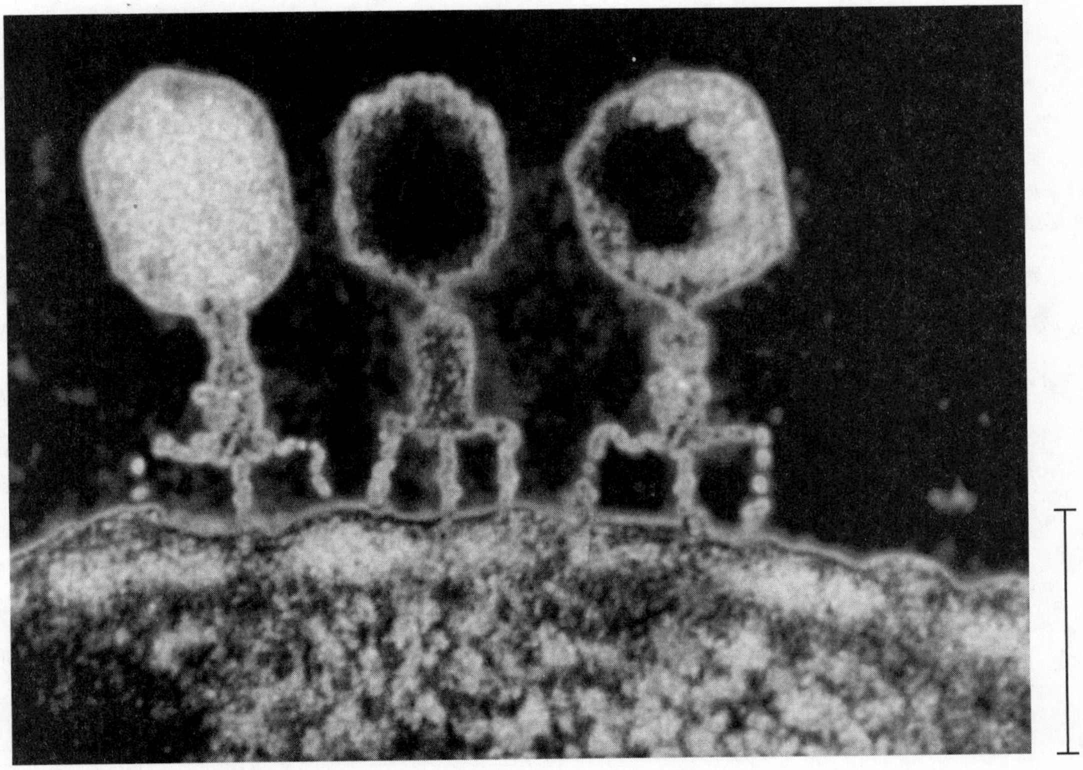

Figure 16.3 Viruses infecting a bacterial cell, page 295

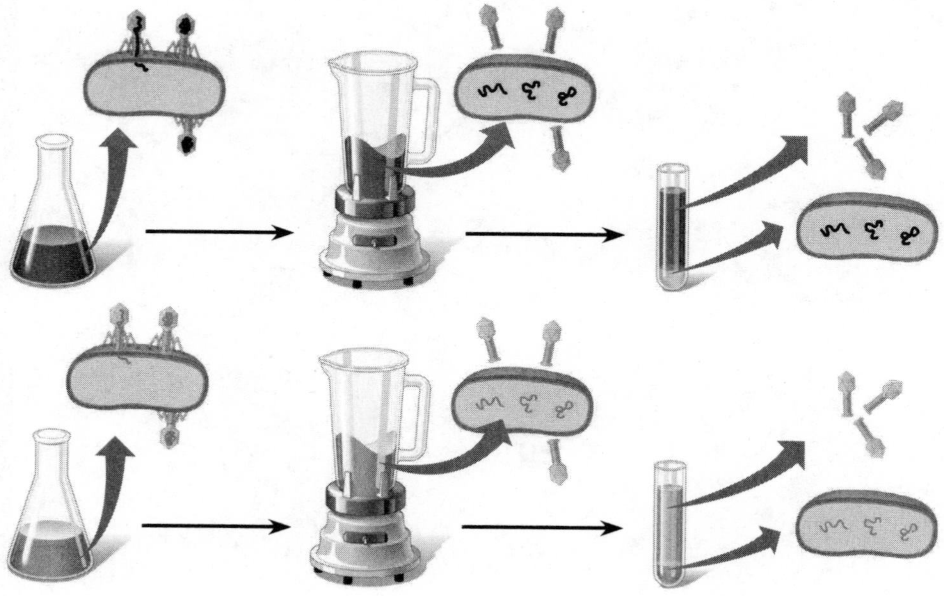

Figure 16.4 Is DNA or protein the genetic material of phage T2?, page 295

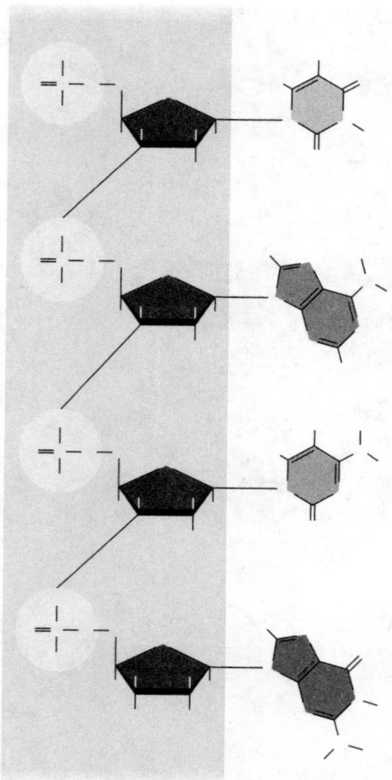

Figure 16.5 The structure of a DNA strand, page 296

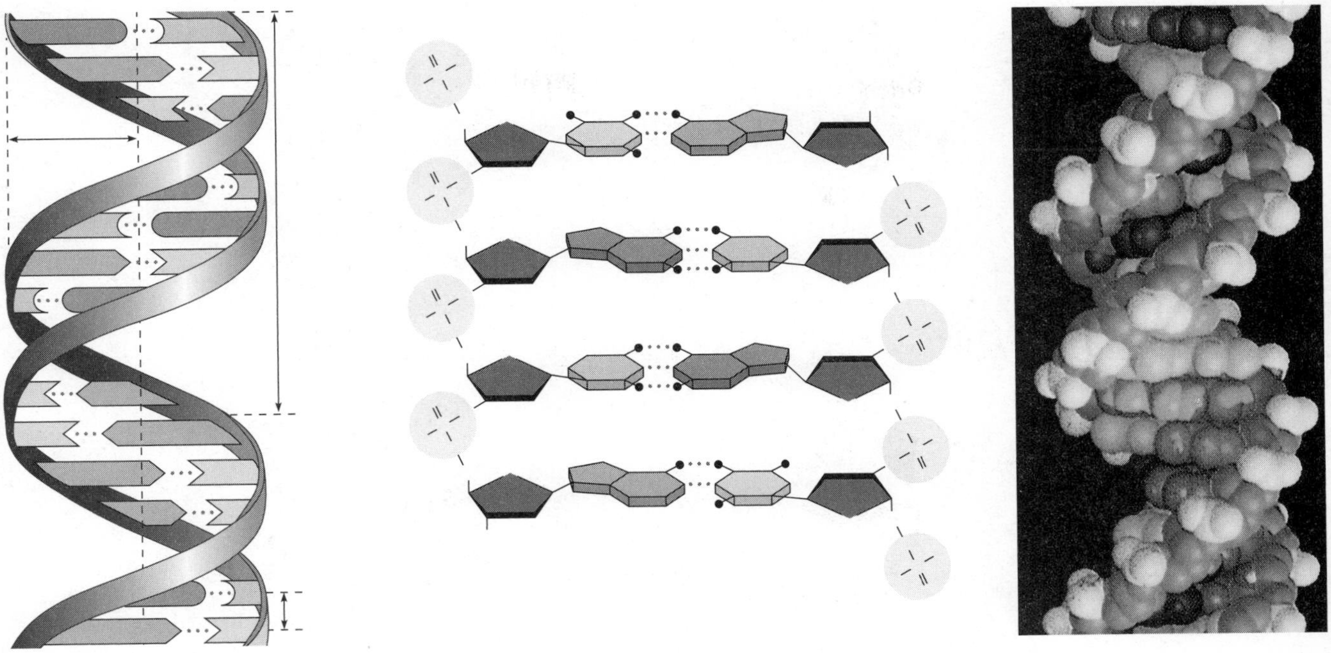

Figure 16.7 The double helix, page 297

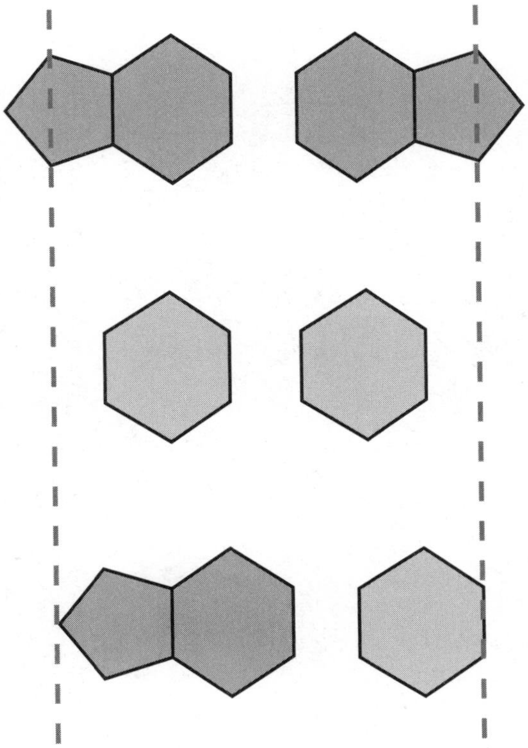

Figure 16.UN1 Purine and pyrimidine, page 298

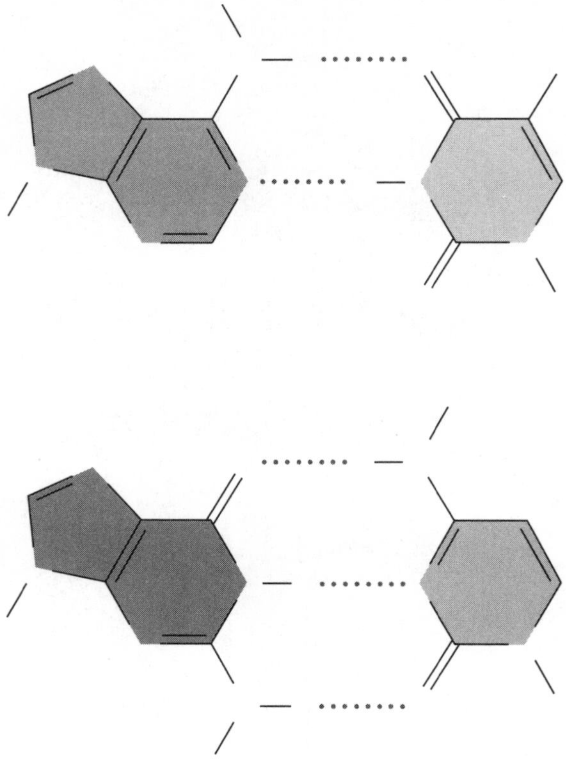

Figure 16.8 Base pairing in DNA, page 298

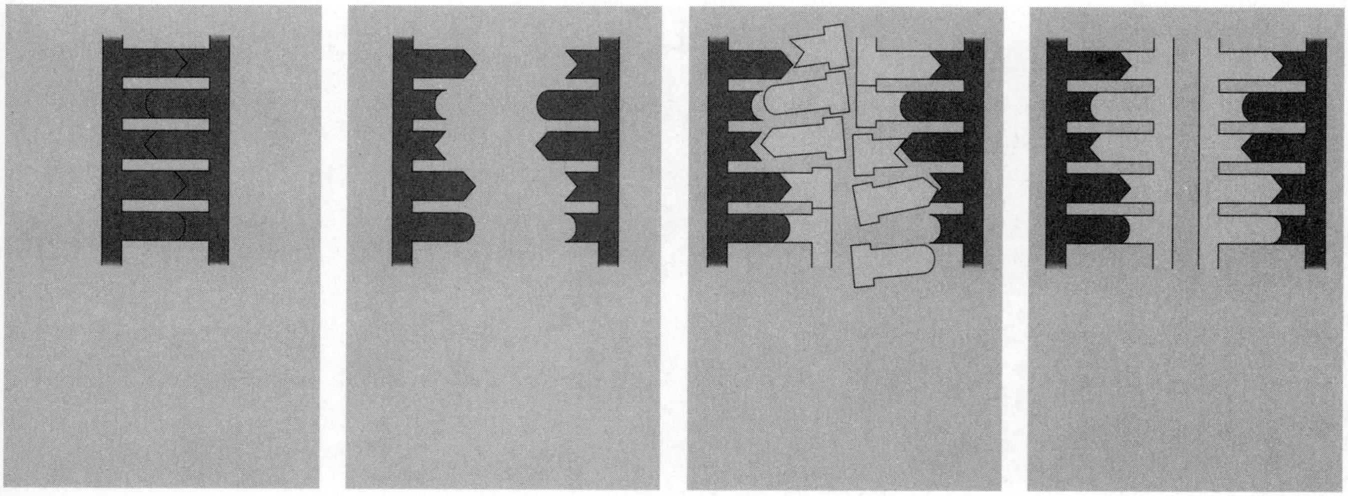

Figure 16.9 A model for DNA replication: the basic concept, page 299

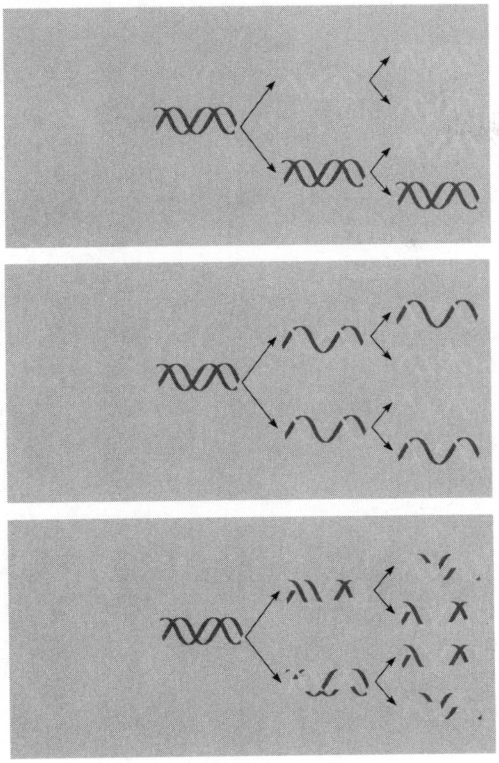

Figure 16.10 Three alternative models of DNA replication, page 300

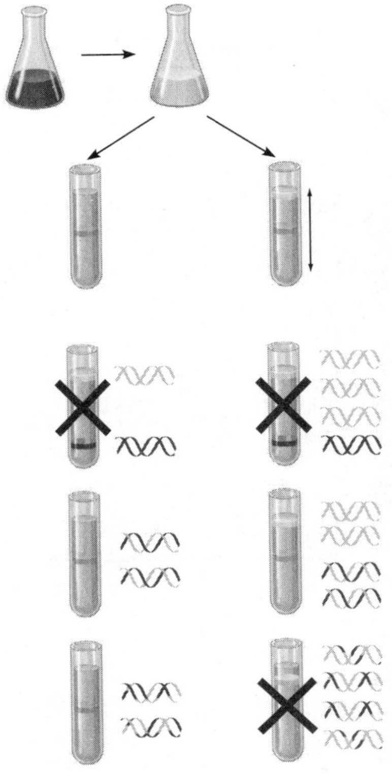

Figure 16.11 Does DNA replication follow the conservative, semiconservative, or dispersive model? page 300

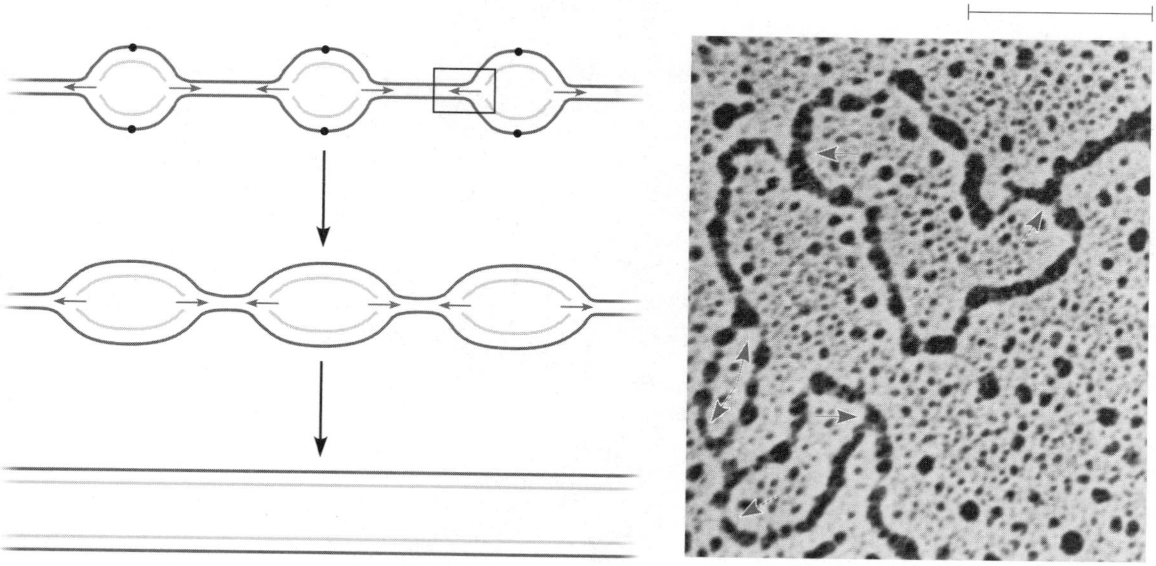

Figure 16.12 Origins of replication to eukaryotes, page 301

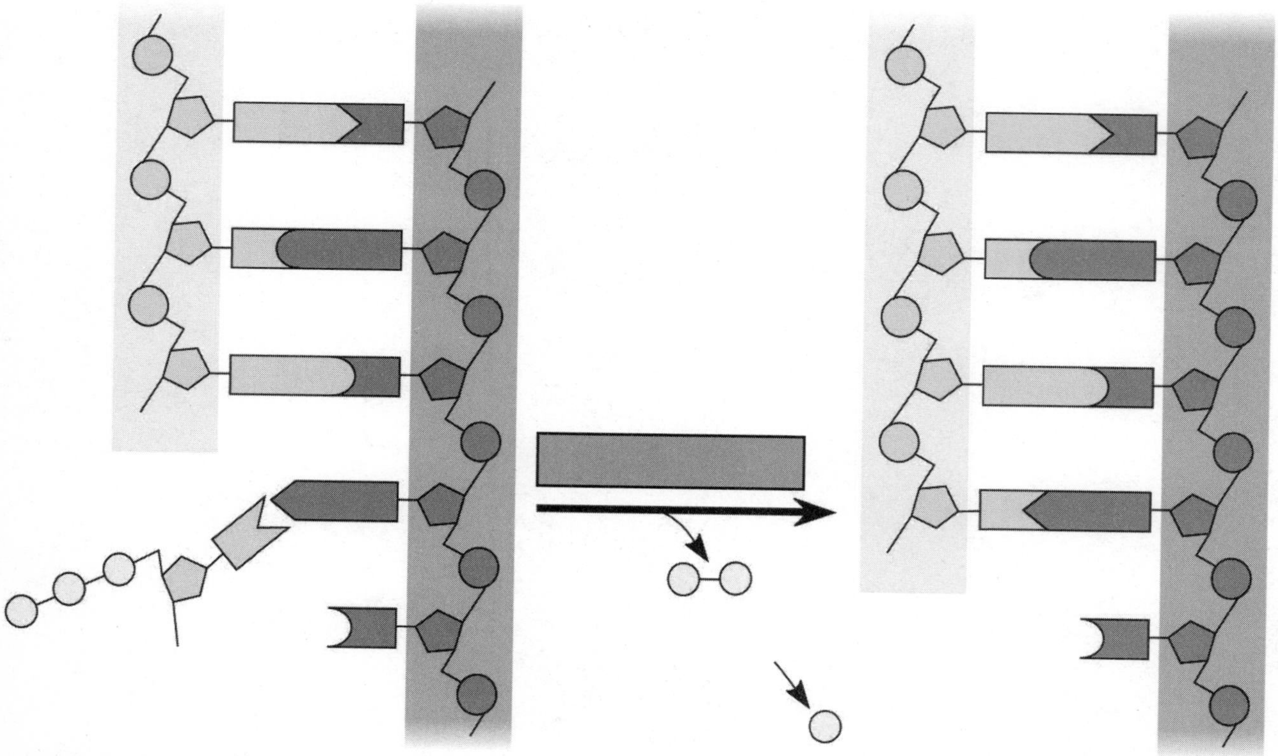

Figure 16.13 Incorporation of a nucleotide into a DNA strand, page 302

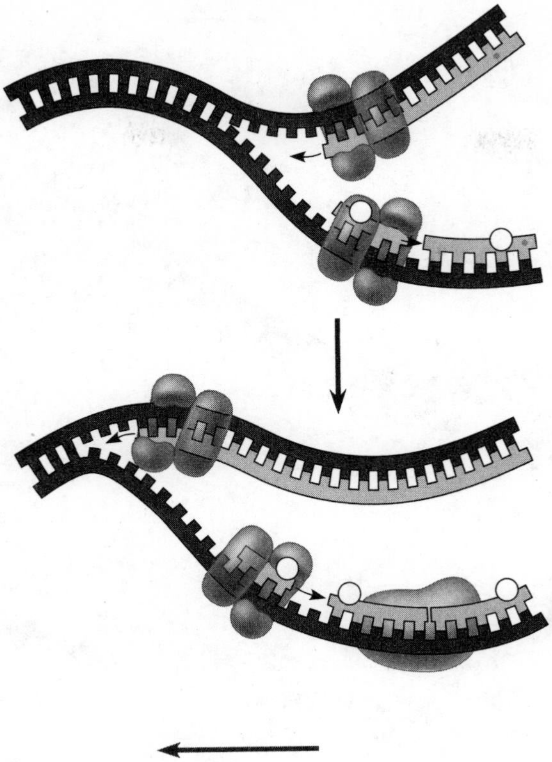

Figure 16.14 Synthesis of leading and lagging strands during DNA replication, page 302

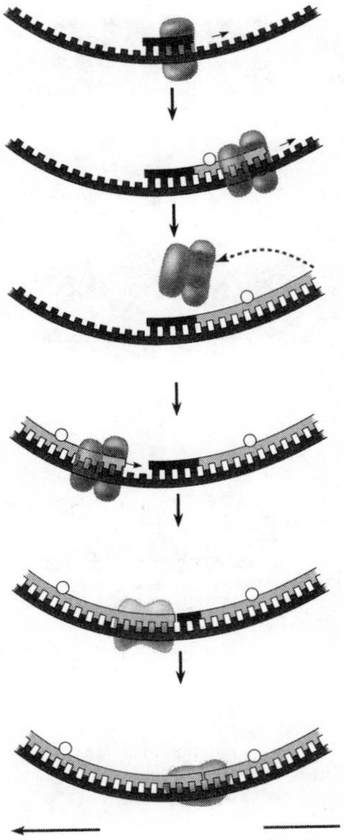

Figure 16.15 Synthesis of the lagging strand, page 303

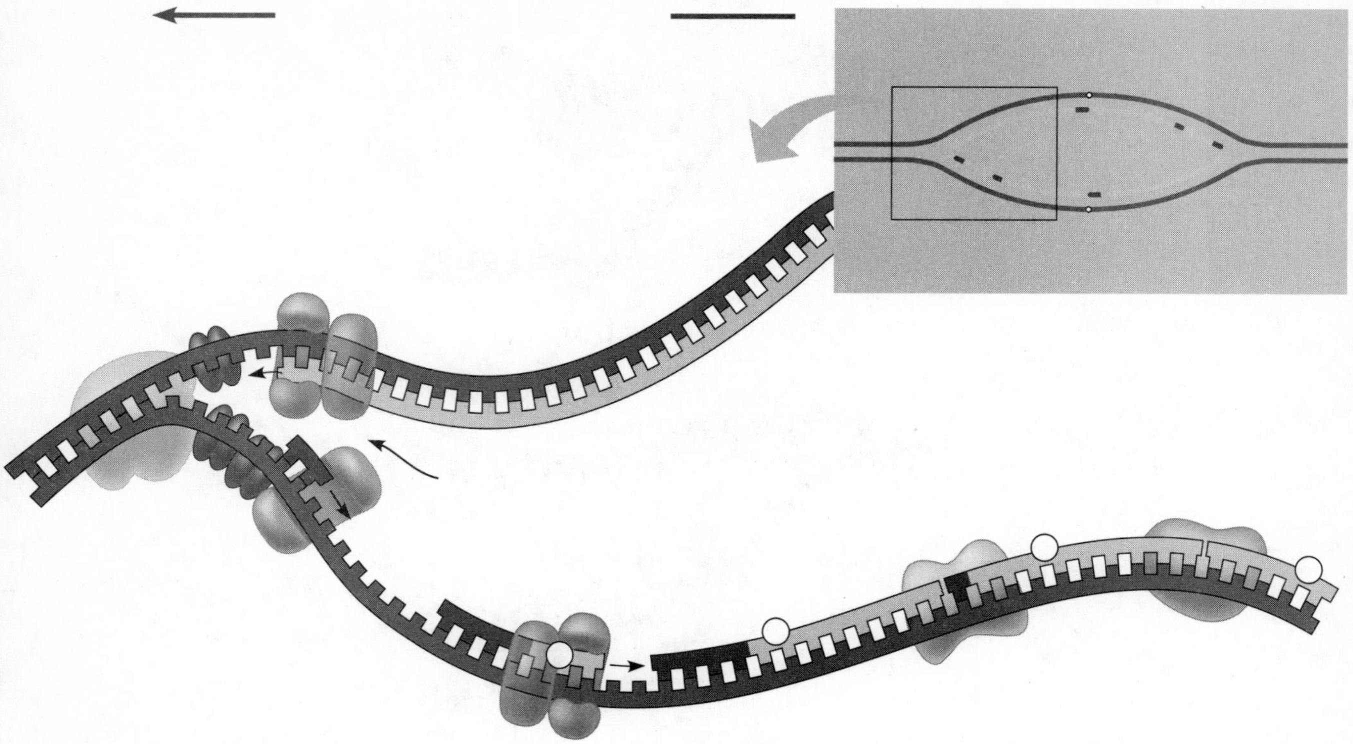

Figure 16.16 A summary of bacterial DNA replication, page 304

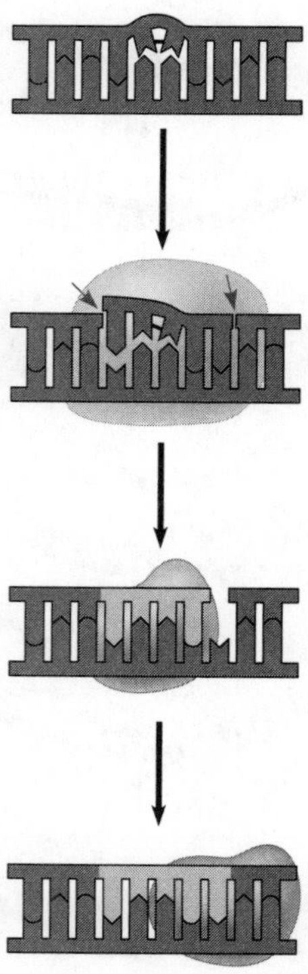

Figure 16.17 Nucleotide excision repair of DNA damage, page 305

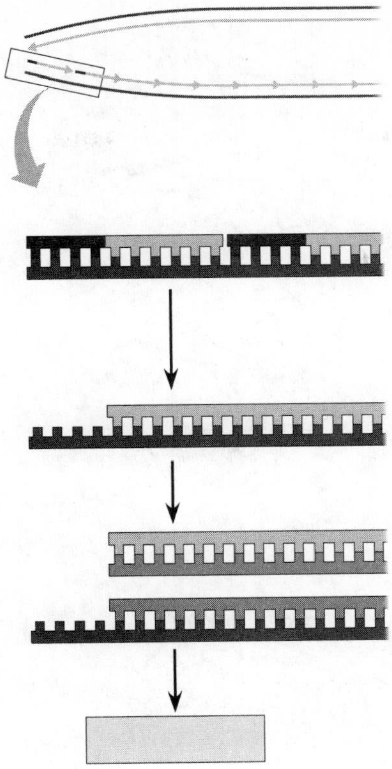

Figure 16.18 Shortening of the ends of linear DNA molecules, page 306

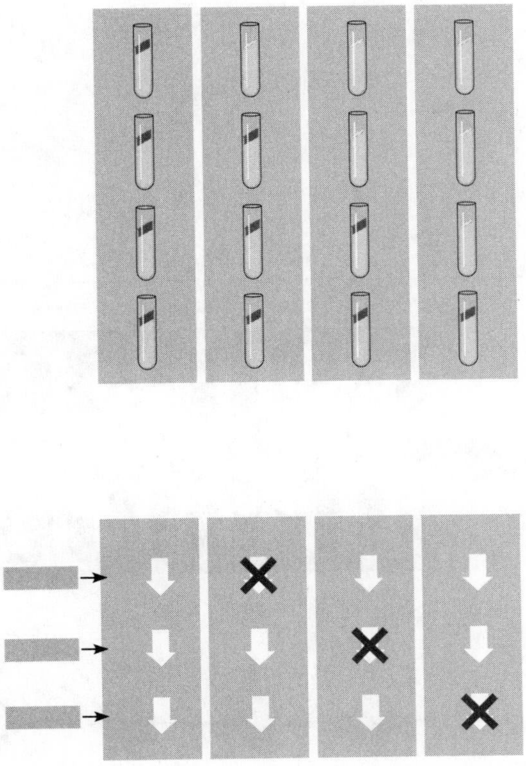

Figure 17.2 Do individual genes specify different enzymes in arginine biosynthesis?, page 311

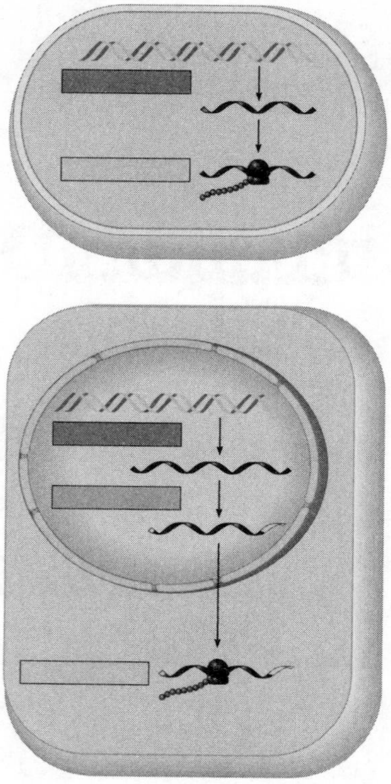

Figure 17.3 Overview: the roles of transcription and translation in the flow of genetic information, page 312

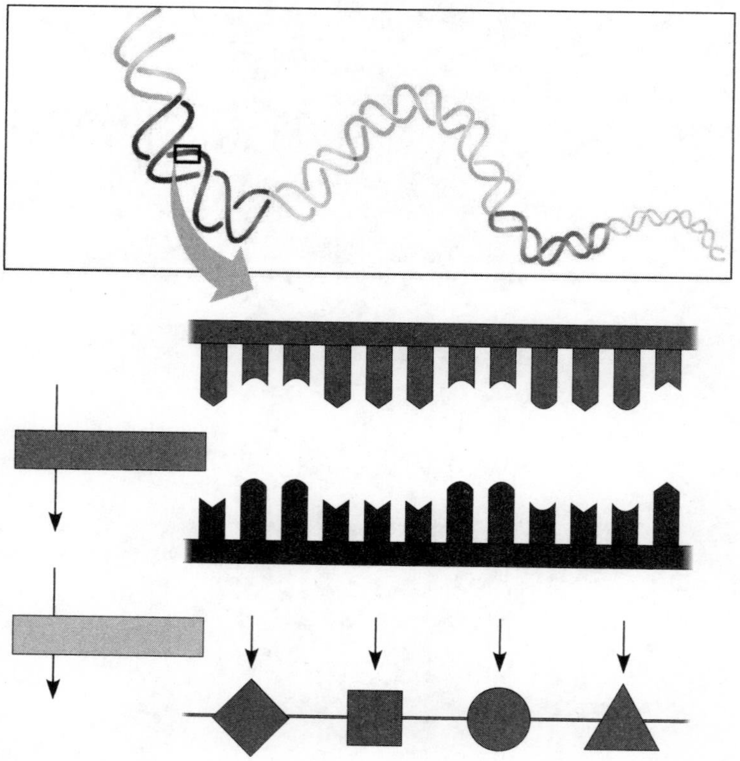

Figure 17.4 The triplet code, page 313

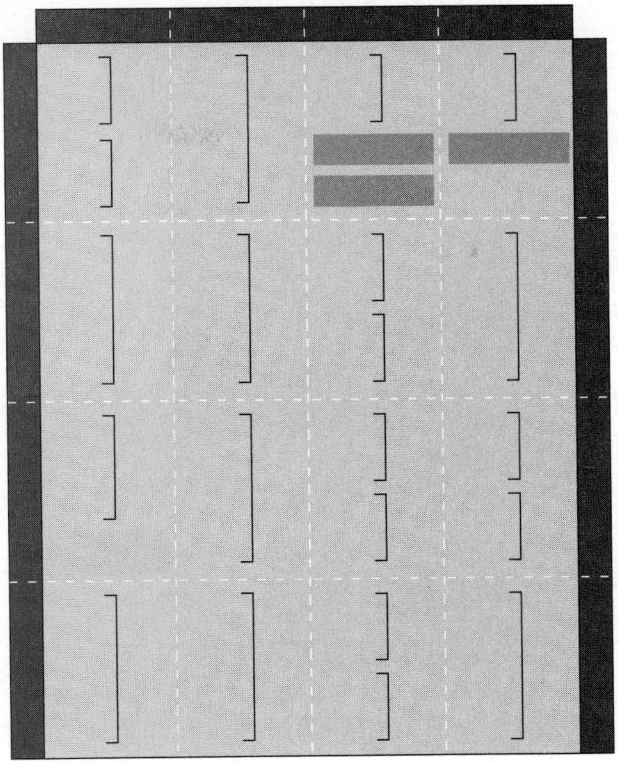

Figure 17.5 The dictionary of the genetic code, page 314

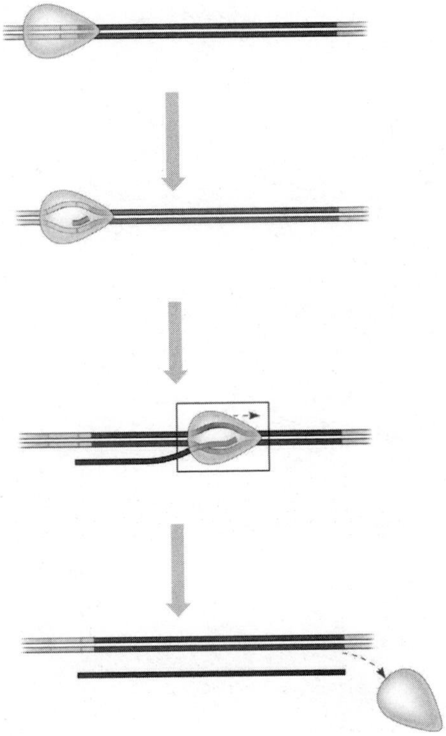

Figure 17.7a The stages of transcription: initiation, elongation, and termination, page 315

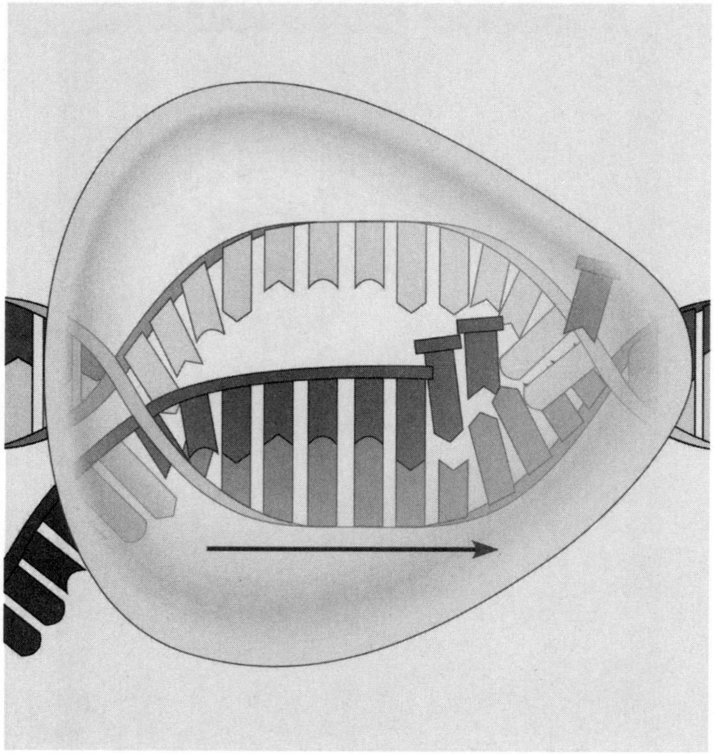

Figure 17.7b The stages of transcription: elongation, page 315

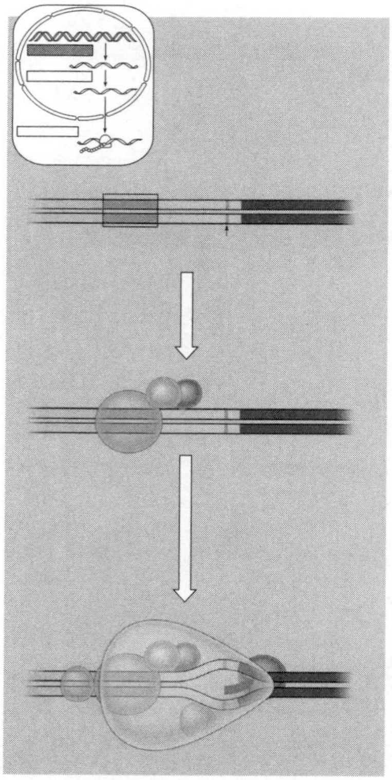

Figure 17.8 The initiation of transcription at a eukaryotic promoter, page 316

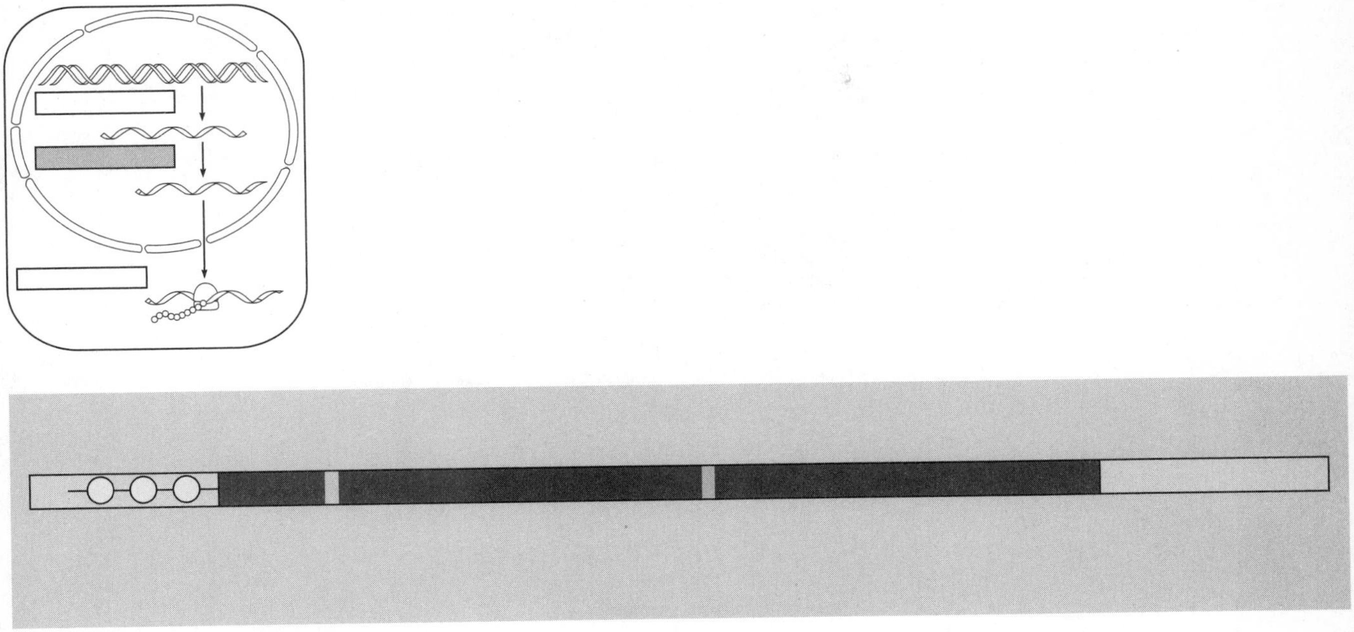

Figure 17.9 RNA processing: addition of the 5′ cap and poly-A tail, page 317

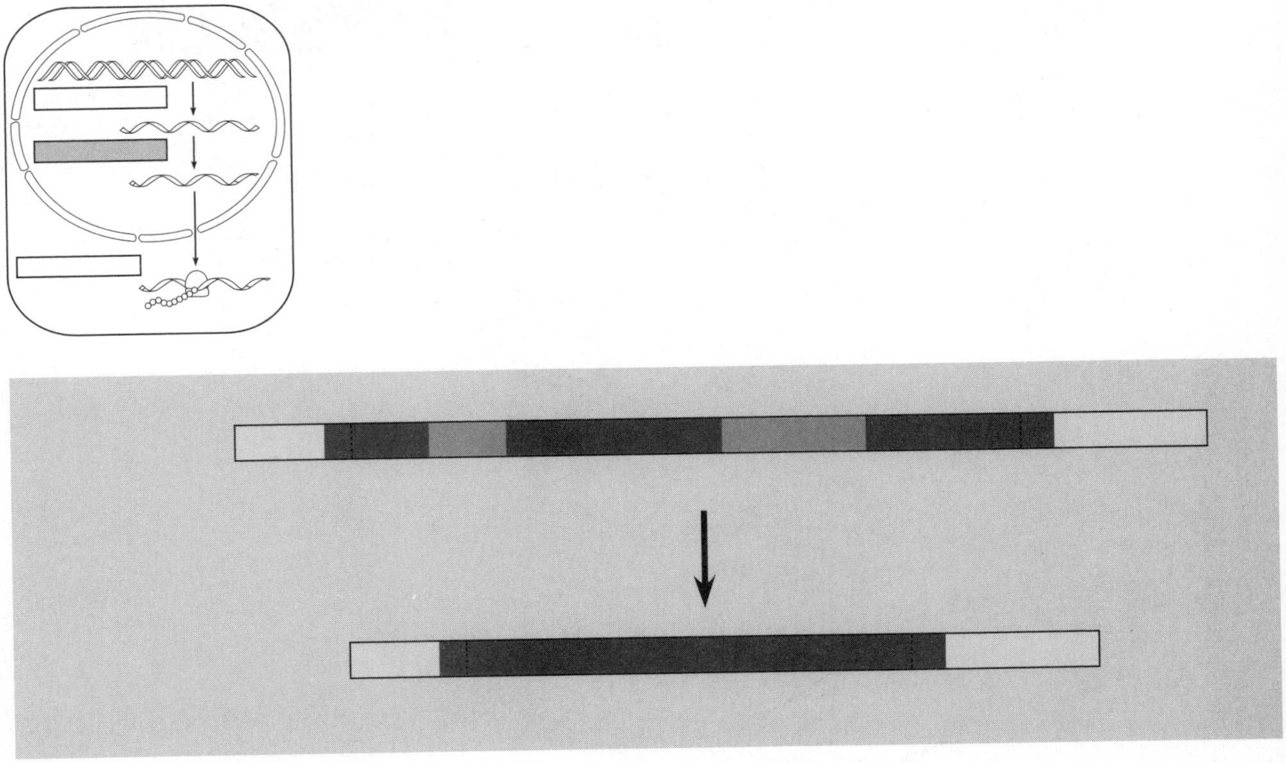

Figure 17.10 RNA processing: RNA splicing, page 318

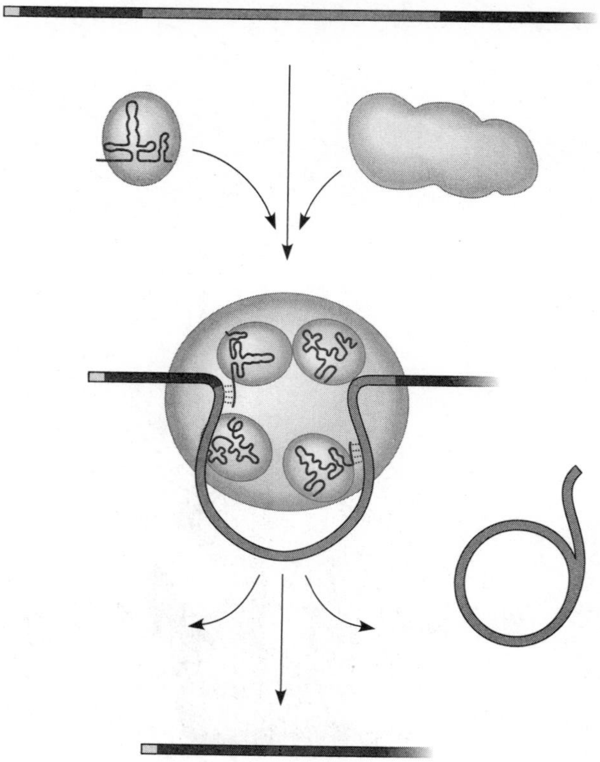

Figure 17.11 The roles of snRNPs and spliceosomes in pre-mRNA splicing, page 319

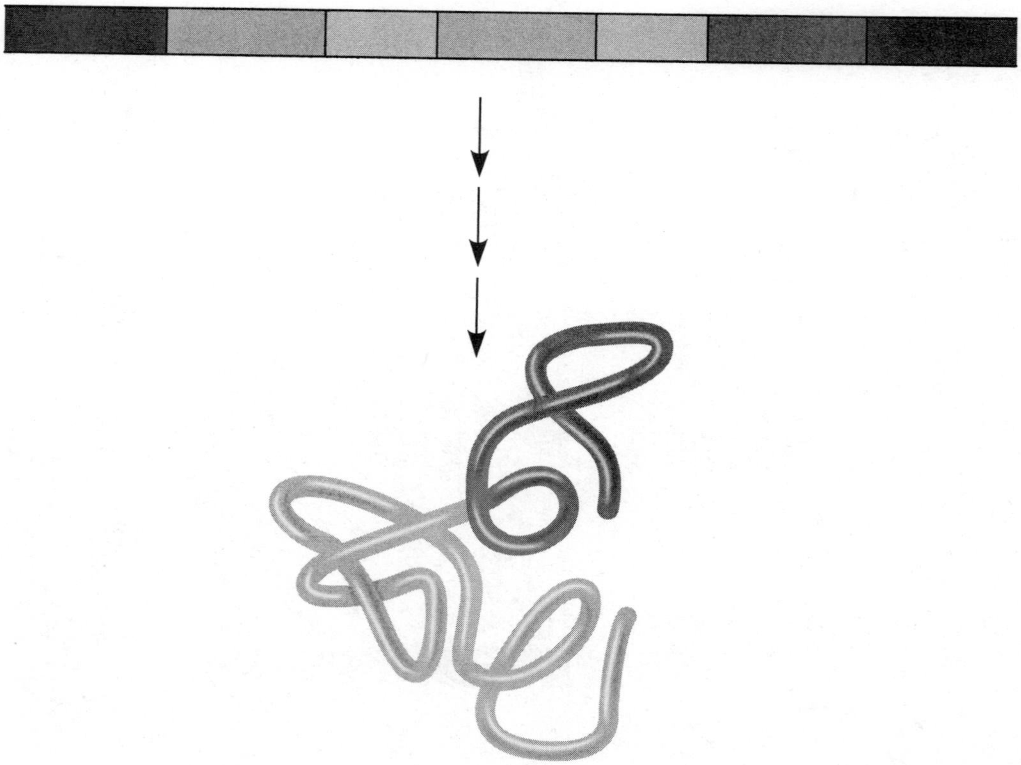

Figure 17.12 Correspondence between exons and protein domains, page 319

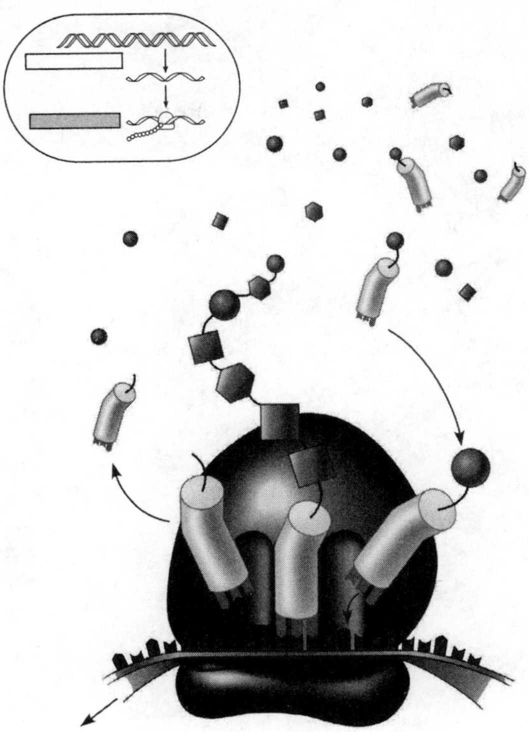

Figure 17.13 Translation: the basic concept, page 320

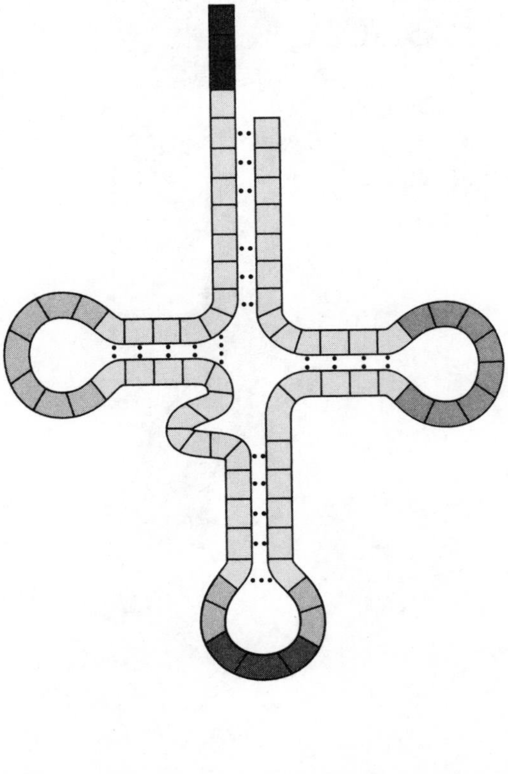

Figure 17.14a The structure of transfer RNA (tRNA): two-dimensional structure, page 321

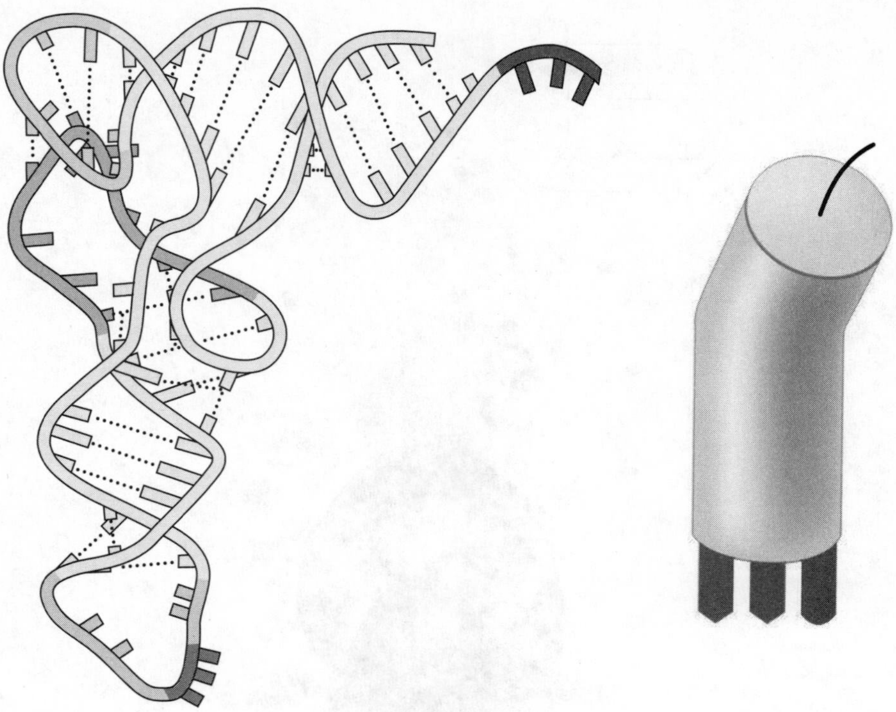

Figure 17.14b The structure of transfer RNA (tRNA): three-dimensional structure, page 321

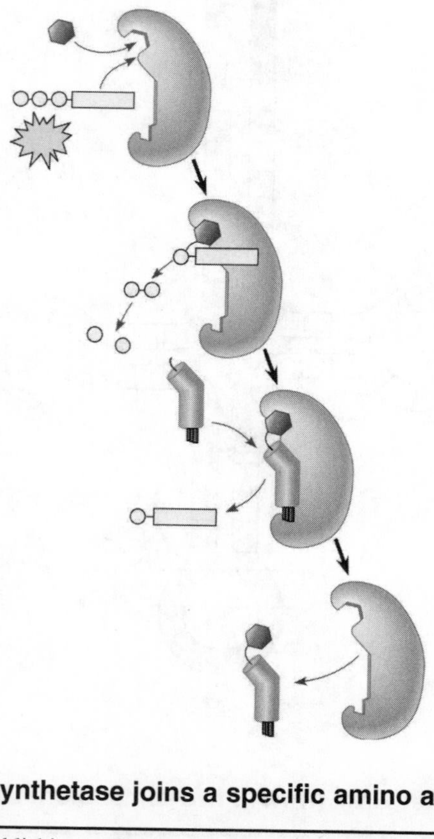

Figure 17.15 An aminoacyl-tRNA synthetase joins a specific amino acid to a tRNA, page 321

Figure 17.16 The anatomy of a functioning ribosome, page 322

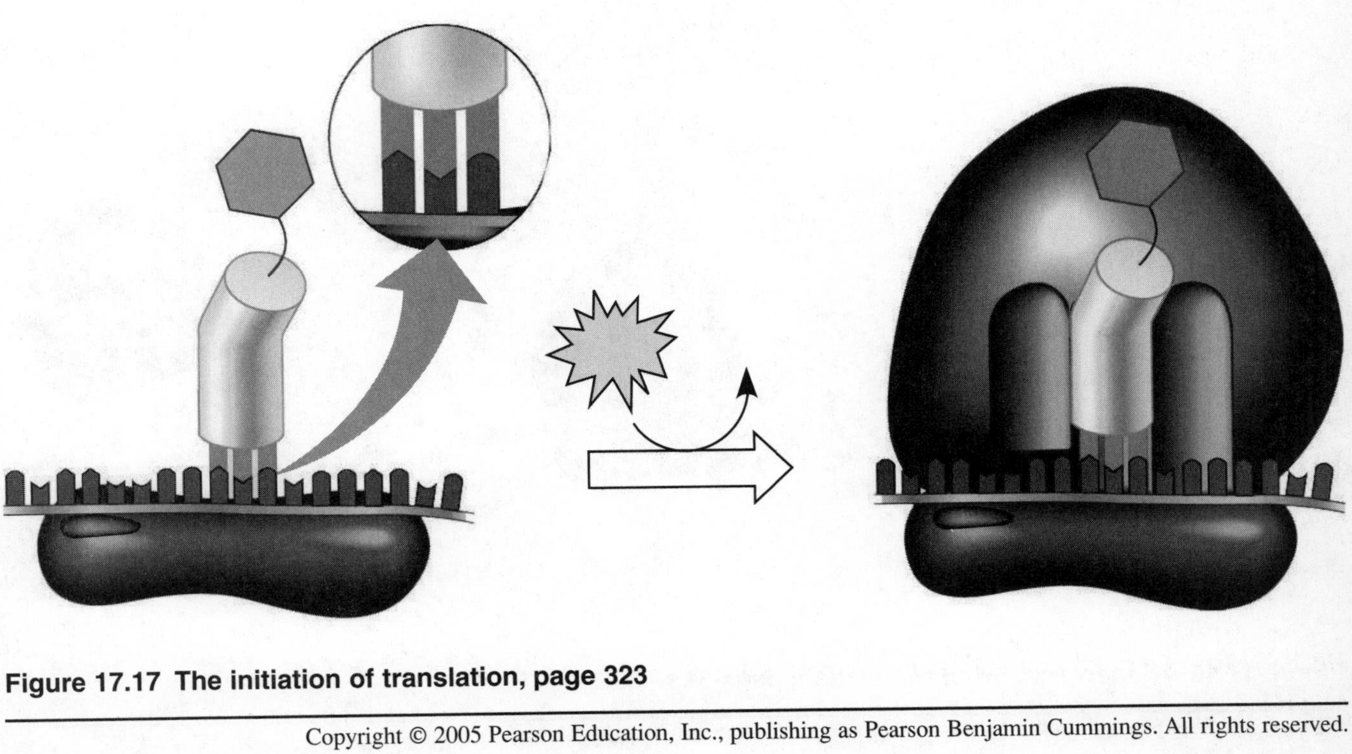

Figure 17.17 The initiation of translation, page 323

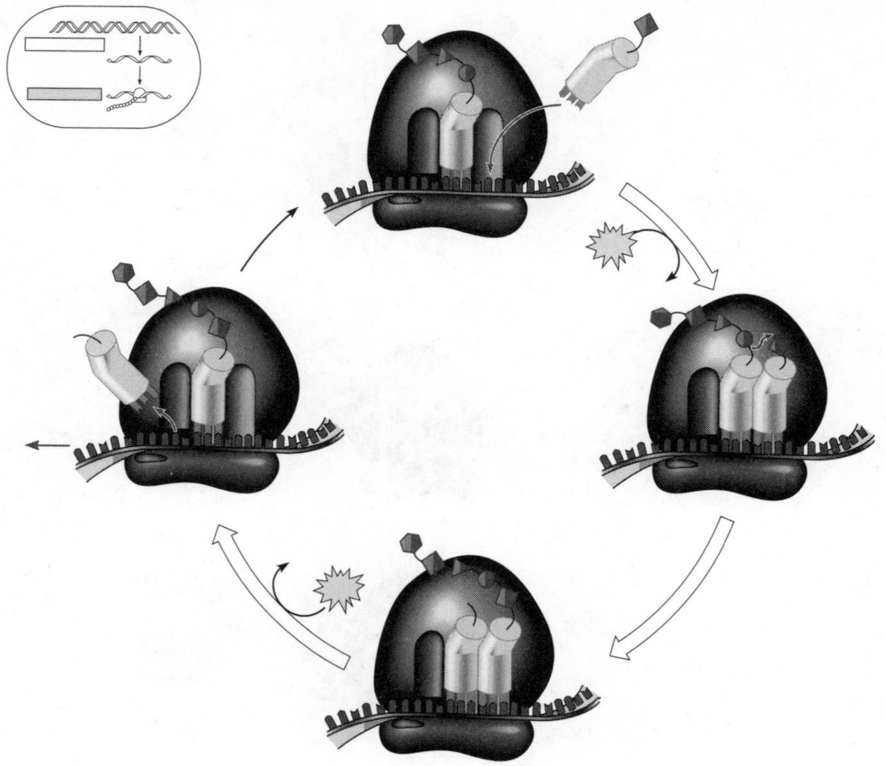

Figure 17.18 The elongation cycle of translation, page 324

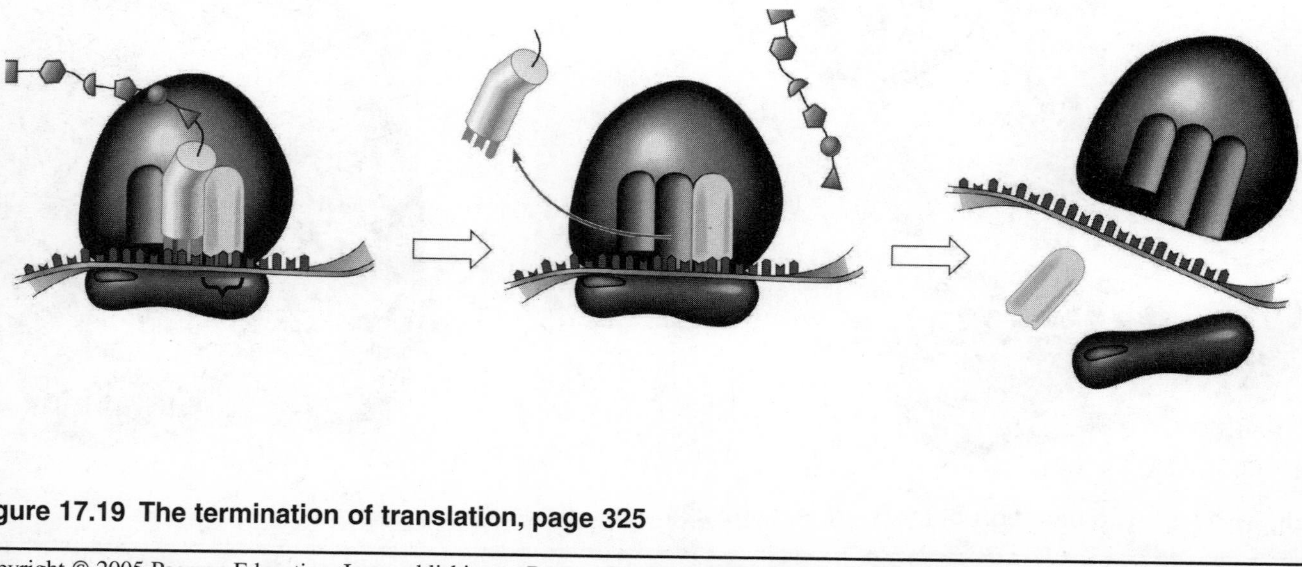

Figure 17.19 The termination of translation, page 325

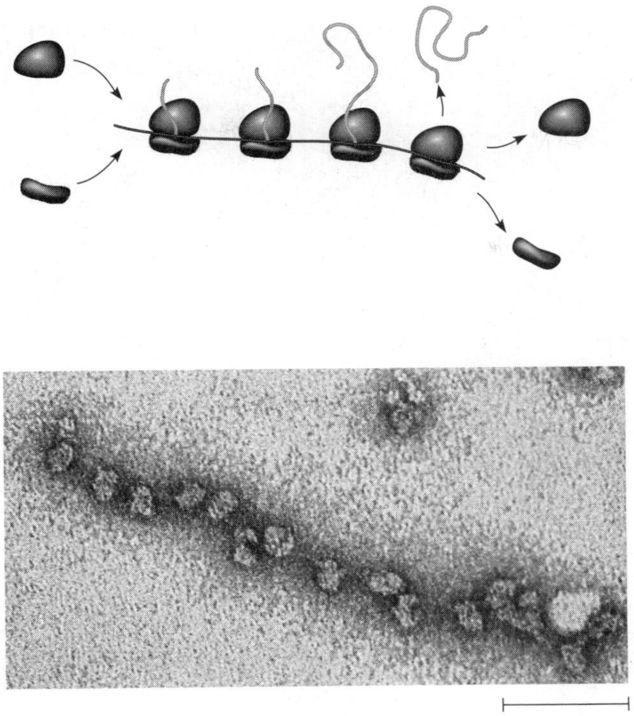

Figure 17.20 Polyribosomes, page 325

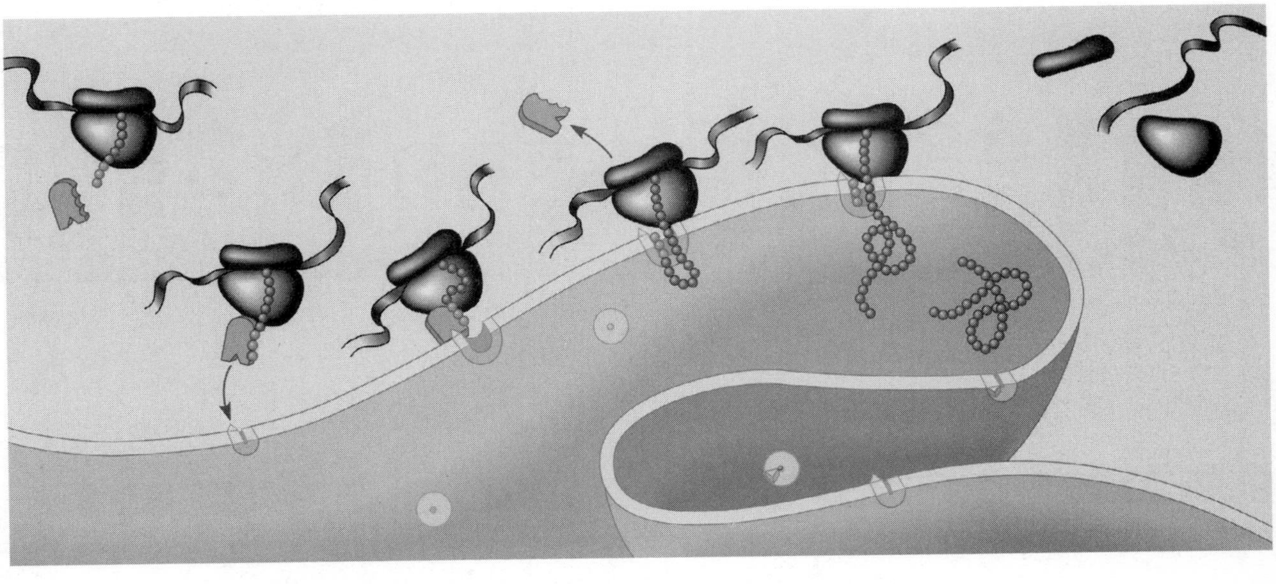

Figure 17.21 The signal mechanism for targeting proteins to the ER, page 326

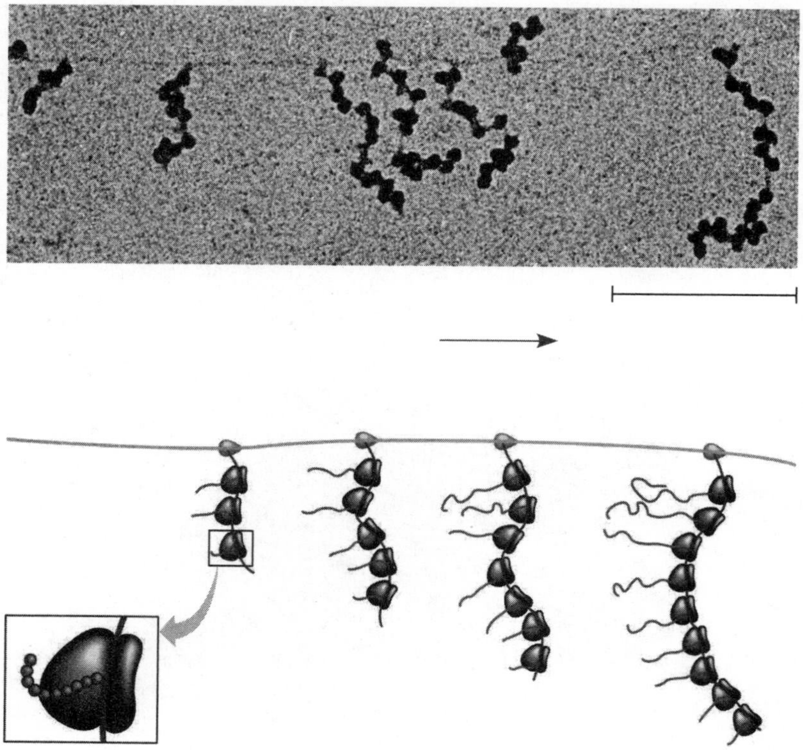

Figure 17.22 Coupled transcription and translation in bacteria, page 328

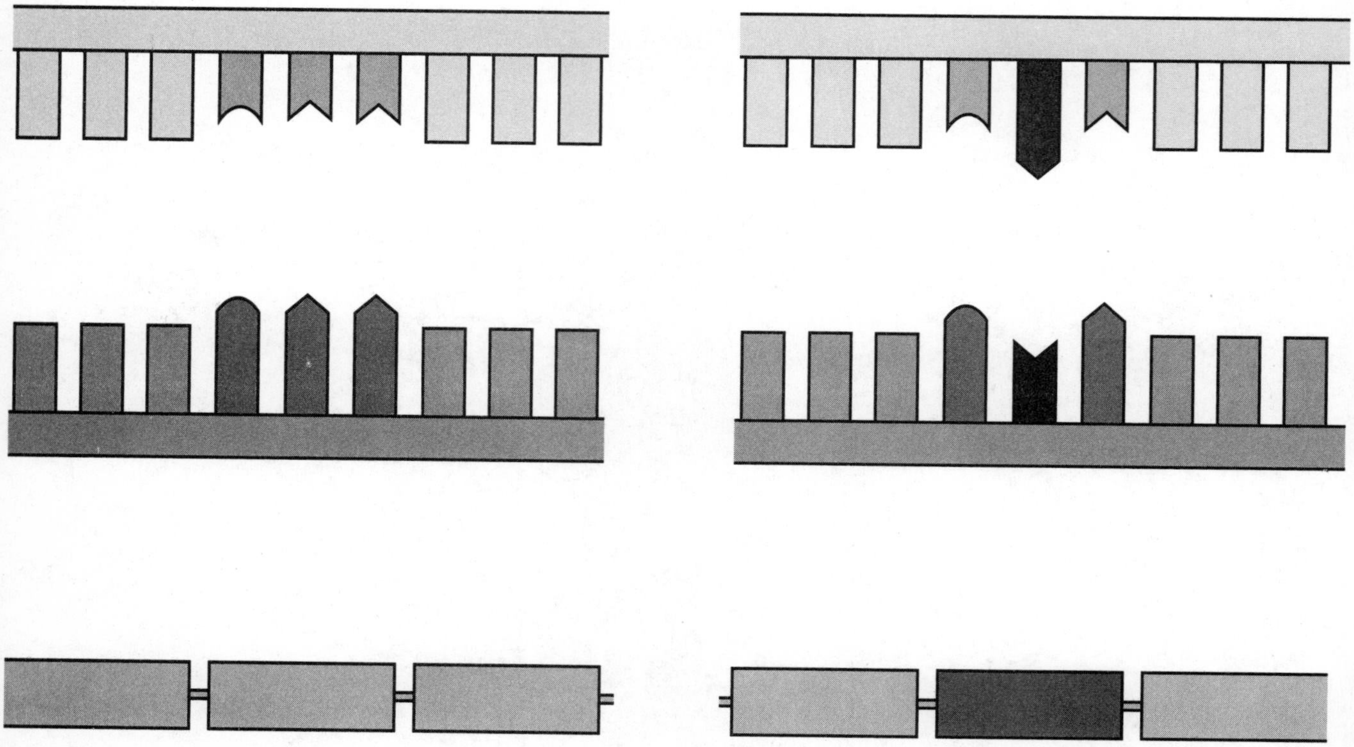

Figure 17.23 The molecular basis of sickle-cell disease: a point mutation, page 329

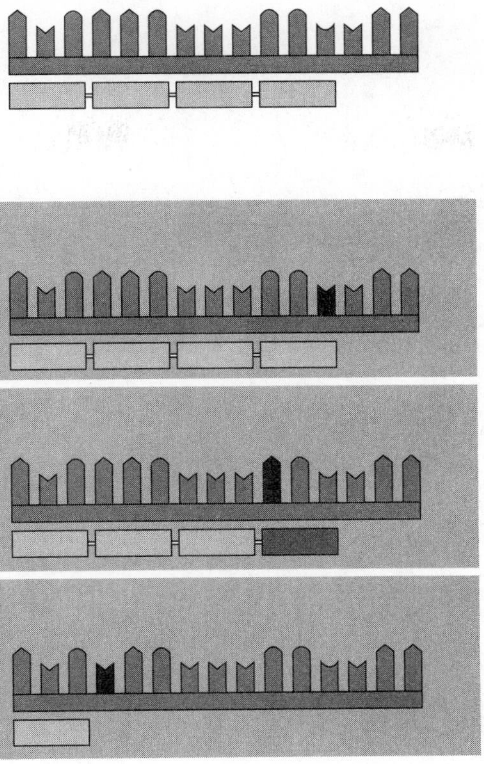

Figure 17.24 Base-pair substitution, page 329

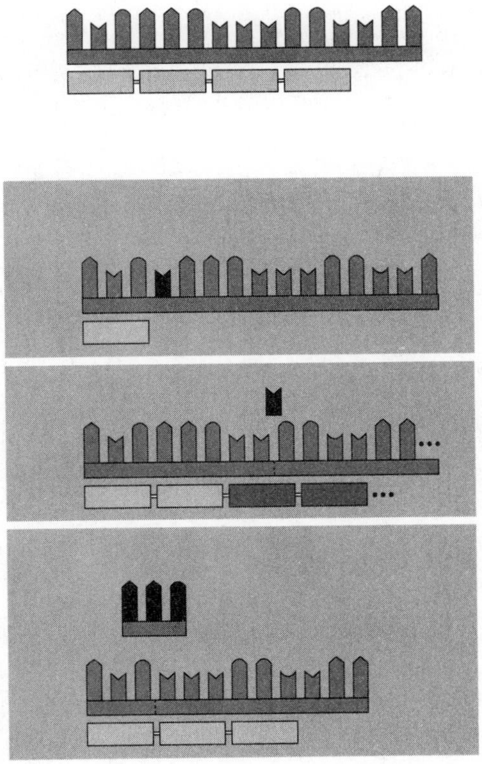

Figure 17.25 Base-pair insertion or deletion, page 330

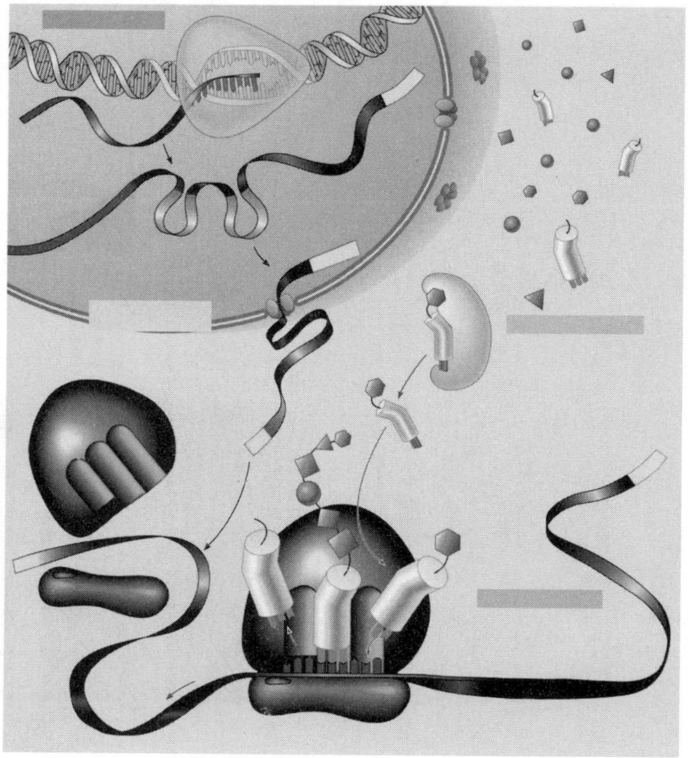

Figure 17.26 A summary of transcription and translation in a eukaryotic cell, page 331

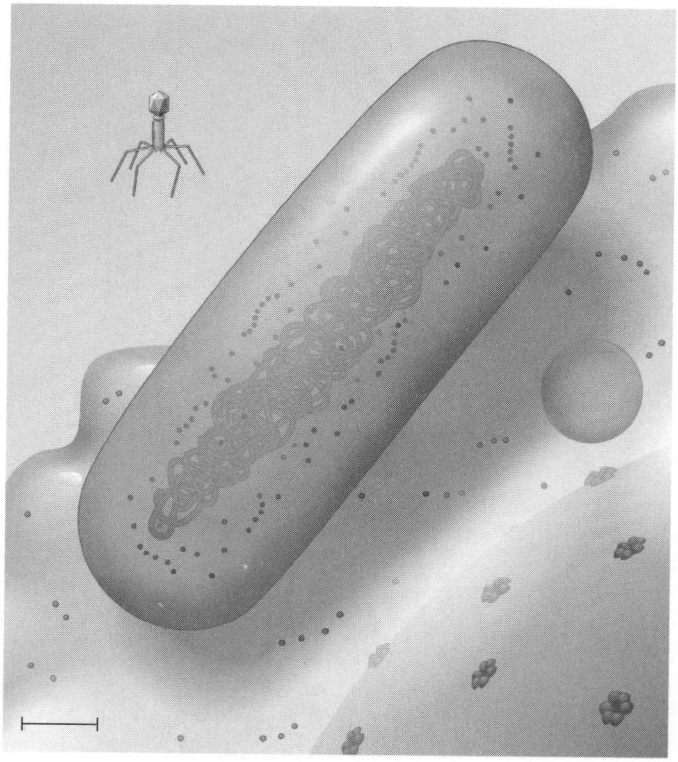

Figure 18.2 Comparing the size of a virus, a bacterium, and an animal cell, page 335

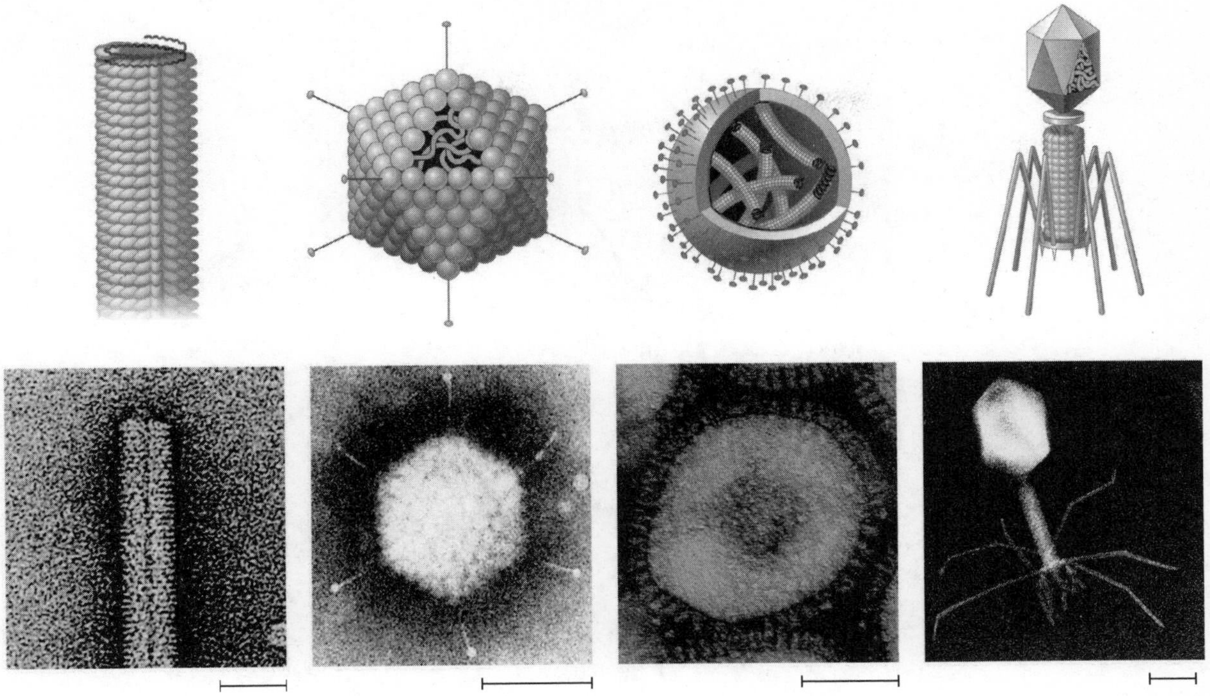

Figure 18.4 Viral structure, page 336

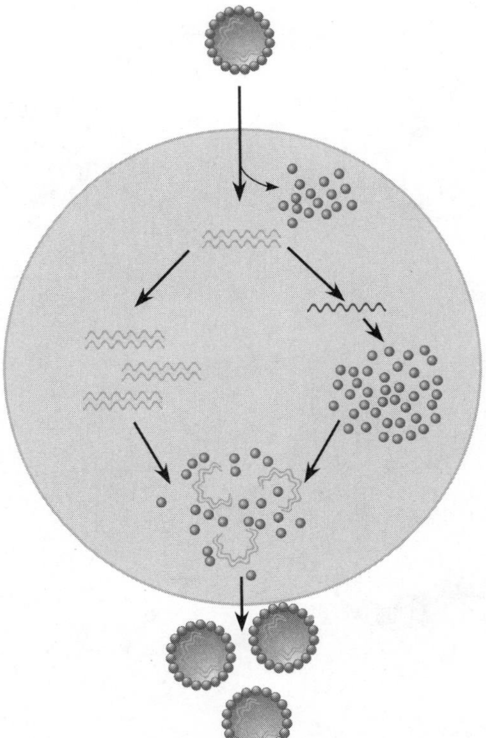

Figure 18.5 A simplified viral reproductive cycle, page 337

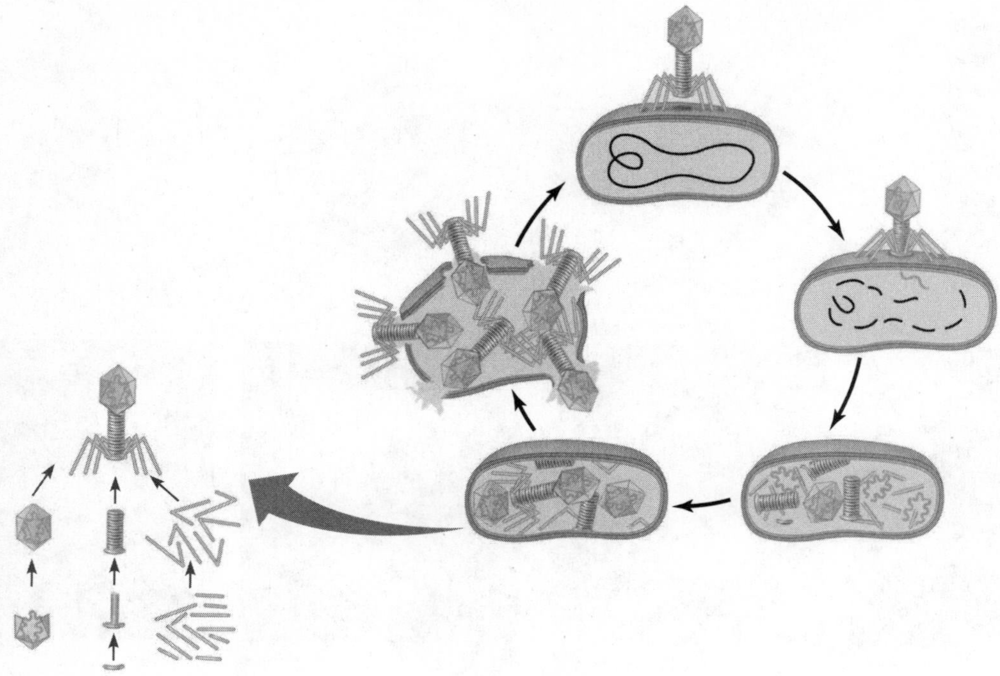

Figure 18.6 The lytic cycle of phage T4, a virulent phage, page 338

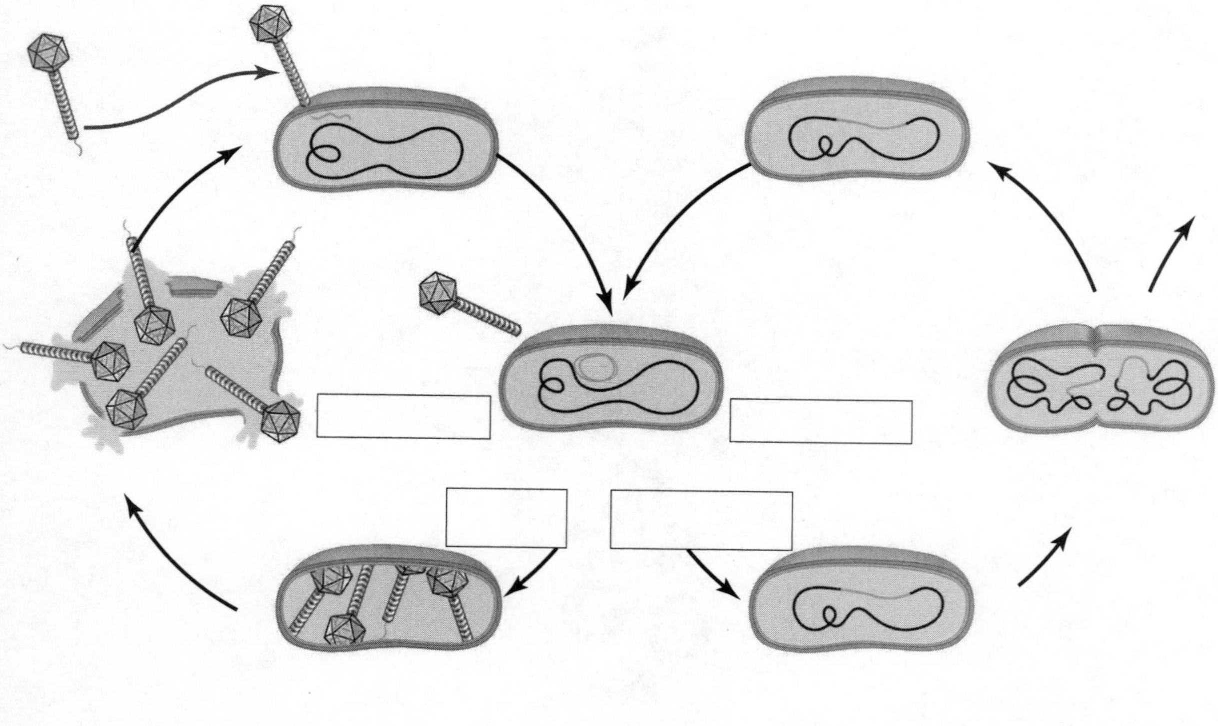

Figure 18.7 The lytic and lysogenic cycles of phage λ, a temperate phage, page 339

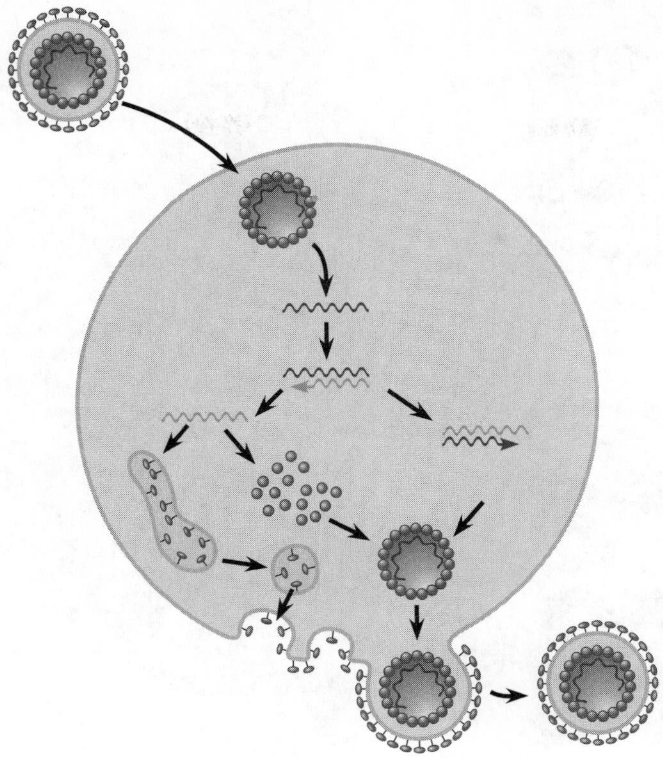

Figure 18.8 The reproductive cycle of an enveloped RNA virus, page 341

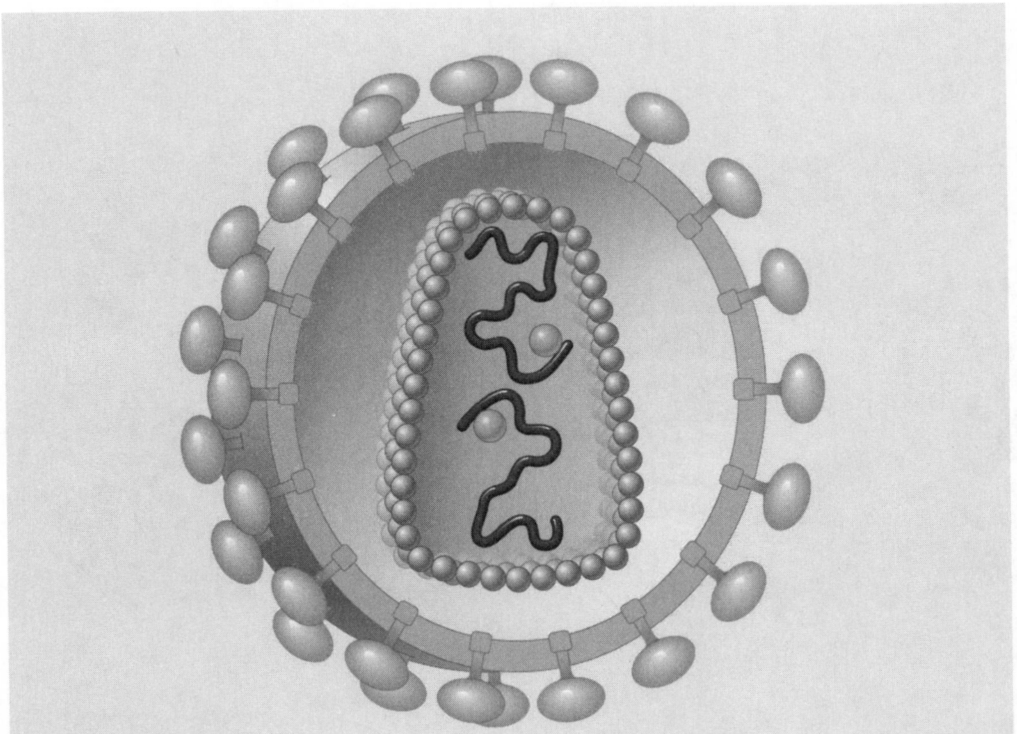

Figure 18.9 The structure of HIV, the retrovirus that causes AIDS, page 341

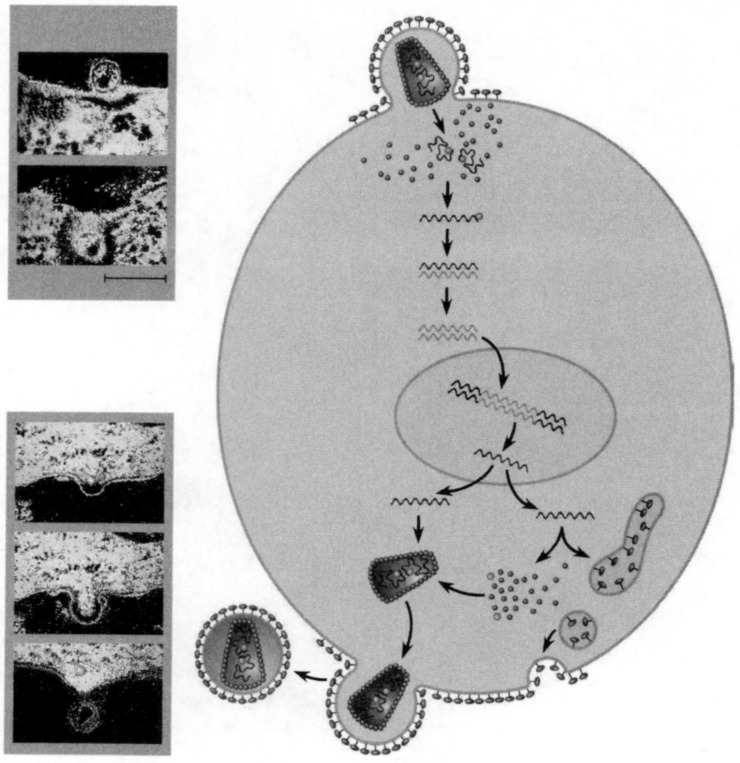

Figure 18.10 The reproductive cycle of HIV, a retrovirus, page 342

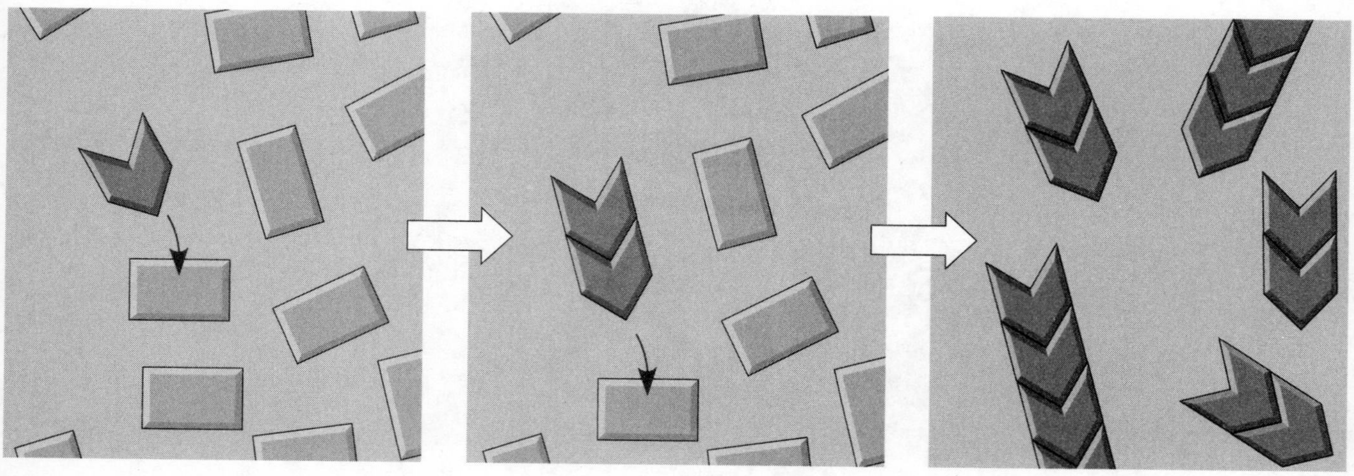

Figure 18.13 Model for how prions propagate, page 346

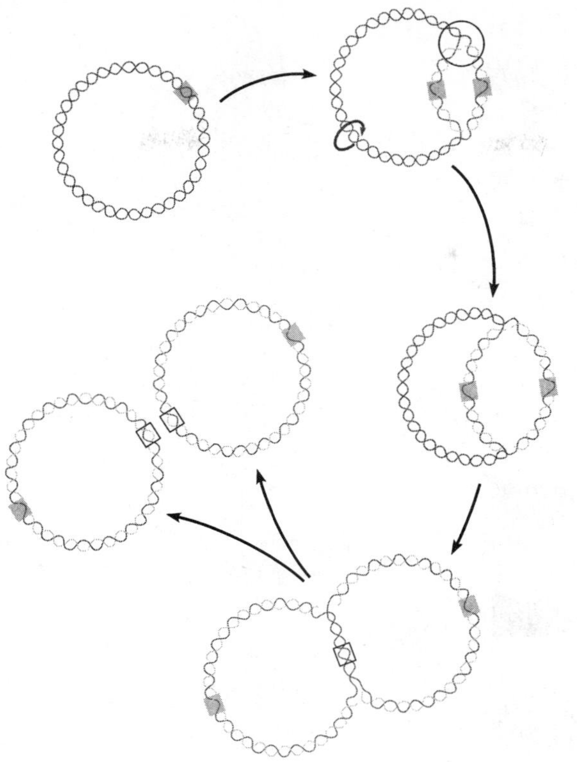

Figure 18.14 Replication of a bacterial chromosome, page 347

Figure 18.15 Can a bacterial cell acquire genes from another bacterial cell?, page 347

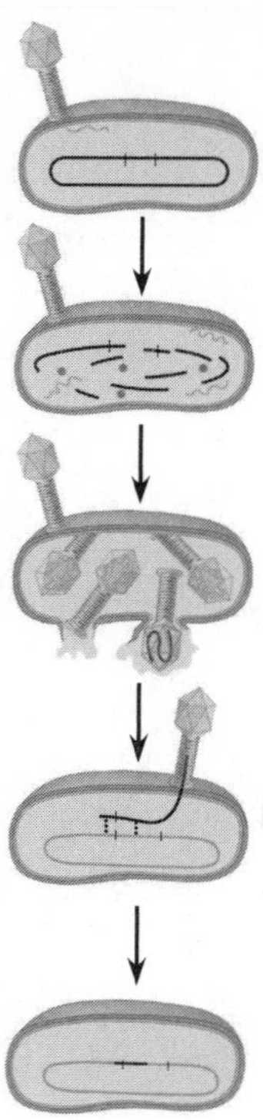

Figure 18.16 Generalized transduction, page 348

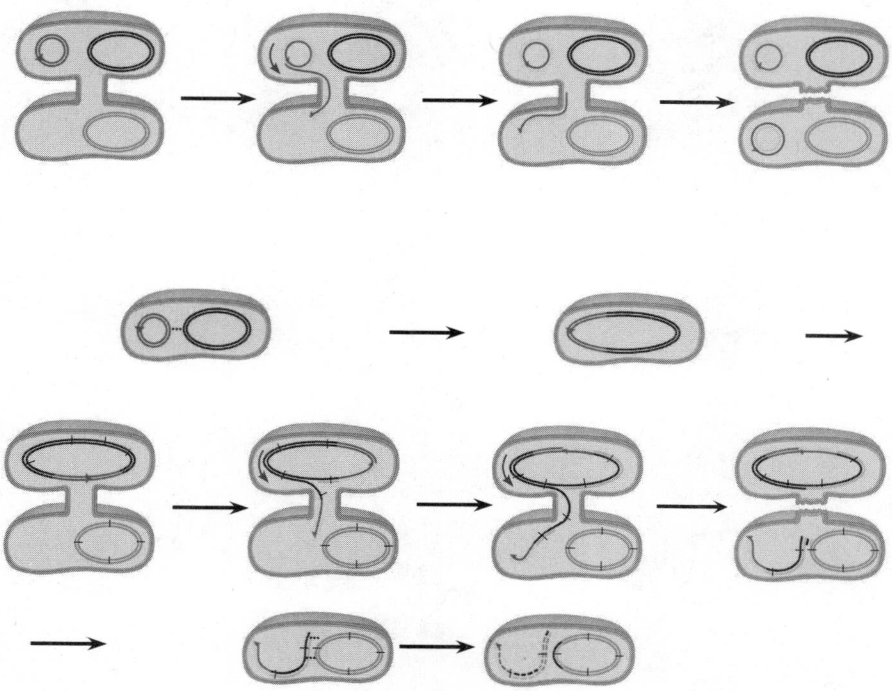

Figure 18.18 Conjugation and recombination in *E. coli*, page 350

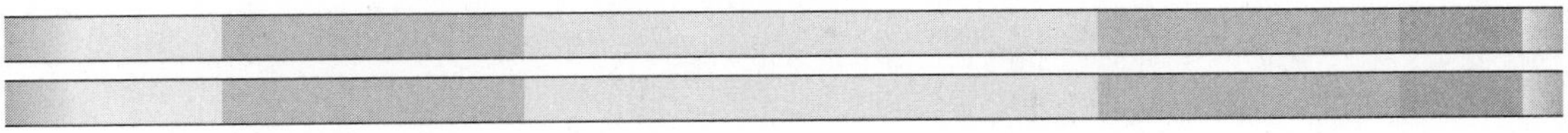

Figure 18.19 Transposable genetic elements in bacteria, page 352

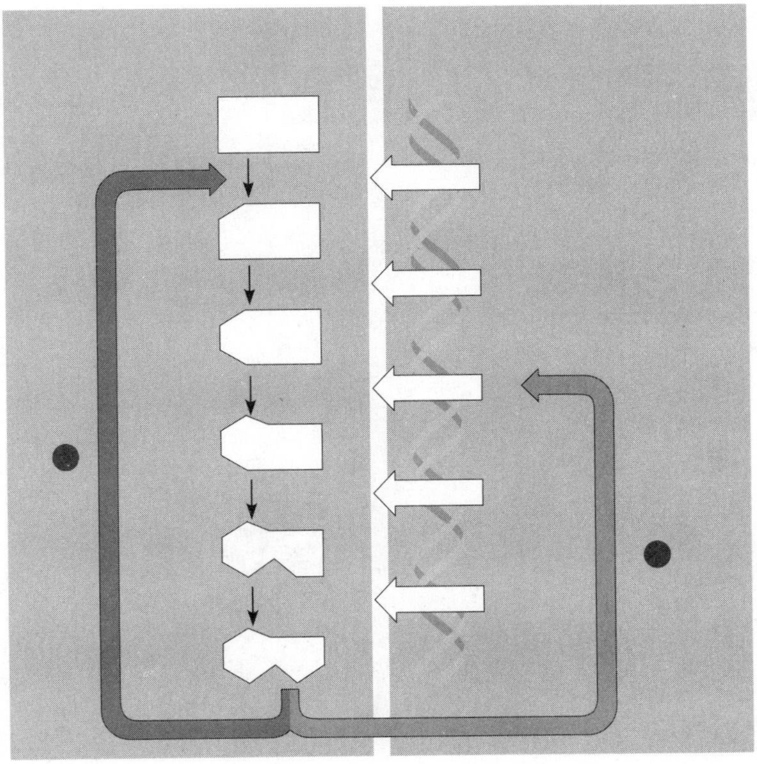

Figure 18.20 Regulation of a metabolic pathway, page 353

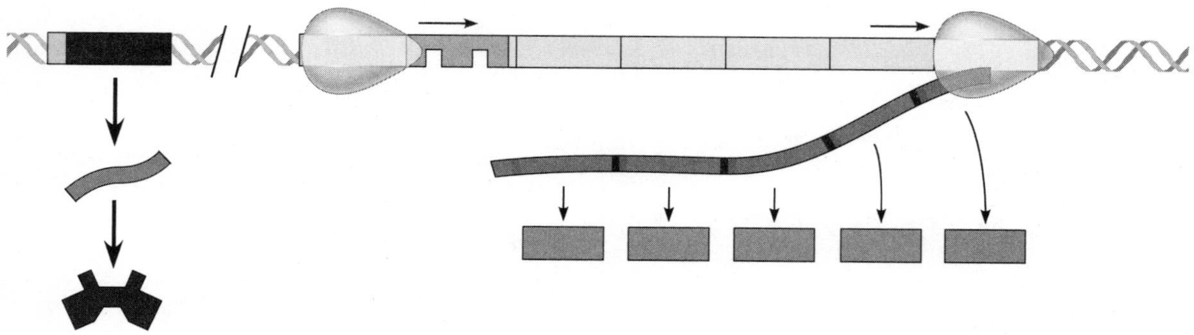

Figure 18.21a The *trp* operon: regulated synthesis of repressible enzymes, page 354

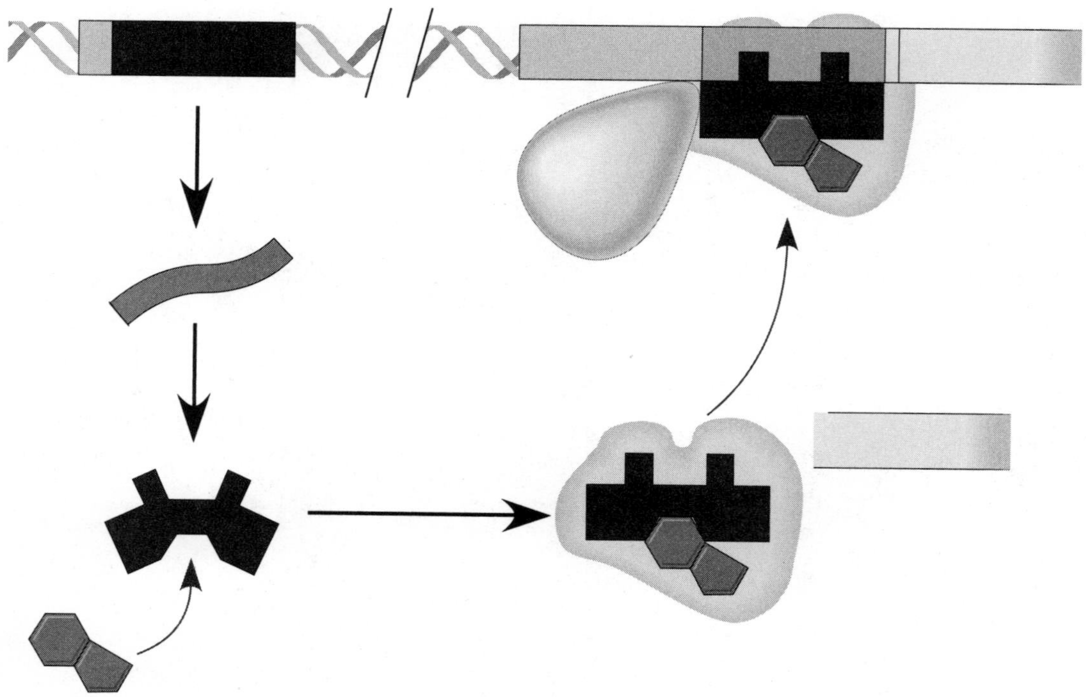

Figure 18.21b The *trp* operon: regulated synthesis of repressible enzymes, page 354

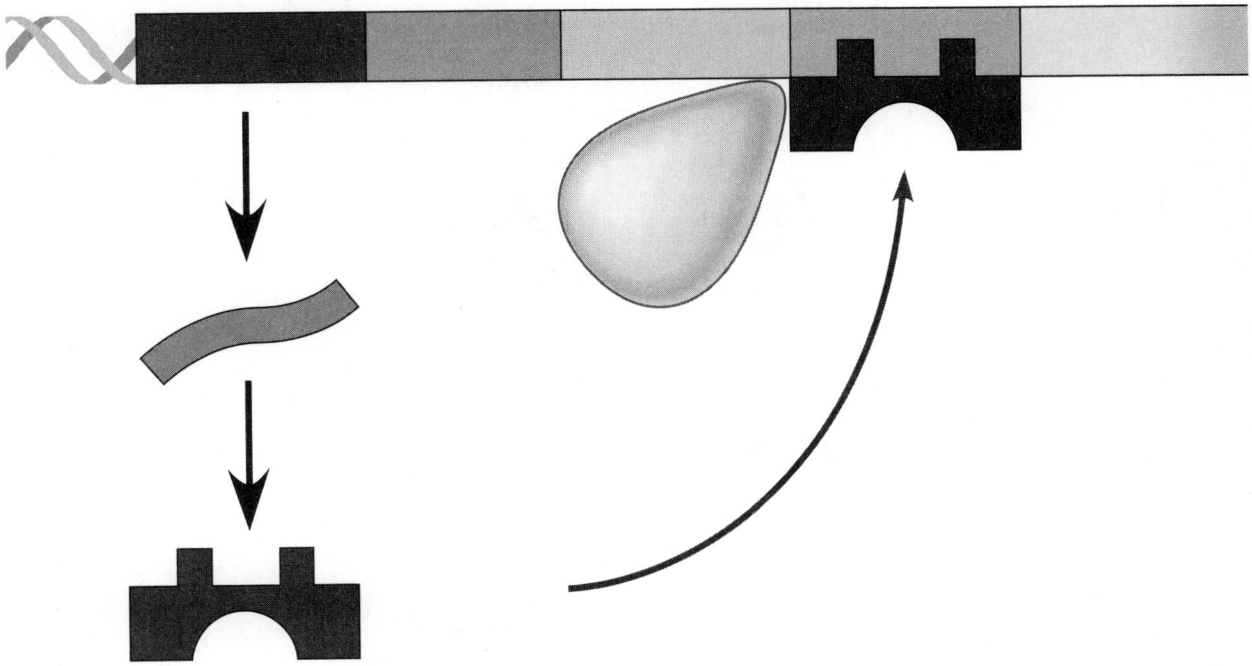

Figure 18.22a The *lac* operon: regulated synthesis of inducible enzymes, page 355

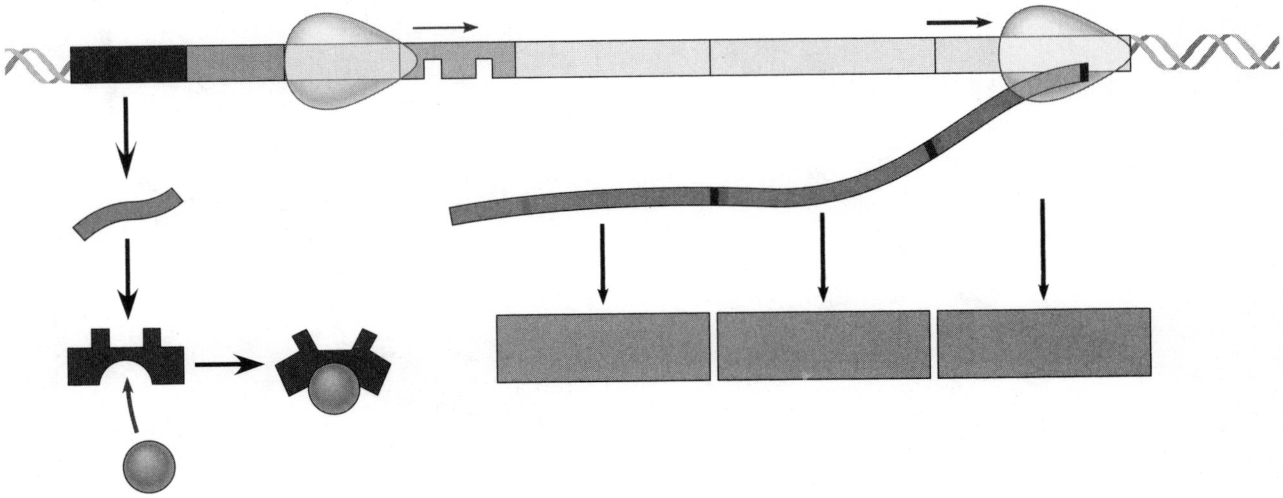

Figure 18.22b The *lac* operon: regulated synthesis of inducible enzymes, page 355

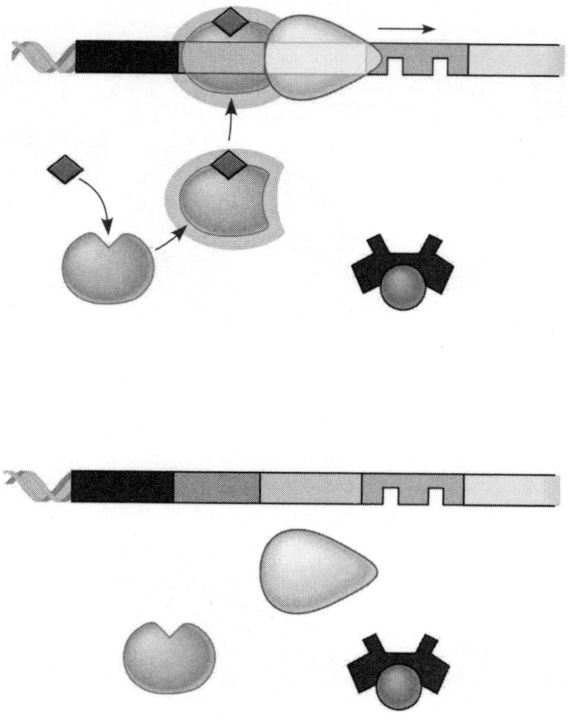

Figure 18.23 Positive control of the *lac* operon by catabolite activator protein (CAP), page 356

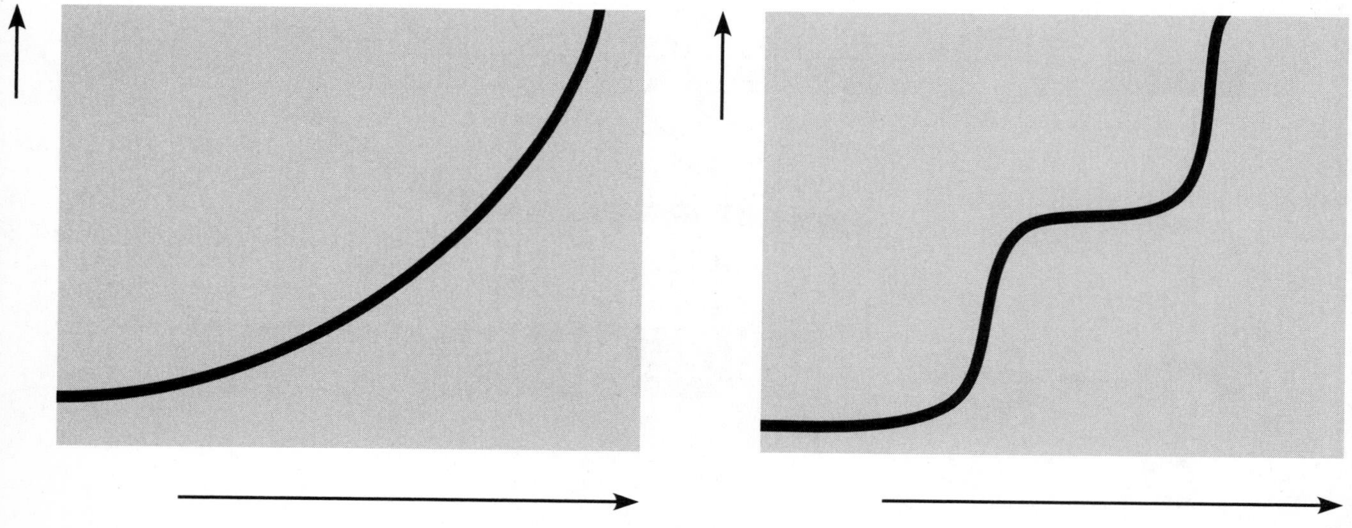

Figure 18.UN1 Bacterial and viral growth curves, page 358

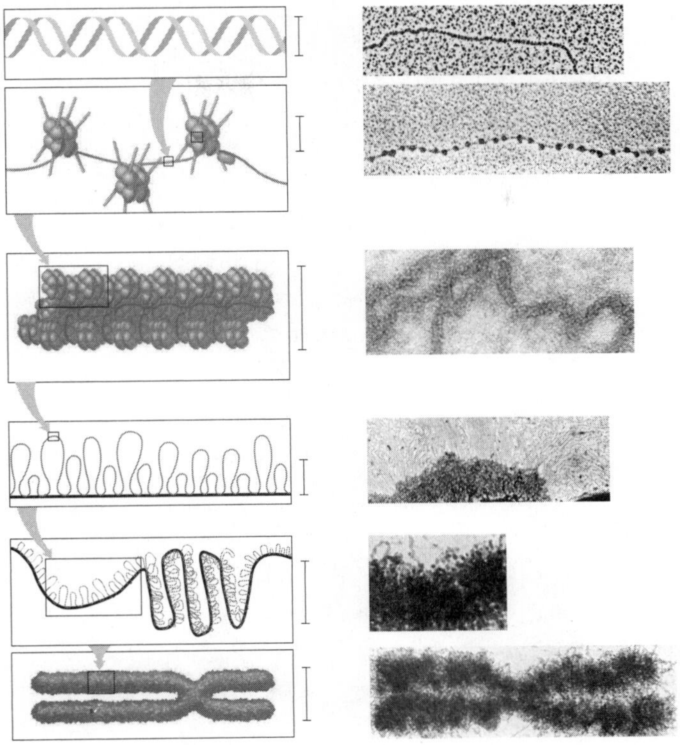

Figure 19.2 Levels of chromatin packing, page 361

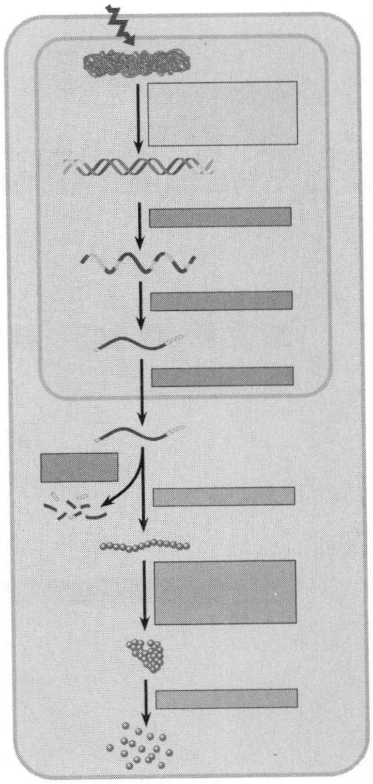

Figure 19.3 Stages in gene expression that can be regulated in eukaryotic cells, page 362

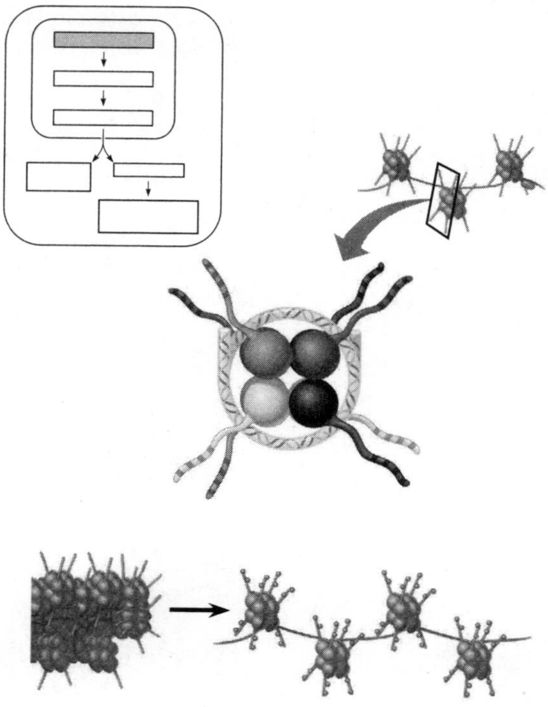

Figure 19.4 A simple model of histone tails and the effect of histone acetylation, page 363

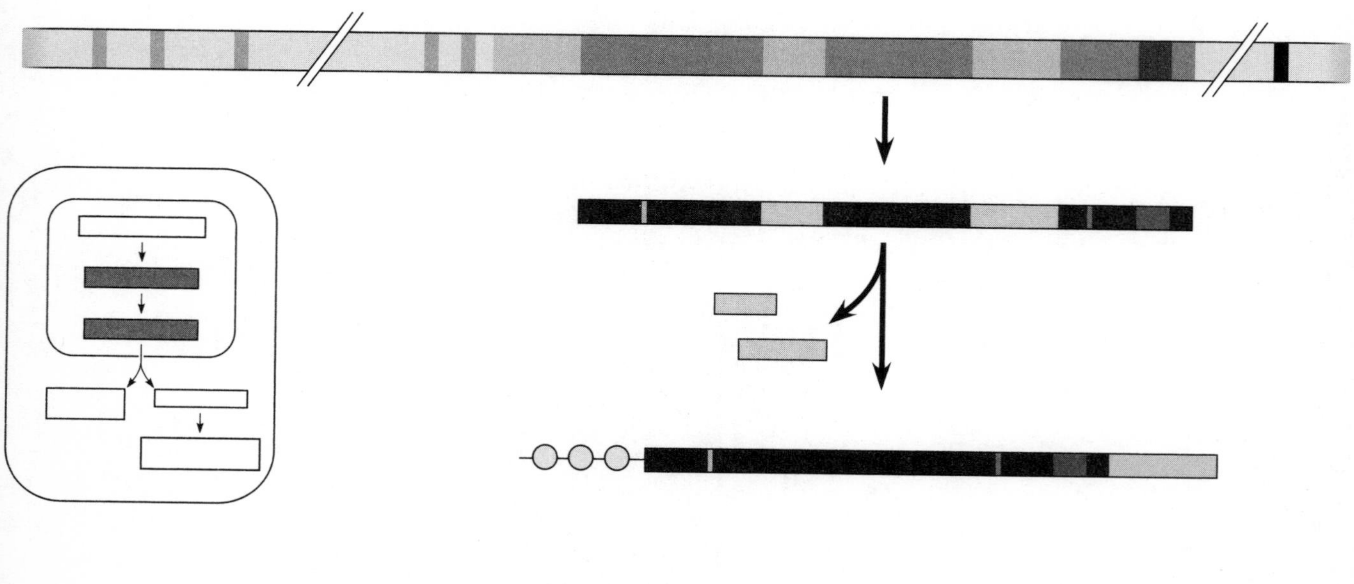

Figure 19.5 A eukaryotic gene and its transcript, page 365

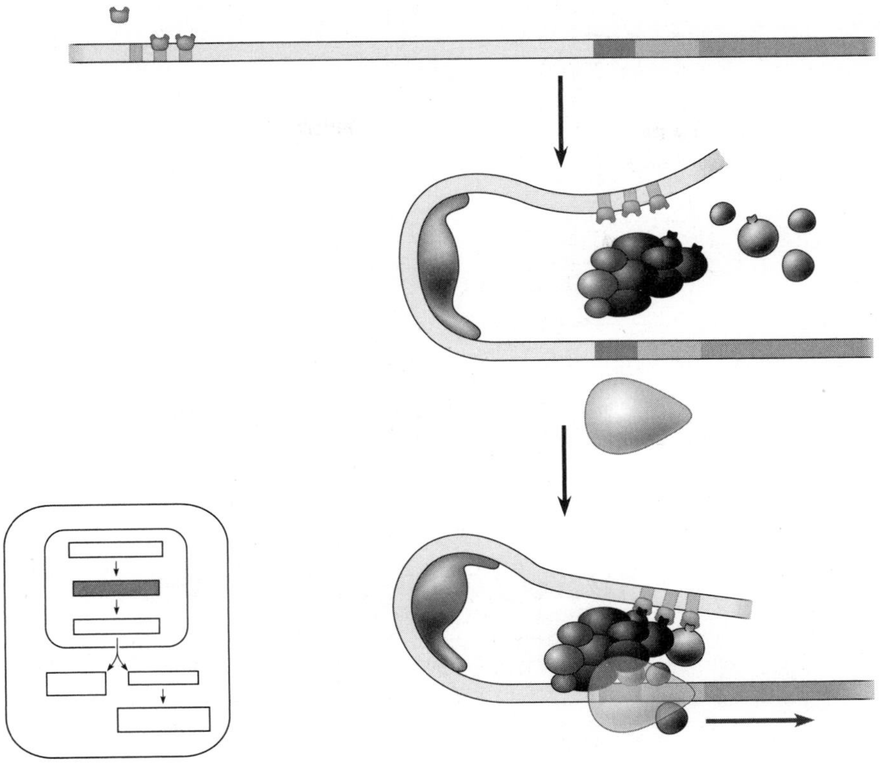

Figure 19.6 A model for the action of enhancers and transcription activators, page 366

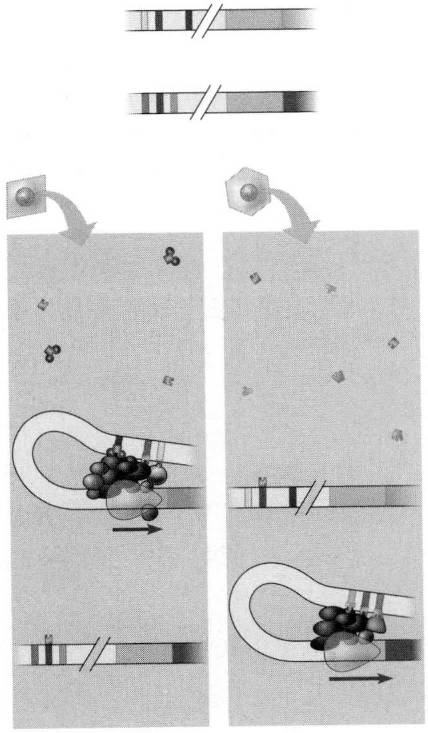

Figure 19.7 Cell type–specific transcription, page 367

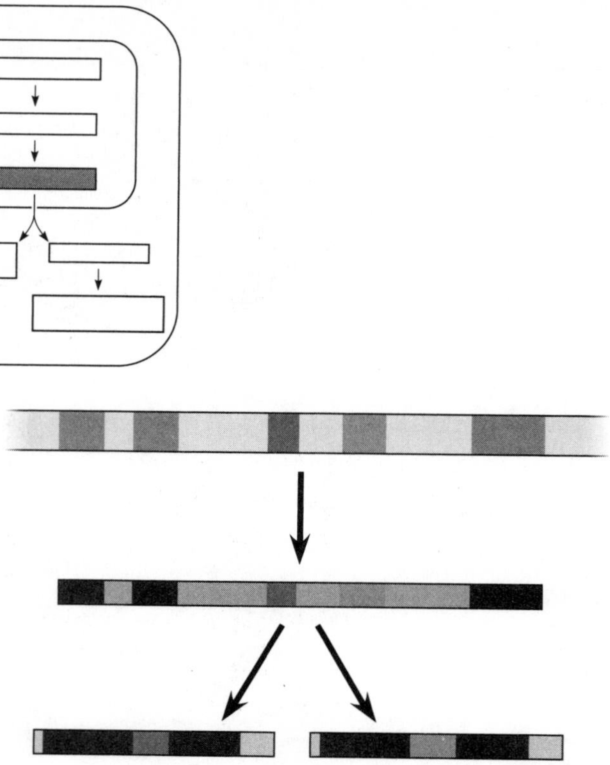

Figure 19.8 Alternative RNA splicing, page 368

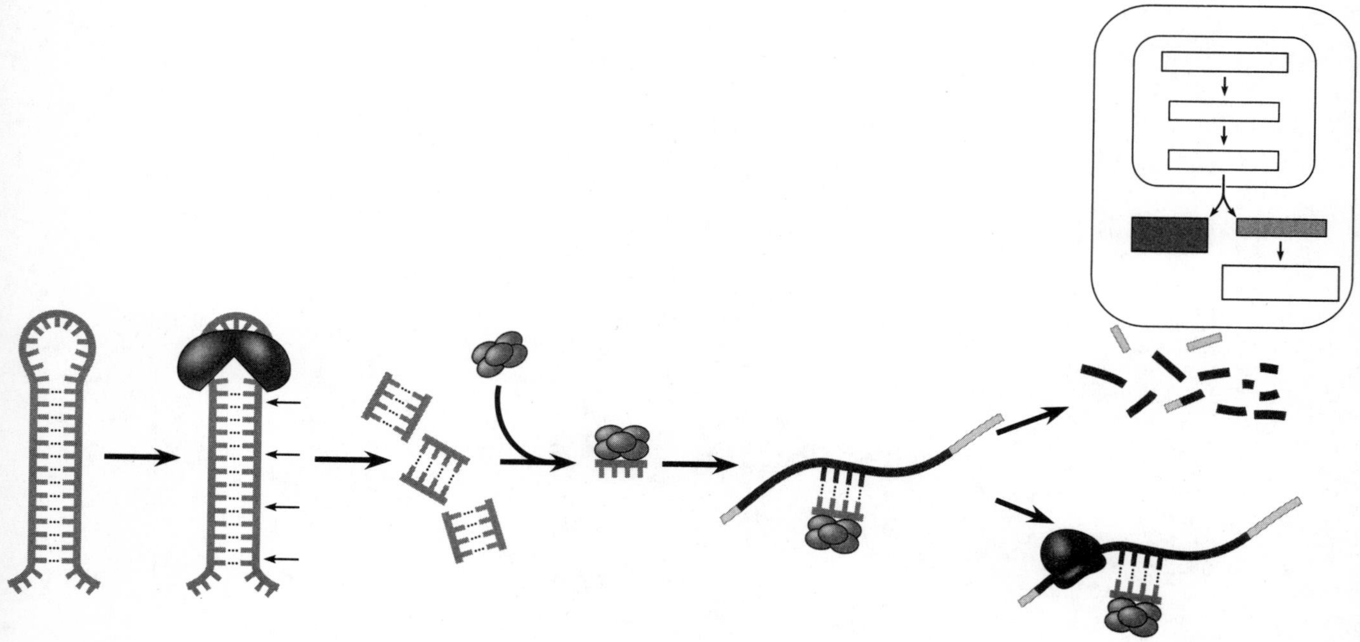

Figure 19.9 Regulation of gene expression by microRNAs (miRNAs), page 369

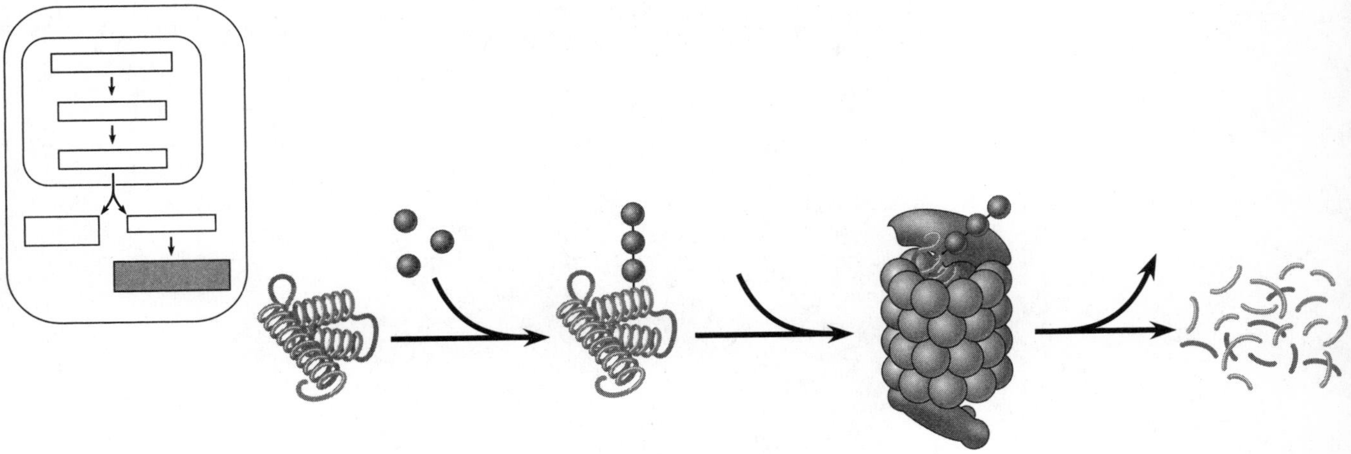

Figure 19.10 Degradation of a protein by a proteasome, page 370

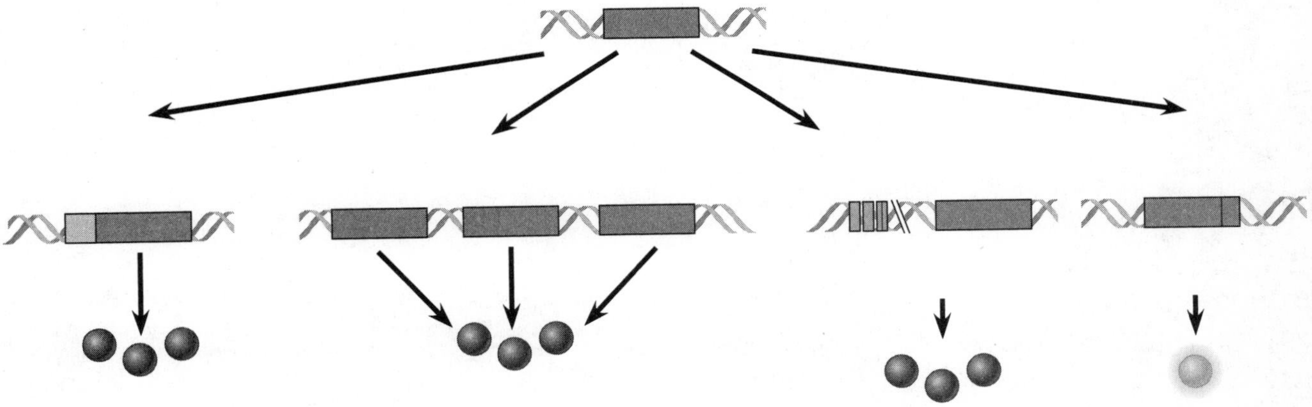

Figure 19.11 Genetic changes that can turn proto-oncogenes into oncogenes, page 371

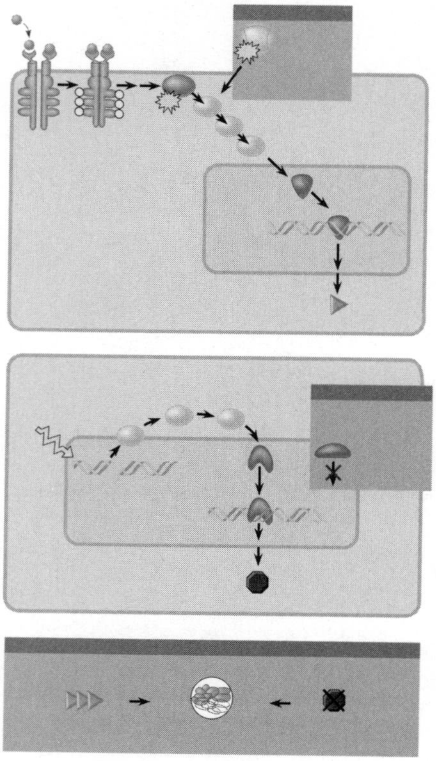

Figure 19.12 Signaling pathways that regulate cell division, page 372

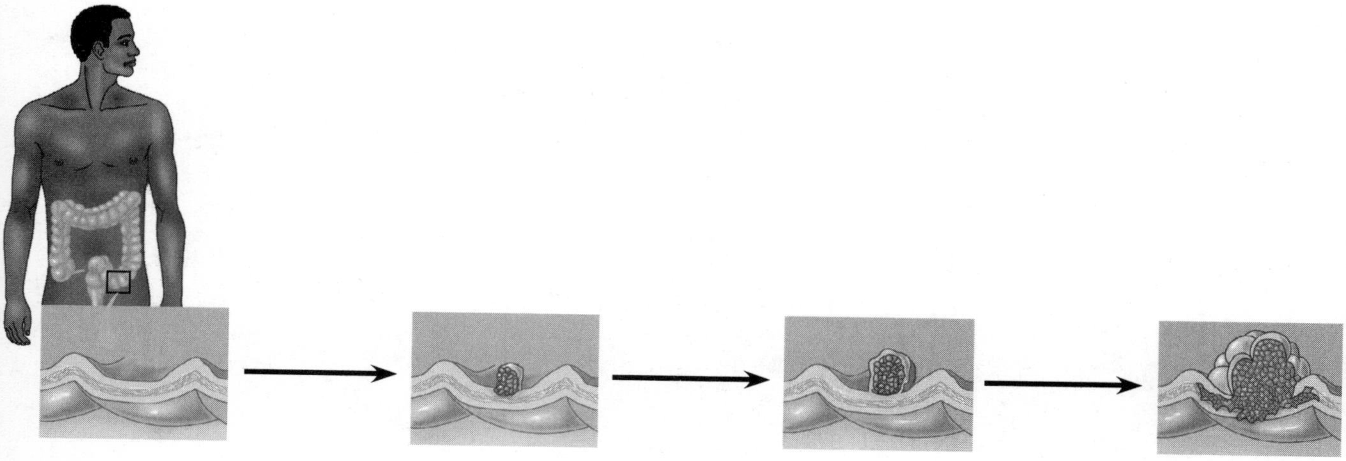

Figure 19.13 A multistep model for the development of colorectal cancer, page 373

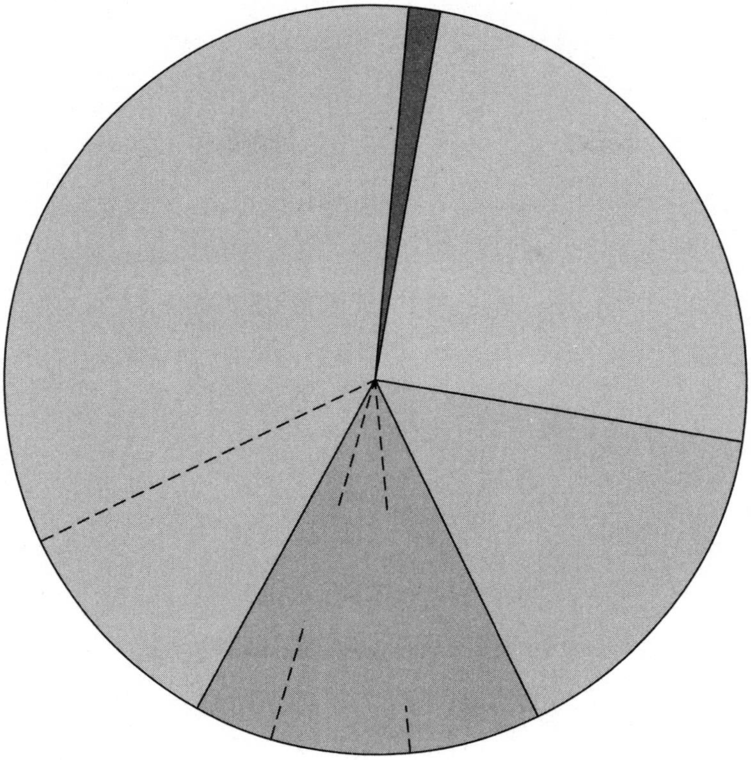

Figure 19.14 Types of DNA sequences in the human genome, page 375

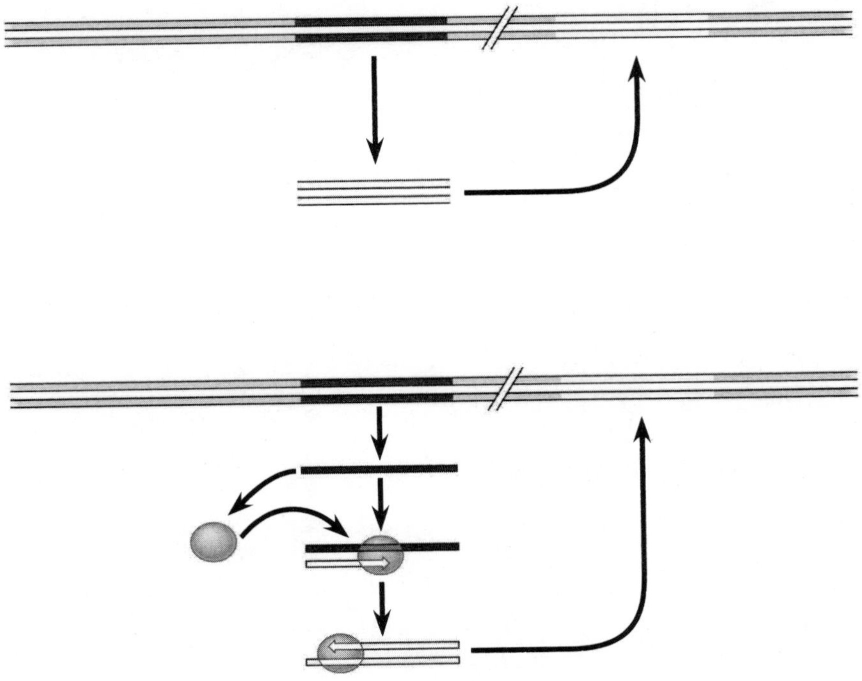

Figure 19.16 Movement of eukaryotic transposable elements, page 376

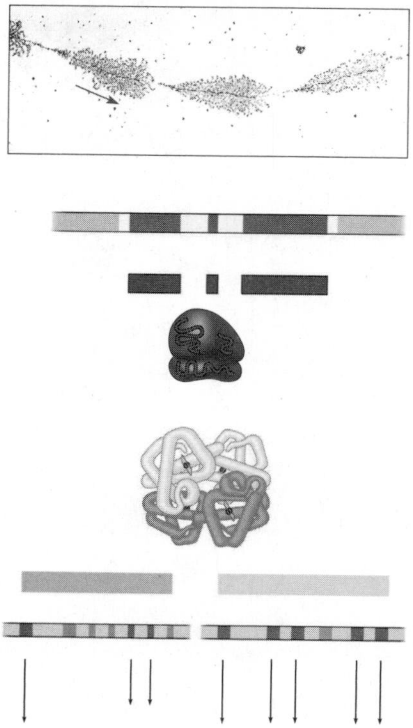

Figure 19.17 Gene families, page 377

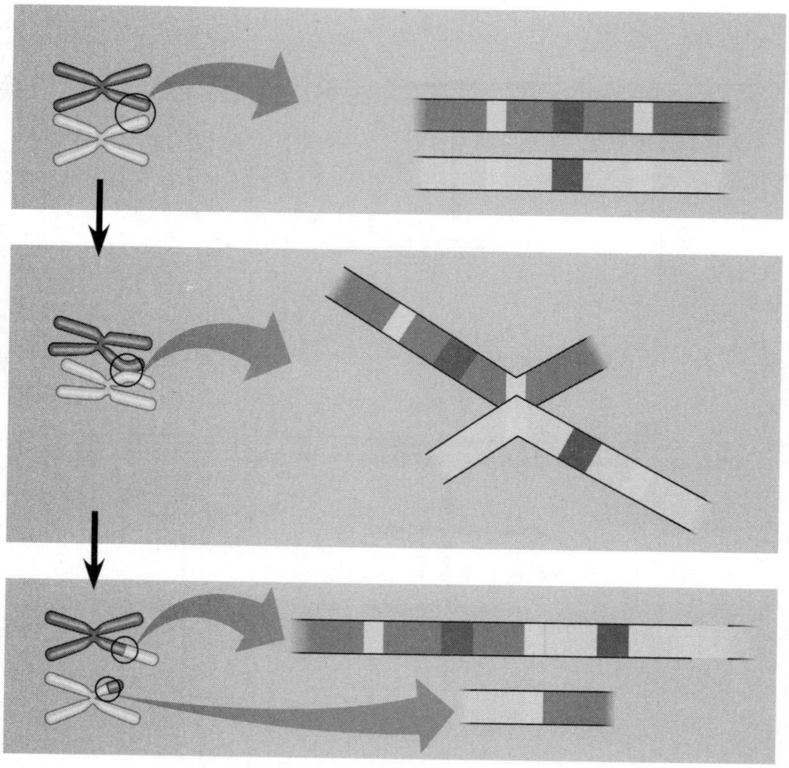

Figure 19.18 Gene duplication due to unequal crossing over, page 378

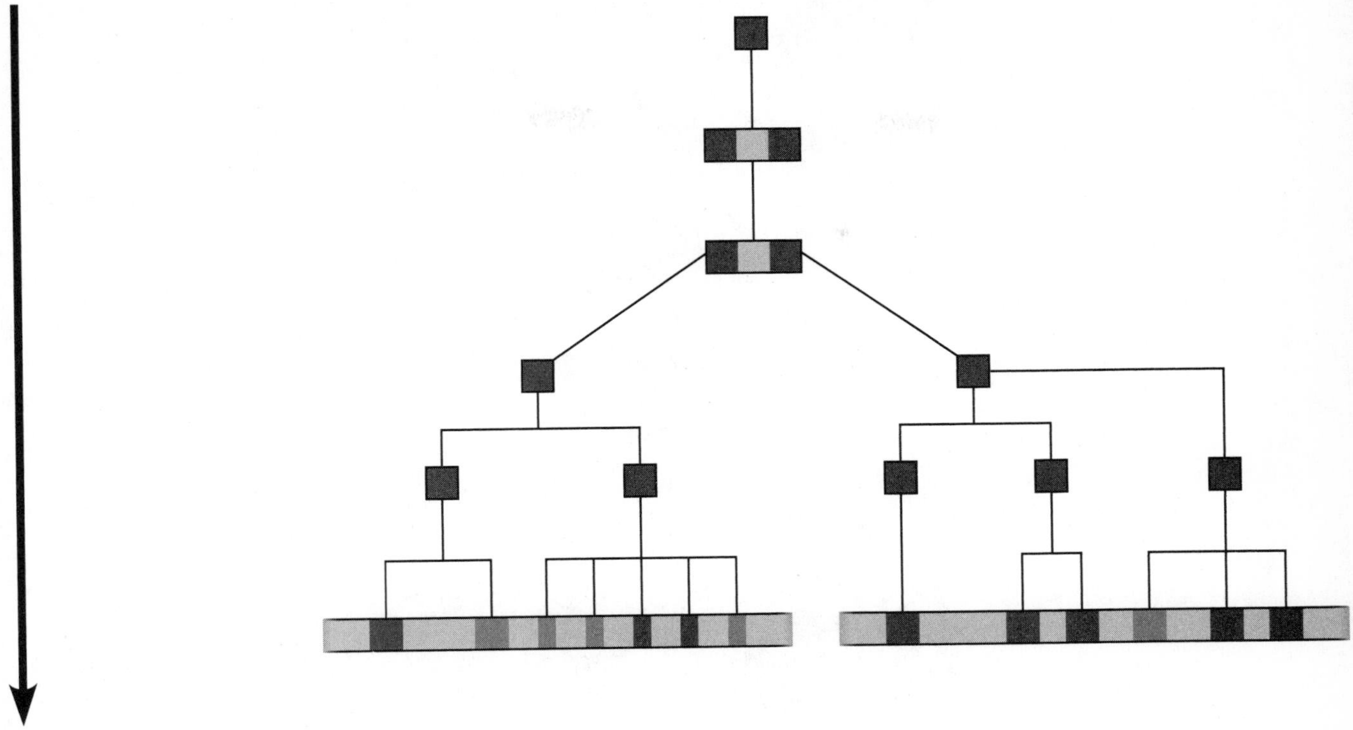

Figure 19.19 Evolution of the human α-globin and β-globin gene families, page 379

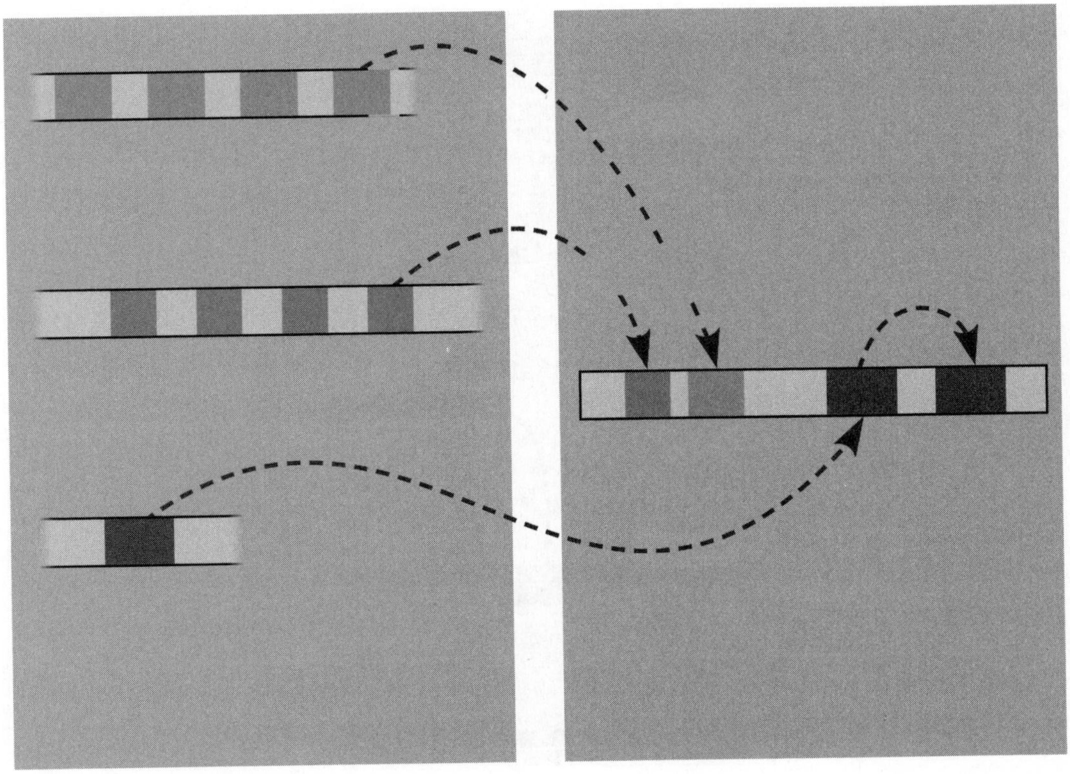

Figure 19.20 Evolution of a new gene by exon shuffling, page 380

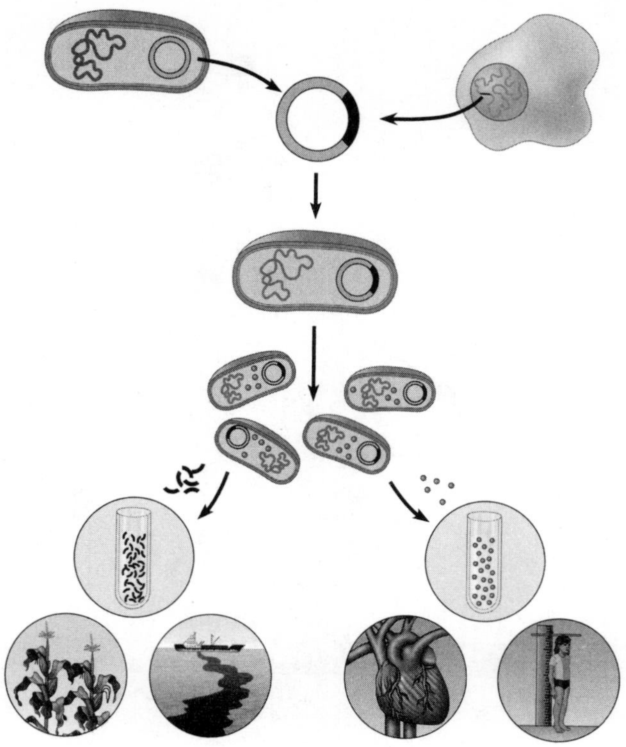

Figure 20.2 Overview of gene cloning with a bacterial plasmid, showing various uses of cloned genes, page 385

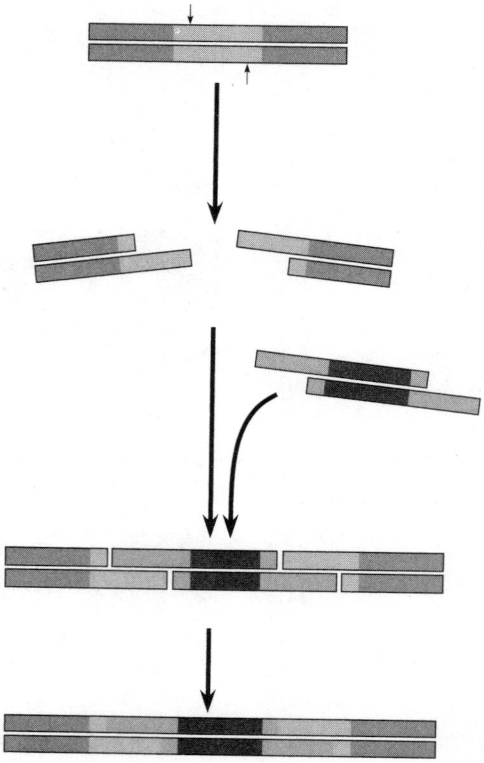

Figure 20.3 Using a restriction enzyme and DNA ligase to make recombinant DNA, page 386

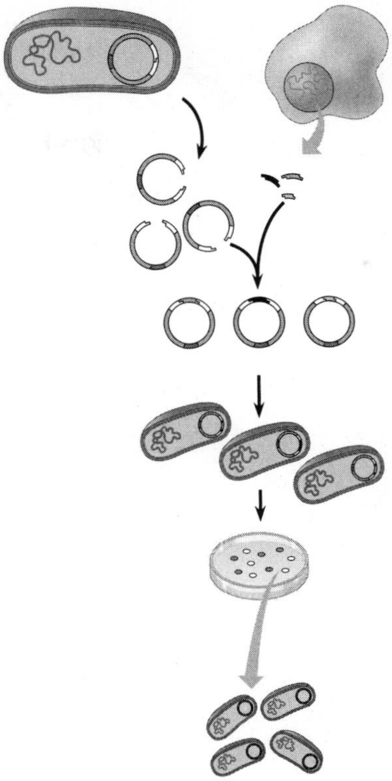

Figure 20.4 Cloning a human gene in a bacterial plasmid, page 387

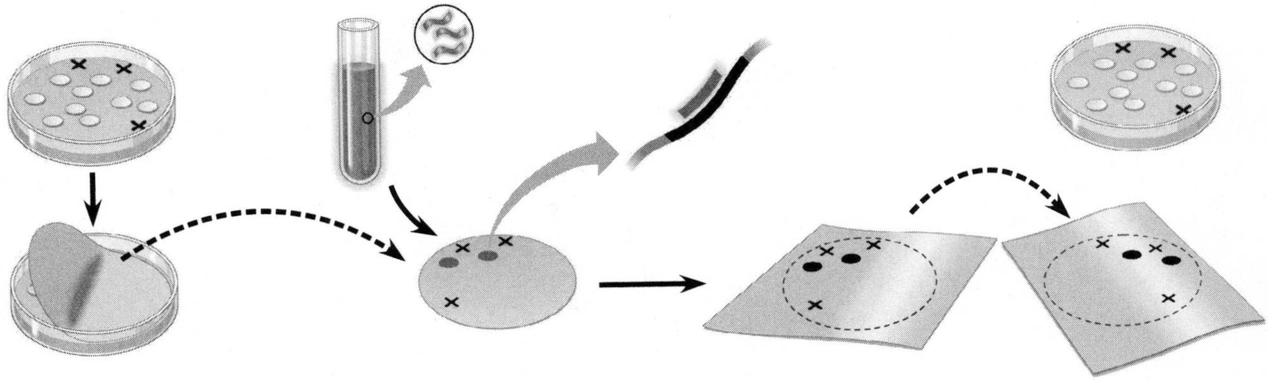

Figure 20.5 Nucleic acid probe hybridization, page 389

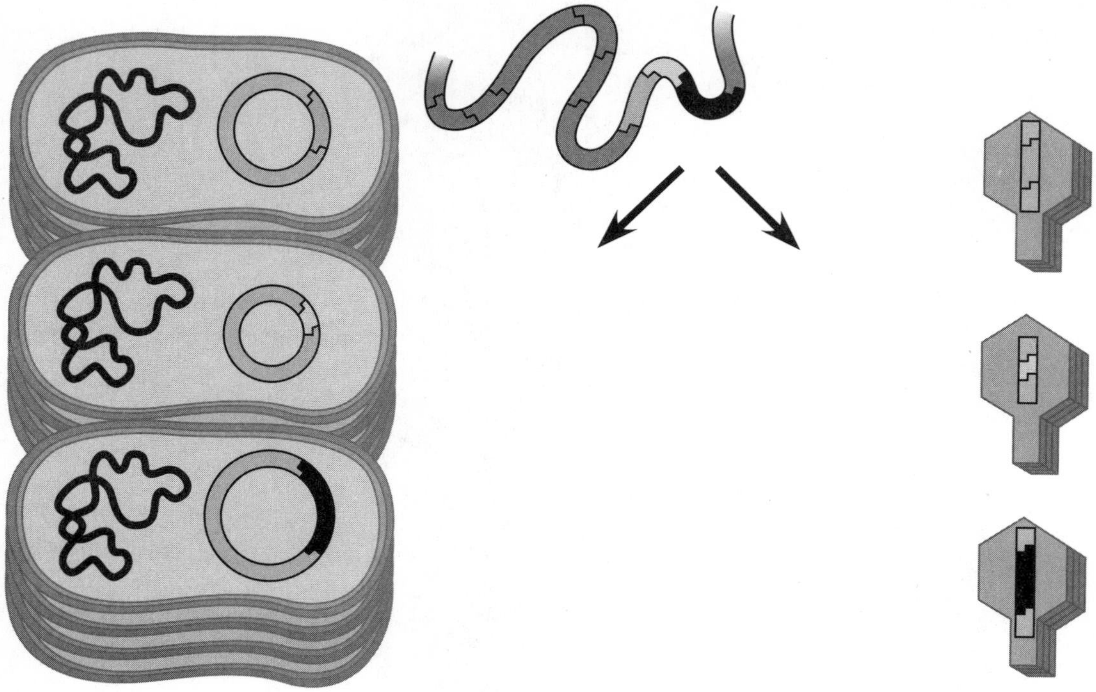

Figure 20.6 Genomic libraries, page 389

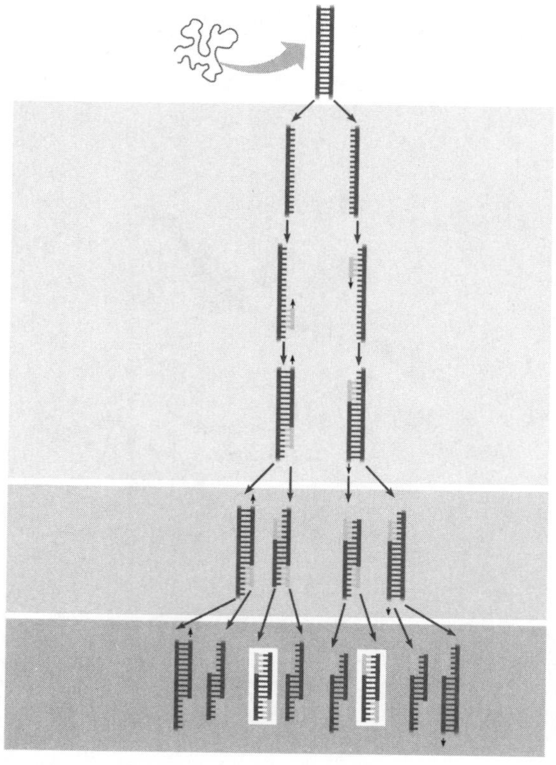

Figure 20.7 The polymerase chain reaction (PCR), page 391

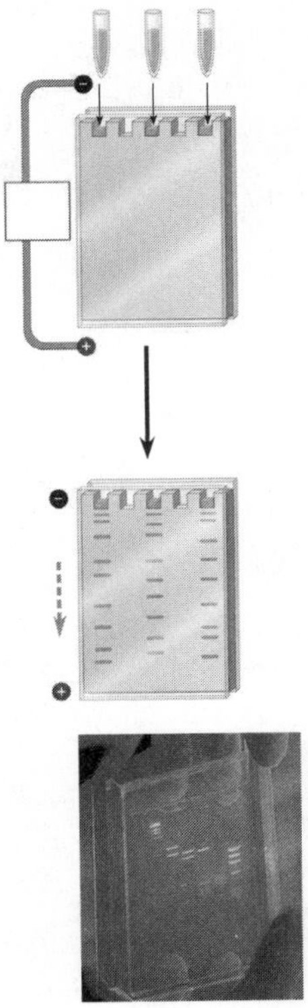

Figure 20.8 Gel electrophoresis, page 393

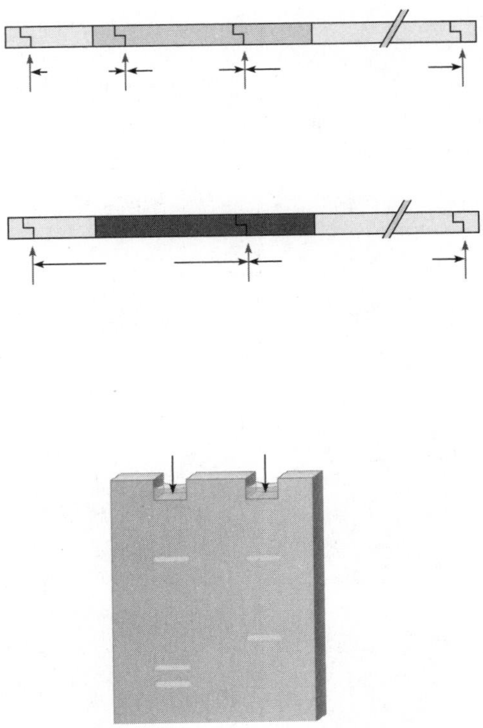

Figure 20.9 Using restriction fragment analysis to distinguish the normal and sickle-cell alleles of the β-globin gene, page 393

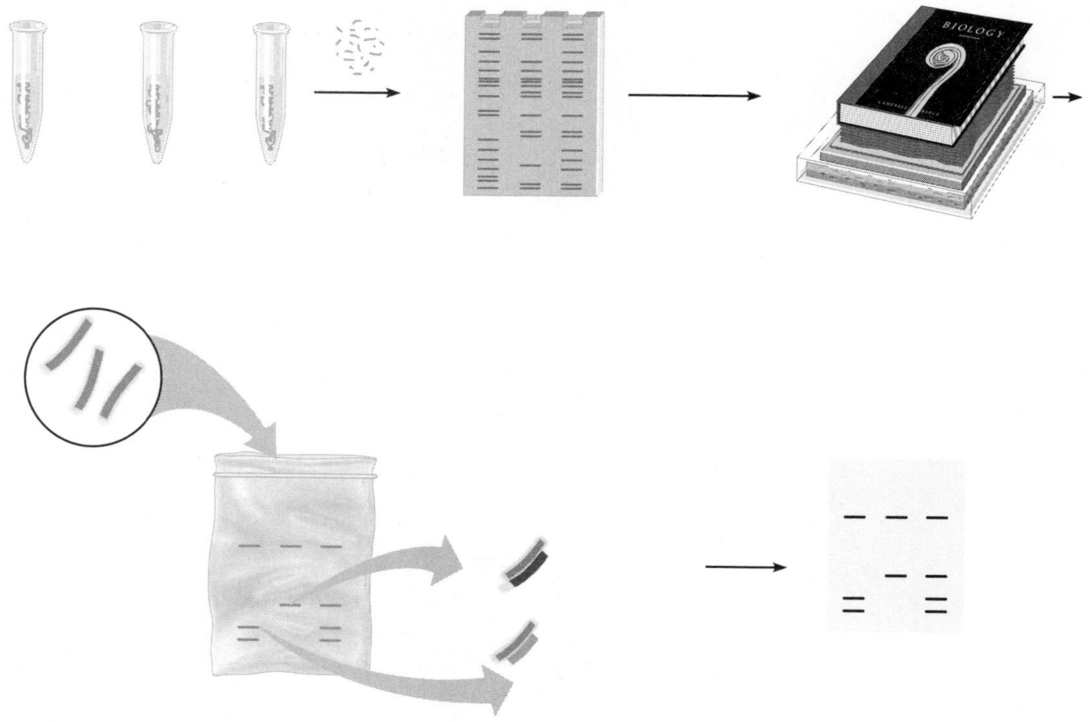

Figure 20.10 Southern blotting of DNA fragments, page 395

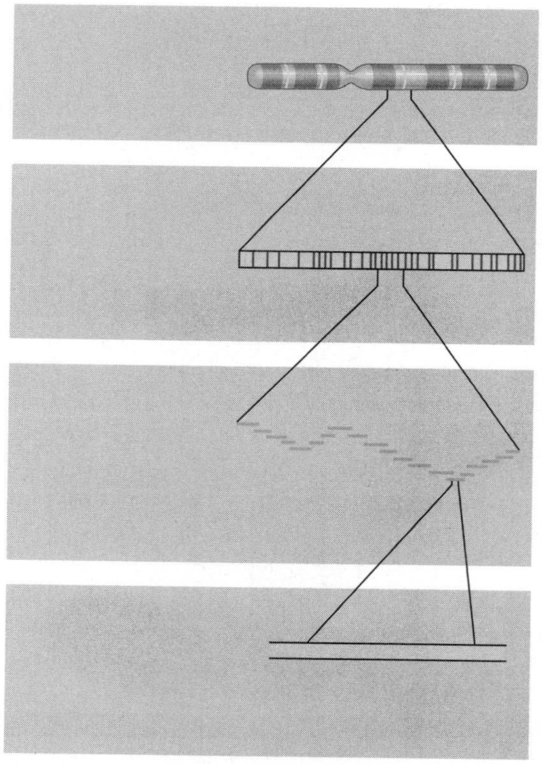

Figure 20.11 Three-stage approach to mapping an entire genome, page 396

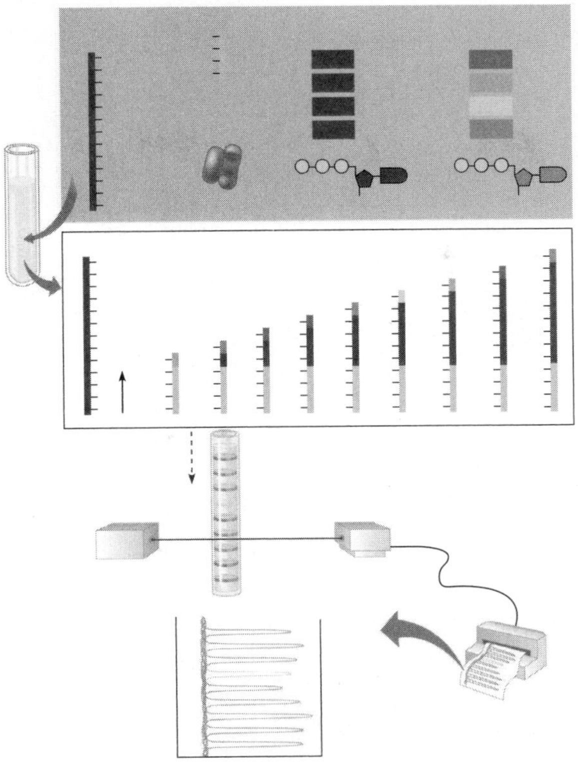

Figure 20.12 Dideoxy chain-termination method for sequencing DNA, page 397

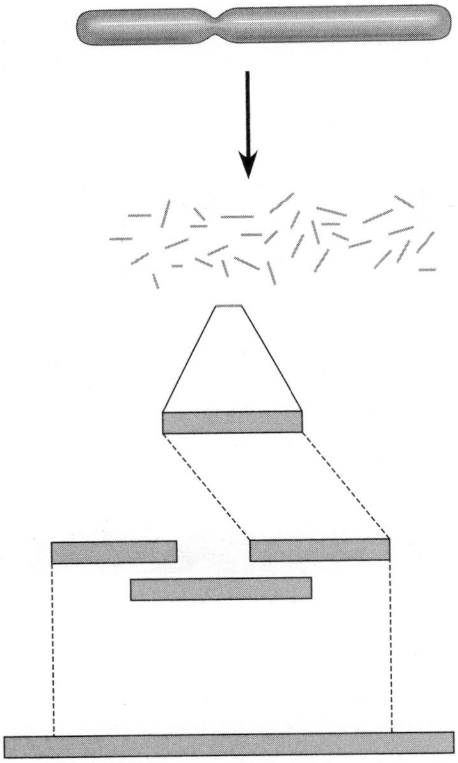

Figure 20.13 Whole-genome shotgun approach to sequencing, page 398

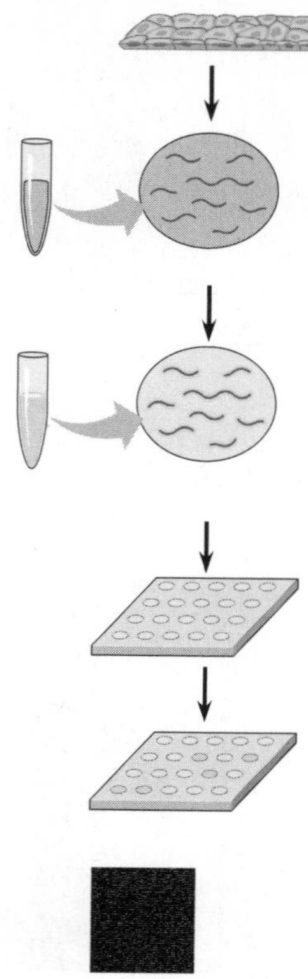

Figure 20.14 DNA microarray assay of gene expression levels, page 401

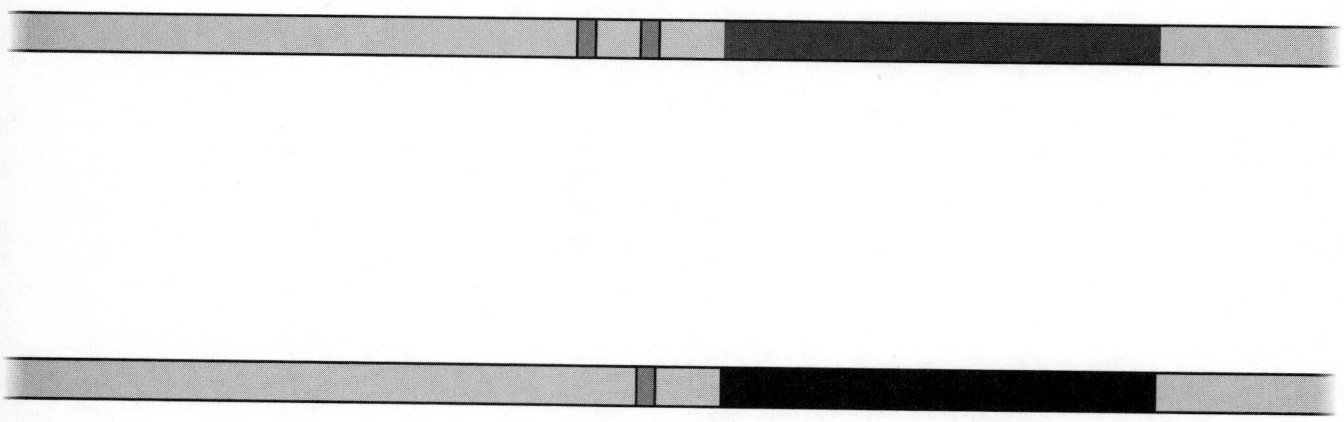

Figure 20.15 RFLPs as markers for disease-causing alleles, page 403

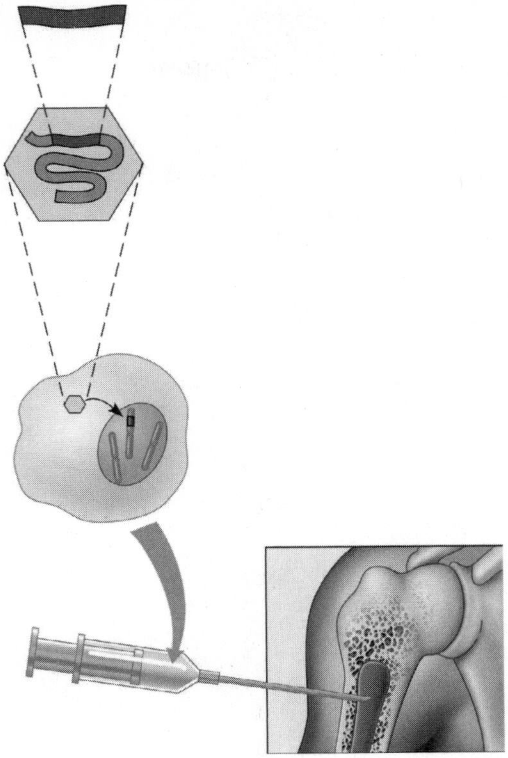

Figure 20.16 Gene therapy using a retroviral vector, page 403

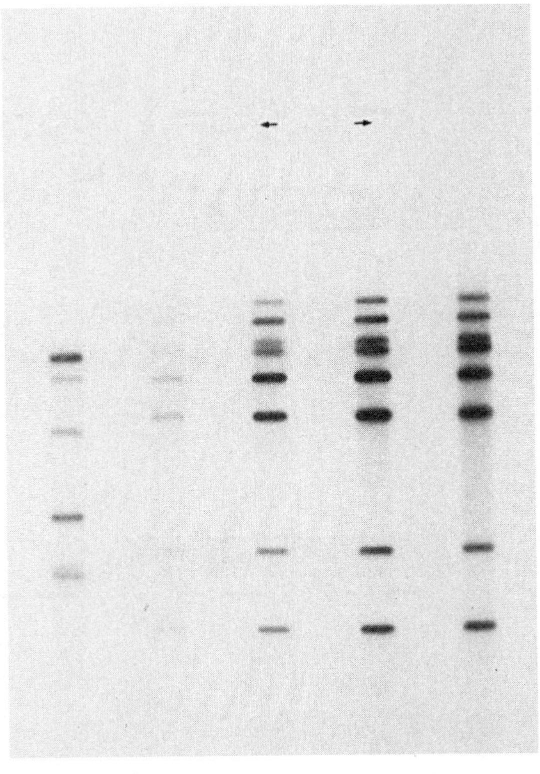

Figure 20.17 DNA fingerprints from a murder case, page 405

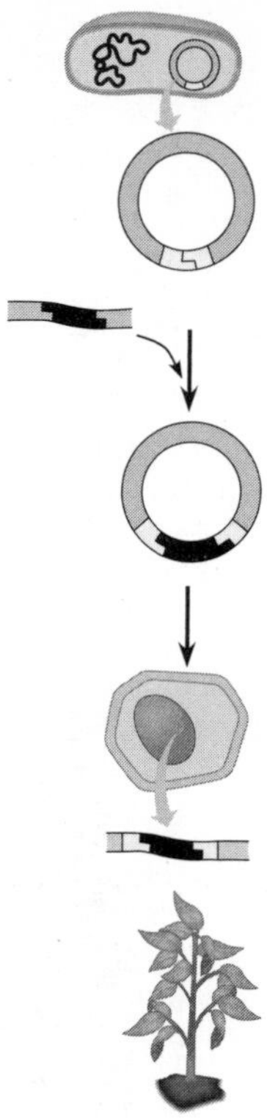

Figure 20.19 Using the Ti plasmid to produce transgenic plants, page 407

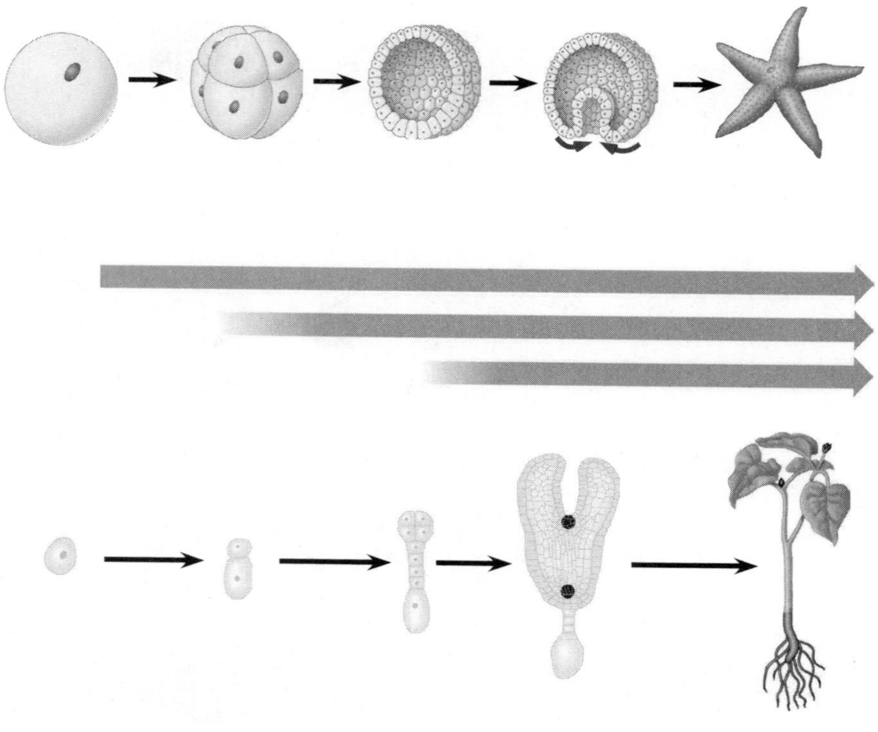

Figure 21.4 Some key stages of development in animals and plants, page 414

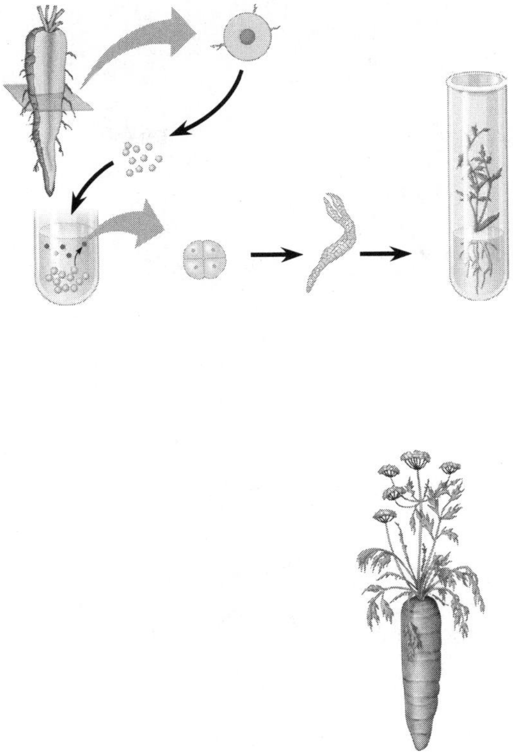

Figure 21.5 Can a differentiated plant cell develop into a whole plant?, page 415

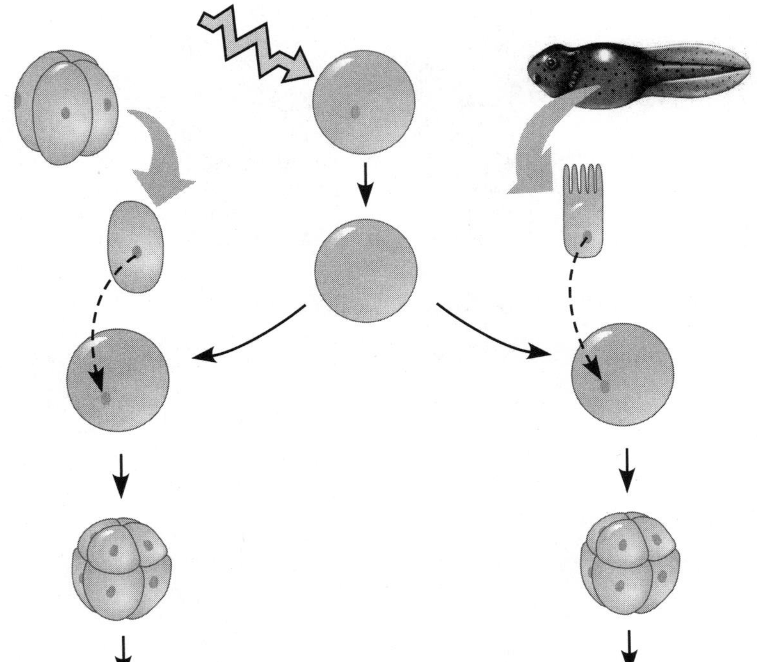

Figure 21.6 Can the nucleus from a differentiated animal cell direct development of an organism?, page 416

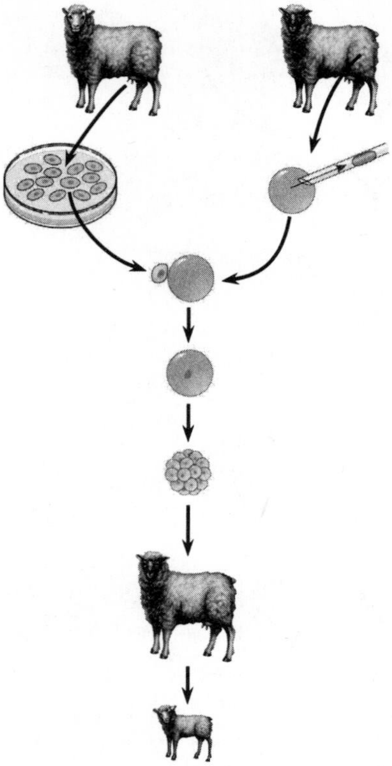

Figure 21.7 Reproductive cloning of a mammal by nuclear transplantation, page 417

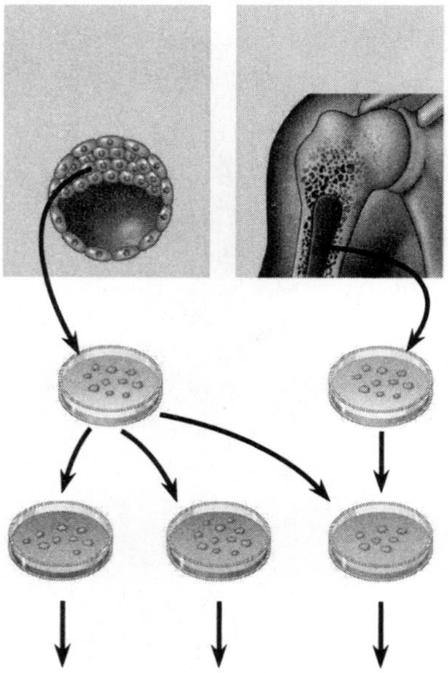

Figure 21.9 Working with stem cells, page 418

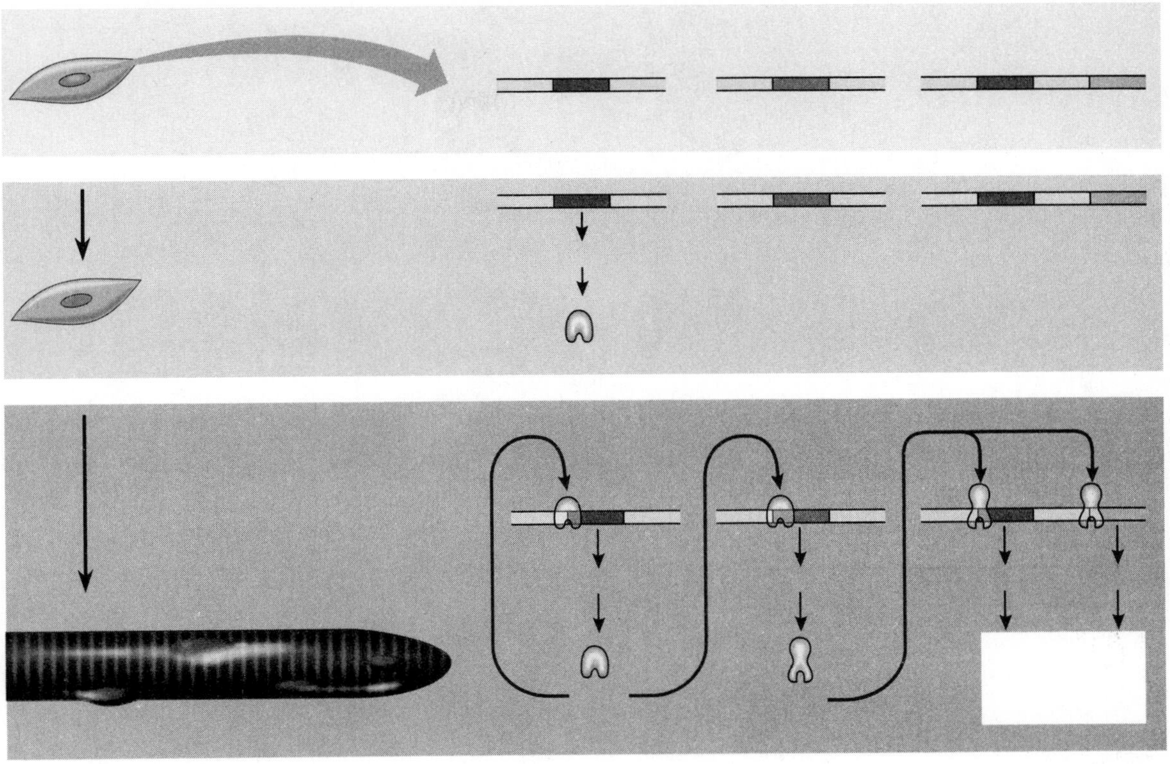

Figure 21.10 Determination and differentiation of muscle cells, page 419

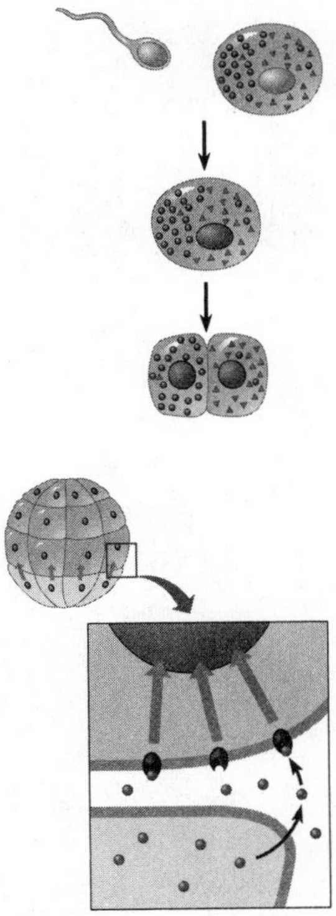

Figure 21.11 Sources of developmental information for the early embryo, page 421

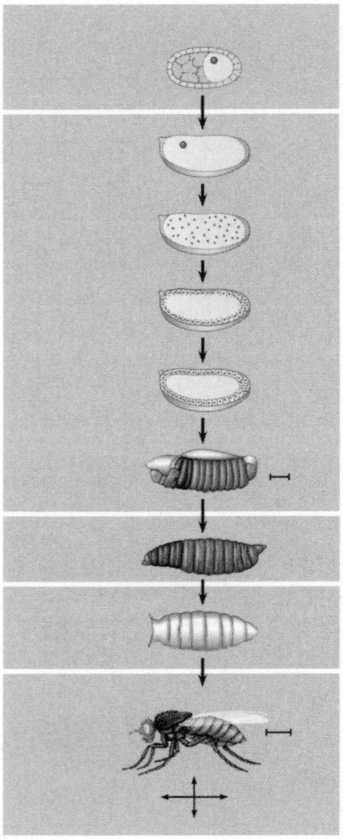

Figure 21.12 Key developmental events in the life cycle of *Drosophila,* page 422

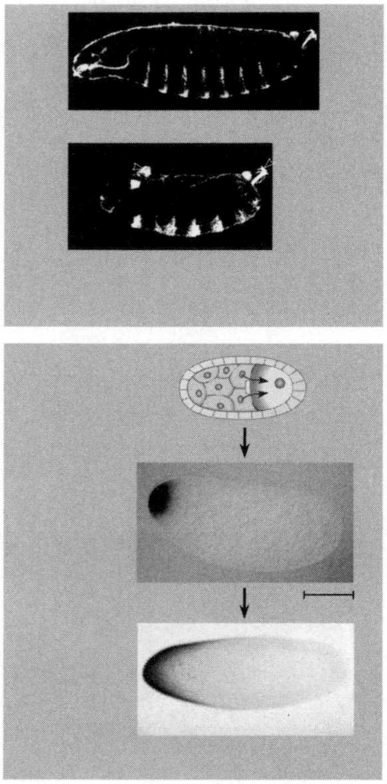

Figure 21.14 The effect of the bicoid gene, a maternal effect (egg-polarity) gene in *Drosophila,* page 424

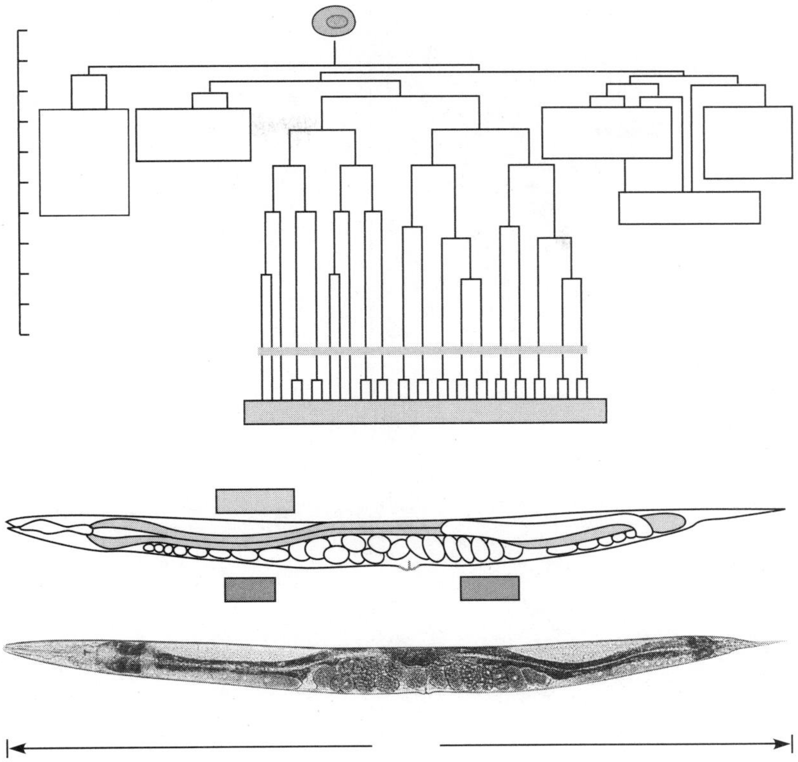

Figure 21.15 Cell lineage in *C. elegans,* page 426

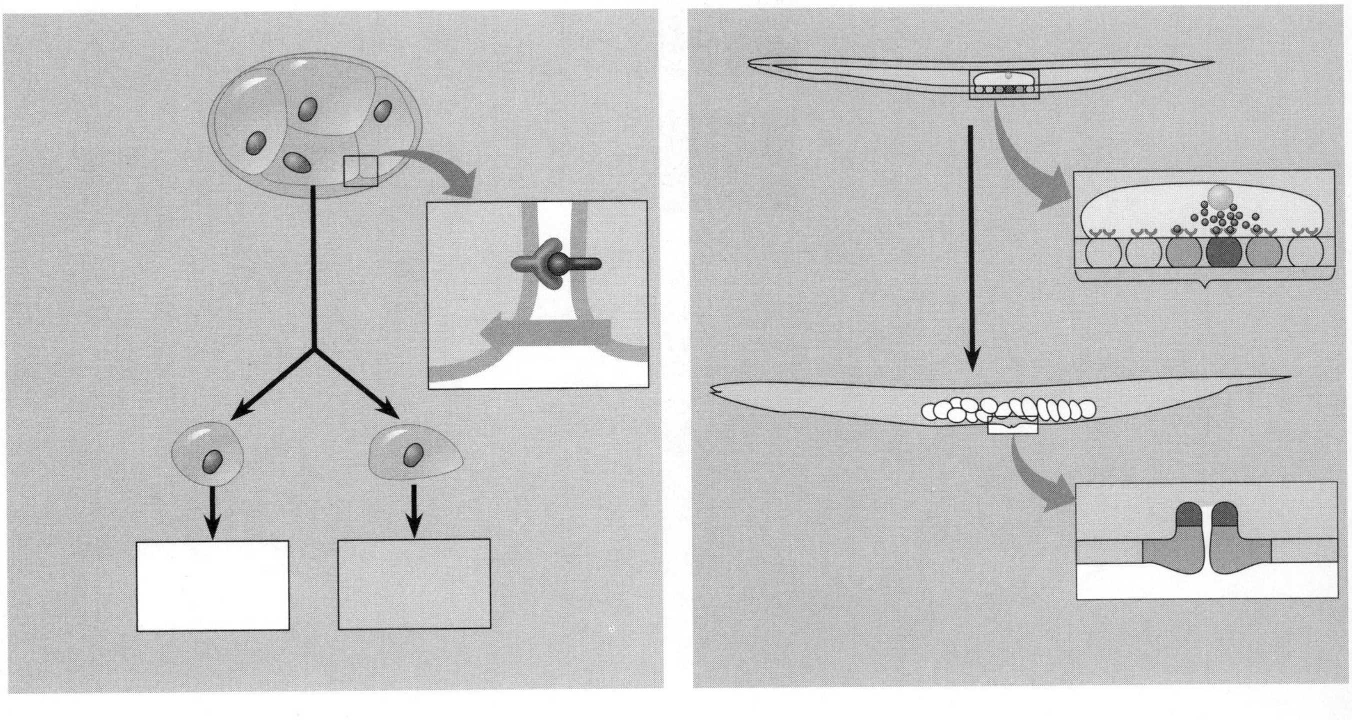

Figure 21.16 Cell signaling and induction during development of the nematode, page 427

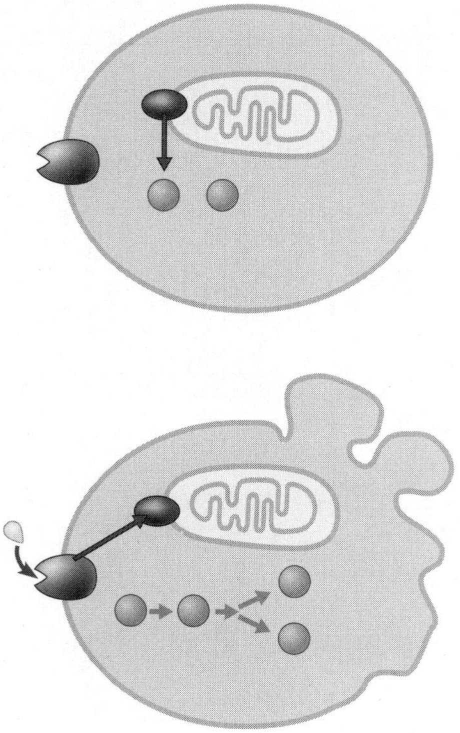

Figure 21.18 Molecular basis of apoptosis in *C. elegans,* page 428

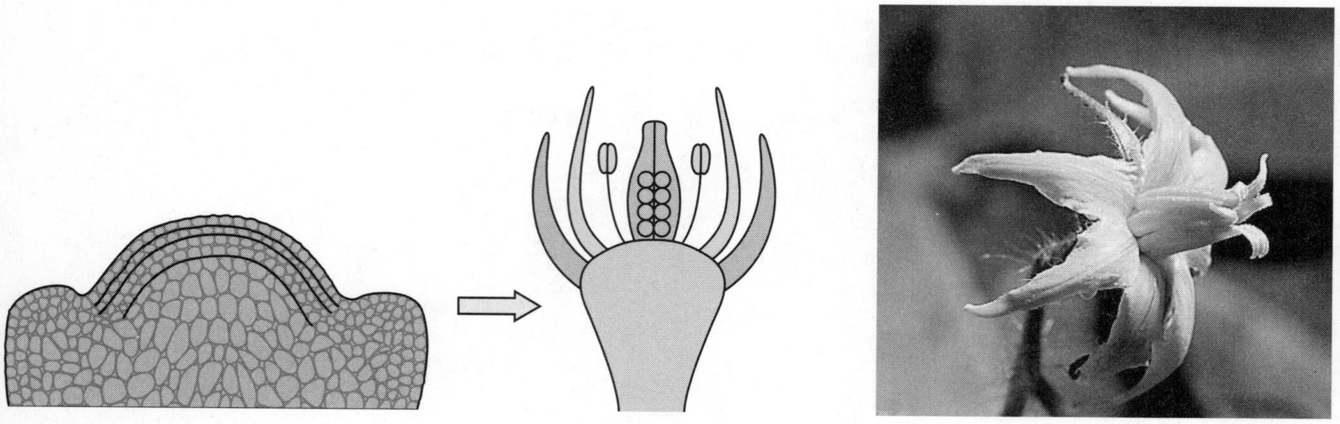

Figure 21.20 Flower development, page 429

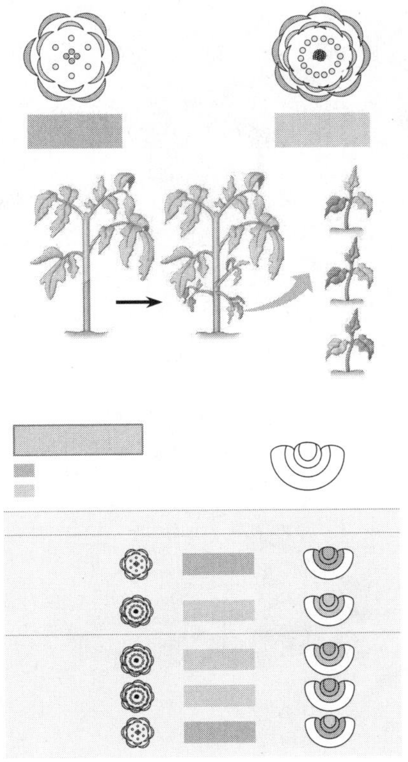

Figure 21.21 Which cell layers in the floral meristem determine the number of floral organs?, page 430

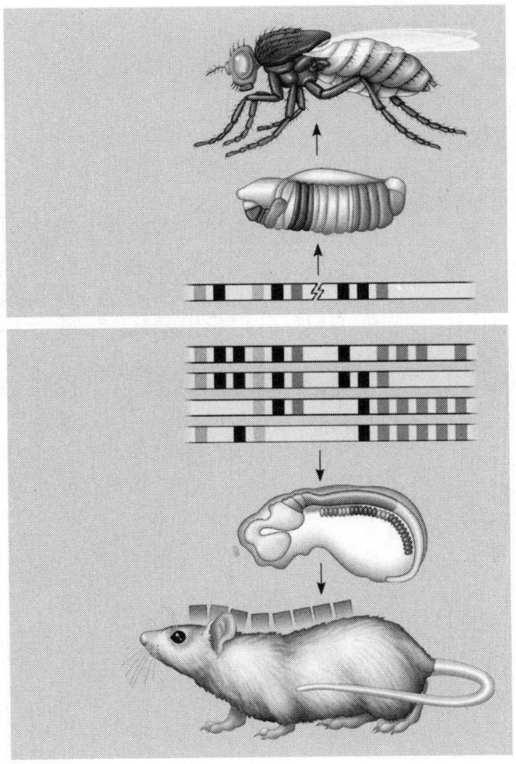

Figure 21.23 Conservation of homeotic genes in a fruit fly and a mouse, page 431

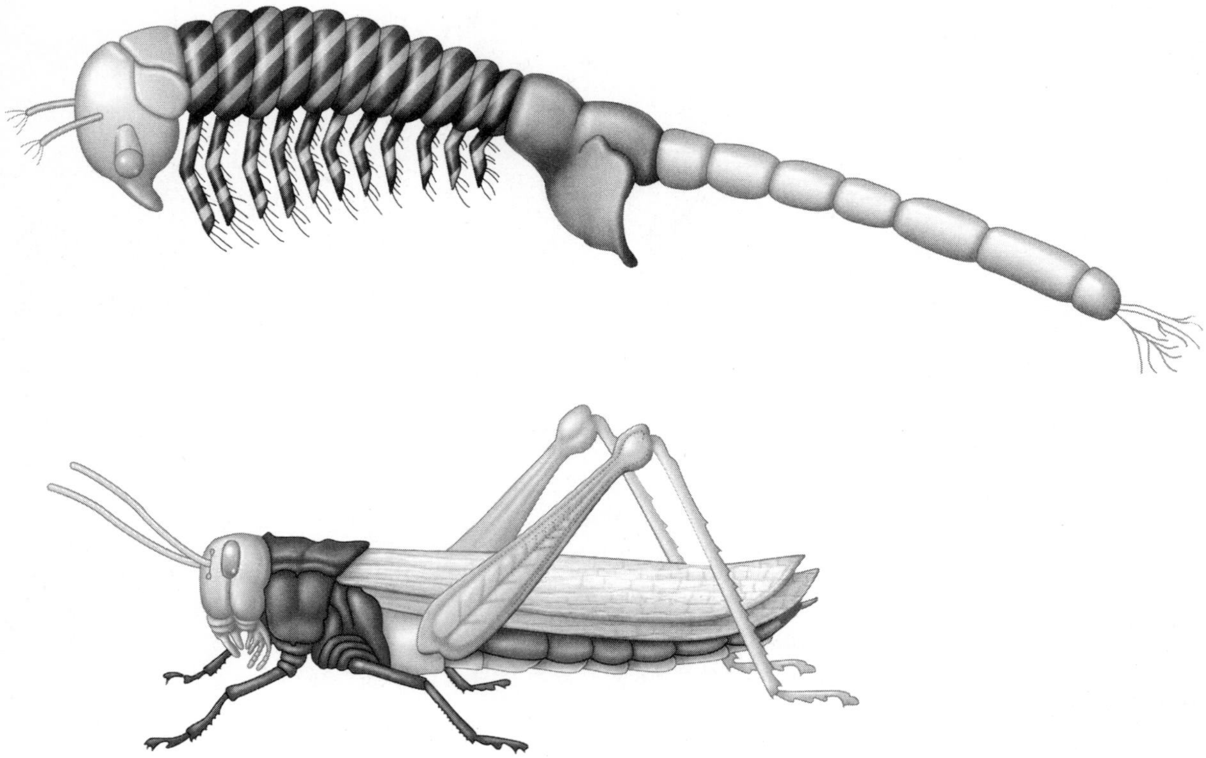

Figure 21.24 Effect of differences in *Hox* gene expression during development in crustaceans and insects, page 432

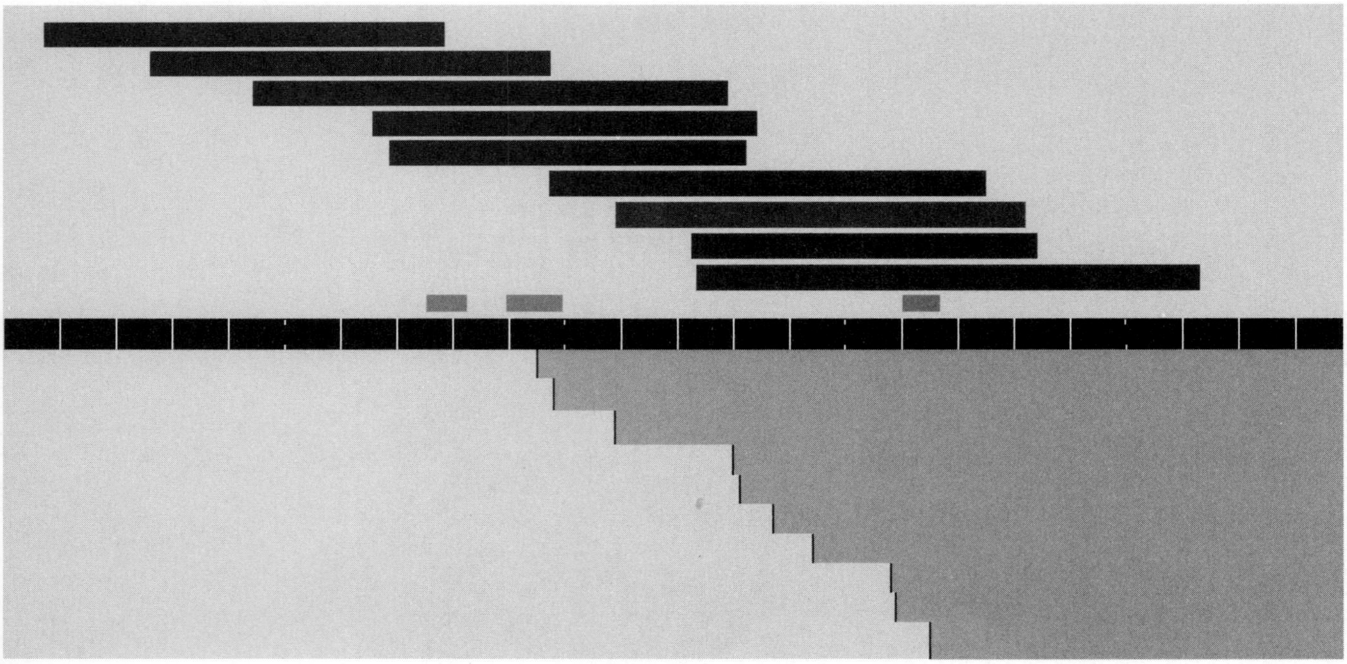

Figure 22.2 The historical context of Darwin's life and ideas, page 439

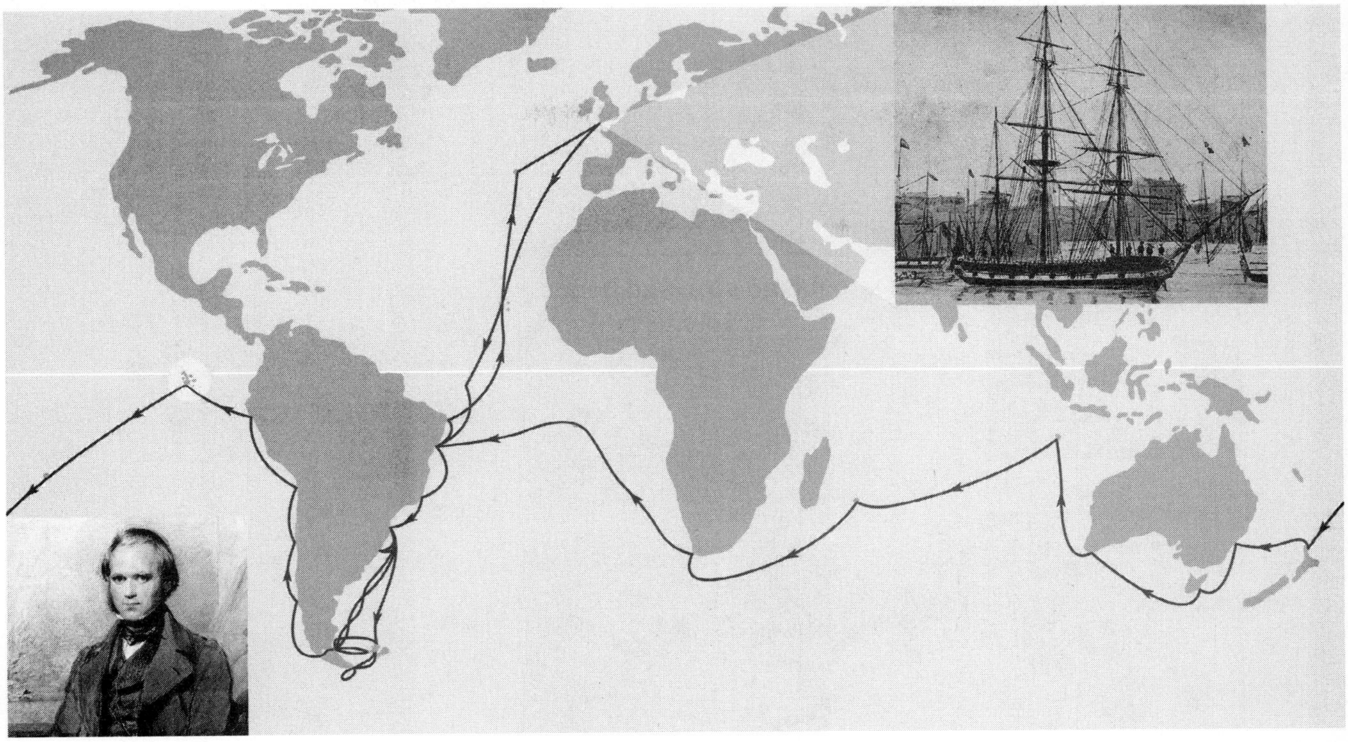

Figure 22.5 The voyage of HMS *Beagle*, page 442

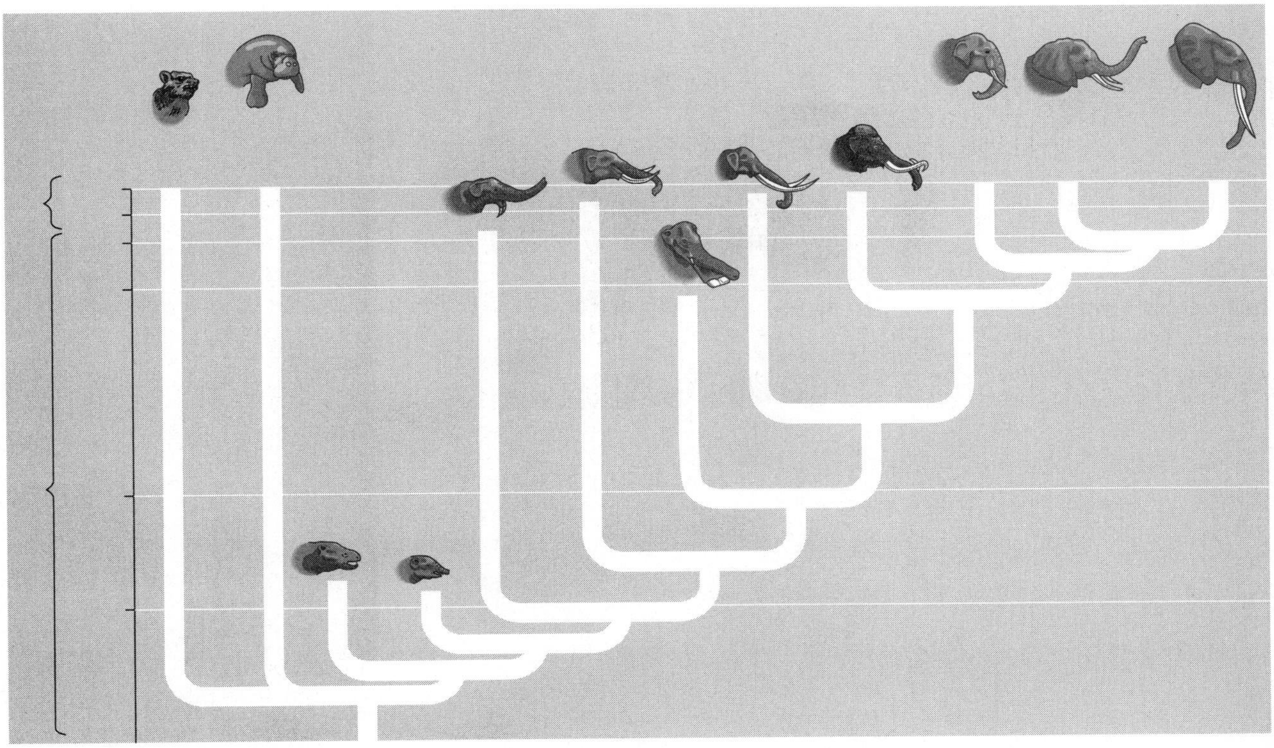

Figure 22.7 Descent with modification, page 444

Figure 22.10 Artificial selection, page 445

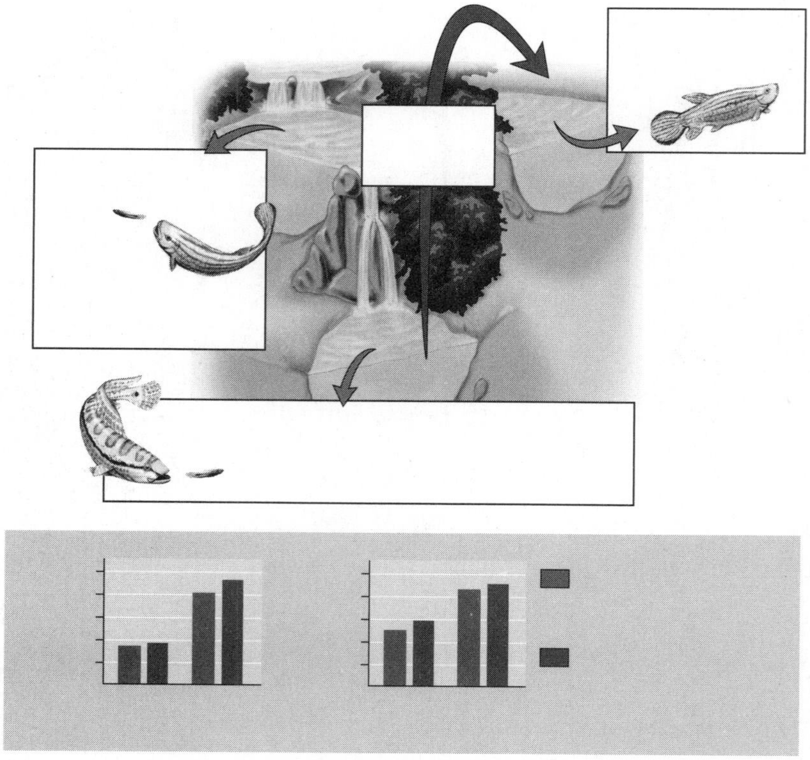

Figure 22.12 Can predation pressure select for size and age at maturity in guppies?, page 447

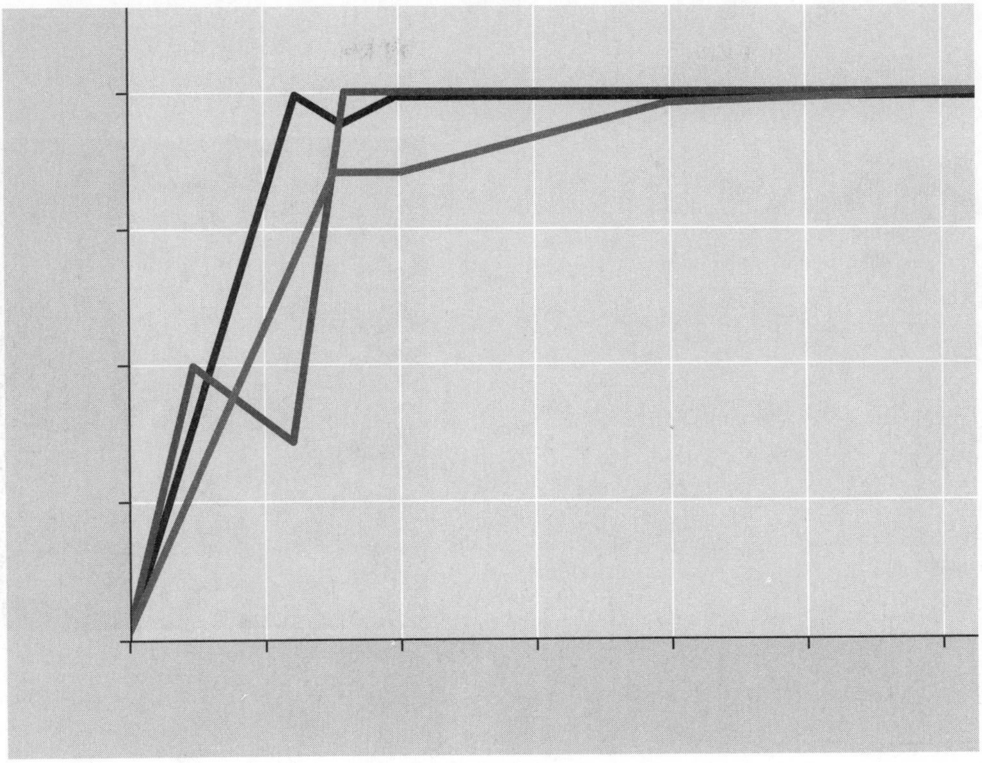

Figure 22.13 Evolution of drug resistance in HIV, page 448

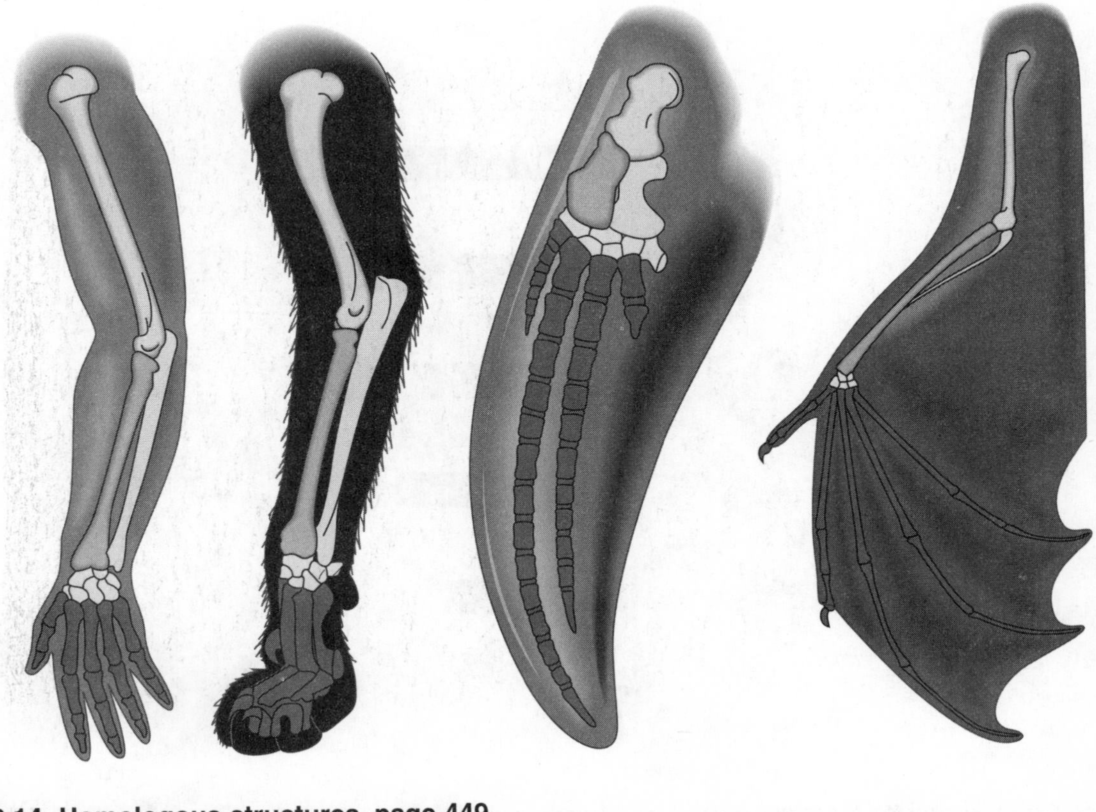

Figure 22.14 Homologous structures, page 449

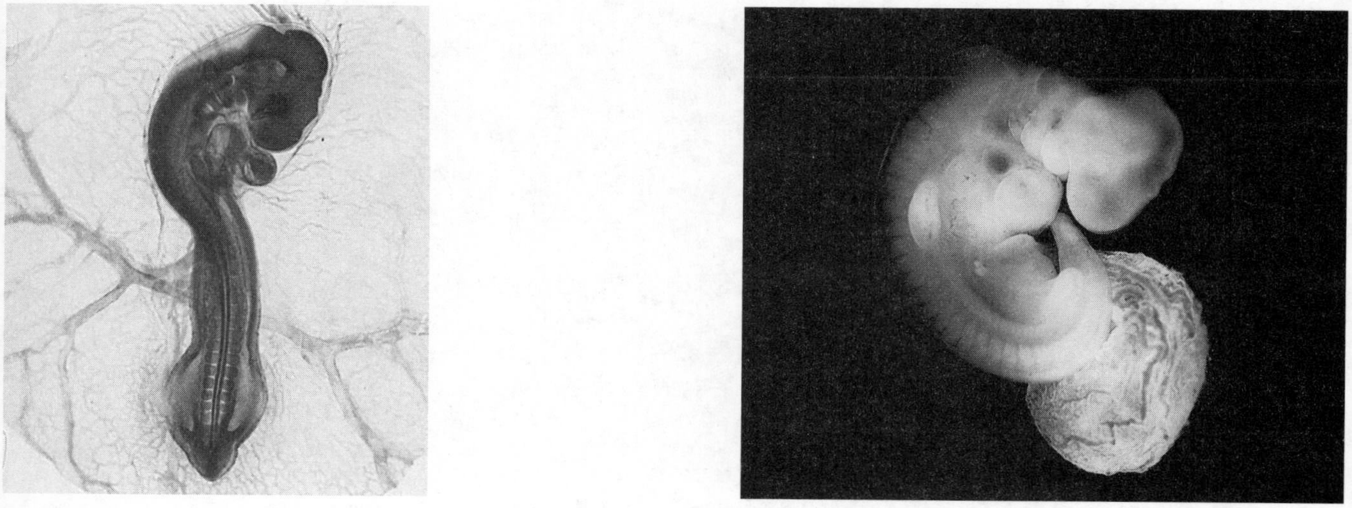

Figure 22.15 Anatomical similarities in vertebrate embryos, page 449

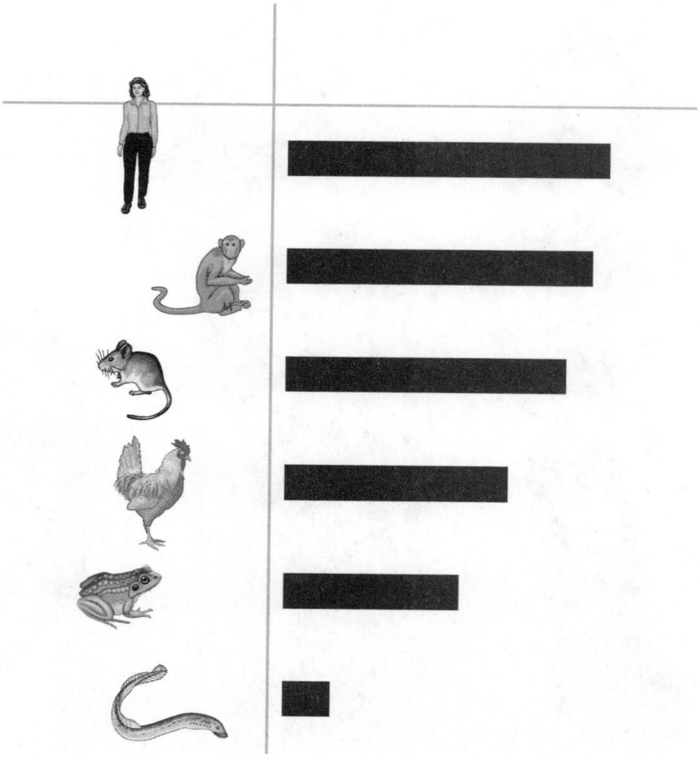

Figure 22.16 Comparison of a protein found in diverse vertebrates, page 450

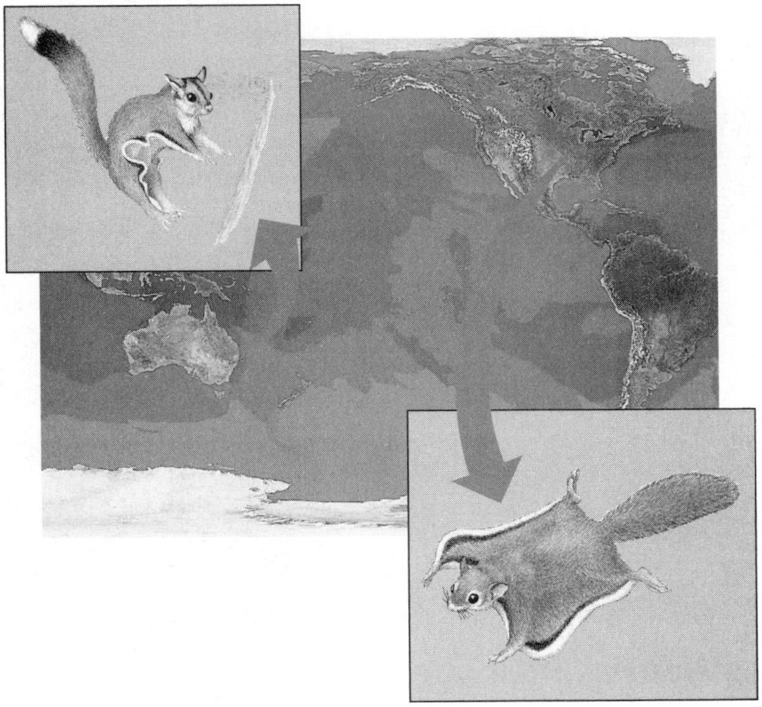

Figure 22.17 Different geographic regions, different mammalian "brands," page 450

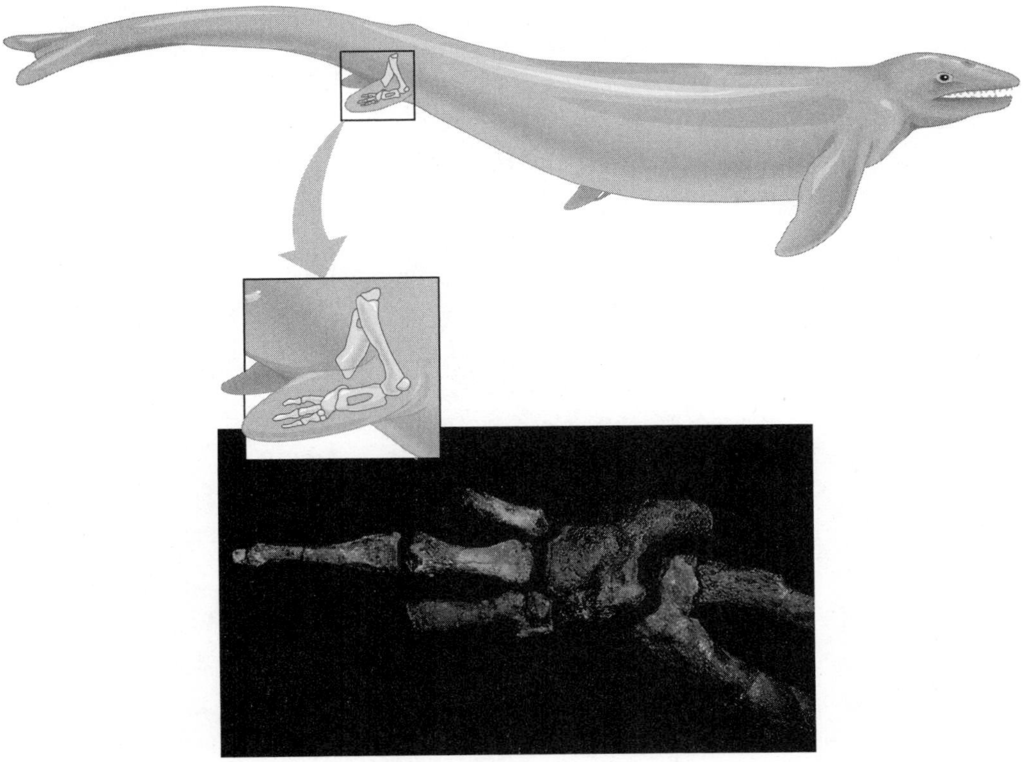

Figure 22.18 A transitional fossil linking past and present, page 451

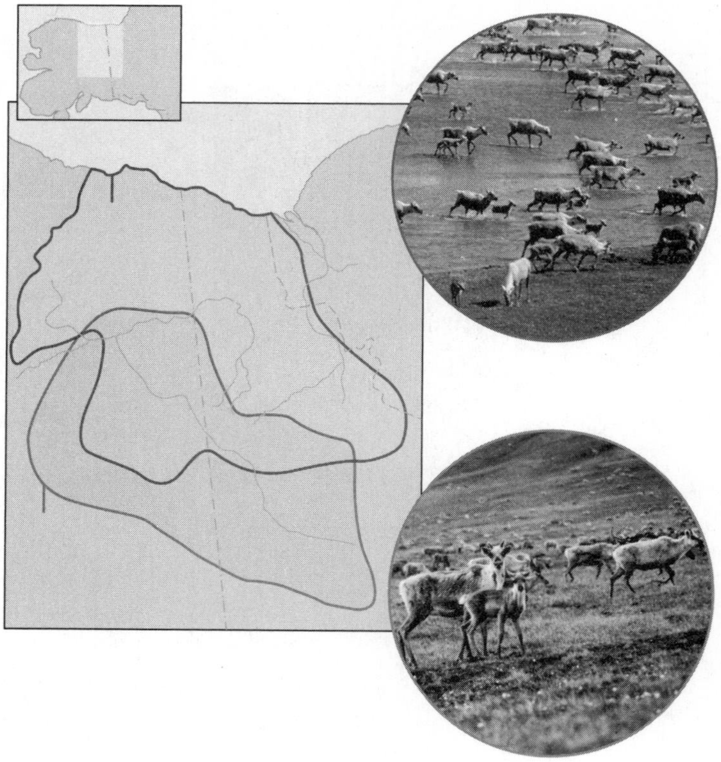

Figure 23.3 One species, two populations, page 455

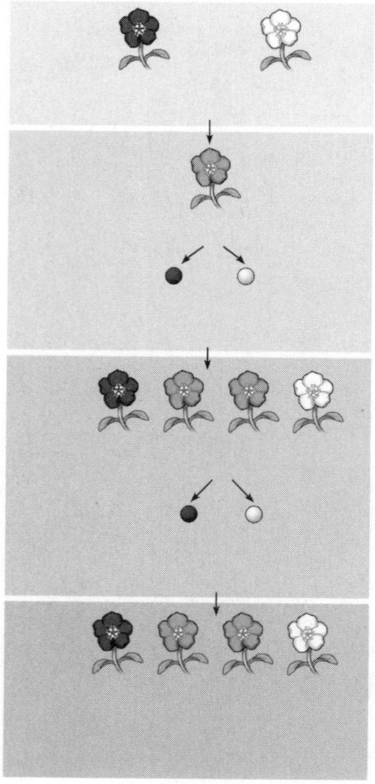

Figure 23.4 Mendelian inheritance preserves genetic variation from one generation to the next, page 456

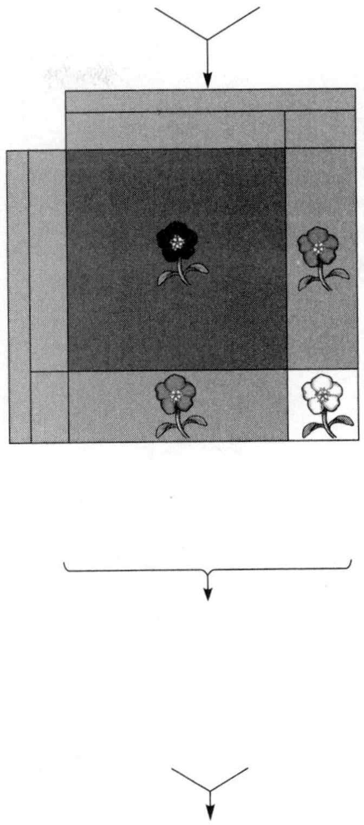

Figure 23.5 The Hardy-Weinberg theorem, page 457

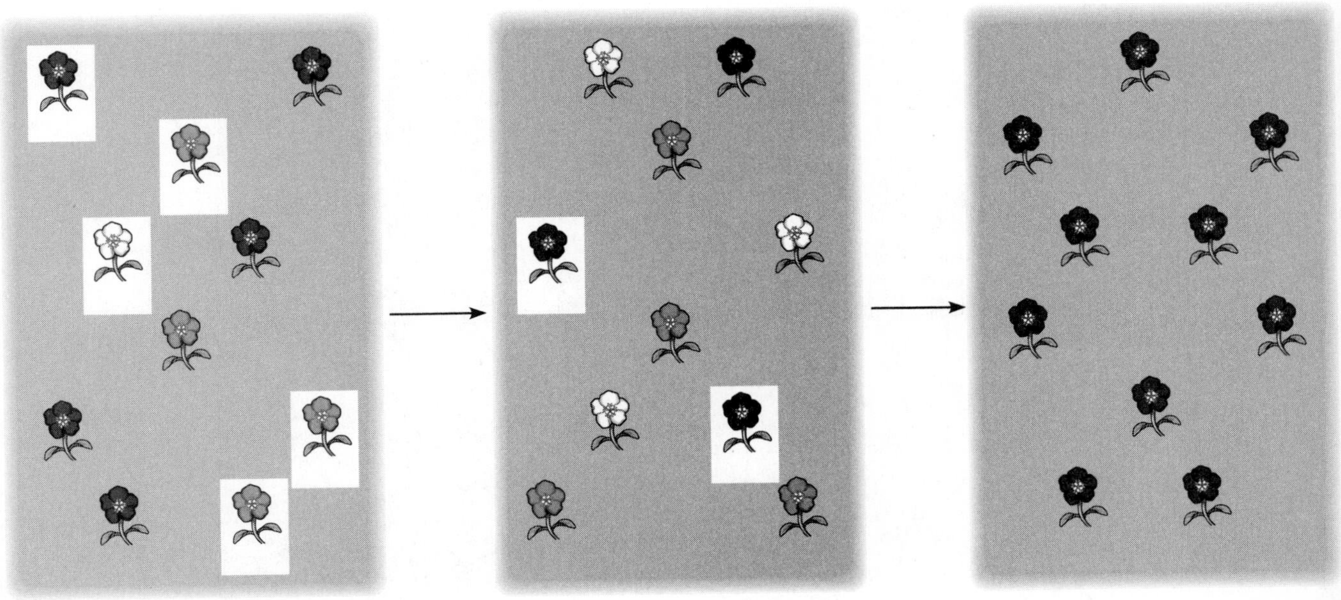

Figure 23.7 Genetic drift, page 461

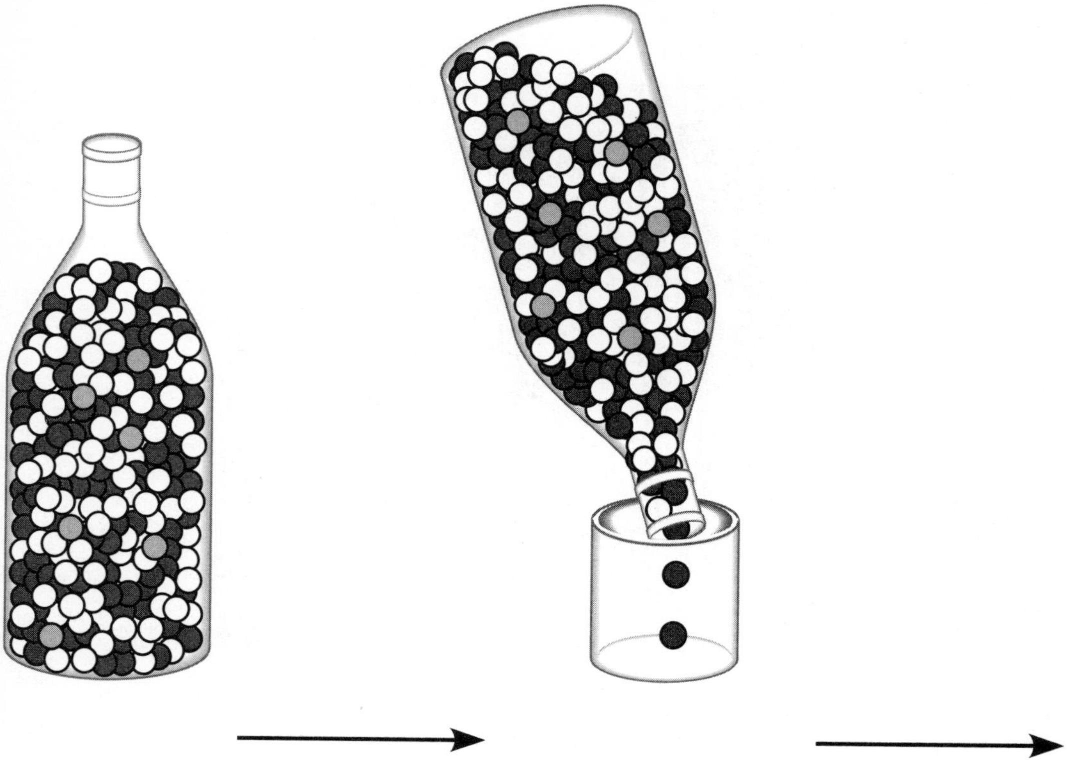

Figure 23.8 The bottleneck effect, page 461

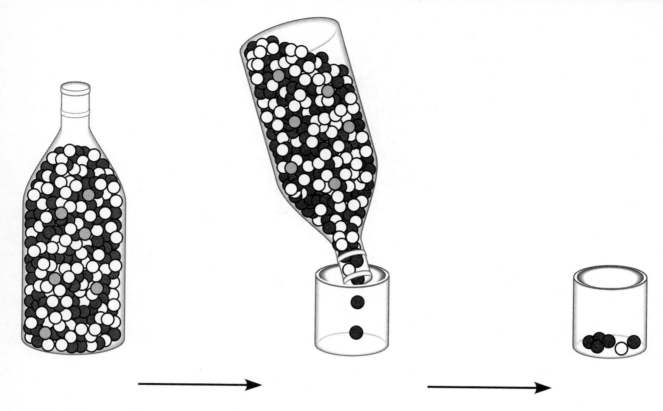

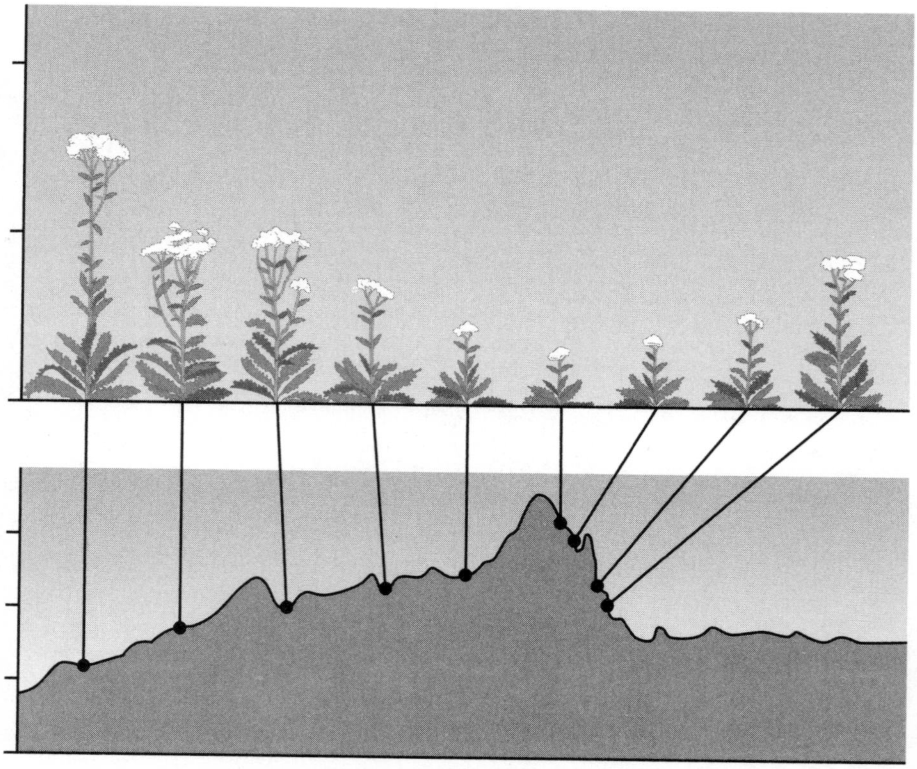

Figure 23.11 Does geographic variation in yarrow plants have a genetic component?, page 464

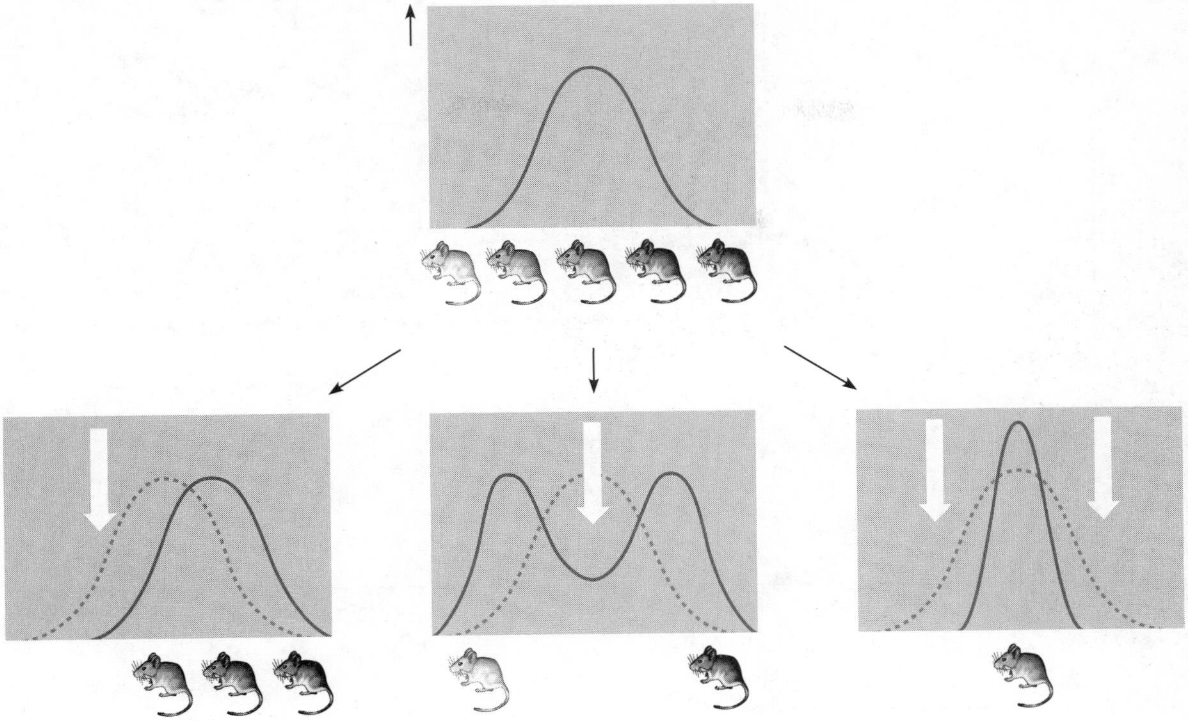

Figure 23.12 Modes of selection, page 465

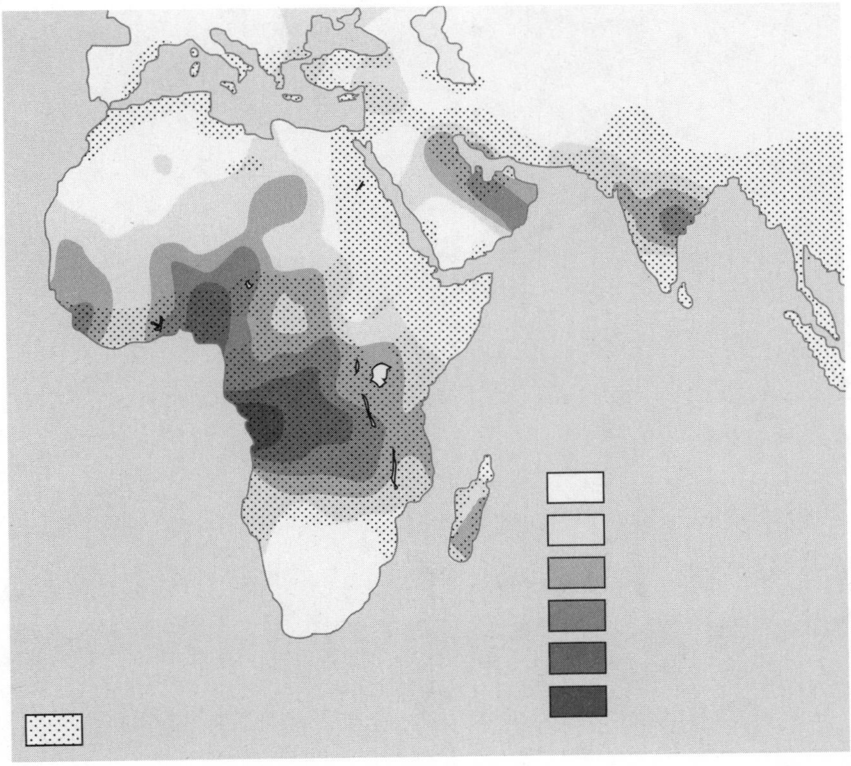

Figure 23.13 Mapping malaria and the sickle-cell allele, page 466

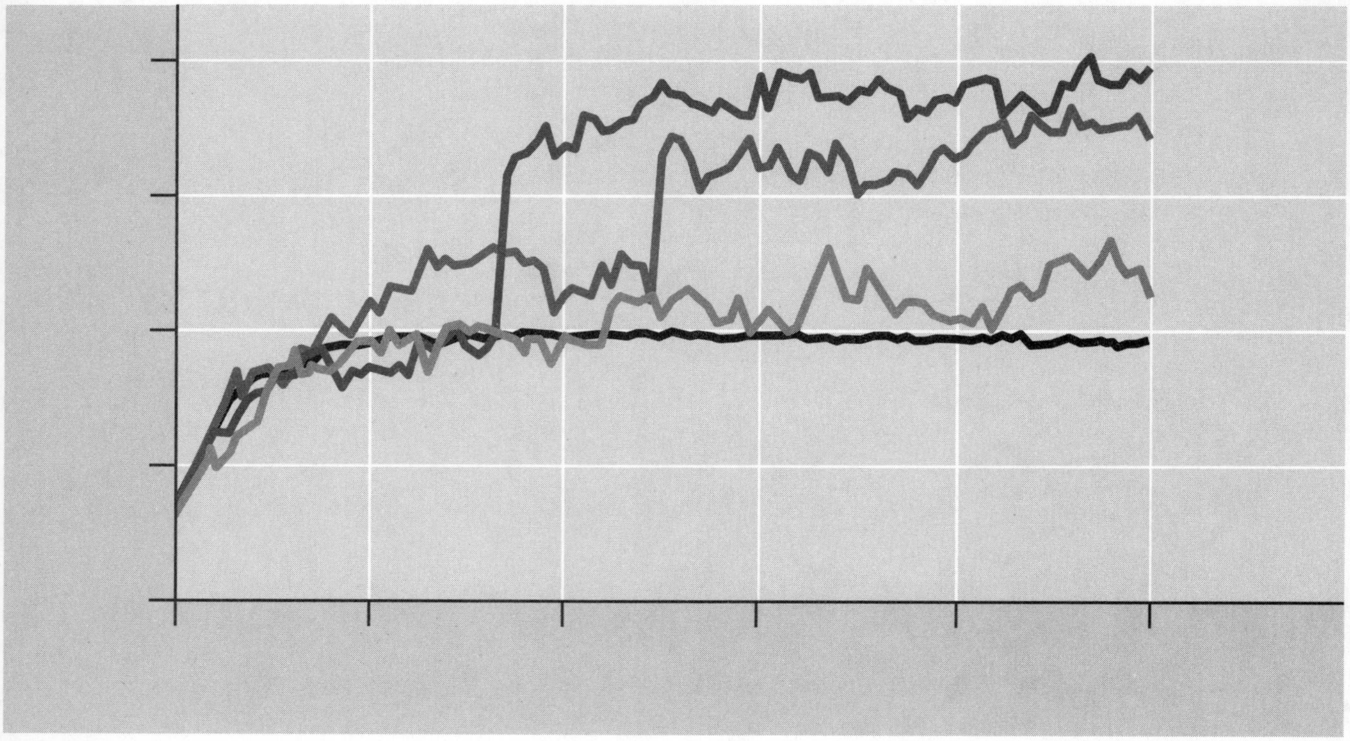

Figure 23.14 Using a virtual population to study the effects of selection, page 467

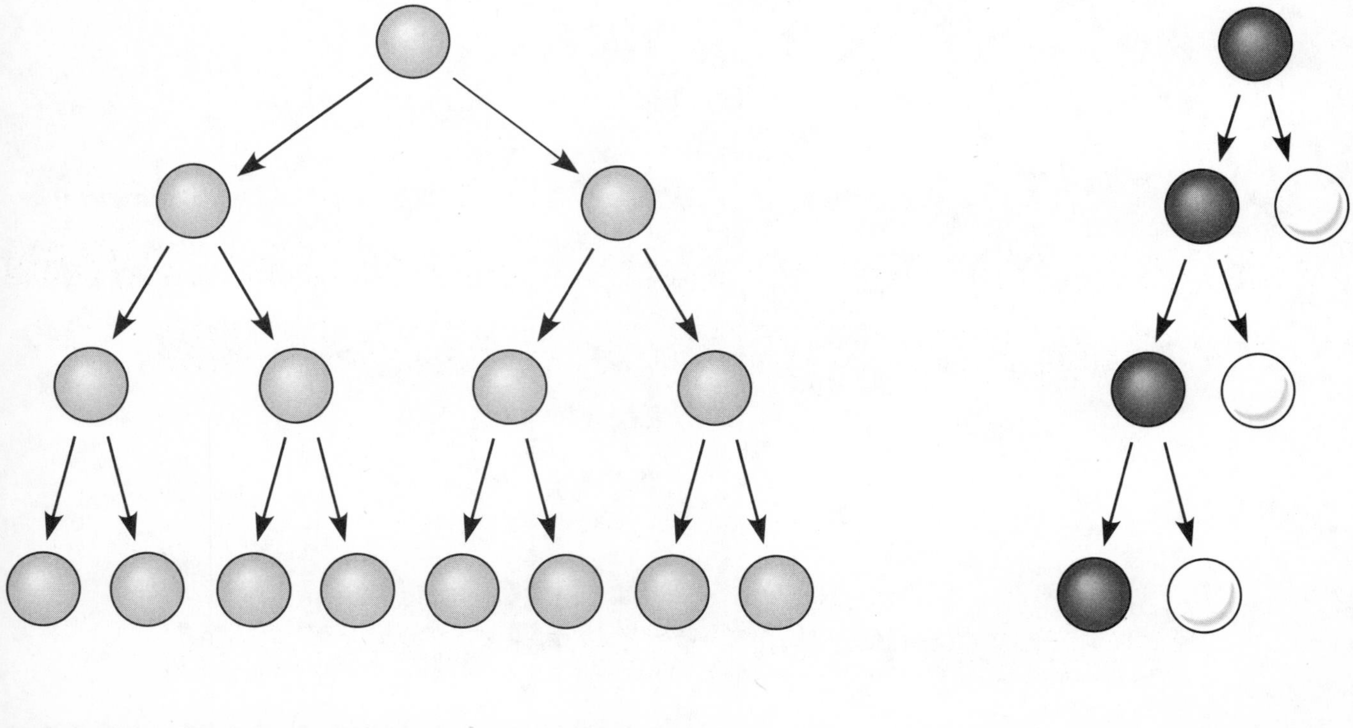

Figure 23.16 The "reproductive handicap" of sex, page 469

Figure 24.2 Two patterns of evolutionary change, page 472

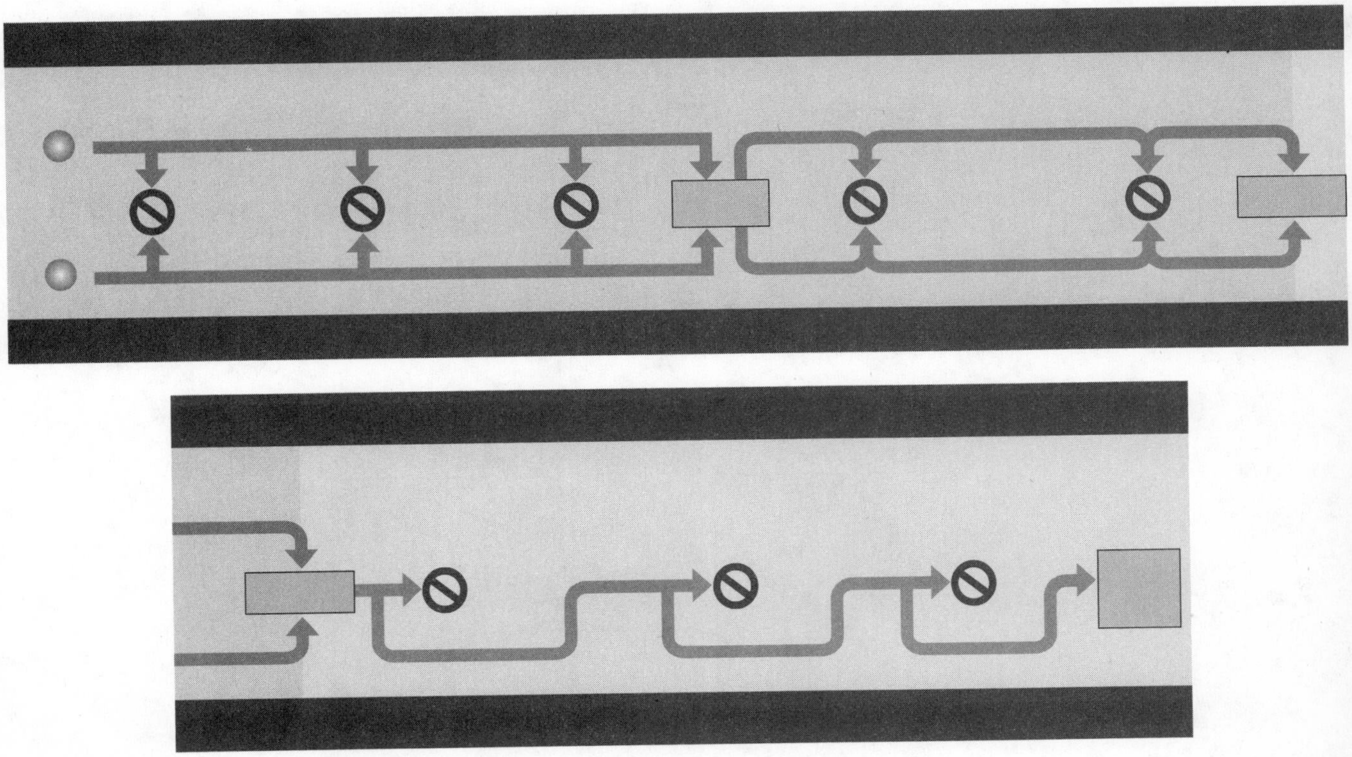

Figure 24.4 Reproductive barriers, pages 474–475

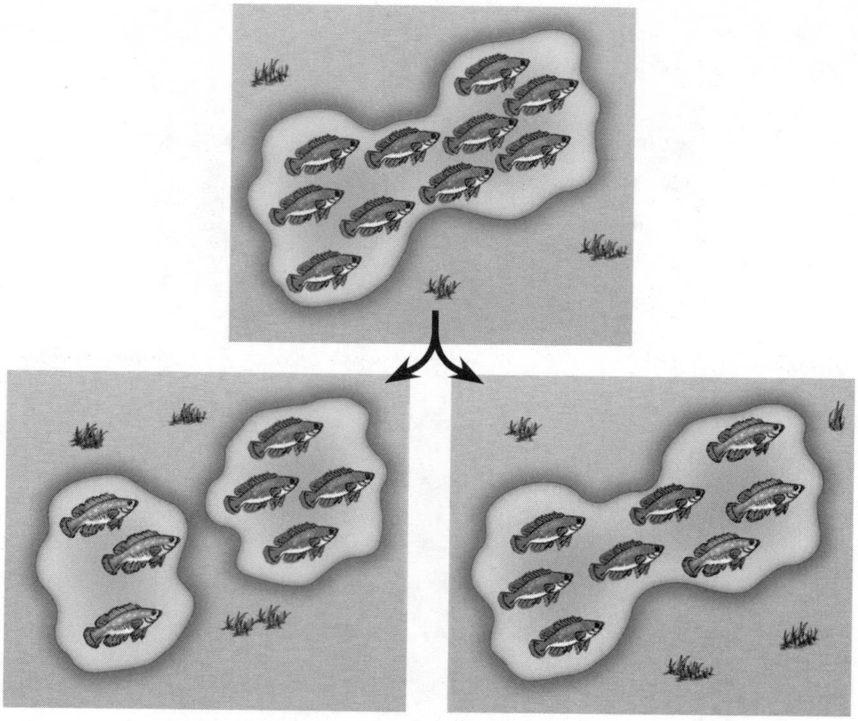

Figure 24.5 Two main modes of speciation, page 476

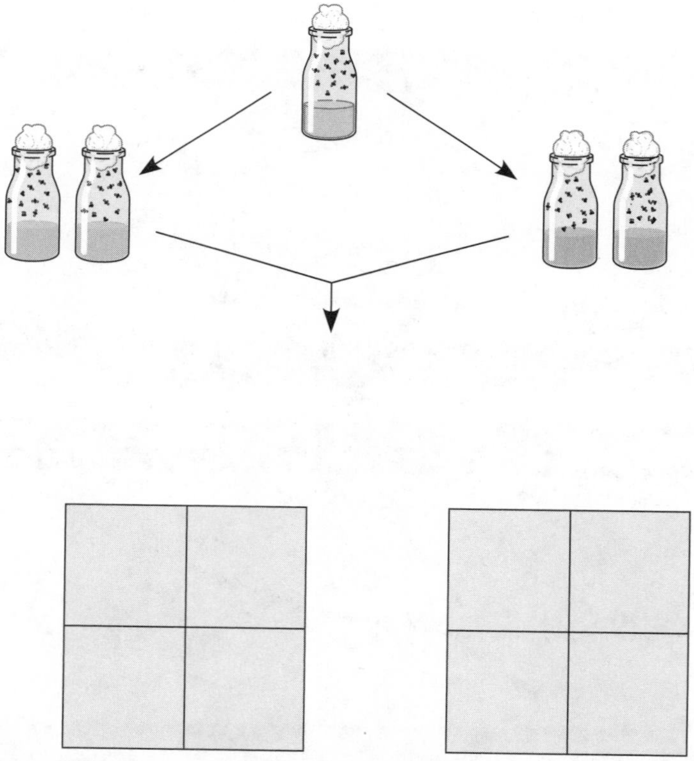

Figure 24.7 Can adaptive divergence of allopatric fruit fly populations lead to reproductive isolation?, page 477

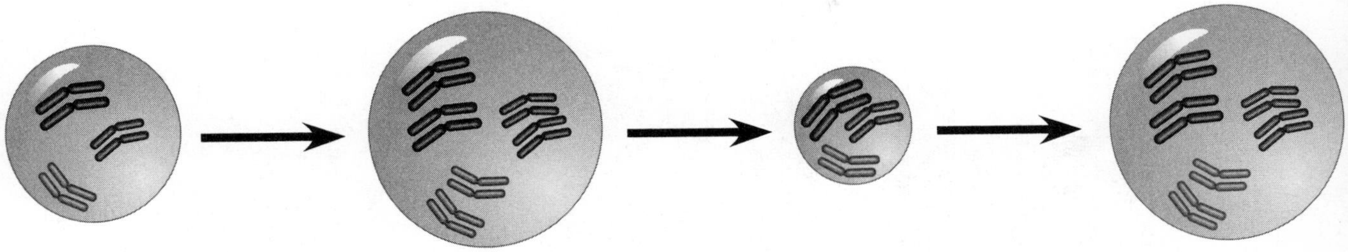

Figure 24.8 Sympatric speciation by autopolyploidy in plants, page 478

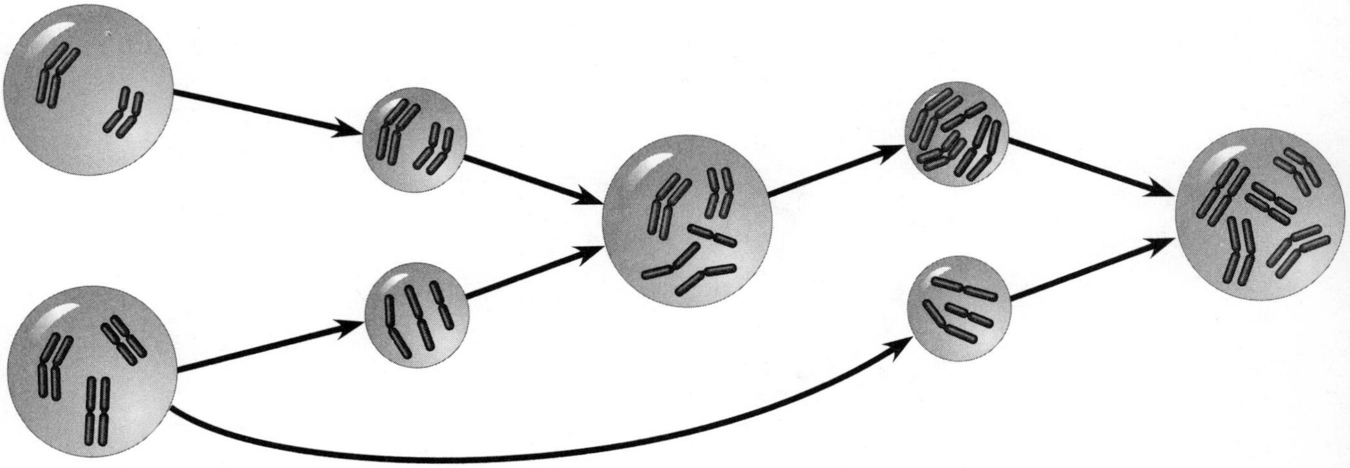

Figure 24.9 One mechanism for allopolyploid speciation in plants, page 479

Figure 24.12 Adaptive radiation, page 481

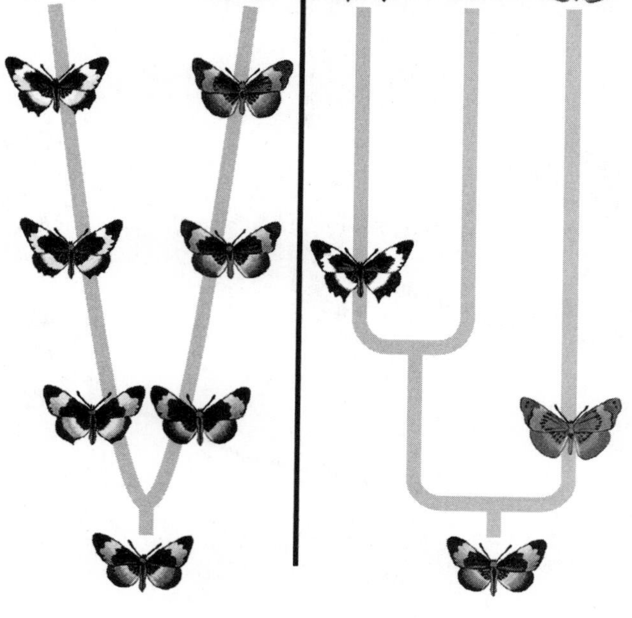

Figure 24.13 Two models for the tempo of speciation, page 482

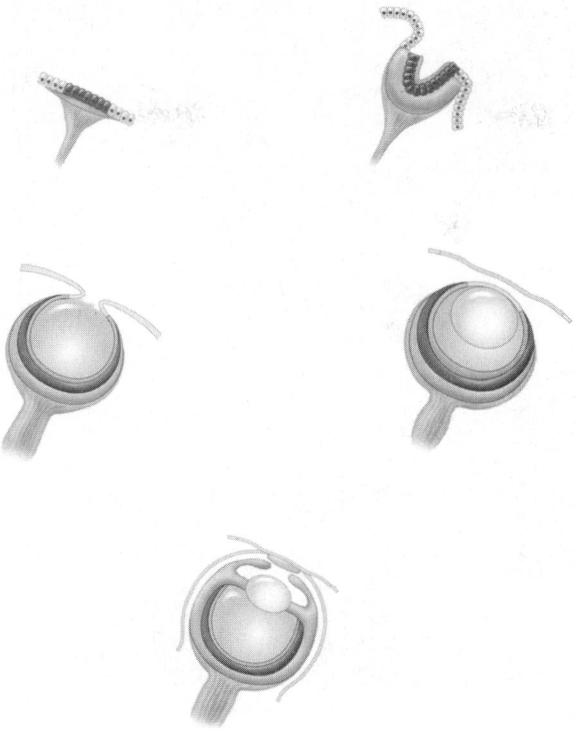

Figure 24.14 A range of eye complexity among molluscs, page 483

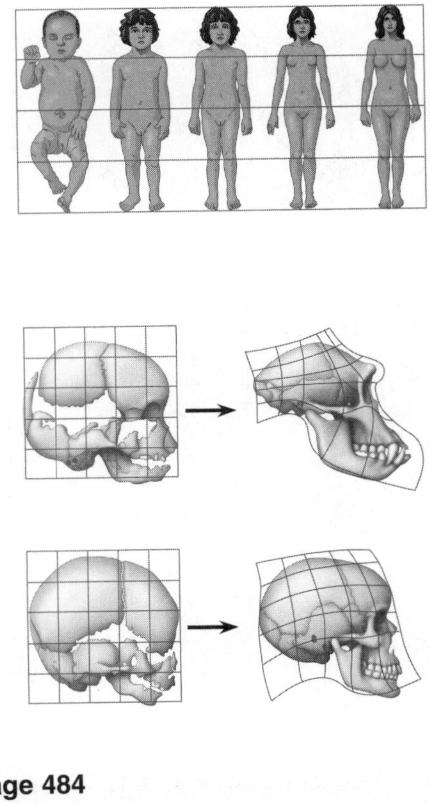

Figure 24.15 Allometric growth, page 484

Figure 24.16 Heterochrony and the evolution of salamander feet in closely related species, page 484

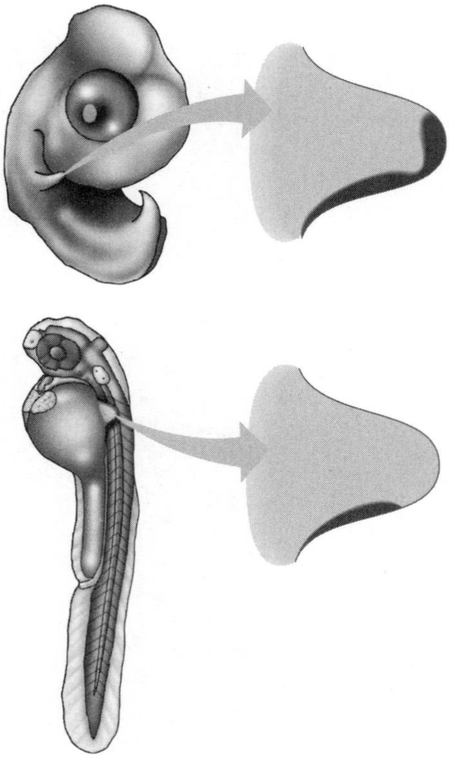

Figure 24.18 *Hox* genes and the evolution of tetrapod limbs, page 485

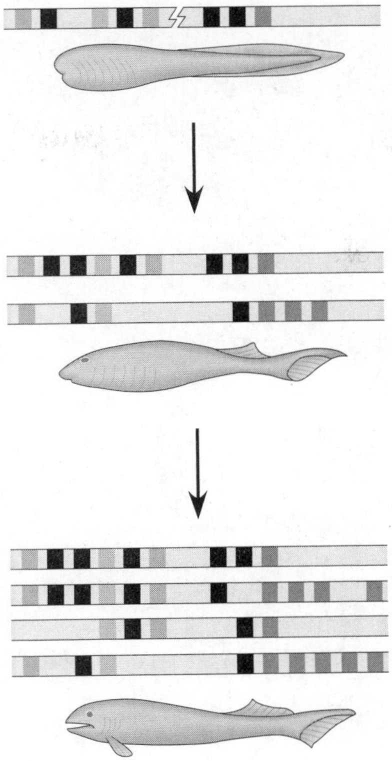

Figure 24.19 *Hox* mutations and the origin of vertebrates, page 486

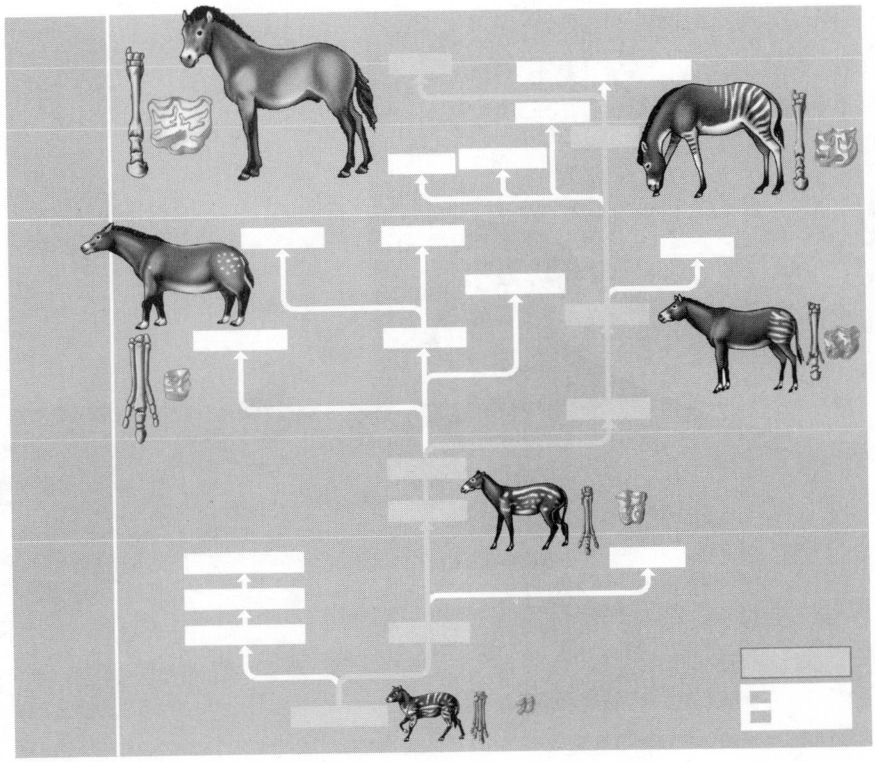

Figure 24.20 The branched evolution of horses, page 487

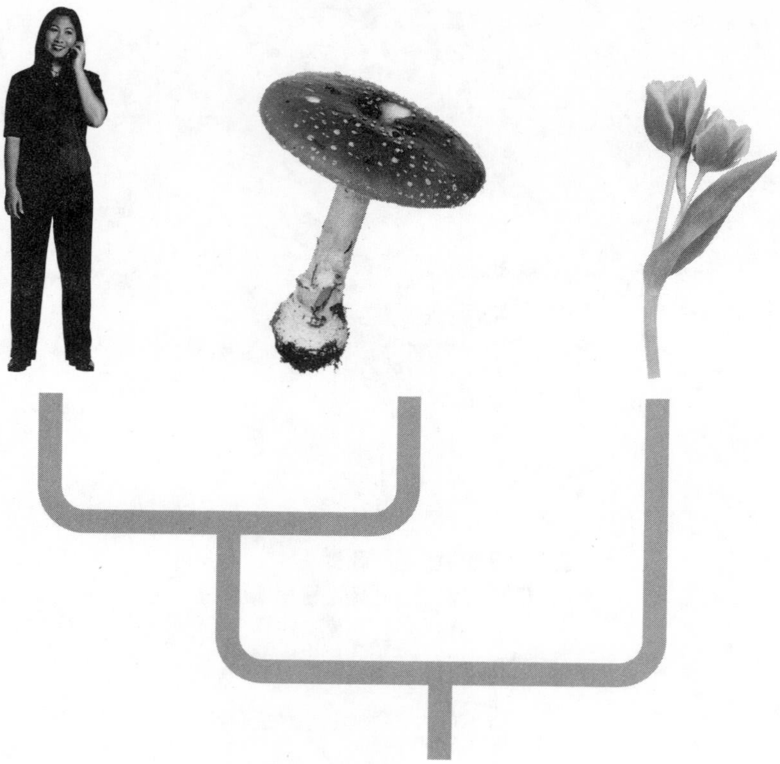

Figure 25.2 An unexpected family tree, page 491

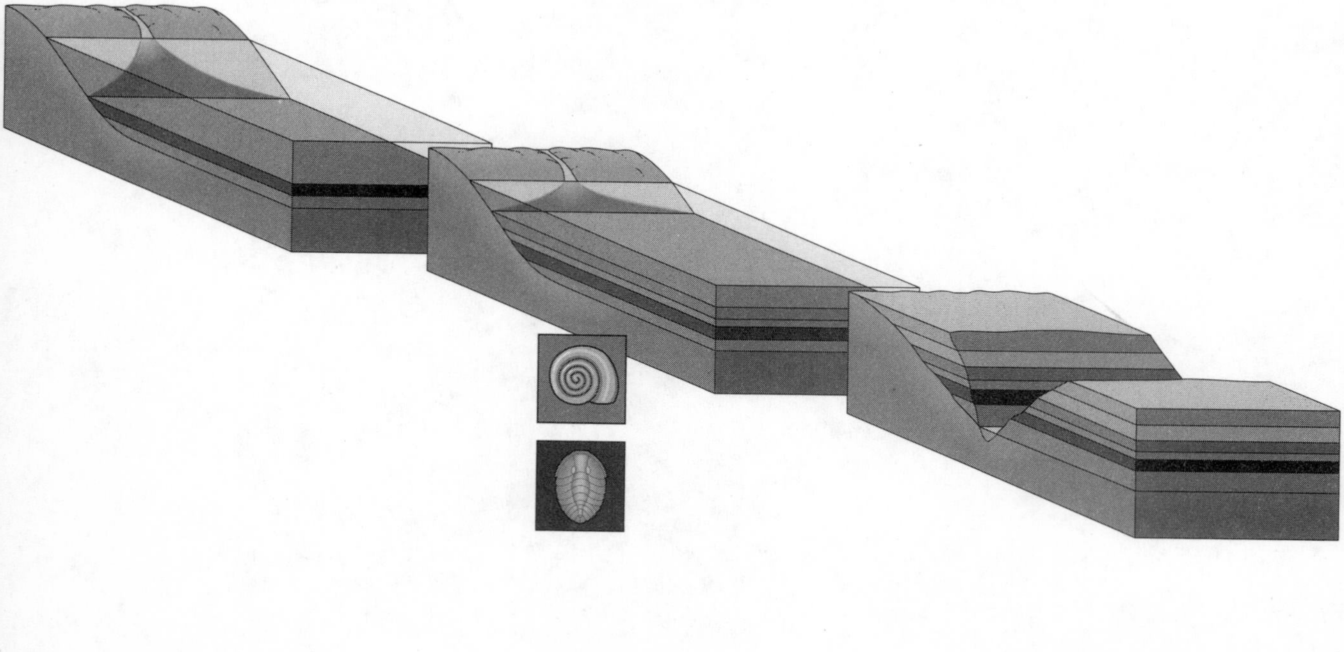

Figure 25.3 Formation of sedimentary strata containing fossils, page 492

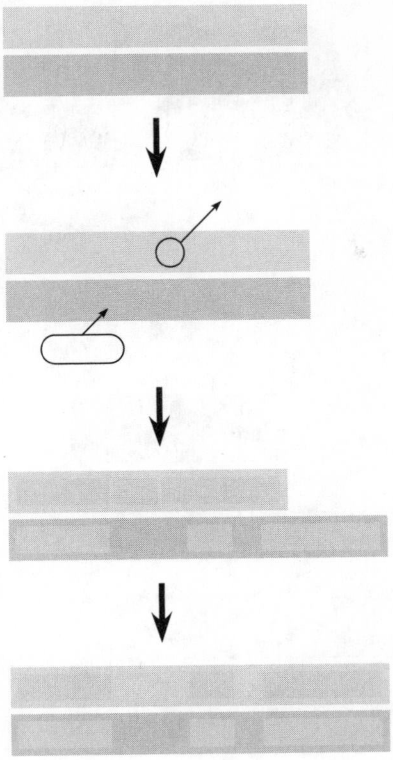

Figure 25.6 Aligning segments of DNA, page 495

Figure 25.7 A molecular homoplasy, page 495

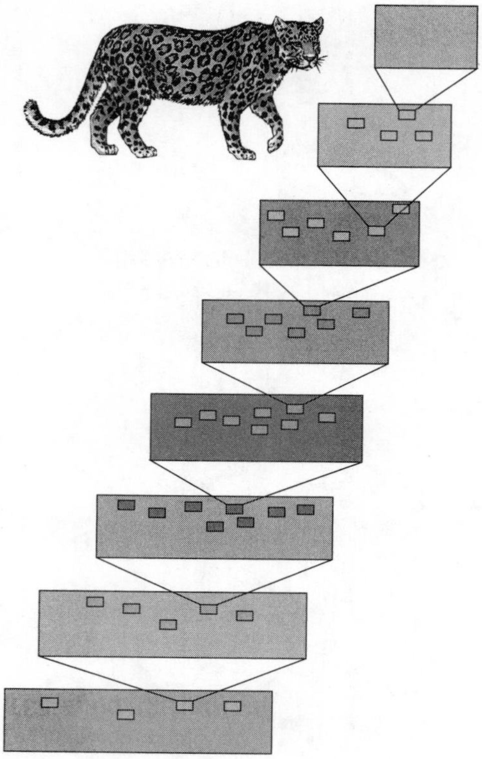

Figure 25.8 Hierarchical classification, page 496

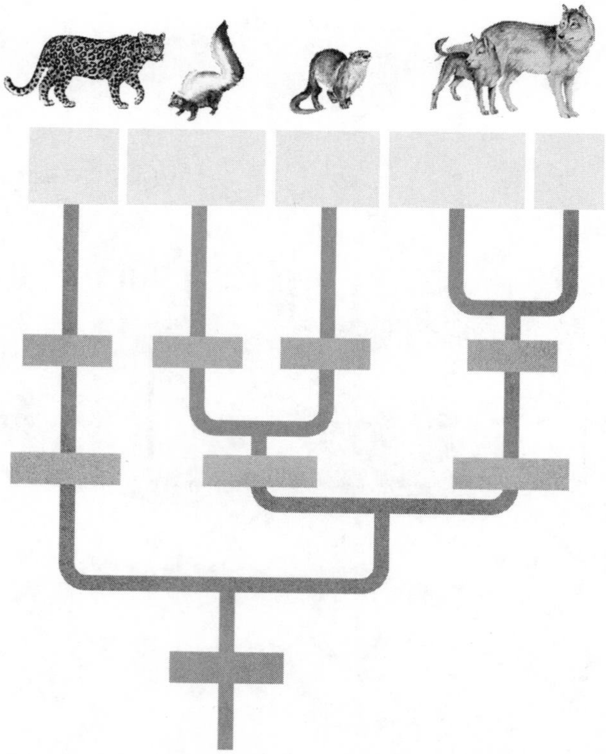

Figure 25.9 The connection between classification and phylogeny, page 497

Figure 25.UN1 Phylogenetic trees, page 497

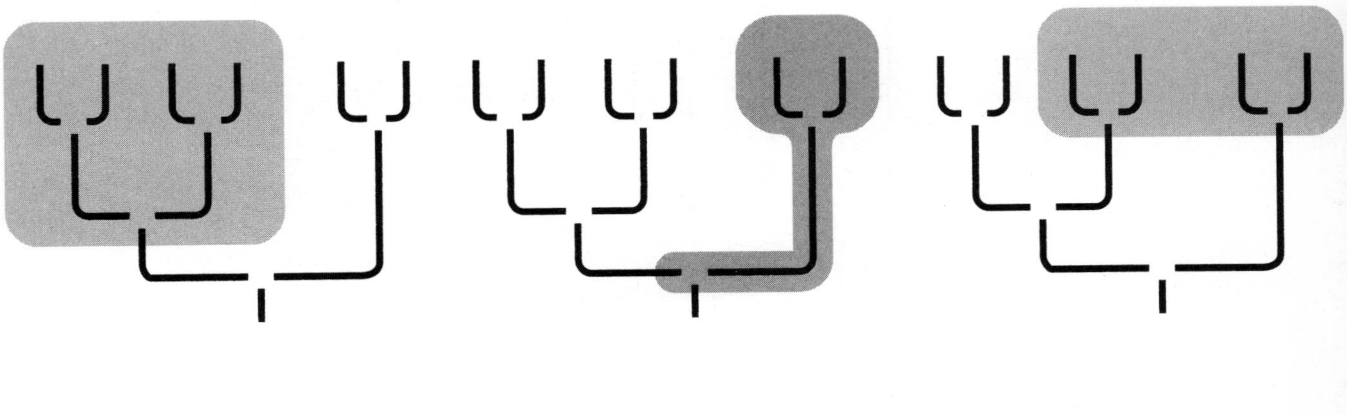

Figure 25.10 Monophyletic, paraphyletic, and polyphyletic groupings, page 498

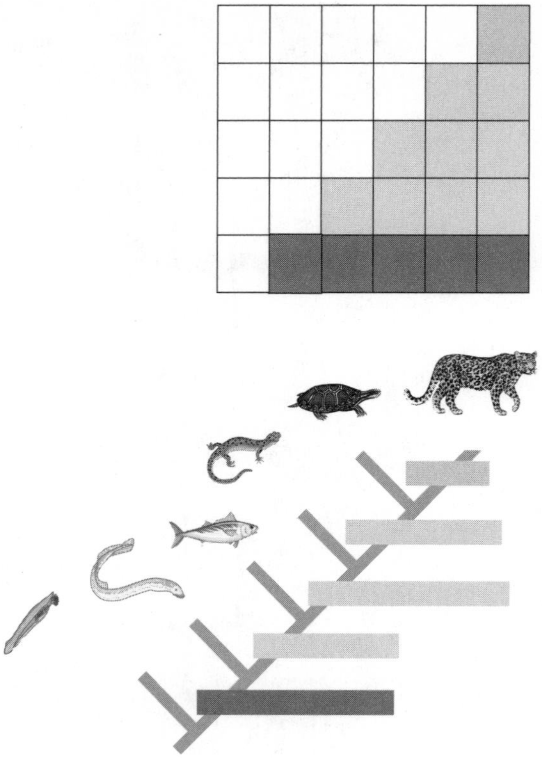

Figure 25.11 Constructing a cladogram, page 499

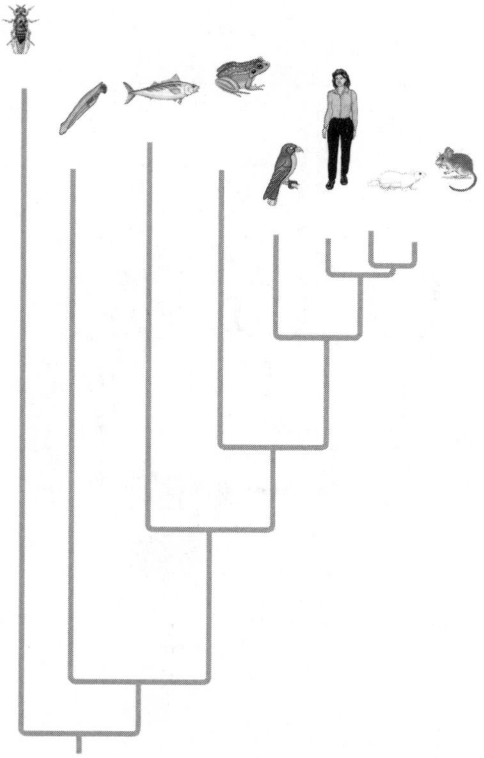

Figure 25.12 Phylogram, page 499

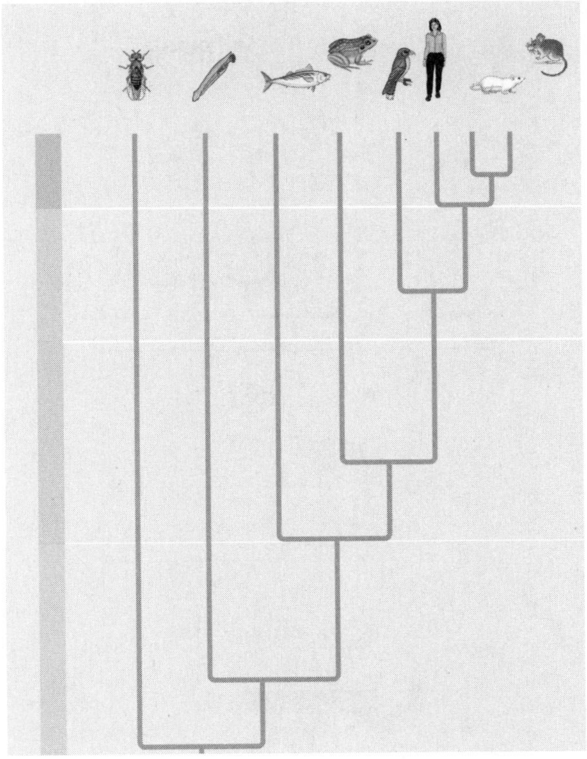

Figure 25.13 Ultrametric tree, page 500

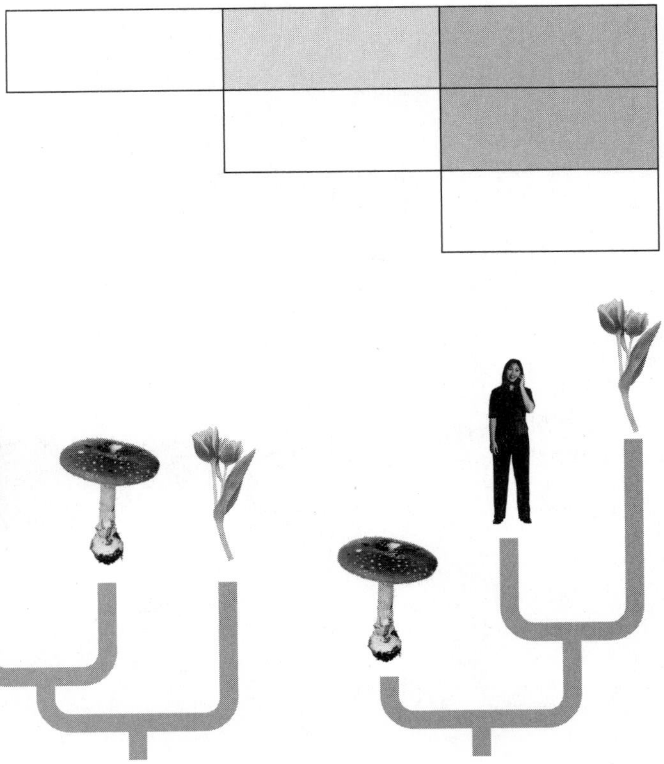

Figure 25.14 Trees with different likelihoods, page 501

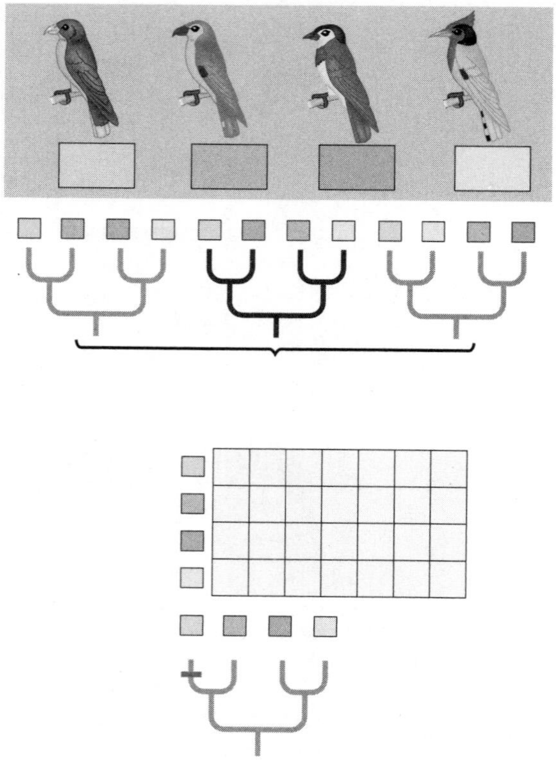

Figure 25.15 Applying parsimony to a problem in molecular systematics (part 1), page 502

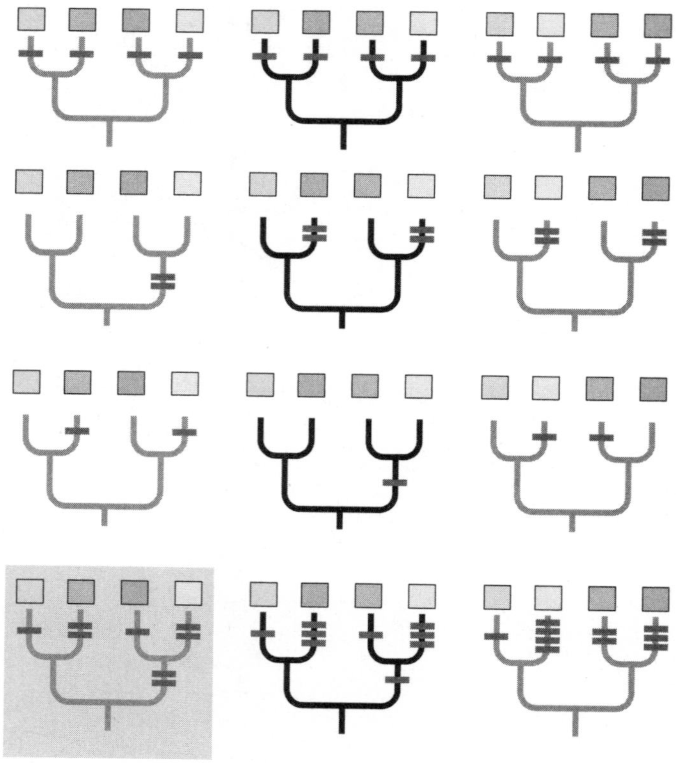

Figure 25.15 Applying parsimony to a problem in molecular systematics (part 2), page 503

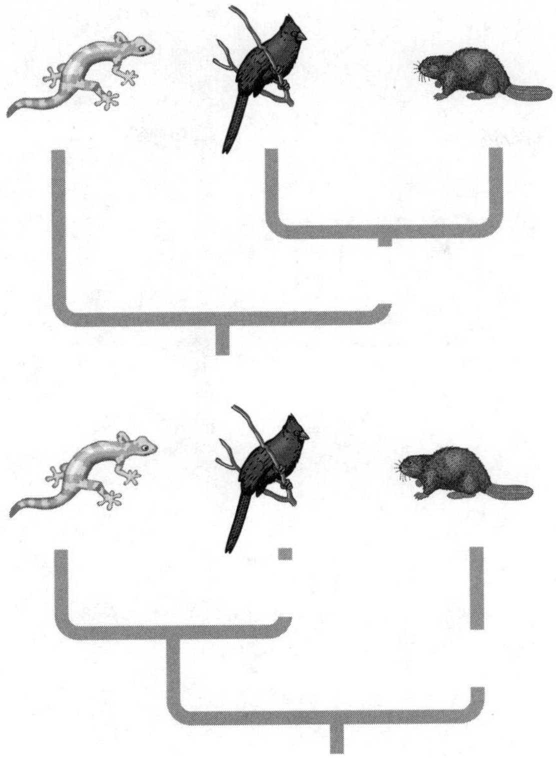

Figure 25.16 Parsimony and the analogy-versus-homology pitfall, page 504

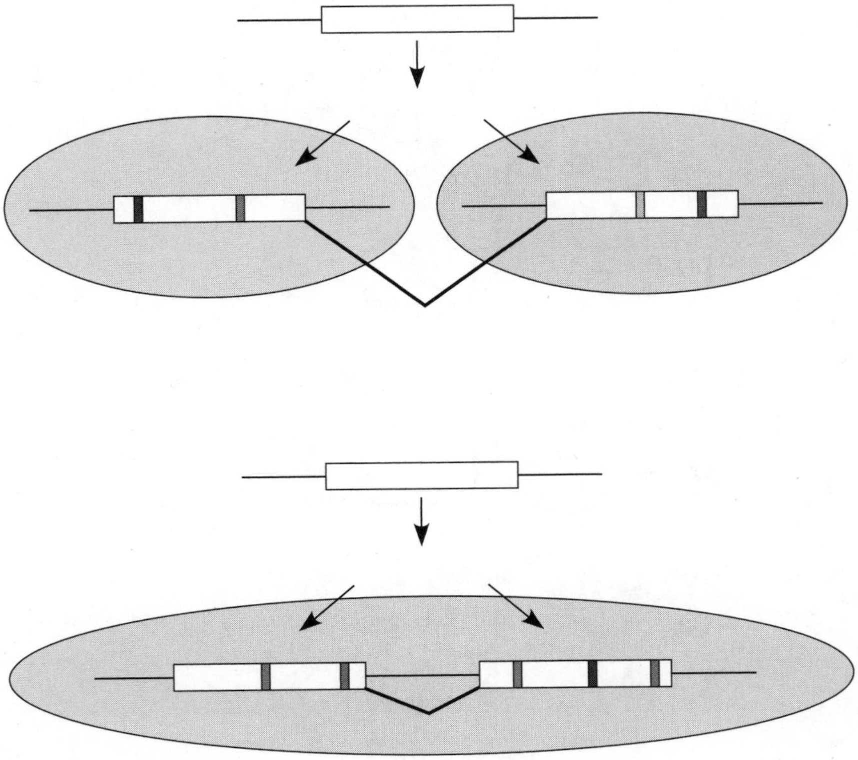

Figure 25.17 Two types of homologous genes, page 505

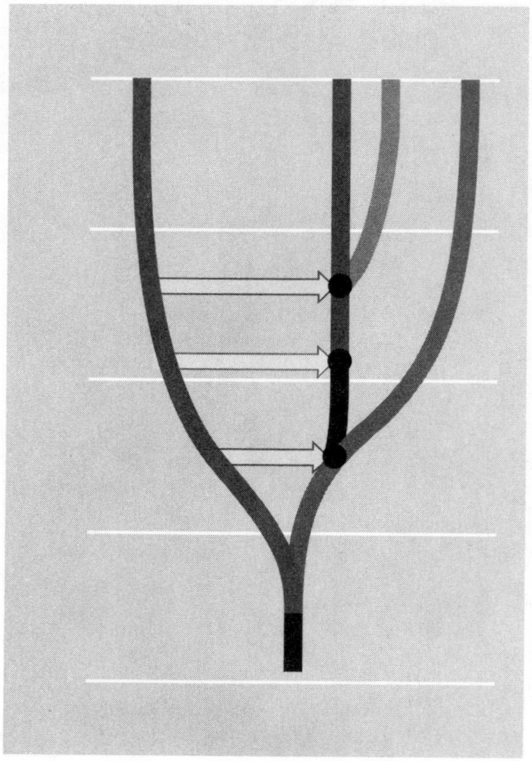

Figure 25.18 The universal tree of life, page 507

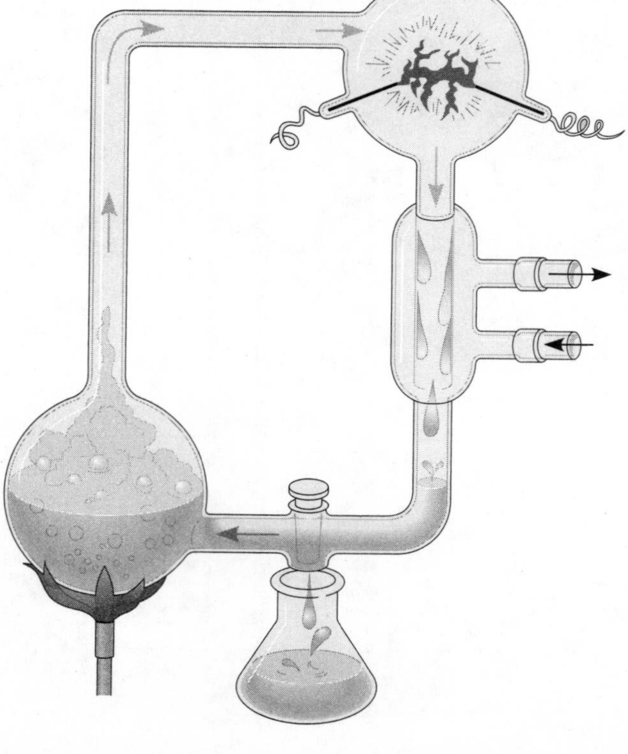

Figure 26.2 Can organic molecules form in a reducing atmosphere? page 513

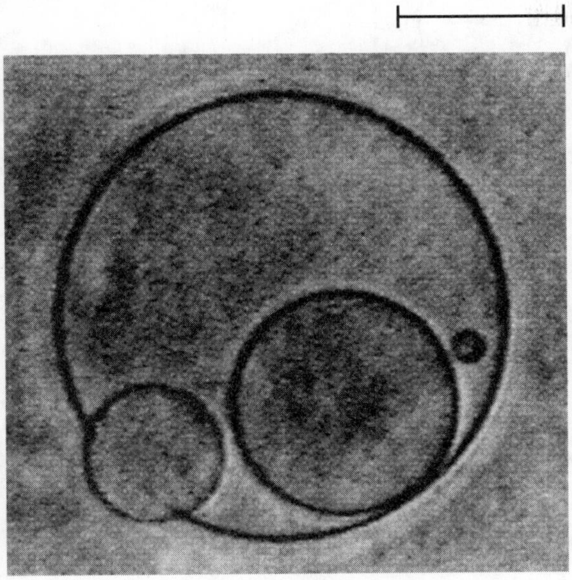

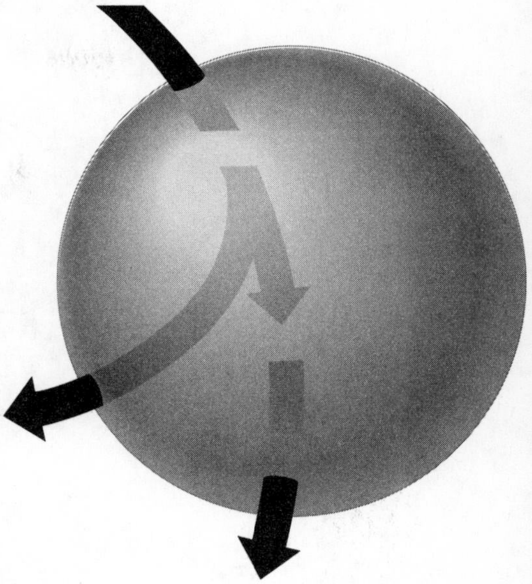

Figure 26.4 Laboratory versions of protobionts, page 515

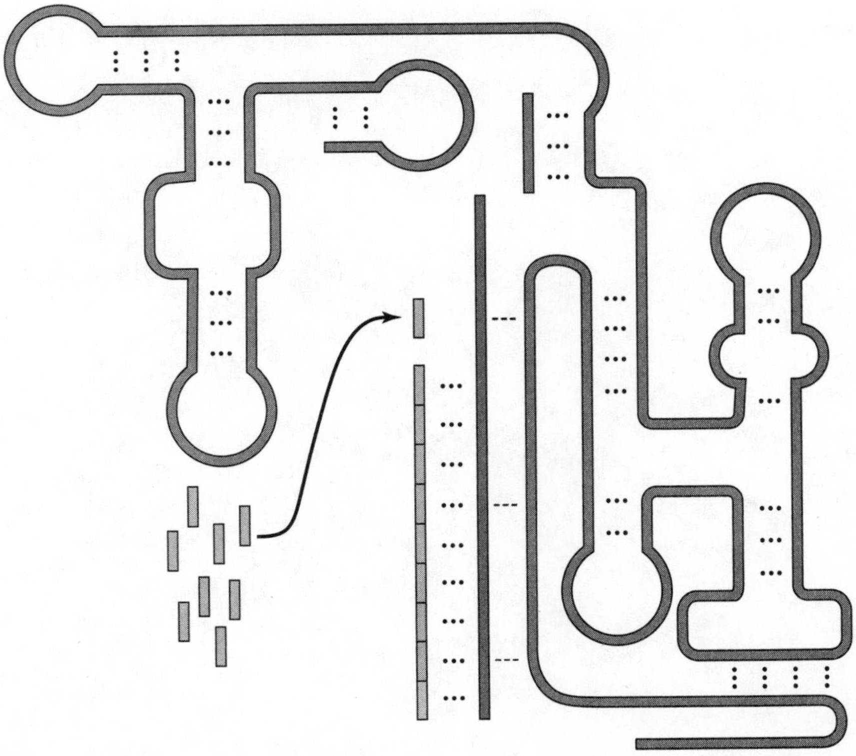

Figure 26.5 A ribozyme capable of replicating RNA, page 515

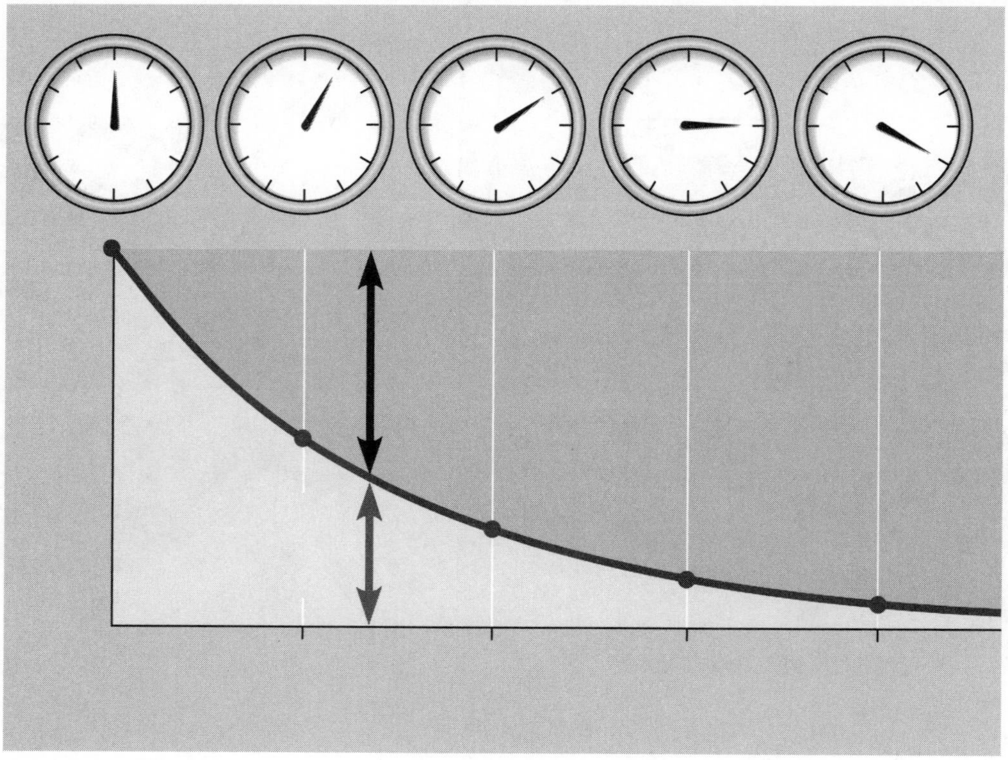

Figure 26.7 Radiometric dating, page 517

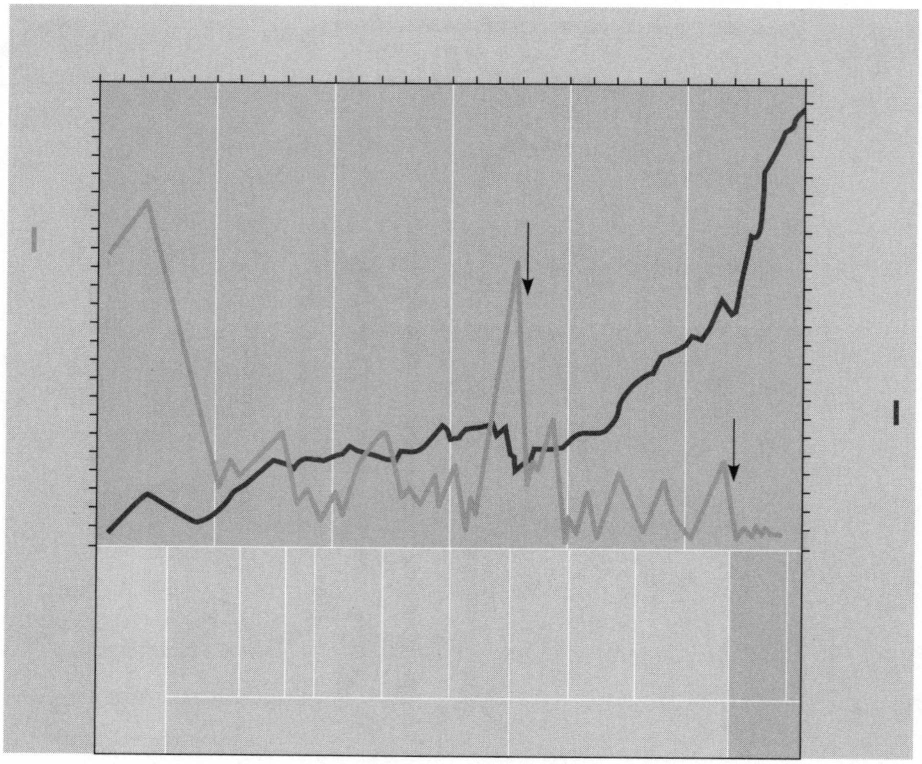

Figure 26.8 Diversity of life and periods of mass extinction, page 518

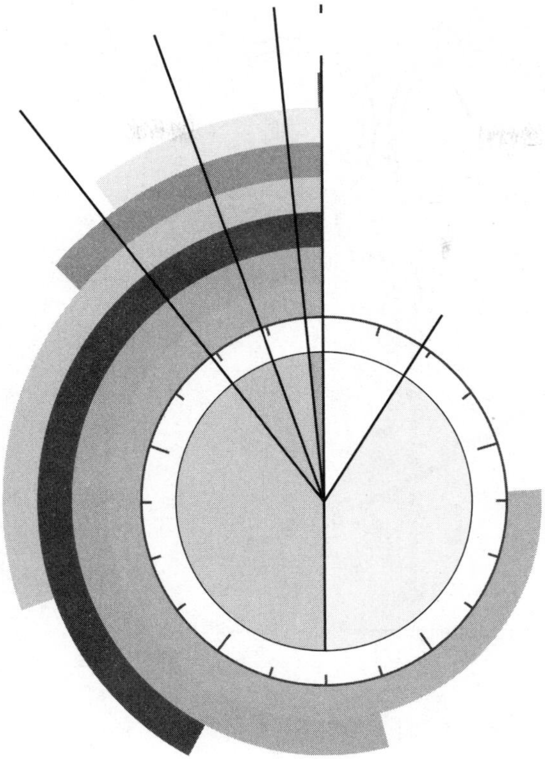

Figure 26.10 Clock analogy for some key events in Earth's history, page 521

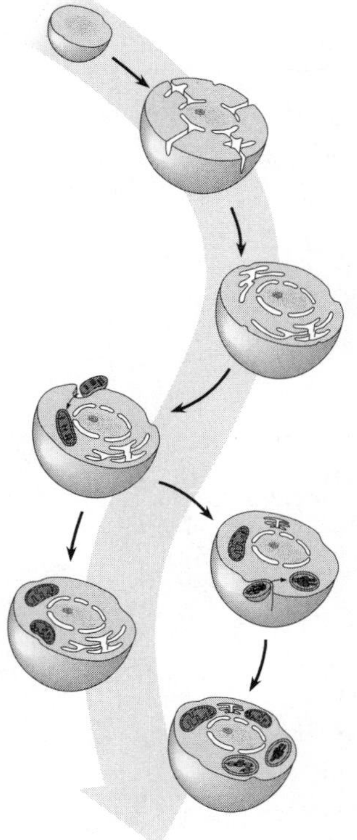

Figure 26.13 A model of the origin of eukaryotes through serial endosymbiosis, page 524

Figure 26.17 The Cambrian radiation of animals, page 527

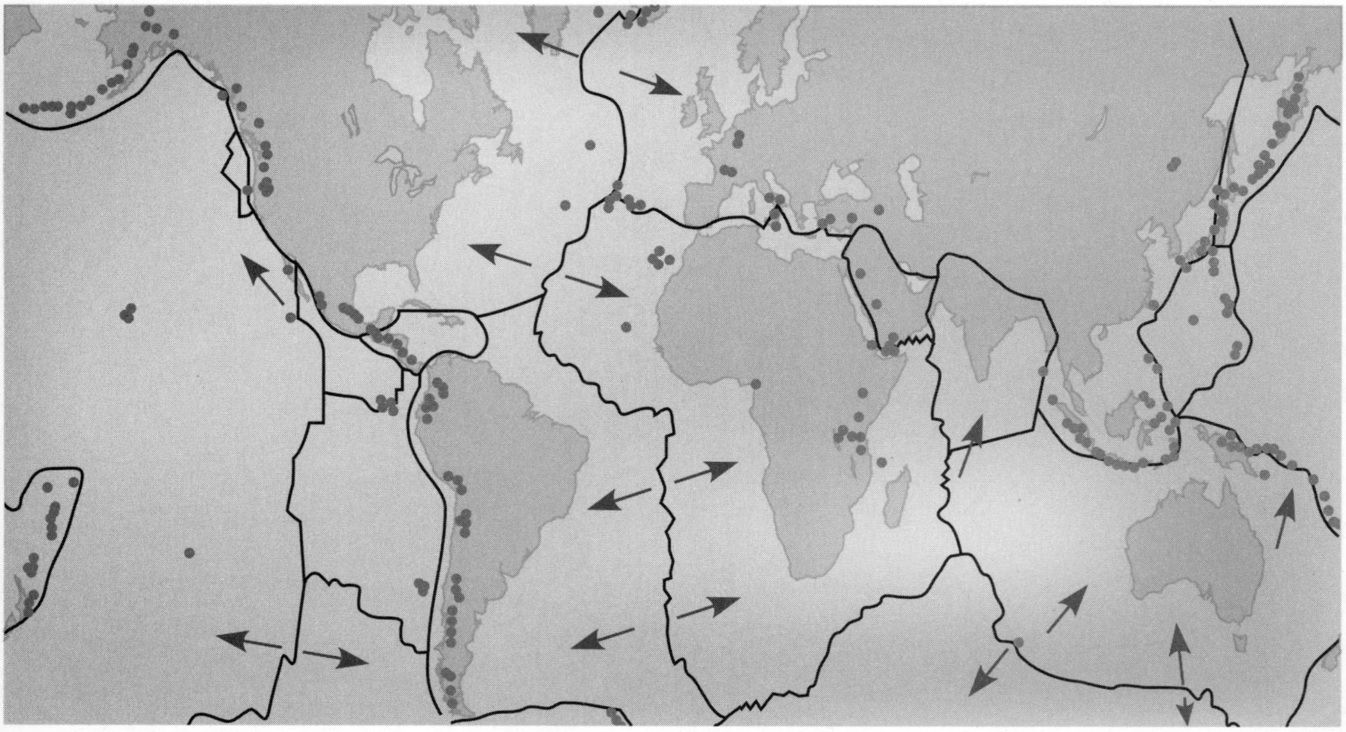

Figure 26.18 Earth's major crustal plates, page 527

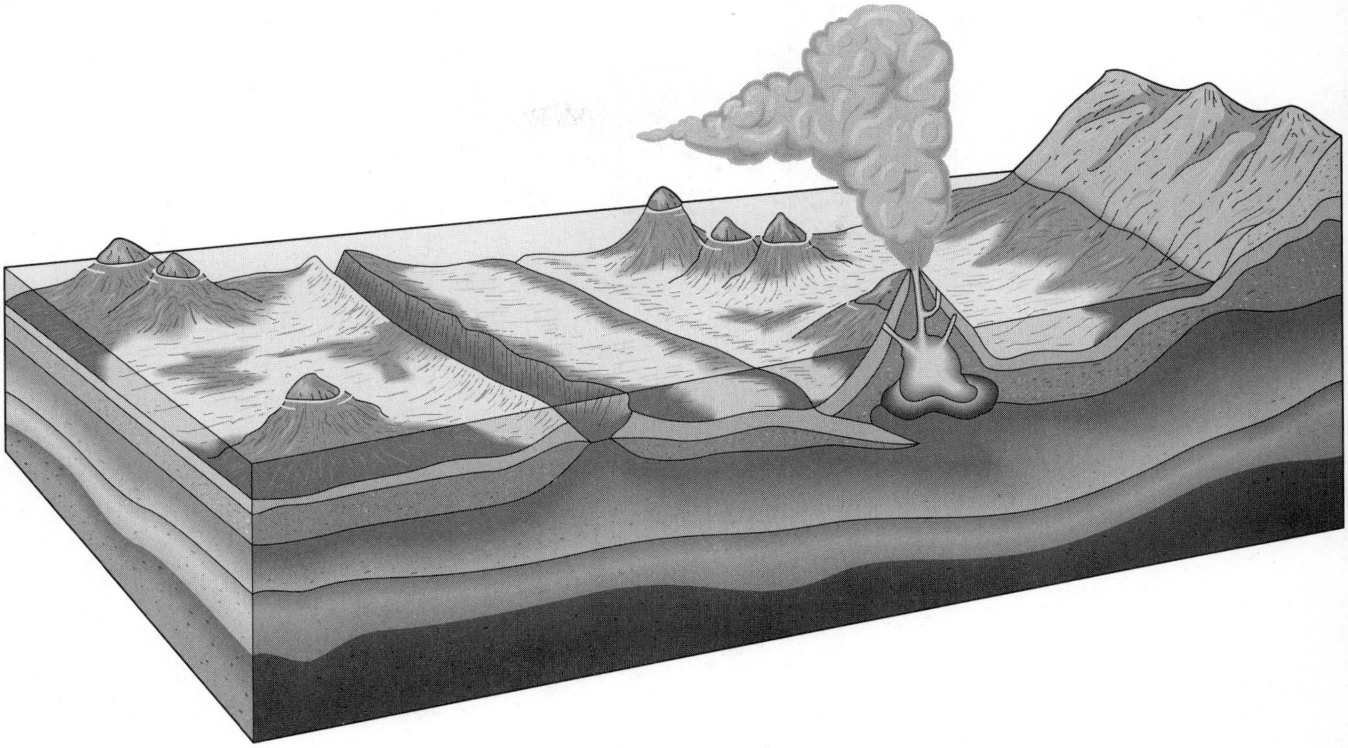

Figure 26.19 Events at plate boundaries, page 528

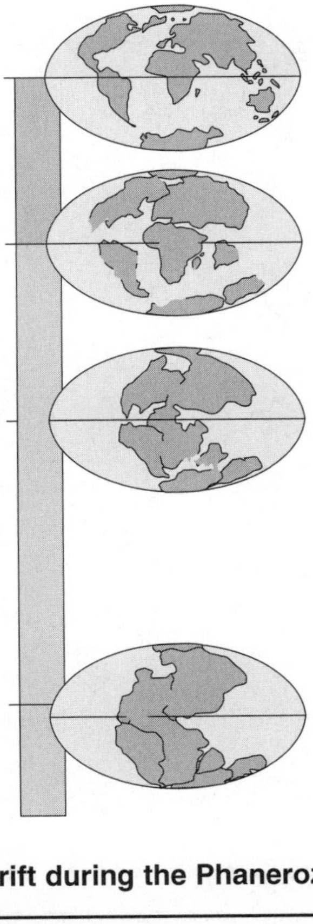

Figure 26.20 The history of continental drift during the Phanerozoic, page 528

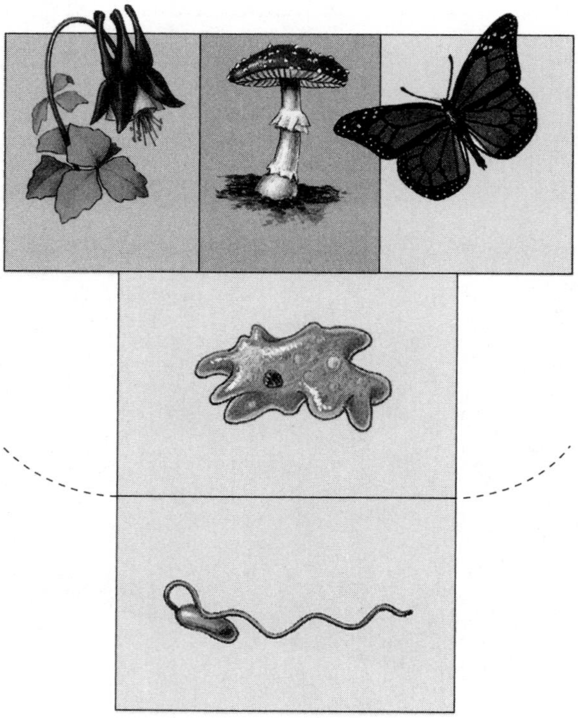

Figure 26.21 Whittaker's five kingdom system, page 529

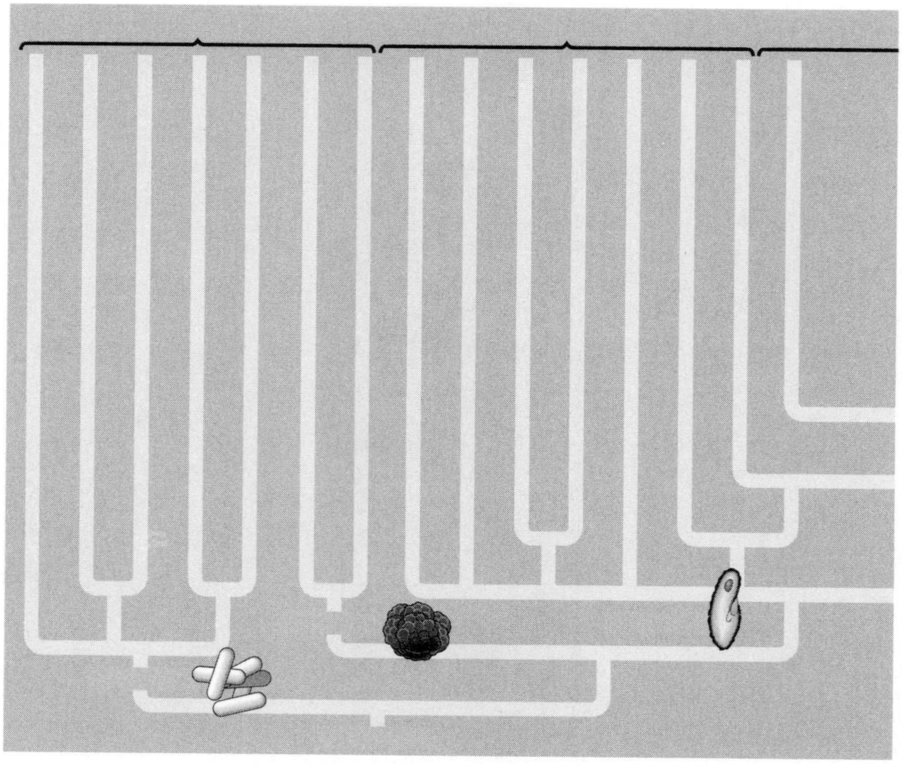

Figure 26.22 One current view of biological diversity (part 1), page 530

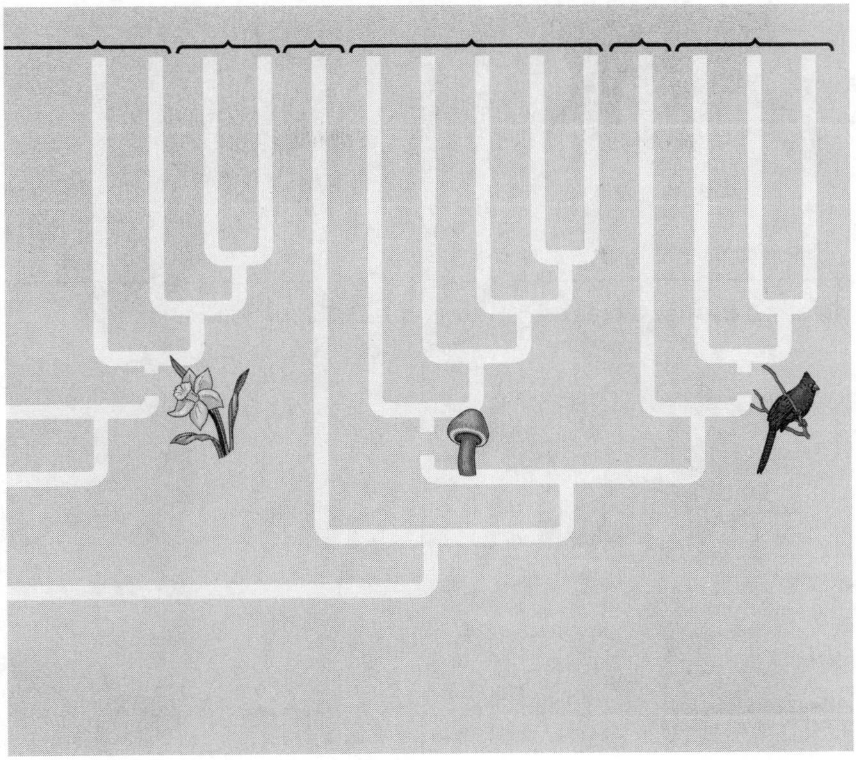

Figure 26.22 One current view of biological diversity (part 2), page 531

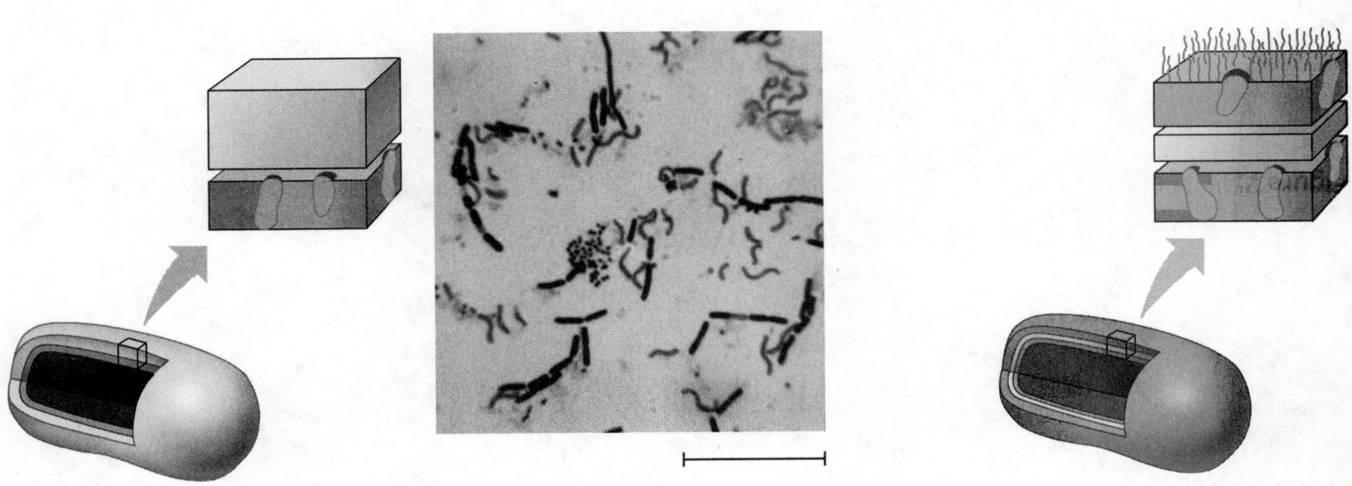

Figure 27.3 Gram staining, page 535

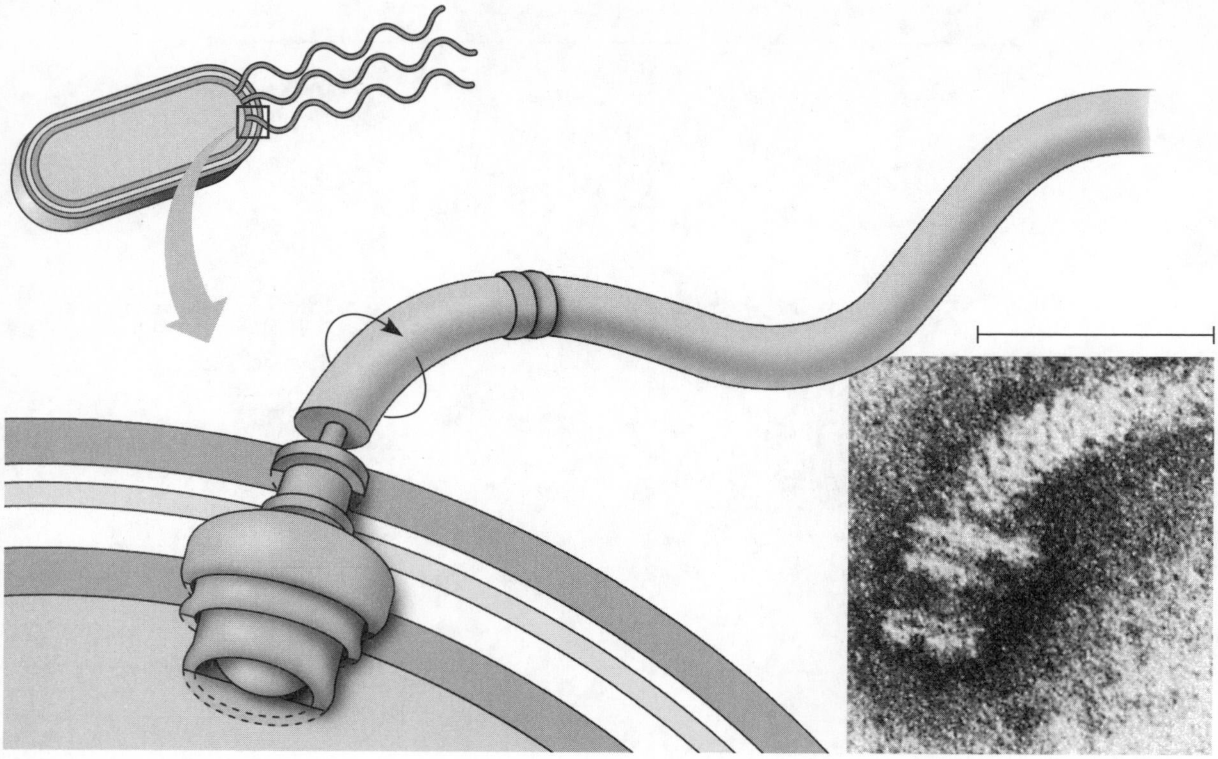

Figure 27.6 Prokaryotic flagellum, page 536

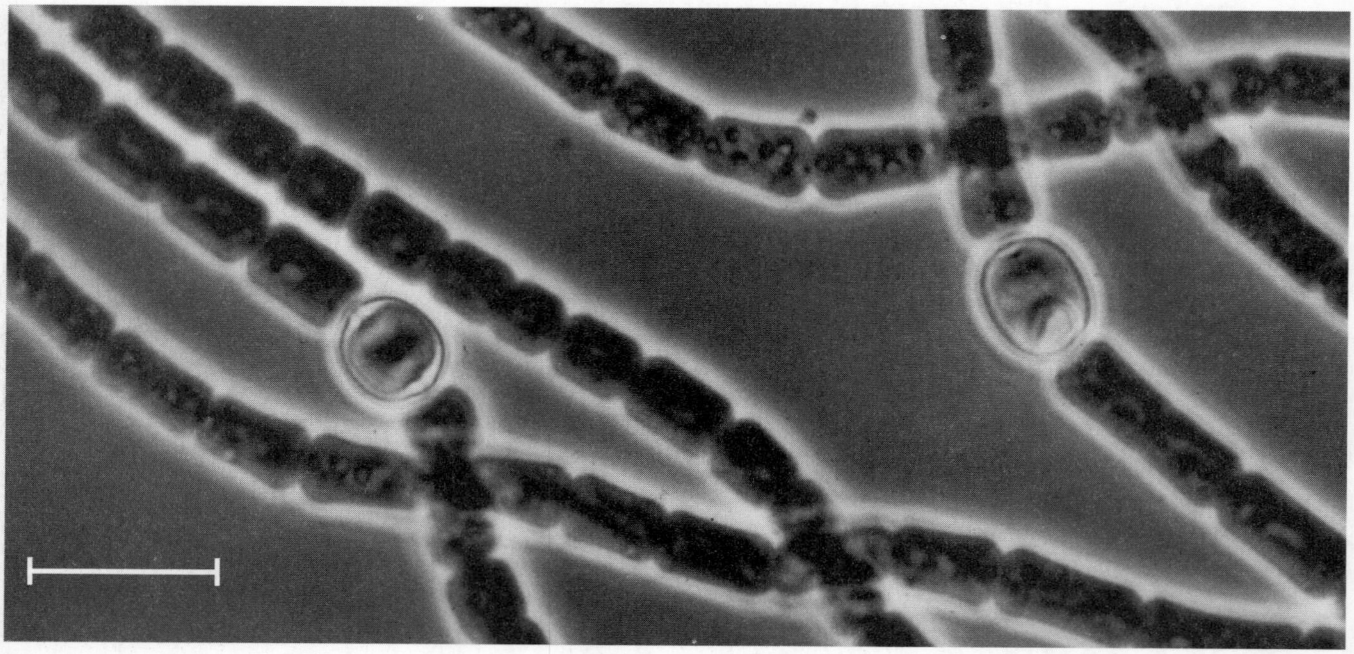

Figure 27.10 Metabolic cooperation in a colonial prokaryote, page 539

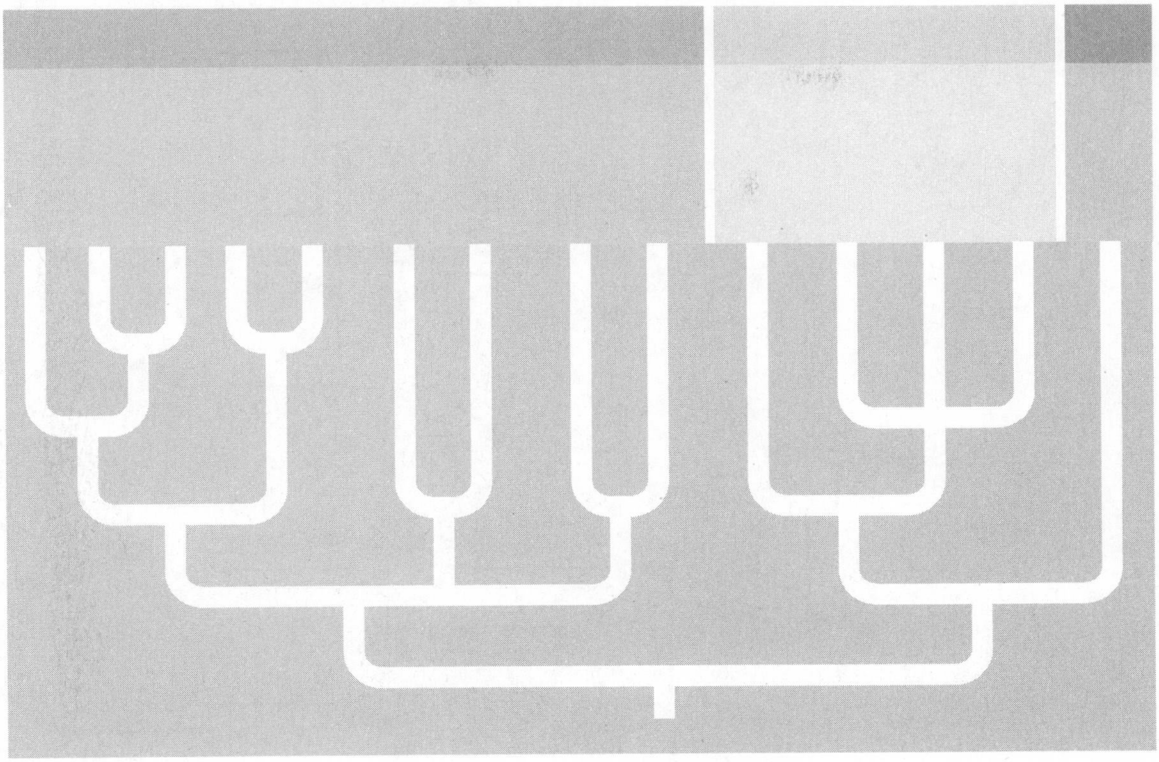

Figure 27.12 A simplified phylogeny of prokaryotes, page 540

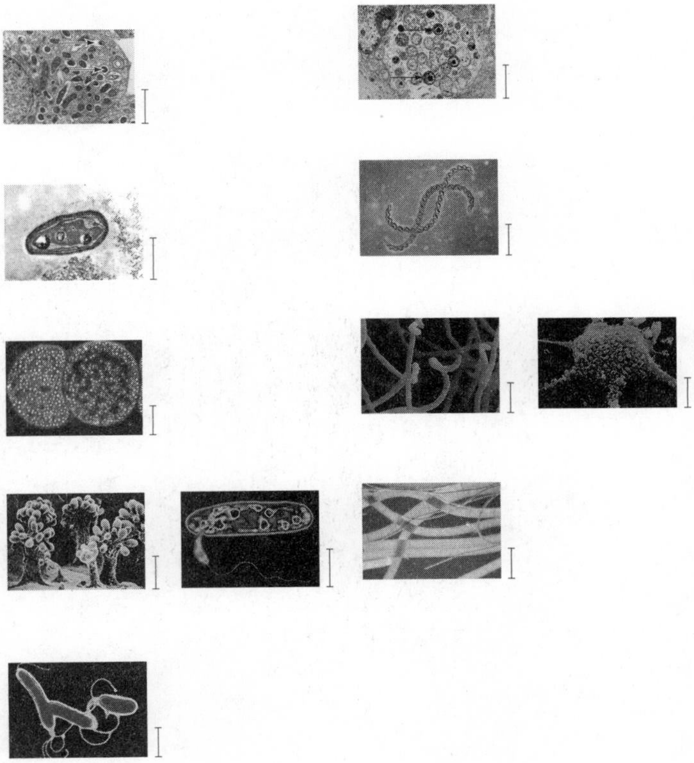

Figure 27.13 Major groups of bacteria, pages 542–543

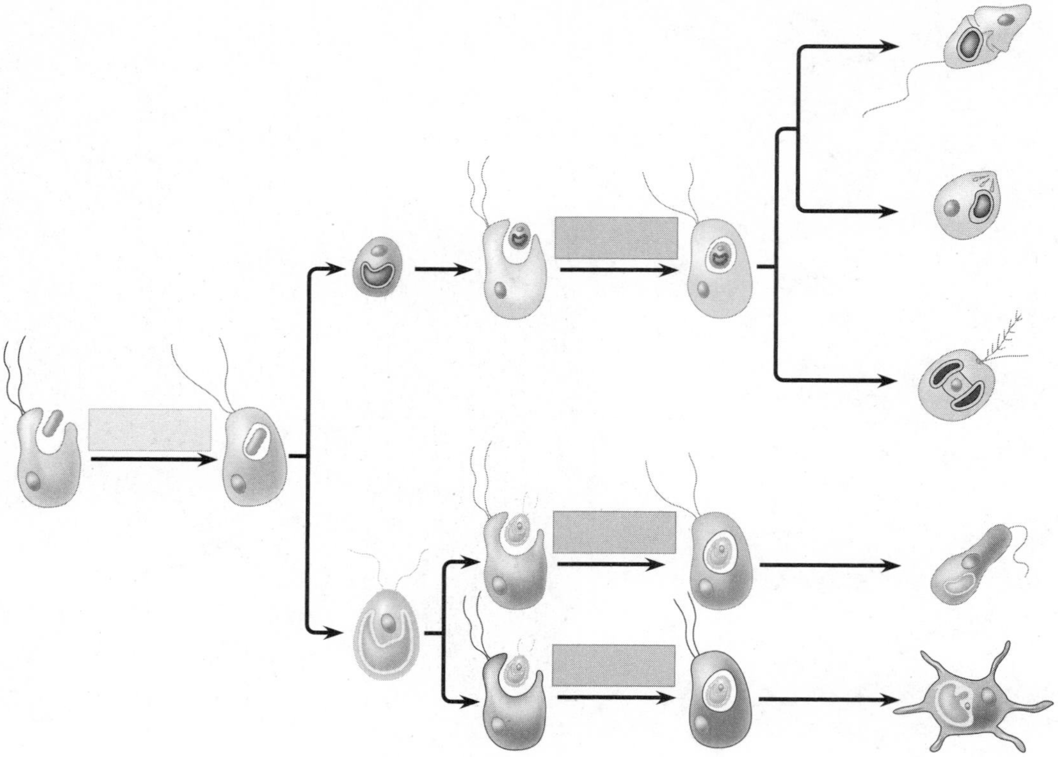

Figure 28.3 Diversity of plastids produced by secondary endosymbiosis, page 551

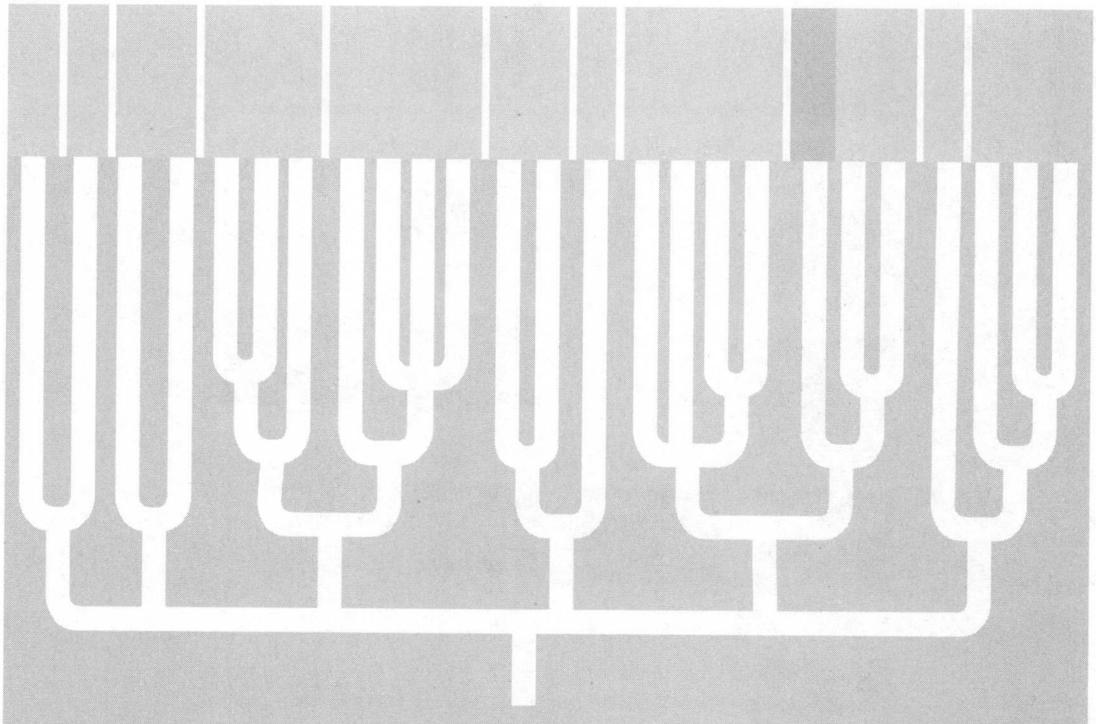

Figure 28.4 A tentative phylogeny of eukaryotes, page 552

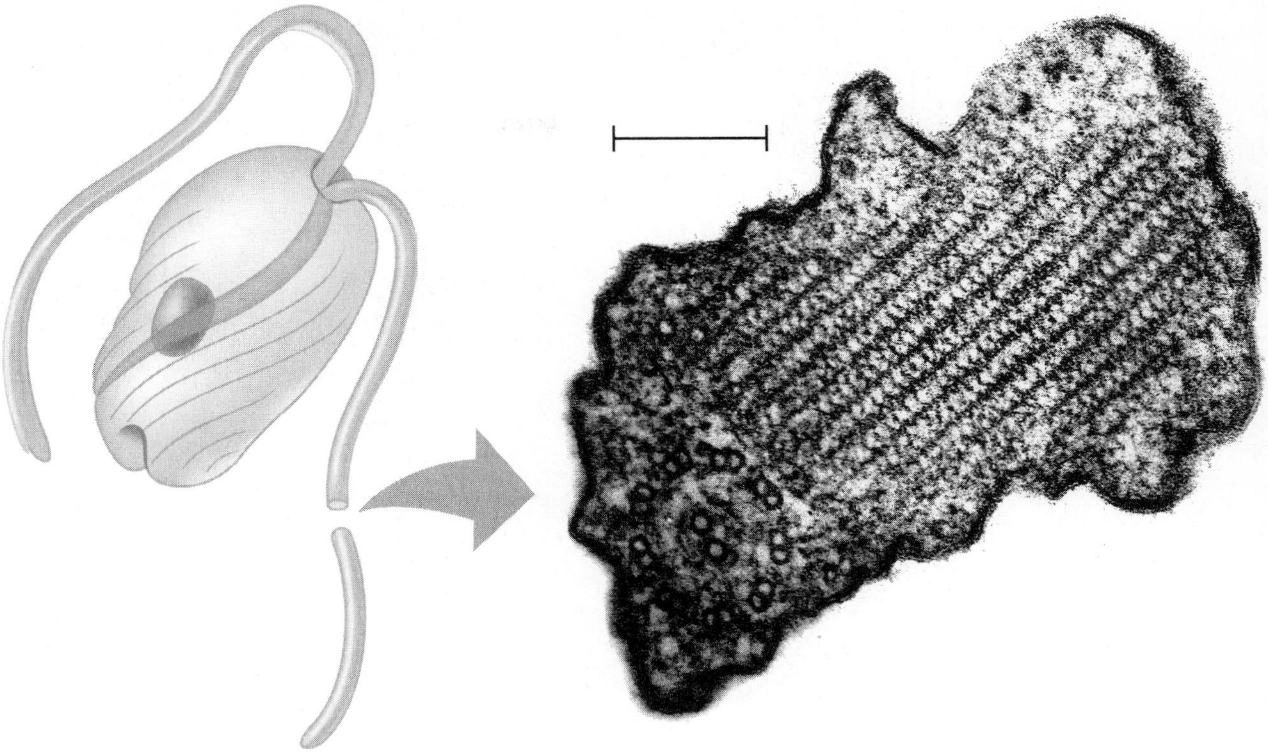

Figure 28.6 Euglenozoan flagellum, page 553

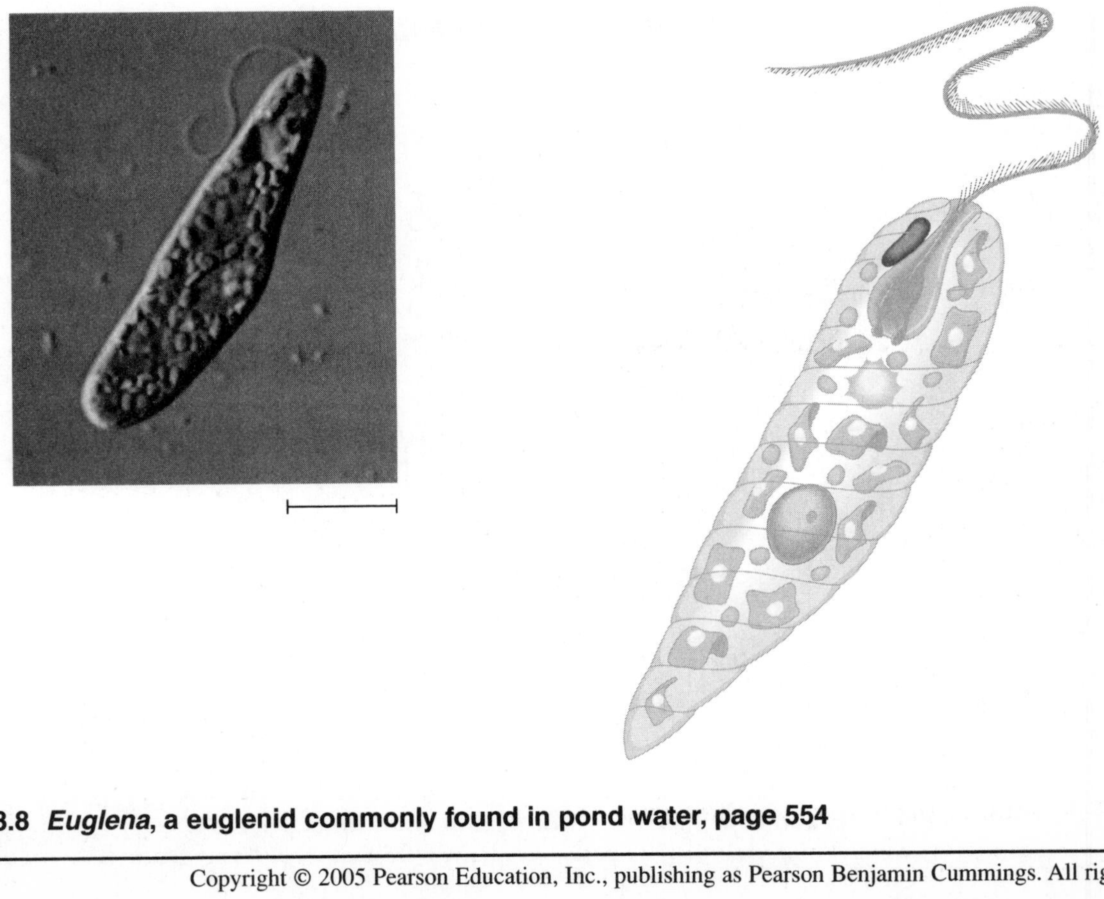

Figure 28.8 *Euglena*, a euglenid commonly found in pond water, page 554

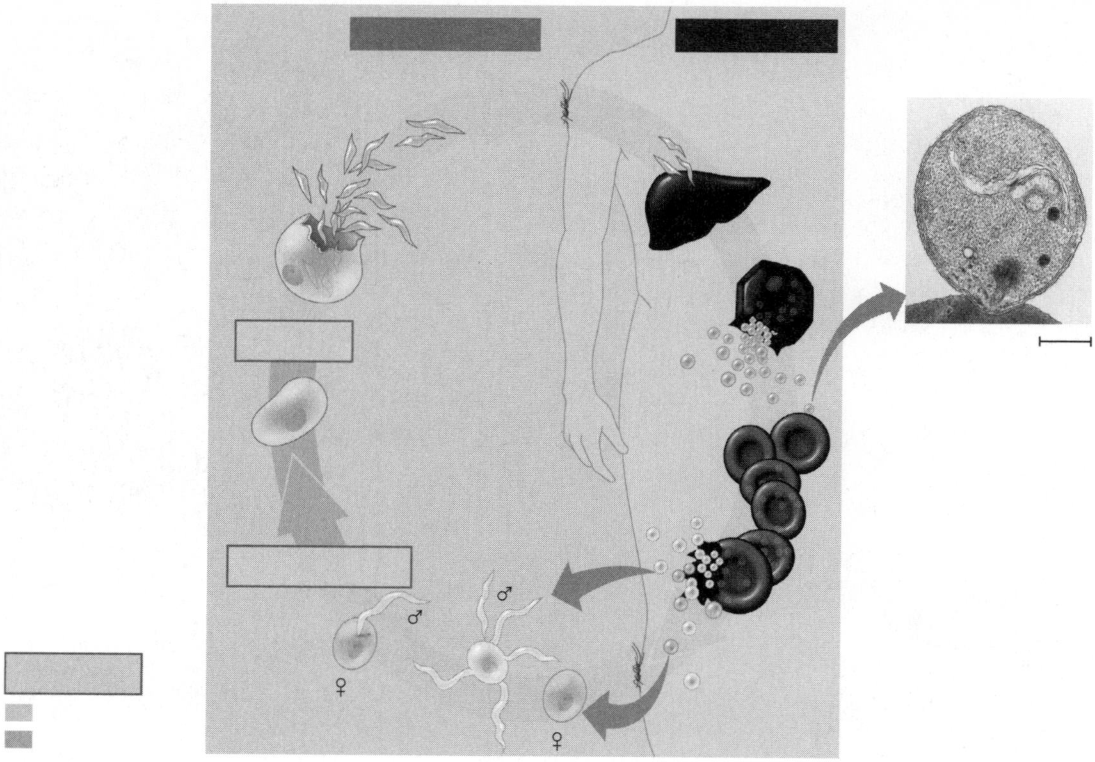

Figure 28.11 The two-host life cycle of *Plasmodium*, the apicomplexan that causes malaria, page 556

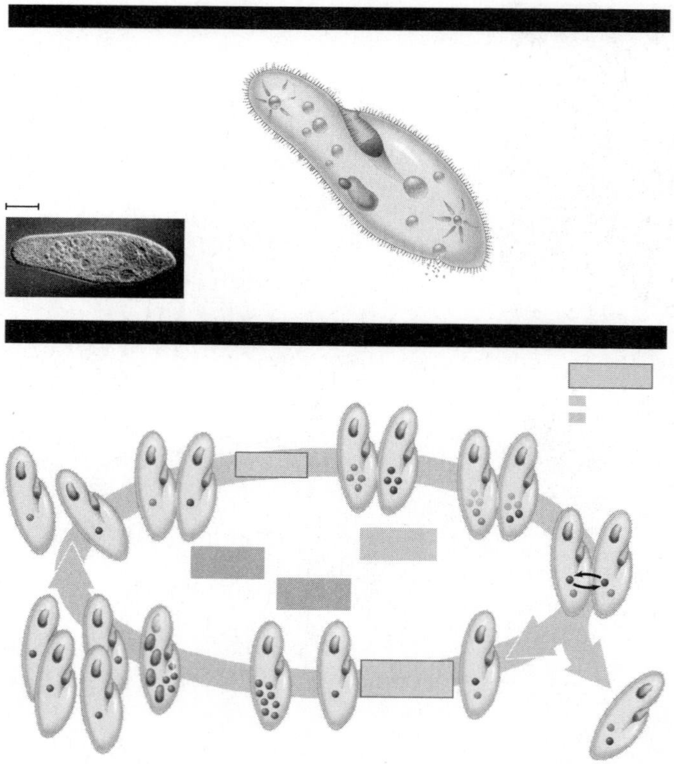

Figure 28.12 Structure and function in the ciliate *Paramecium caudatum*, page 557

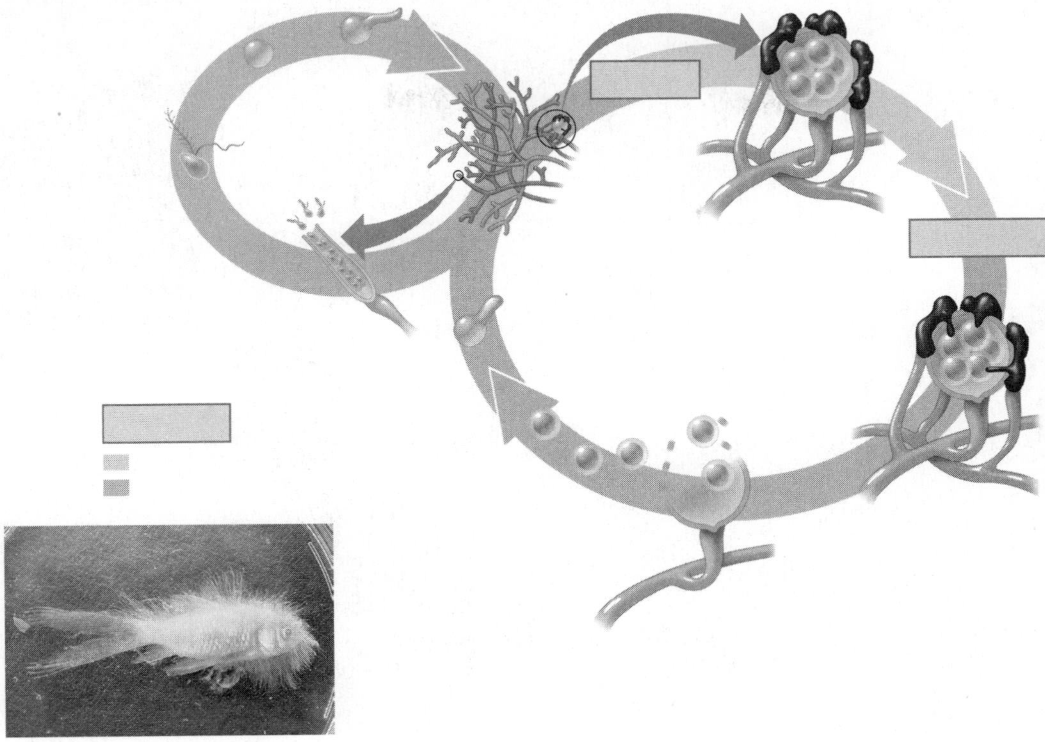

Figure 28.14 The life cycle of a water mold, page 559

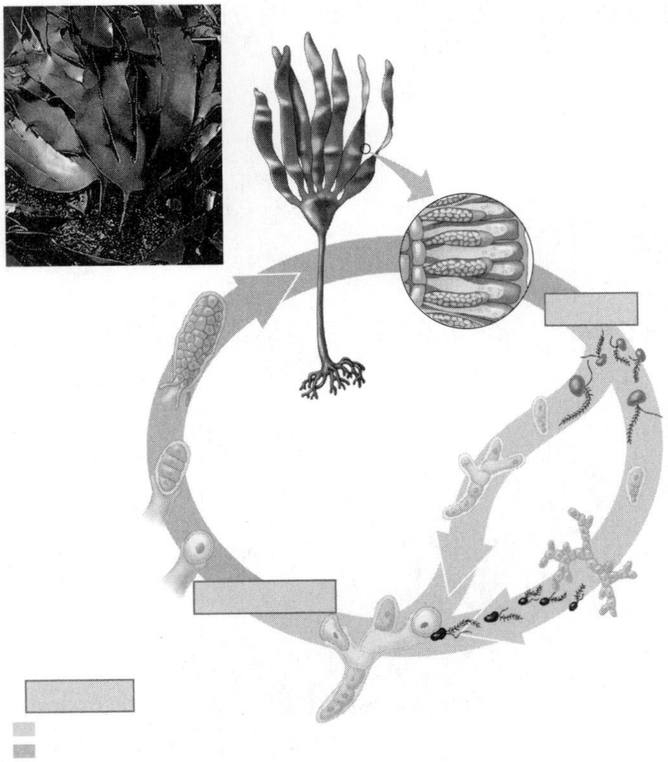

Figure 28.21 The life cycle of *Laminaria*: an example of alternation of generations, page 562

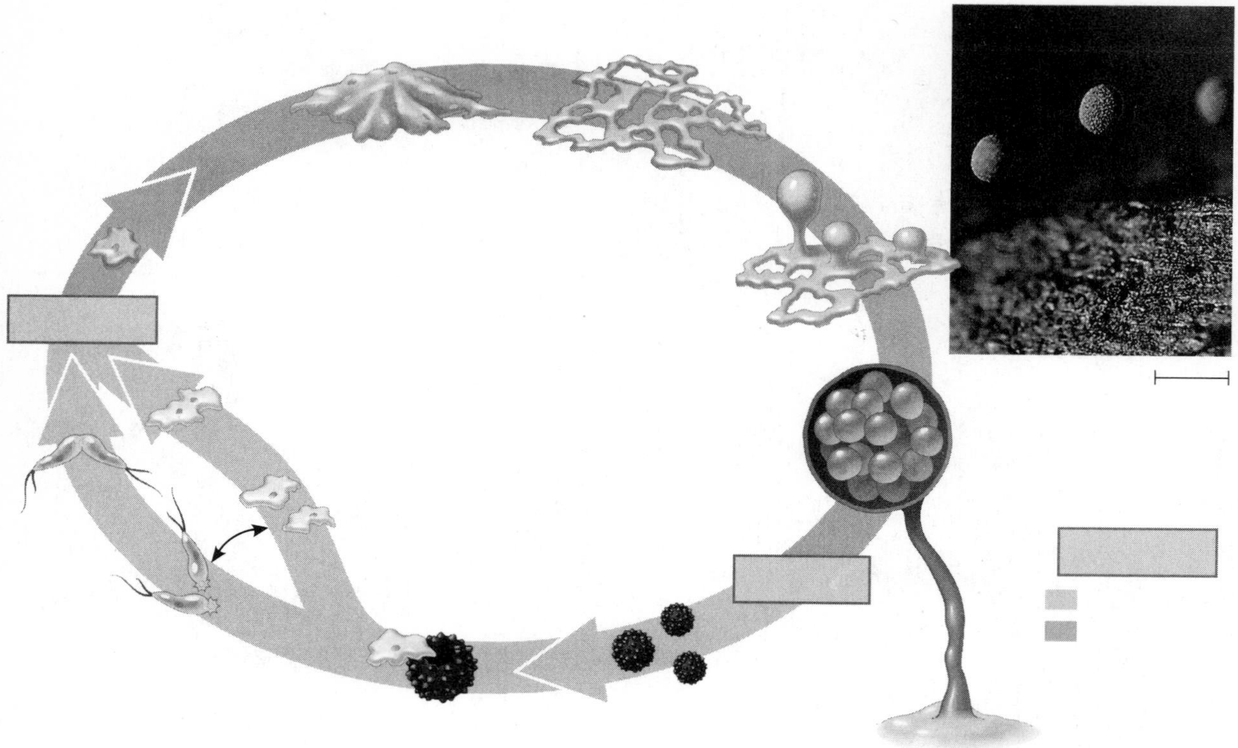

Figure 28.26 The life cycle of a plasmodial slime mold, page 565

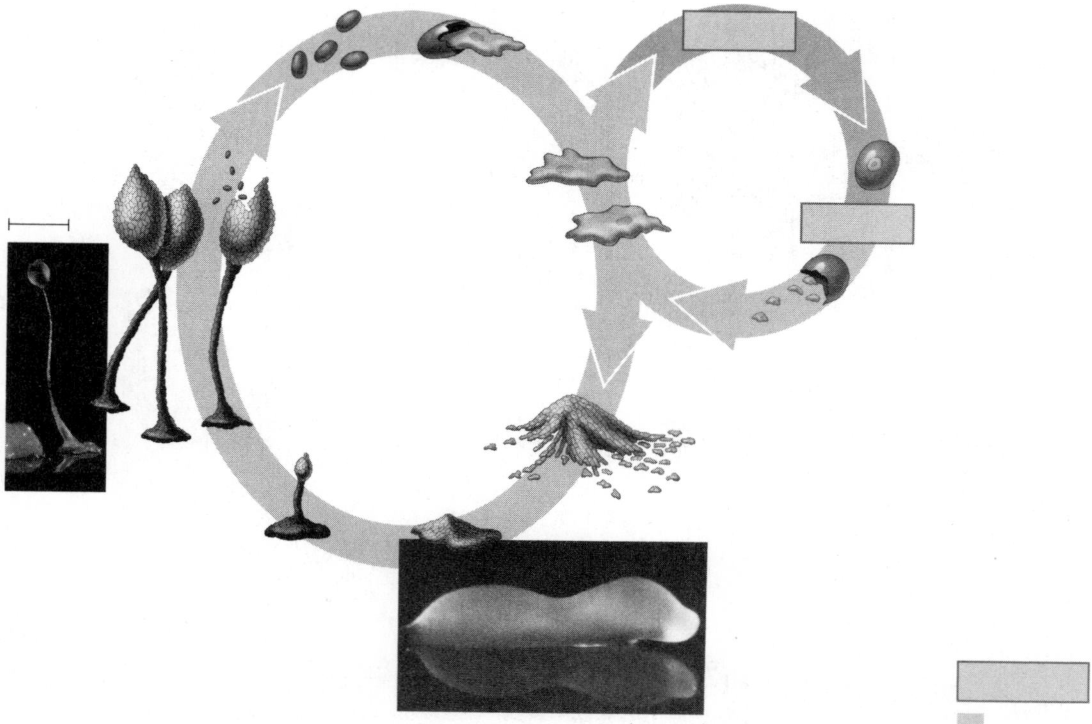

Figure 28.27 The life cycle of *Dictyostelium*, a cellular slime mold, page 566

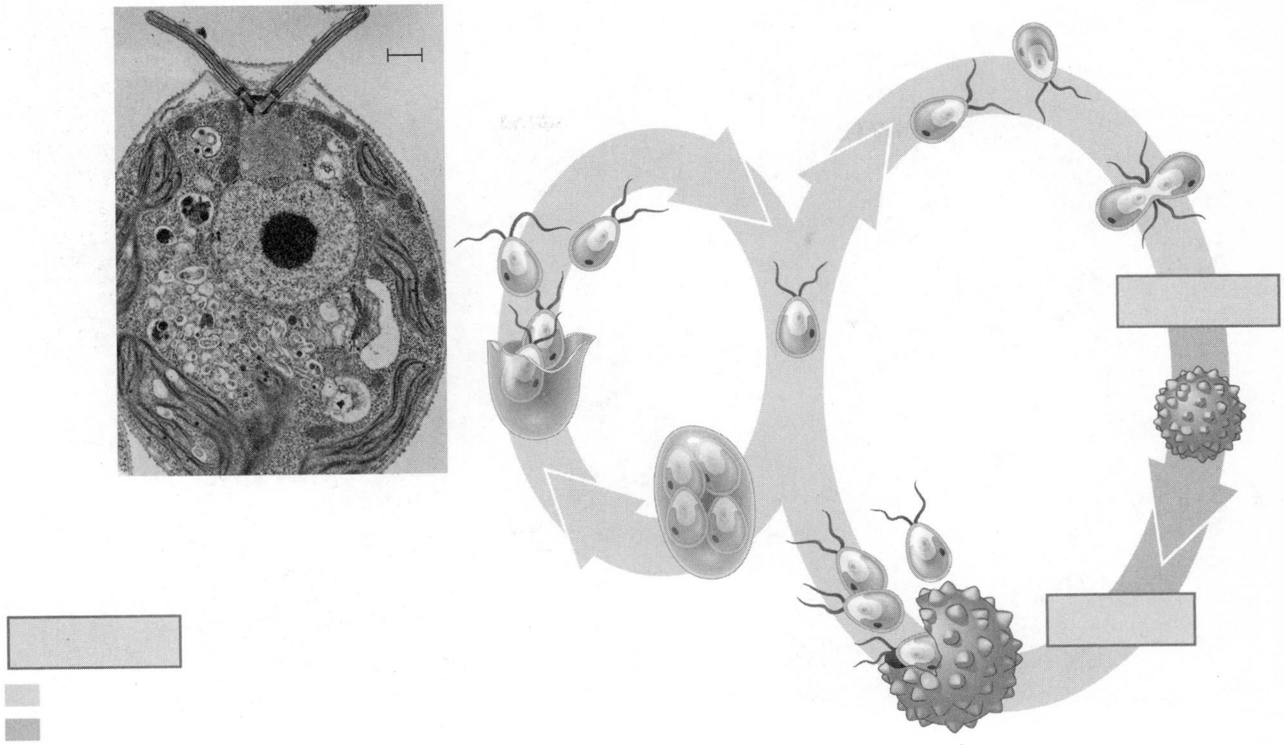

Figure 28.31 The life cycle of *Chlamydomonas,* a unicellular chlorophyte, page 569

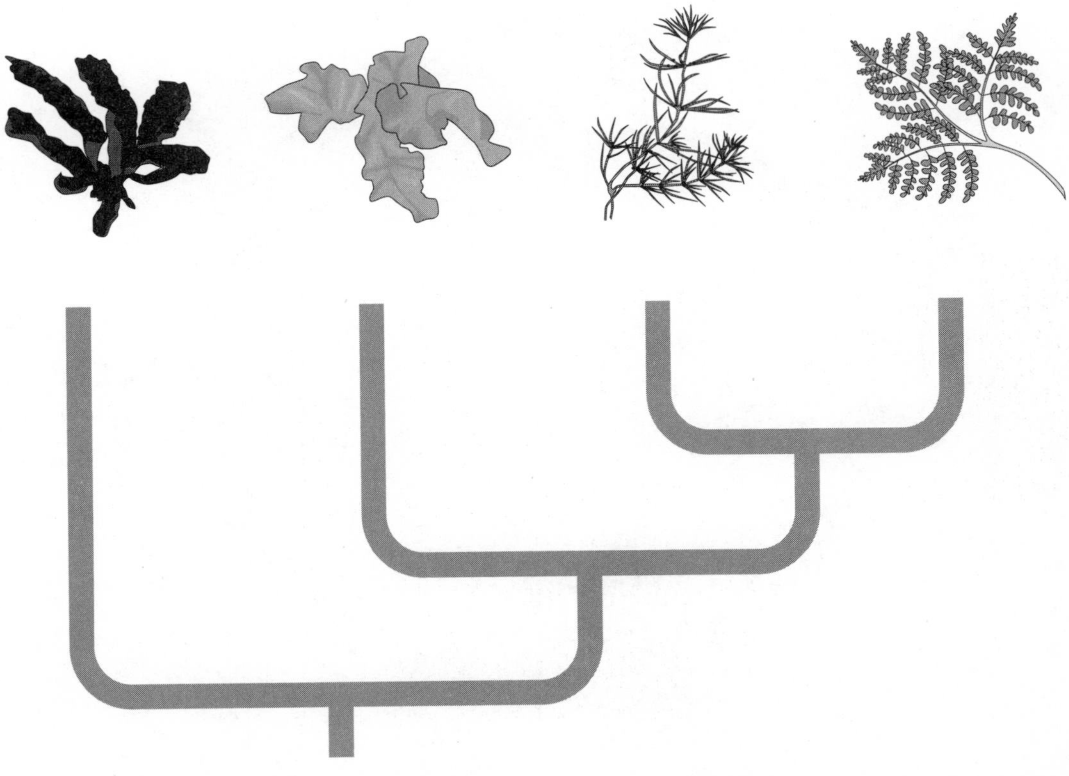

Figure 29.4 Three clades that are candidates for designation as the plant kingdom, page 575

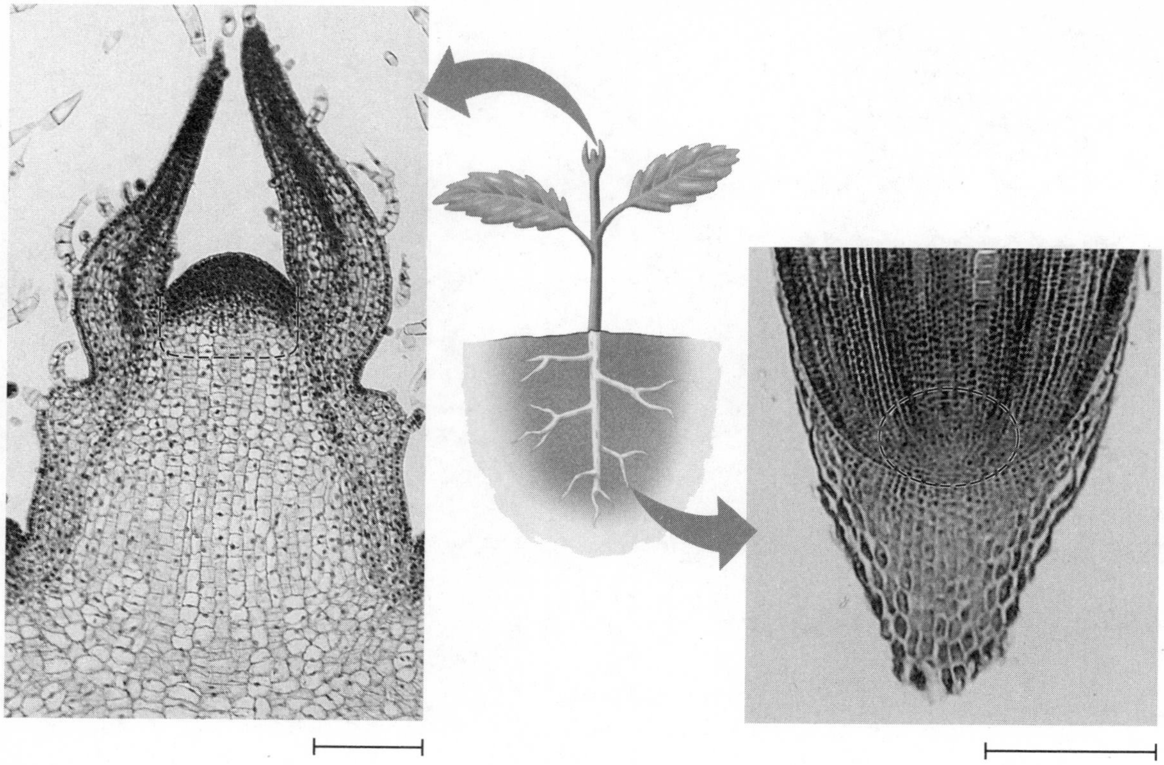

Figure 29.5 Derived traits of land plants (part 1), page 576

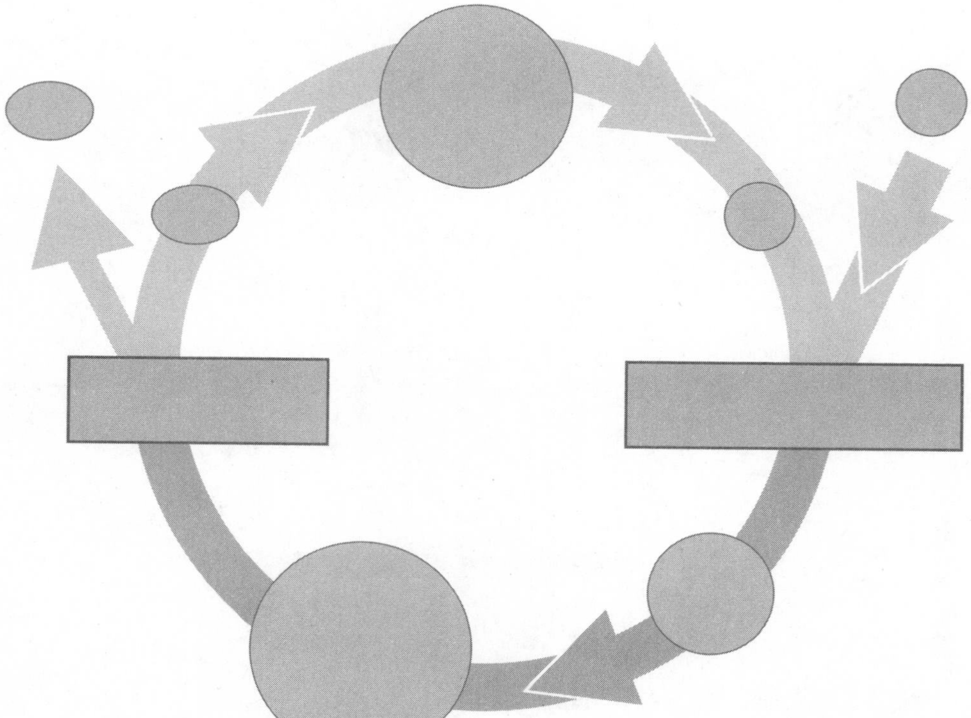

Figure 29.5 Derived traits of land plants (part 2), page 576

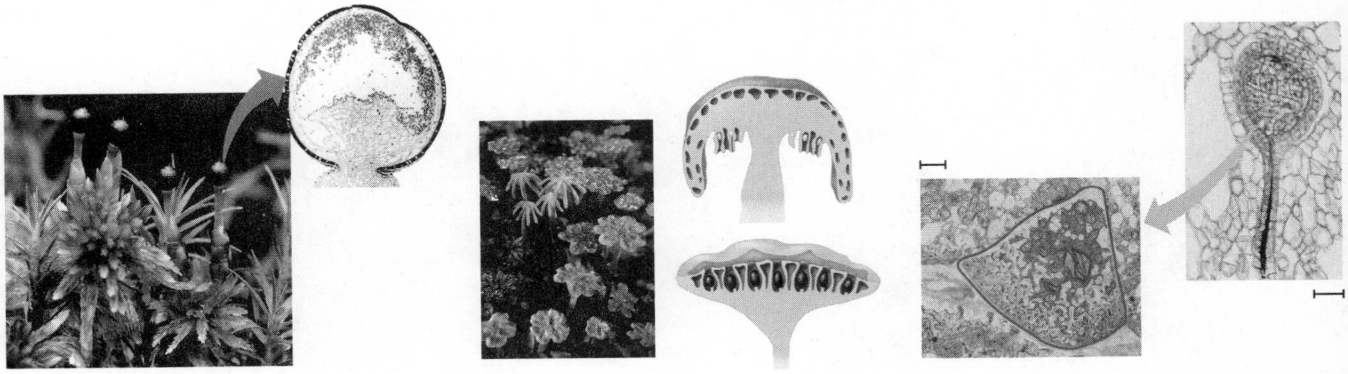

Figure 29.5 Derived traits of land plants (part 3), page 577

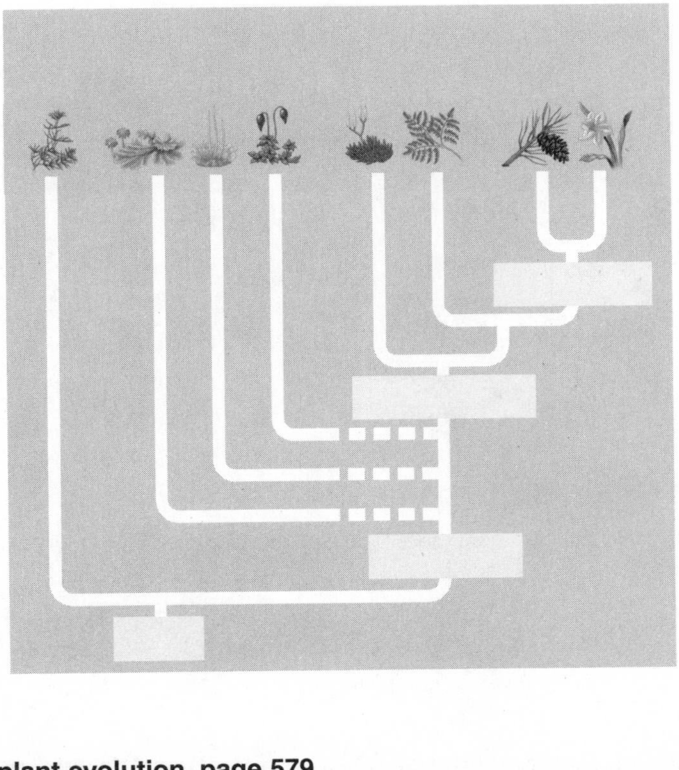

Figure 29.7 Highlights of plant evolution, page 579

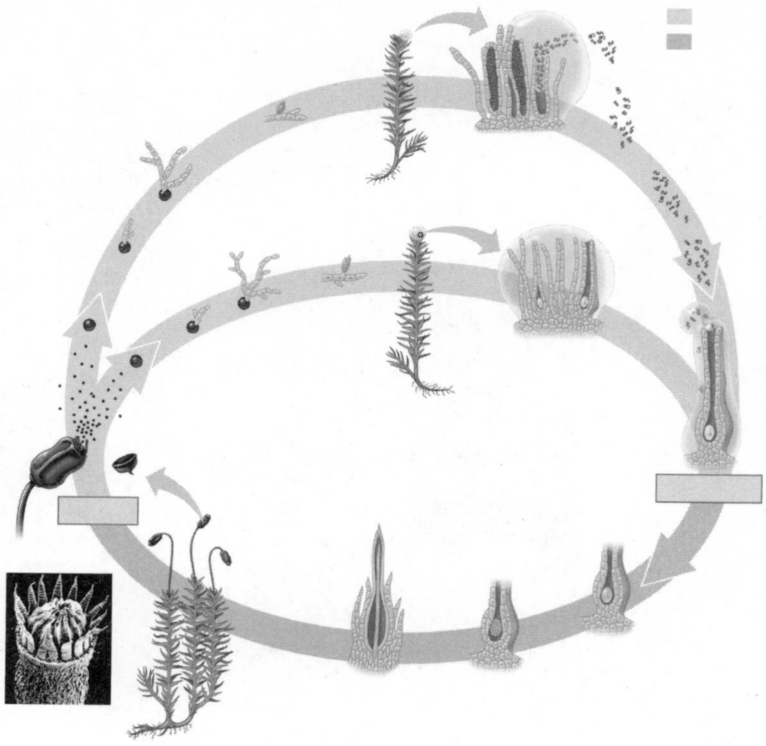

Figure 29.8 The life cycle of a *Polytrichum* moss, page 581

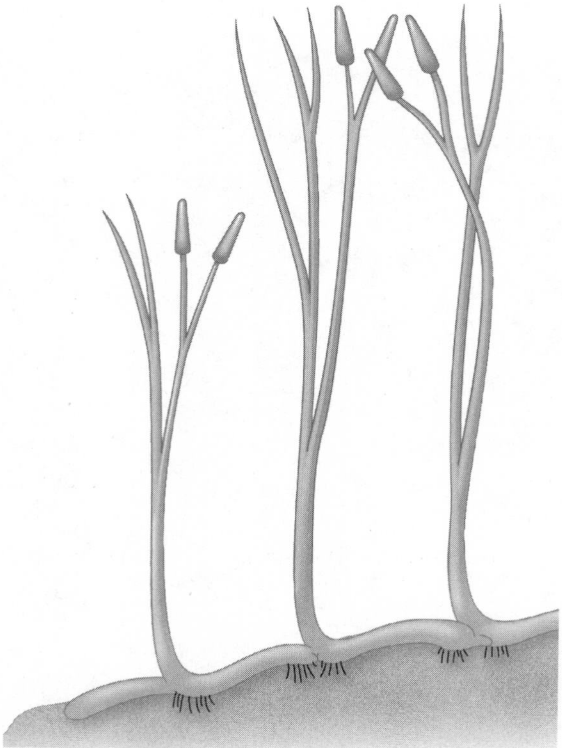

Figure 29.11 *Aglaophyton major*, an ancient relative of modern vascular plants, page 584

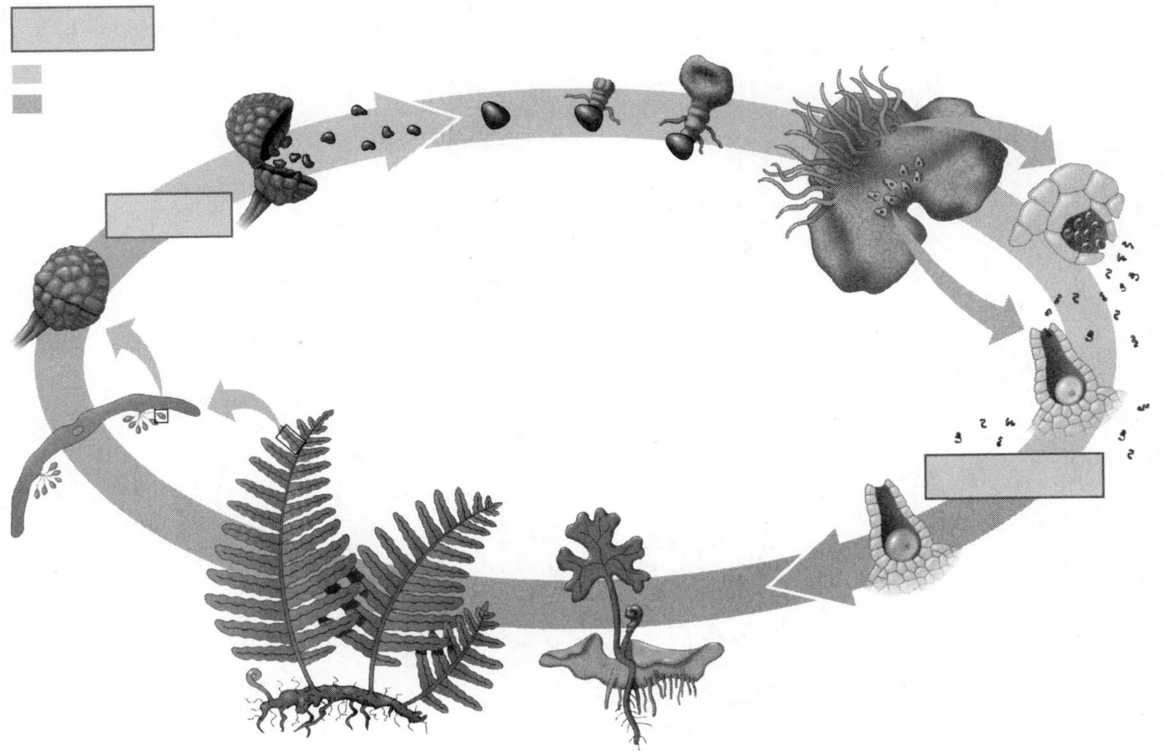

Figure 29.12 The life cycle of a fern, page 585

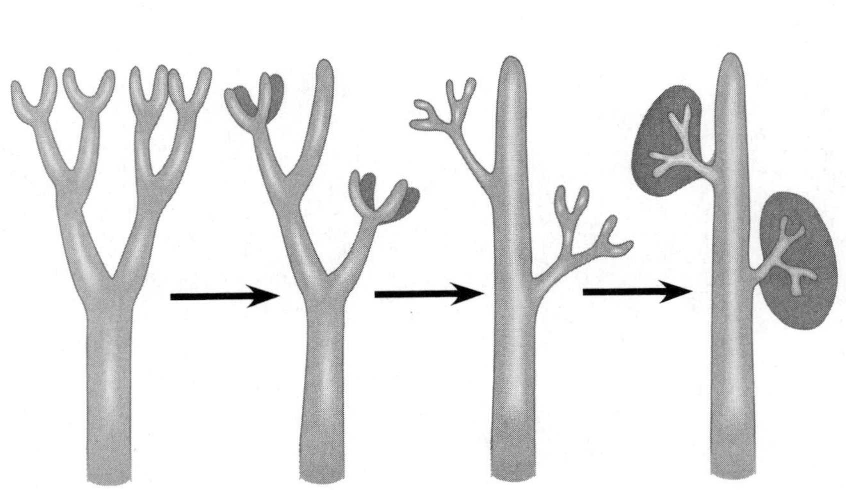

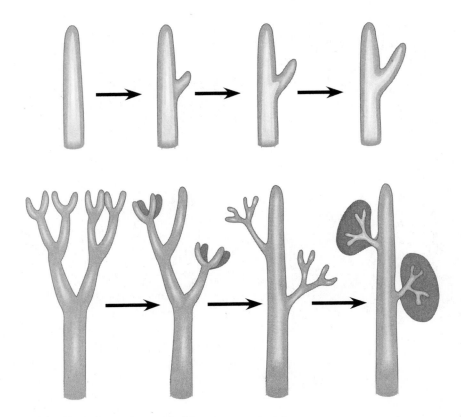

Figure 29.13 Hypotheses for the evolution of leaves, page 586

Figure 30.2 Gametophyte/sporophyte relationships, page 592

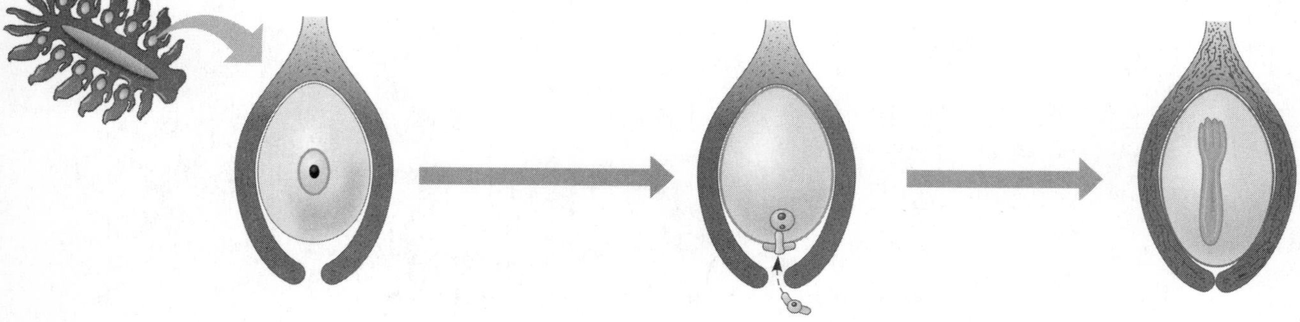

Figure 30.3 From ovule to seed, page 593

Figure 30.5 A progymnosperm, page 596

Figure 30.6 The life cycle of a pine, page 597

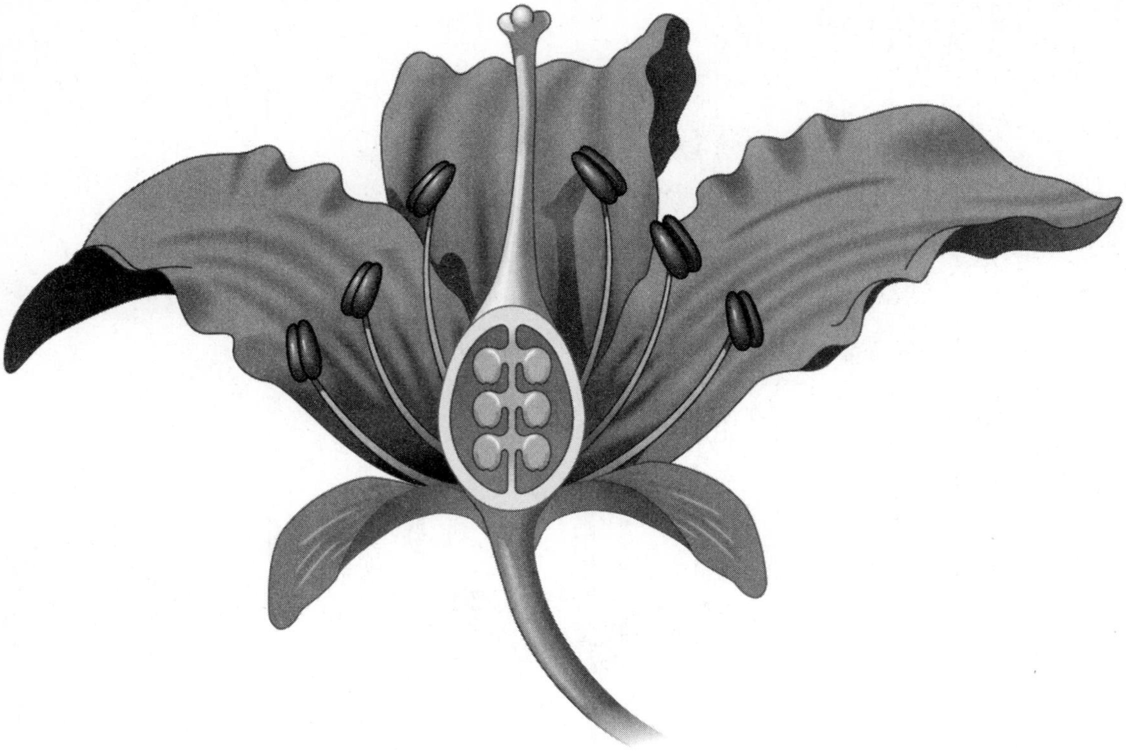

Figure 30.7 The structure of an idealized flower, page 598

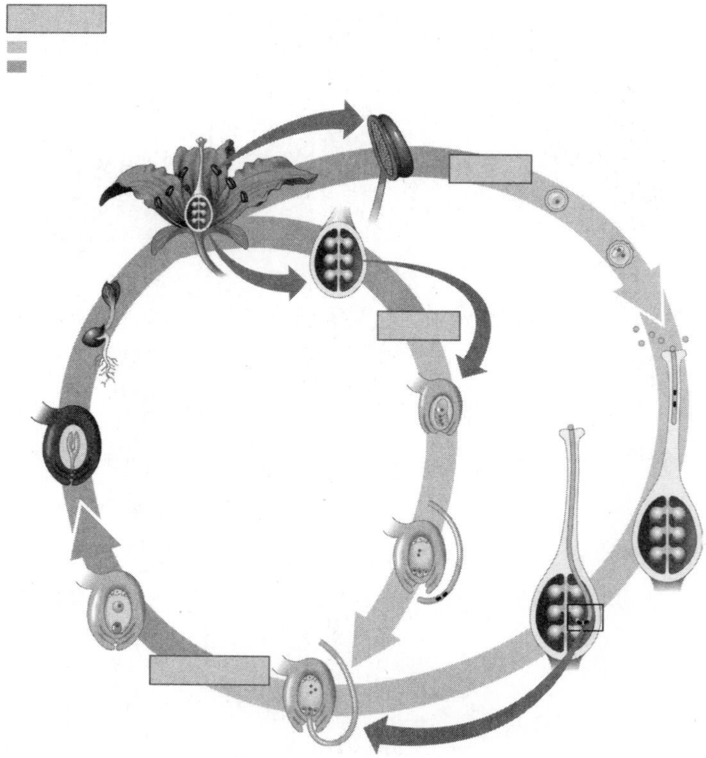

Figure 30.10 The life cycle of an angiosperm, page 600

Figure 30.11 A primitive flowering plant?, page 601

Figure 30.12 Angiosperm diversity, page 602

Figure 30.12 Angiosperm diversity, page 603

Figure 31.2 Structure of a multicellular fungus, page 609

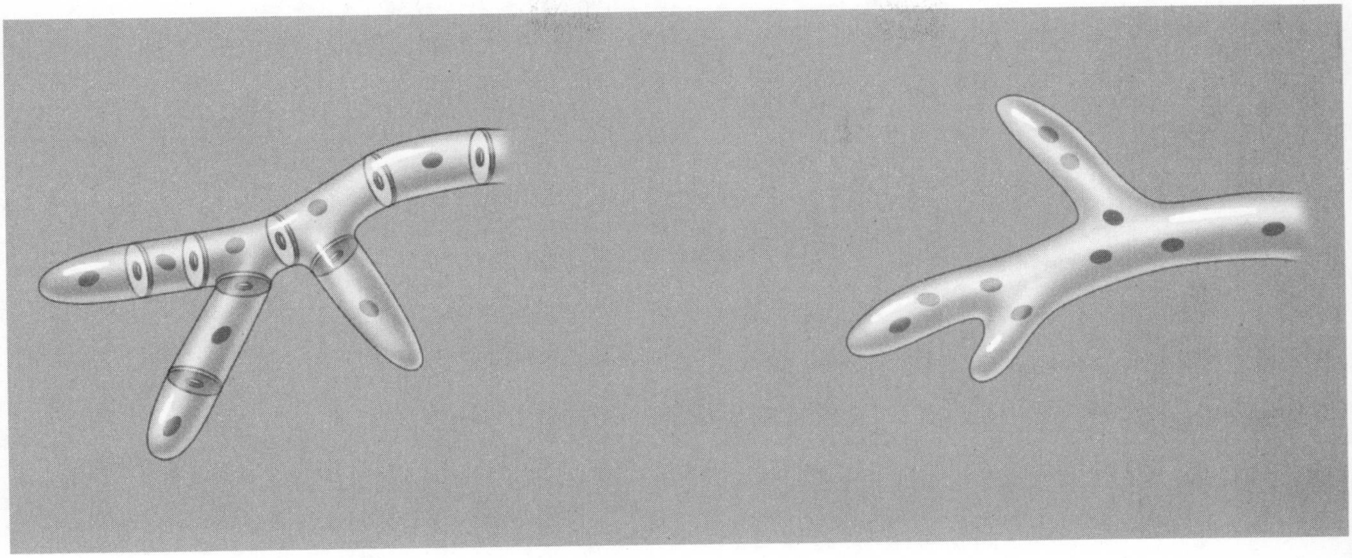

Figure 31.3 Structure of hyphae, page 609

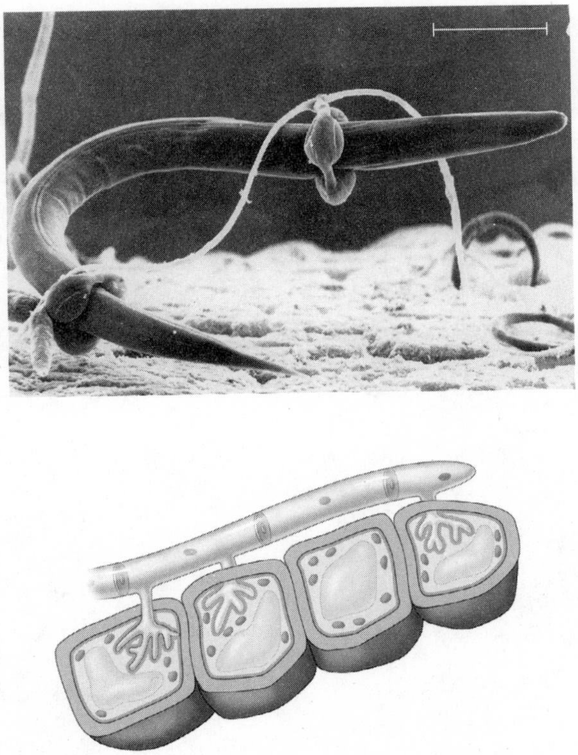

Figure 31.4 Specialized hyphae, page 610

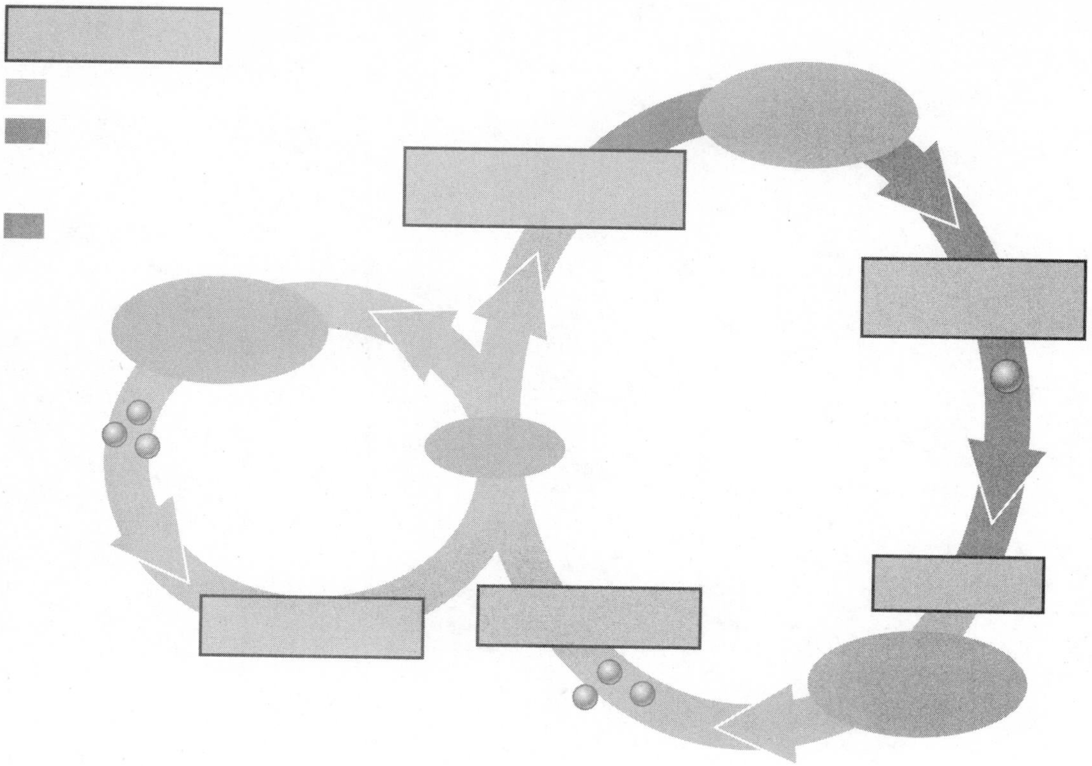

Figure 31.5 Generalized life cycle of fungi, page 611

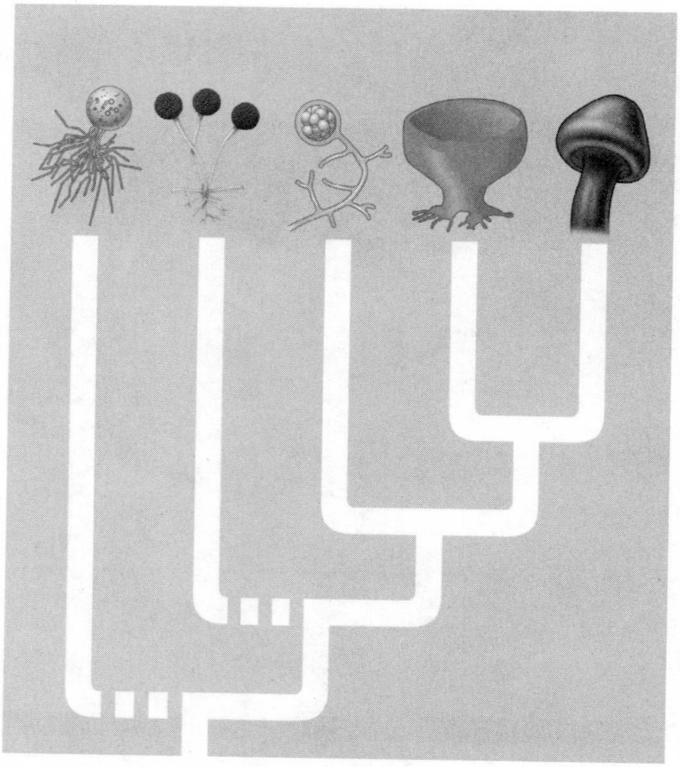

Figure 31.9 Phylogeny of fungi, page 613

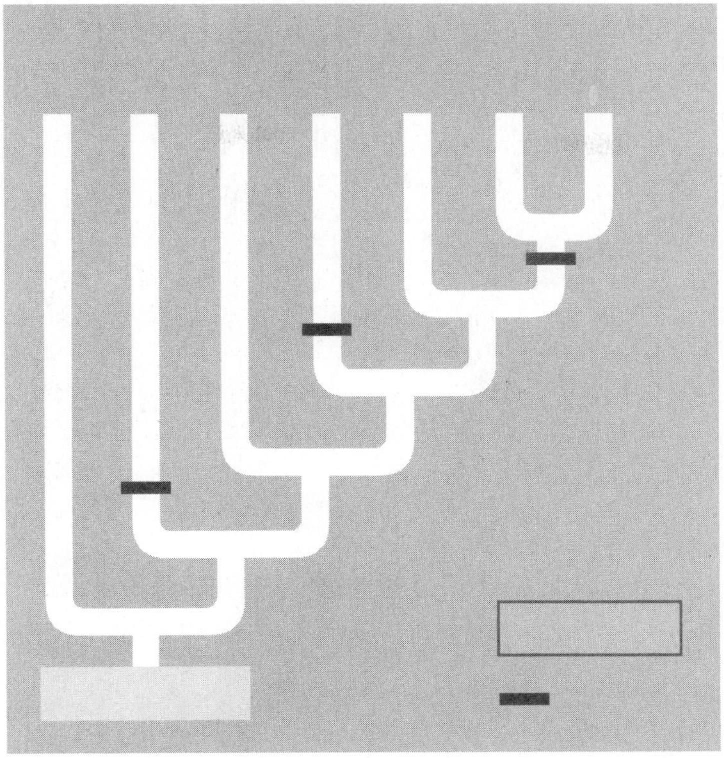

Figure 31.11 Multiple evolutionary losses of flagella, page 613

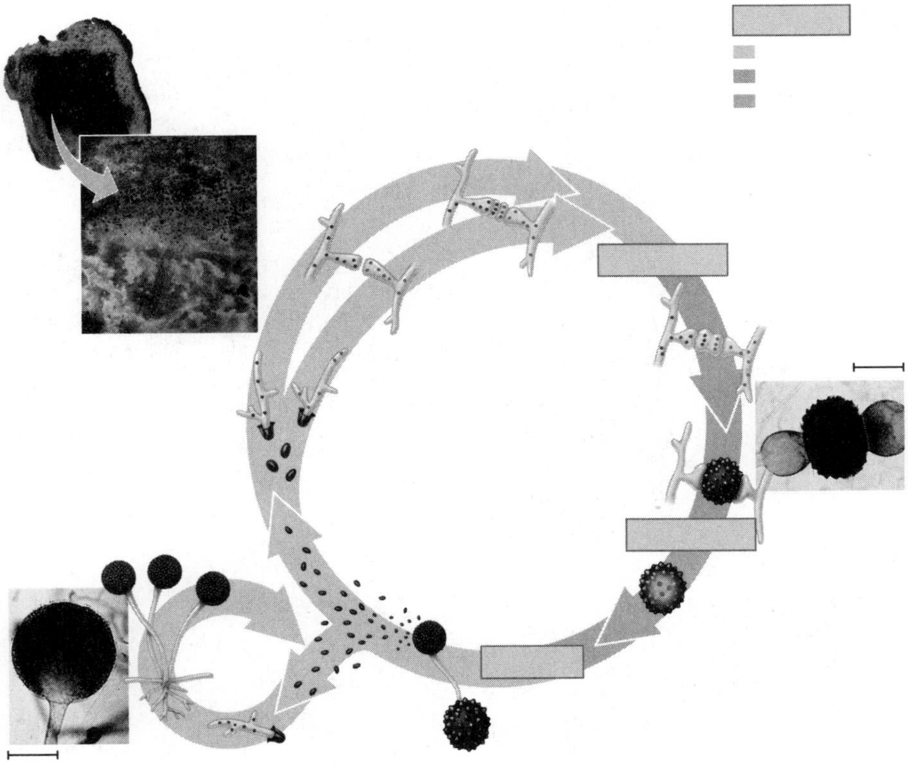

Figure 31.12 The life cycle of the zygomycete *Rhizopus stolonifer* (black bread mold), page 614

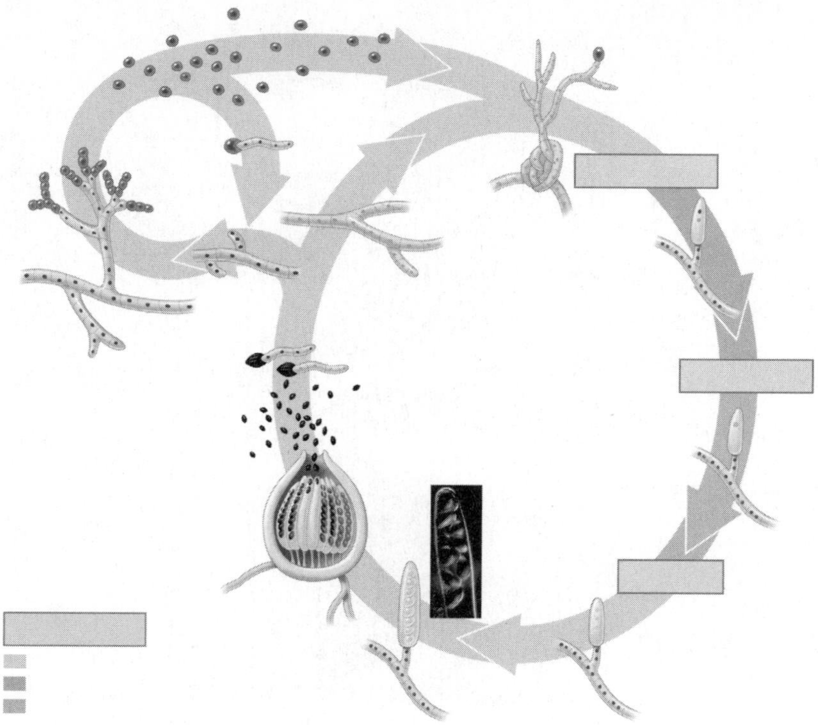

Figure 31.17 The life cycle of *Neurospora crassa*, an ascomycete, page 617

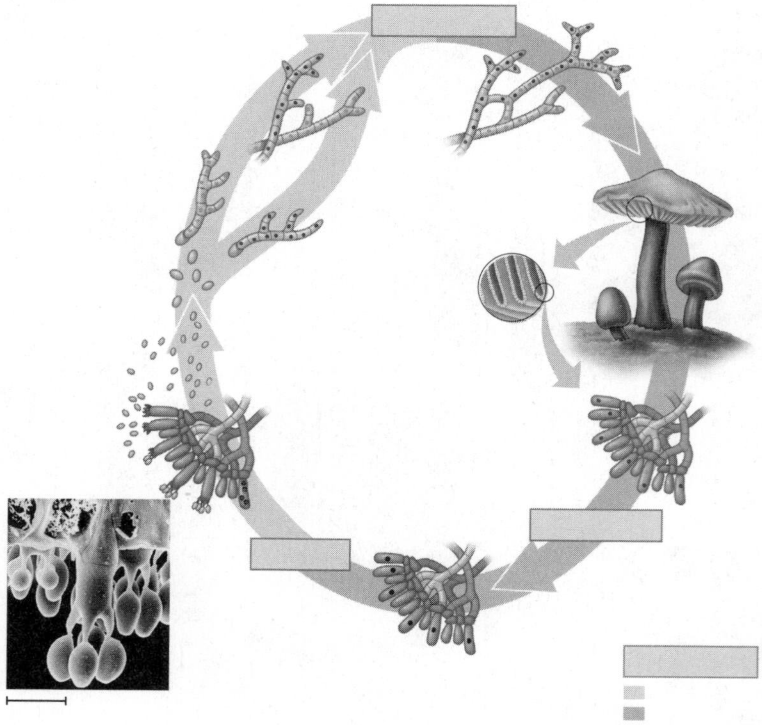

Figure 31.20 The life cycle of a mushroom-forming basidiomycete, page 619

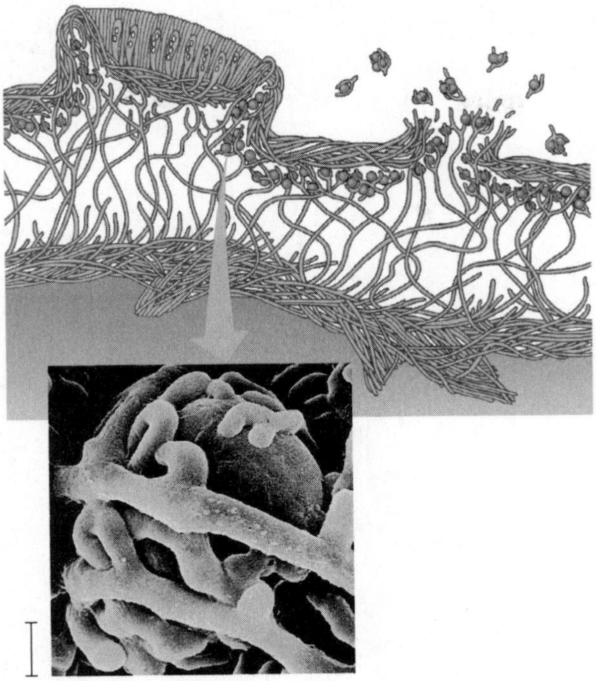

Figure 31.24 Anatomy of an ascomycete lichen, page 621

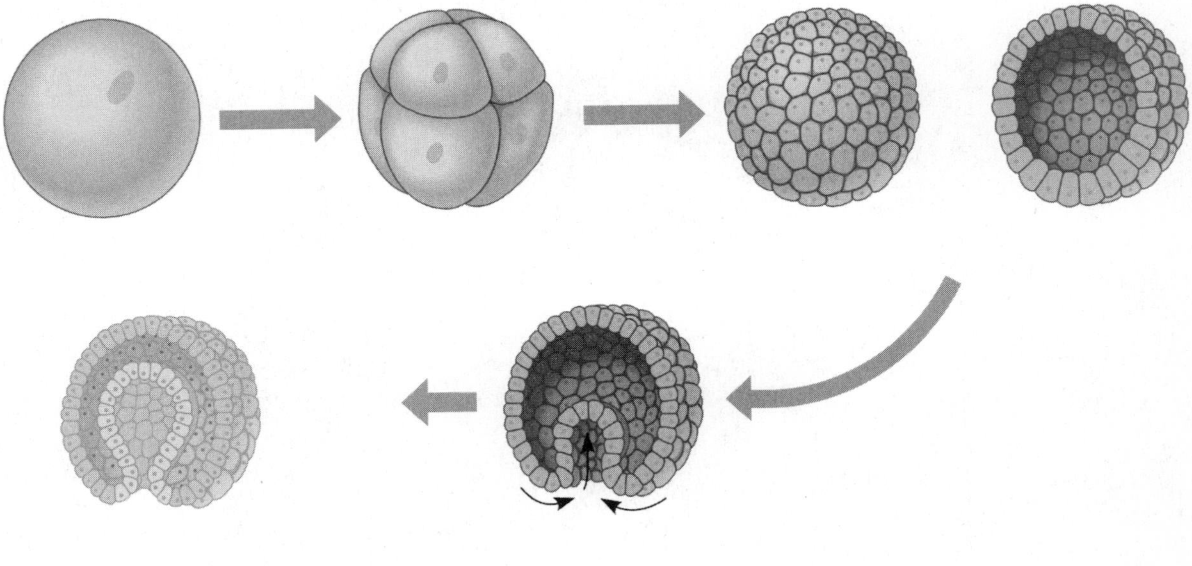

Figure 32.2 Early embryonic development in animals, page 627

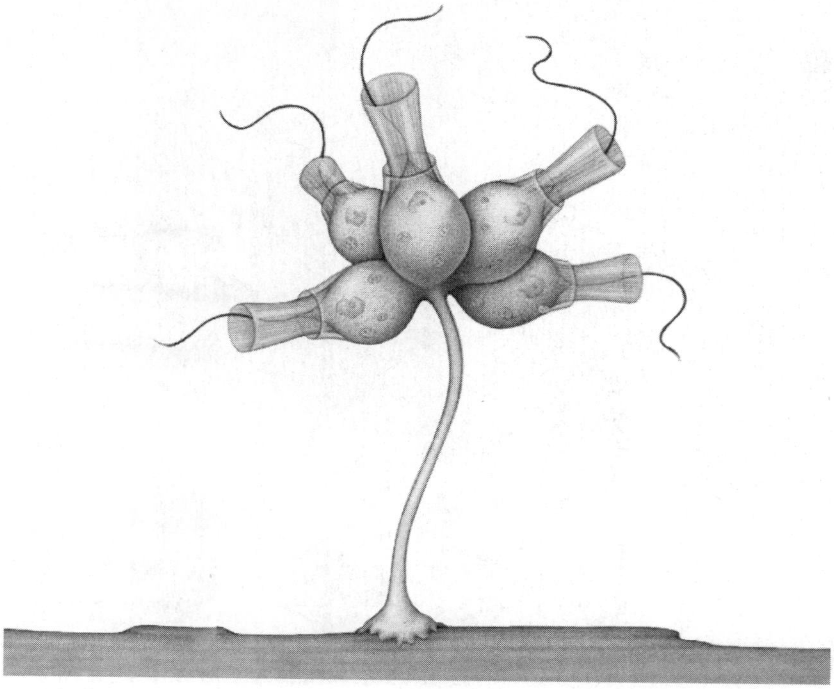

Figure 32.3 A choanoflagellate colony, page 628

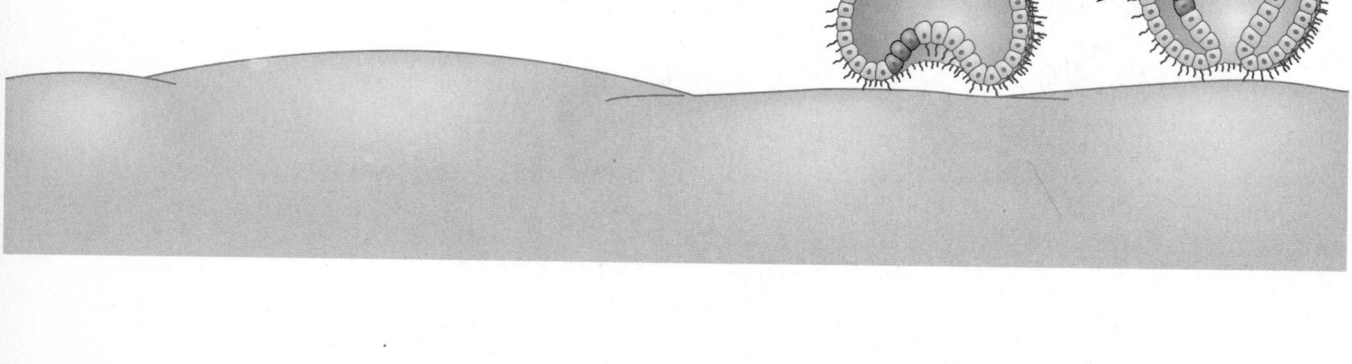

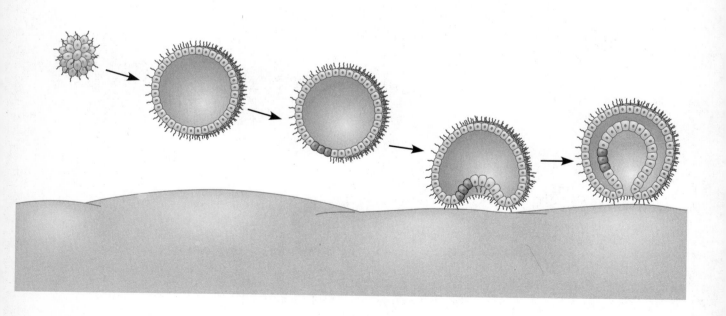

Figure 32.4 One hypothesis for the origin of animals from a flagellated protist, page 628

Figure 32.6 A Cambrian seascape, page 629

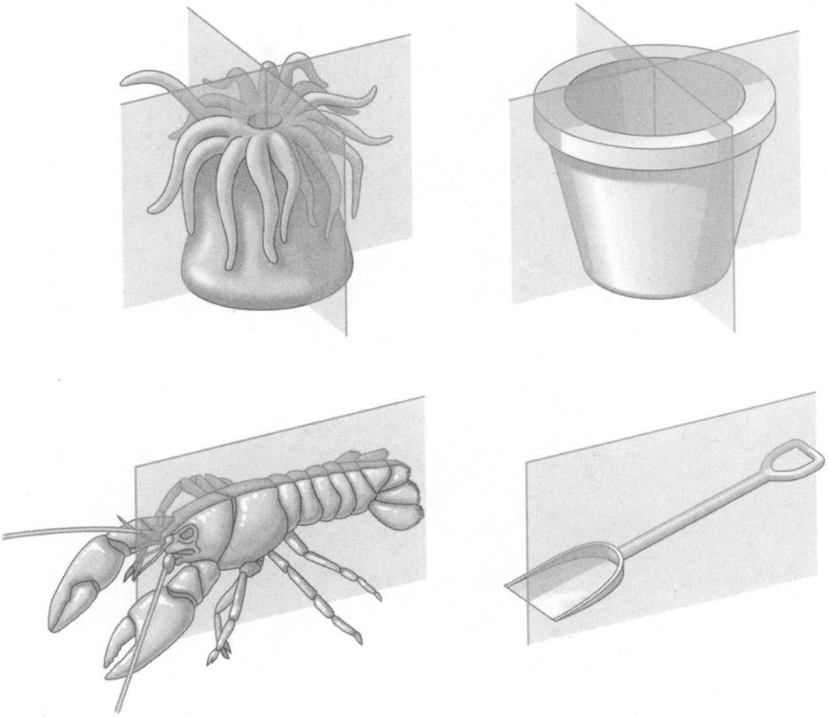

Figure 32.7 Body symmetry, page 630

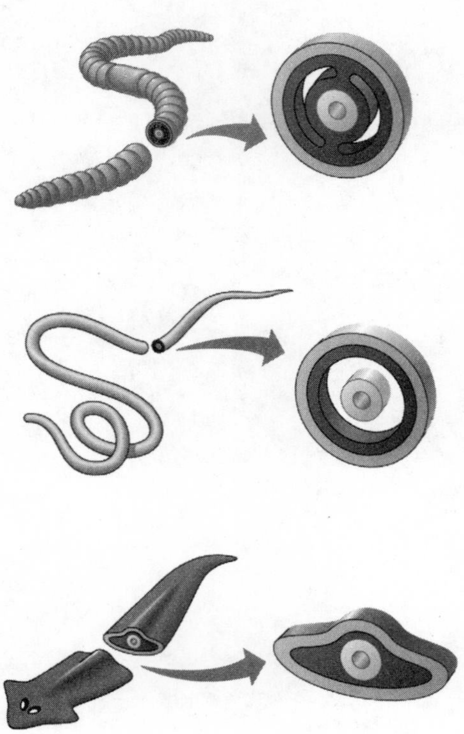

Figure 32.8 Body plans of triploblastic animals, page 631

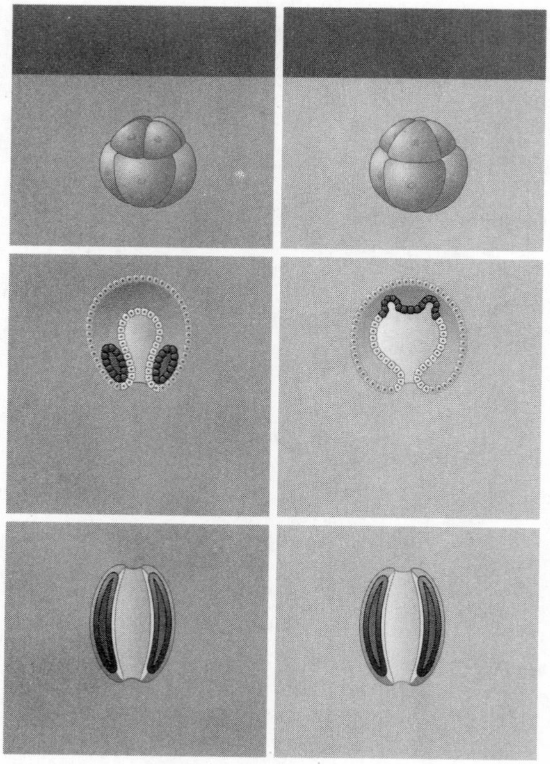

Figure 32.9 A comparison of protostome and deuterostome development, page 632

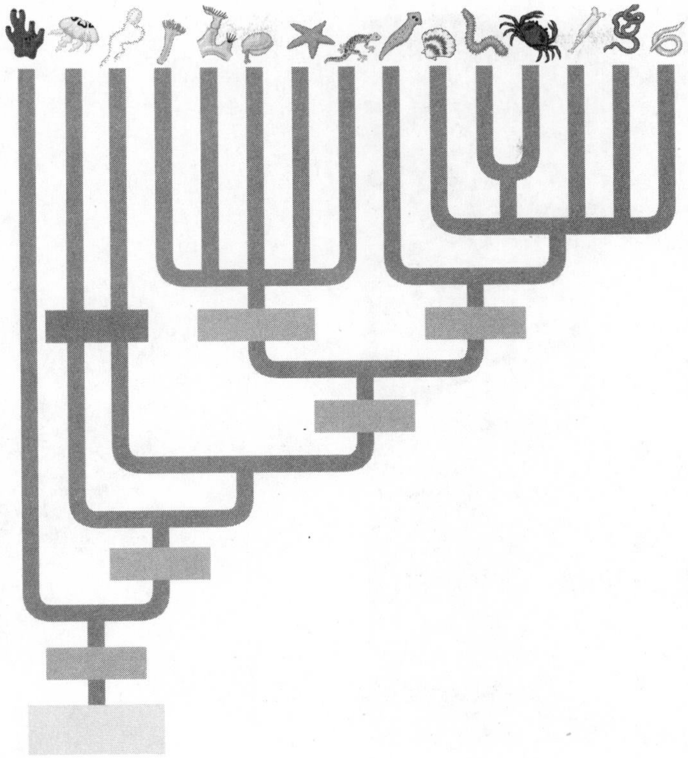

Figure 32.10 One hypothesis of animal phylogeny based mainly on morphological and developmental comparisons, page 634

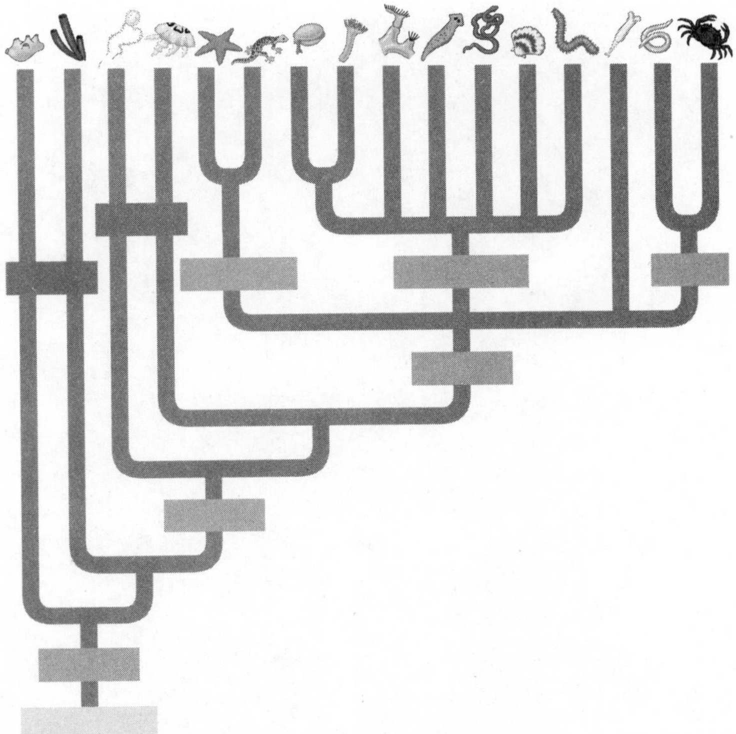

Figure 32.11 One hypothesis of animal phylogeny based mainly on molecular data, page 635

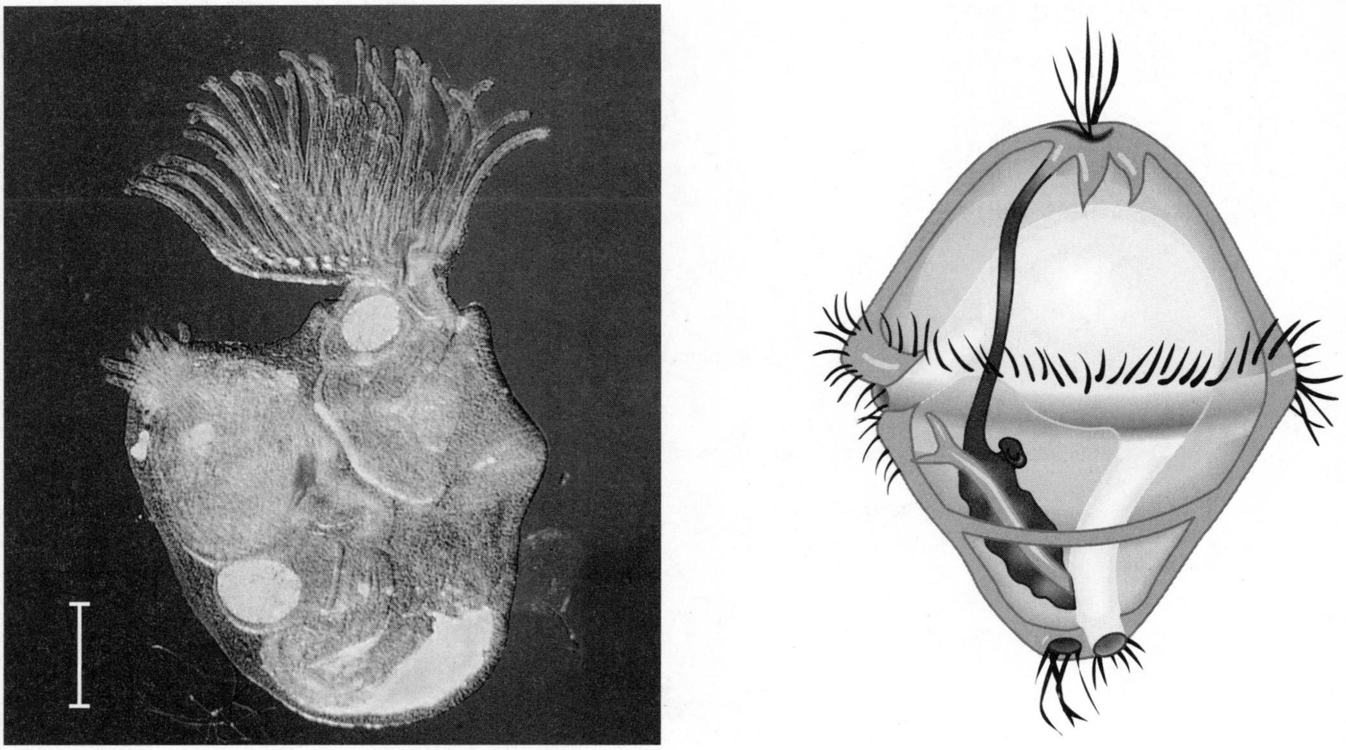

Figure 32.13 Characteristics of lophotrochozoans, page 635

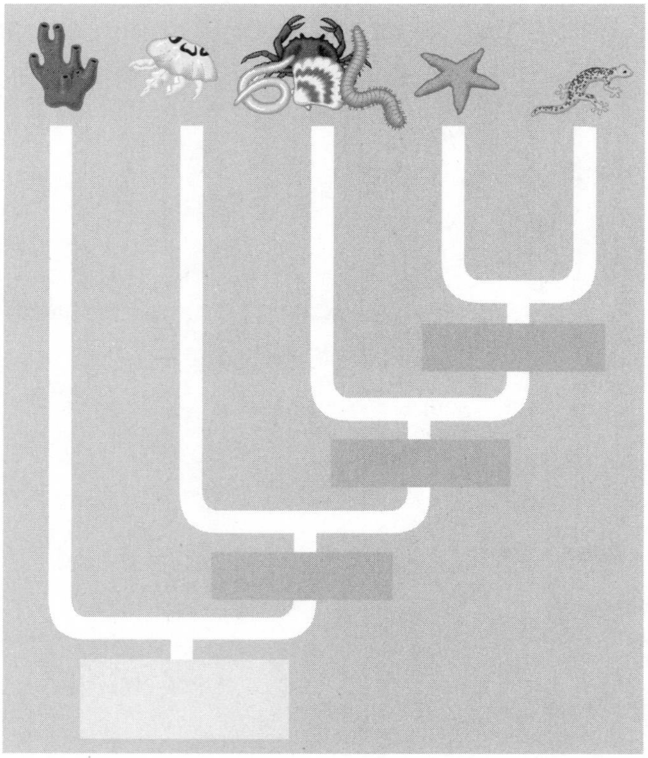

Figure 33.2 Review of animal phylogeny, page 638

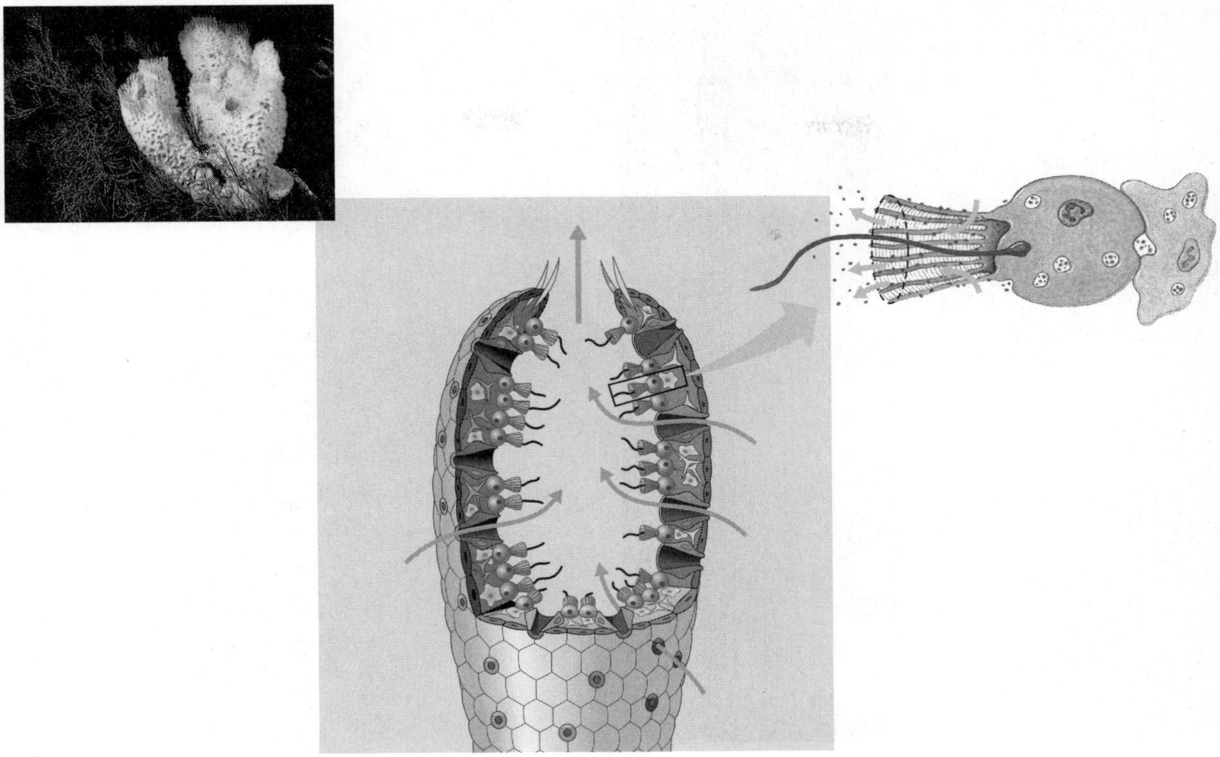

Figure 33.4 Anatomy of a sponge, page 642

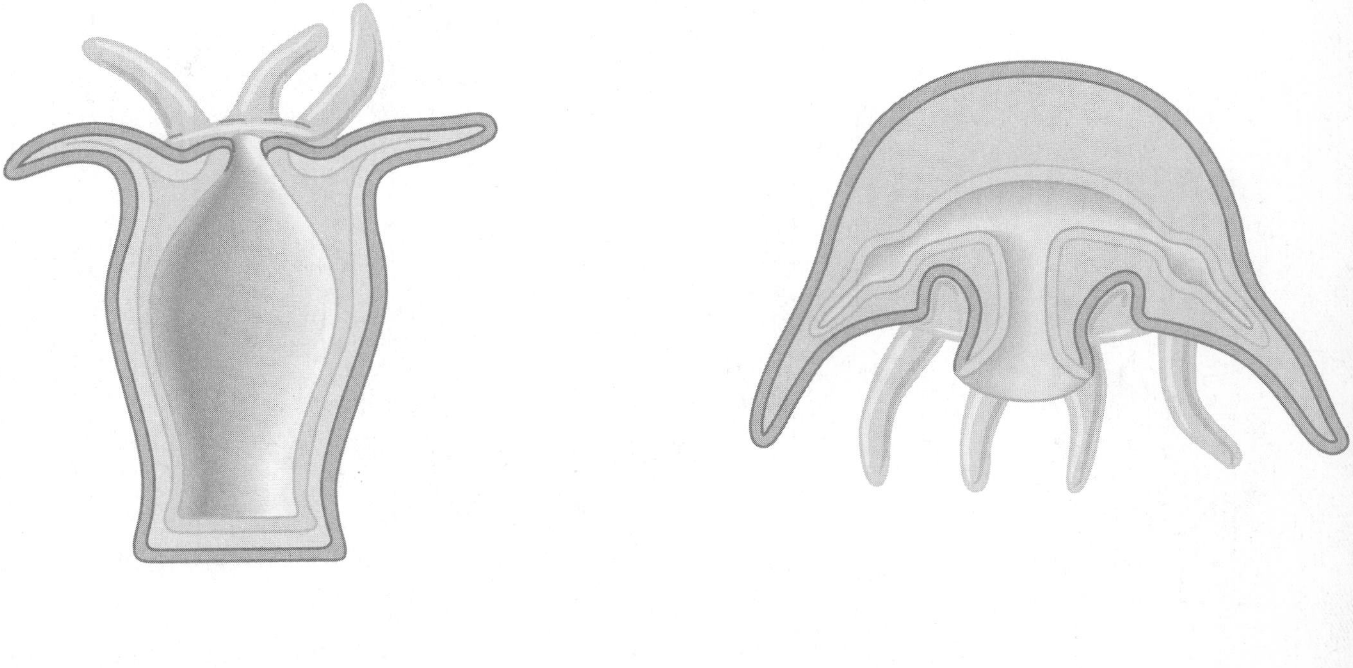

Figure 33.5 Polyp and medusa forms of cnidarians, page 643

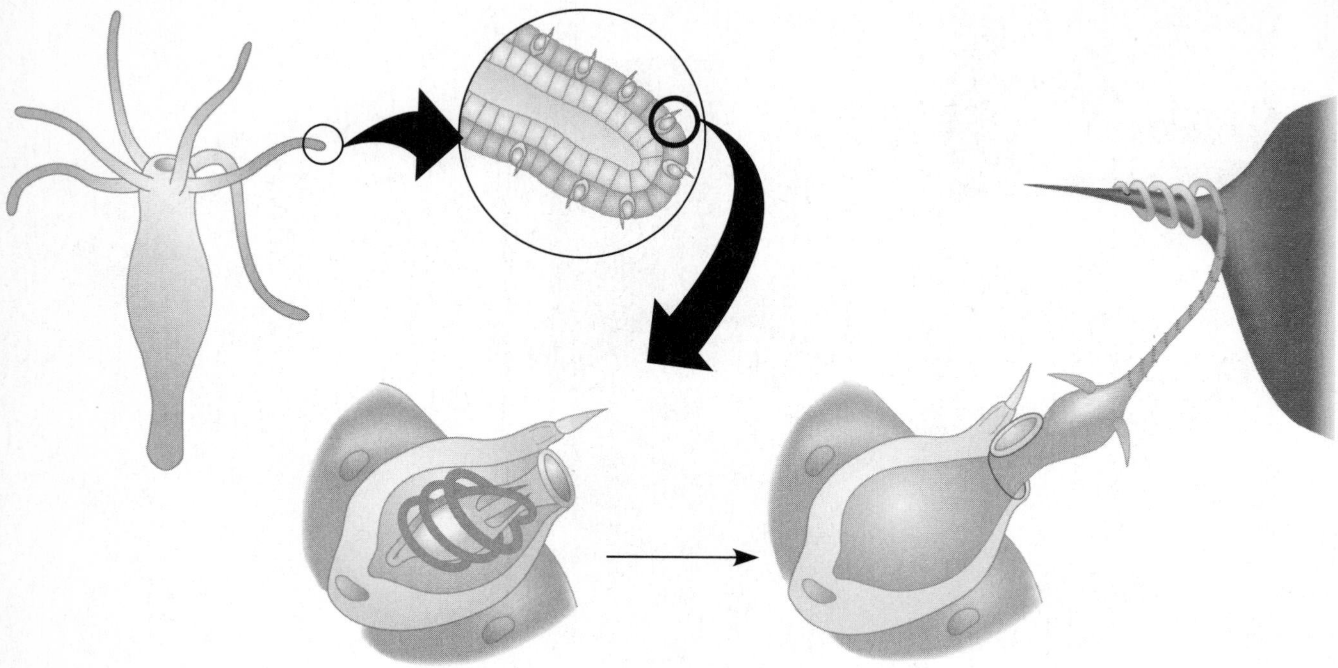

Figure 33.6 A cnidocyte of a hydra, page 643

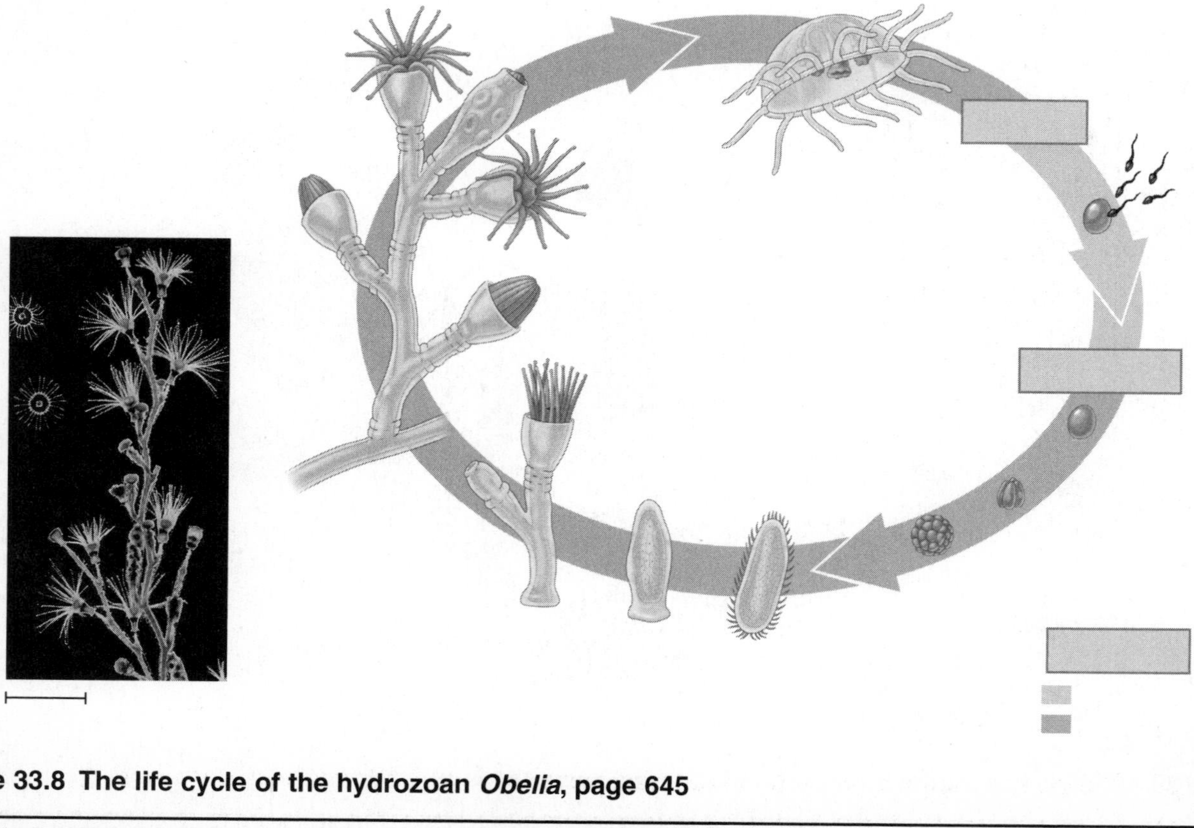

Figure 33.8 The life cycle of the hydrozoan *Obelia*, page 645

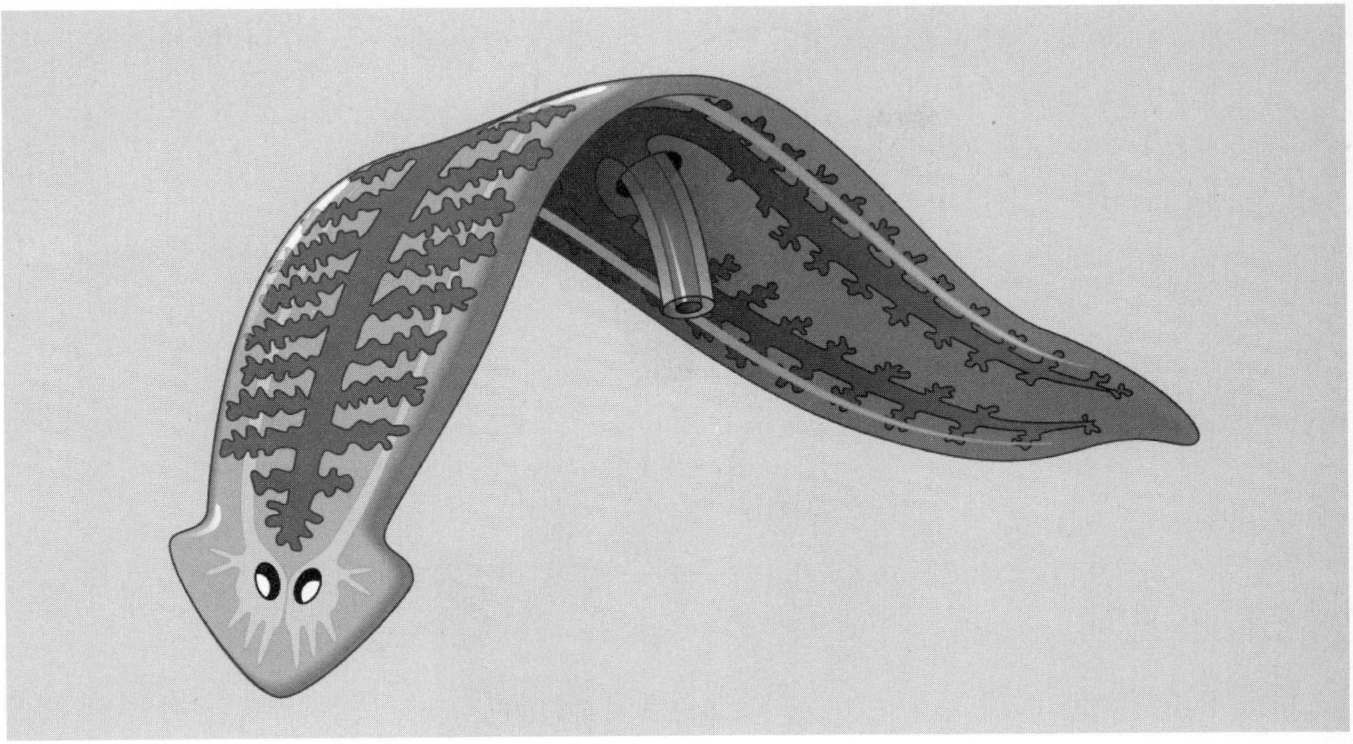

Figure 33.10 Anatomy of a planarian, a turbellarian, page 647

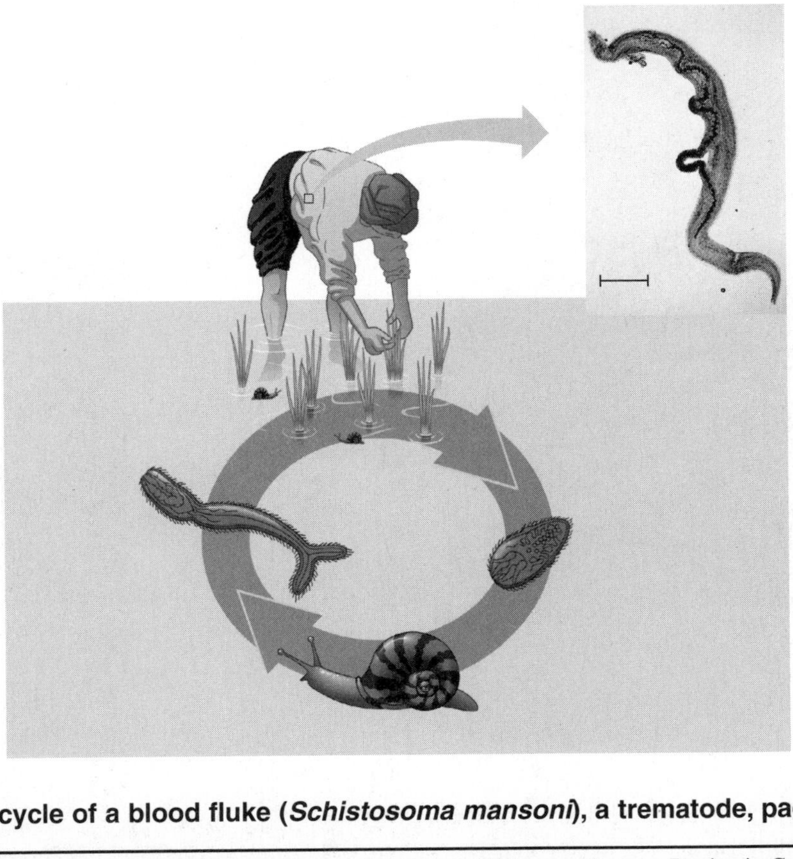

Figure 33.11 The life cycle of a blood fluke (*Schistosoma mansoni*), a trematode, page 647

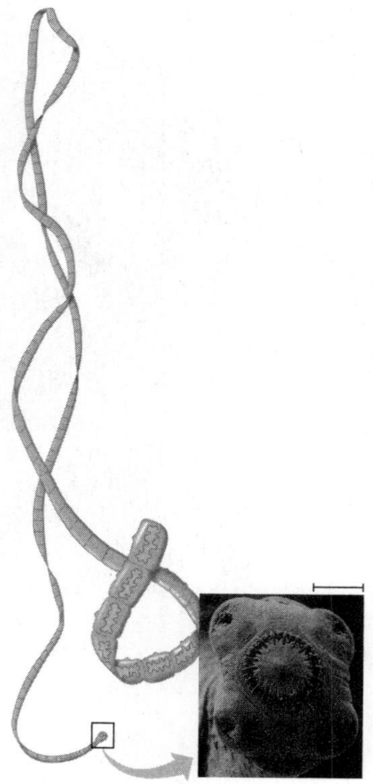

Figure 33.12 Anatomy of a tapeworm, page 648

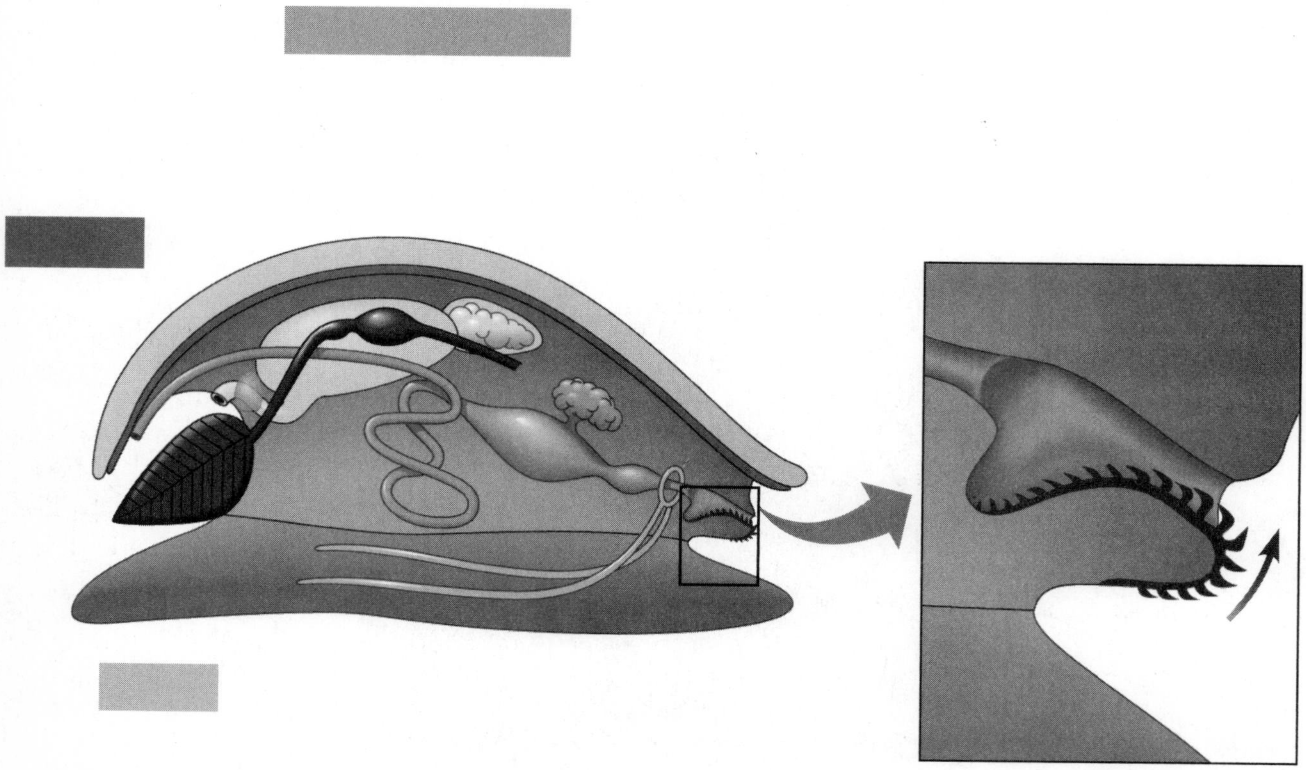

Figure 33.16 The basic body plan of a mollusc, page 650

Figure 33.19 The results of torsion in a gastropod, page 651

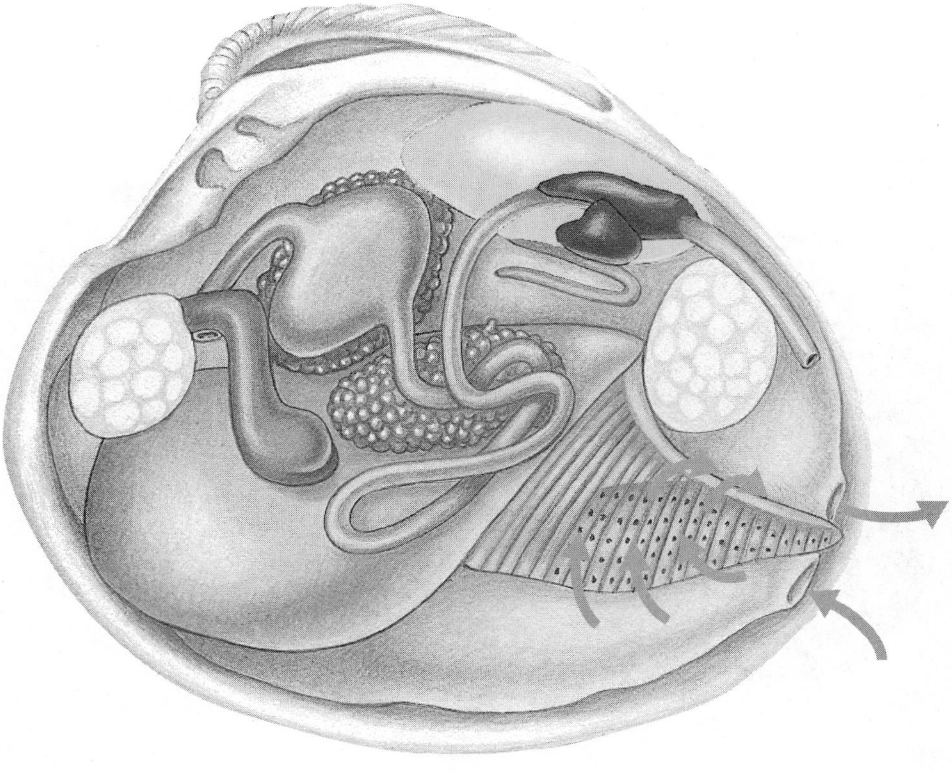

Figure 33.21 Anatomy of a clam, page 652

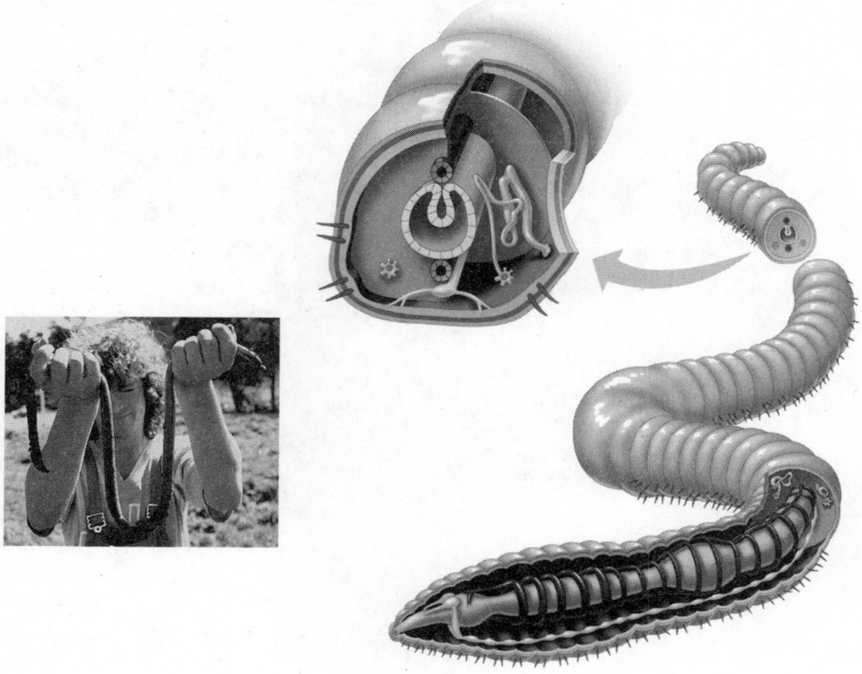

Figure 33.23 Anatomy of an earthworm, page 654

Figure 33.29 External anatomy of an arthropod, page 657

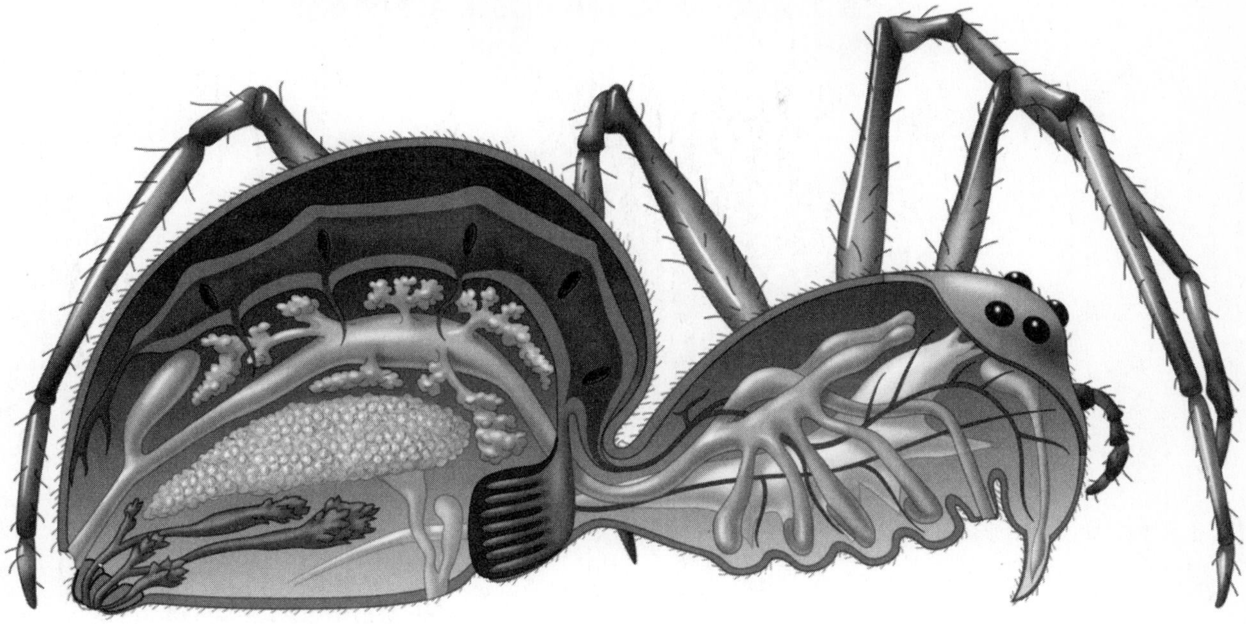

Figure 33.32 Anatomy of a spider, page 659

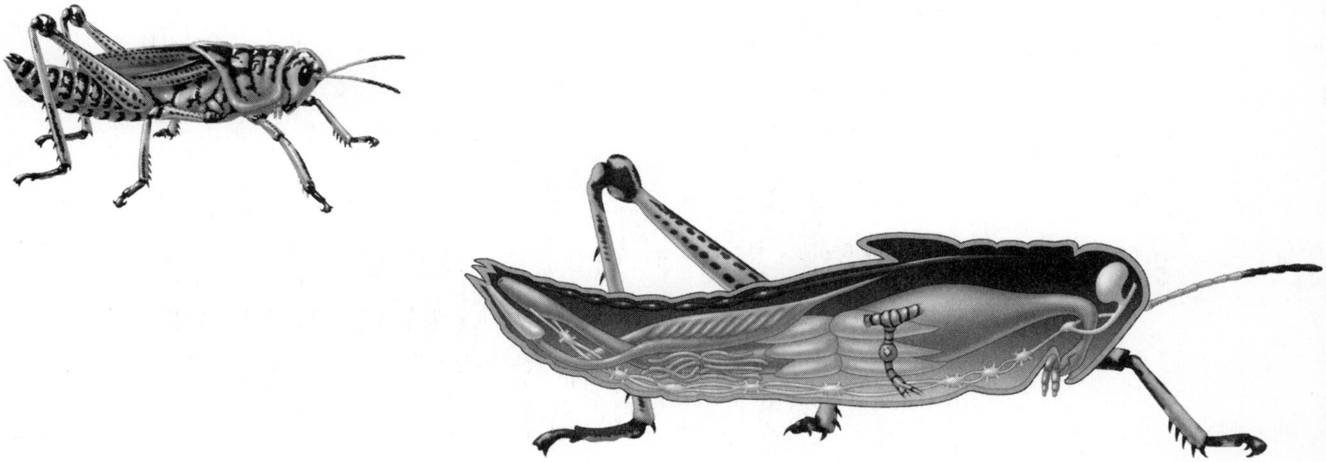

Figure 33.35 Anatomy of a grasshopper, an insect, page 660

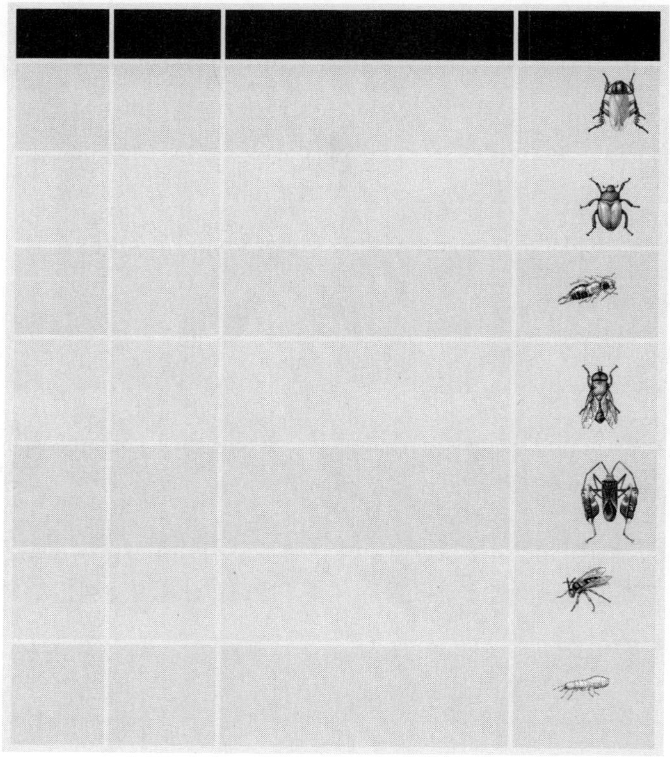

Figure 33.37 Insect diversity, page 662

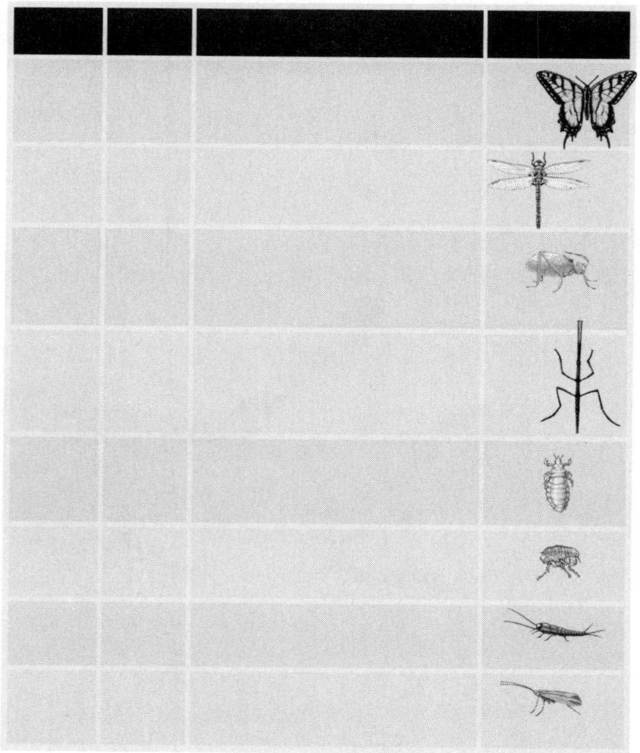

Figure 33.37 Insect diversity, page 663

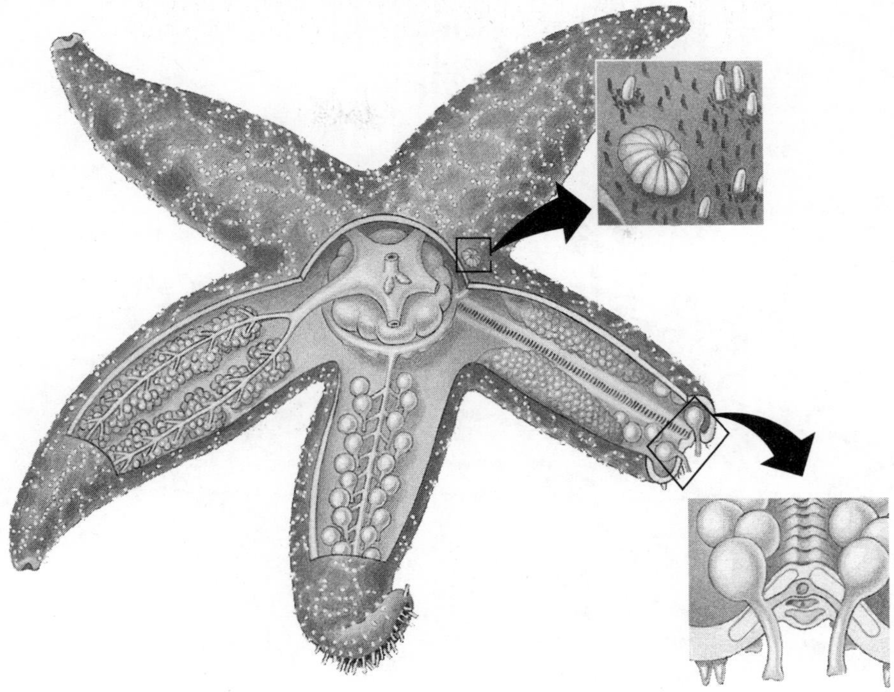

Figure 33.39 Anatomy of a sea star, an echinoderm, page 665

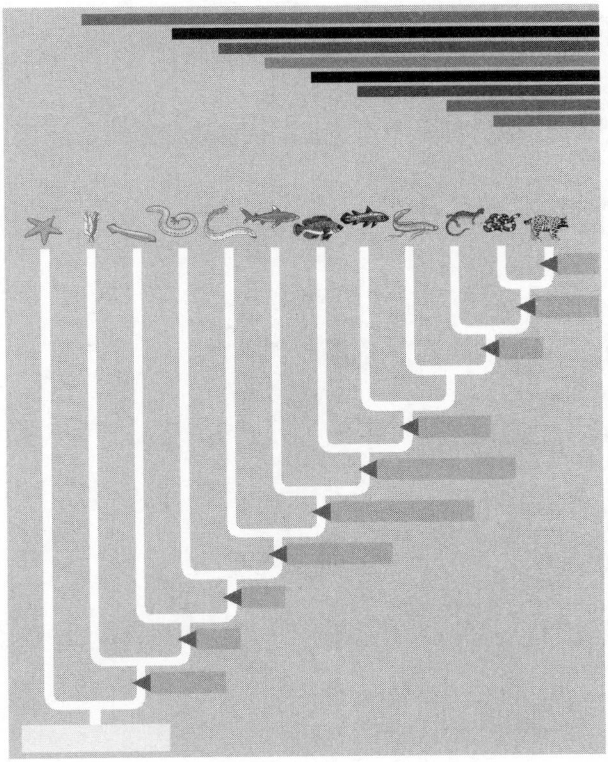

Figure 34.2 Hypothetical phylogeny of chordates, page 672

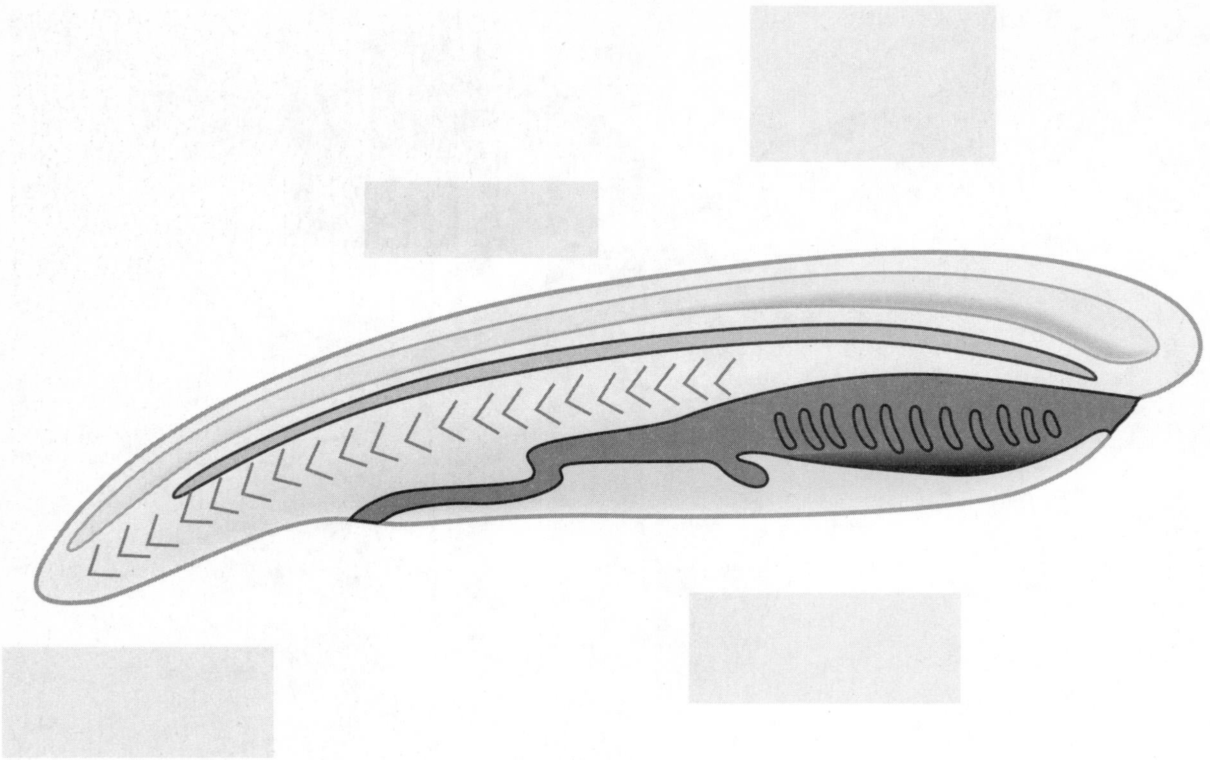

Figure 34.3 Chordate characteristics, page 673

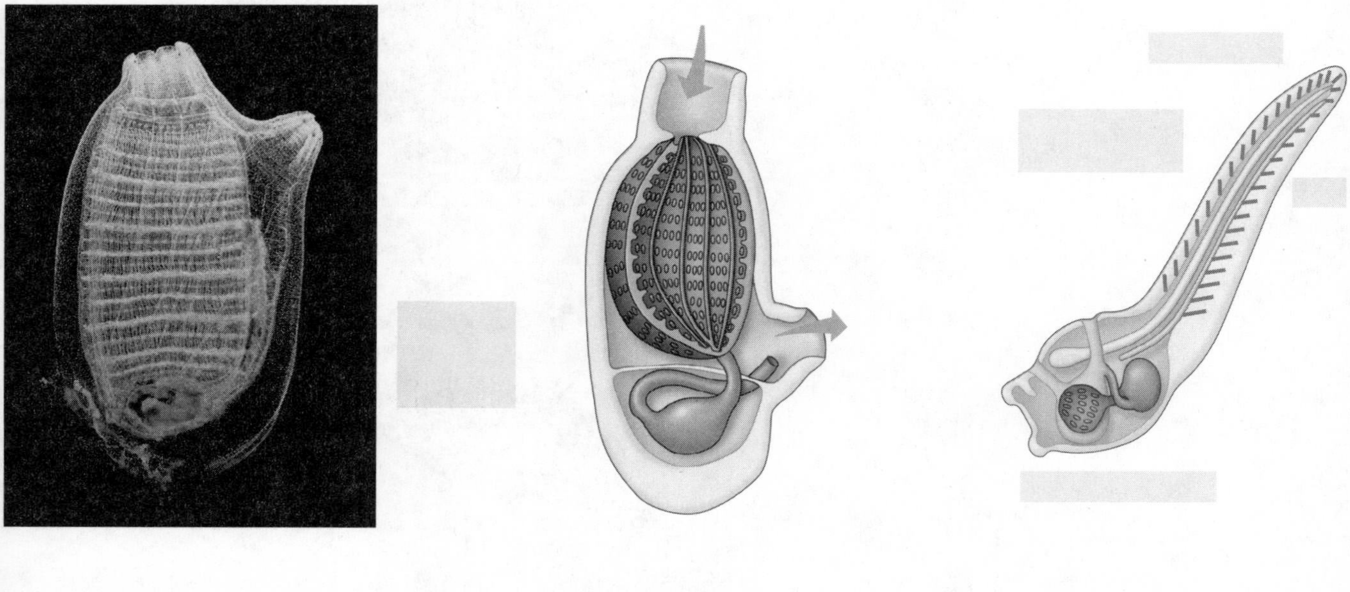

Figure 34.4 A tunicate, a urochordate, page 674

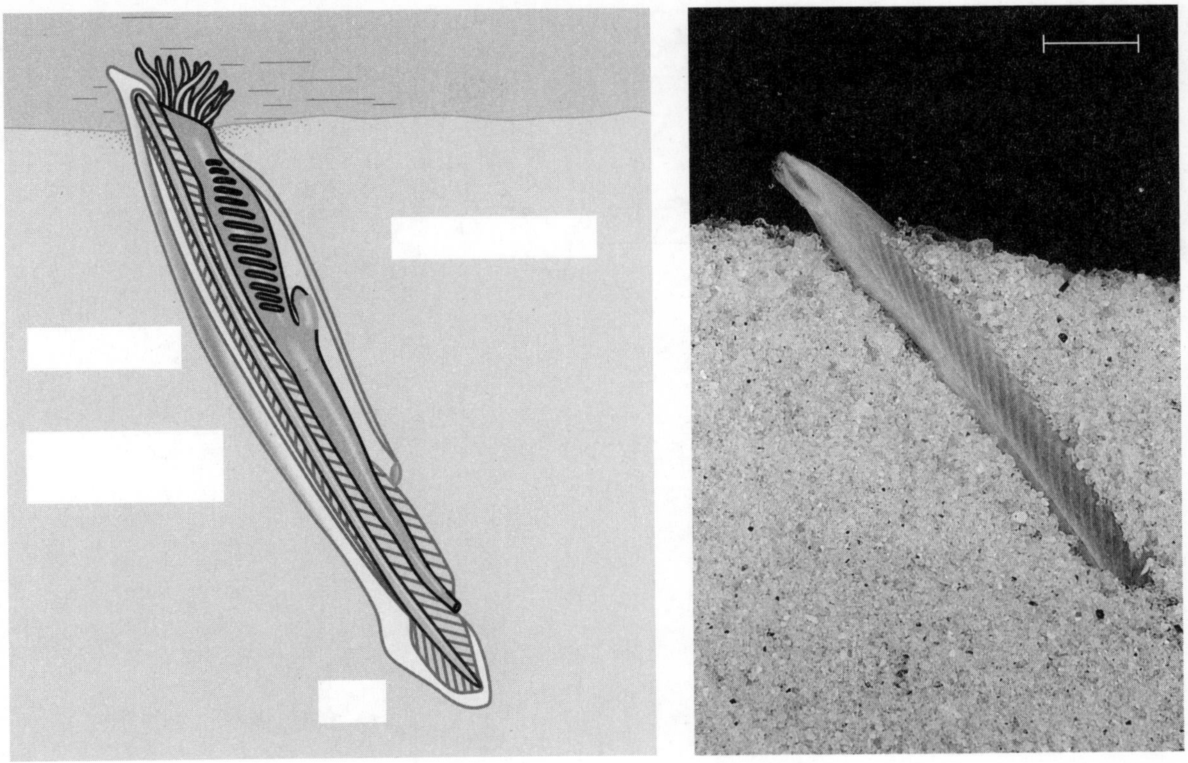

Figure 34.5 The lancelet *Branchiostoma*, a cephalochordate, page 675

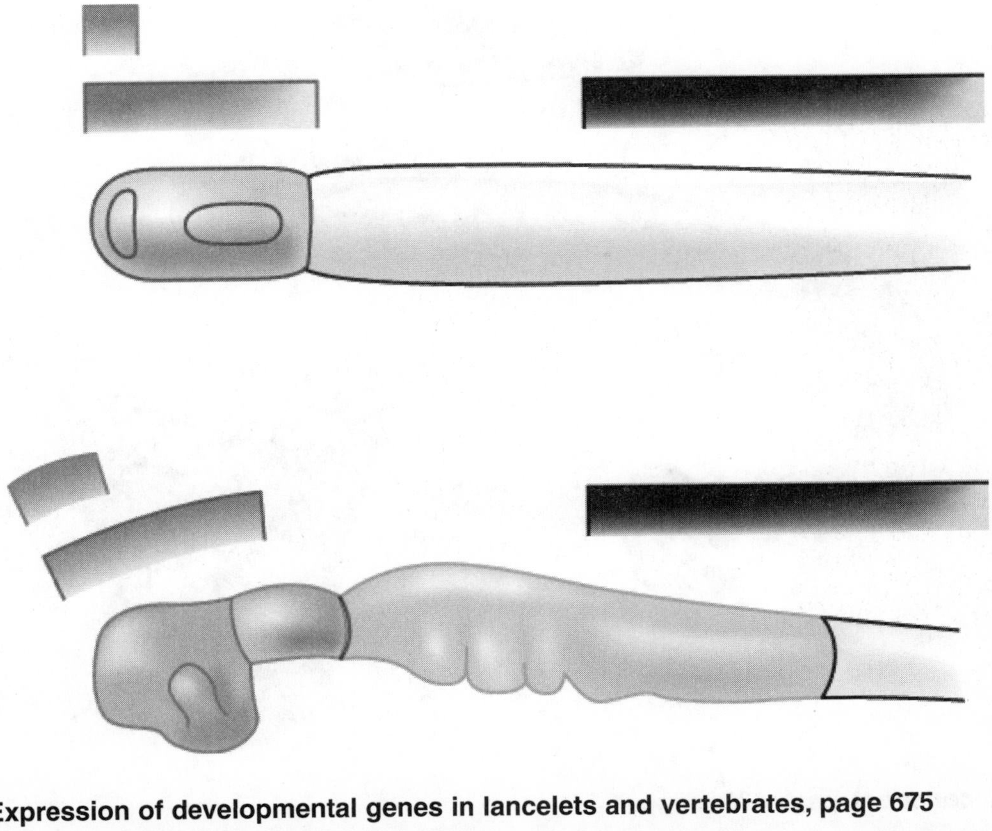

Figure 34.6 Expression of developmental genes in lancelets and vertebrates, page 675

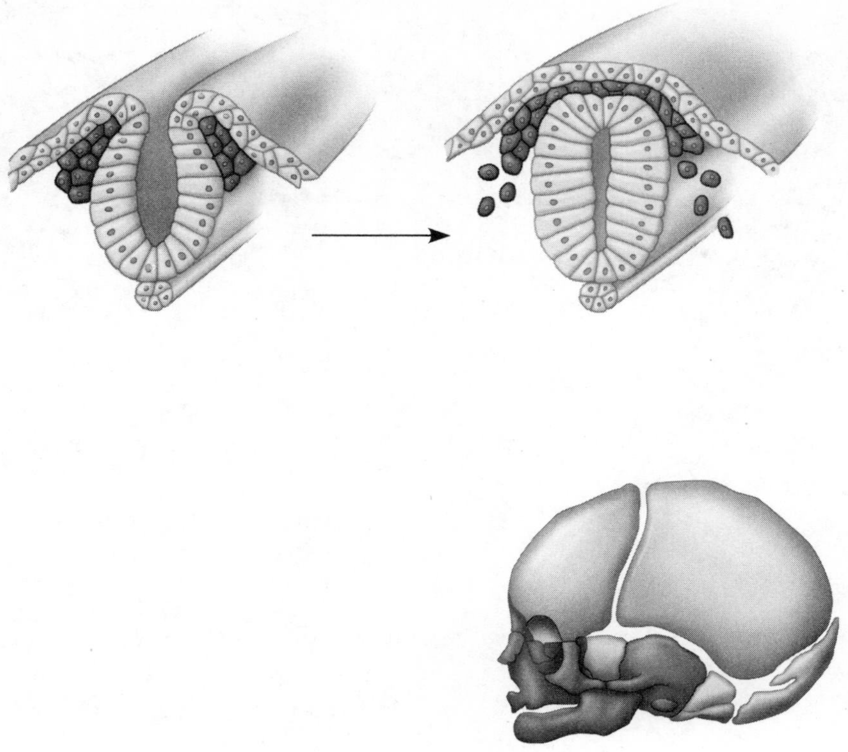

Figure 34.7 The neural crest, embryonic source of many unique vertebrate characters, page 676

Figure 34.11 A conodont, page 679

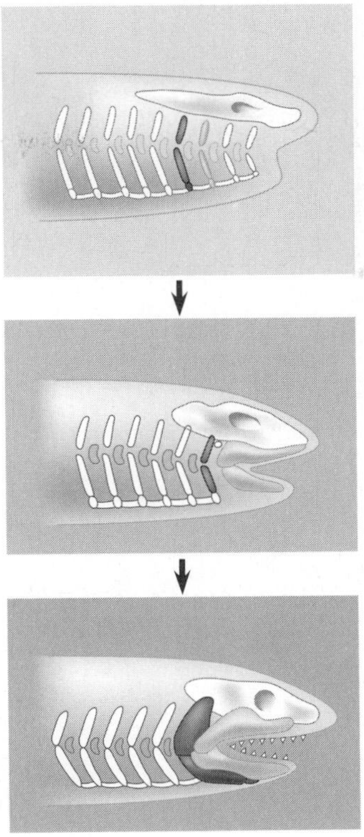

Figure 34.13 Hypothesis for the evolution of vertebrate jaws, page 680

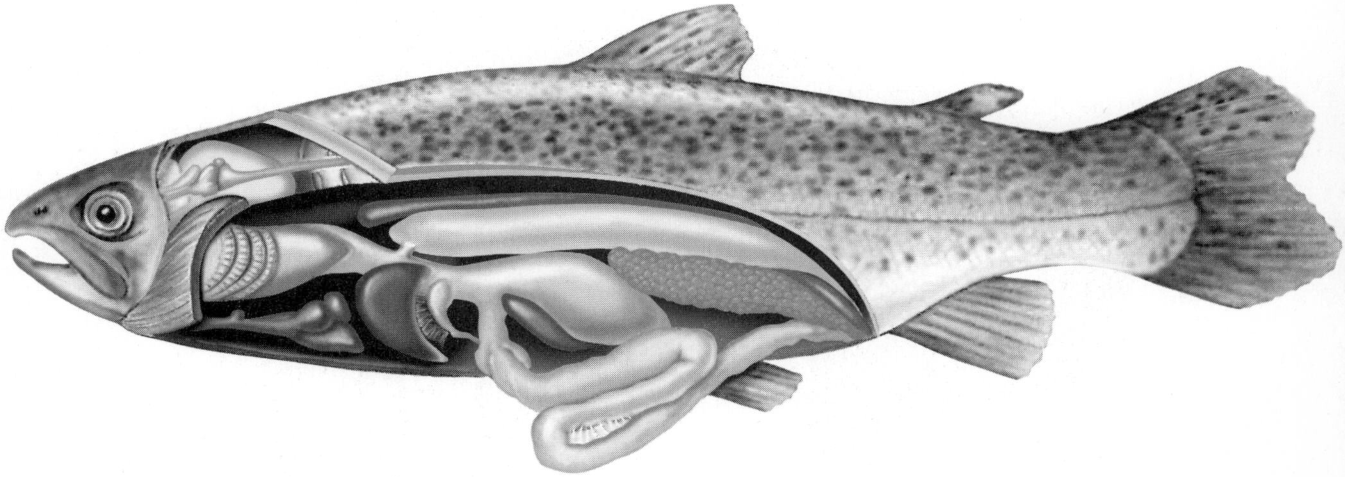

Figure 34.16 Anatomy of a trout, an aquatic osteichthyan, page 682

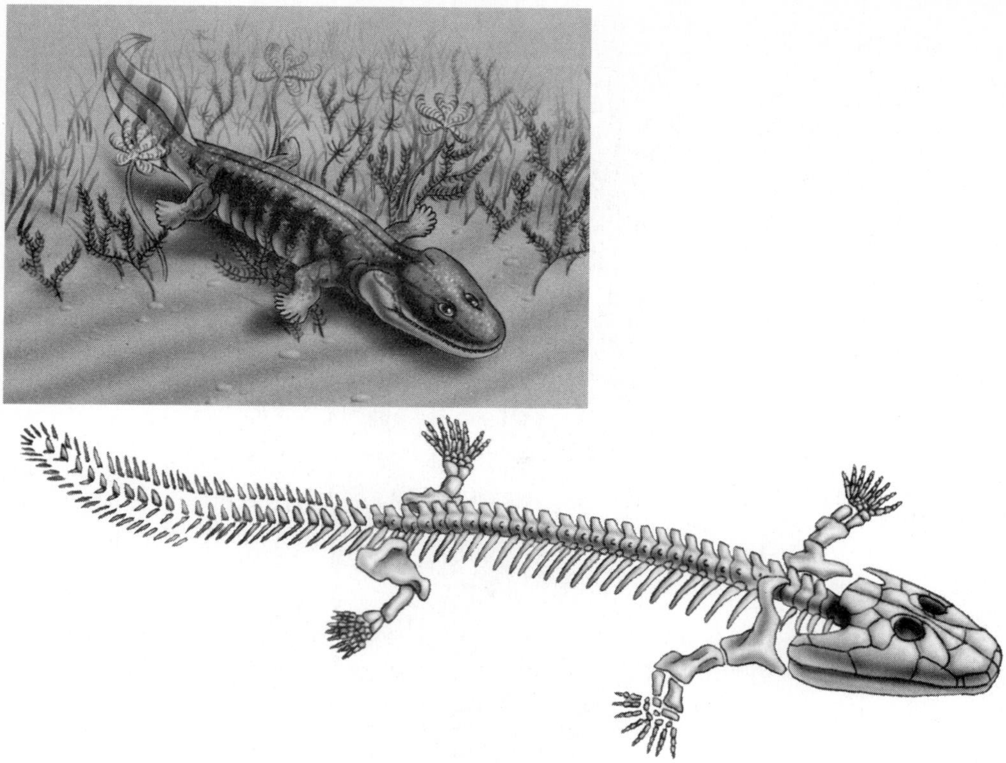

Figure 34.19 ***Acanthostega*, a Devonian relative of tetrapods, page 684**

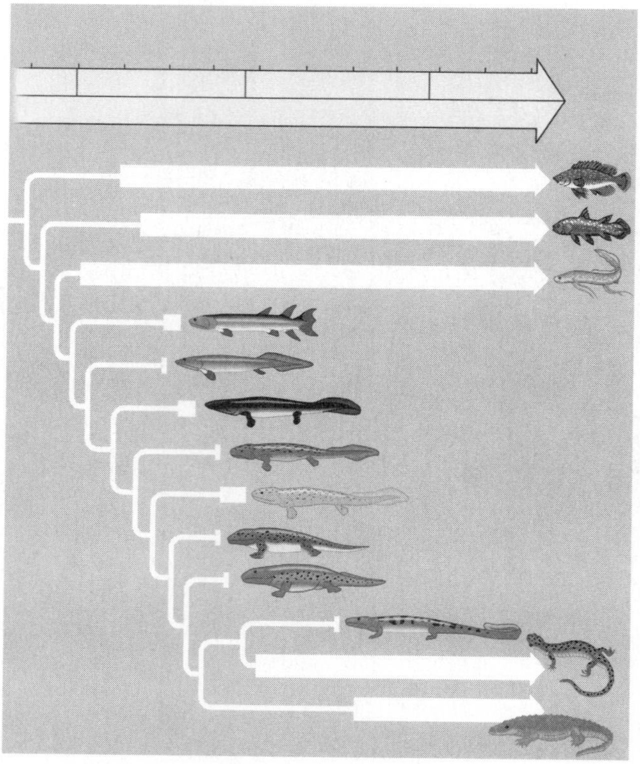

Figure 34.20 The origin of tetrapods, page 685

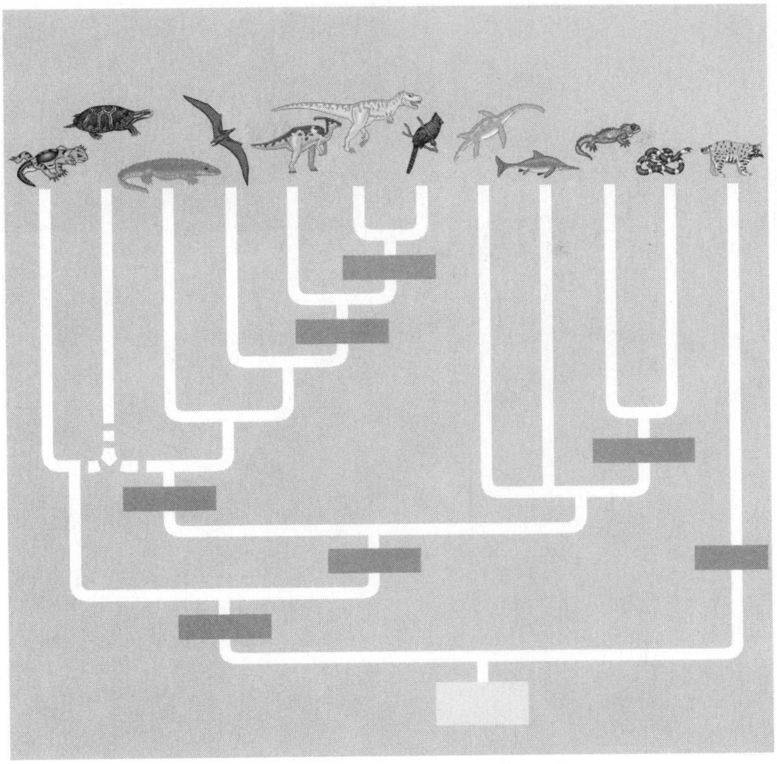

Figure 34.23 A phylogeny of amniotes, page 687

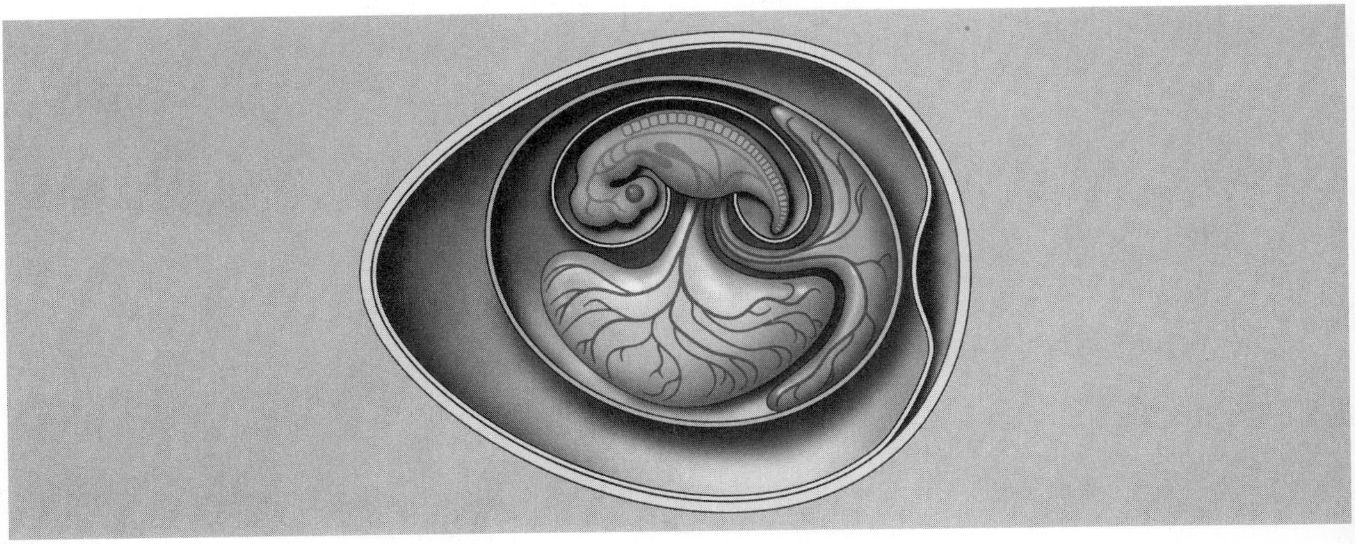

Figure 34.24 The amniotic egg, page 688

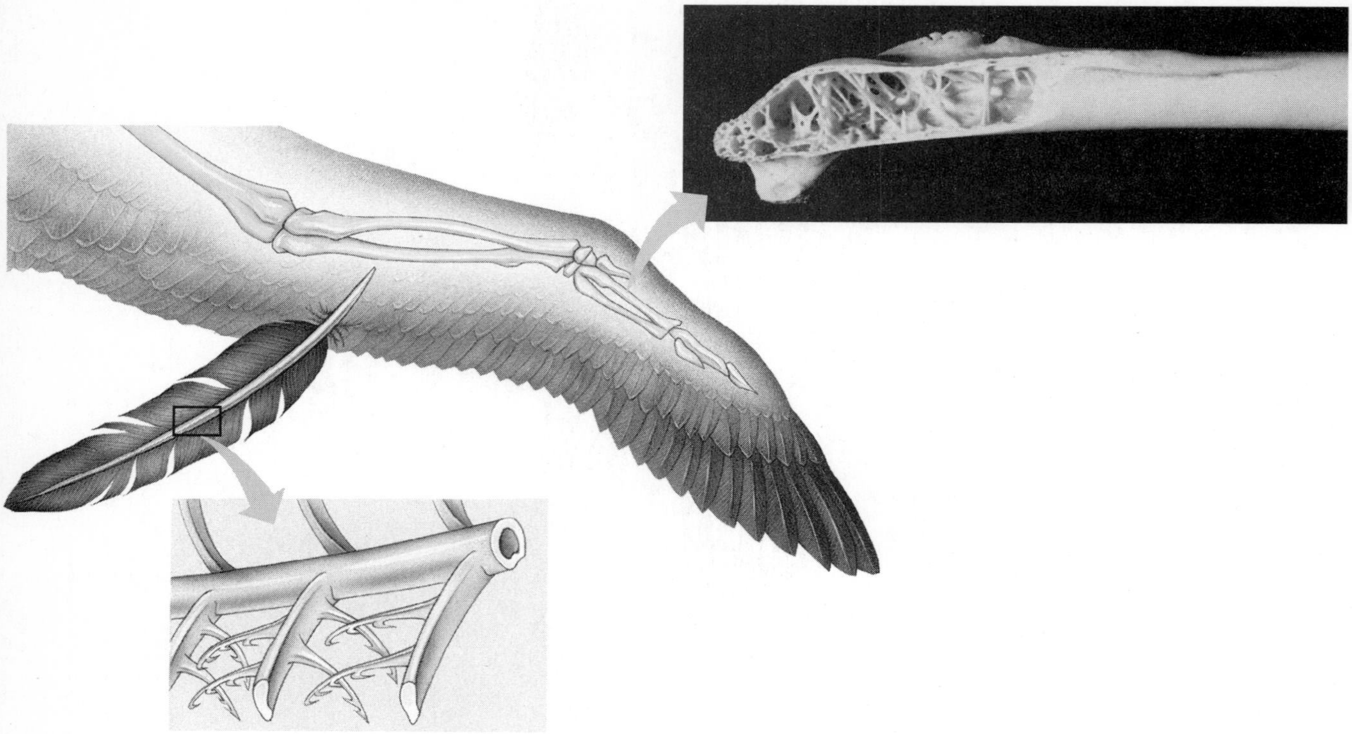

Figure 34.28 Form fits function: the avian wing and feather, page 692

Figure 34.29 Artist's reconstruction of *Archaeopteryx*, the earliest known bird, page 693

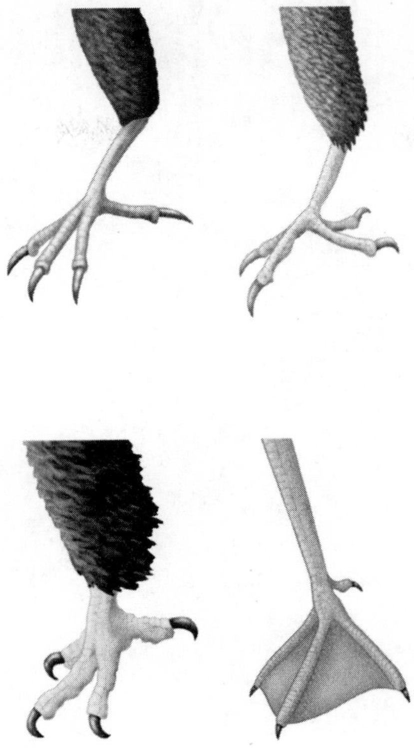

Figure 34.31 Diversity of form and function in bird feet, page 694

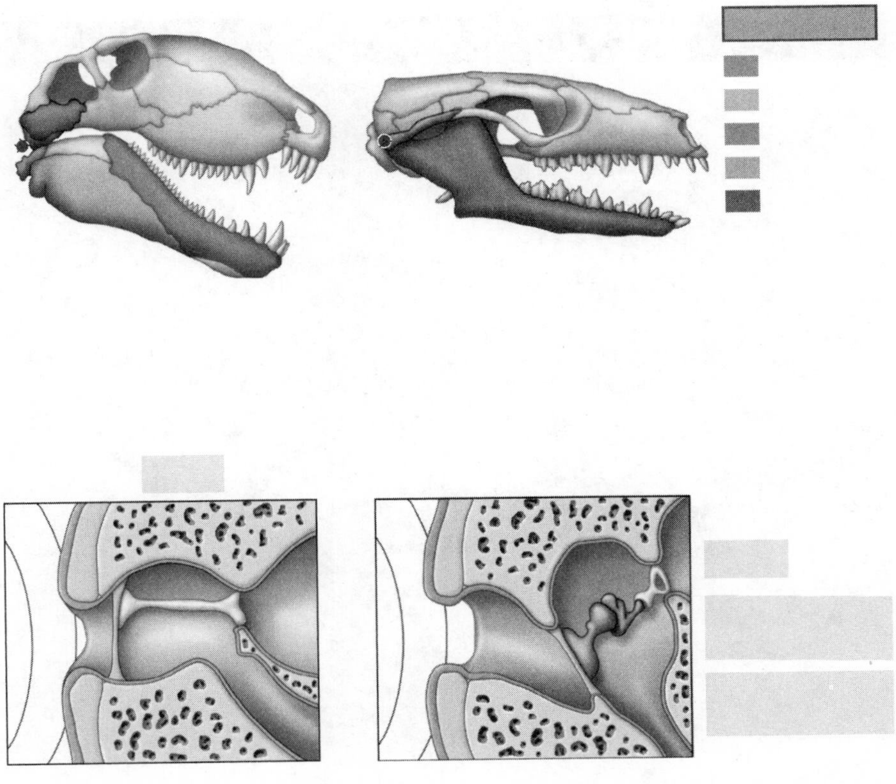

Figure 34.32 The evolution of the mammalian jaw and ear bones, page 695

Figure 34.35 Evolutionary convergence of marsupials and eutherians (placental mammals), page 696

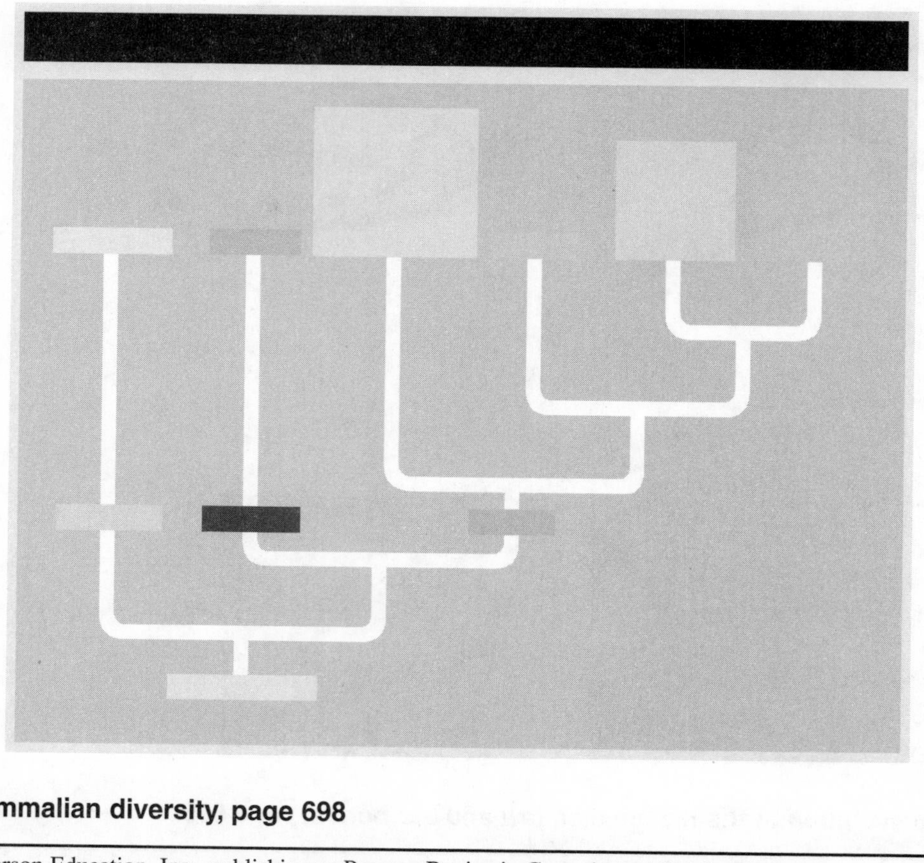

Figure 34.36 Mammalian diversity, page 698

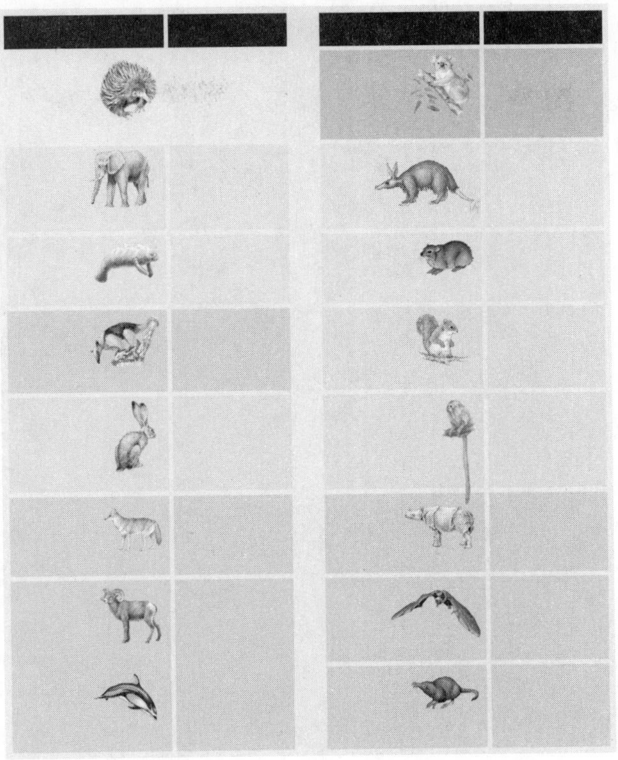

Figure 34.36 Mammalian diversity, page 699

Figure 34.38 A phylogenetic tree of primates, page 700

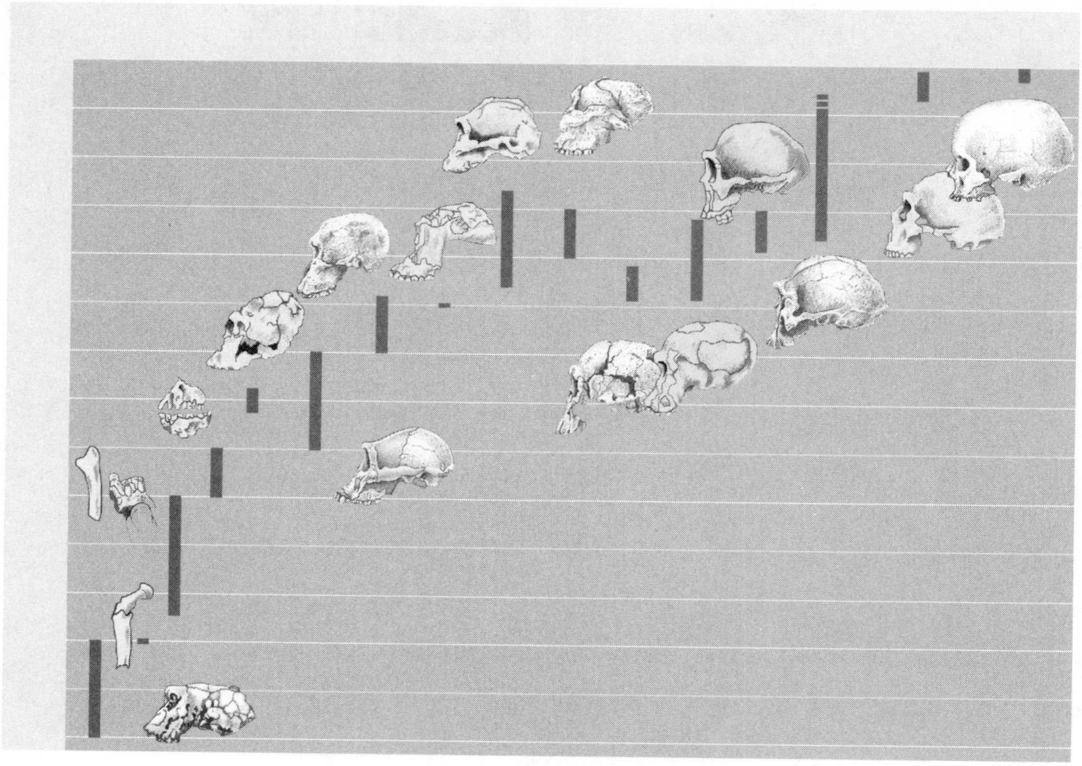

Figure 34.41 A timeline for some hominid species, page 702

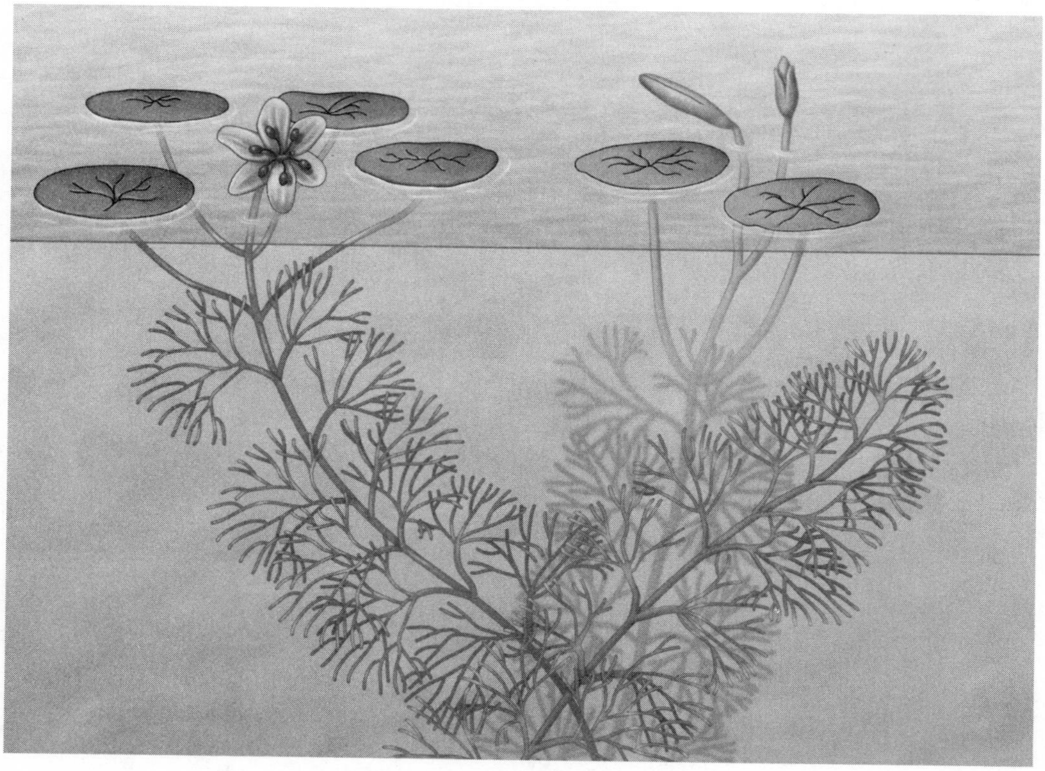

Figure 35.1 Fanwort (*Cabomba caroliniana*), page 712

Figure 35.2 An overview of a flowering plant, page 713

Figure 35.6 Simple versus compound leaves, page 716

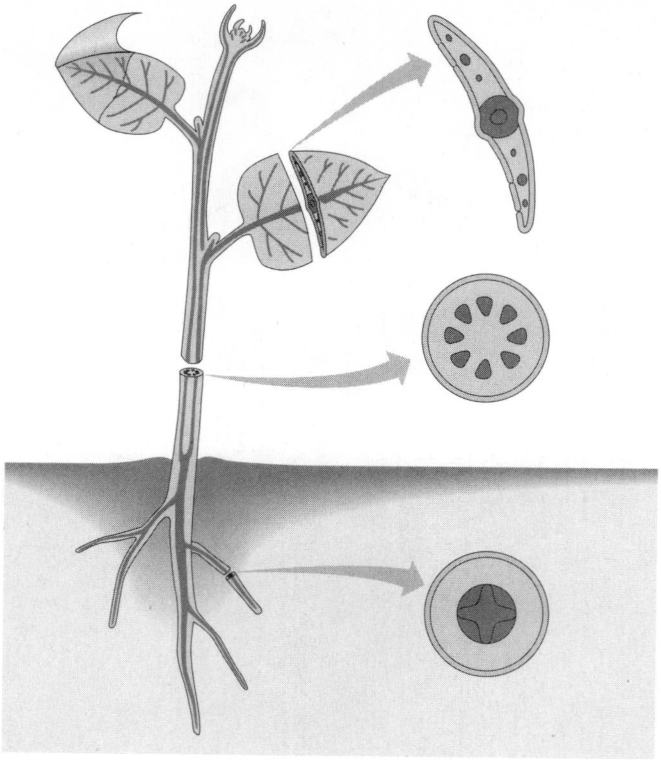

Figure 35.8 The three tissue systems, page 717

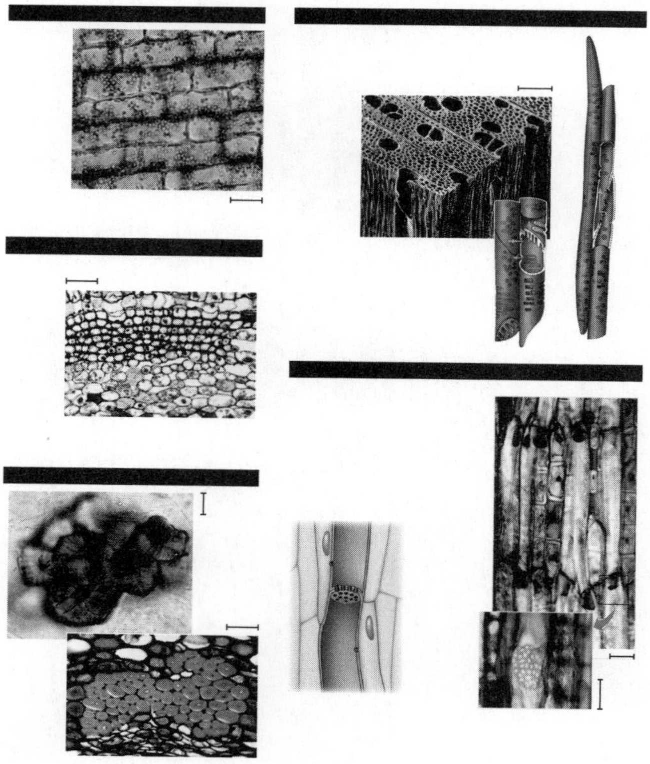

Figure 35.9 Examples of differentiated plant cells, pages 718–719

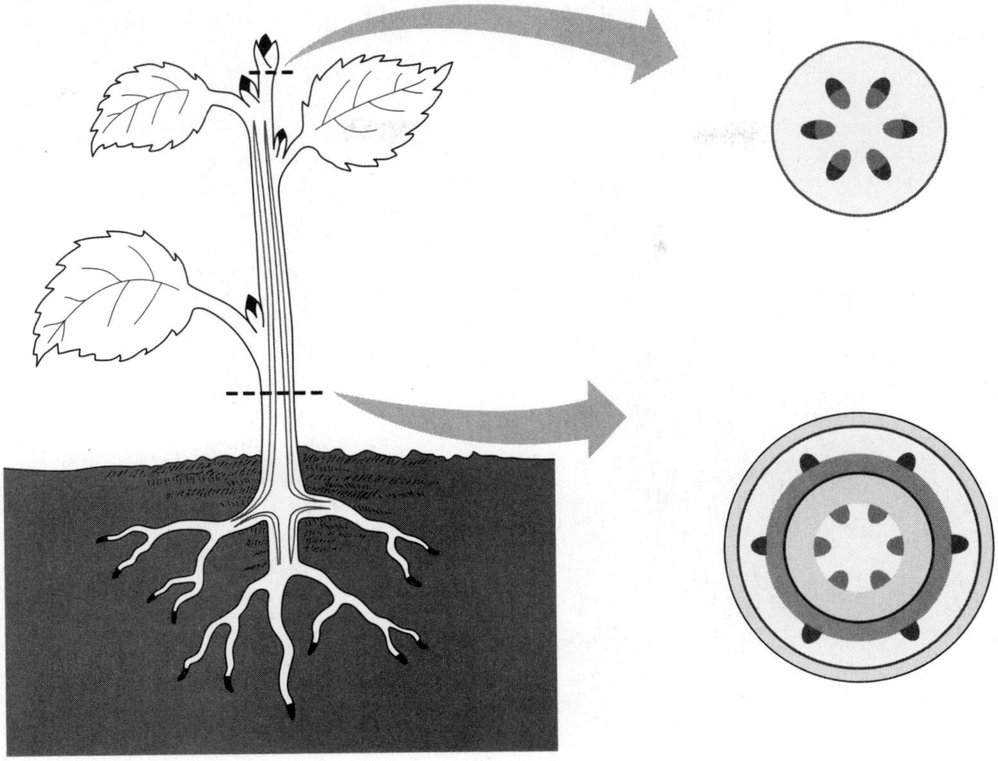

Figure 35.10 An overview of primary and secondary growth, page 720

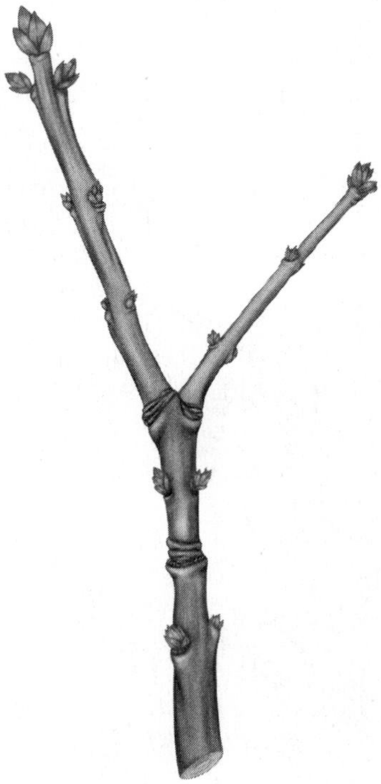

Figure 35.11 Three years' past growth evident in a winter twig, page 721

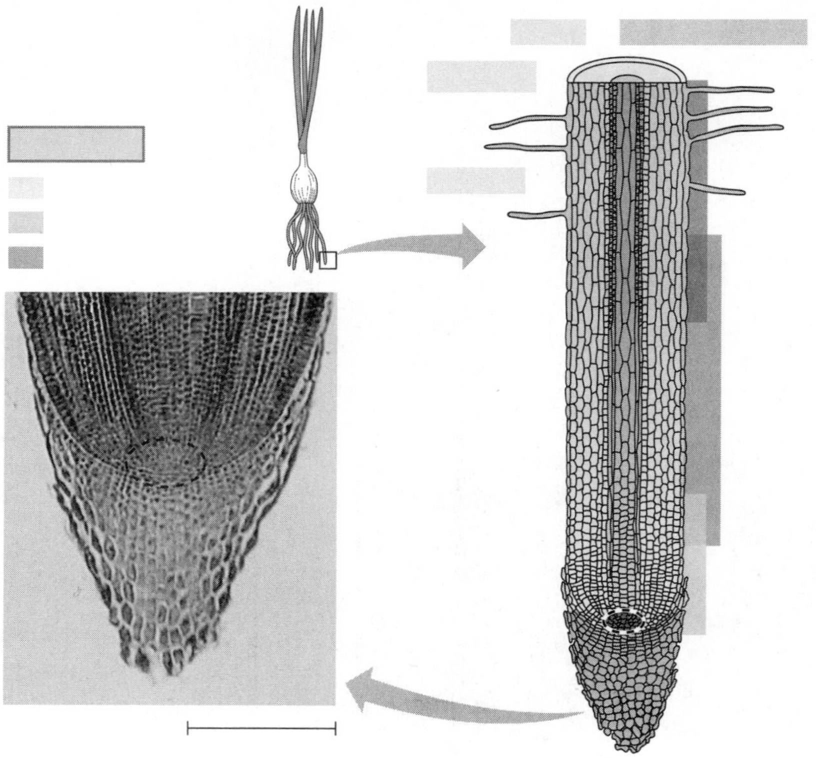

Figure 35.12 Primary growth of a root, page 721

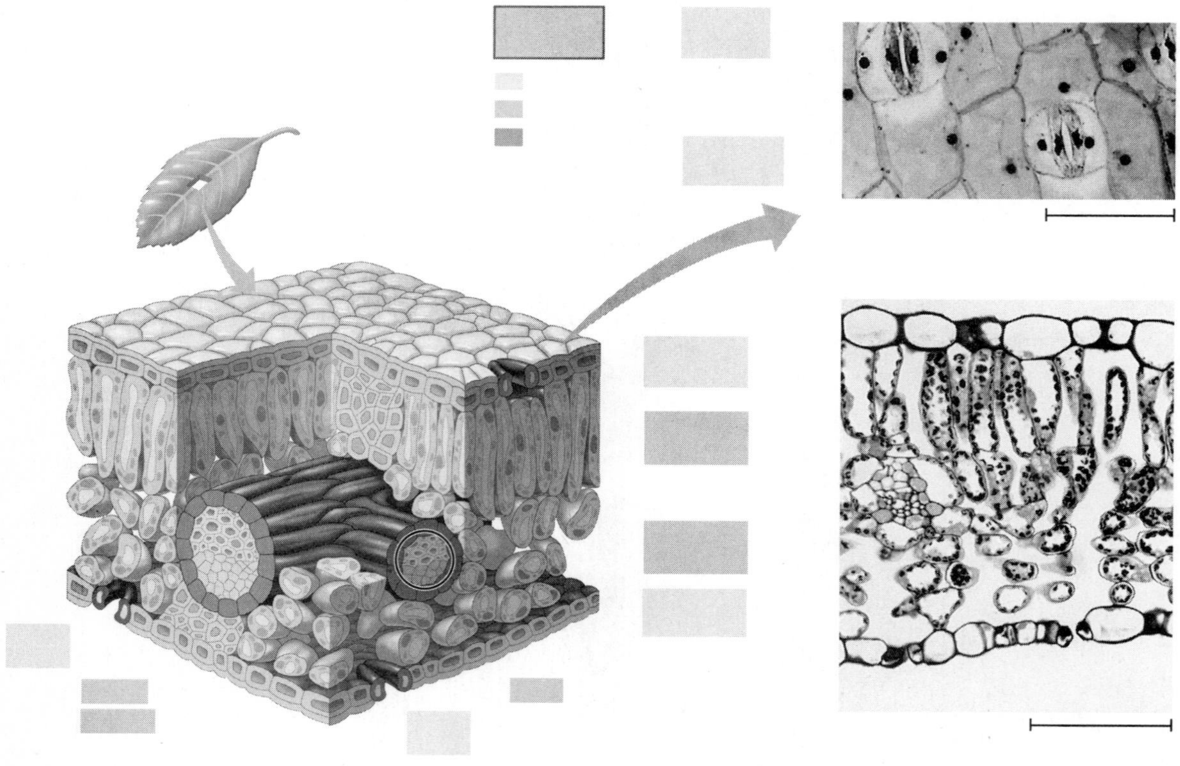

Figure 35.17 Leaf anatomy, page 725

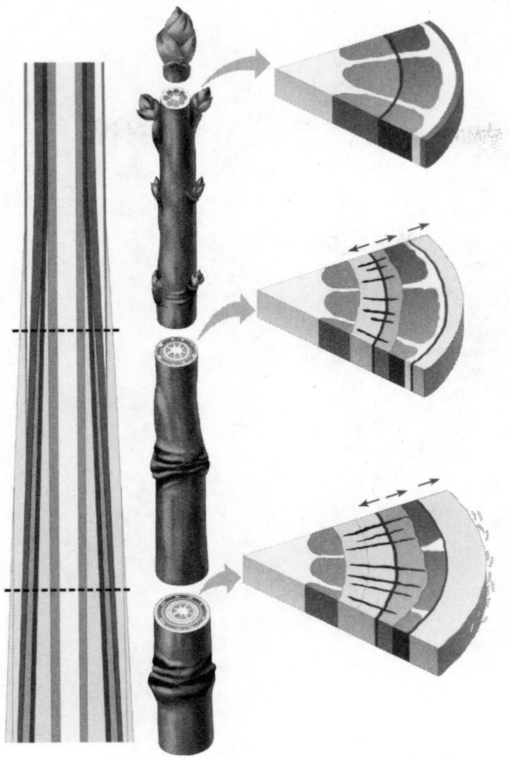

Figure 35.18 Primary and secondary growth of a stem, page 726

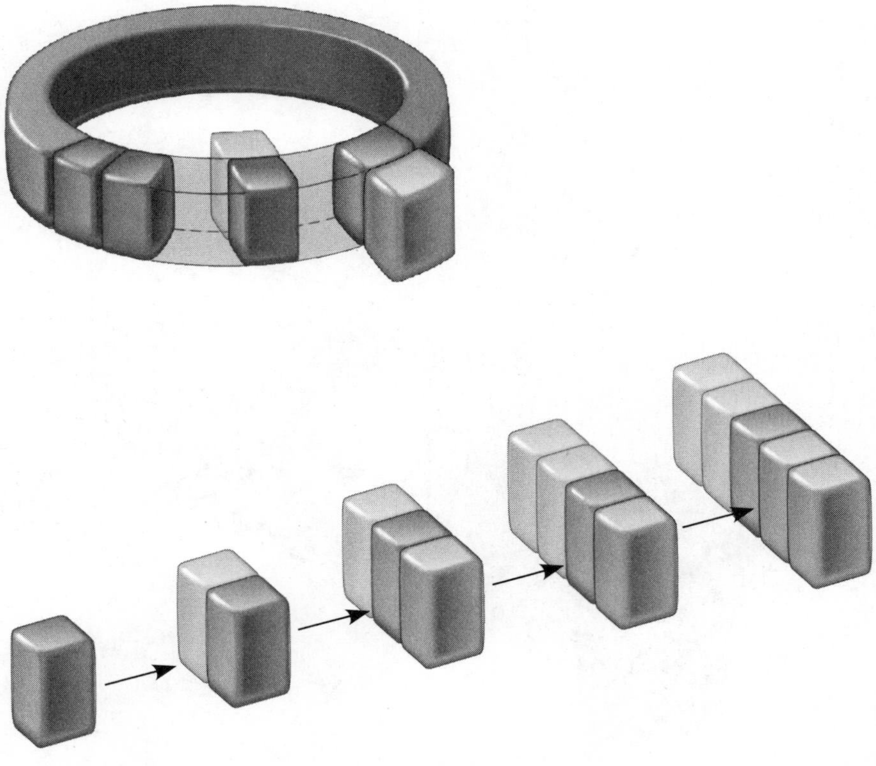

Figure 35.19 Cell division in the vascular cambium, page 727

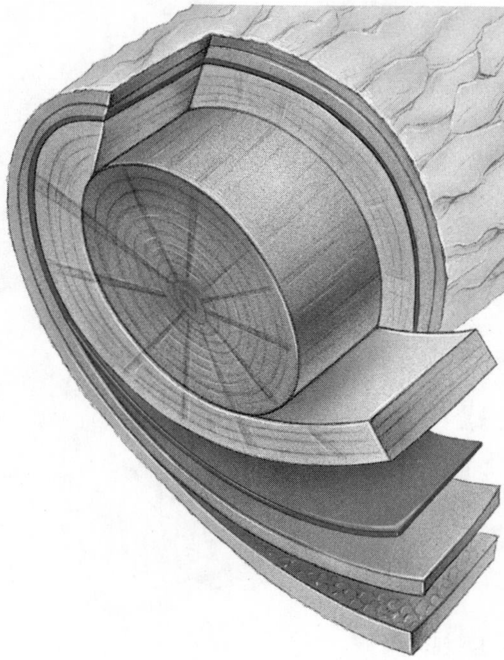

Figure 35.20 Anatomy of a tree trunk, page 727

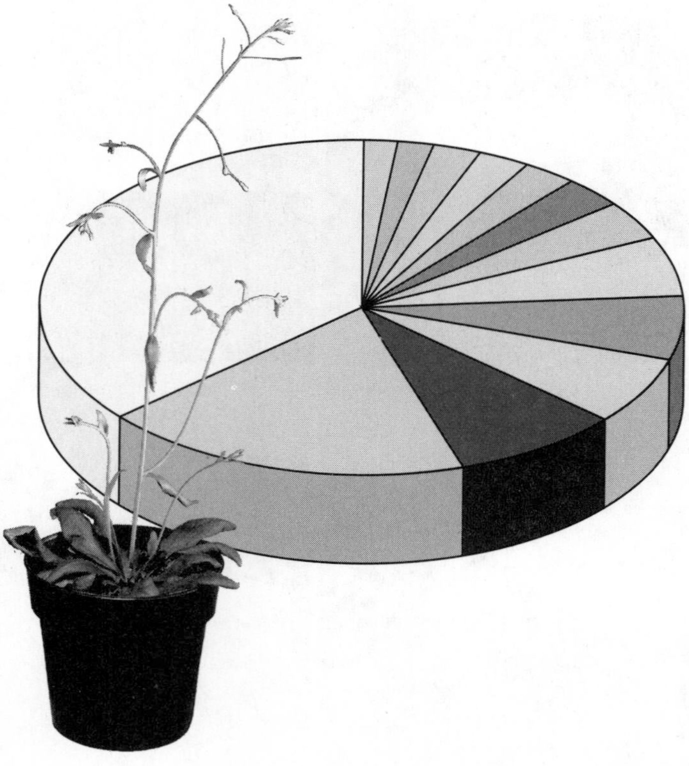

Figure 35.21 *Arabidopsis thaliana*, page 729

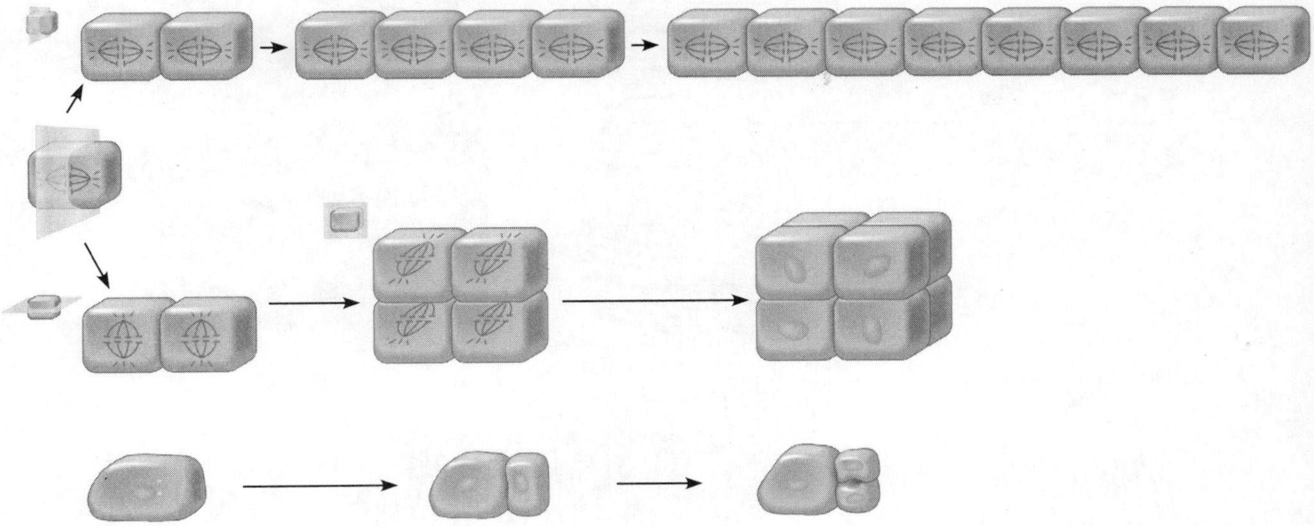

Figure 35.22 The plane and symmetry of cell division influence development of form, page 729

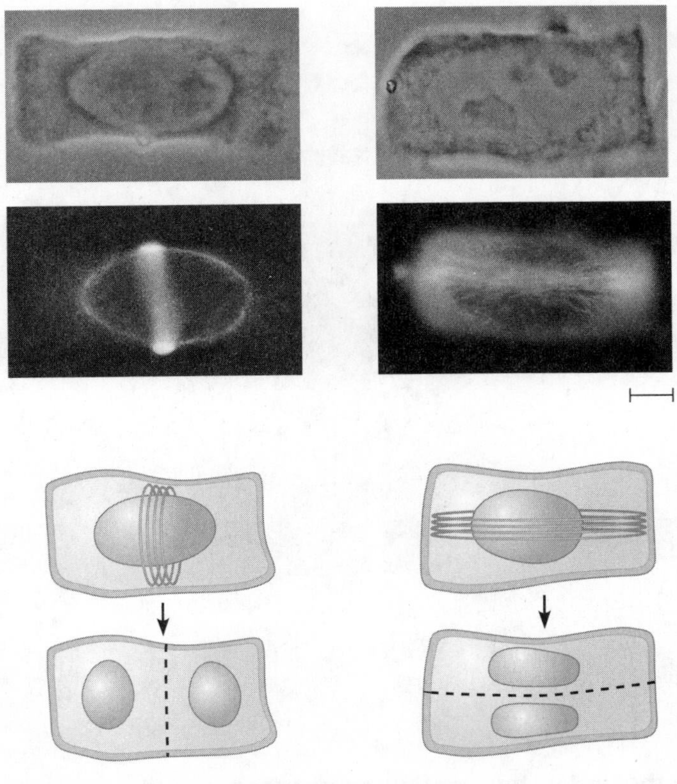

Figure 35.23 The preprophase band and the plane of cell division, page 730

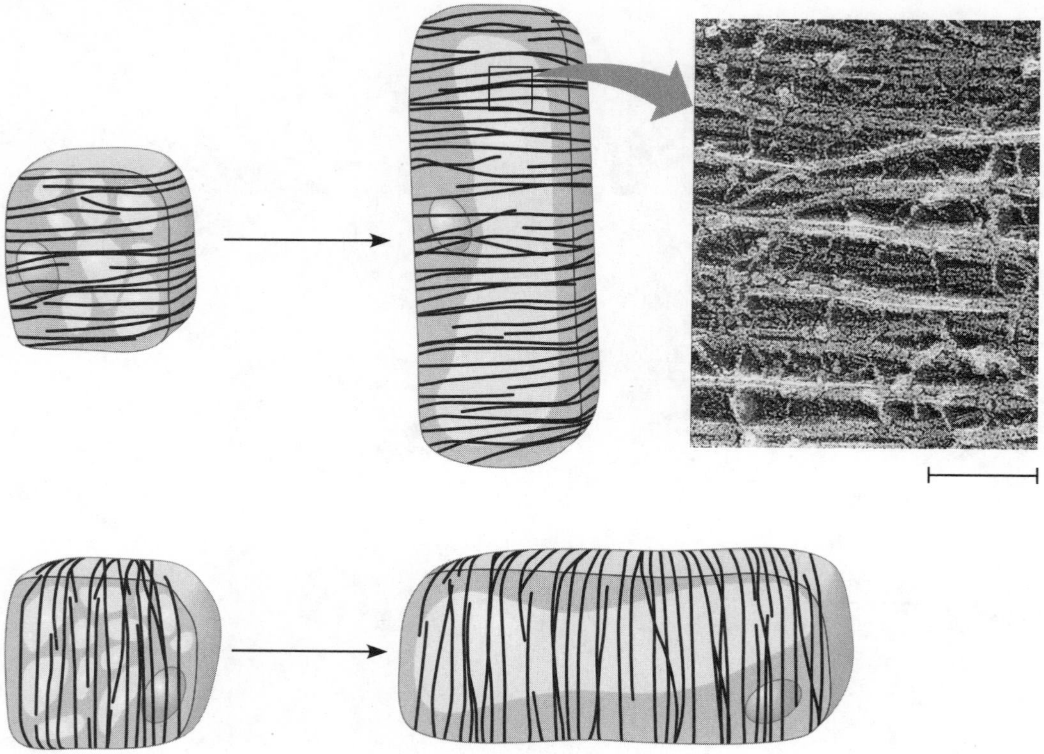

Figure 35.24 The orientation of plant cell expansion, page 731

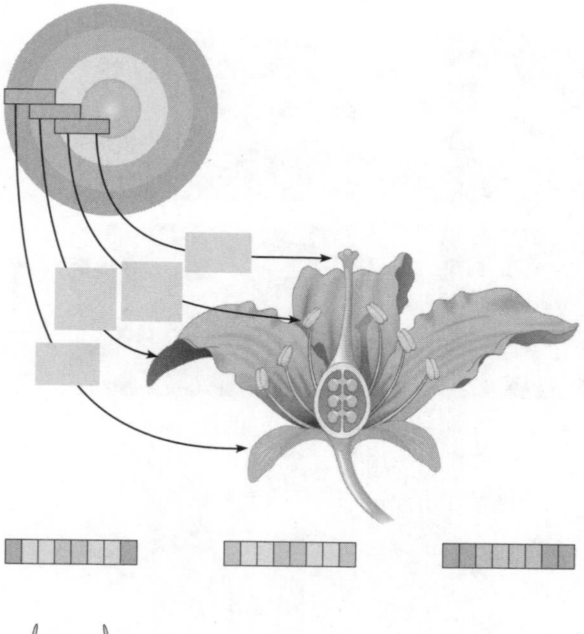

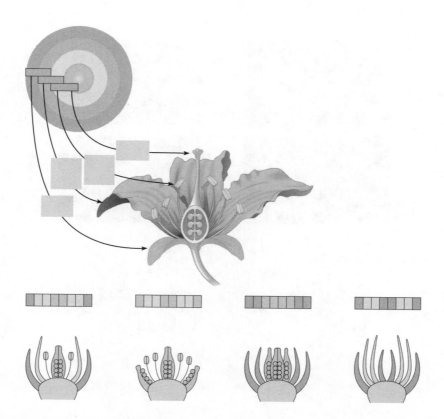

Figure 35.31 The ABC hypothesis for the functioning of organ identity genes in flower development, page 735

Figure 36.2 An overview of transport in a vascular plant, page 739

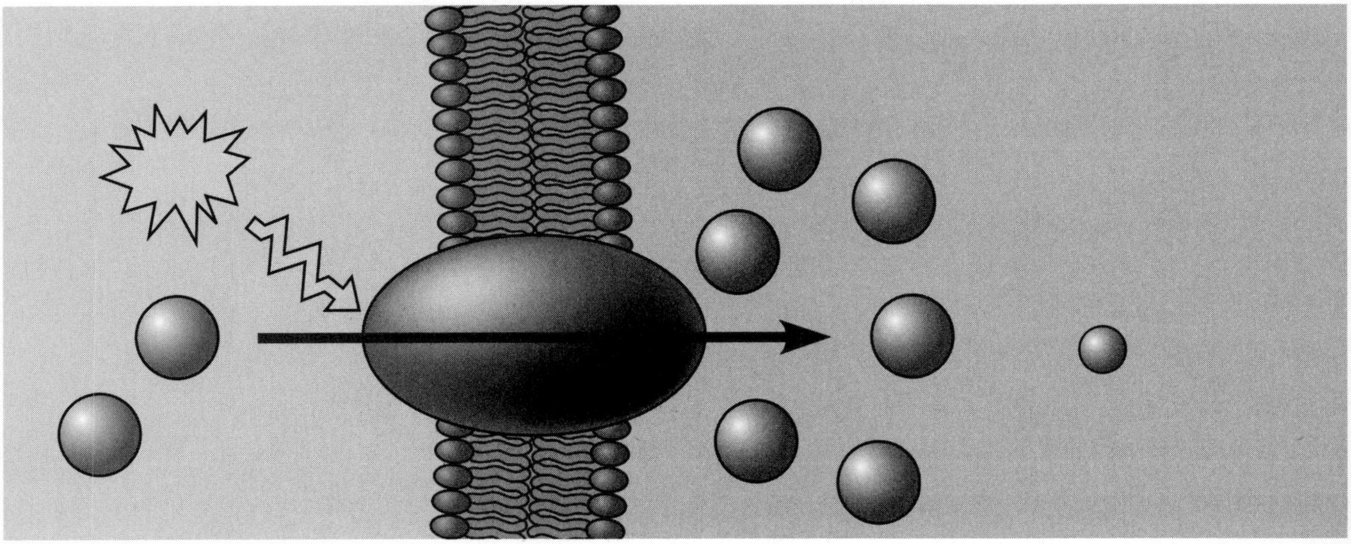

Figure 36.3 Proton pumps provide energy for solute transport, page 739

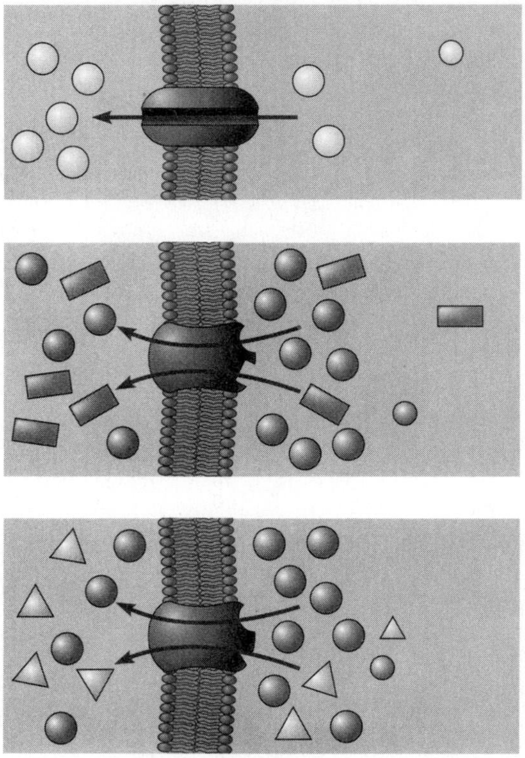

Figure 36.4 Solute transport in plant cells, page 740

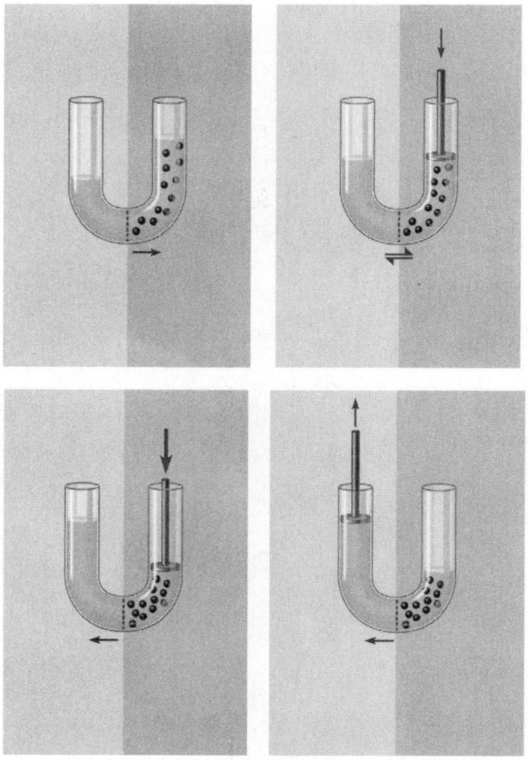

Figure 36.5 Water potential and water movement: an artificial model, page 741

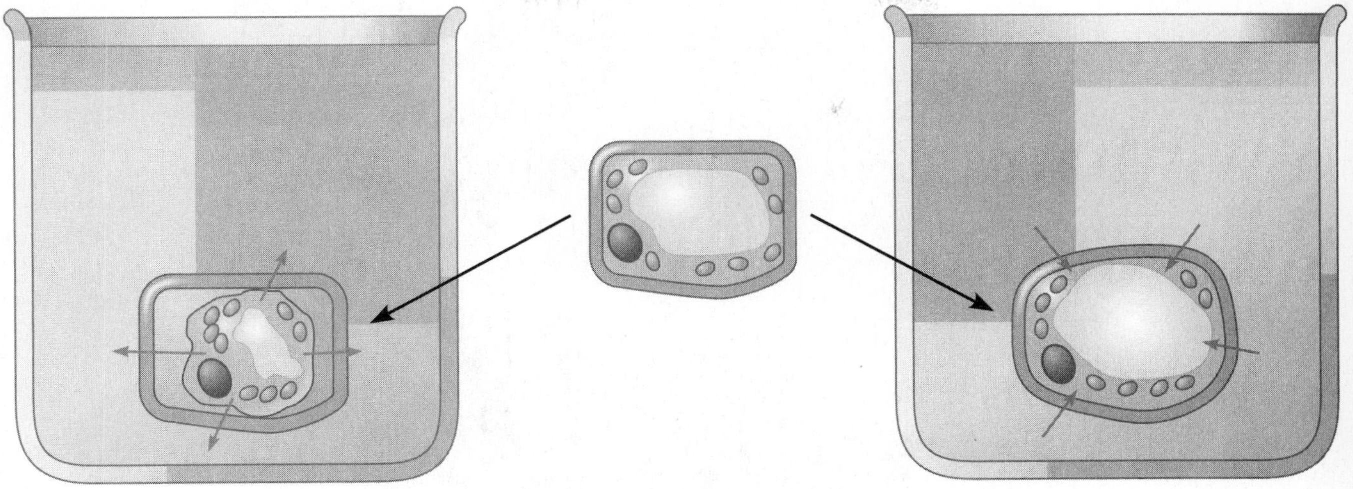

Figure 36.6 Water relations in plant cells, page 742

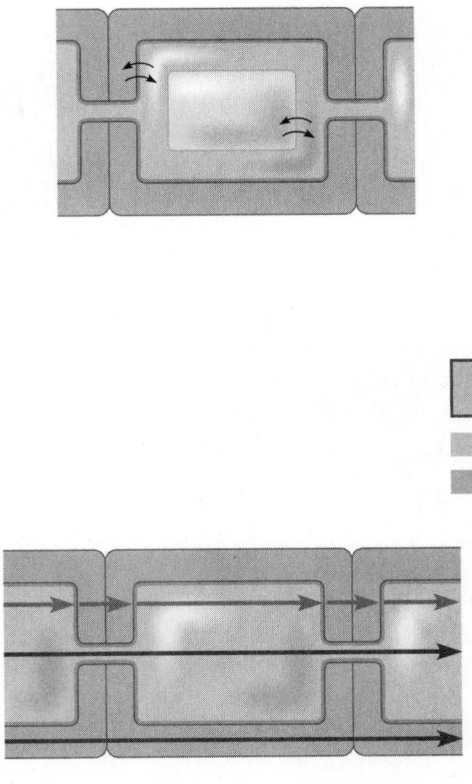

Figure 36.8 Cell compartments and routes for short-distance transport, page 743

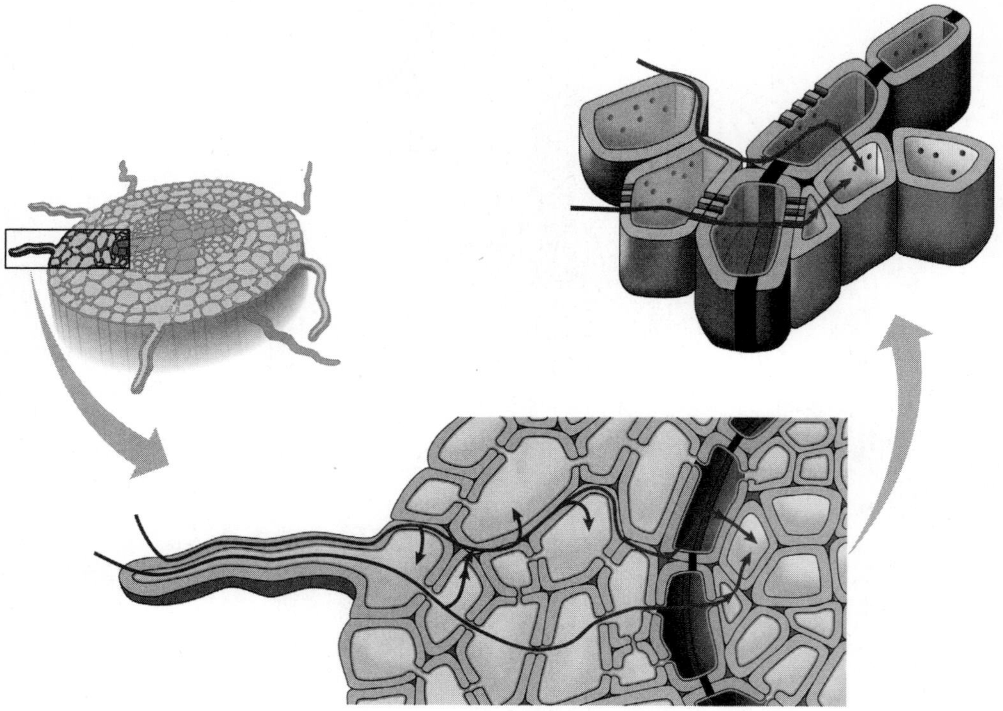

Figure 36.9 Lateral transport of minerals and water in roots, page 745

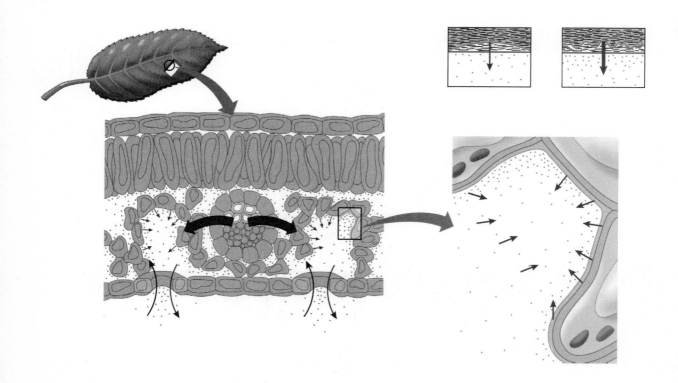

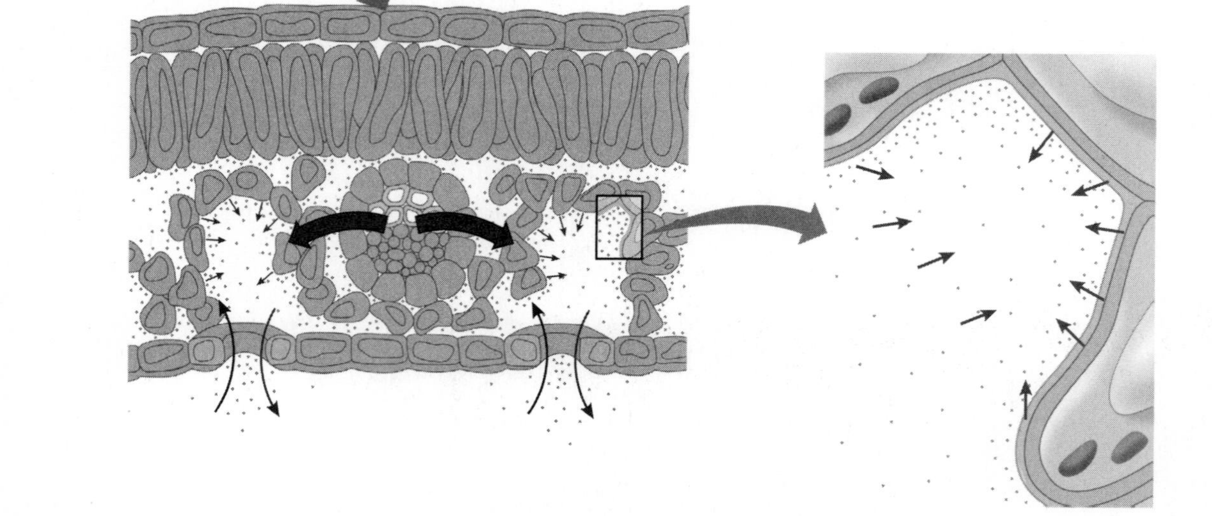

Figure 36.12 The generation of transpirational pull in a leaf, page 747

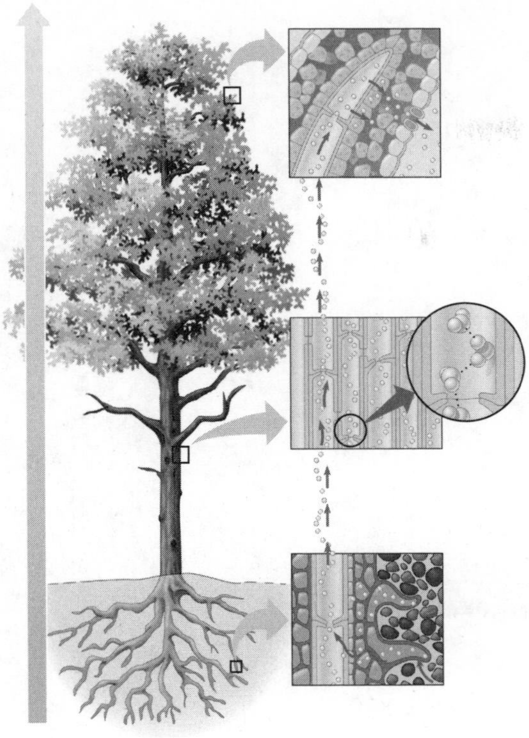

Figure 36.13 Ascent of xylem sap, page 748

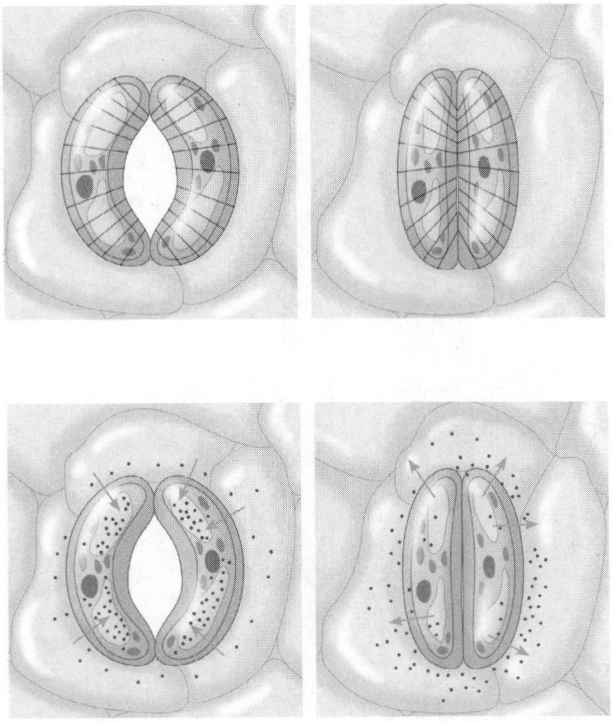

Figure 36.15 The mechanism of stomatal opening and closing, page 750

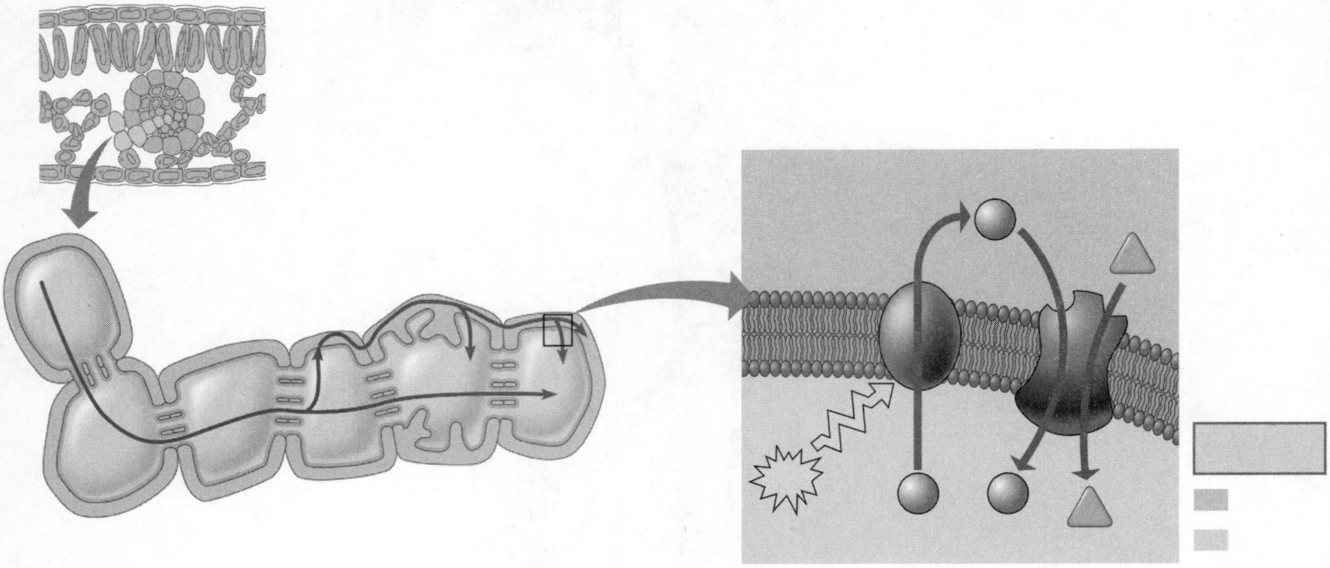

Figure 36.17 Loading of sucrose into phloem, page 752

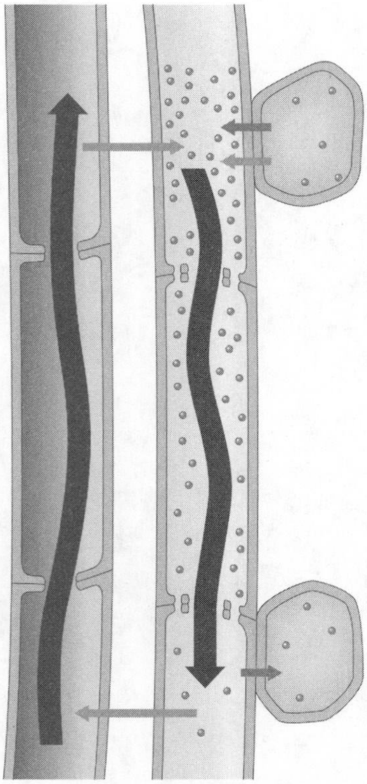

Figure 36.18 Pressure flow in a sieve tube, page 753

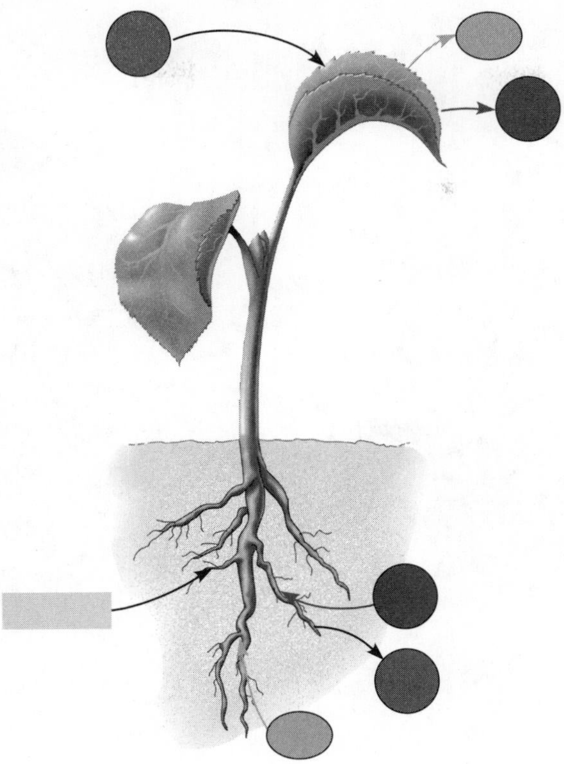

Figure 37.2 The uptake of nutrients by a plant: a review, page 757

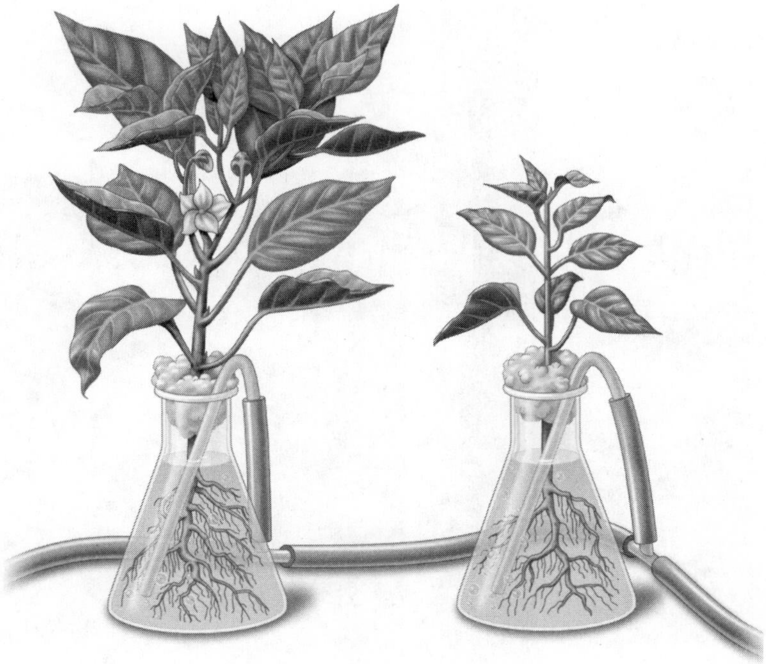

Figure 37.3 Hydroponic culture, page 757

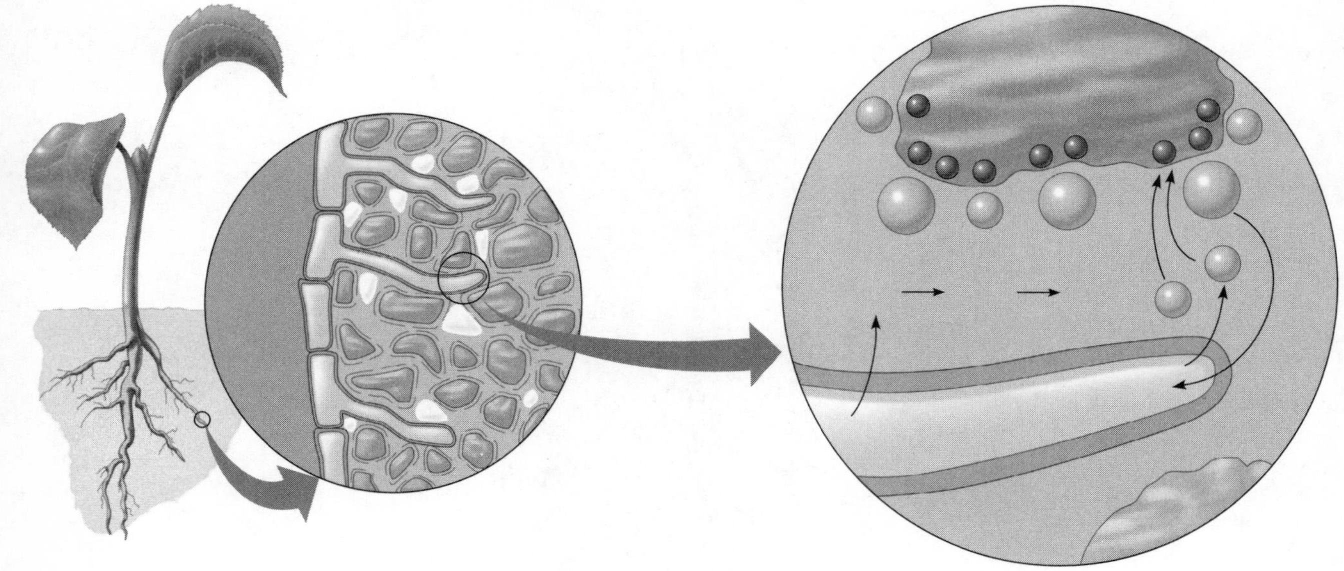

Figure 37.6 The availability of soil water and minerals, page 761

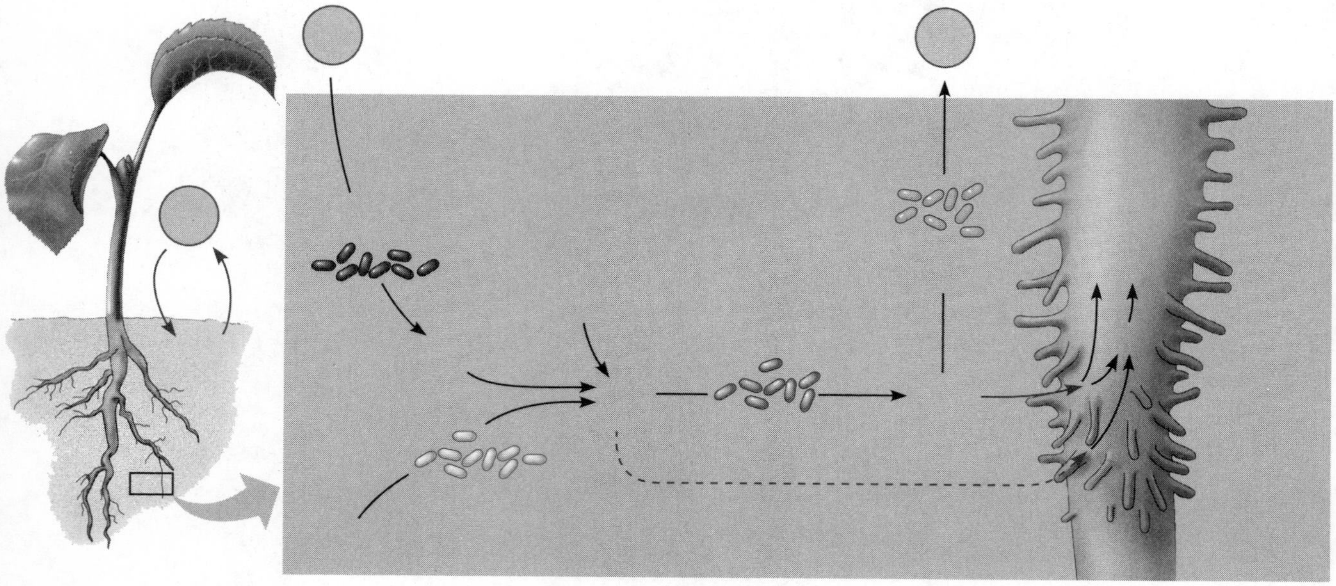

Figure 37.9 The role of soil bacteria in the nitrogen nutrition of plants, page 763

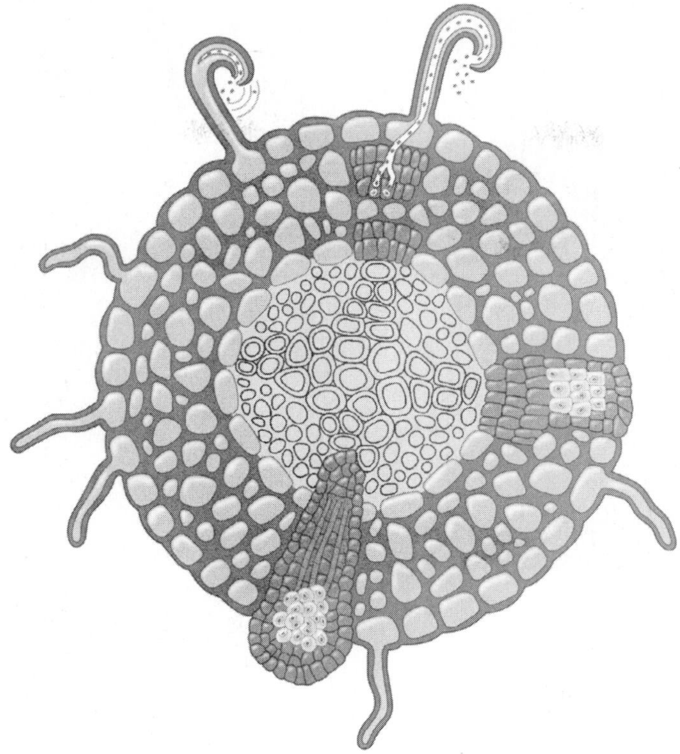

Figure 37.11 Development of a soybean root nodule, page 765

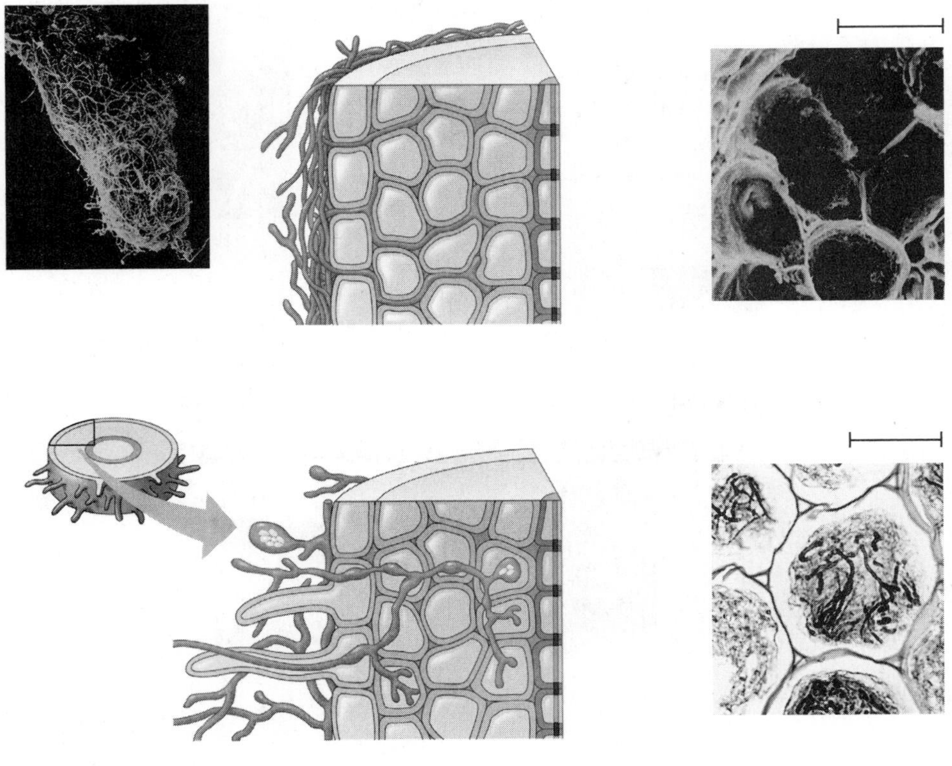

Figure 37.12 Mycorrhizae, page 767

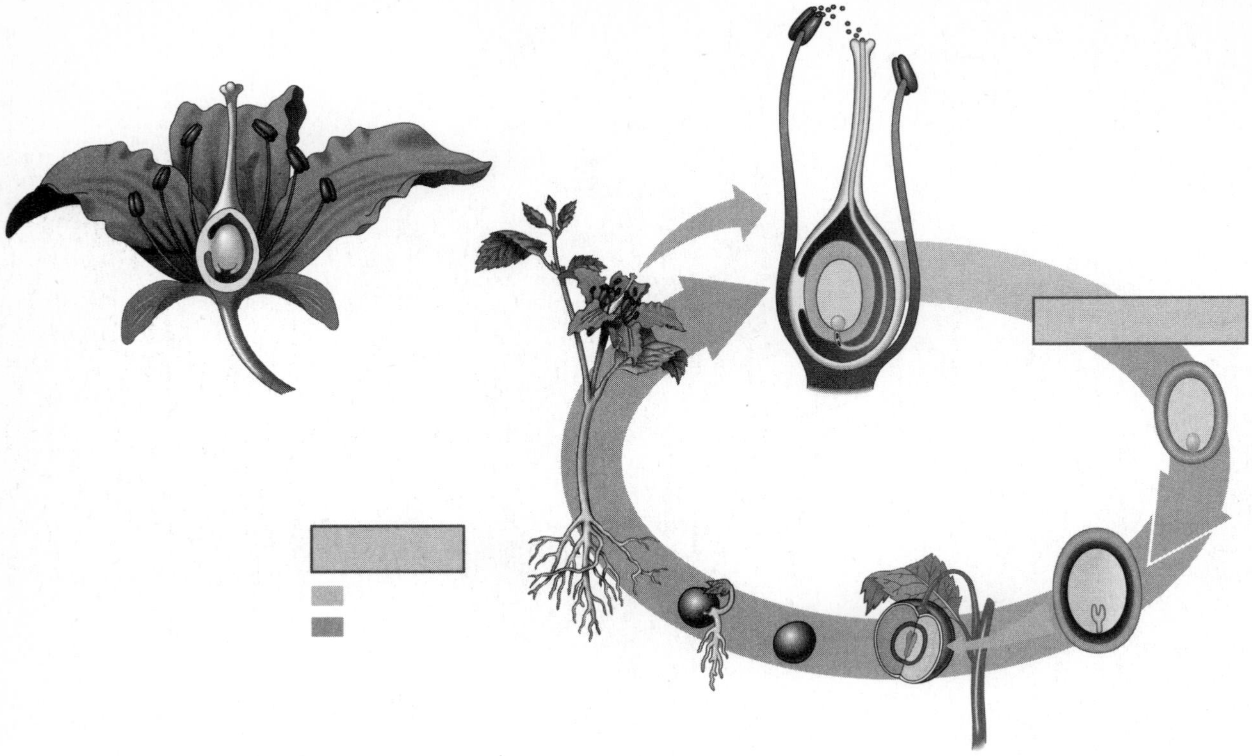

Figure 38.2 An overview of angiosperm reproduction, page 772

Figure 38.3 Floral variations, page 773

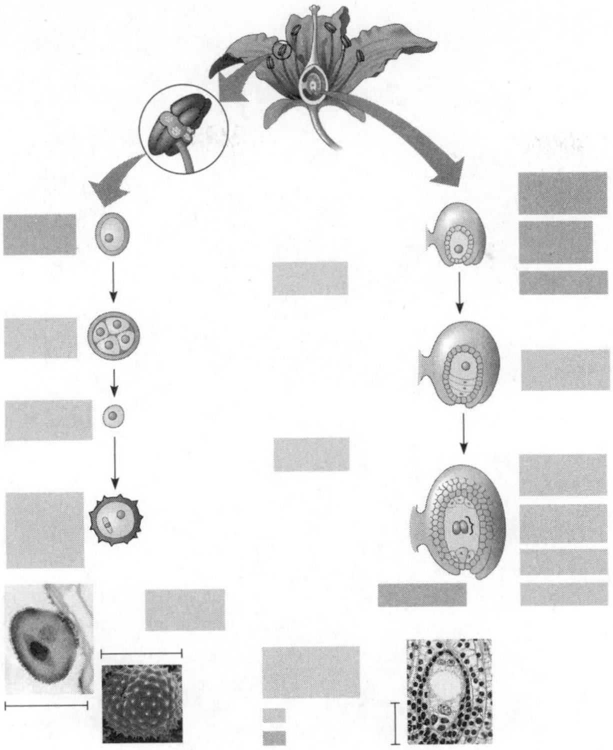

Figure 38.4 The development of angiosperm gametophytes (pollen grains and embryo sacs), page 774

Figure 38.5 "Pin" and "thrum" flower types reduce self-fertilization, page 775

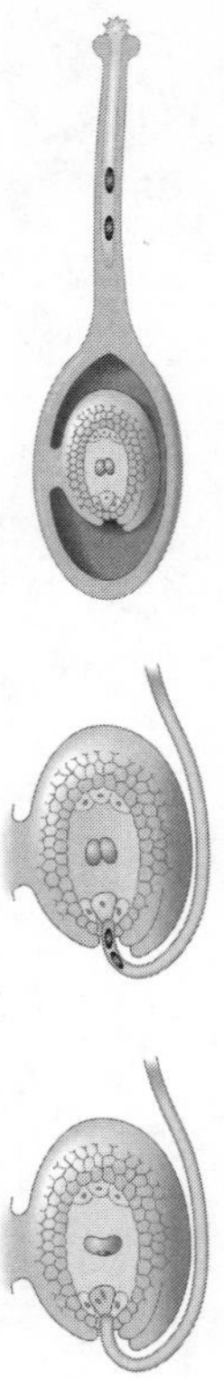

Figure 38.6 Growth of the pollen tube and double fertilization, page 776

Figure 38.7 The development of a eudicot plant embryo, page 777

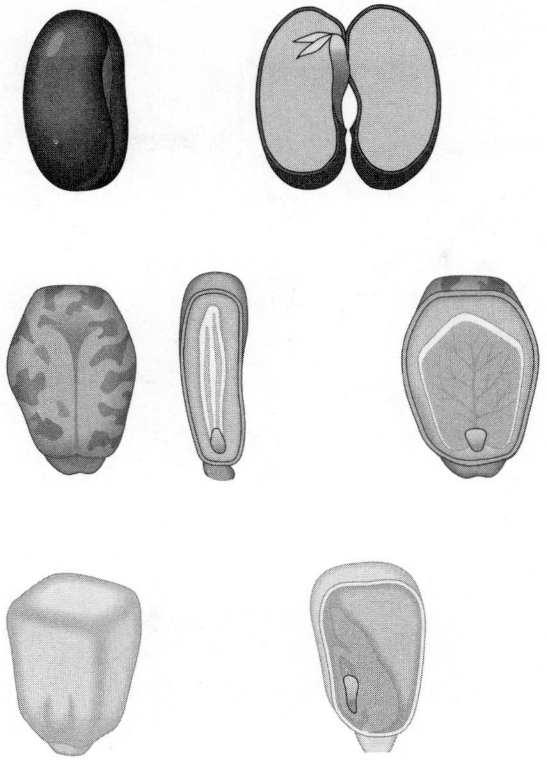

Figure 38.8 Seed structure, page 778

Figure 38.9 Developmental origin of fruits, page 779

Figure 38.10 Two common types of seed germination, page 780

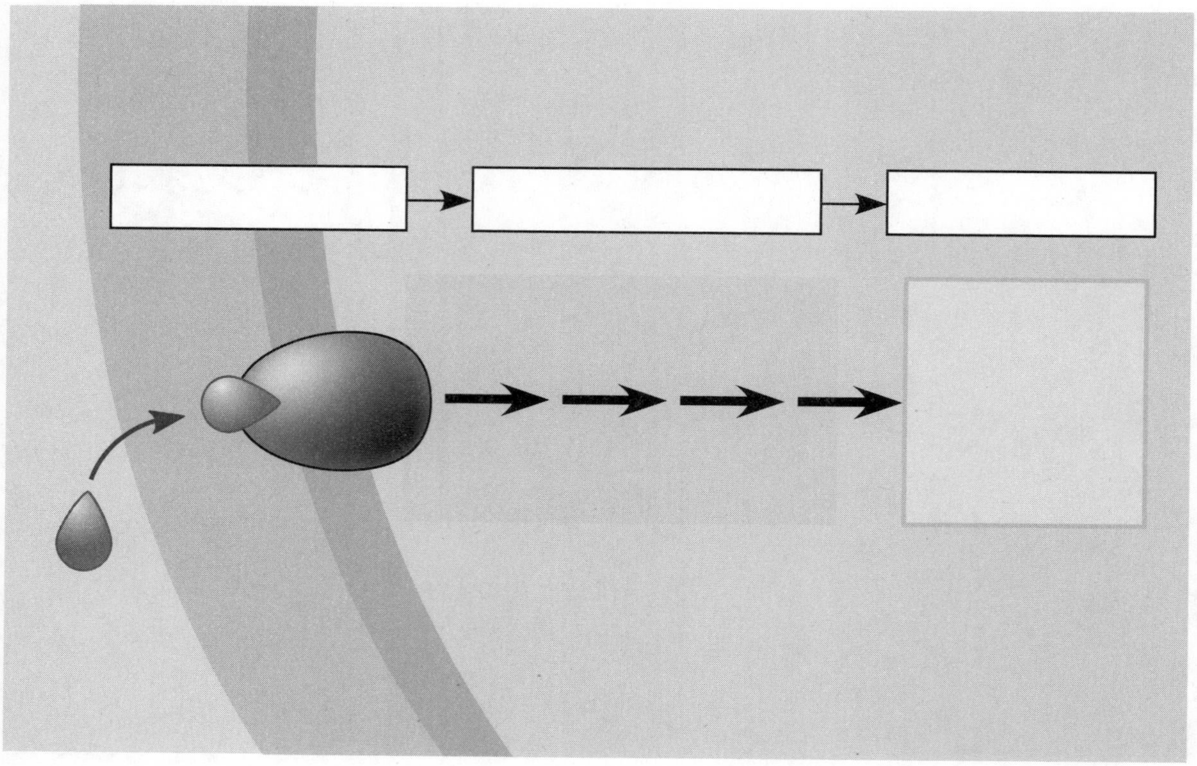

Figure 39.3 Review of a general model for signal transduction pathways, page 789

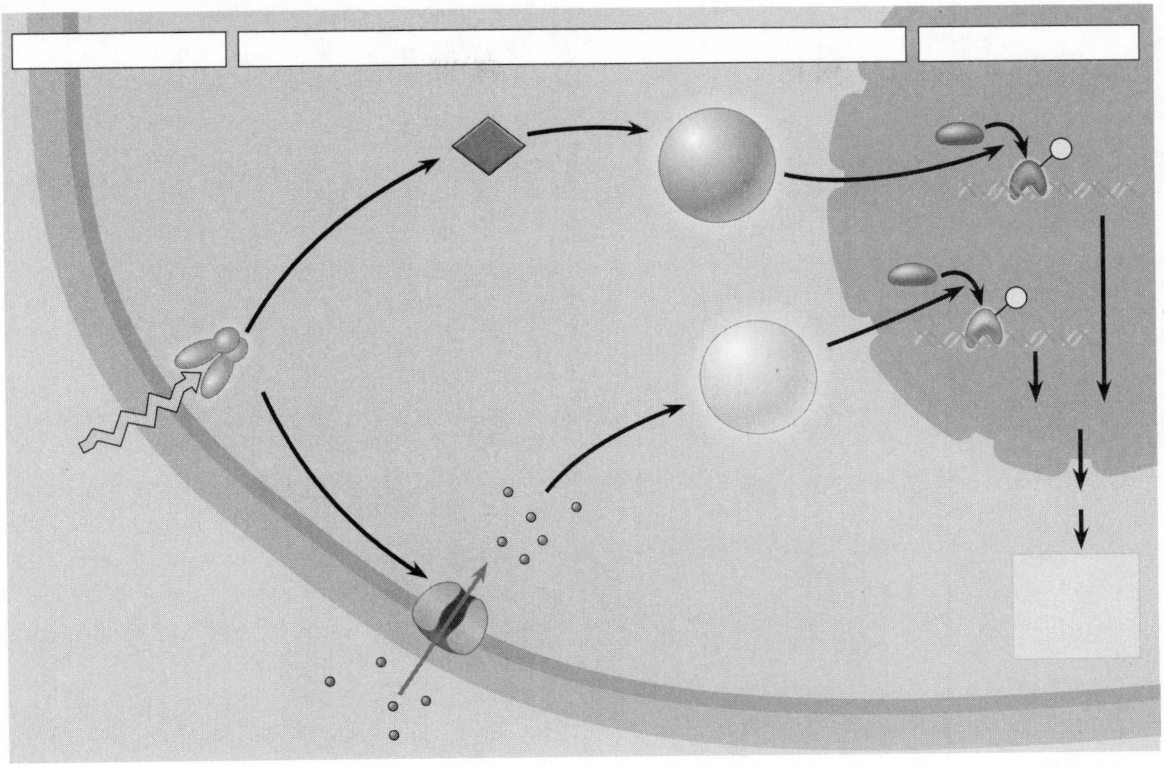

Figure 39.4 An example of signal transduction in plants: the role of phytochrome in the de-etiolation (greening) response, page 790

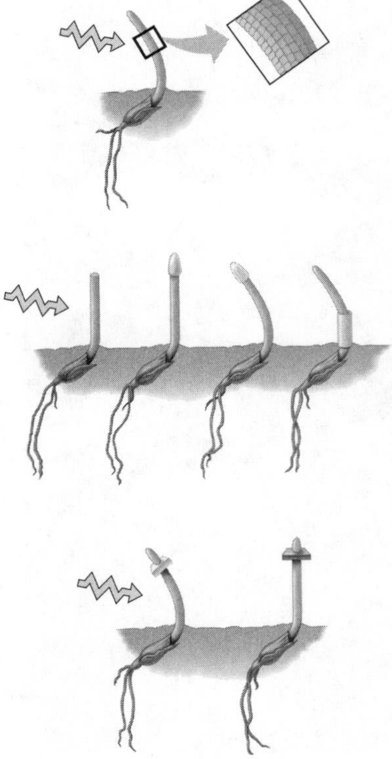

Figure 39.5 What part of a coleoptile senses light, and how is the signal transmitted?,page 792

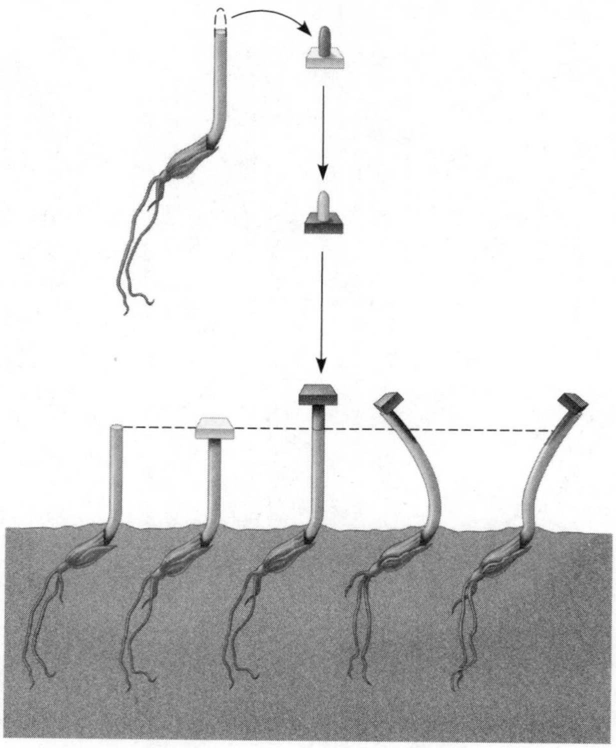

Figure 39.6 Does asymmetric distribution of a growth-promoting chemical cause a coleoptile to grow toward the light? page 793

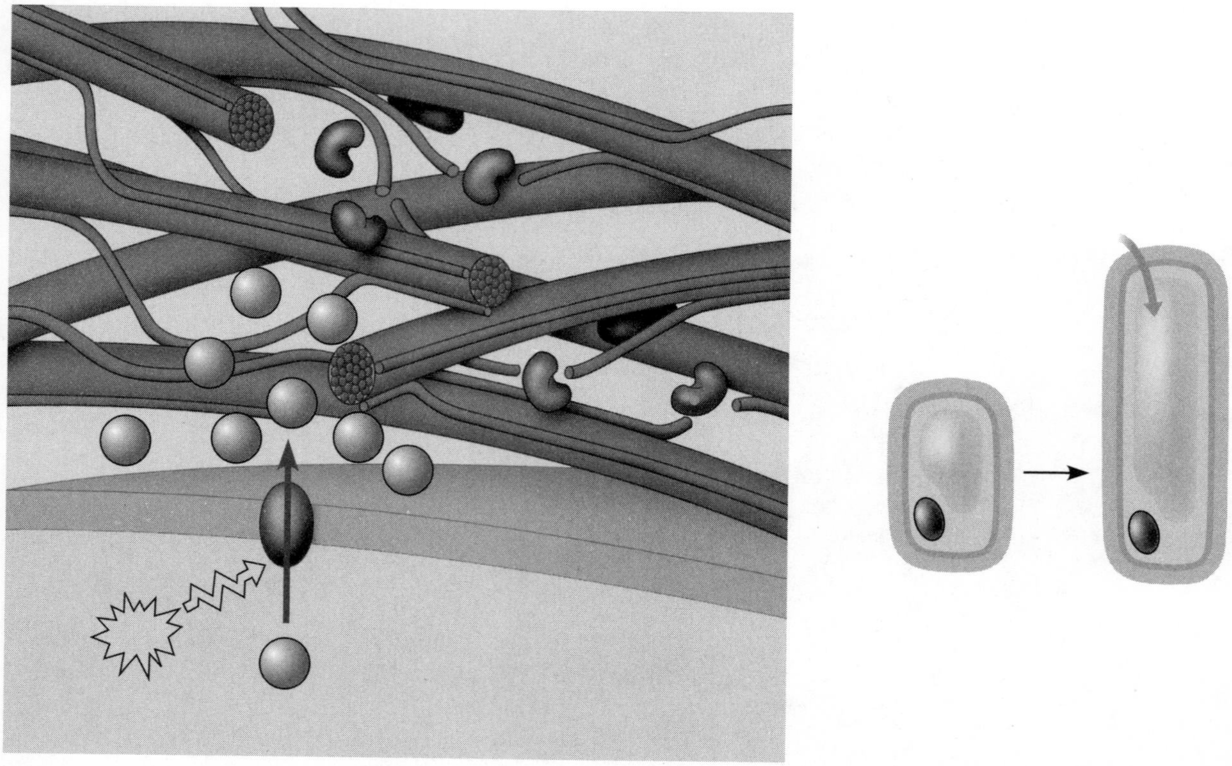

Figure 39.8 Cell elongation in response to auxin: the acid growth hypothesis, page 795

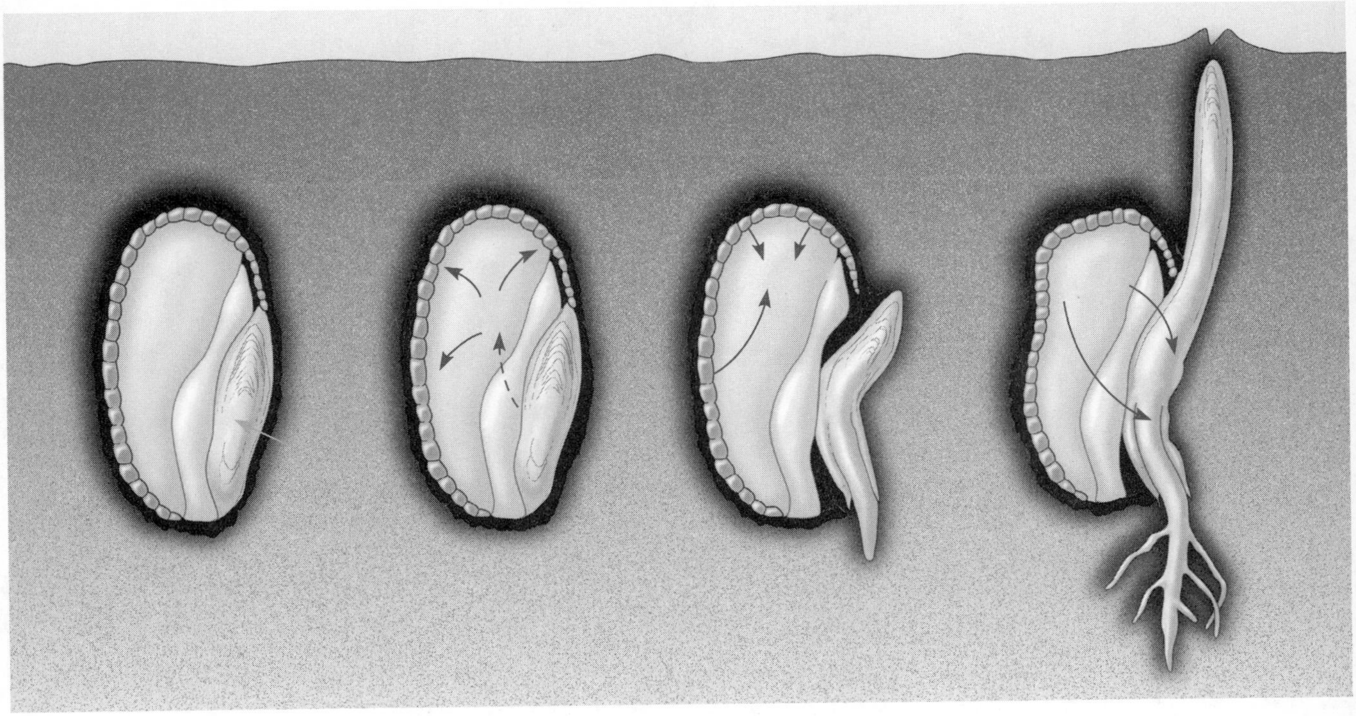

Figure 39.11 Gibberellins mobilize nutrients during the germination of grain seeds, page 798

Figure 39.13 How does ethylene concentration affect the triple response in seedlings? page 799

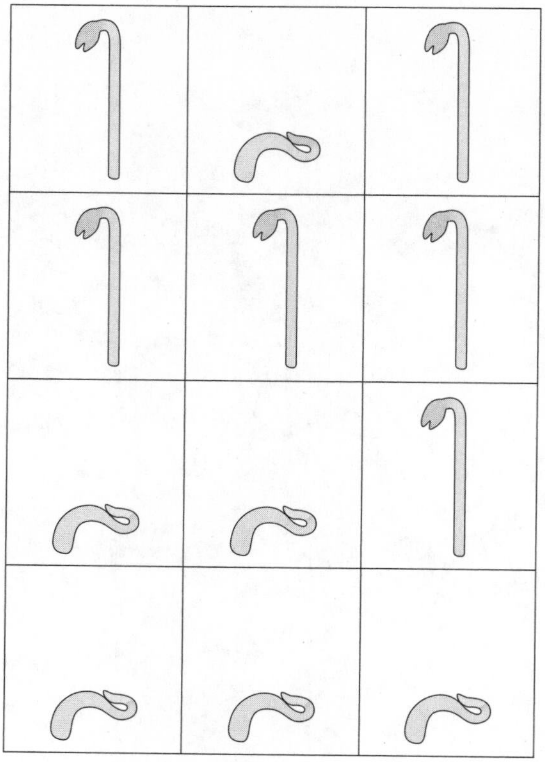

Figure 39.15 Ethylene signal transduction mutants can be distinguished by their different responses to experimental treatments, page 800

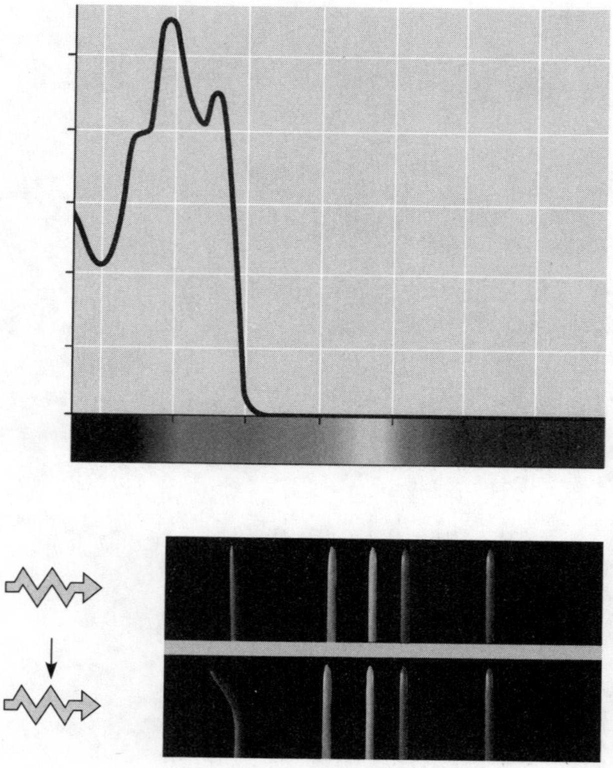

Figure 39.17 What wavelengths stimulate phototropic bending toward light? page 803

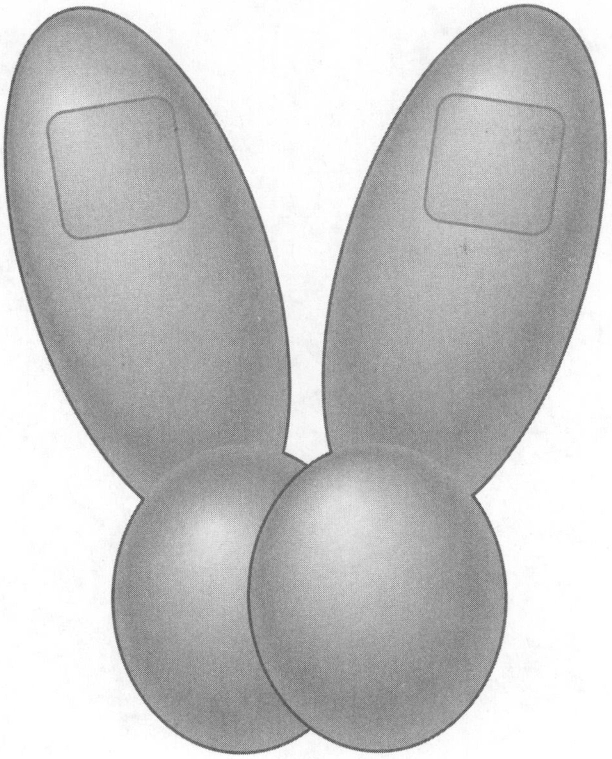

Figure 39.19 Structure of a phytochrome, page 804

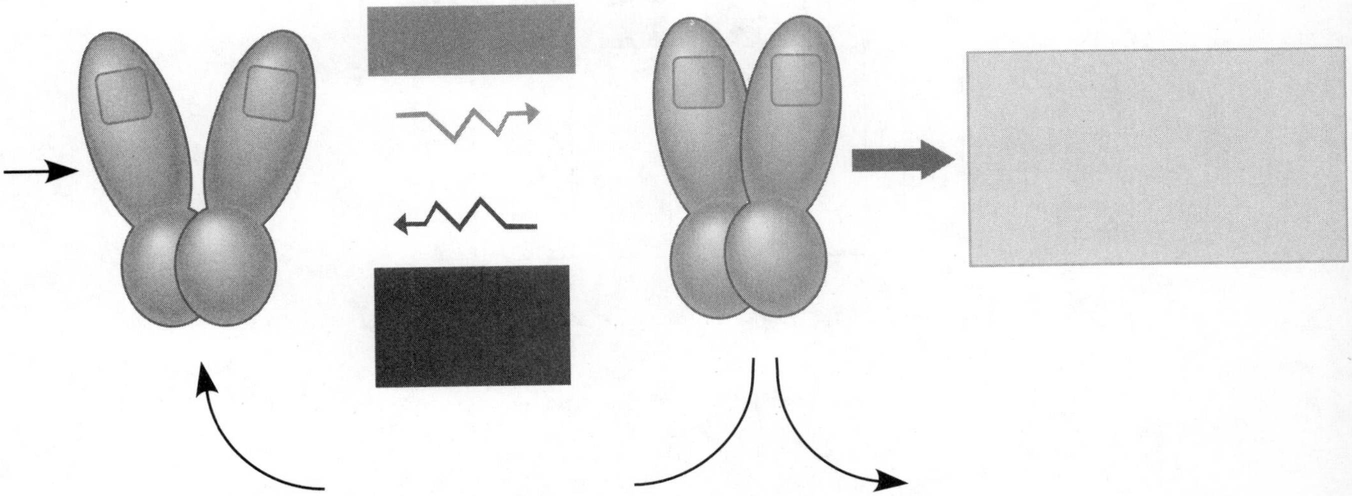

Figure 39.20 Phytochrome: a molecular switching mechanism, page 804

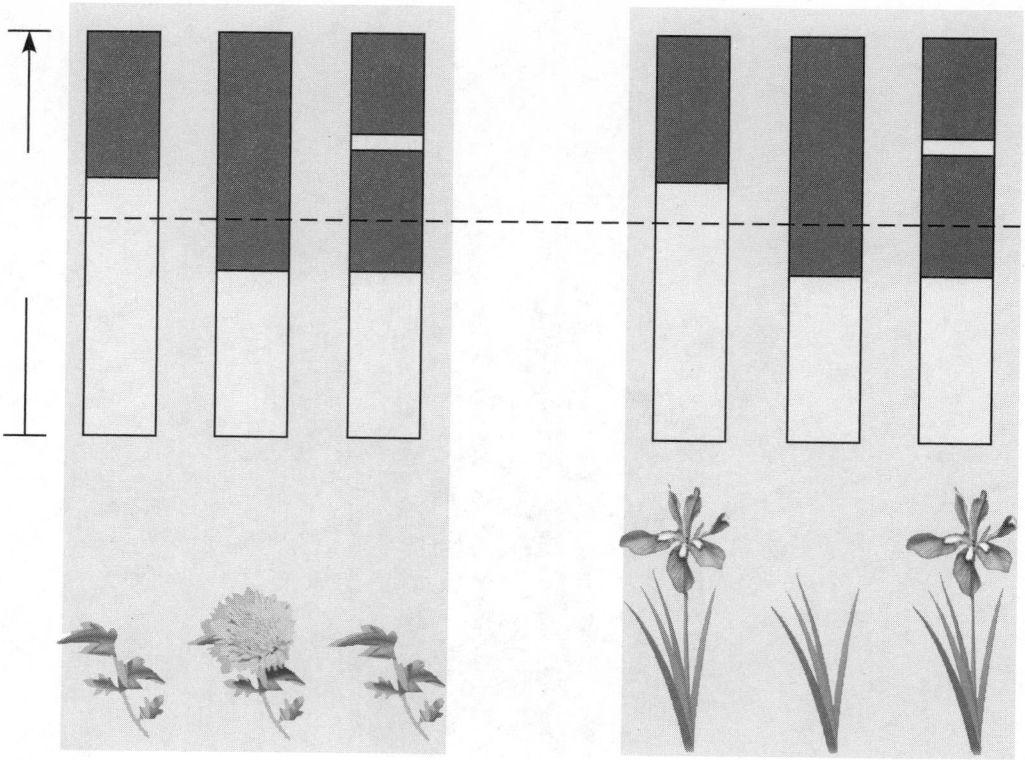

Figure 39.22 How does interrupting the dark period with a brief exposure to light affect flowering? page 807

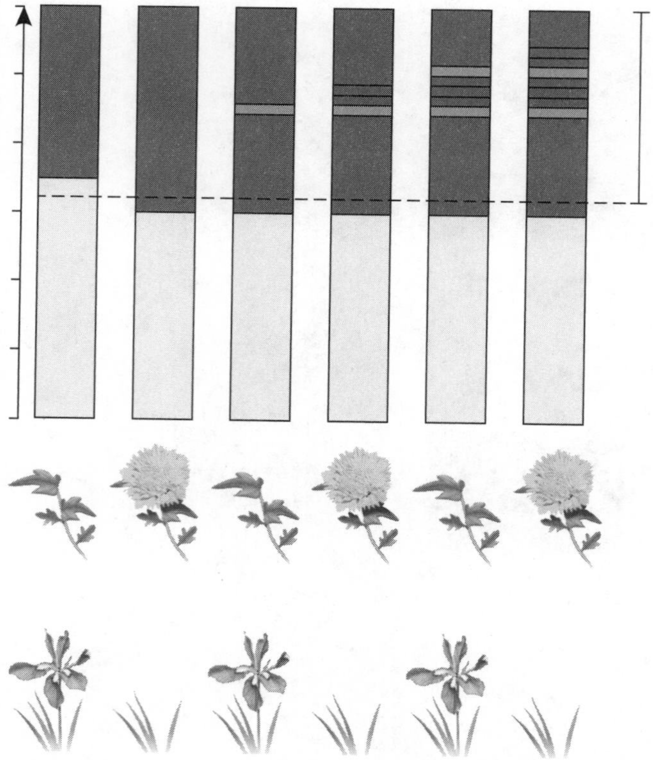

Figure 39.23 Is phytochrome the pigment that measures the interruption of dark periods in photoperiodic response? page 807

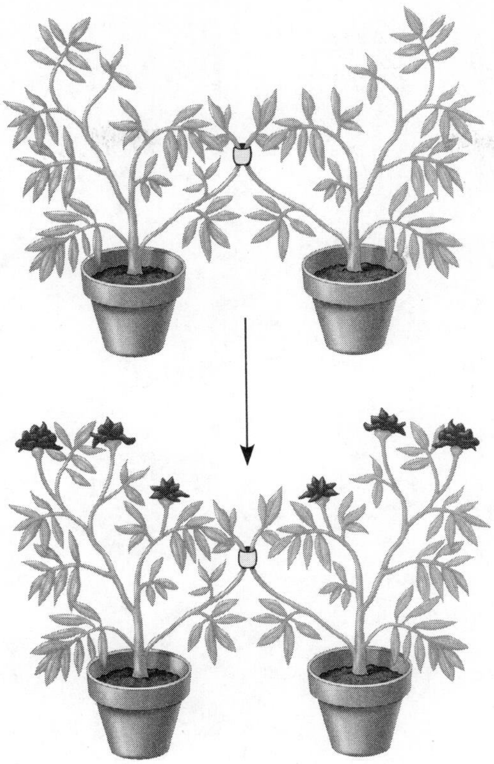

Figure 39.24 Is there a flowering hormone? page 808

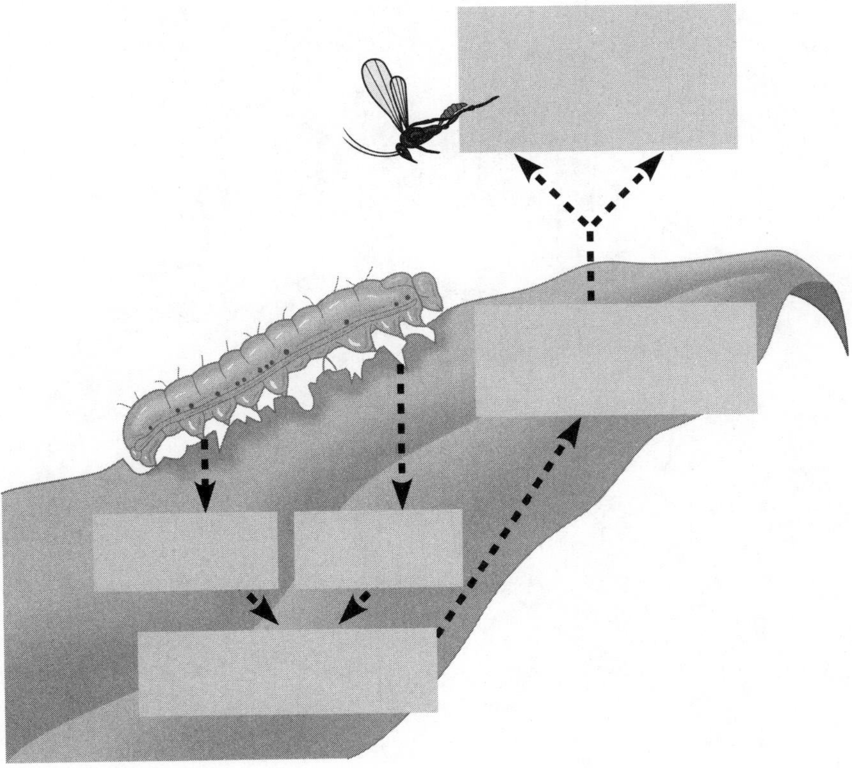

Figure 39.29 A maize leaf "recruits" a parasitoid wasp as a defensive response to an herbivore, an army-worm caterpillar, page 813

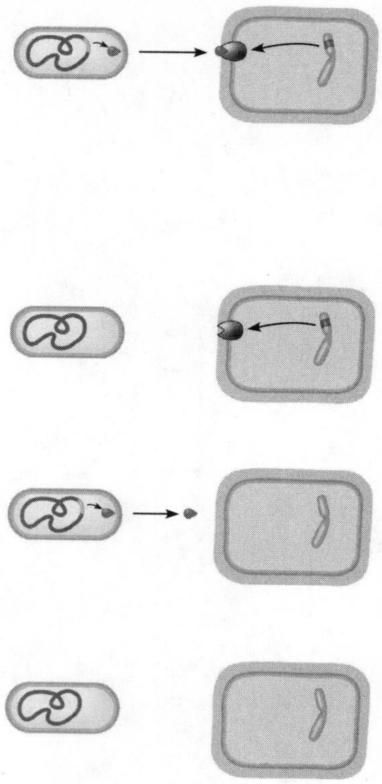

Figure 39.30 Gene-for-gene resistance of plants to pathogens: the receptor-ligand model, page 814

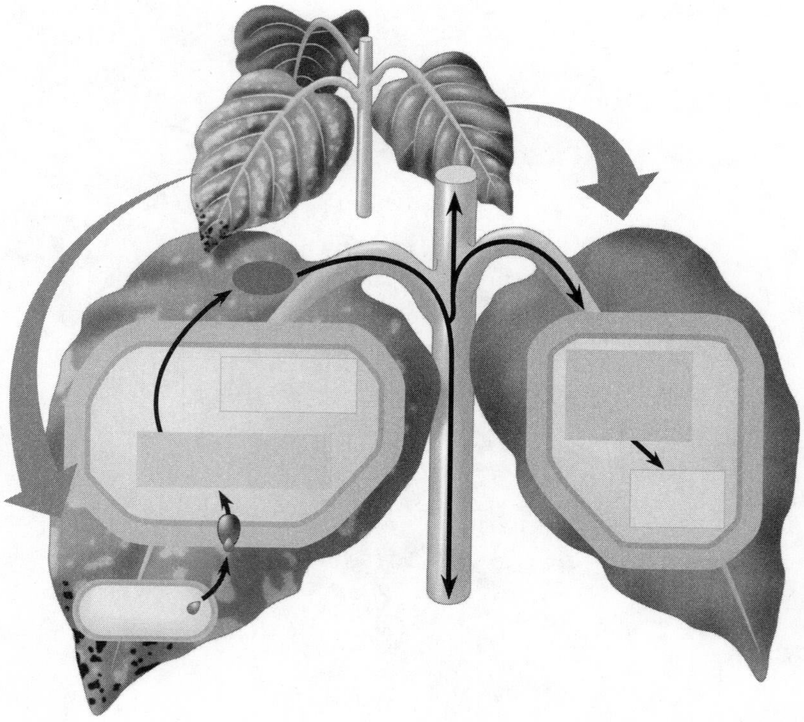

Figure 39.31 Defense responses against an avirulent pathogen, page 815

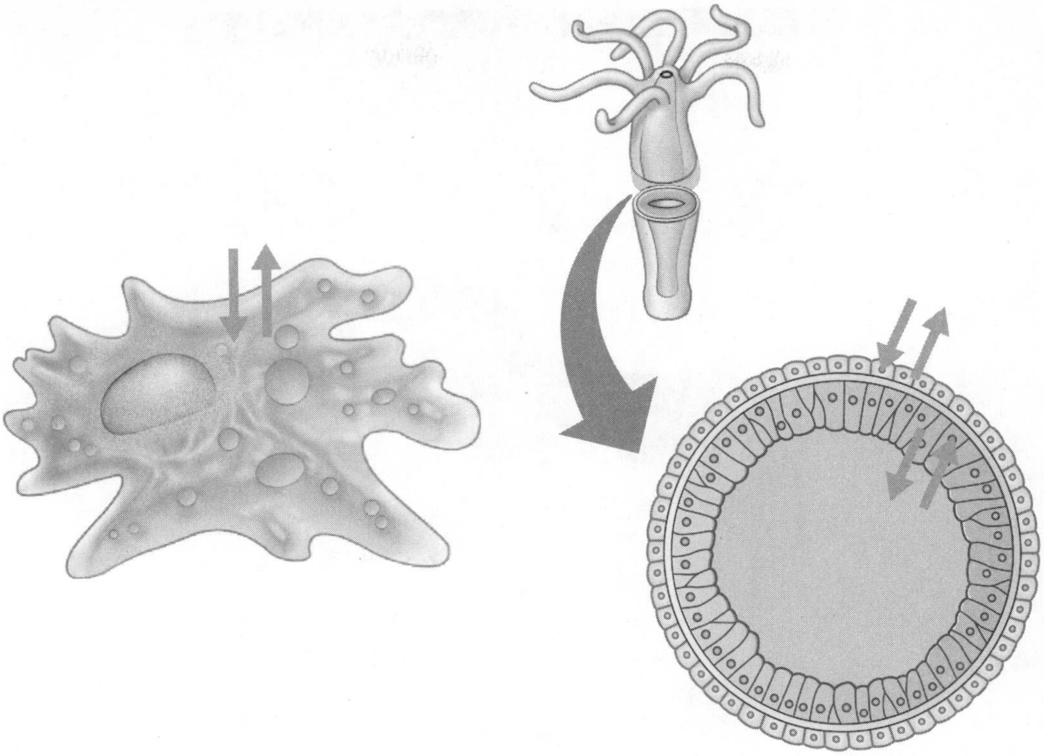

Figure 40.3 Contact with the environment, page 821

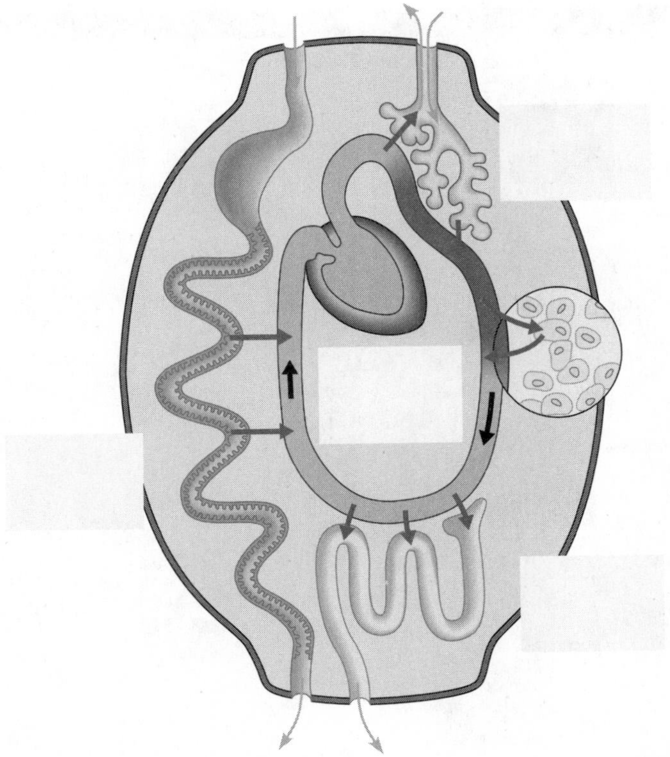

Figure 40.4 Internal exchange surfaces of complex animals, page 822

[redacted]

Figure 40.5 Structure and function in animal tissues, page 824

[redacted]

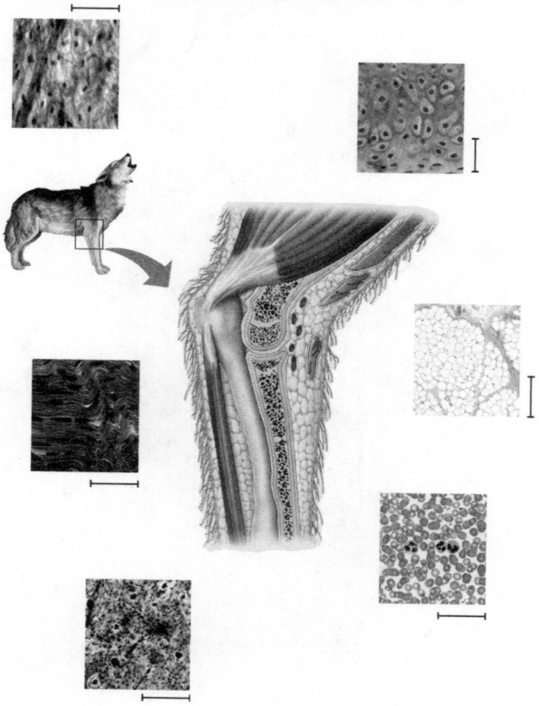

Figure 40.5 Structure and function in animal tissues, page 825

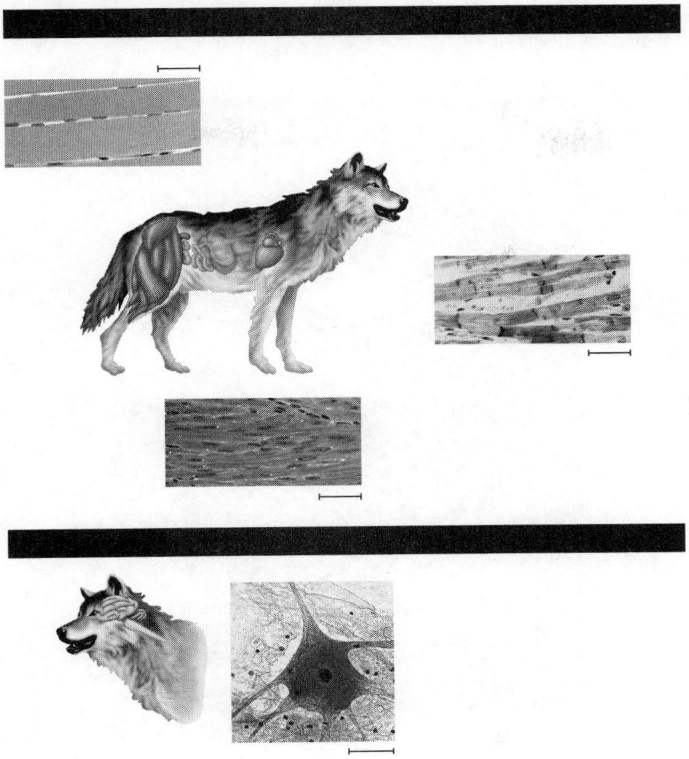

Figure 40.5 Structure and function in animal tissues, page 826

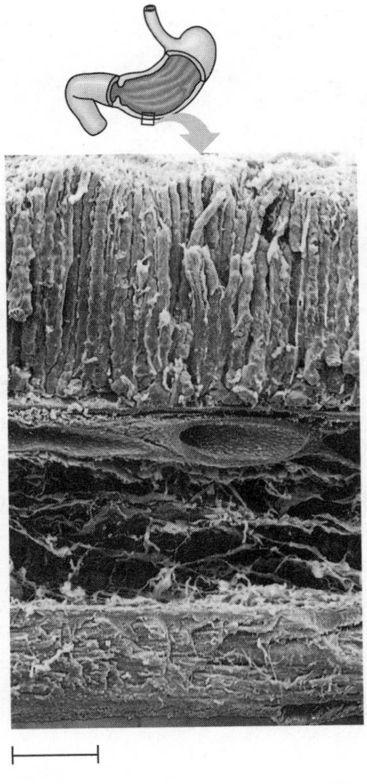

Figure 40.6 Tissue layers of the stomach, a digestive organ, page 827

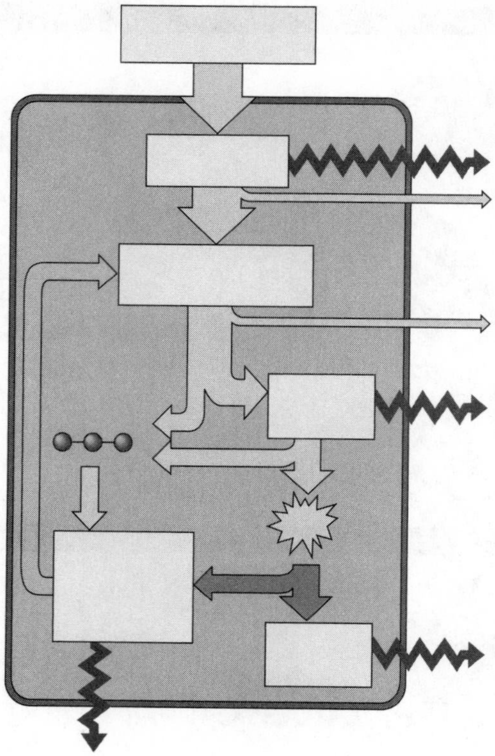

Figure 40.7 Bioenergetics of an animal: an overview, page 828

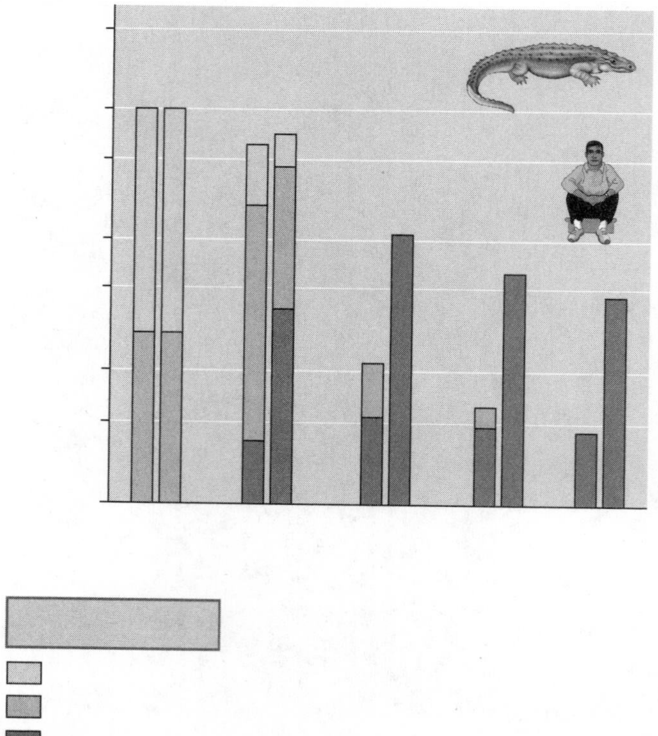

Figure 40.9 Maximum metabolic rates over different time spans, page 830

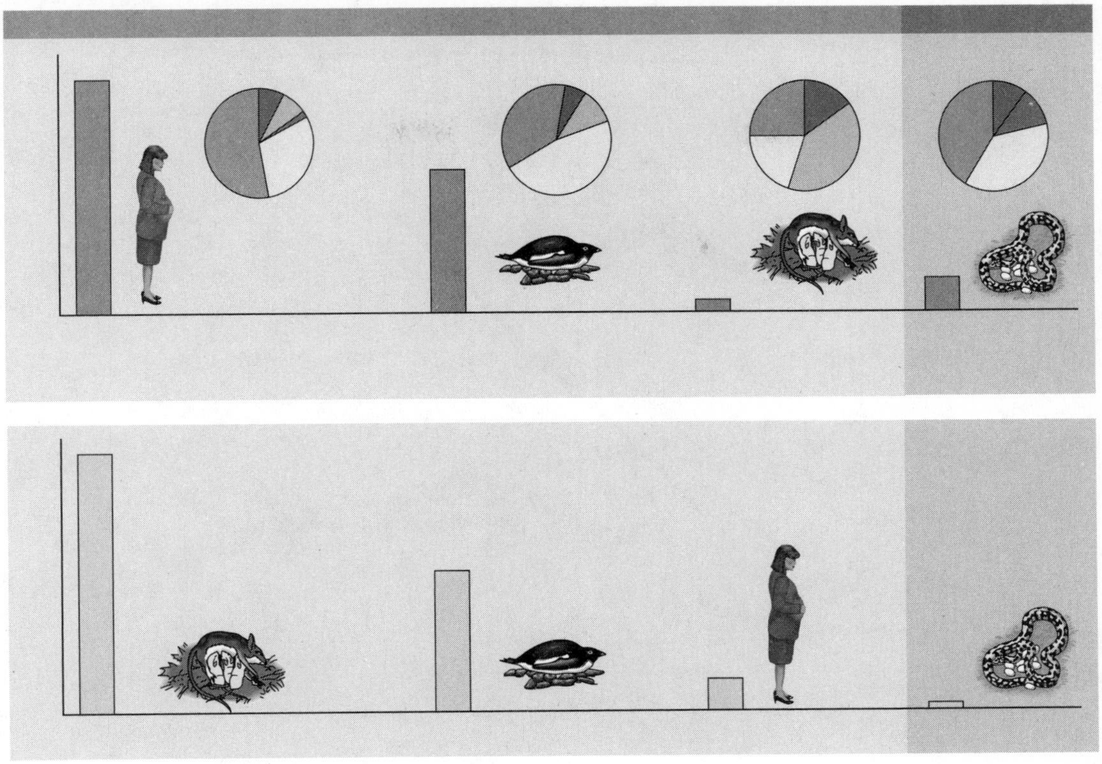

Figure 40.10 Energy budgets for four animals, page 831

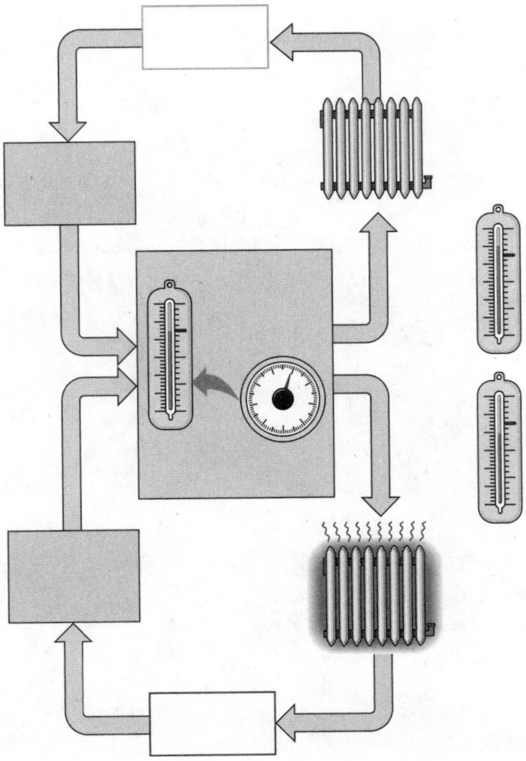

Figure 40.11 A nonliving example of negative feedback: control of room temperature, page 832

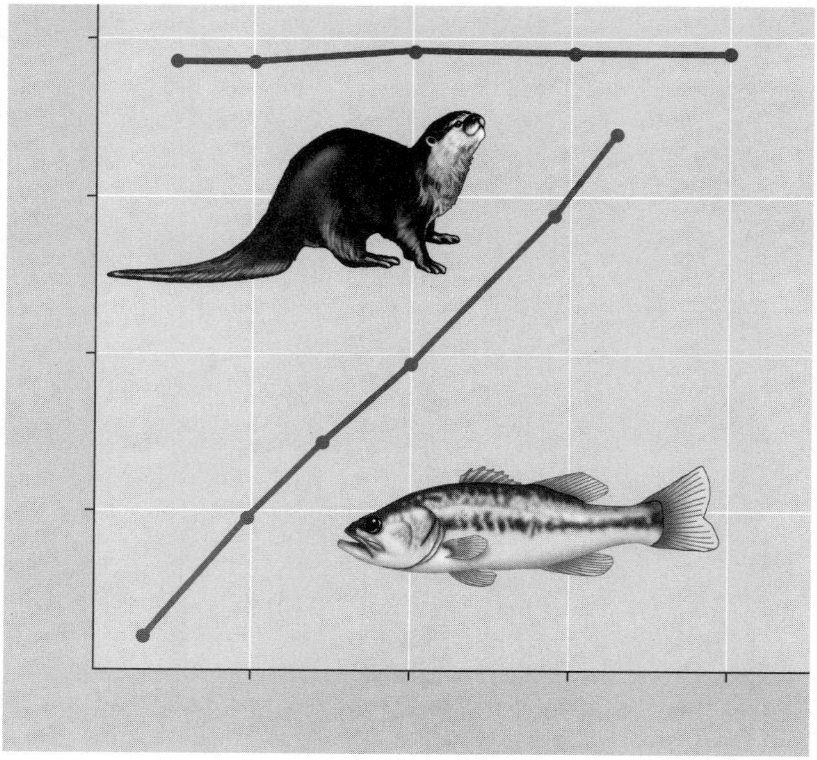

Figure 40.12 The relationship between body temperature and environmental temperature in an aquatic endotherm and ecotherm, page 834

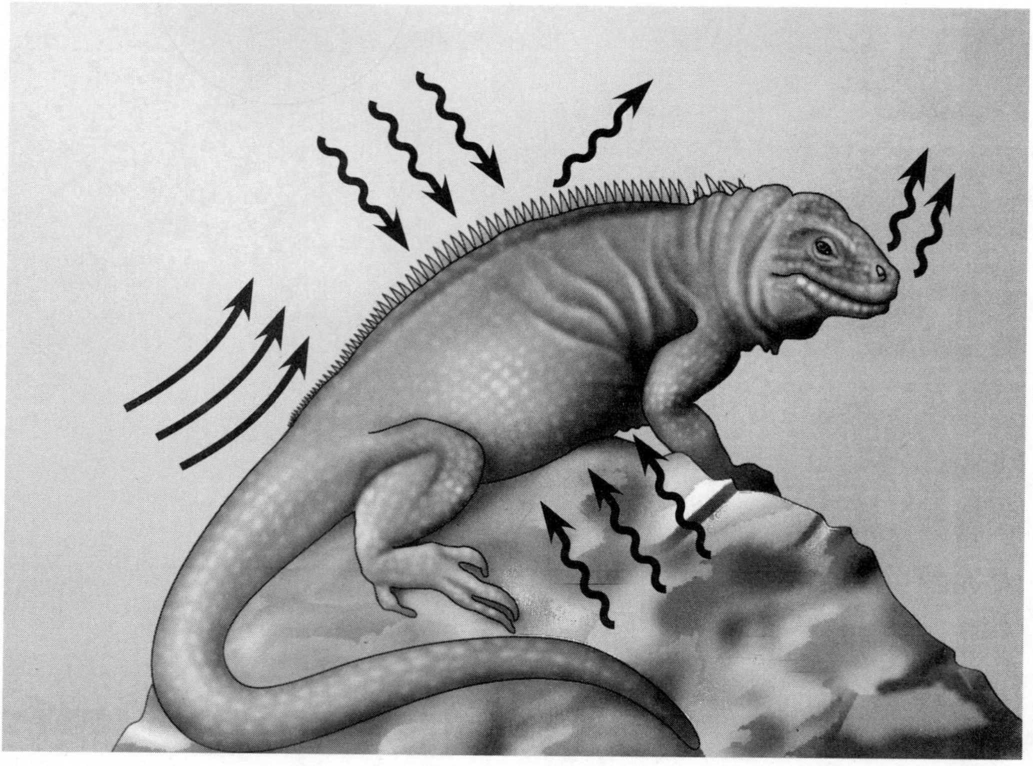

Figure 40.13 Heat exchange between an organism and its environment, page 835

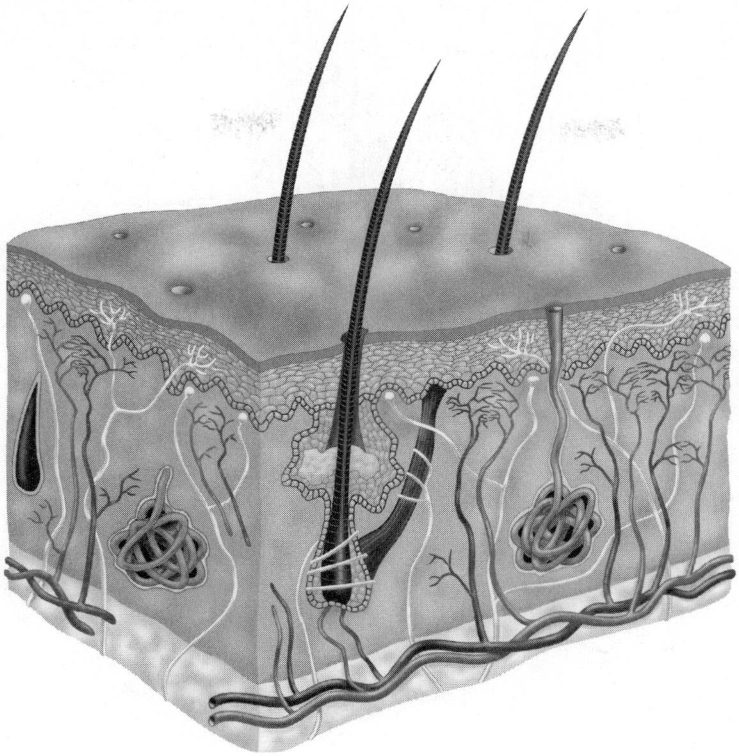

Figure 40.14 Mammalian integumentary system, page 835

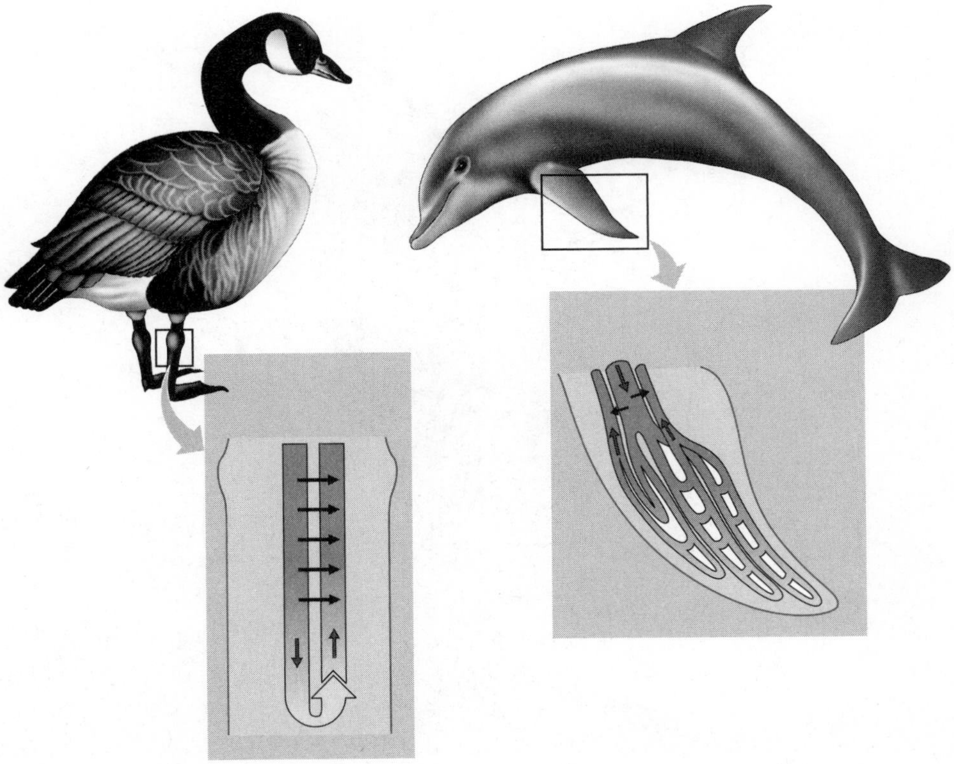

Figure 40.15 Countercurrent heat exchangers, page 836

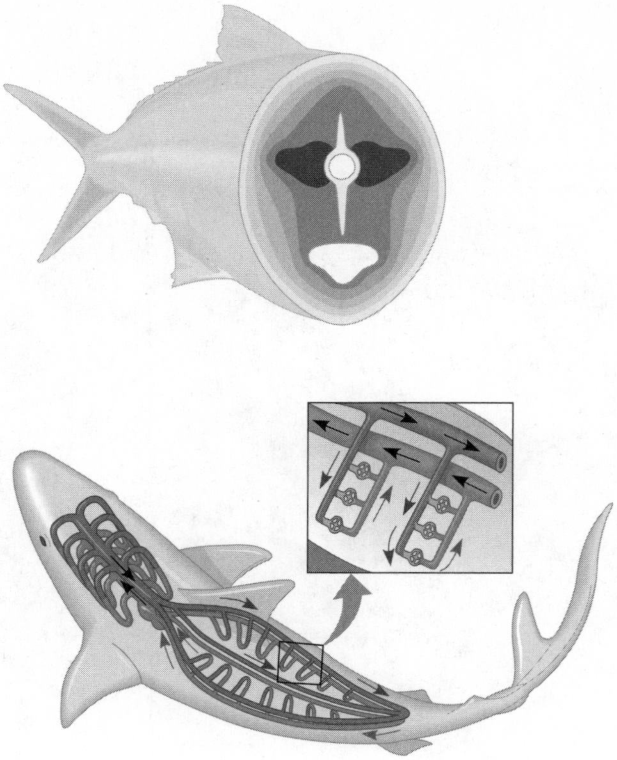

Figure 40.16 Thermoregulation in large, active bony fishes and sharks, page 837

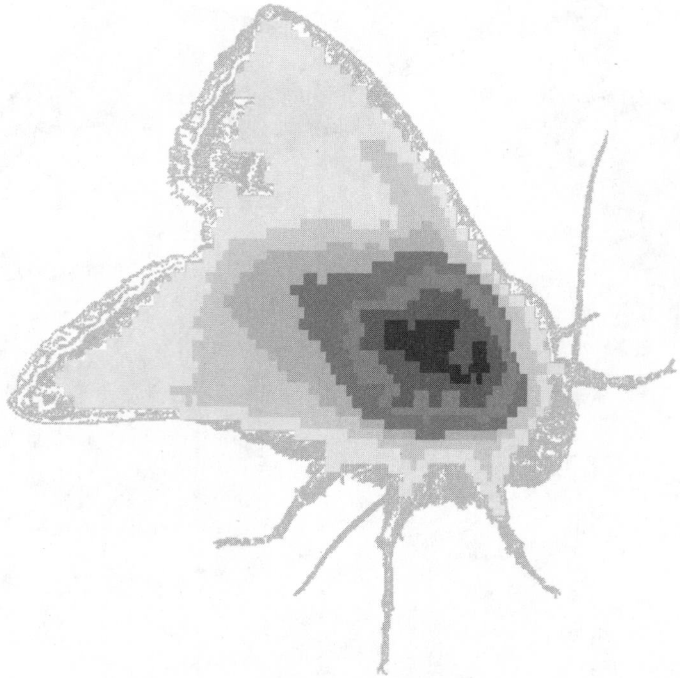

Figure 40.17 Internal temperature in the winter moth, page 837

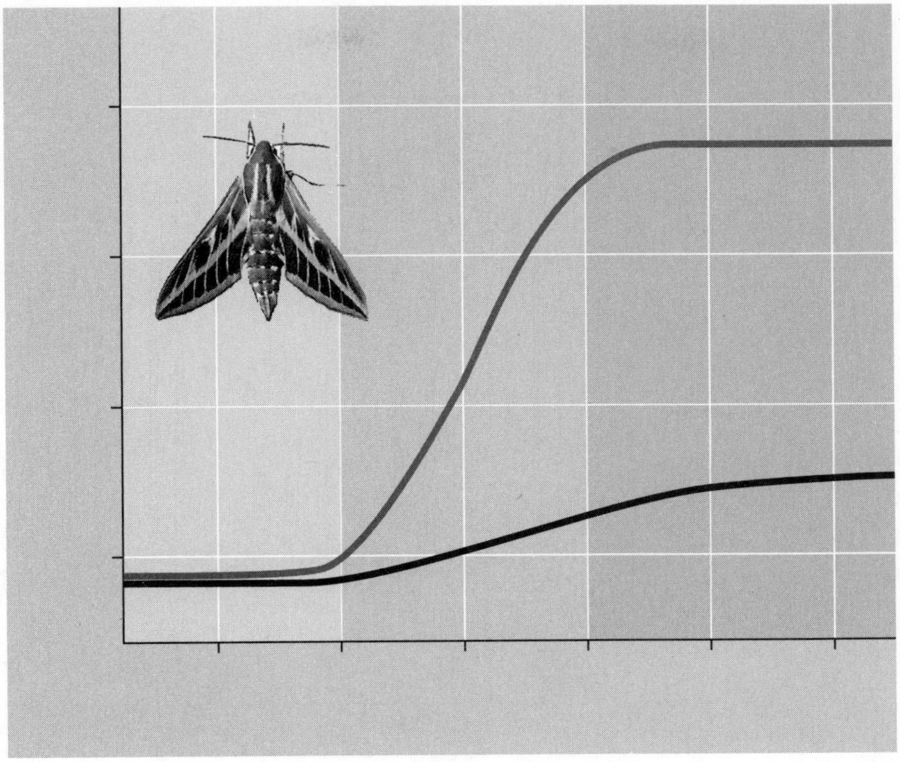

Figure 40.20 Preflight warmup in the hawkmoth, page 839

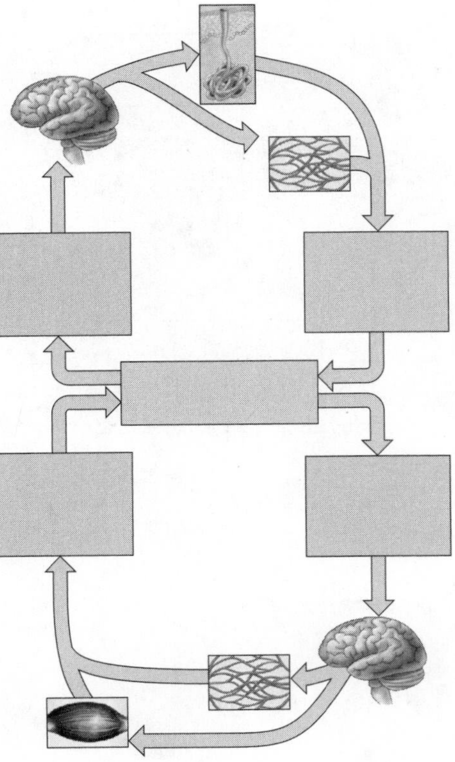

Figure 40.21 The thermostat function of the hypothalamus in human thermoregulation, page 839

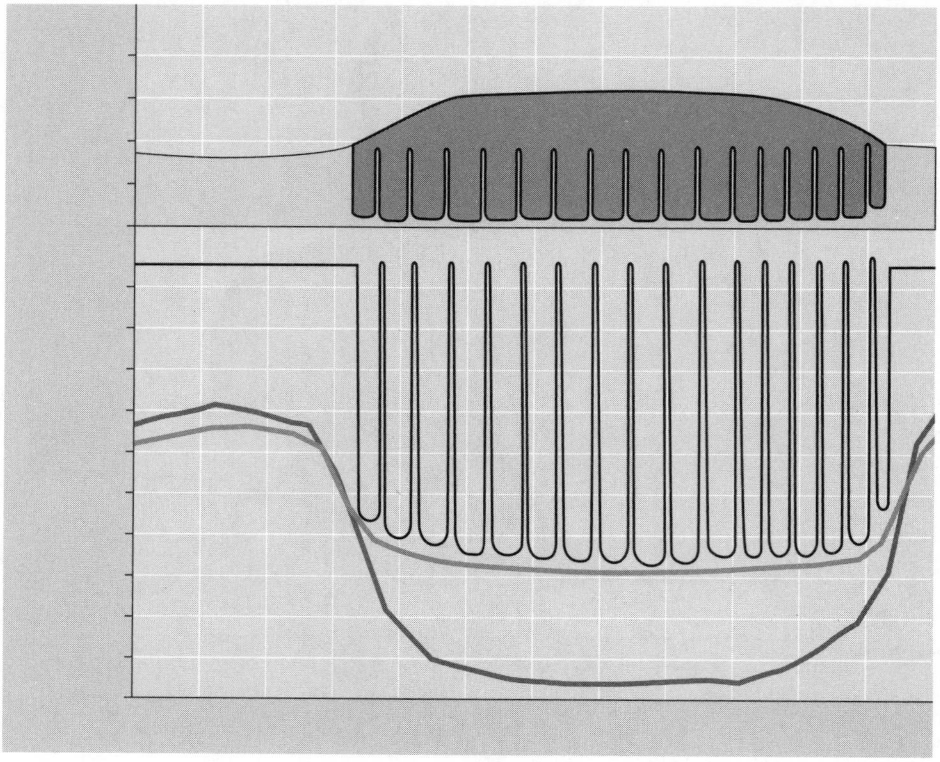

Figure 40.22 Body temperature and metabolism during hibernation in Belding's ground squirrels, page 840

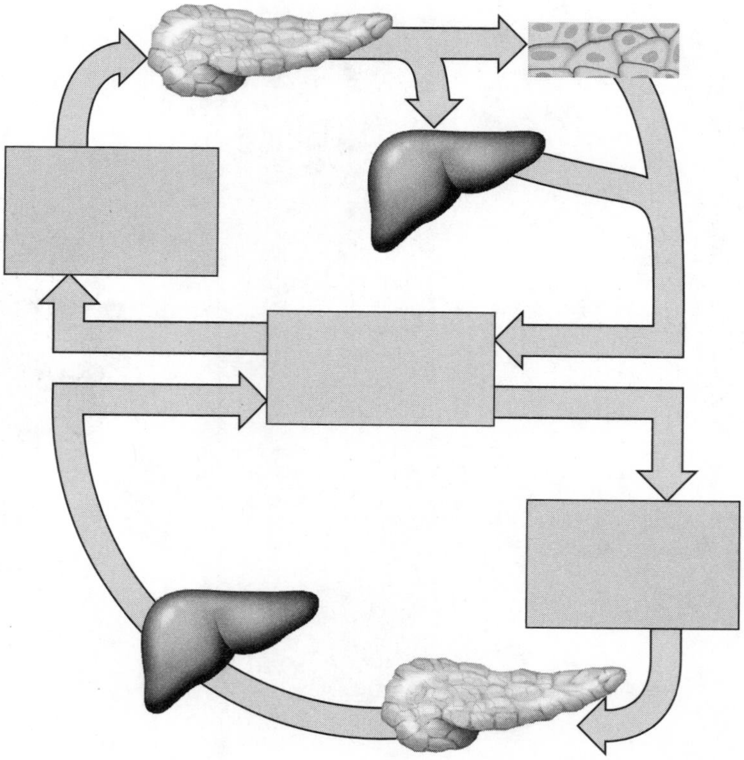

Figure 41.3 Homeostatic regulation of cellular fuel, page 846

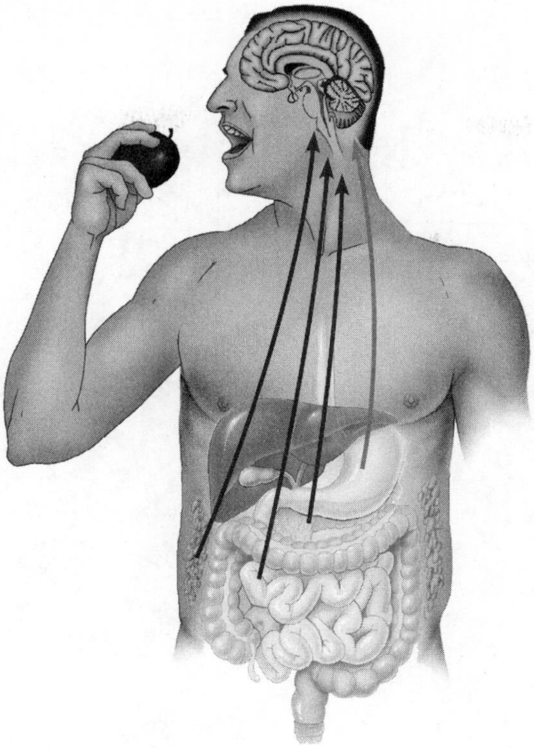

Figure 41.5 A few of the appetite-regulating hormones, page 847

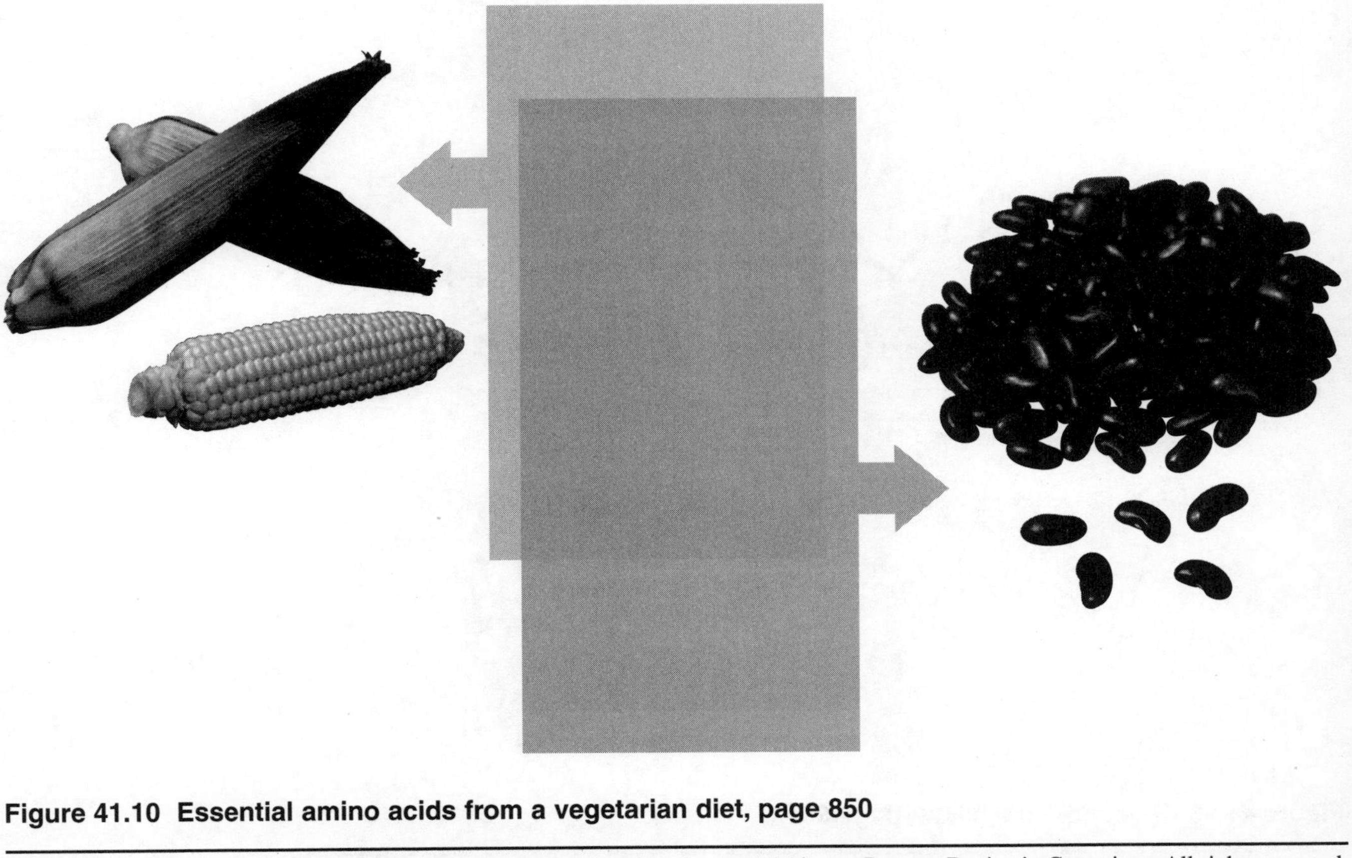

Figure 41.10 Essential amino acids from a vegetarian diet, page 850

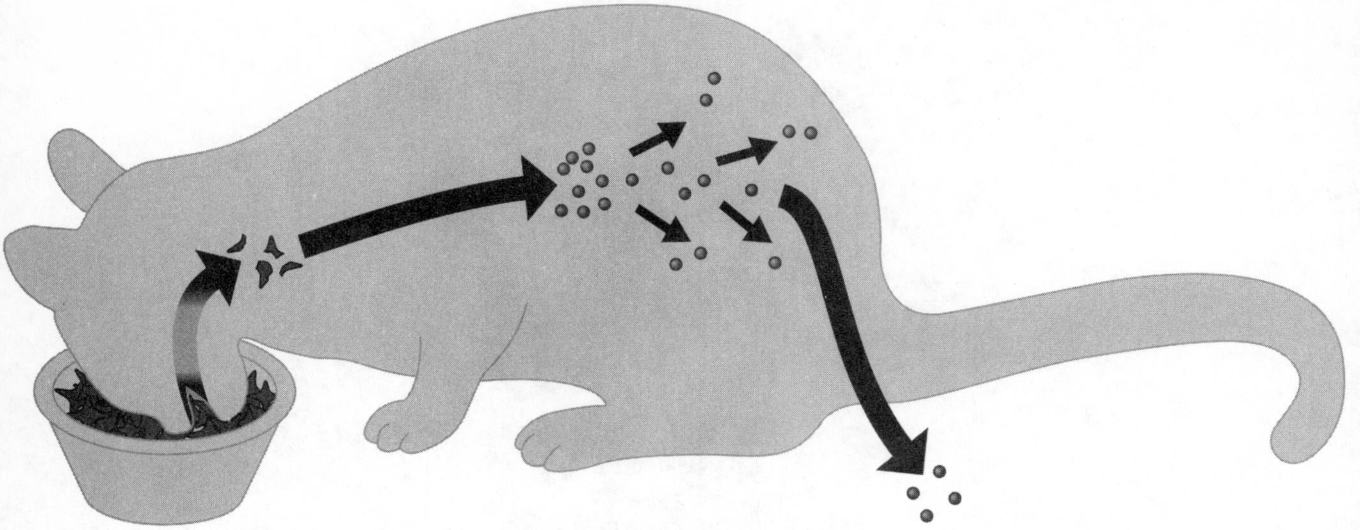

Figure 41.12 The four stages of food processing, page 853

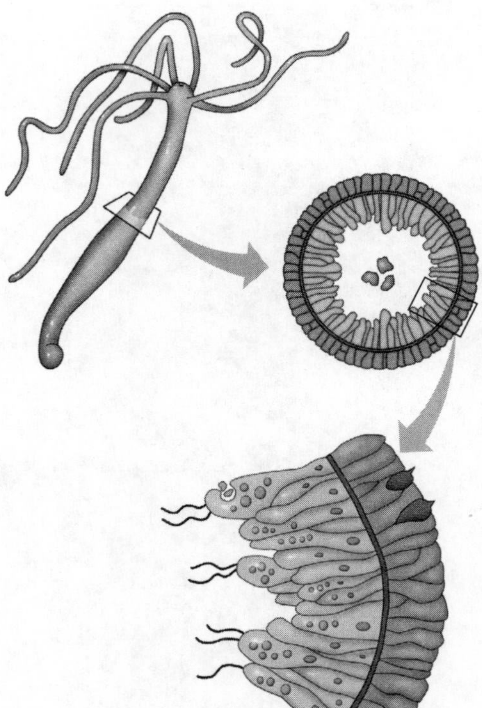

Figure 41.13 Digestion in a hydra, page 854

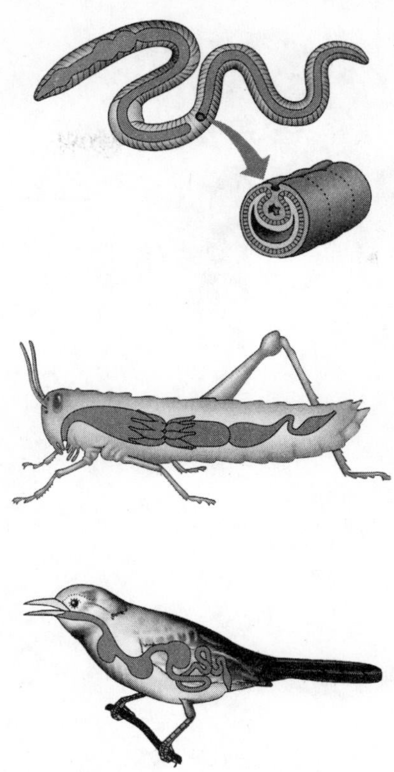

Figure 41.14 Variation in alimentary canals, page 854

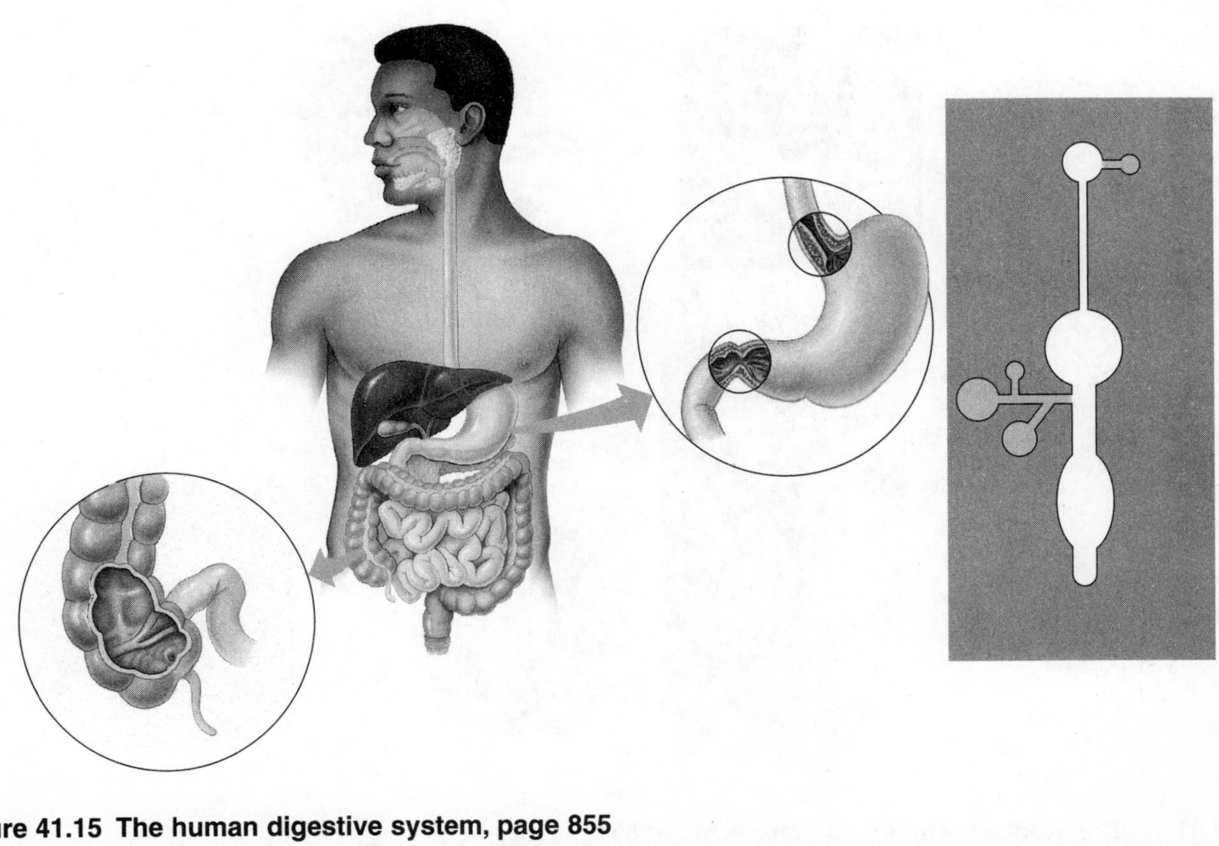

Figure 41.15 The human digestive system, page 855

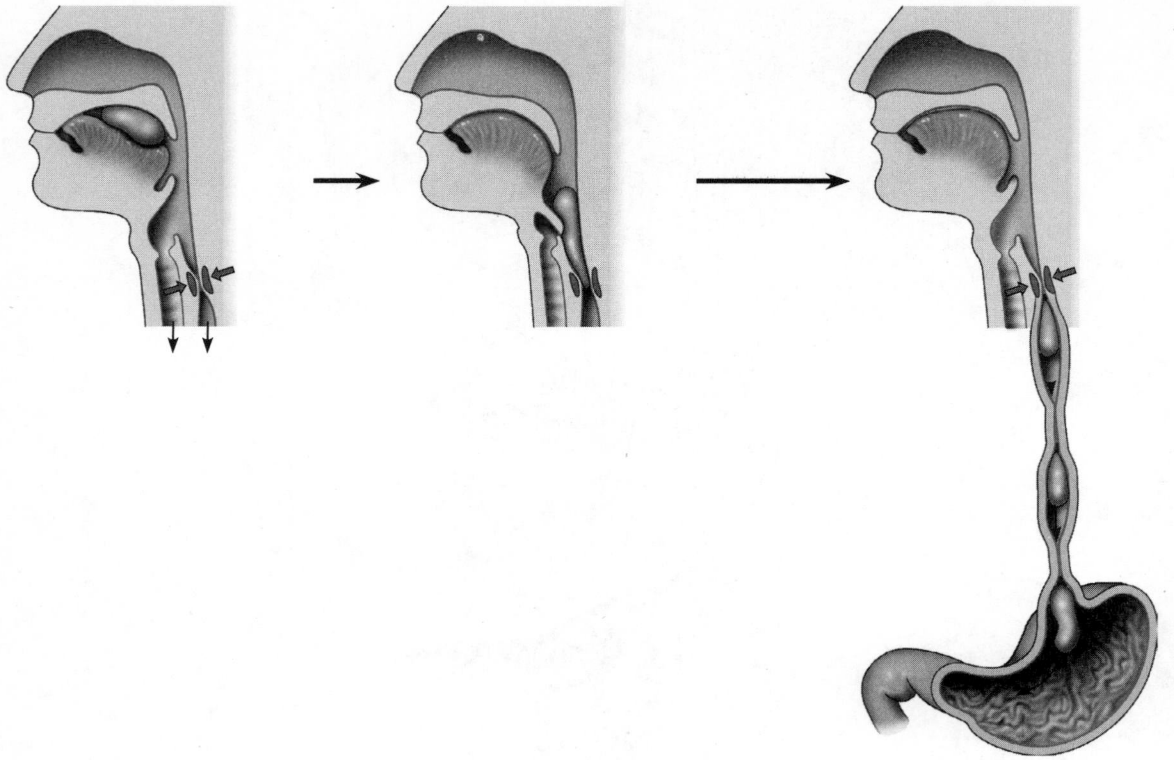

Figure 41.16 From mouth to stomach: the swallowing reflex and esophageal peristalsis, page 856

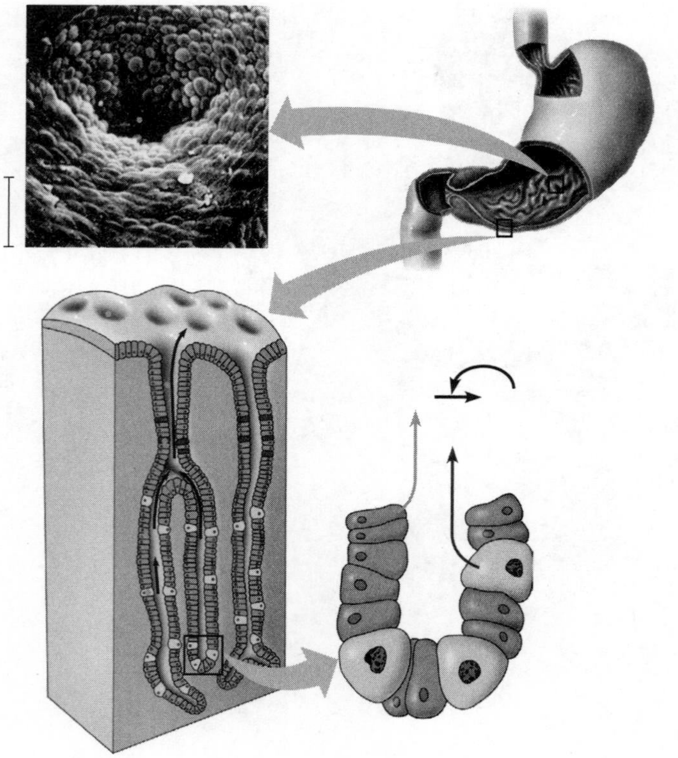

Figure 41.17 The stomach and its secretions, page 857

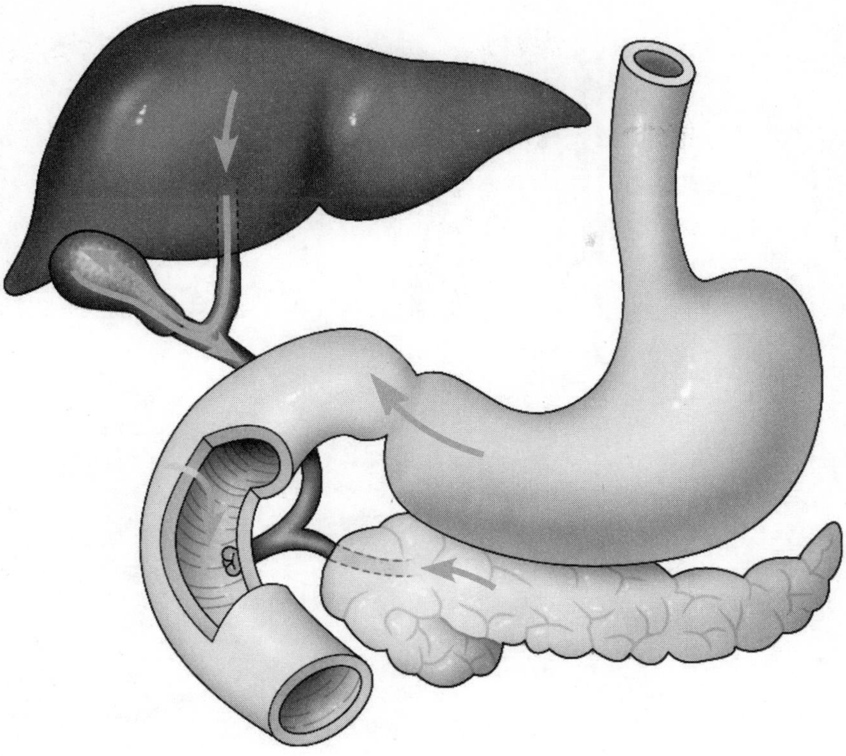

Figure 41.19 The duodenum, page 858

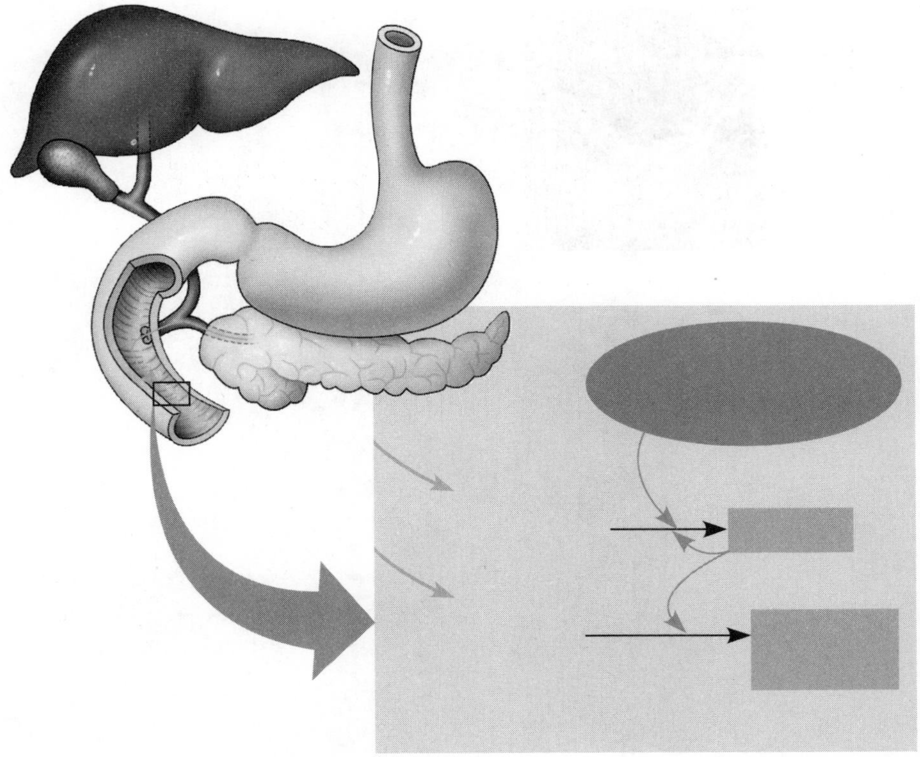

Figure 41.20 Protease activation, page 859

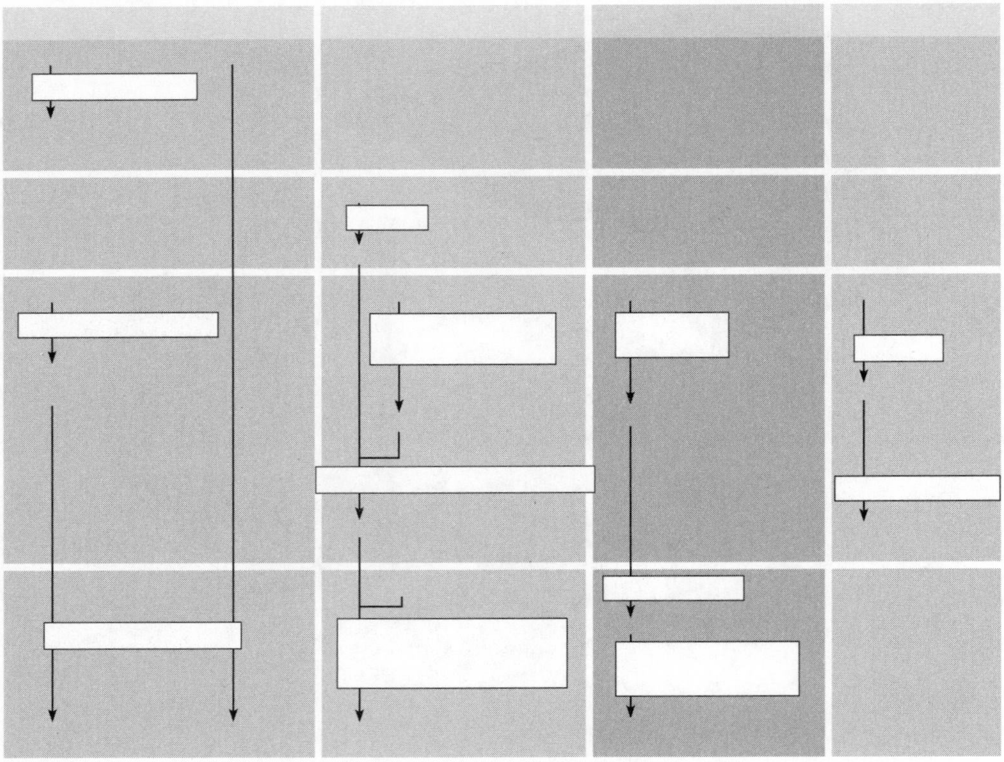

Figure 41.21 Flowchart of enzymatic digestion in the human digestive system, page 859

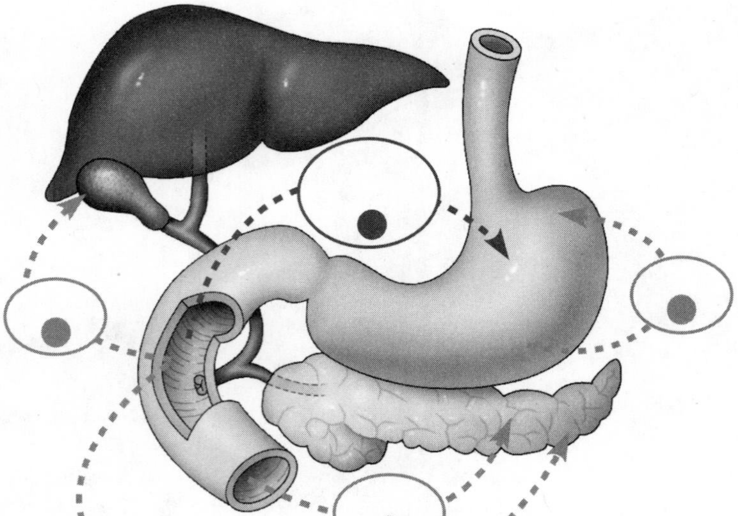

Figure 41.22 Hormonal control of digestion, page 860

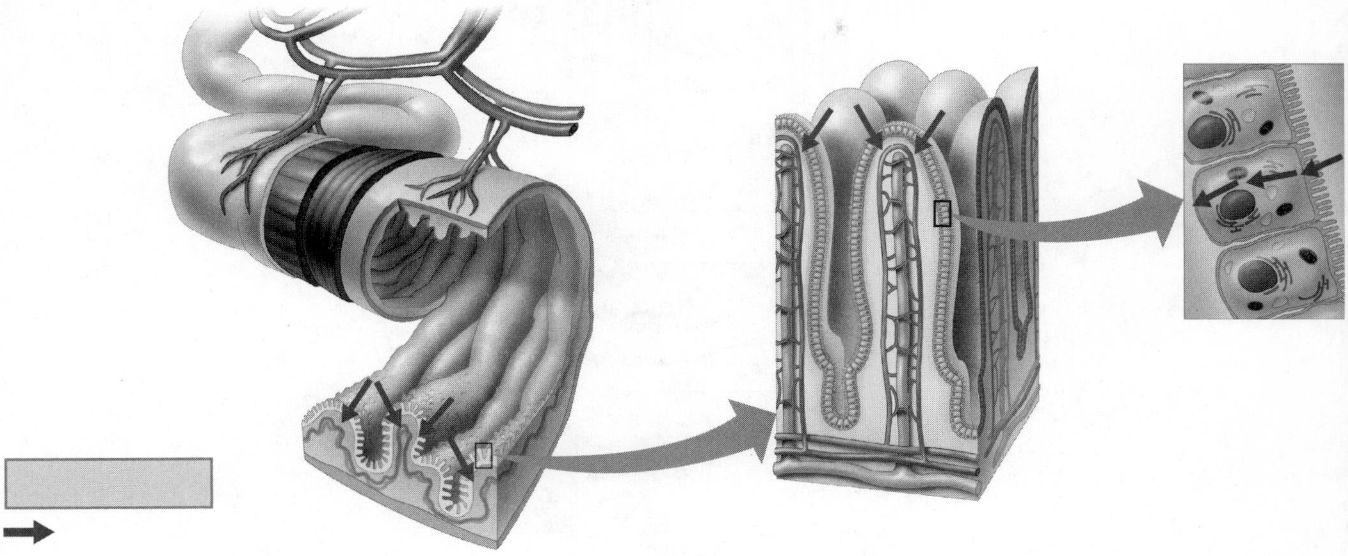

Figure 41.23 The structure of the small intestine, page 860

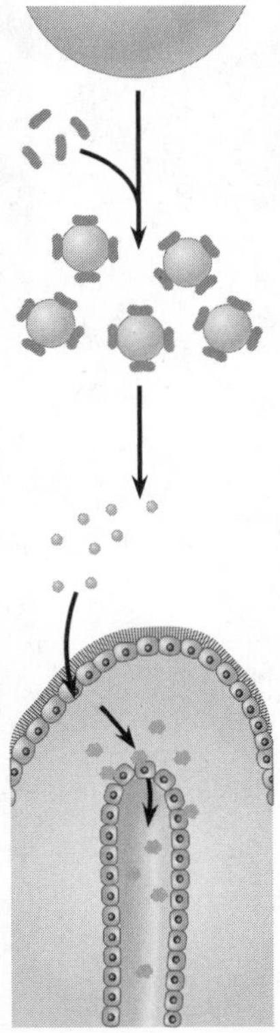

Figure 41.24 Digestion and absorption of fats, page 861

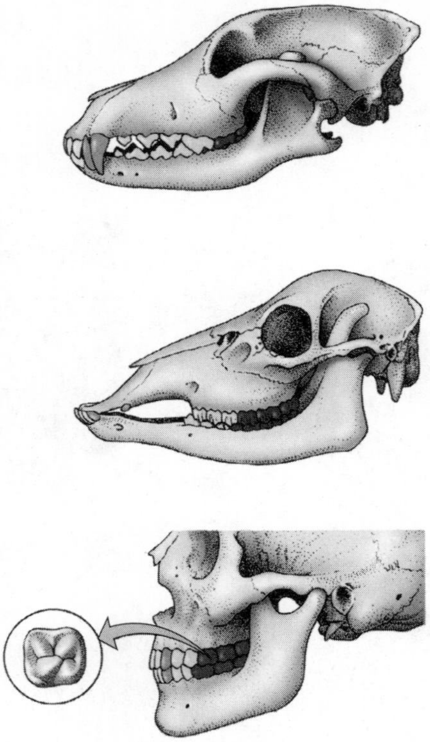

Figure 41.26 Dentition and diet, page 863

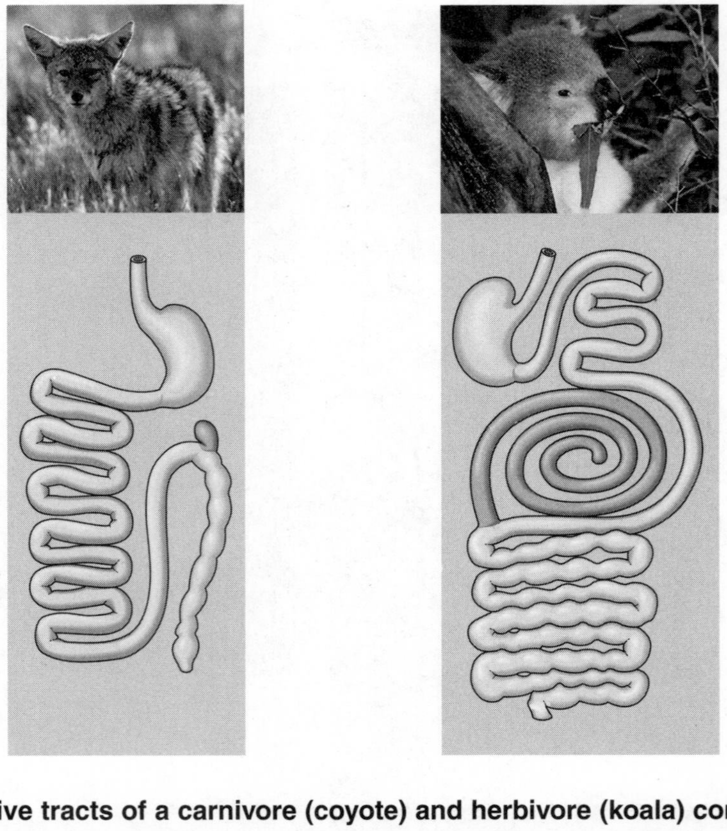

Figure 41.27 The digestive tracts of a carnivore (coyote) and herbivore (koala) compared, page 863

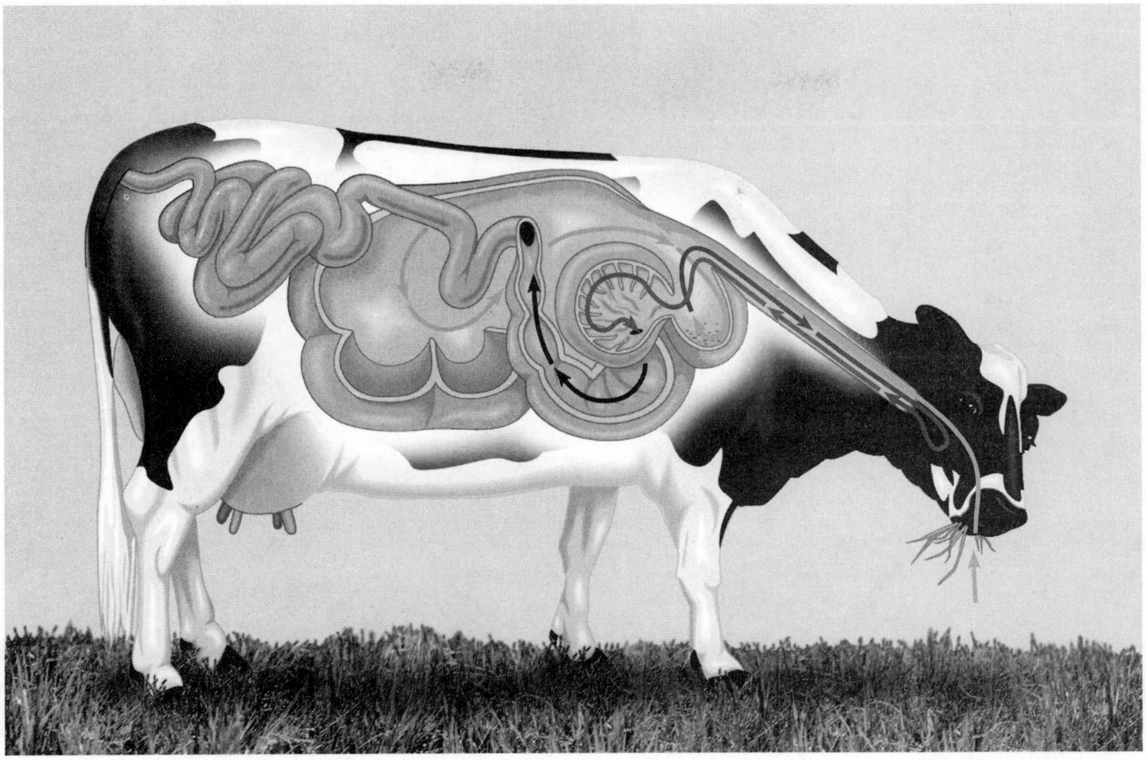

Figure 41.28 Ruminant digestion, page 864

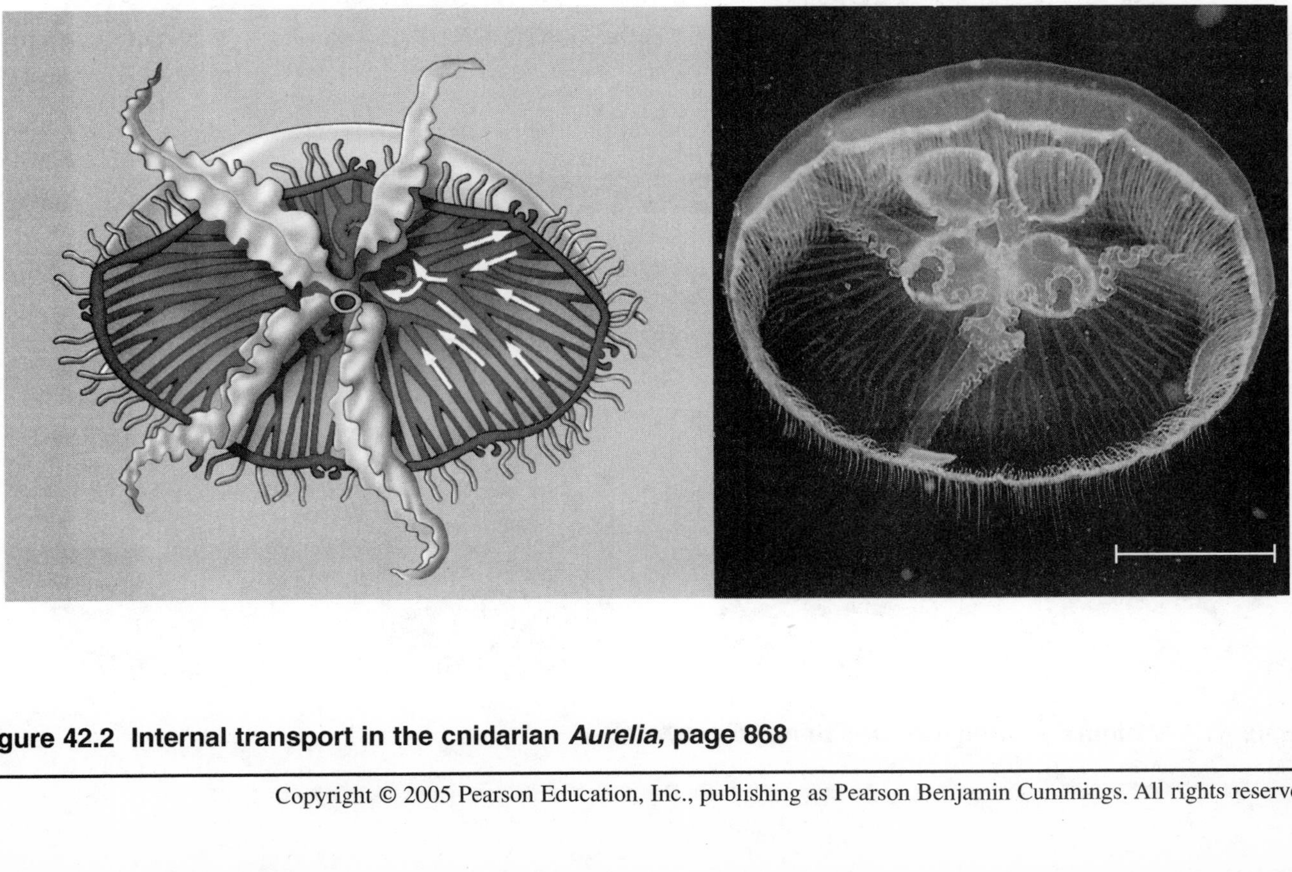

Figure 42.2 Internal transport in the cnidarian *Aurelia,* page 868

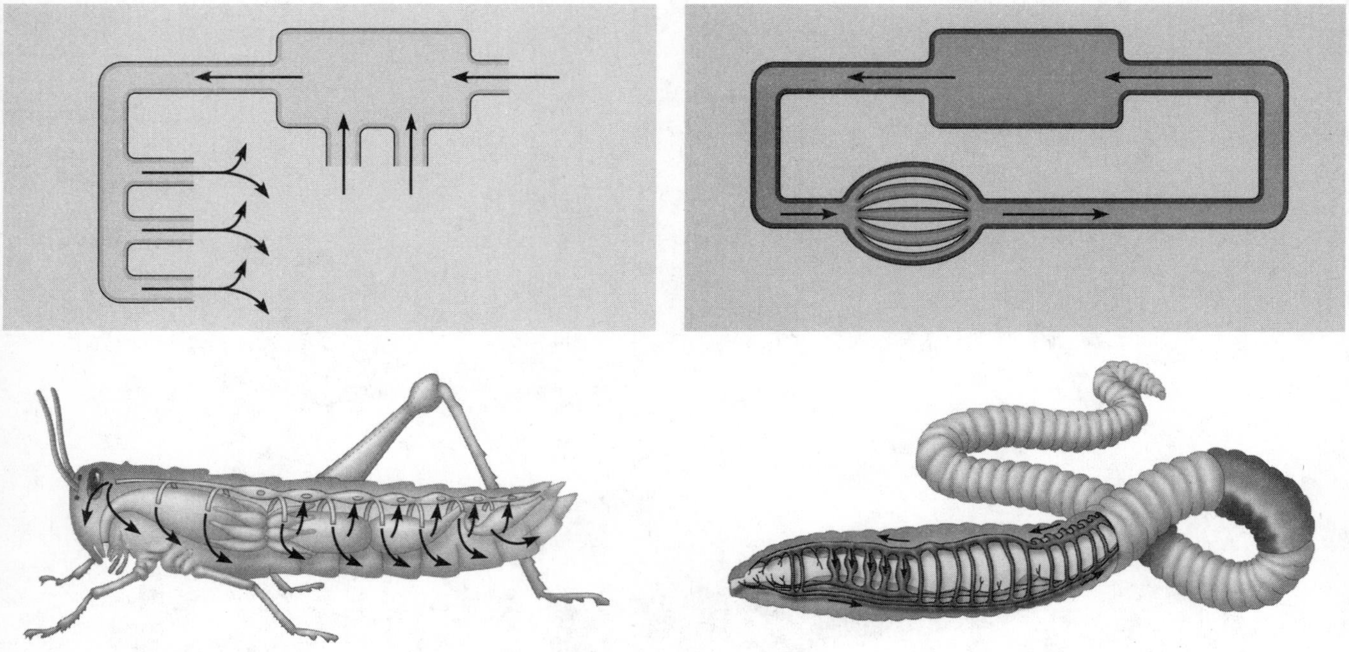

Figure 42.3 Open and closed circulatory systems, page 869

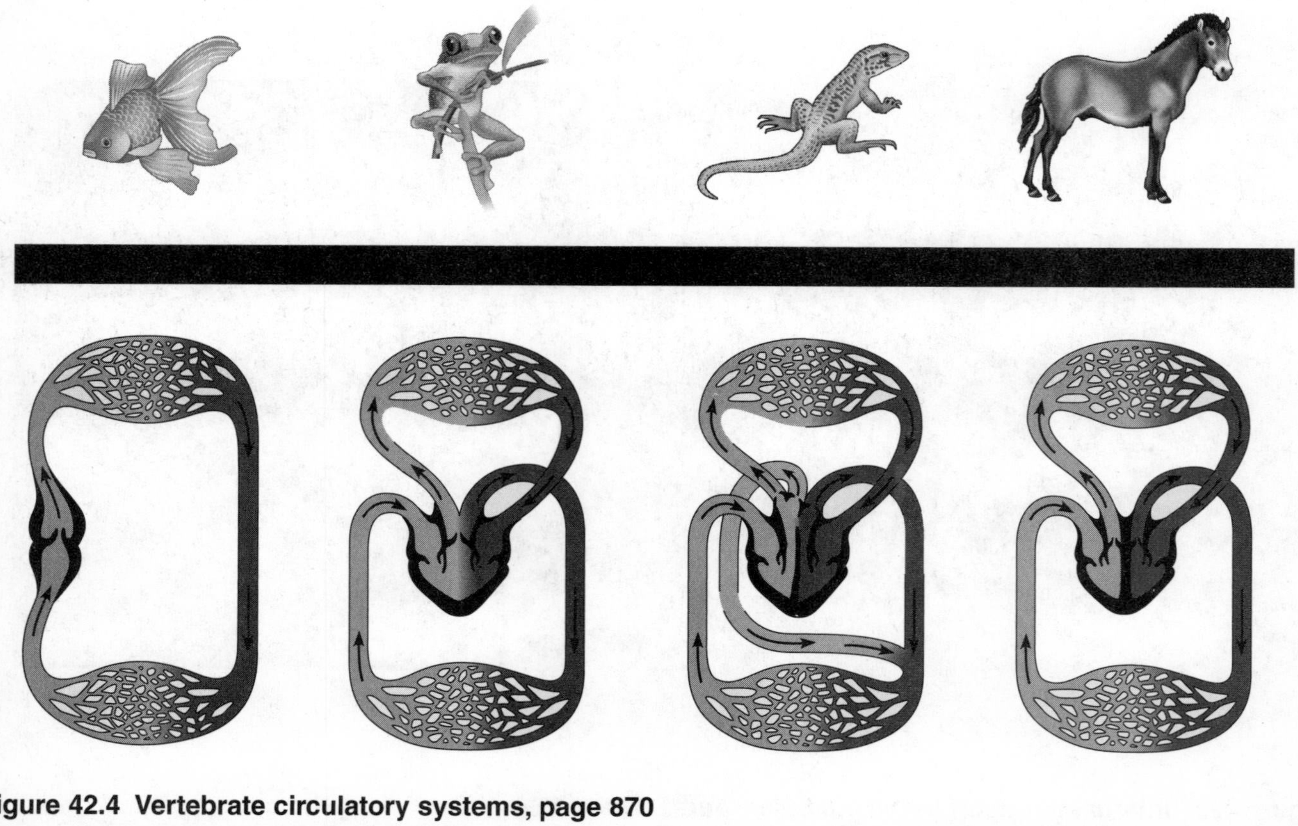

Figure 42.4 Vertebrate circulatory systems, page 870

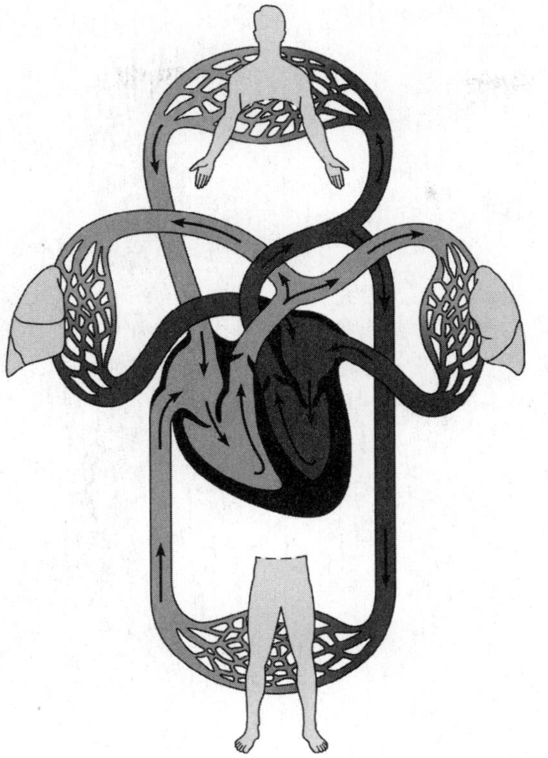

Figure 42.5 The mammalian cardiovascular system: an overview, page 872

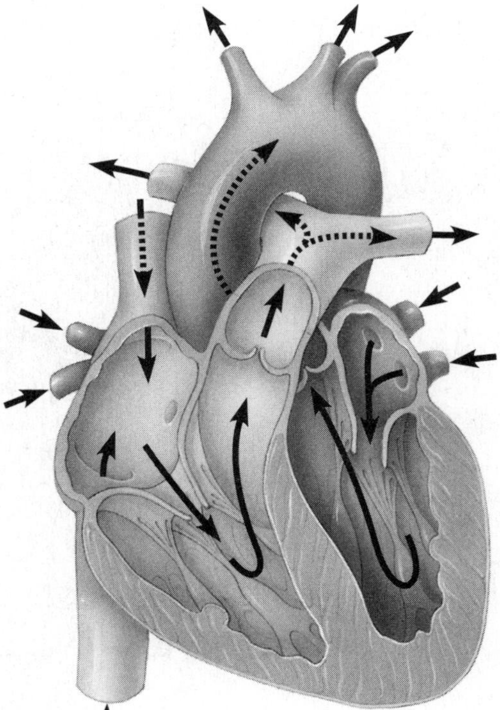

Figure 42.6 The mammalian heart: a closer look, page 872

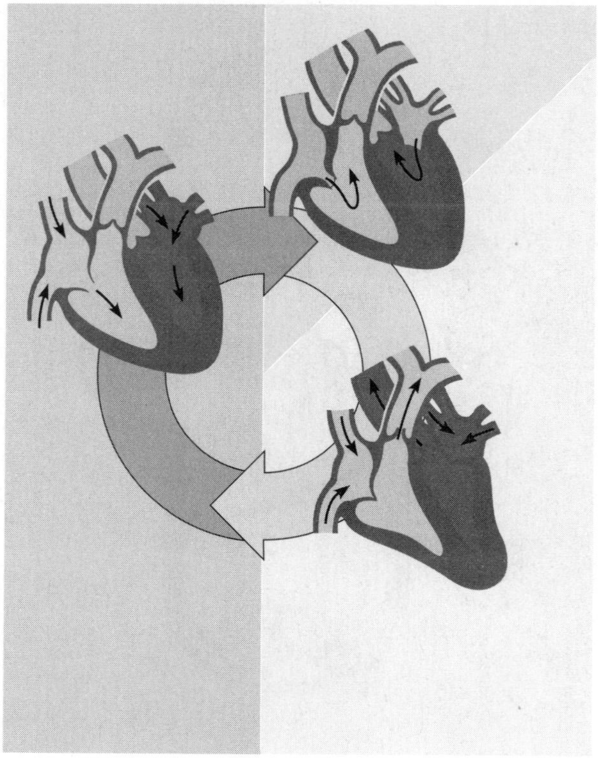

Figure 42.7 The cardiac cycle, page 873

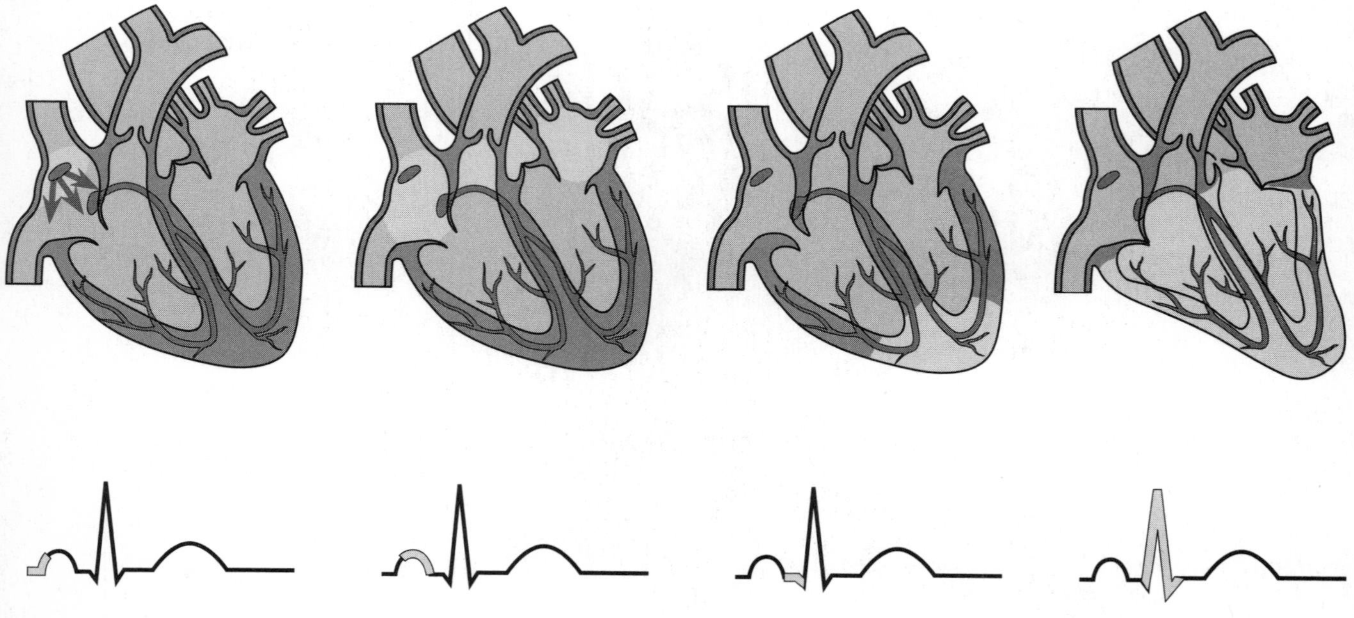

Figure 42.8 The control of heart rhythm, page 874

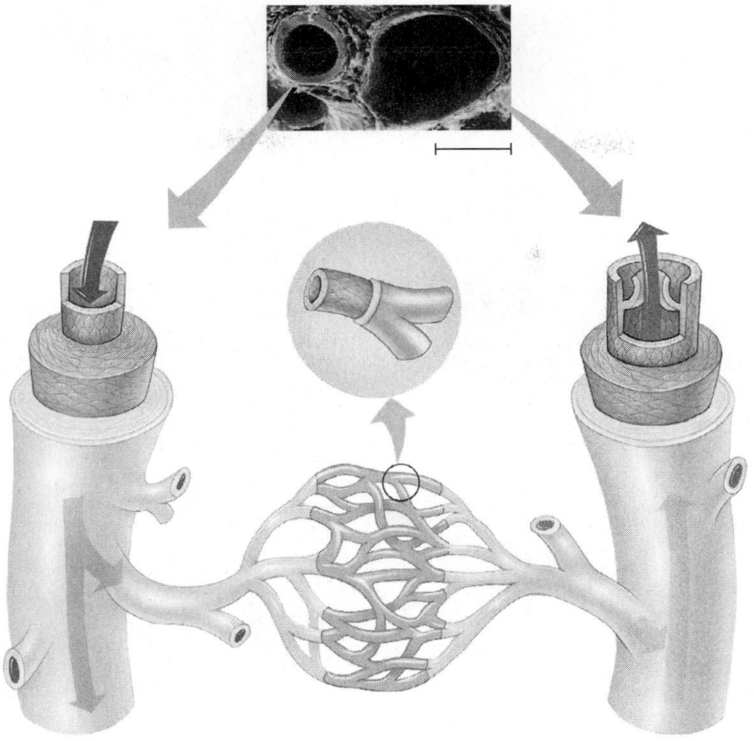

Figure 42.9 The structure of blood vessels, page 875

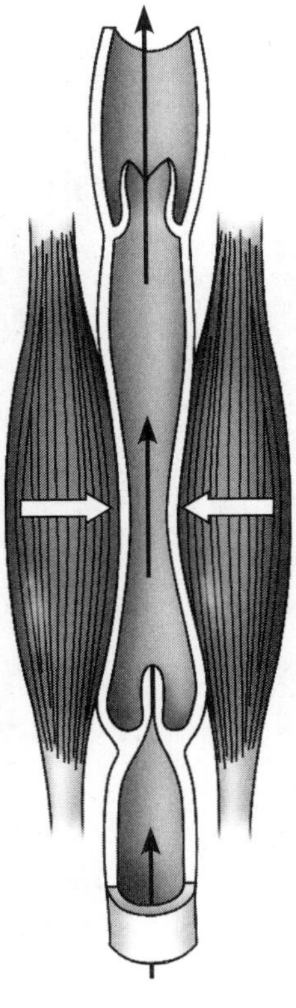

Figure 42.10 Blood flow in veins, page 875

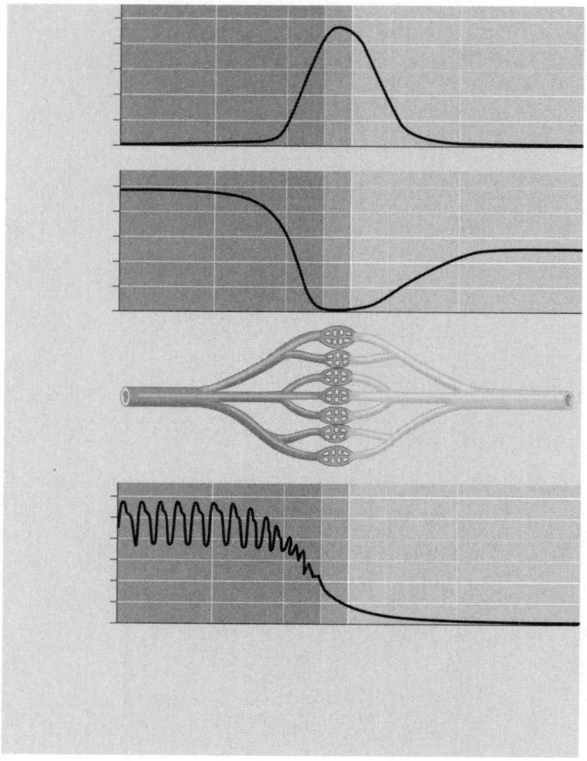

Figure 42.11 The interrelationship of blood flow velocity, cross-sectional area of blood vessels, and blood pressure, page 876

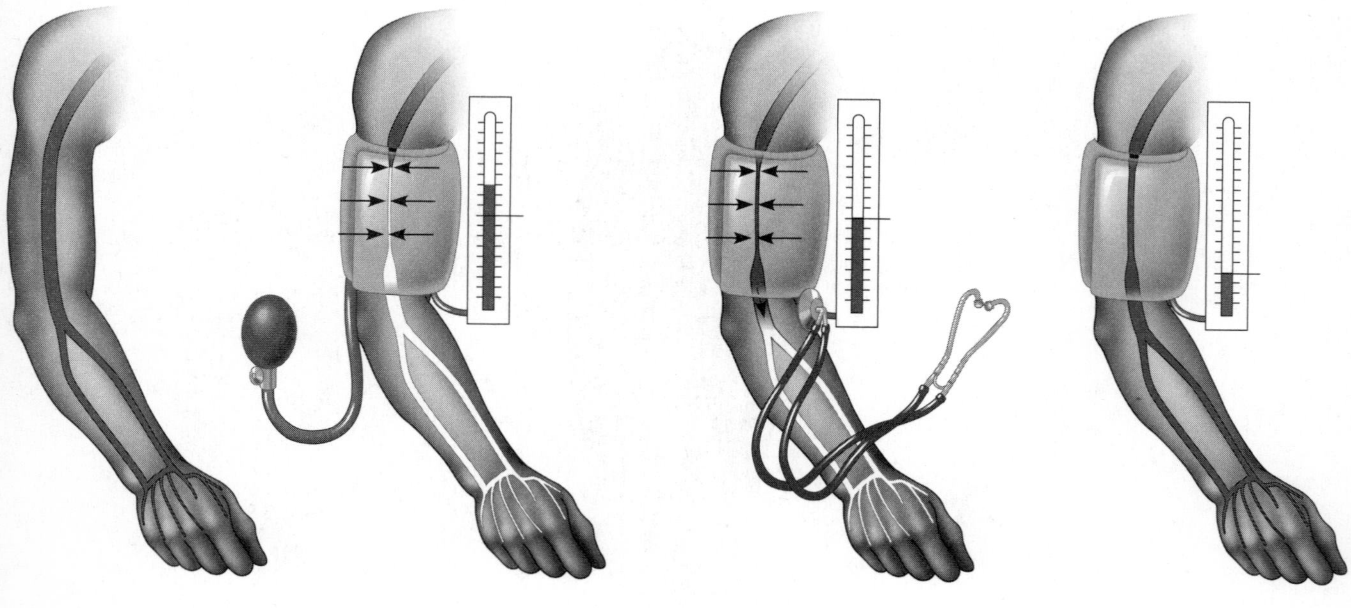

Figure 42.12 Measurement of blood pressure, page 877

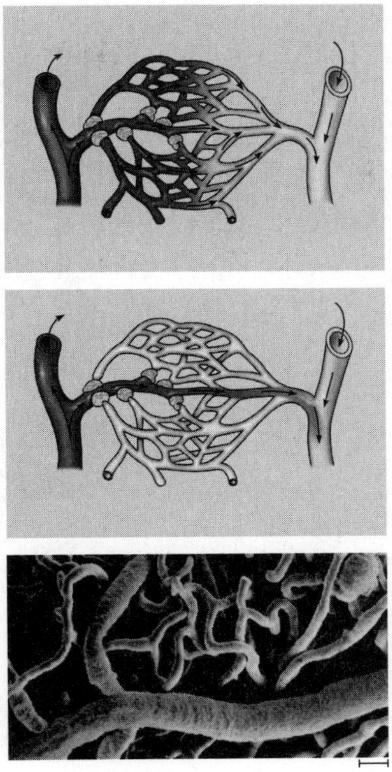

Figure 42.13 Blood flow in capillary beds, page 878

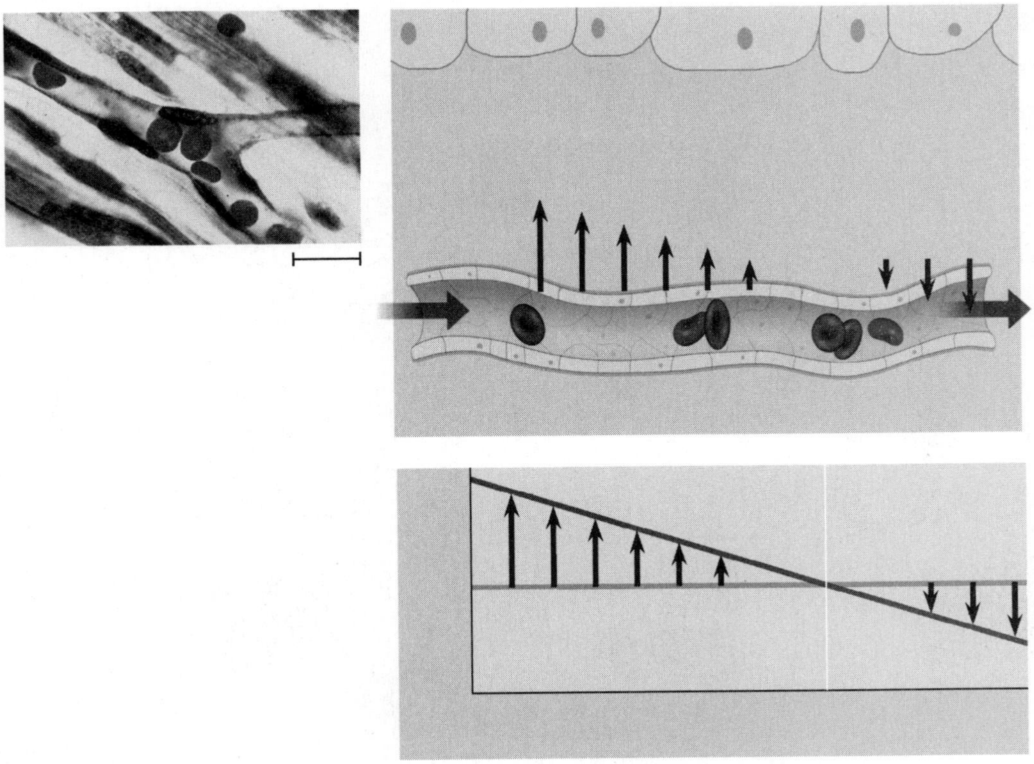

Figure 42.14 Fluid exchange between capillaries and the interstitial fluid, page 879

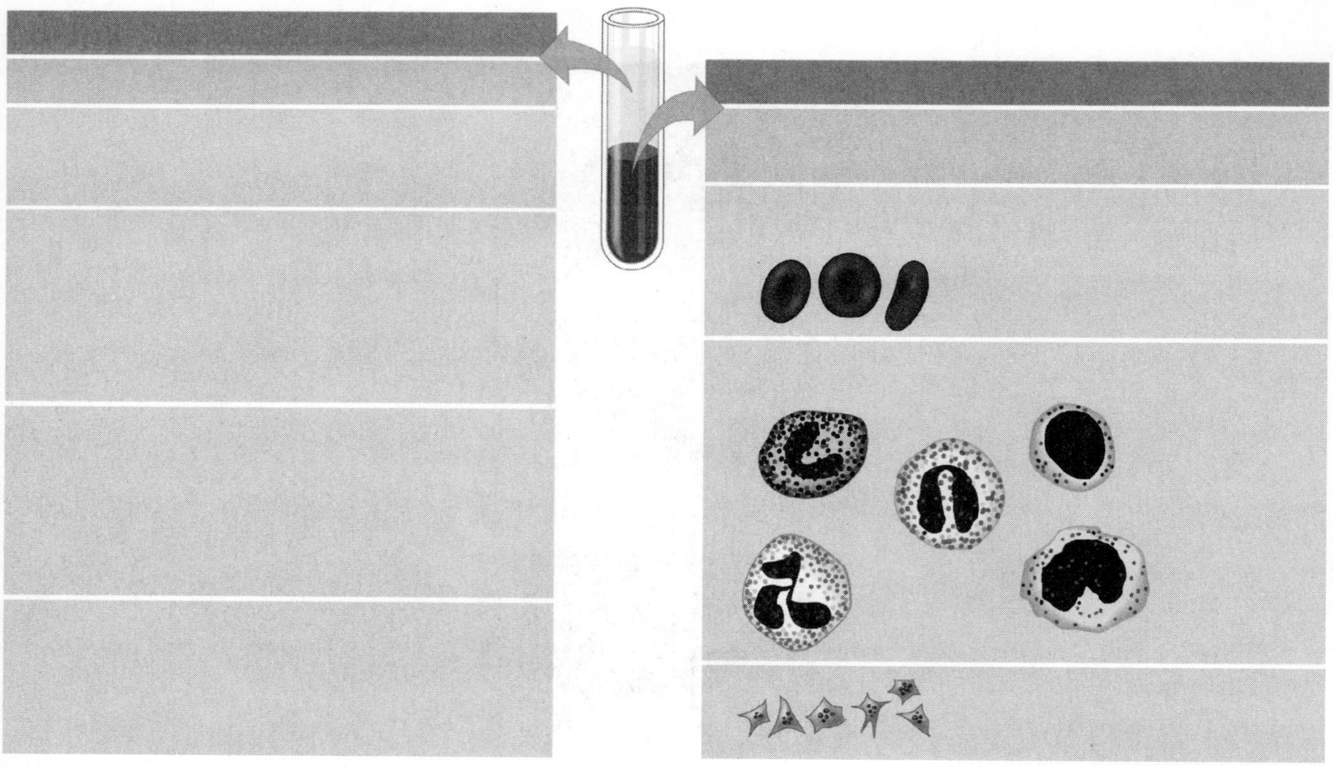

Figure 42.15 The composition of mammalian blood, page 880

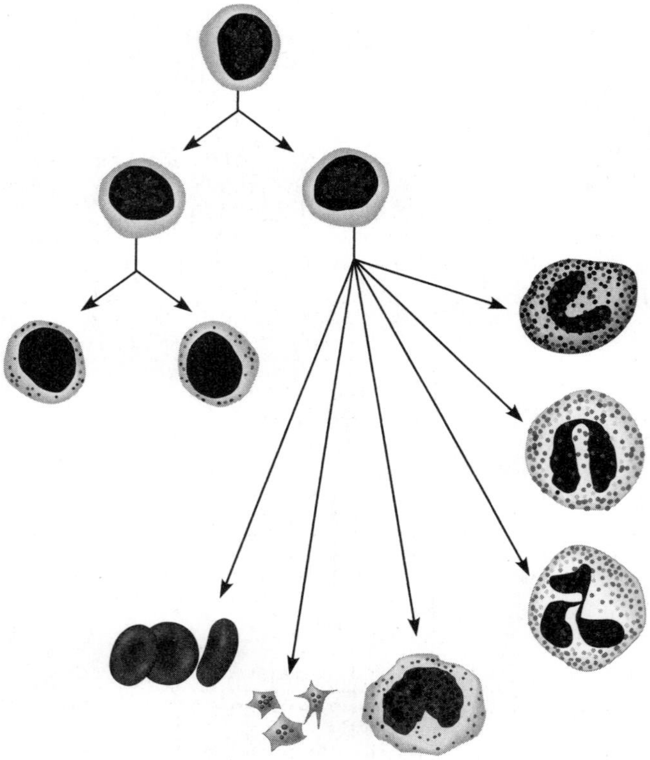

Figure 42.16 Differentiation of blood cells, page 881

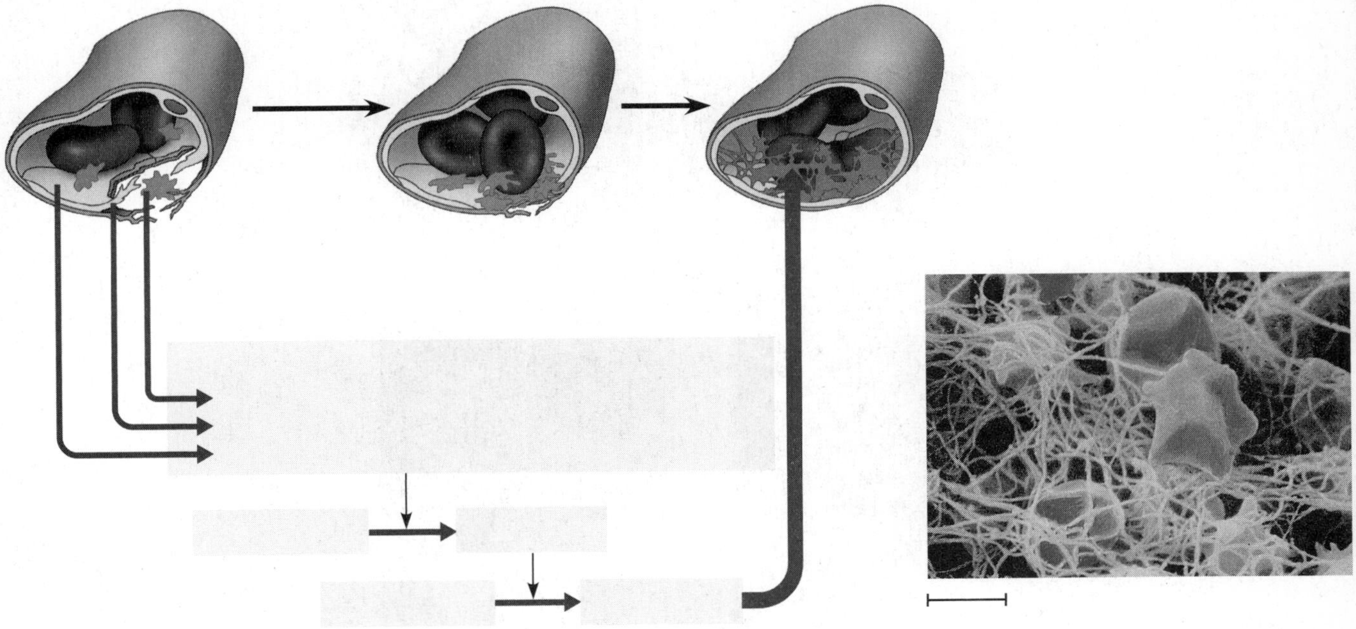

Figure 42.17 Blood clotting, page 882

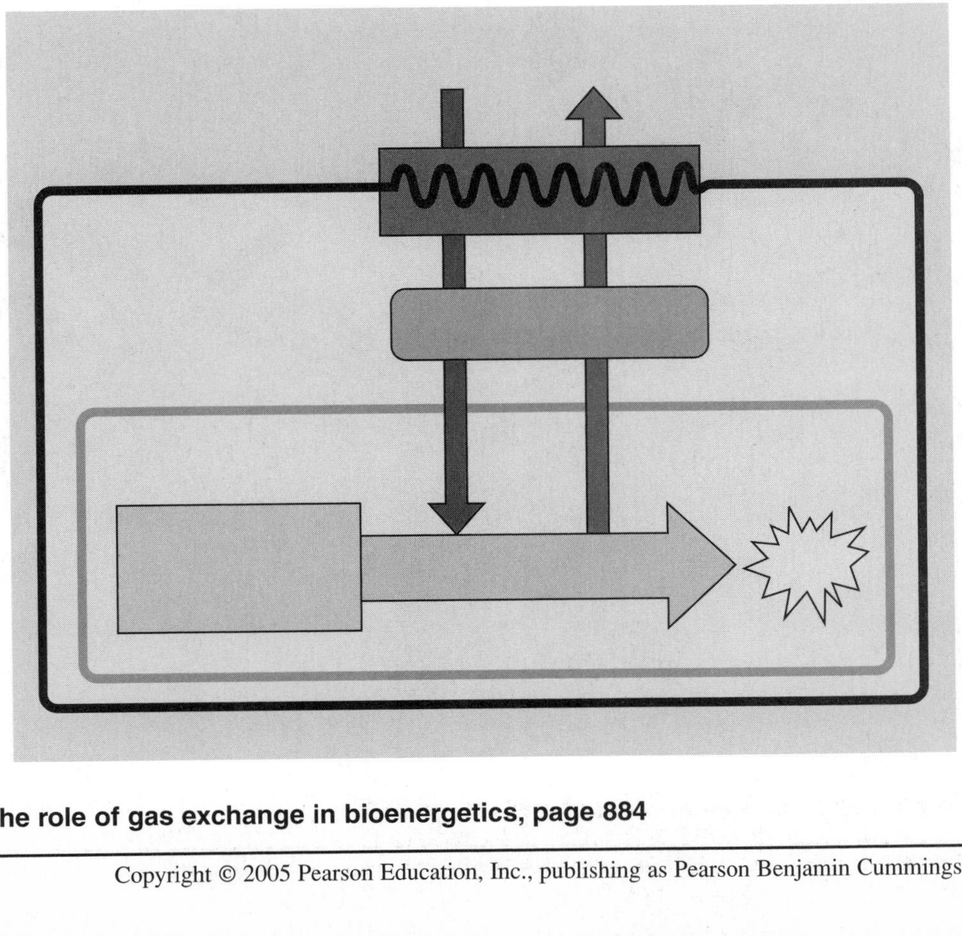

Figure 42.19 The role of gas exchange in bioenergetics, page 884

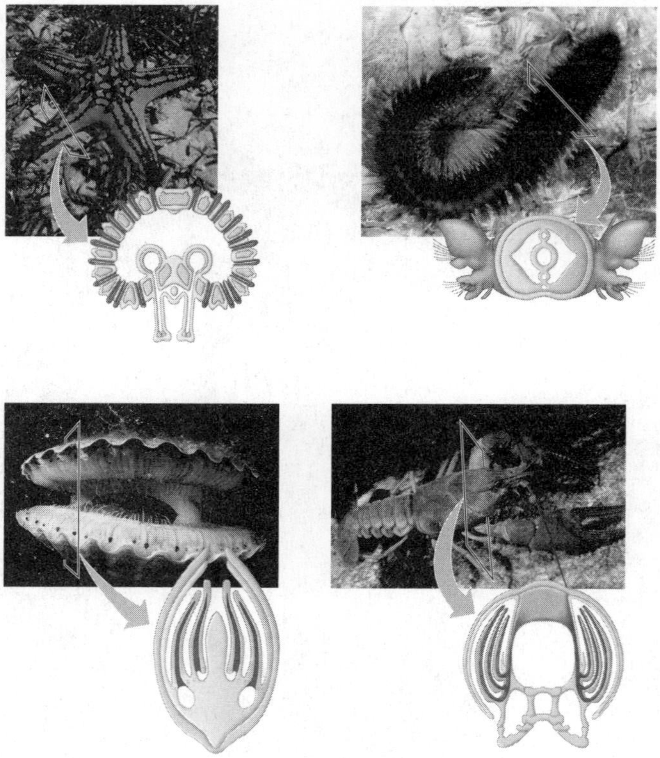

Figure 42.20 Diversity in the structure of gills, external body surfaces functioning in gas exchange, page 885

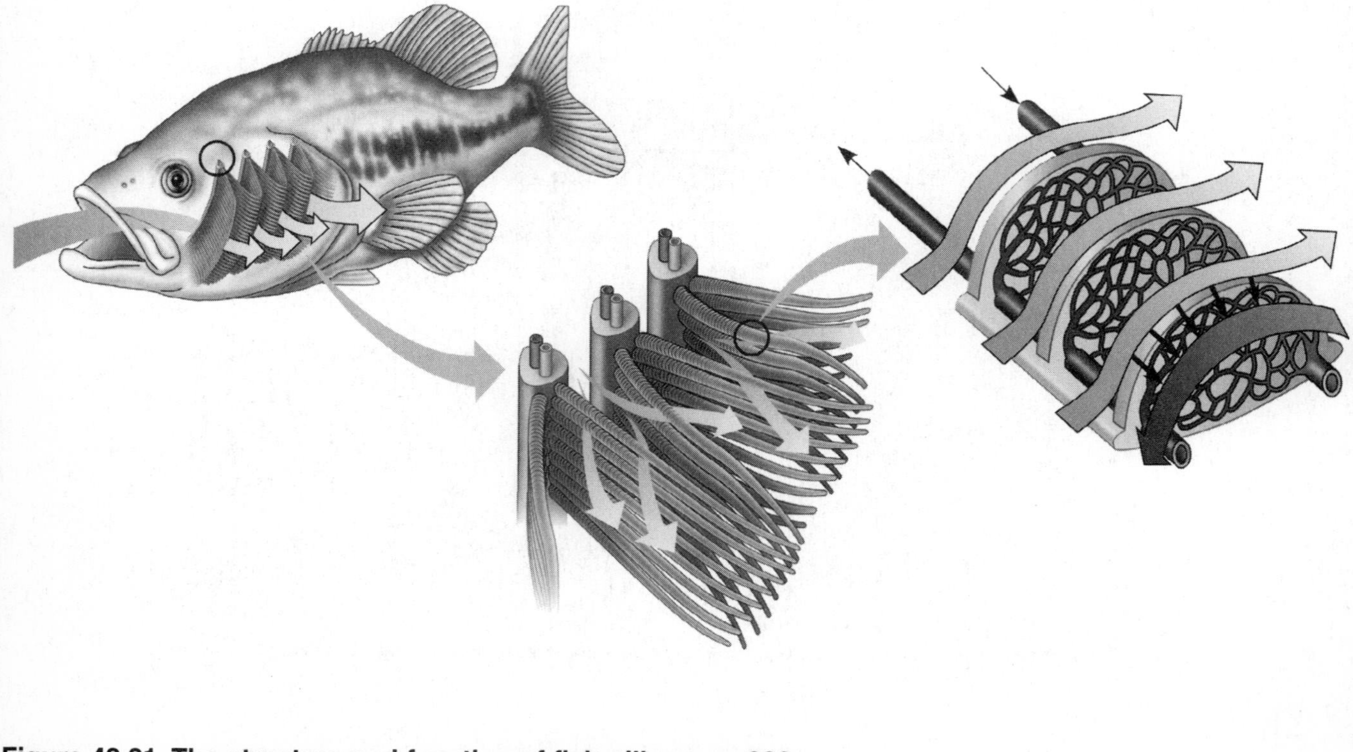

Figure 42.21 The structure and function of fish gills, page 886

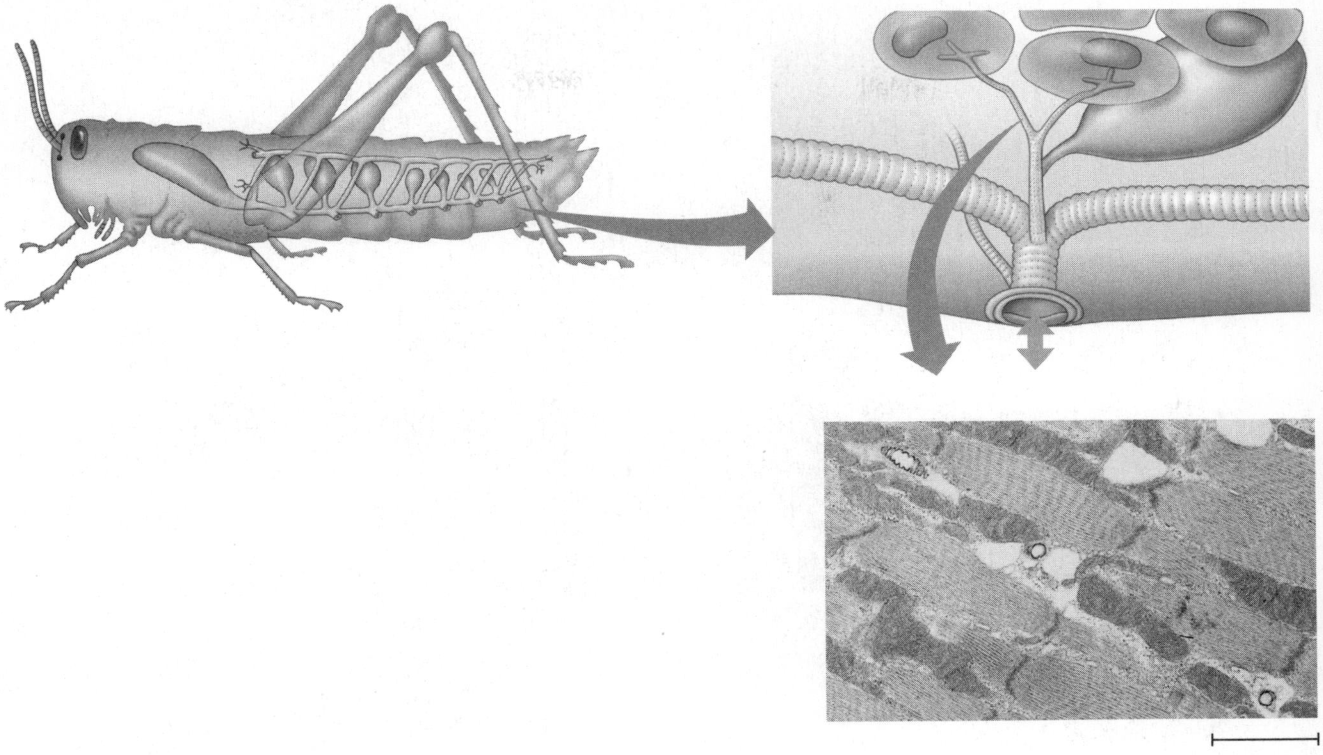

Figure 42.22 Tracheal systems, page 887

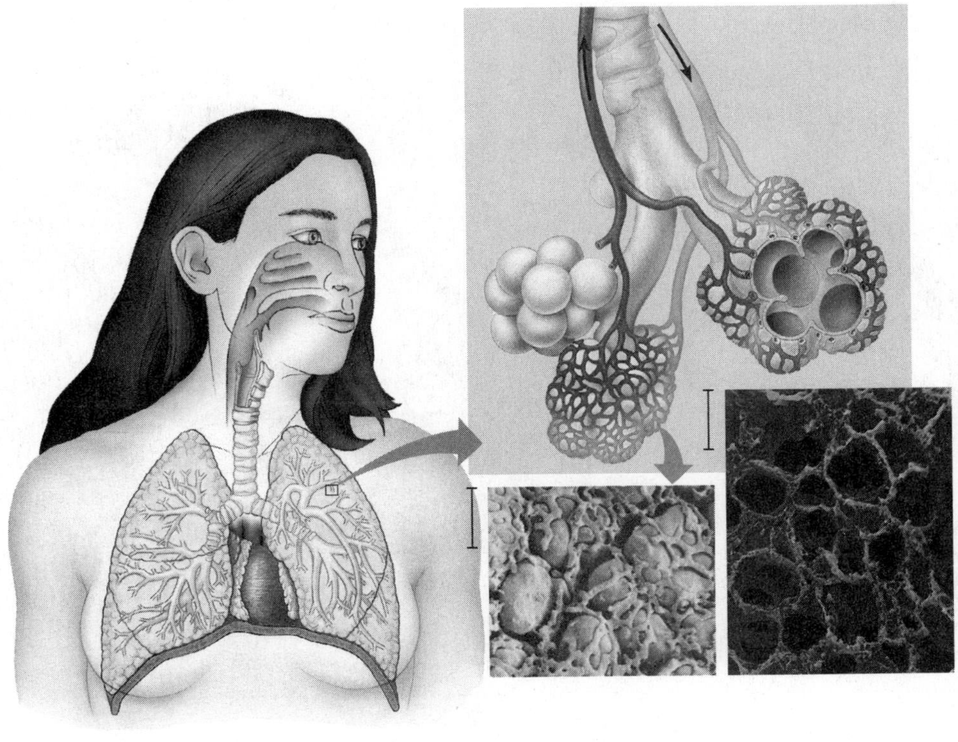

Figure 42.23 The mammalian respiratory system, page 888

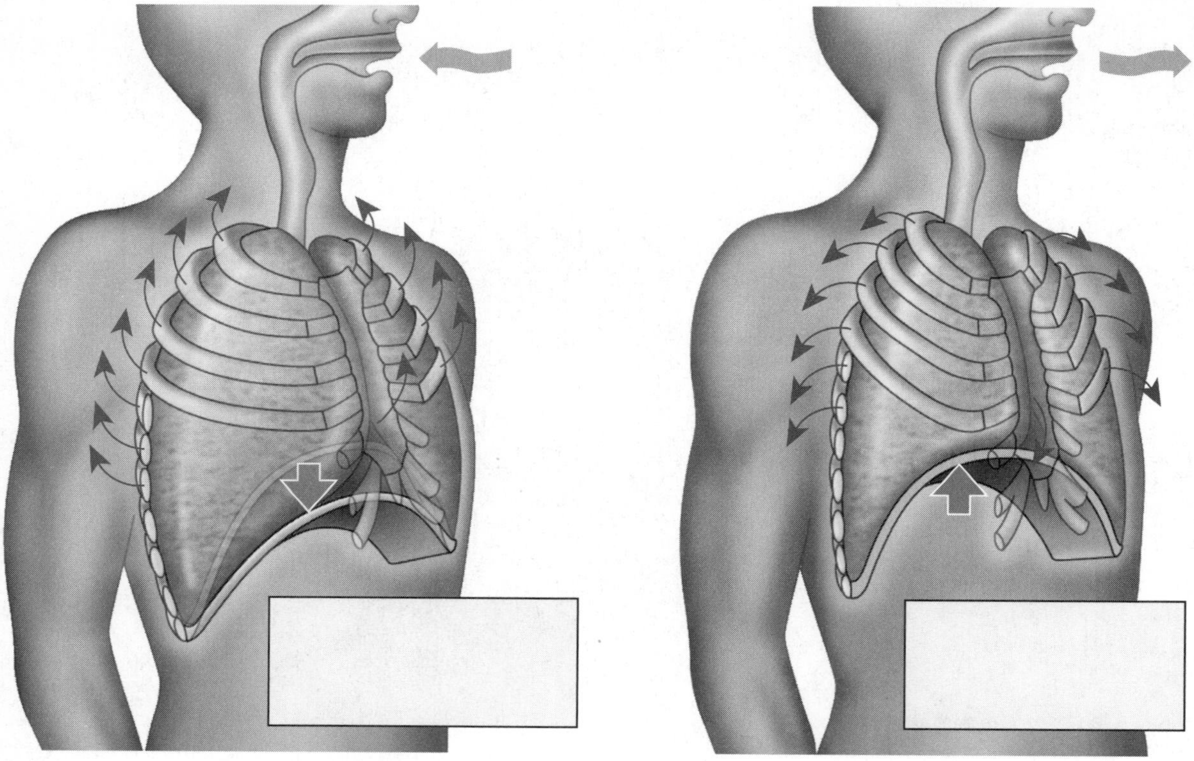

Figure 42.24 Negative pressure breathing, page 889

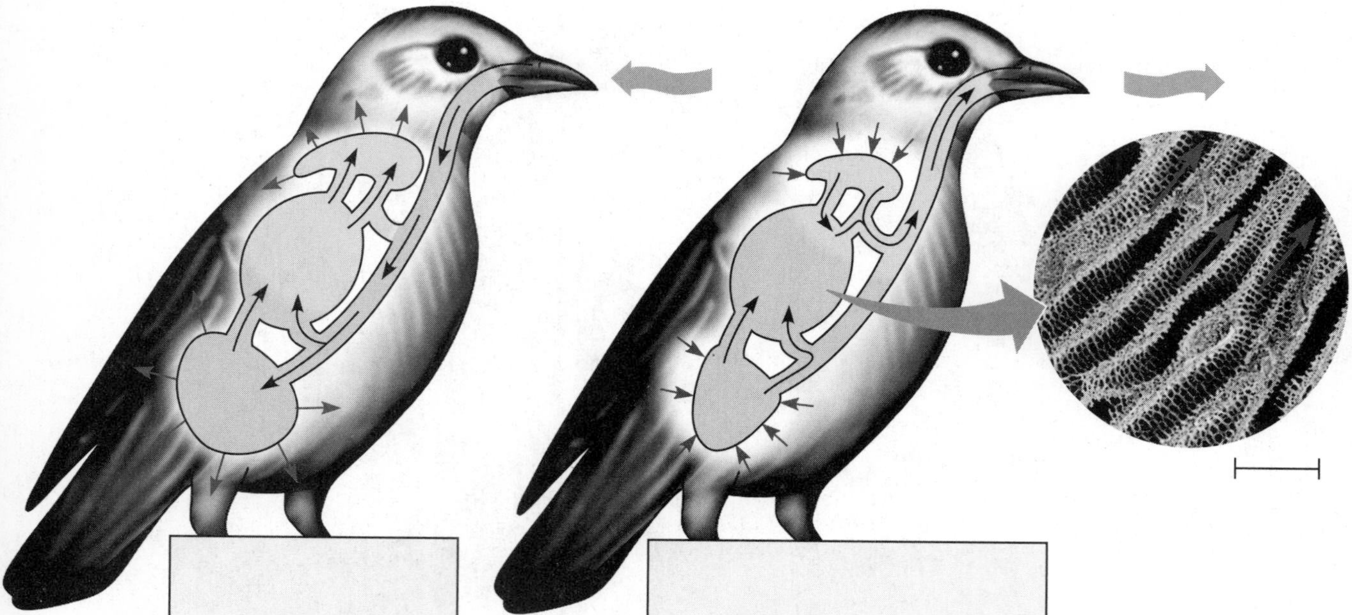

Figure 42.25 The avian respiratory system, page 889

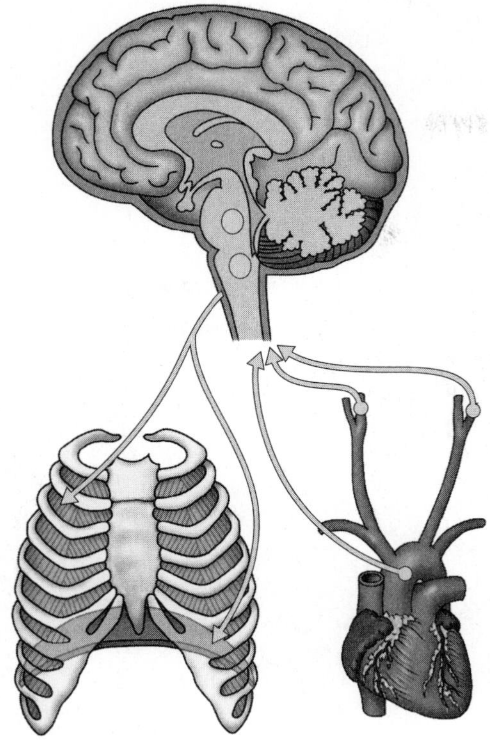

Figure 42.26 Automatic control of breathing, page 890

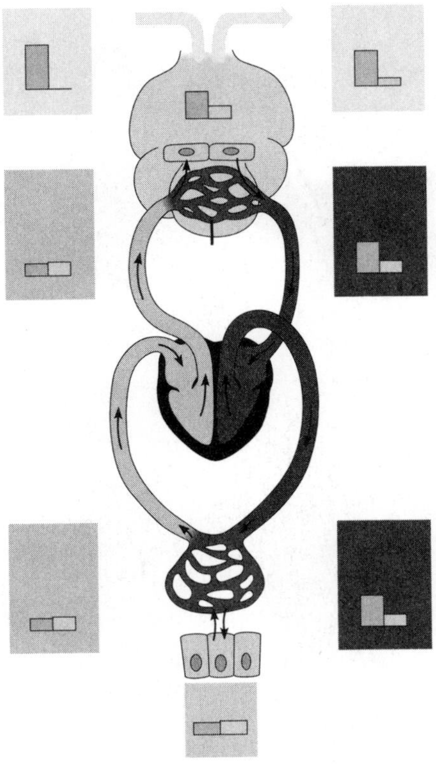

Figure 42.27 Loading and unloading of respiratory gases, page 891

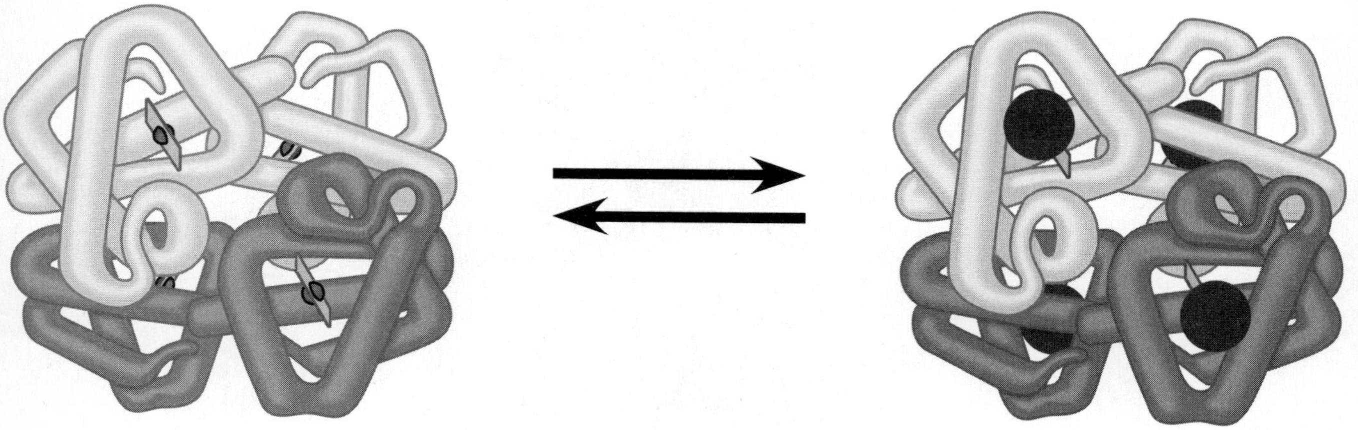

Figure 42.28 Hemoglobin loading and unloading O_2, page 892

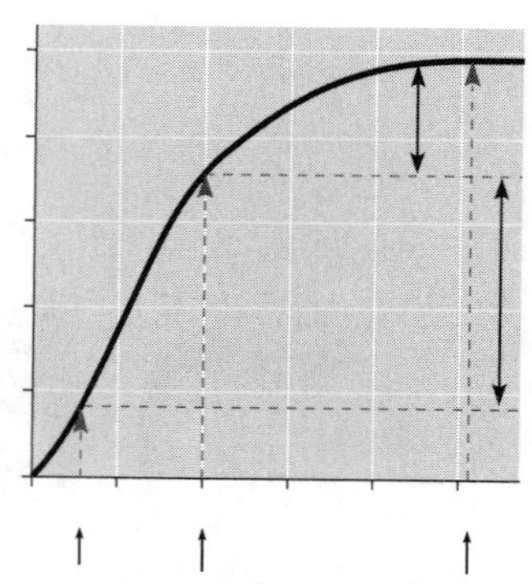

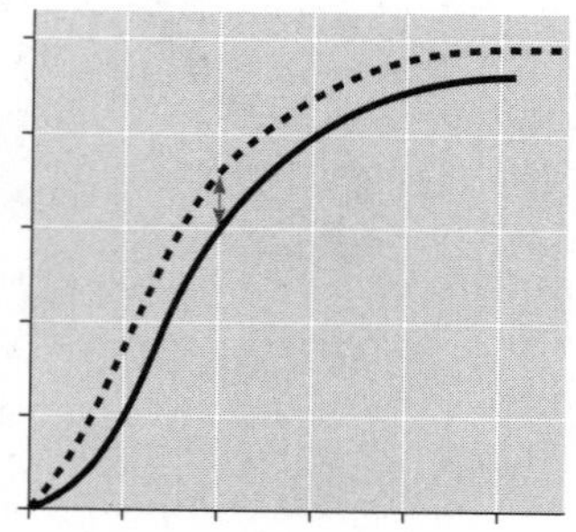

Figure 42.29 Oxygen dissociation curves for hemoglobin, page 892

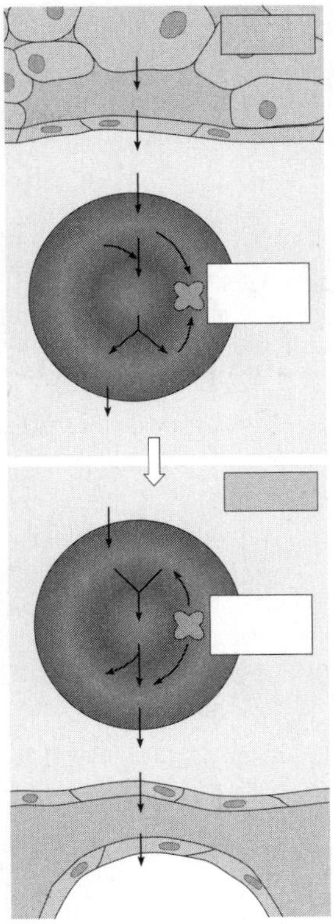

Figure 42.30 Carbon dioxide transport in the blood, page 893

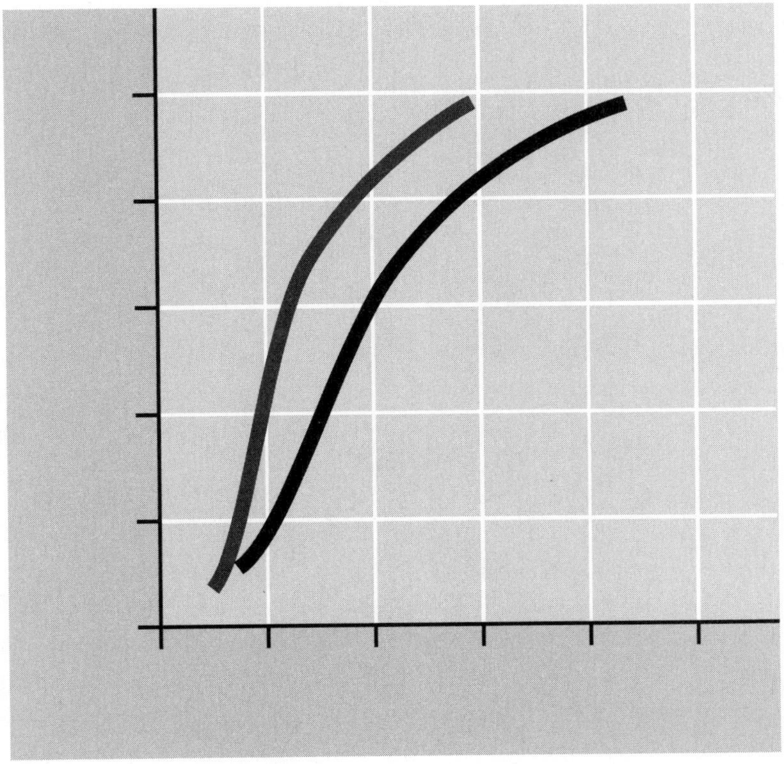

Figure UN1 Dissociation curves for two hemoglobins, page 897

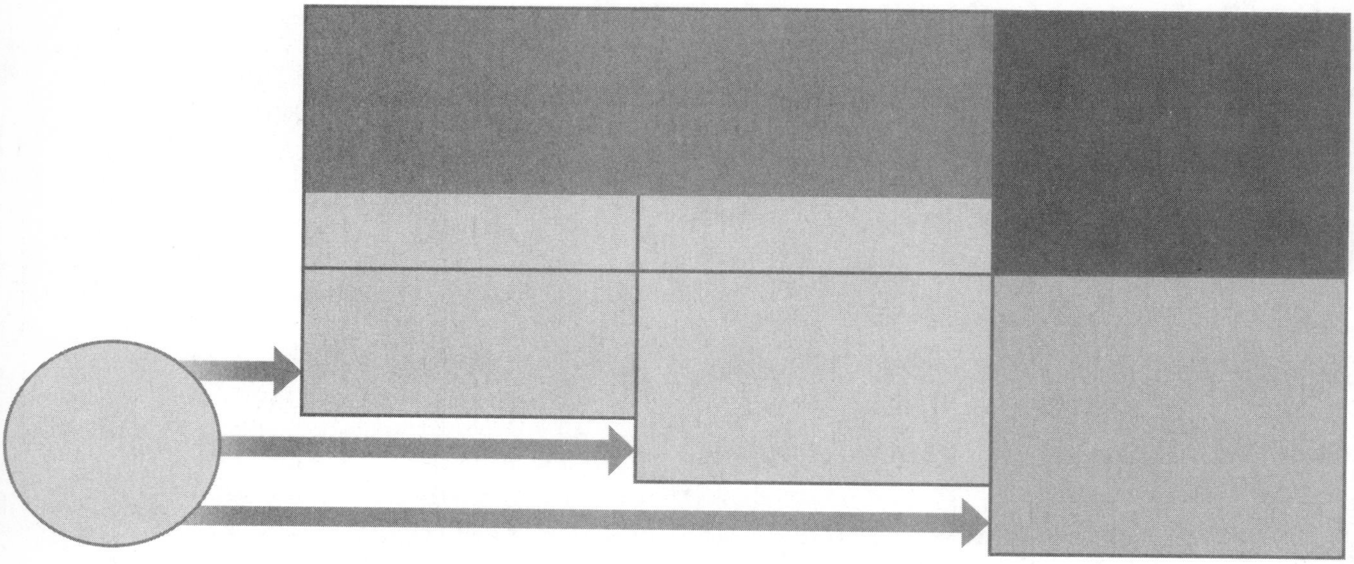

Figure 43.2 Overview of vertebrate defenses against bacteria, viruses, and other pathogens, page 899

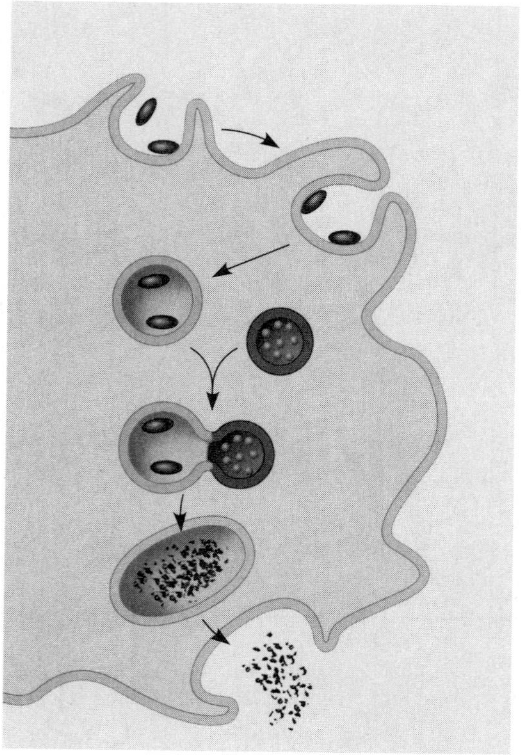

Figure 43.4 Phagocytosis, page 900

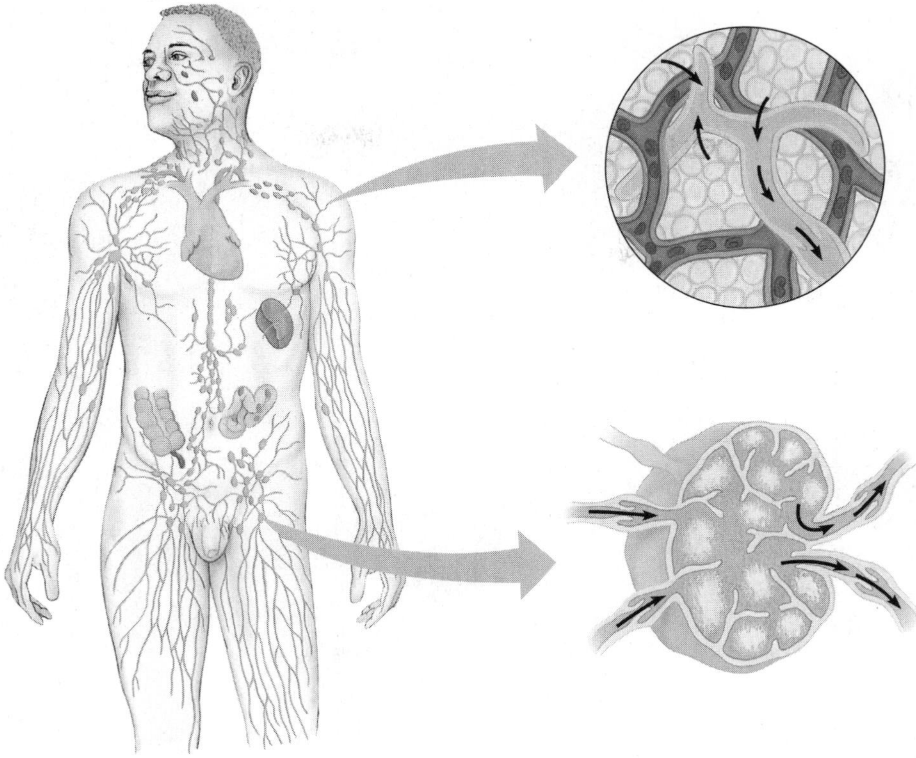

Figure 43.5 The human lymphatic system, page 901

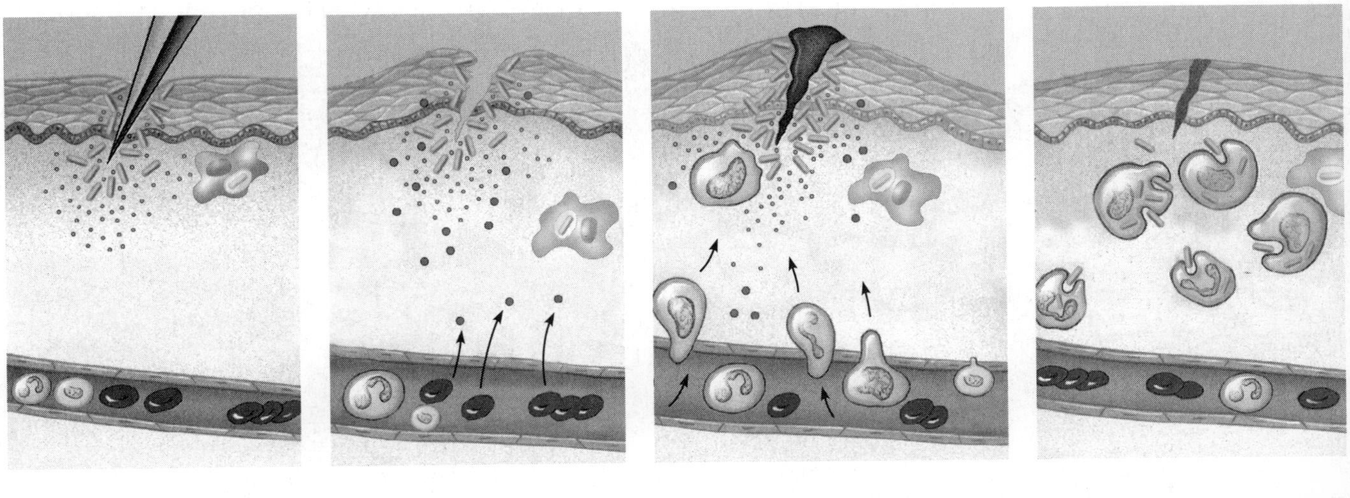

Figure 43.6 Major events in the local inflammatory response, page 902

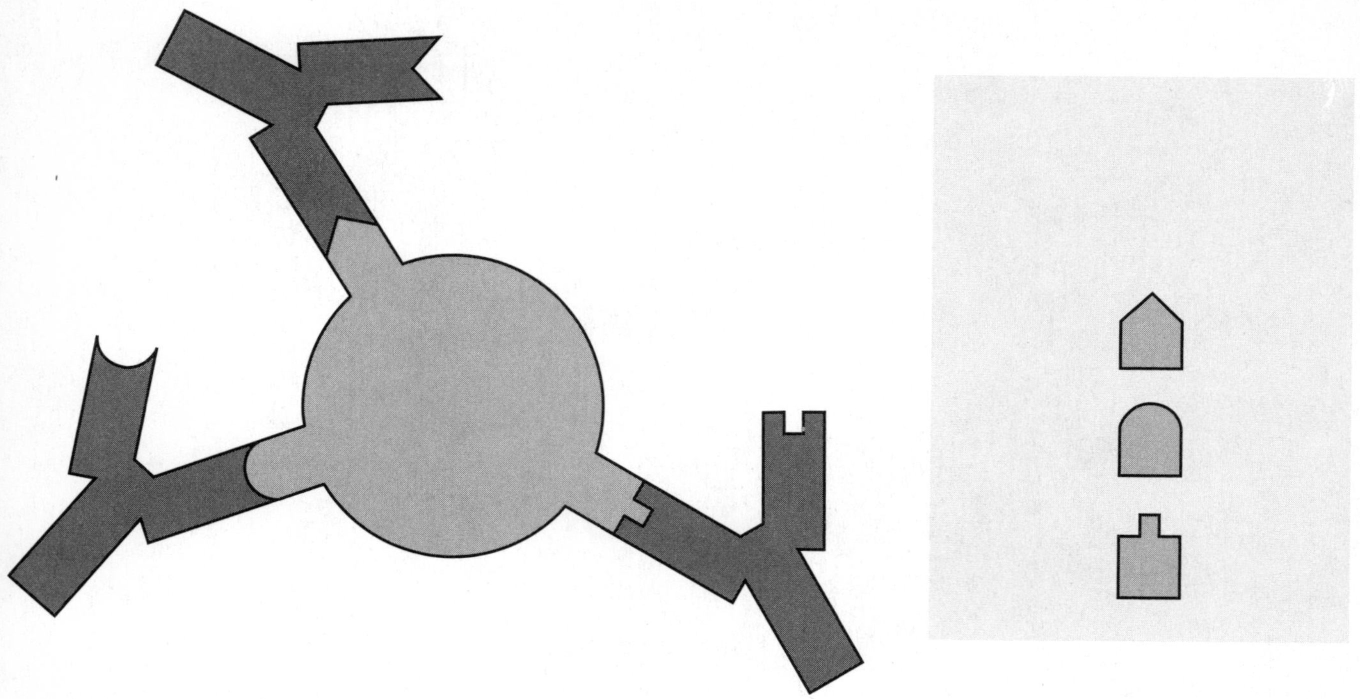

Figure 43.7 Epitopes (antigenic determinants), page 903

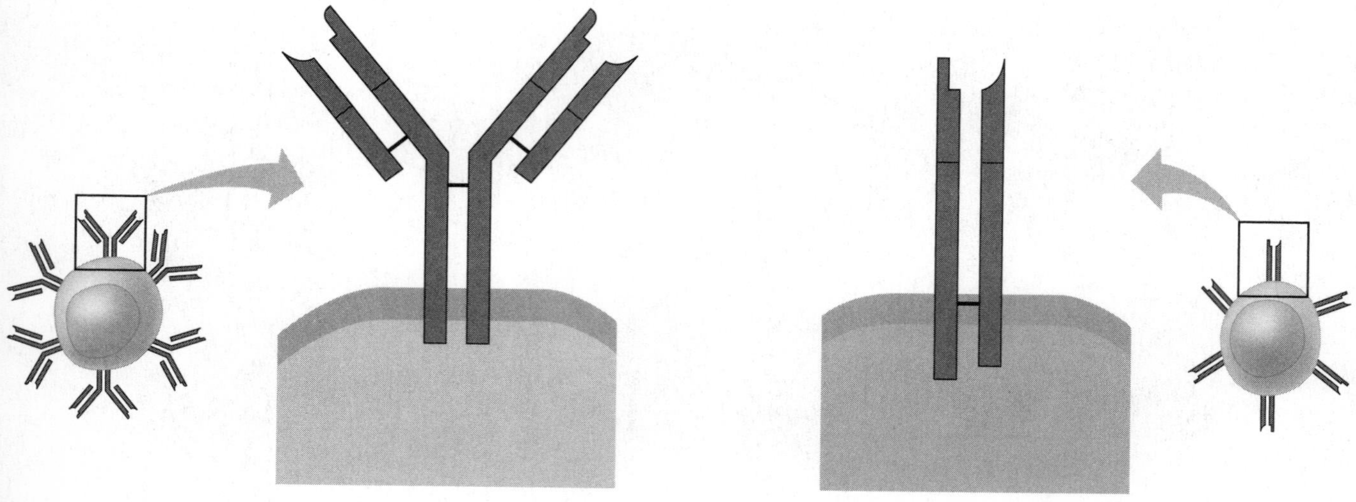

Figure 43.8 Antigen receptors on lymphocytes, page 904

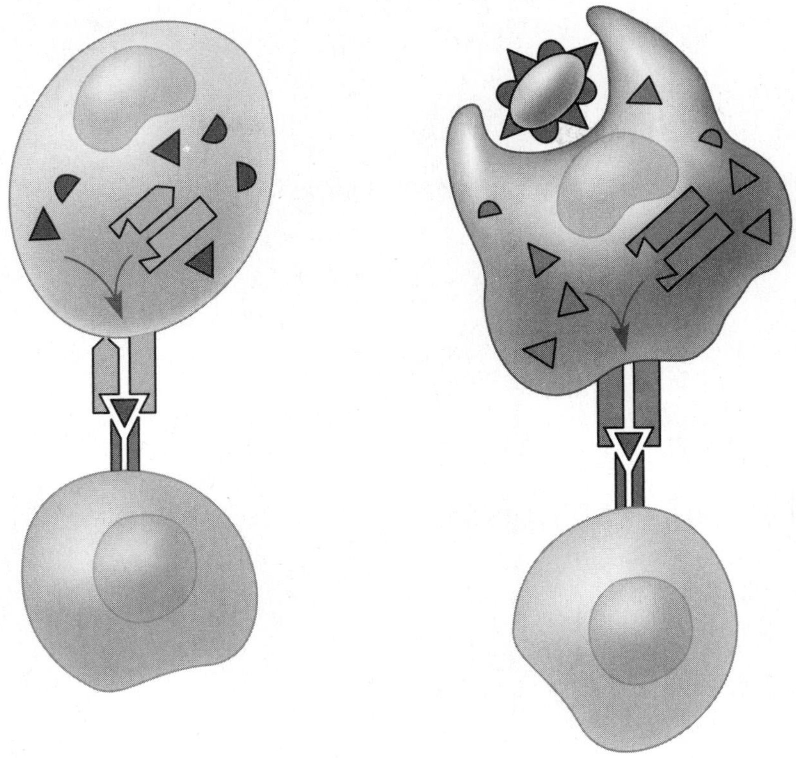

Figure 43.9 The interaction of T cells with MHC molecules, page 905

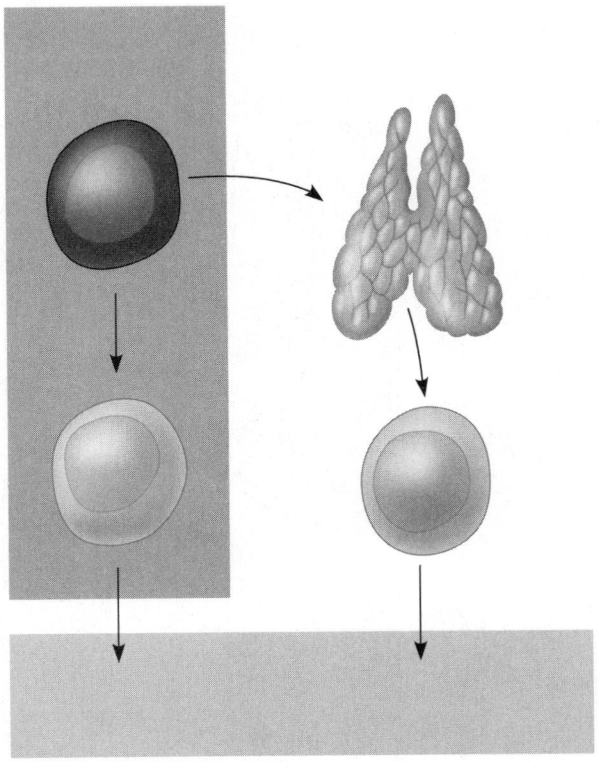

Figure 43.10 Overview of lymphocyte development, page 905

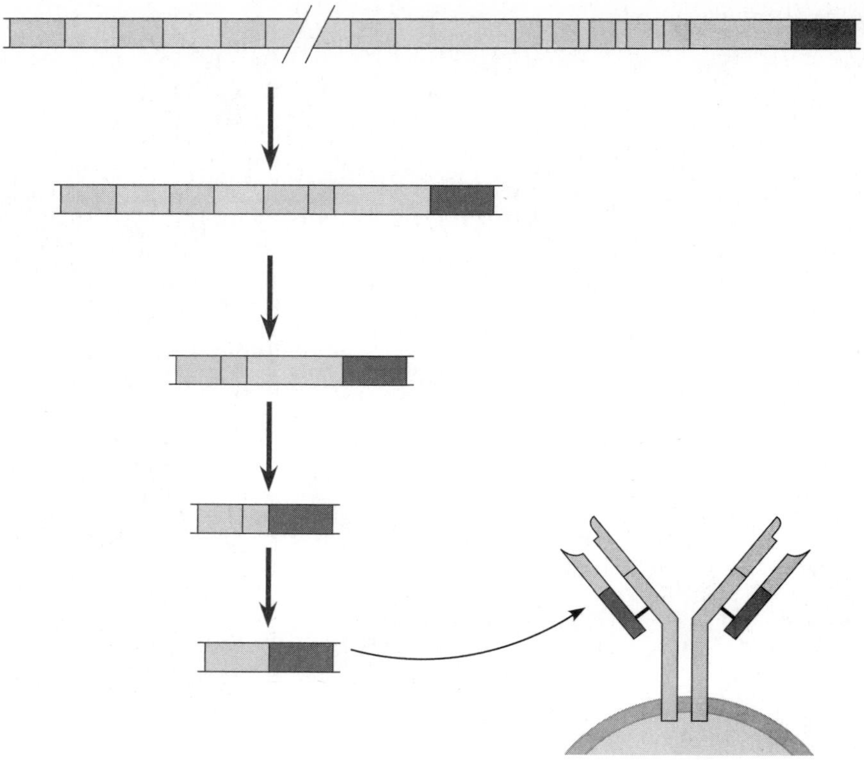

Figure 43.11 Immunoglobulin gene rearrangement, page 906

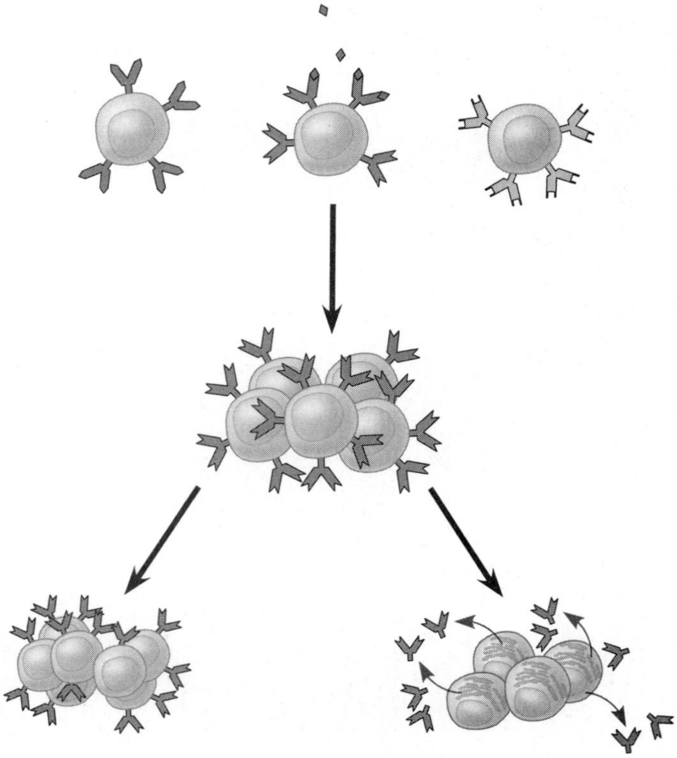

Figure 43.12 Clonal selection of B cells, page 907

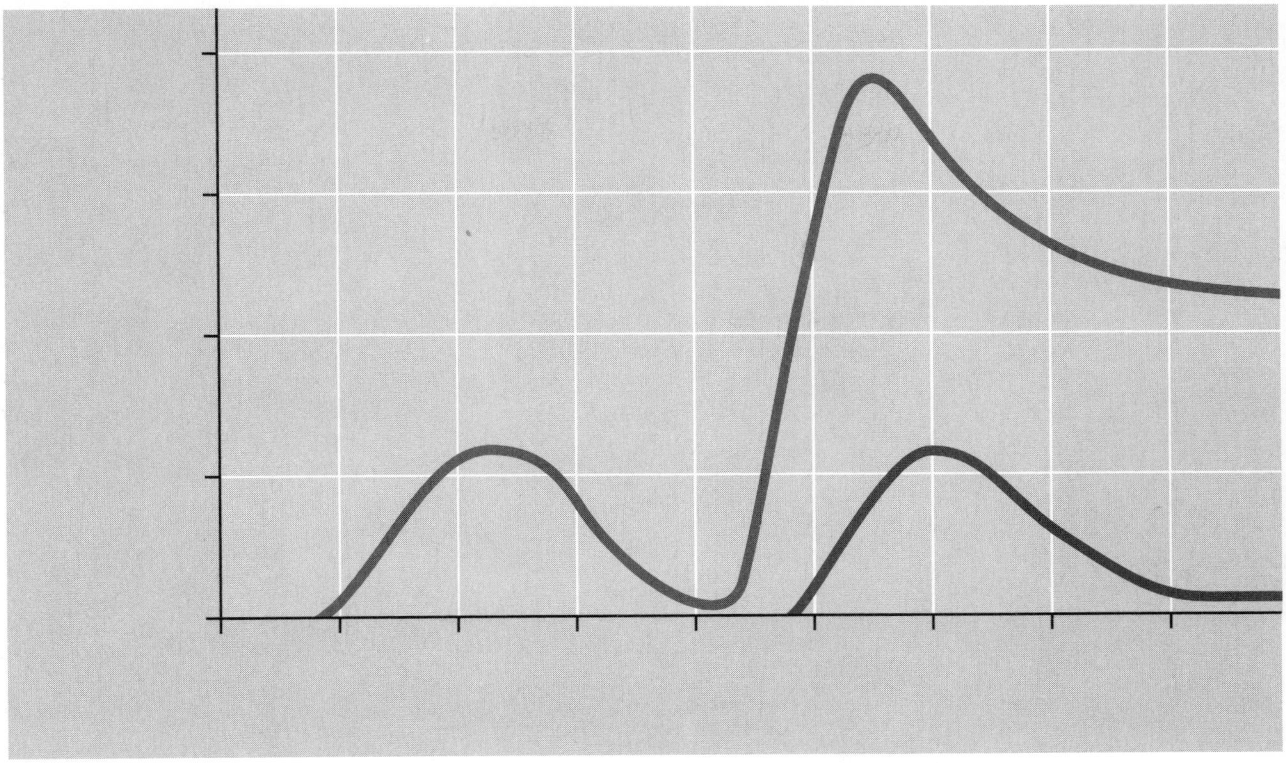

Figure 43.13 The specificity of immunological memory, page 908

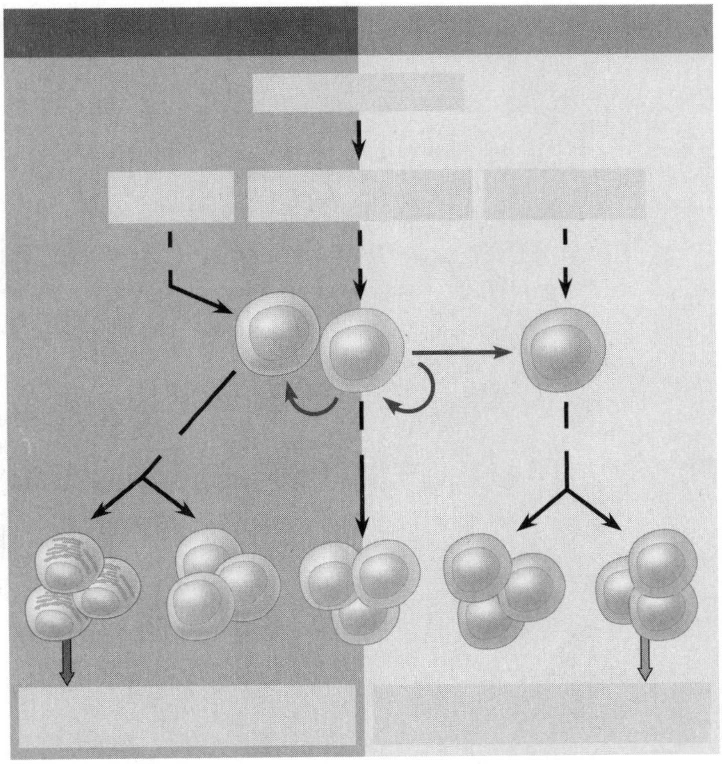

Figure 43.14 An overview of the acquired immune response, page 909

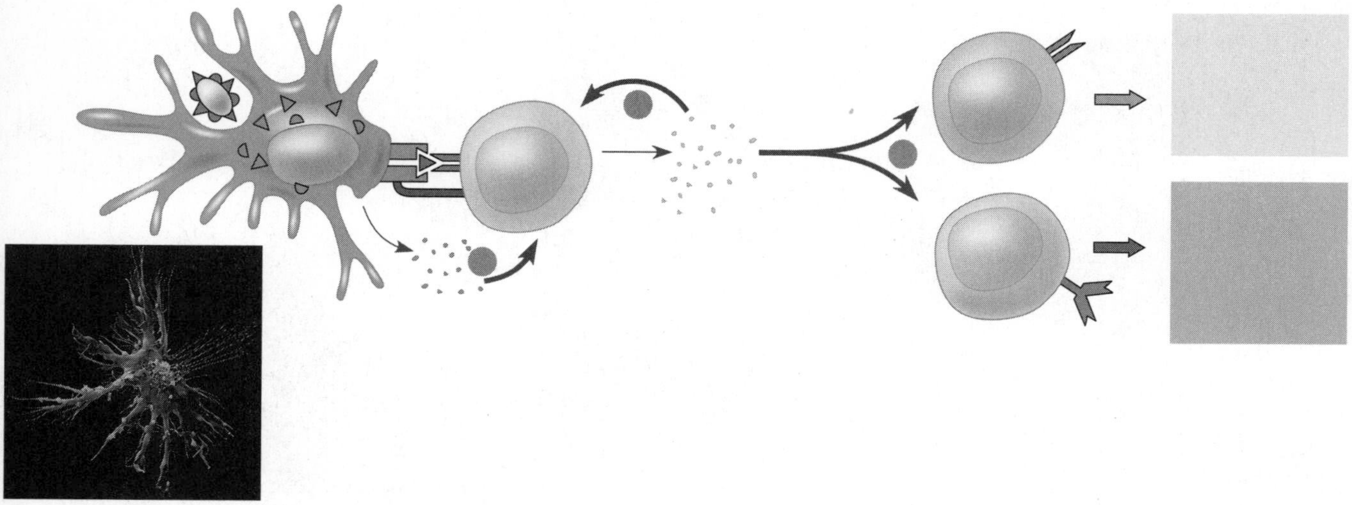

Figure 43.15 The central role of helper T cells in humoral and cell-mediated immune responses, page 910

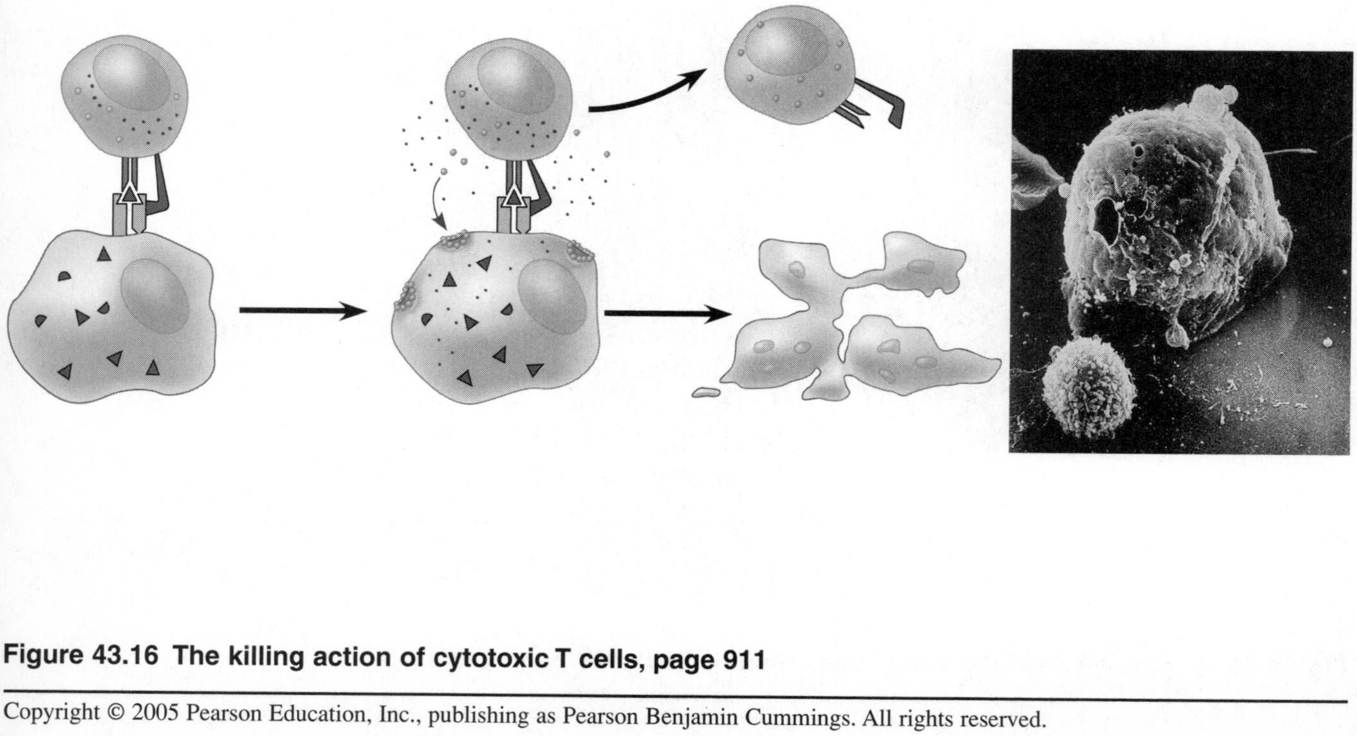

Figure 43.16 The killing action of cytotoxic T cells, page 911

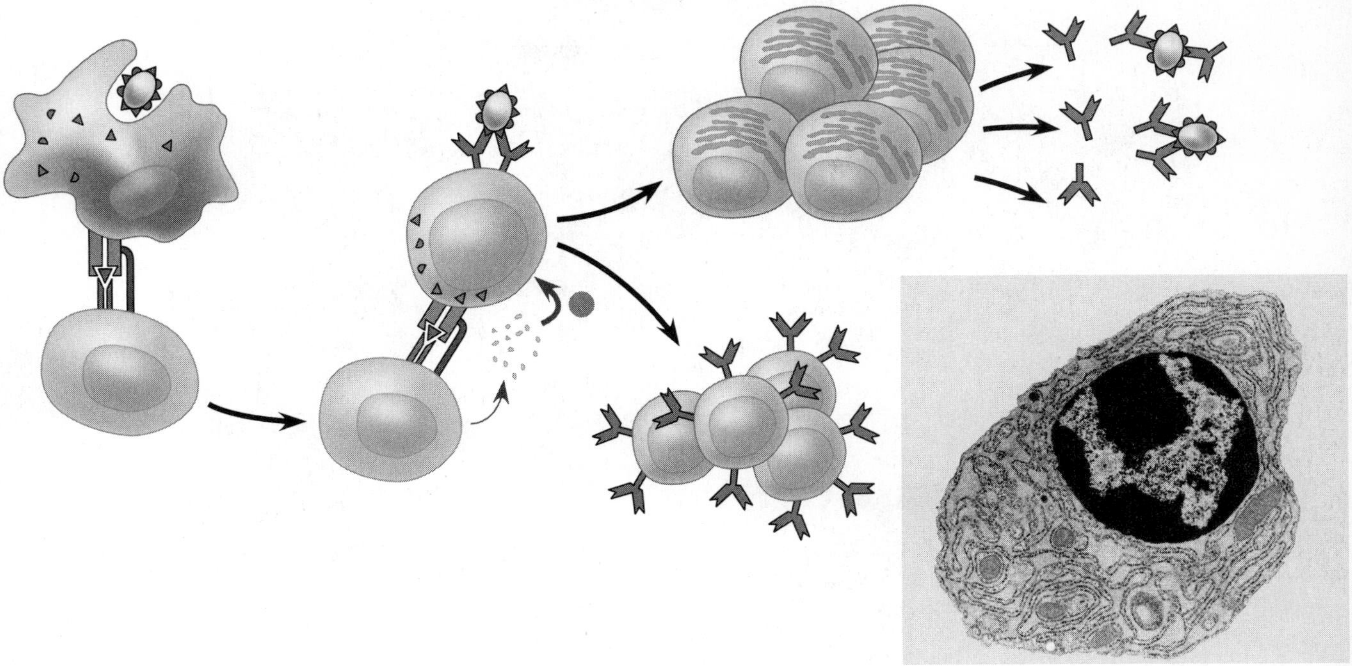

Figure 43.17 Humoral immune response, page 911

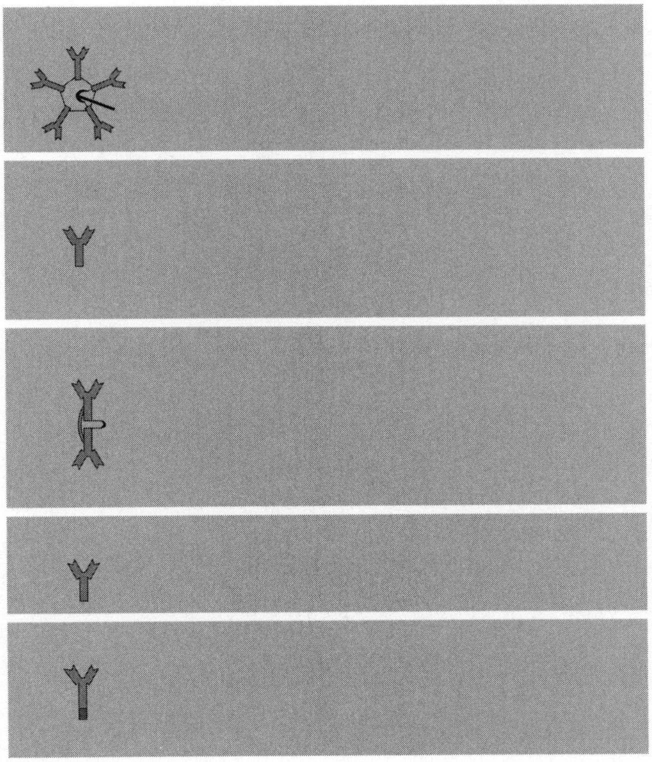

Figure 43.18 The five classes of immunoglobulins, page 912

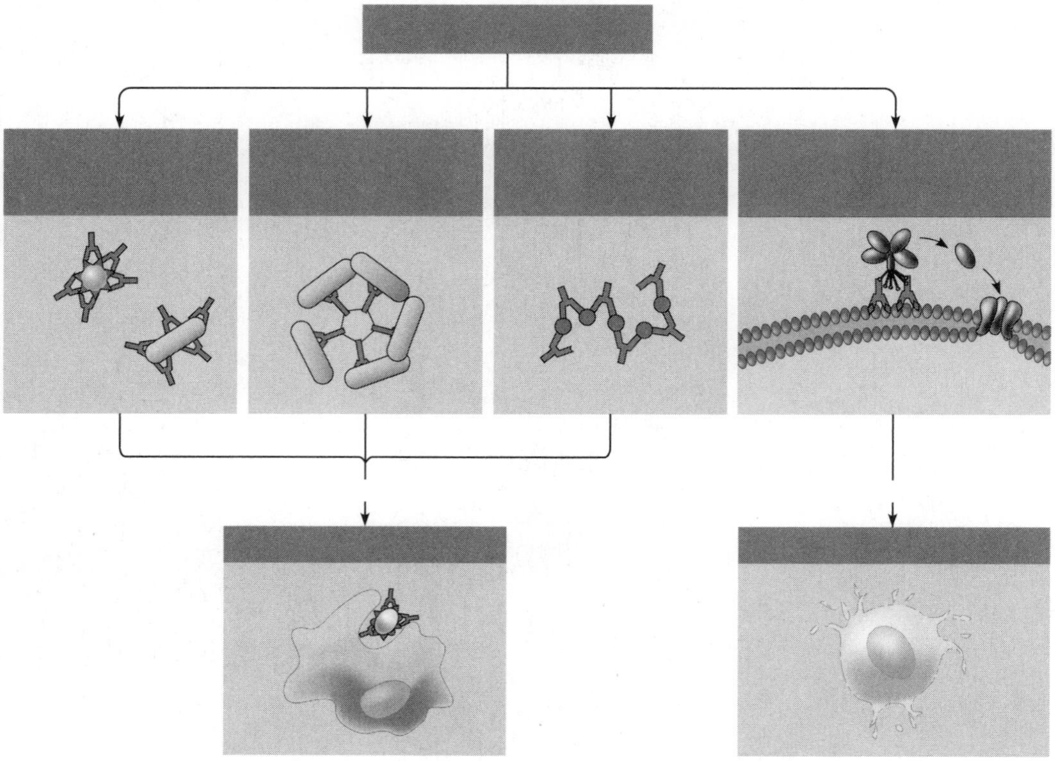

Figure 43.19 Antibody-mediated mechanisms of antigen disposal, page 913

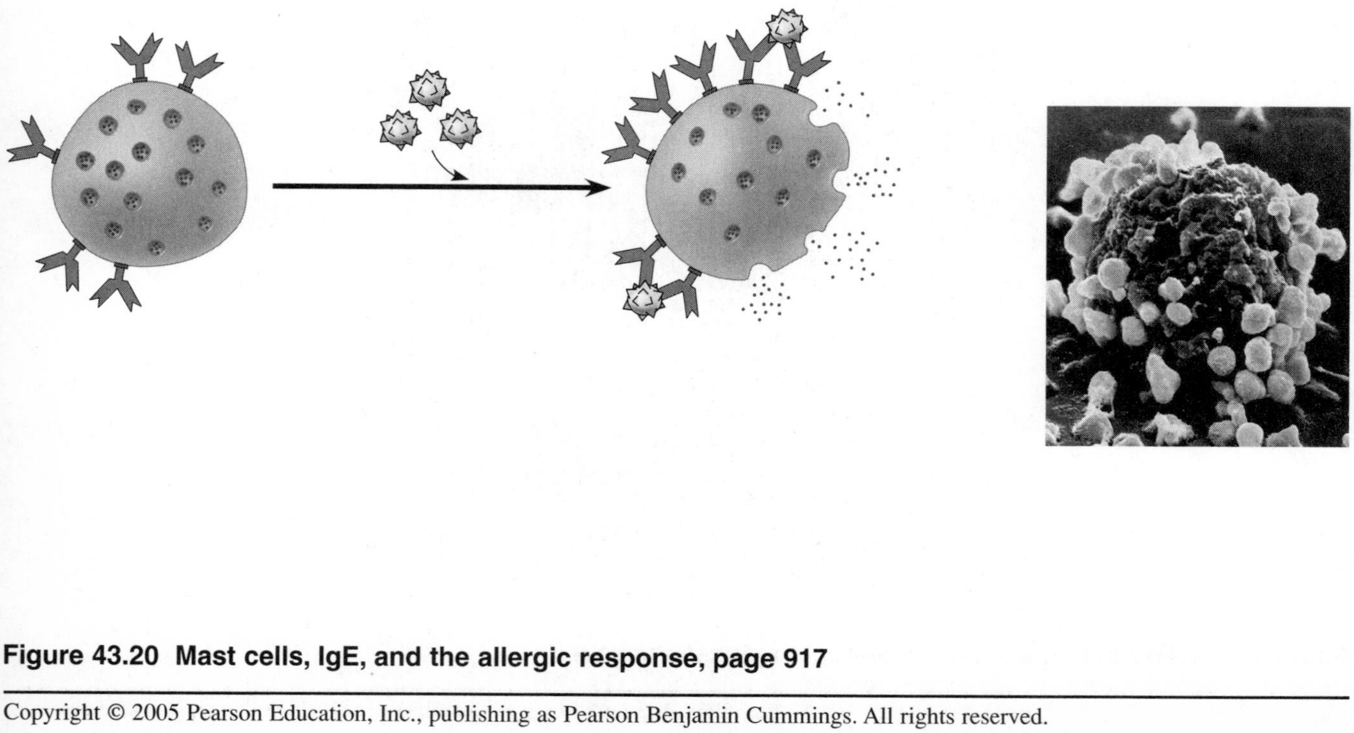

Figure 43.20 Mast cells, IgE, and the allergic response, page 917

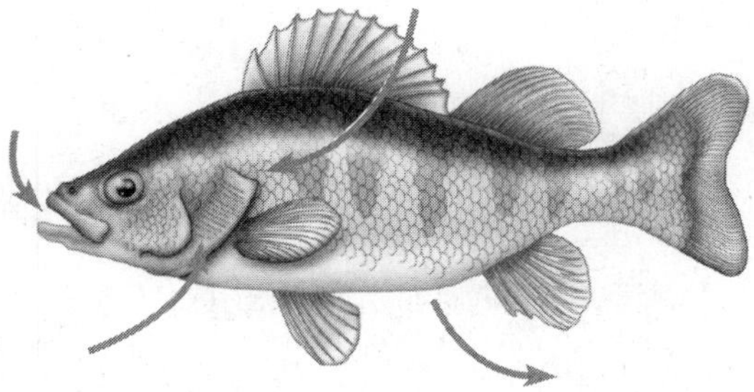

Figure 44.3 Osmoregulation in marine and freshwater bony fishes: a comparison, page 924

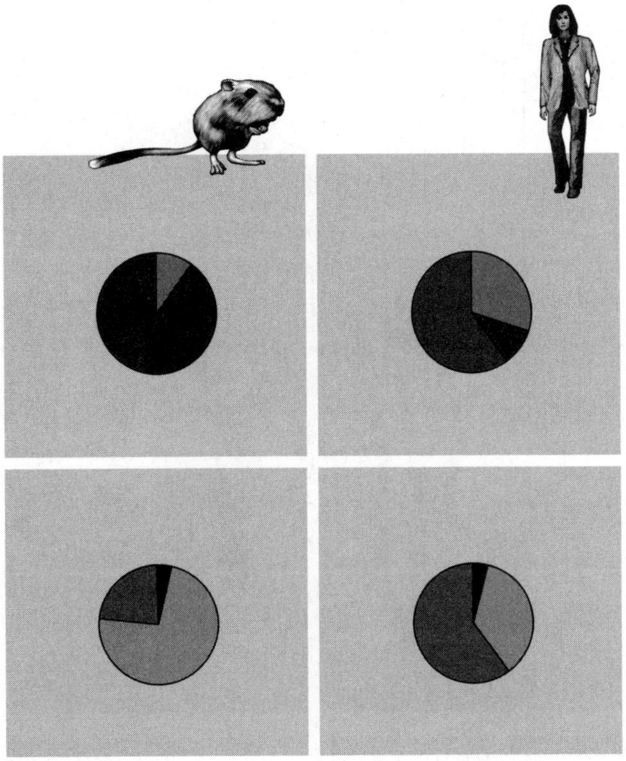

Figure 44.5 Water balance in two terrestrial mammals, page 925

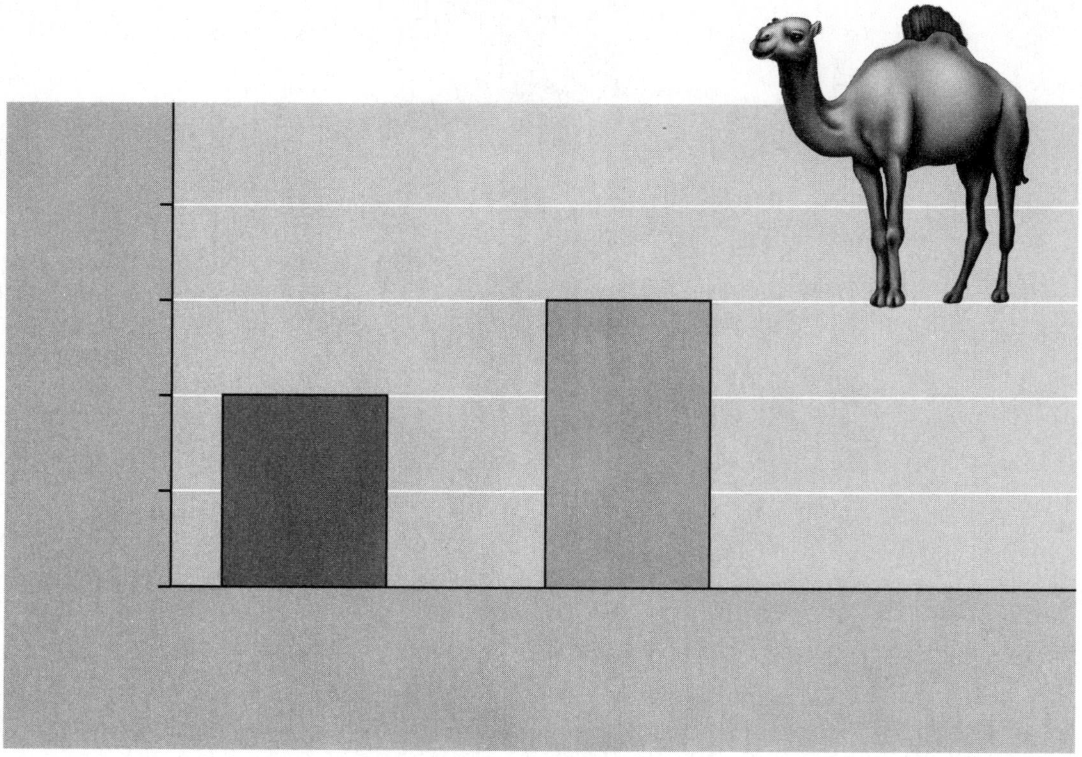

Figure 44.6 What role does fur play in water conservation by camels? page 926

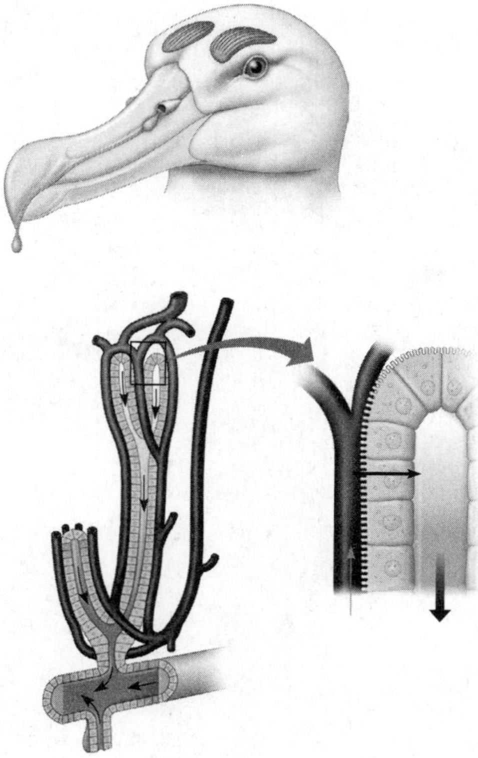

Figure 44.7 Salt-excreting glands in birds, page 926

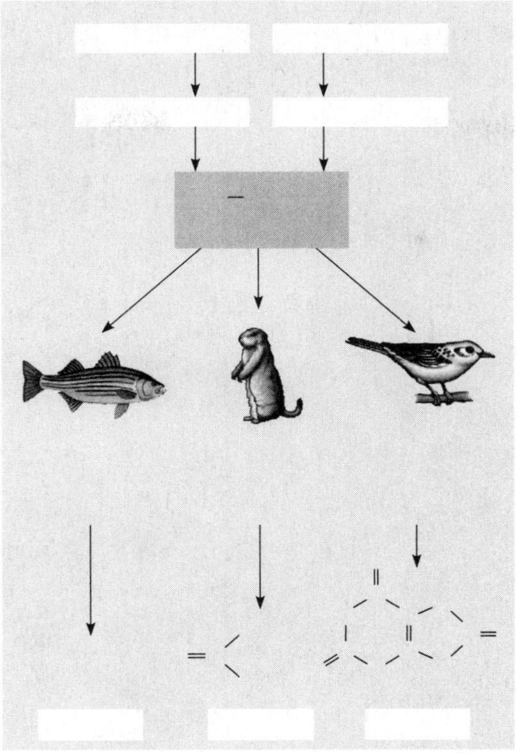

Figure 44.8 Nitrogenous wastes, page 927

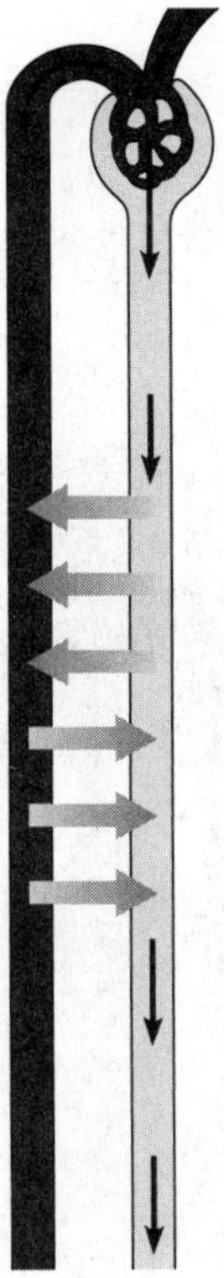

Figure 44.9 Key functions of excretory systems: an overview, page 929

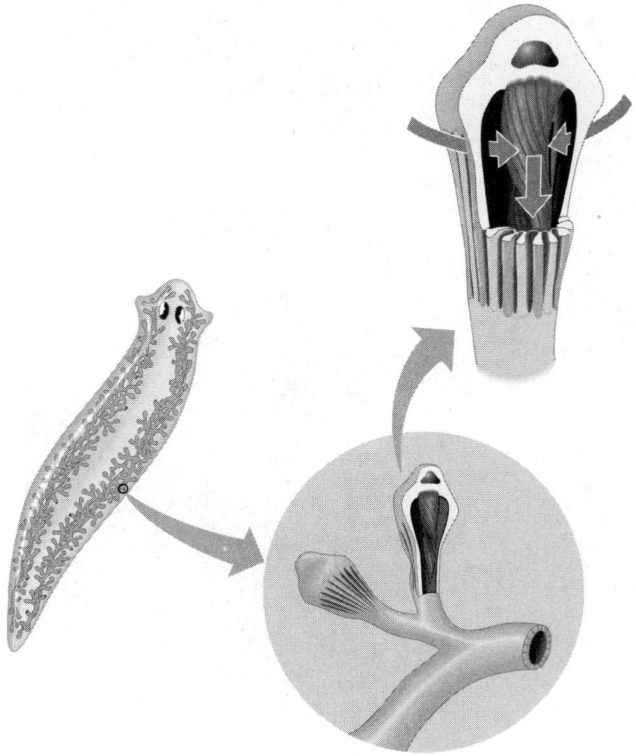

Figure 44.10 Protonephridia: the flame-bulb system of a planarian, page 929

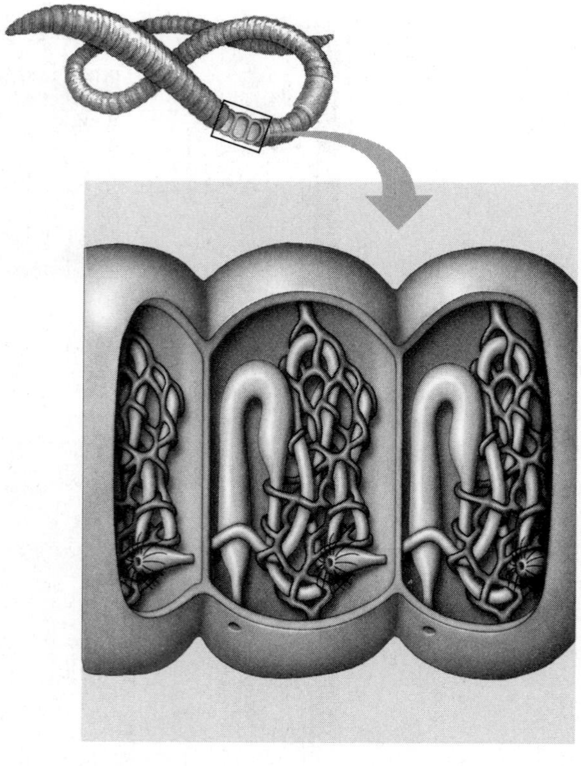

Figure 44.11 Metanephridia of an earthworm, page 930

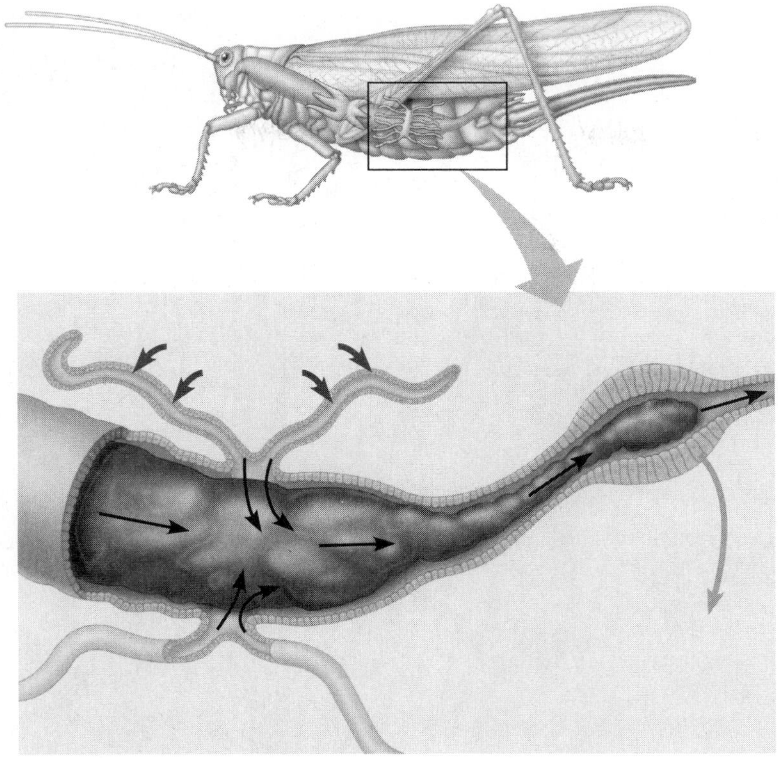

Figure 44.12 Malpighian tubules of insects, page 930

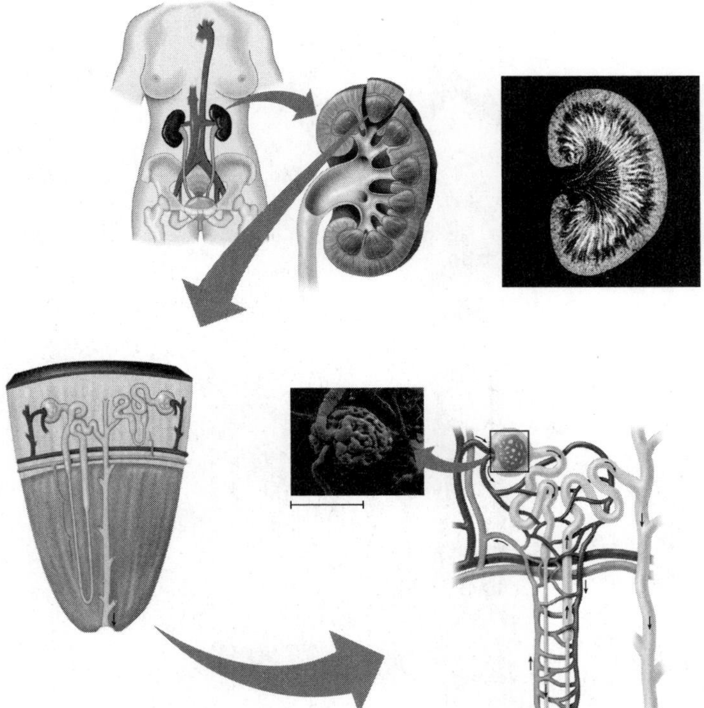

Figure 44.13 The mammalian excretory system, page 932

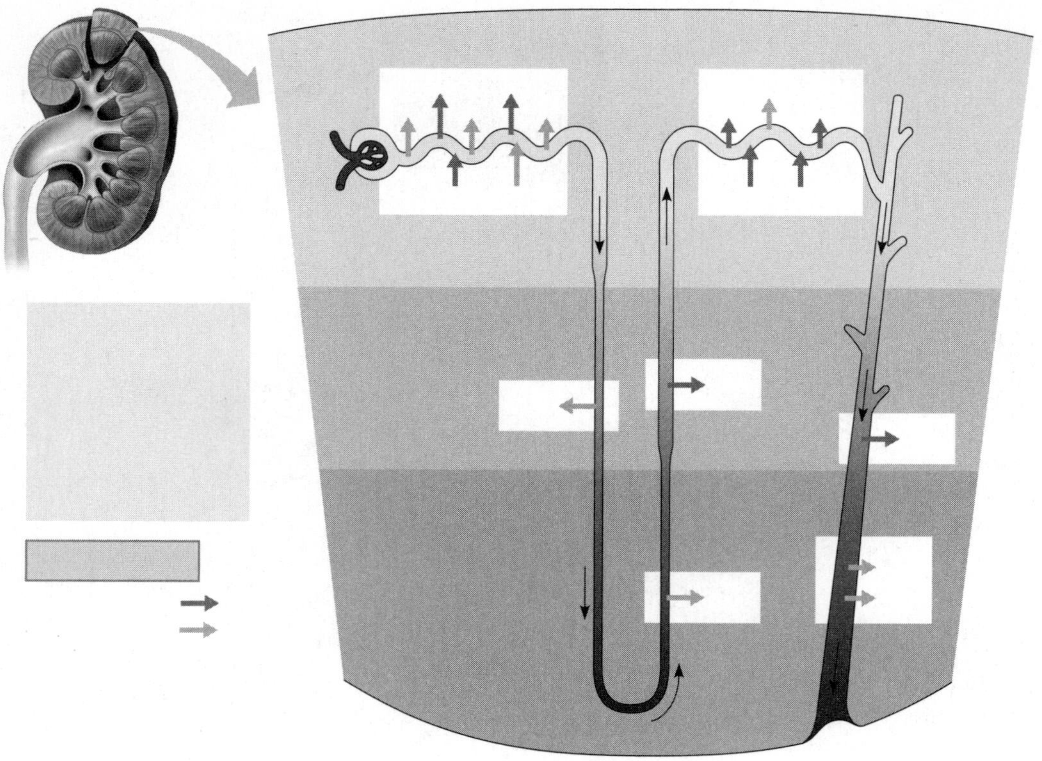

Figure 44.14 The nephron and collecting duct: regional functions of the transport epithelium, page 933

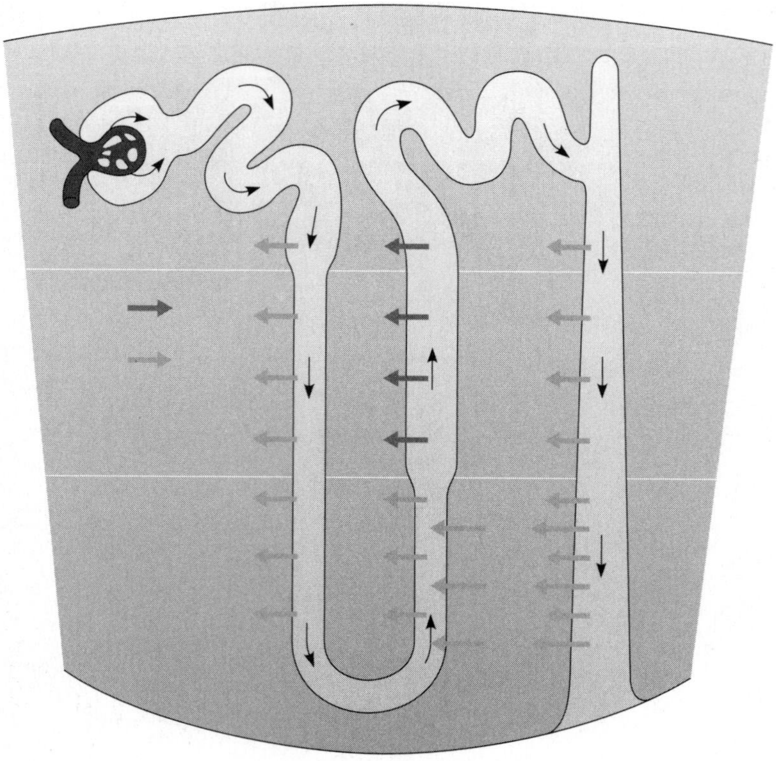

Figure 44.15 How the human kidney concentrates urine: the two-solute model, page 935

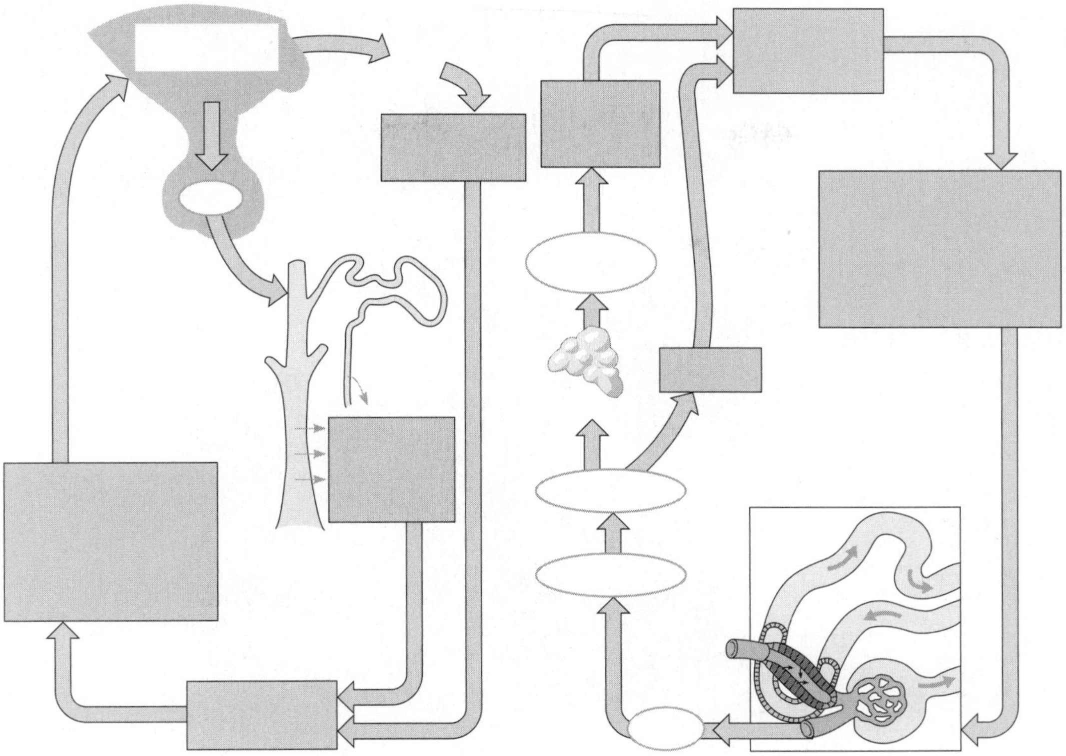

Figure 44.16 Hormonal control of the kidney by negative feedback circuits, page 937

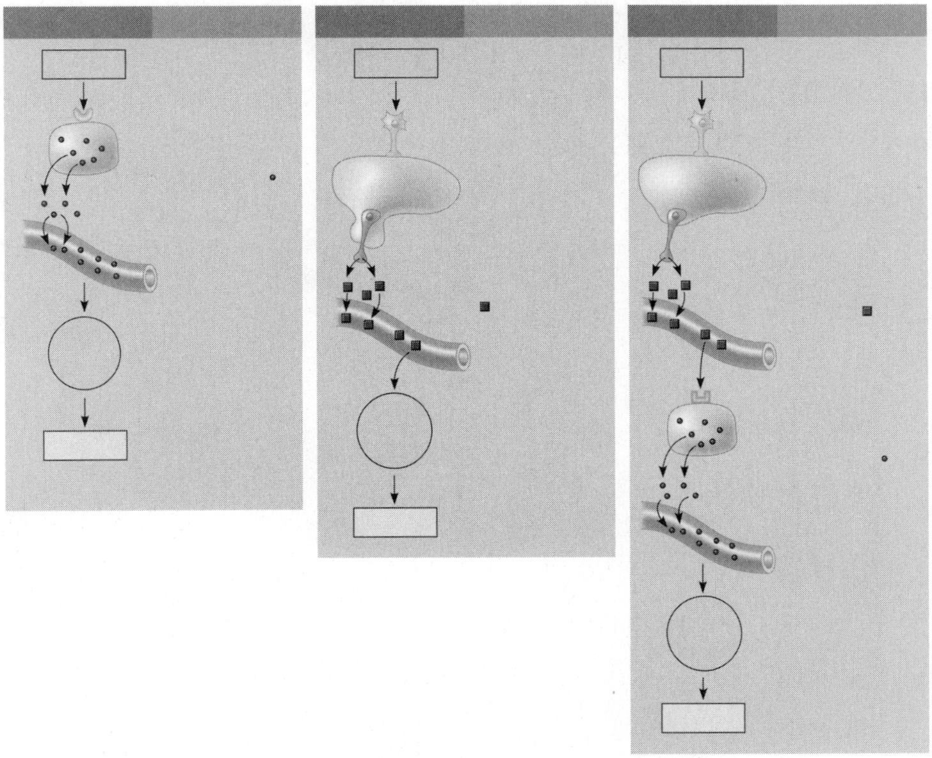

Figure 45.2 Basic patterns of simple hormonal control pathways, page 945

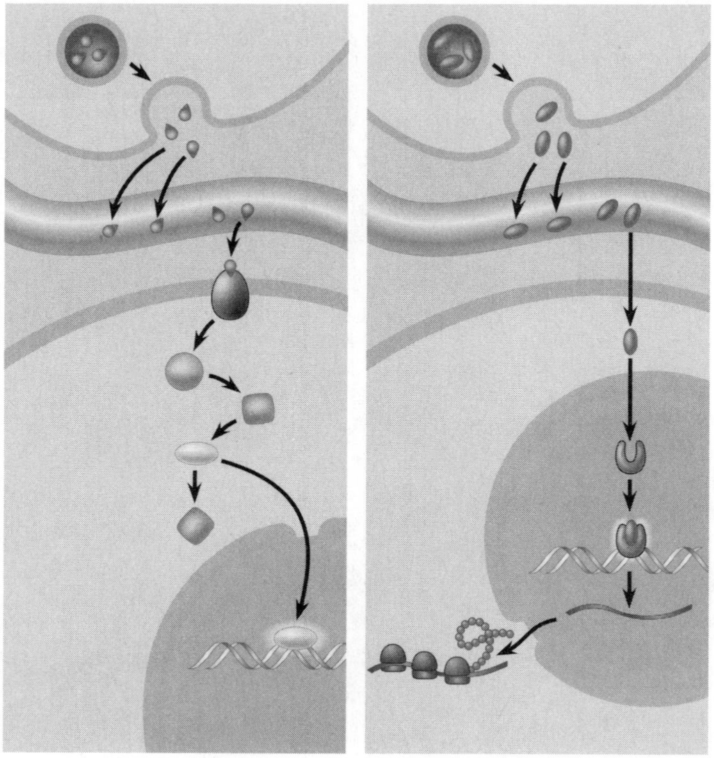

Figure 45.3 Mechanisms of hormonal signaling: a review, page 946

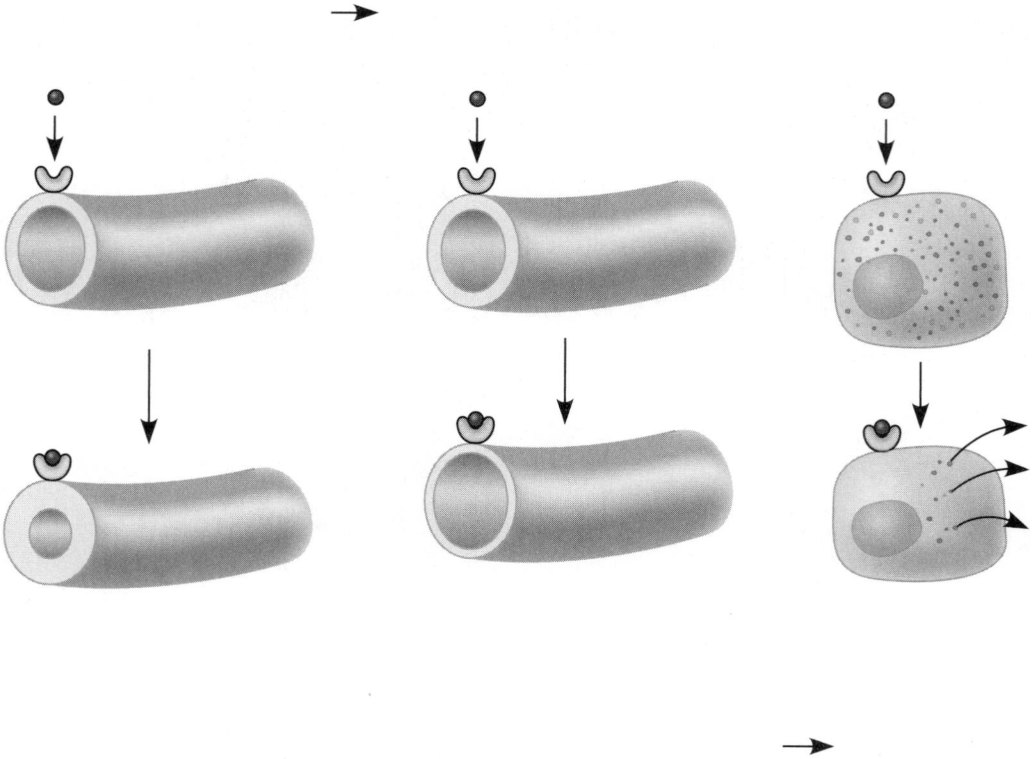

Figure 45.4 One chemical signal, different effects, page 947

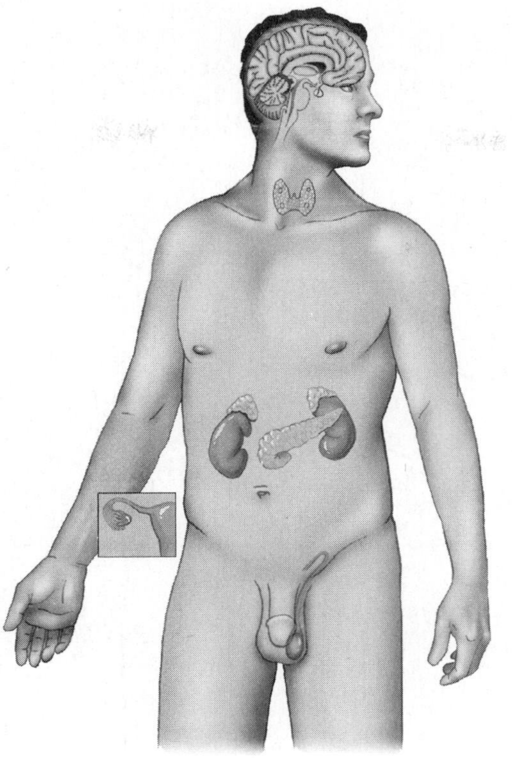

Figure 45.6 Human endocrine glands surveyed in this chapter, page 950

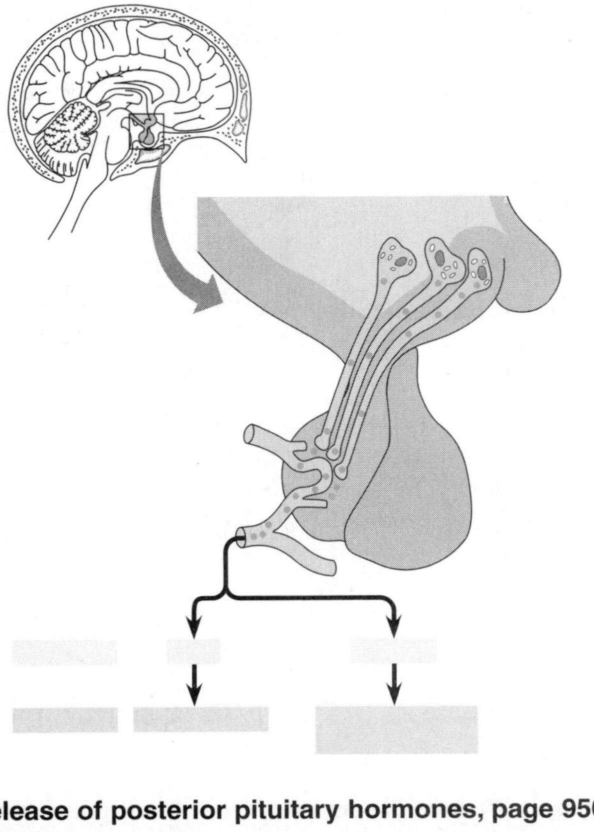

Figure 45.7 Production and release of posterior pituitary hormones, page 950

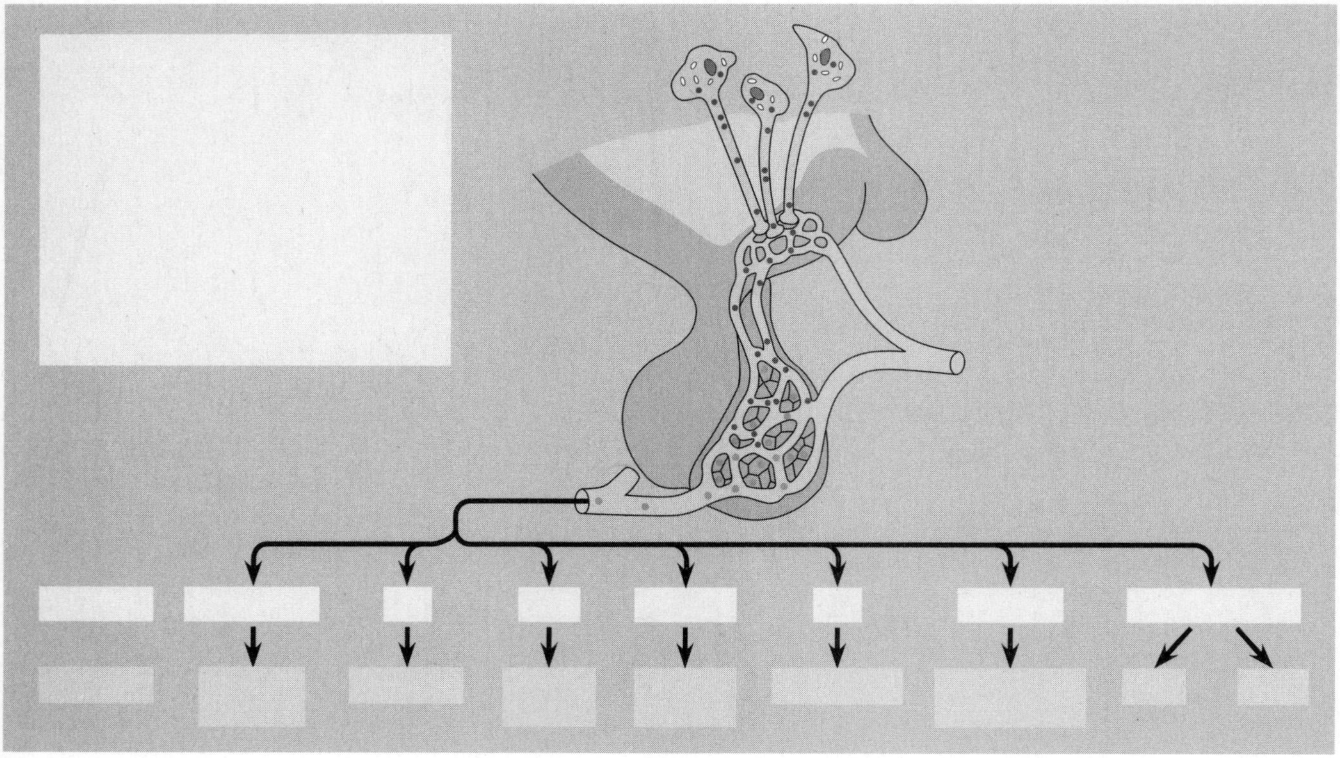

Figure 45.8 Production and release of anterior pituitary hormones, page 951

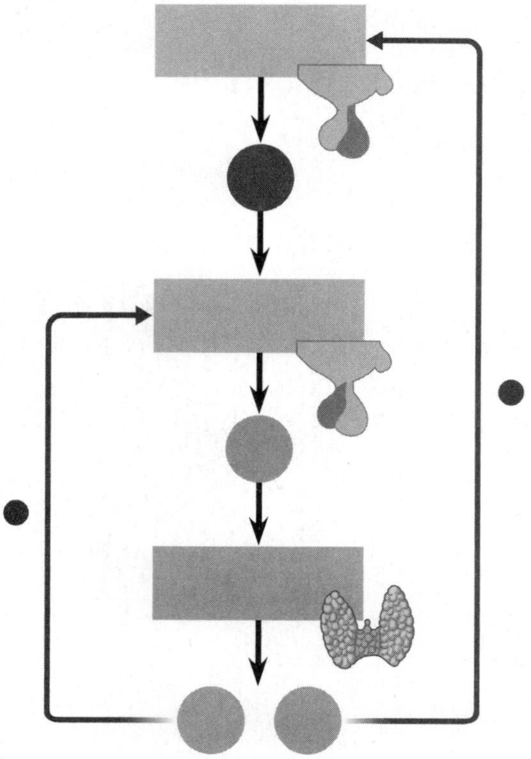

Figure 45.9 Feedback regulation of T_3 and T_4 secretion from the thyroid gland, page 953

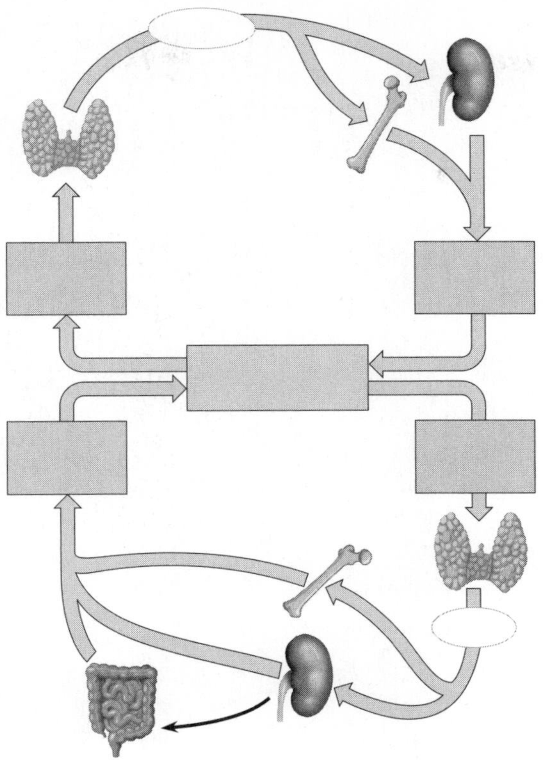

Figure 45.11 Hormonal control of calcium homeostasis in mammals, page 954

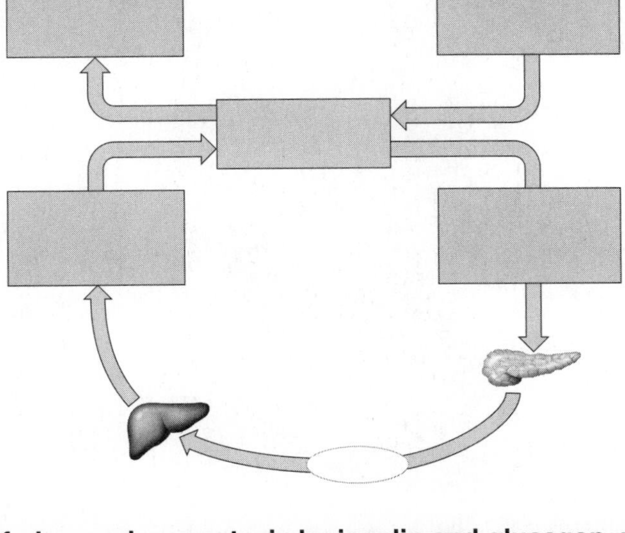

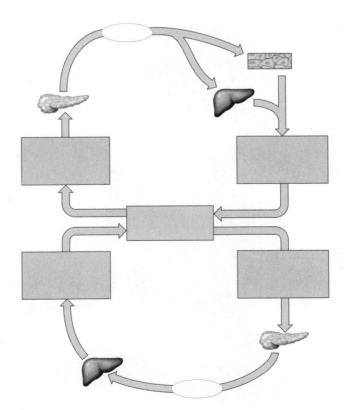

Figure 45.12 Maintenance of glucose homeostasis by insulin and glucagon, page 956

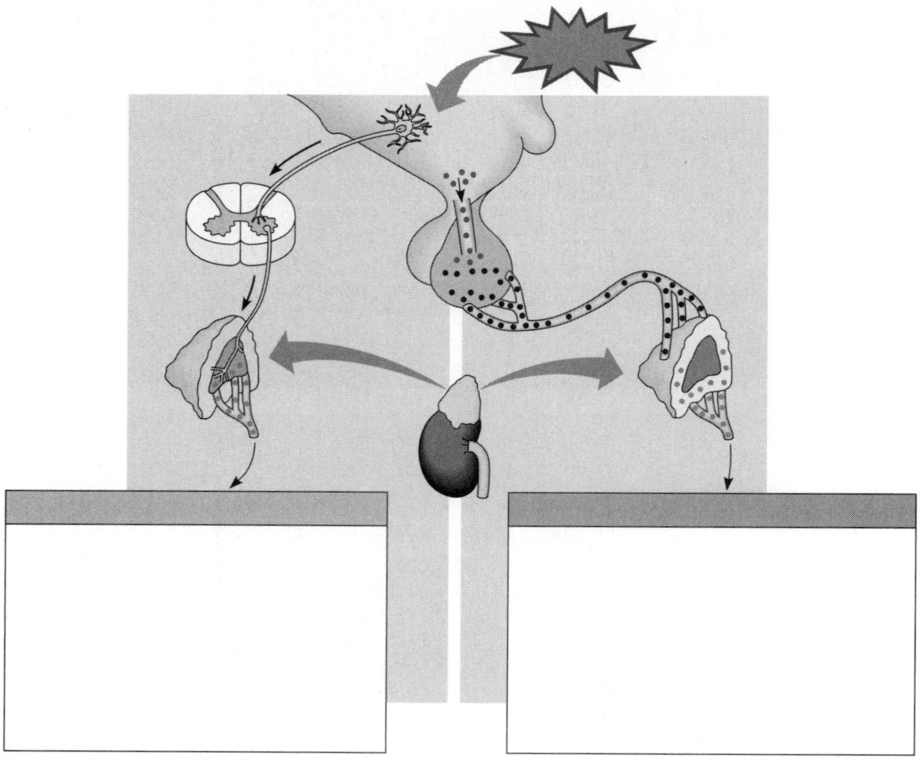

Figure 45.13 Stress and the adrenal gland, page 957

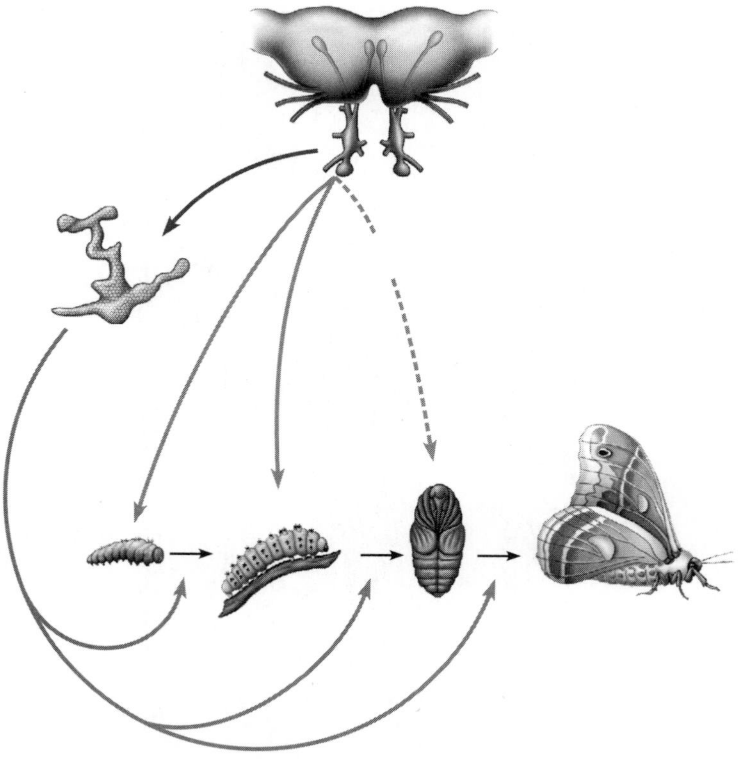

Figure 45.15 Hormonal regulation of insect development, page 960

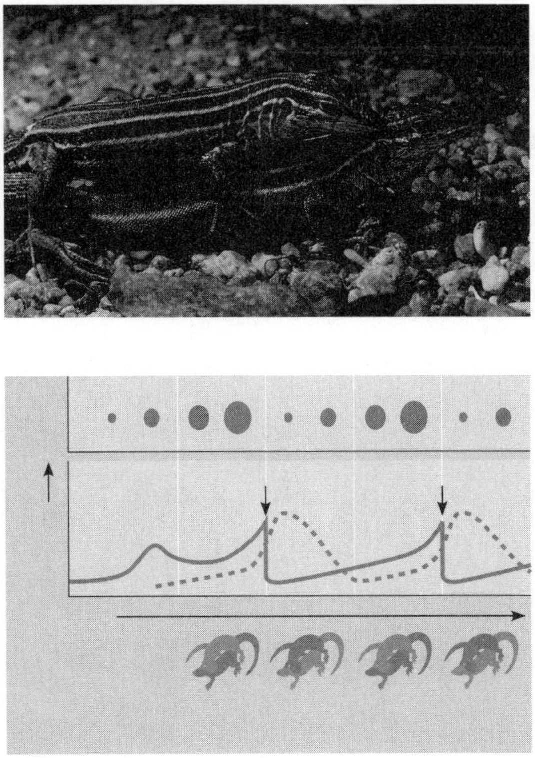

Figure 46.3 Sexual behavior in parthenogenetic lizards, page 966

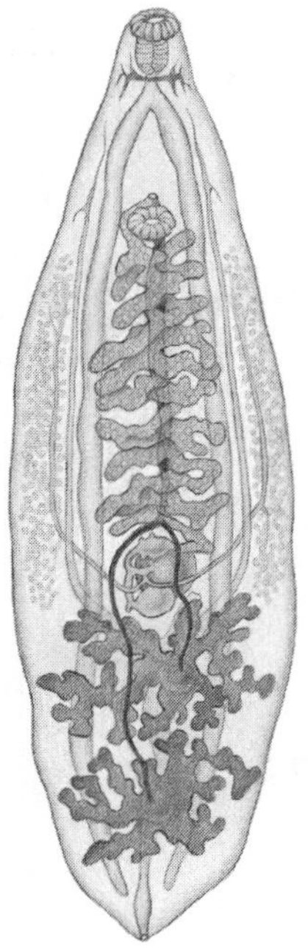

Figure 46.7 Reproductive anatomy of a parasitic flatworm, page 968

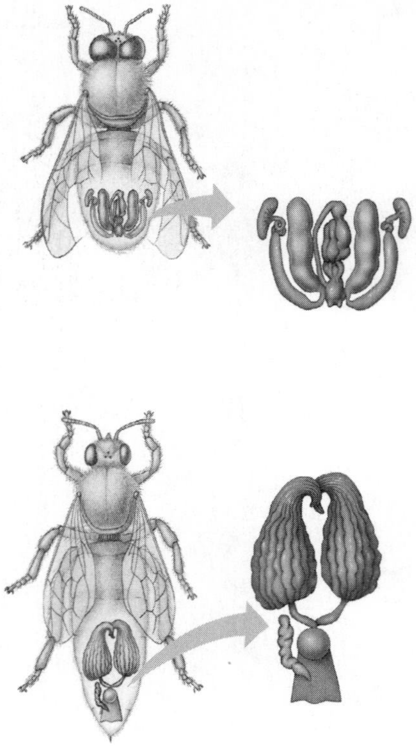

Figure 46.8 Insect reproductive anatomy, page 969

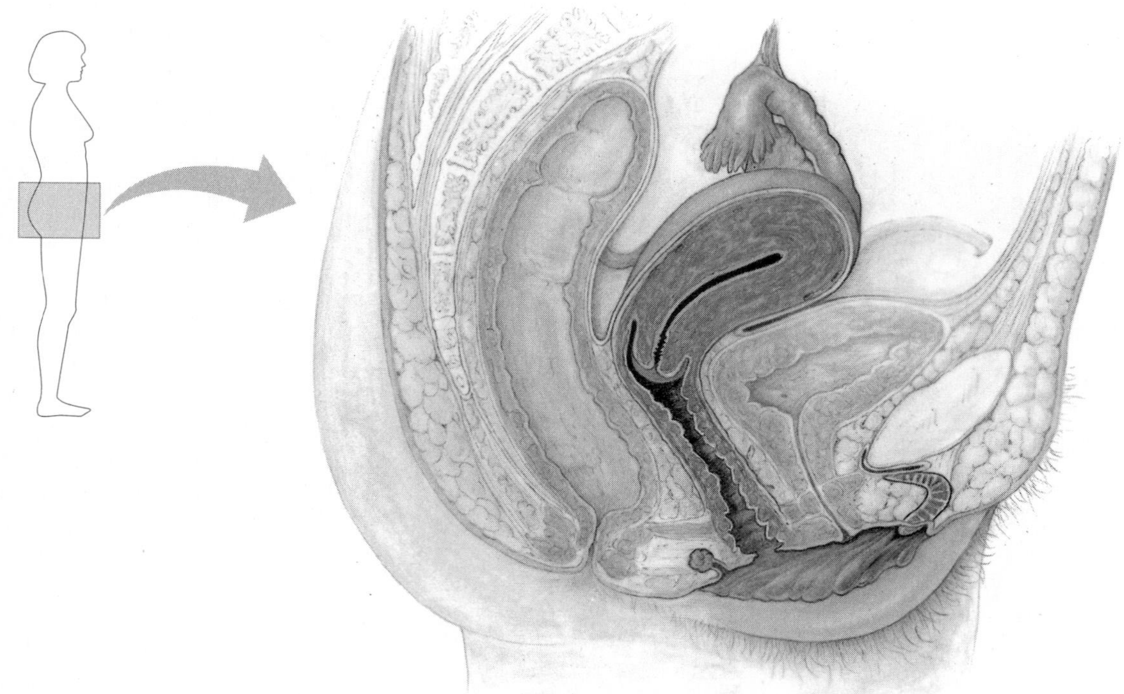

Figure 46.9 Reproductive anatomy of the human female (part 1), page 970

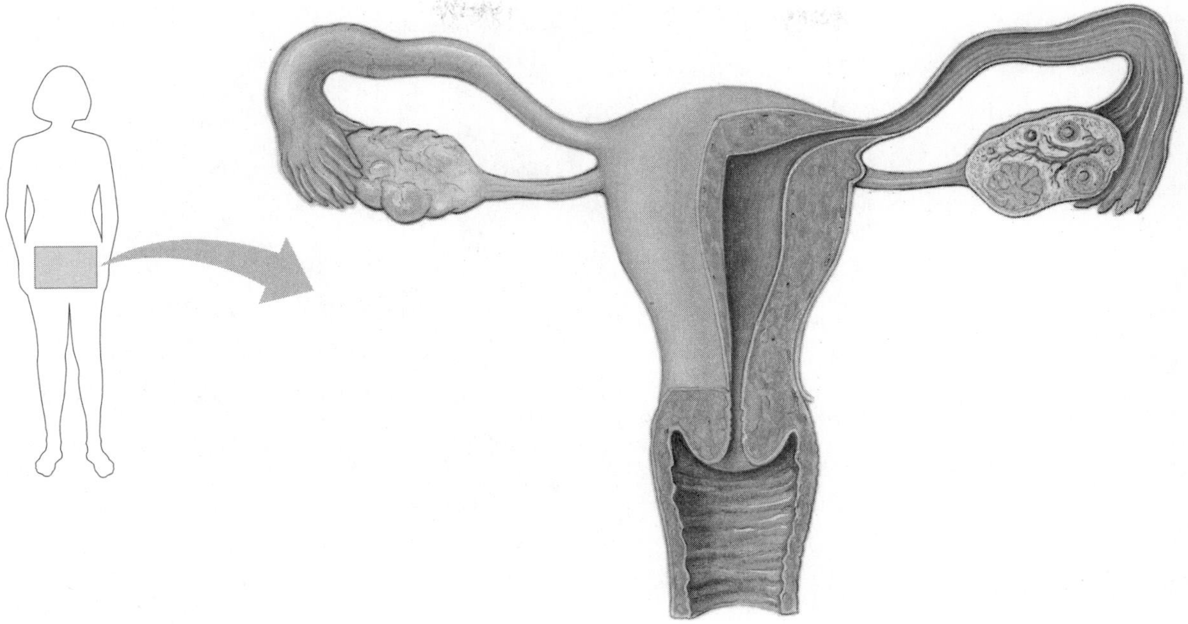

Figure 46.9 Reproductive anatomy of the human female (part 2), page 970

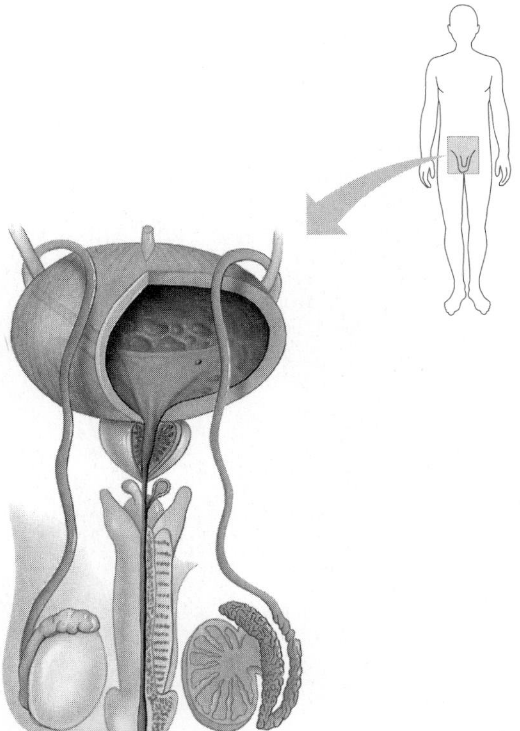

Figure 46.10 Reproductive anatomy of the human male (part 1), page 971

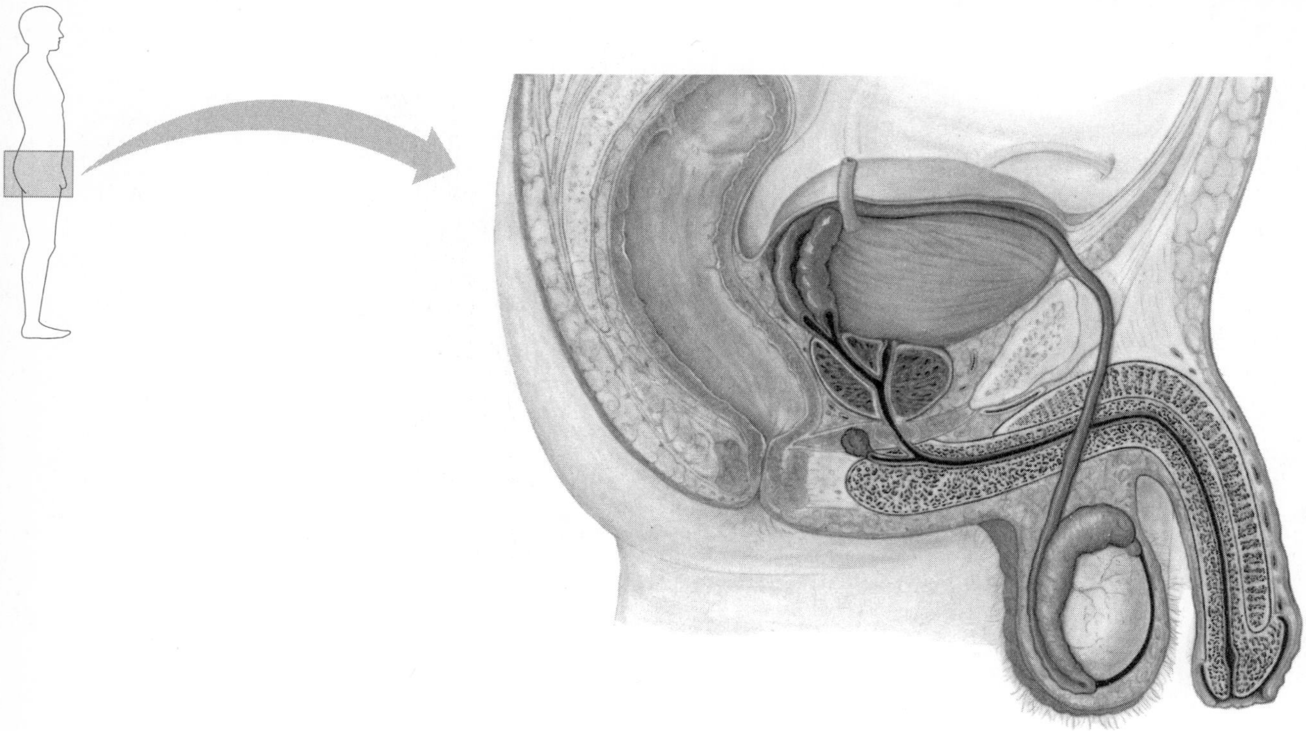

Figure 46.10 Reproductive anatomy of the human male (part 2), page 971

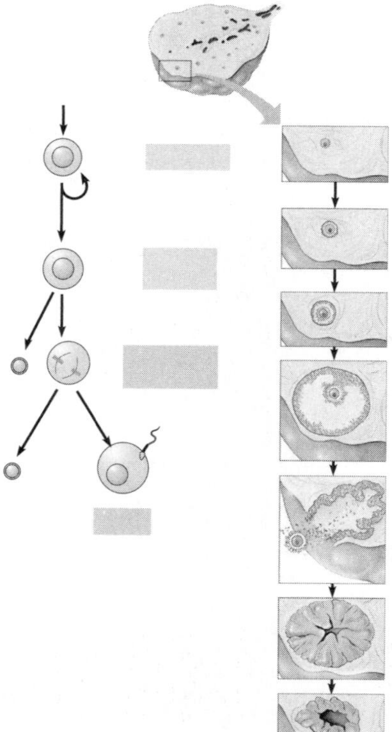

Figure 46.11 Human oogenesis, page 974

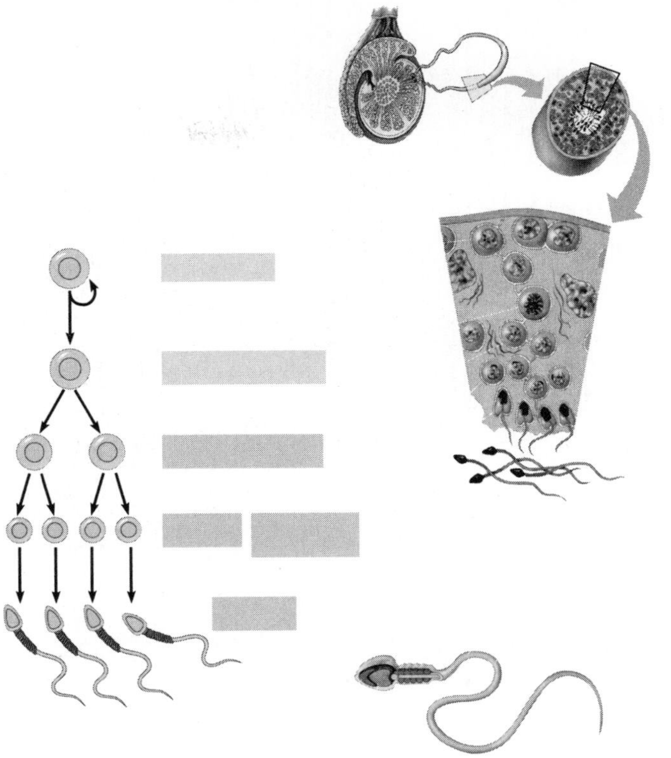

Figure 46.12 Human spermatogenesis, page 975

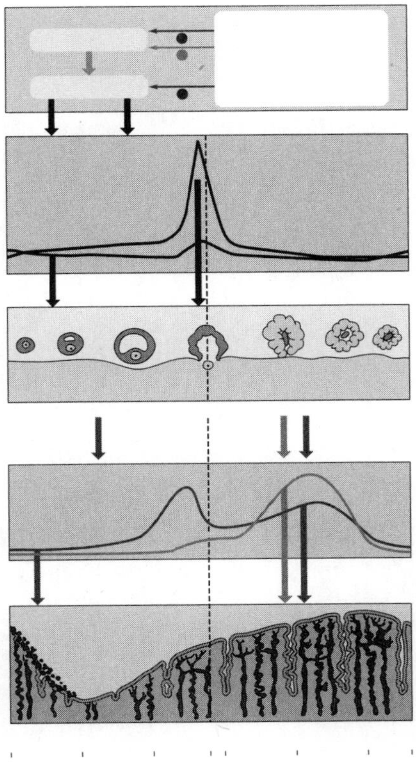

Figure 46.13 The reproductive cycle of the human female, page 976

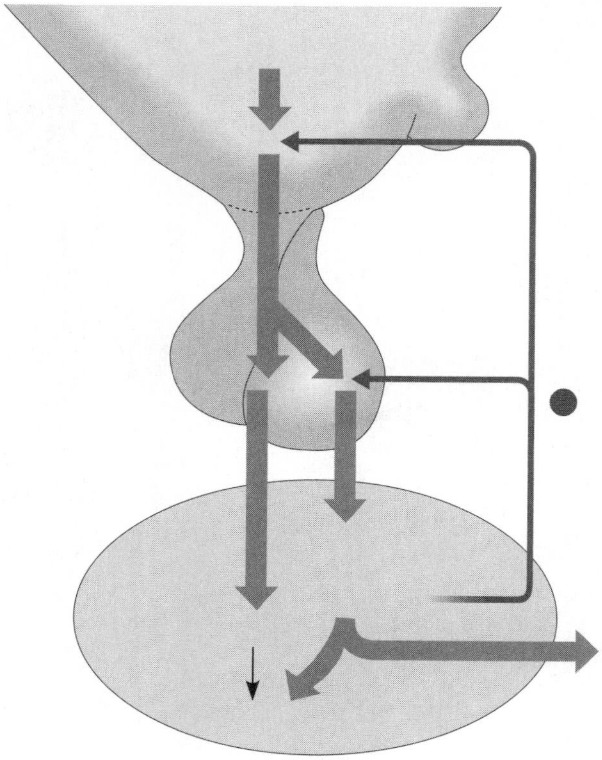

Figure 46.14 Hormonal control of the testes, page 978

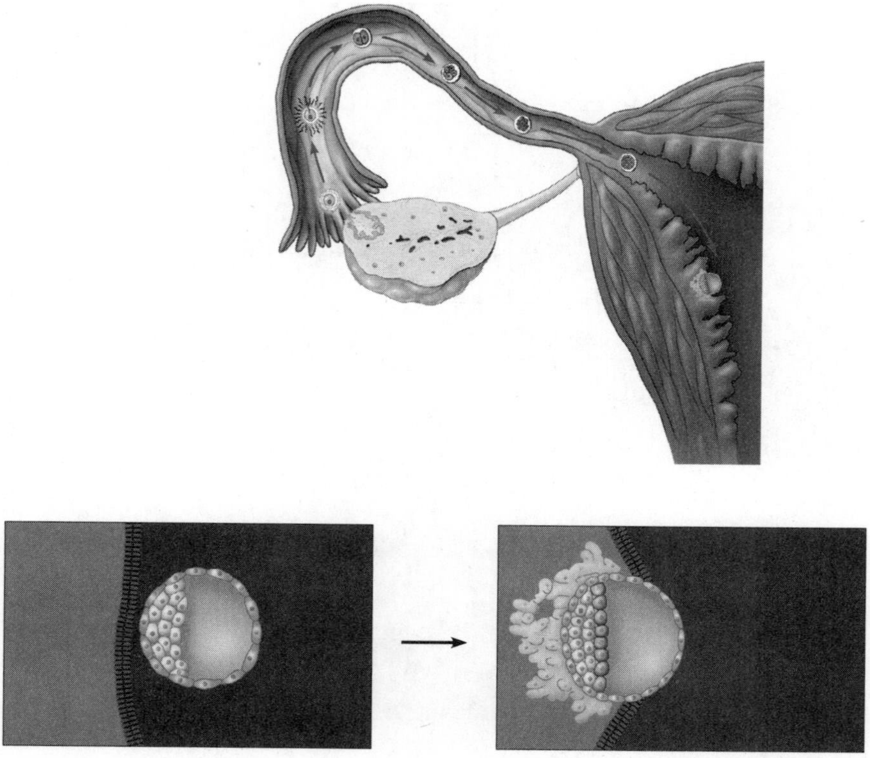

Figure 46.15 Formation of the zygote and early postfertilization events, page 979

Figure 46.16 Placental circulation, page 980

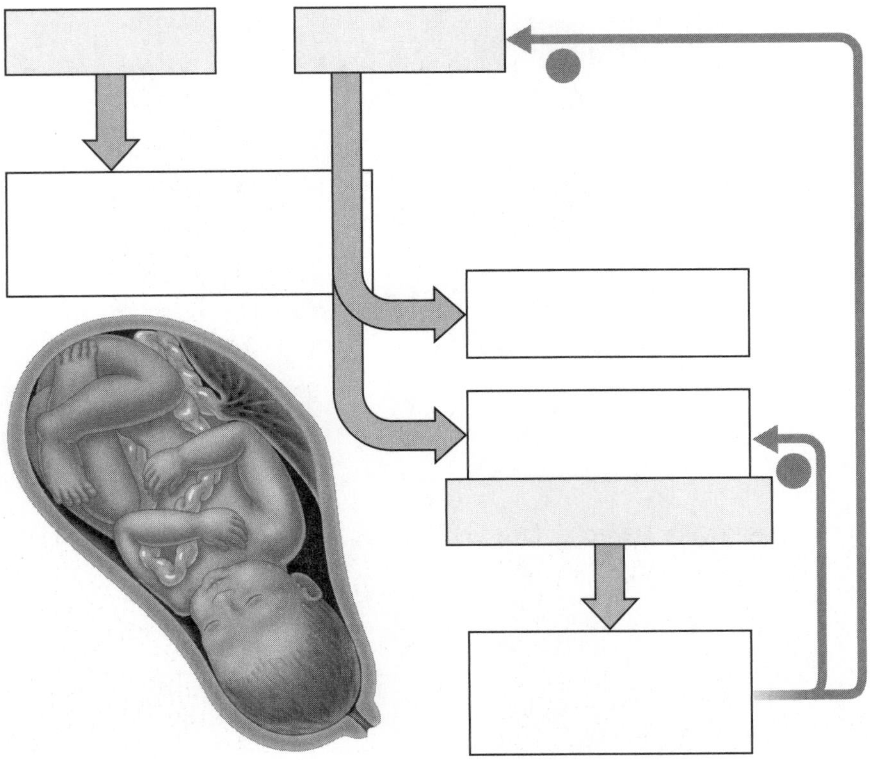

Figure 46.18 A model for the induction of labor, page 981

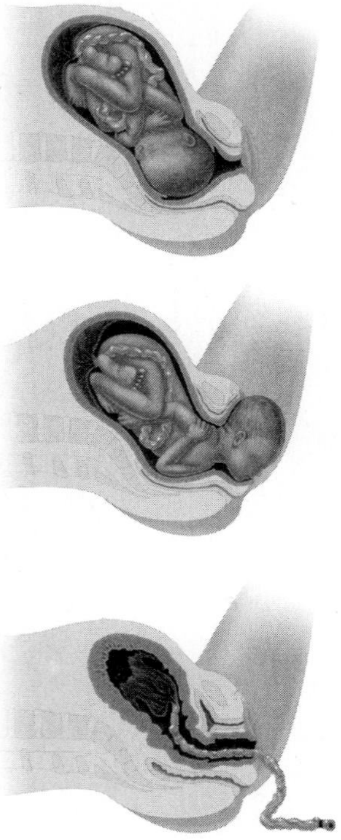

Figure 46.19 The three stages of labor, page 981

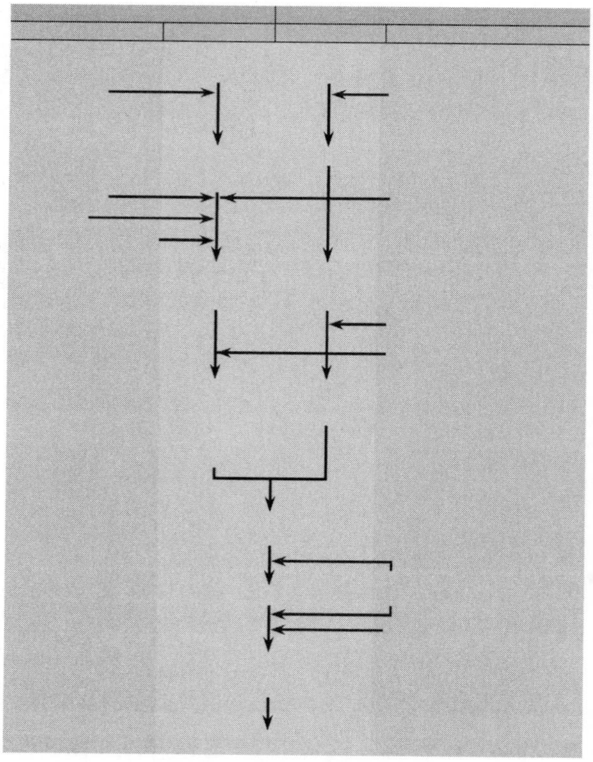

Figure 46.20 Mechanisms of some contraceptive methods, page 982

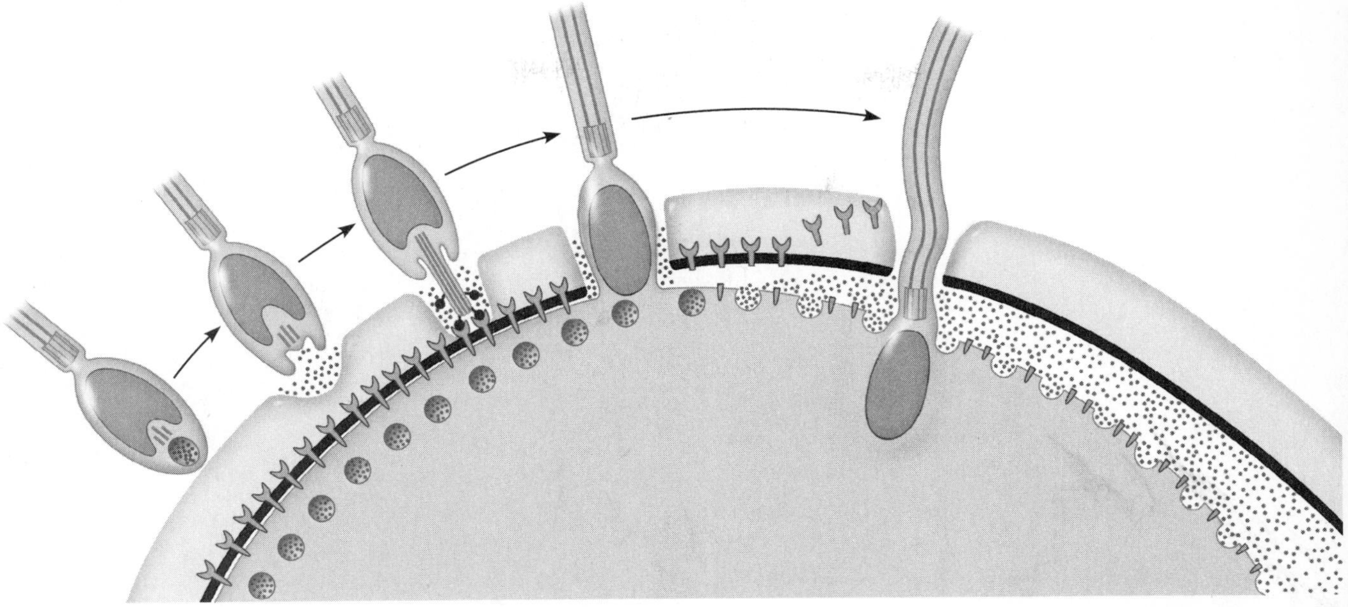

Figure 47.3 The acrosomal and cortical reactions during sea urchin fertilization, page 989

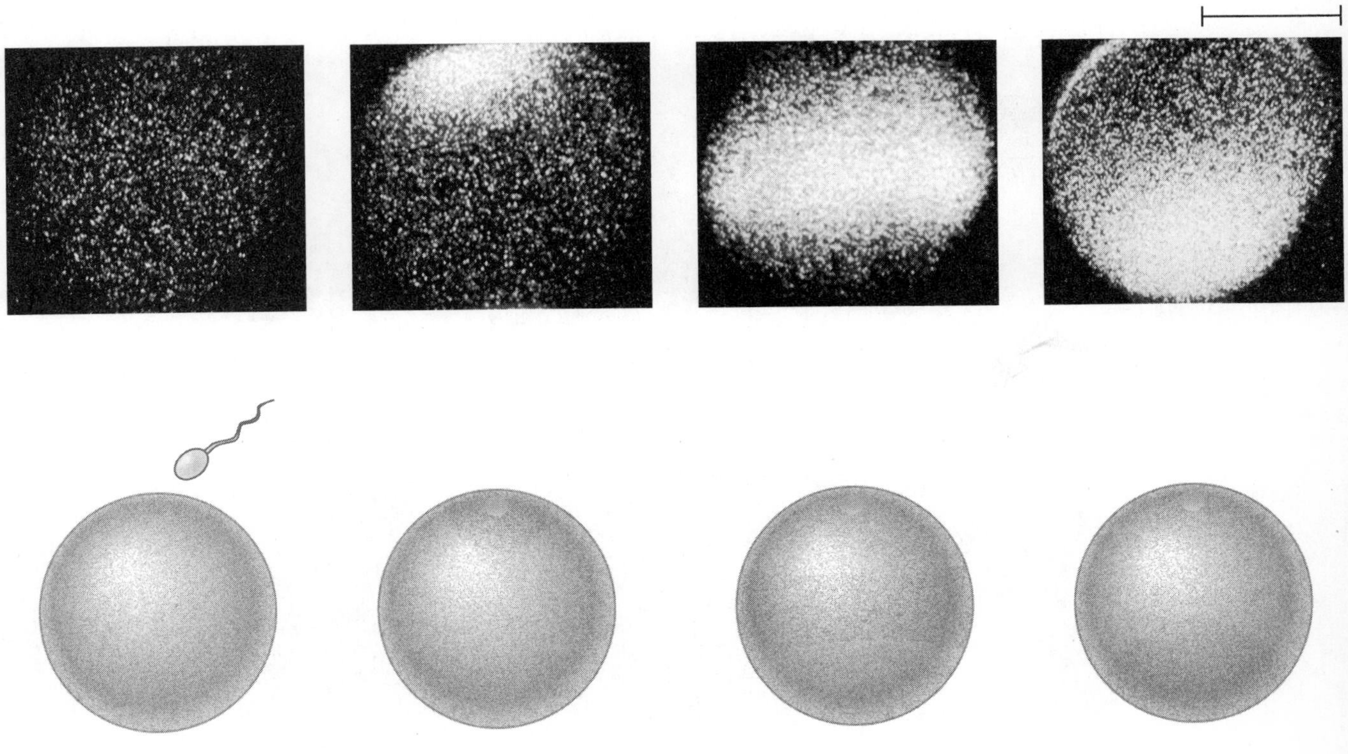

Figure 47.4 What is the effect of sperm binding on Ca^{2+} distribution in the egg? page 990

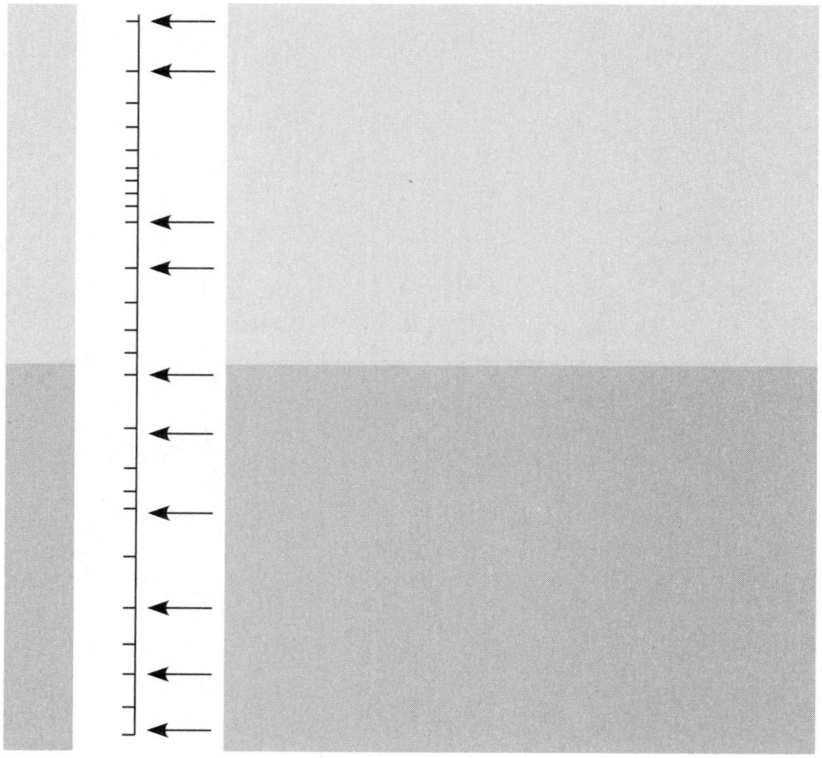

Figure 47.5 Timeline for the fertilization of sea urchin eggs, page 990

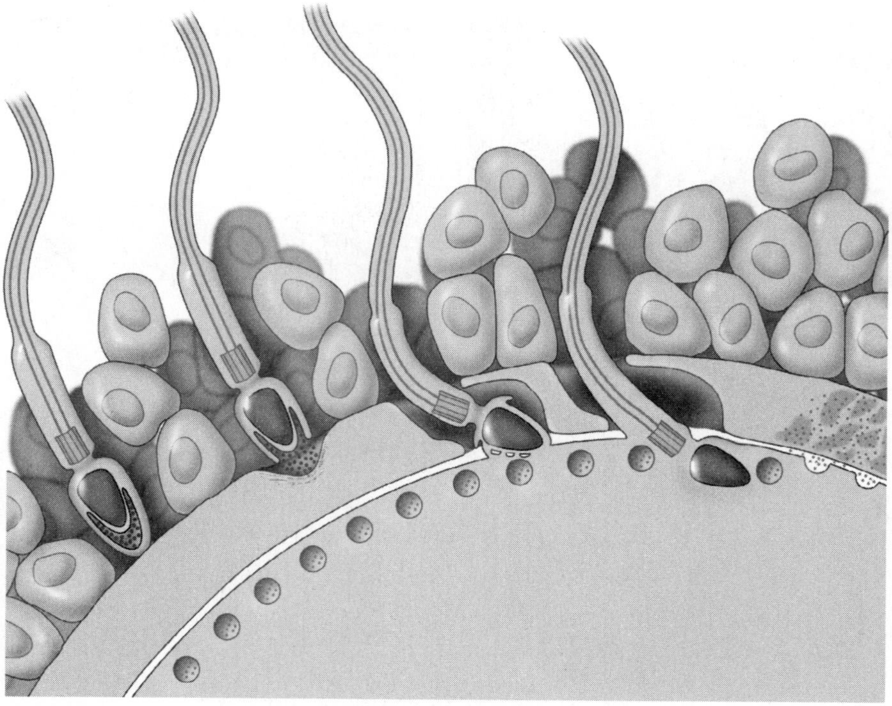

Figure 47.6 Early events of fertilization in mammals, page 991

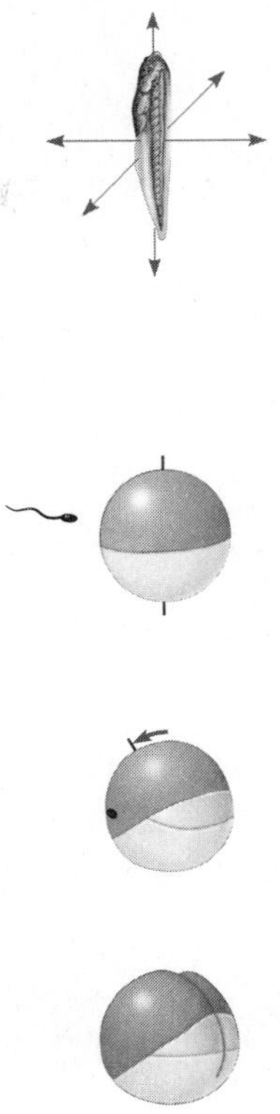

Figure 47.8 The body axes and their establishment in an amphibian, page 993

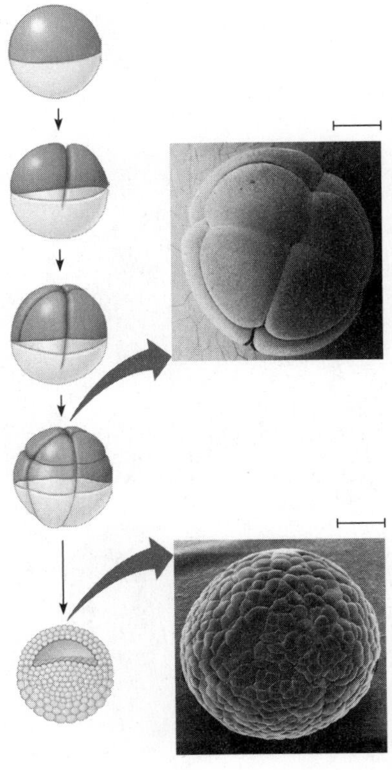

Figure 47.9 Cleavage in a frog embryo, page 993

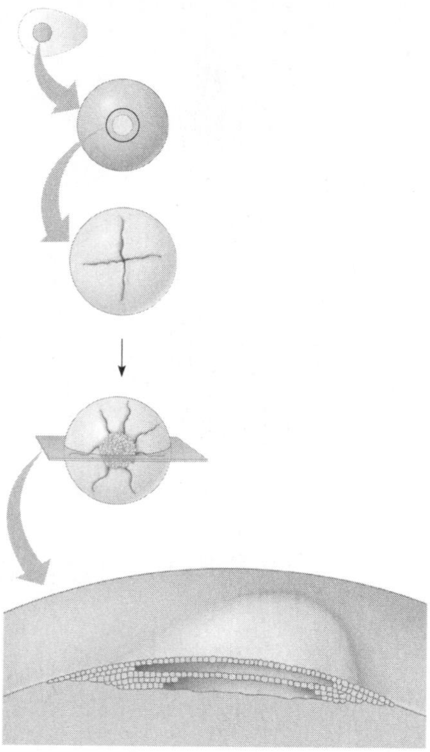

Figure 47.10 Cleavage in a chick embryo, page 994

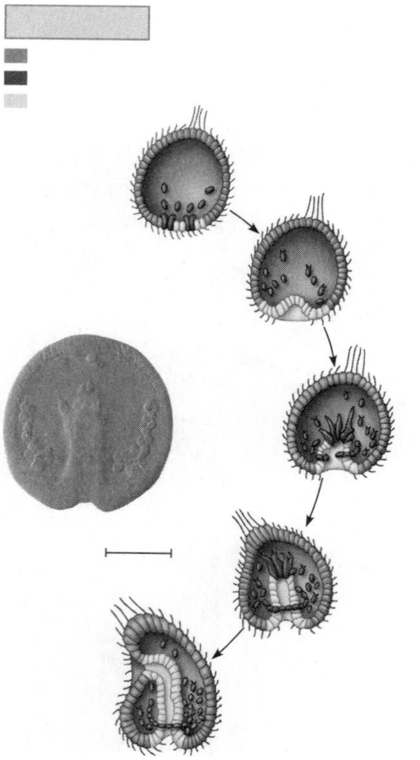

Figure 47.11 Gastrulation in a sea urchin embryo, page 995

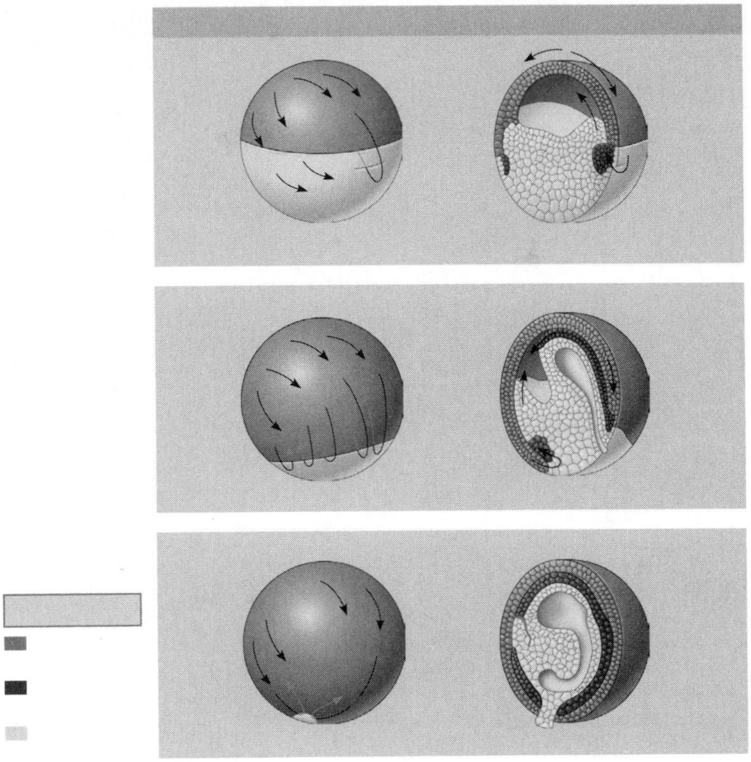

Figure 47.12 Gastrulaton in a frog embryo, page 996

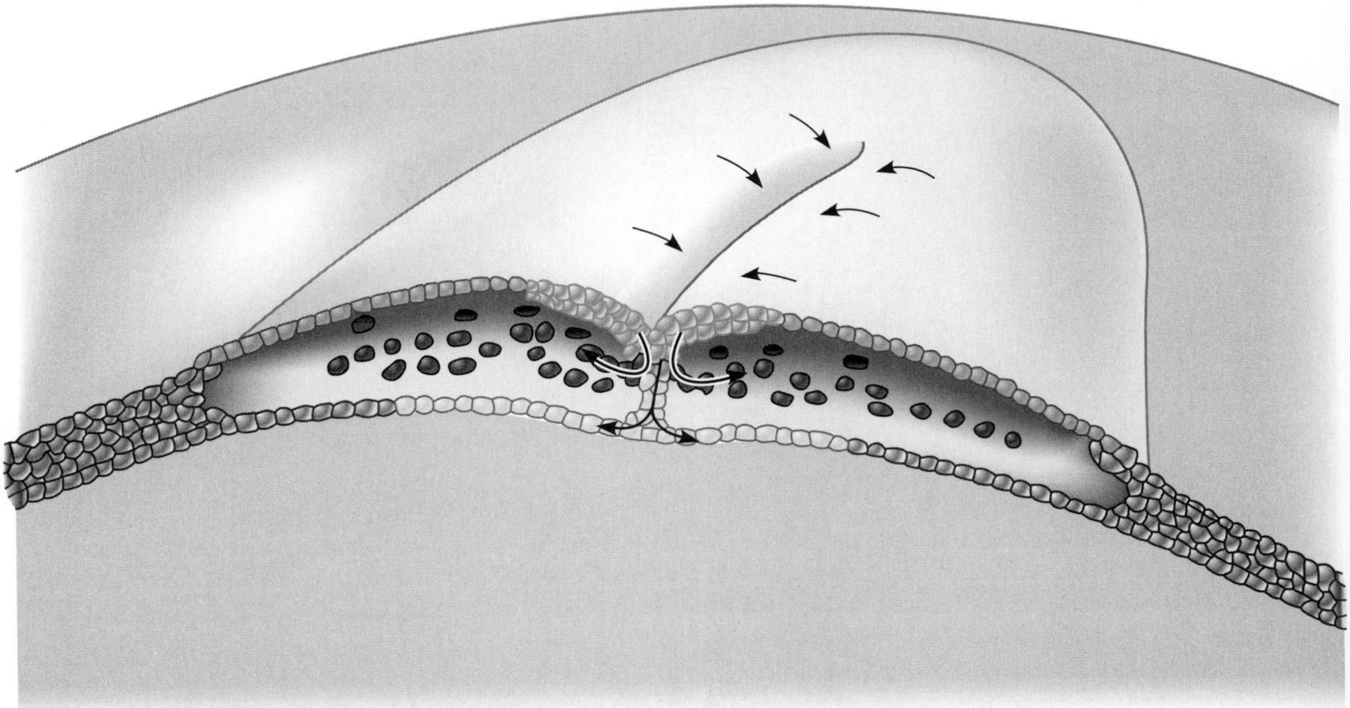

Figure 47.13 Gastrulation in a chick embryo, page 997

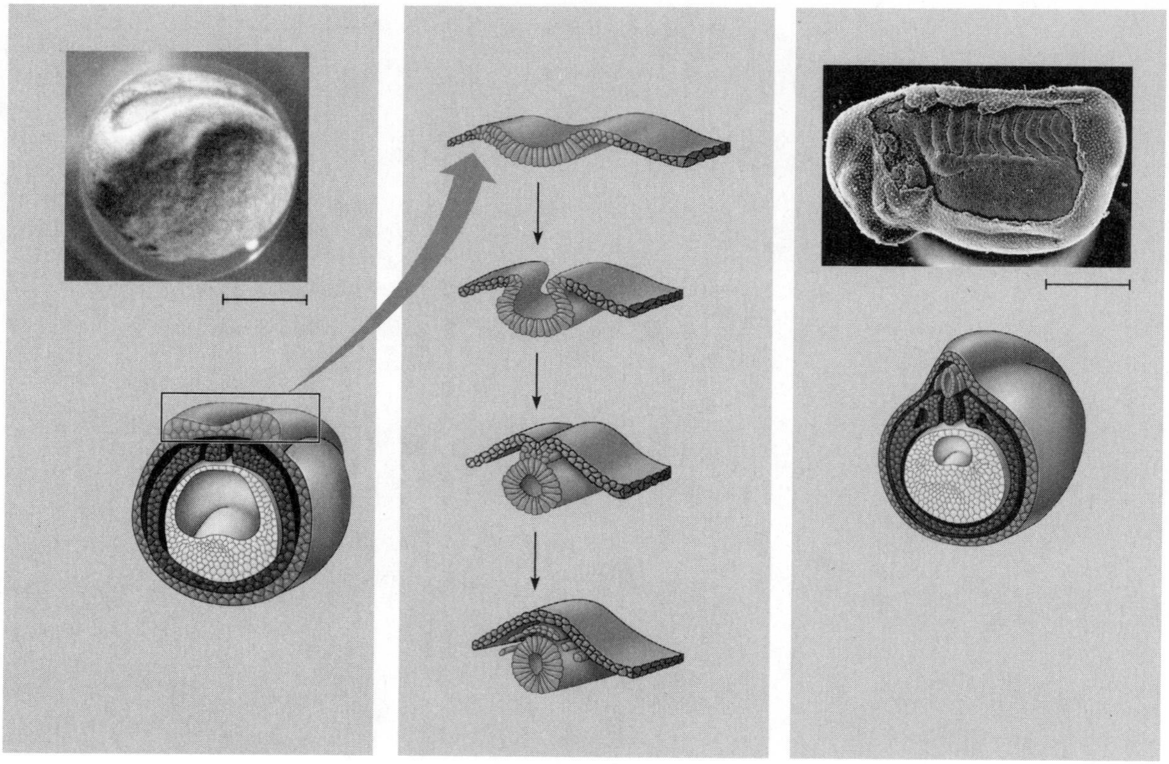

Figure 47.14 Early organogenesis in a frog embryo, page 997

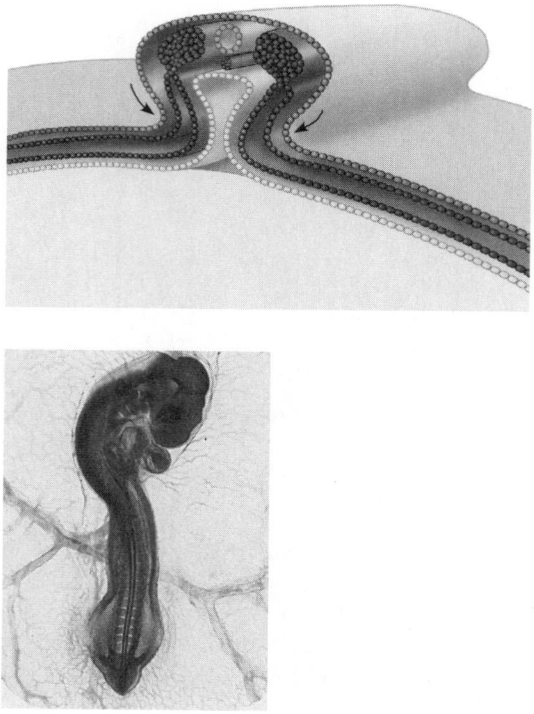

Figure 47.15 Organogenesis in a chick embryo, page 998

Figure 47.16 Adult derivatives of the three embryonic germ layers in vertebrates, page 999

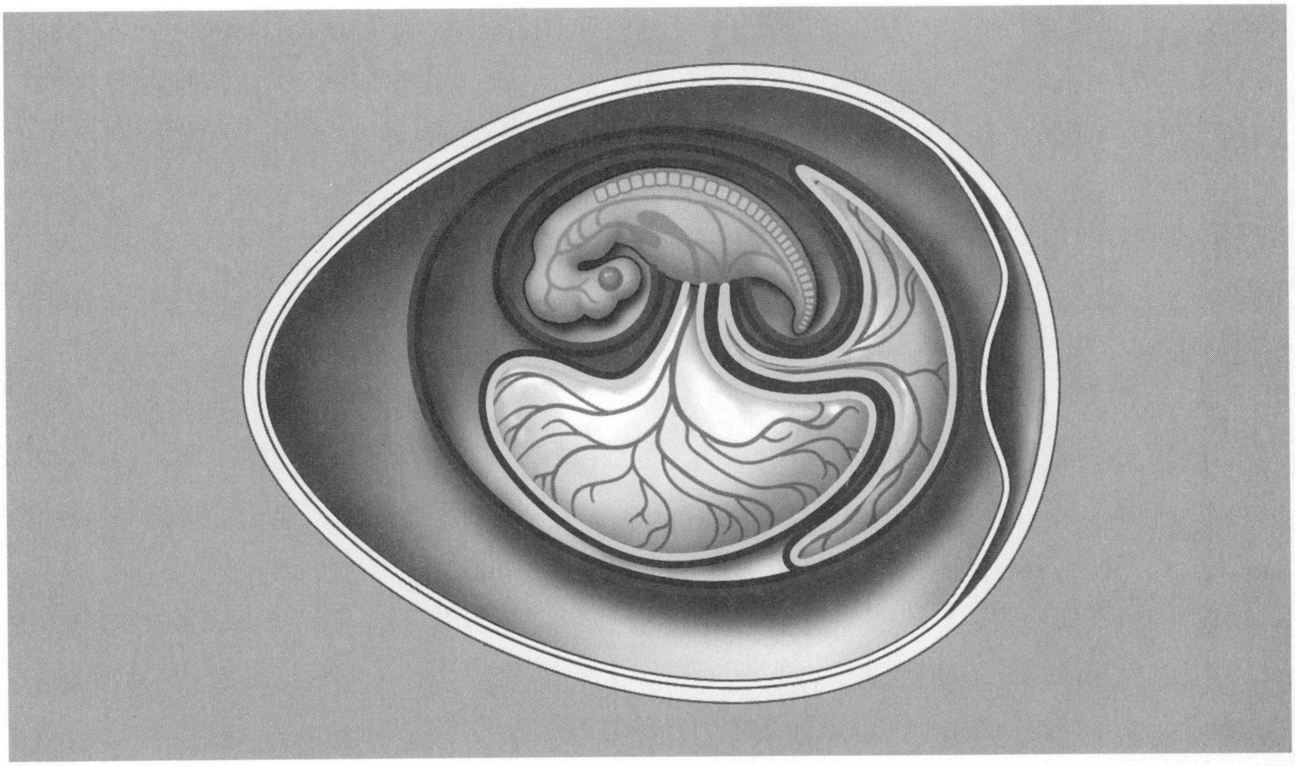

Figure 47.17 Extraembryonic membranes in birds and other reptiles, page 999

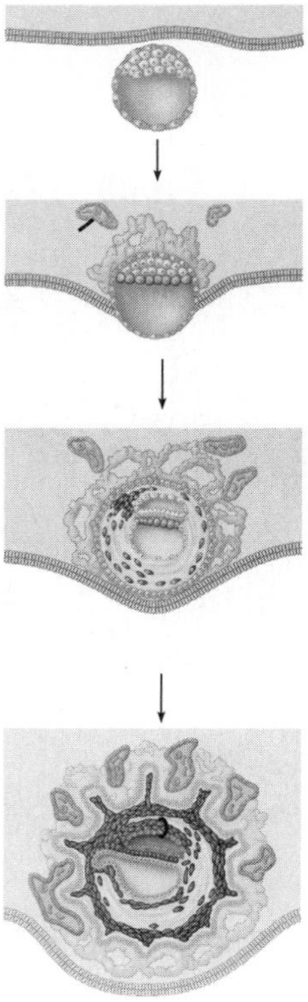

Figure 47.18 Four stages in early embryonic development of a human, page 1000

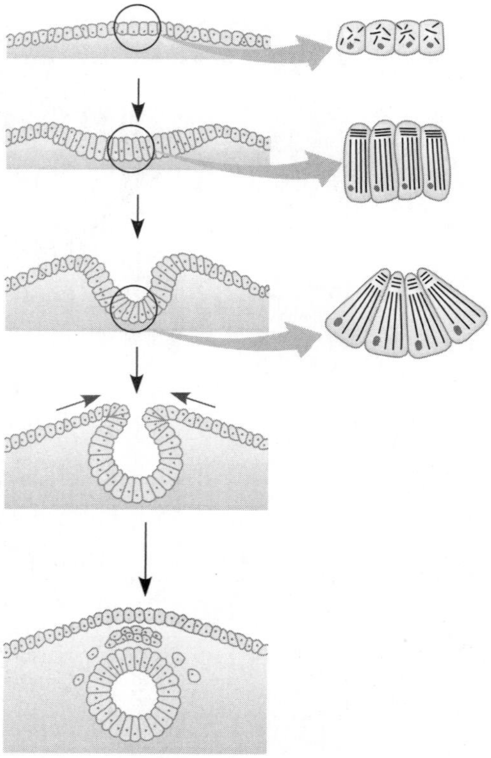

Figure 47.19 Change in cellular shape during morphogenesis, page 1001

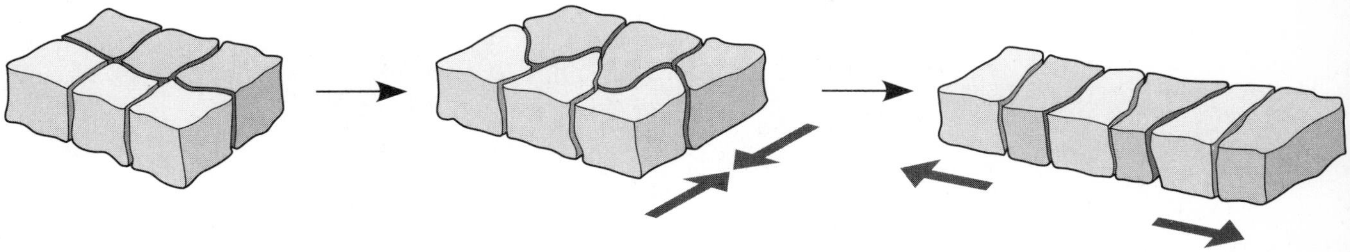

Figure 47.20 Convergent extension of a sheet of cells, page 1002

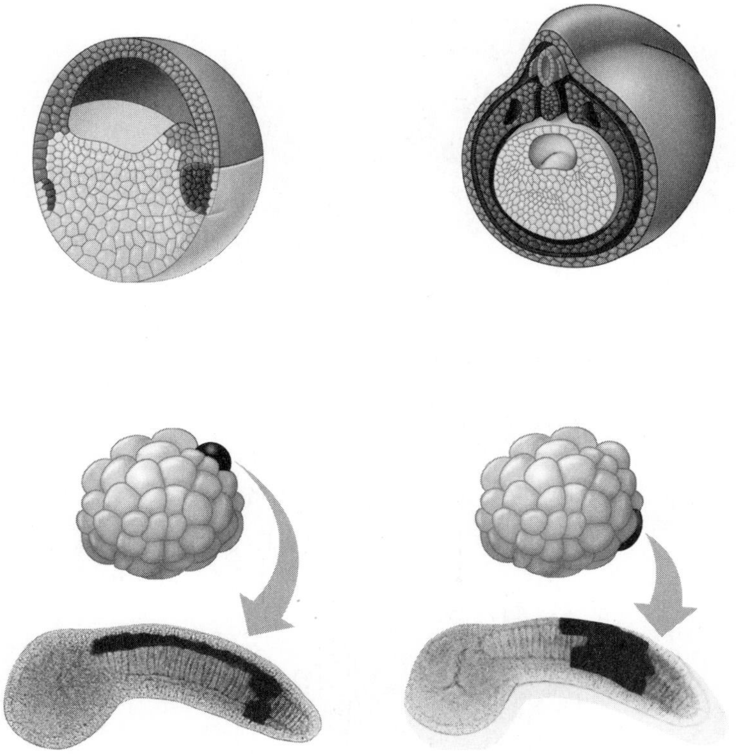

Figure 47.23 Fate mapping for two chordates, page 1004

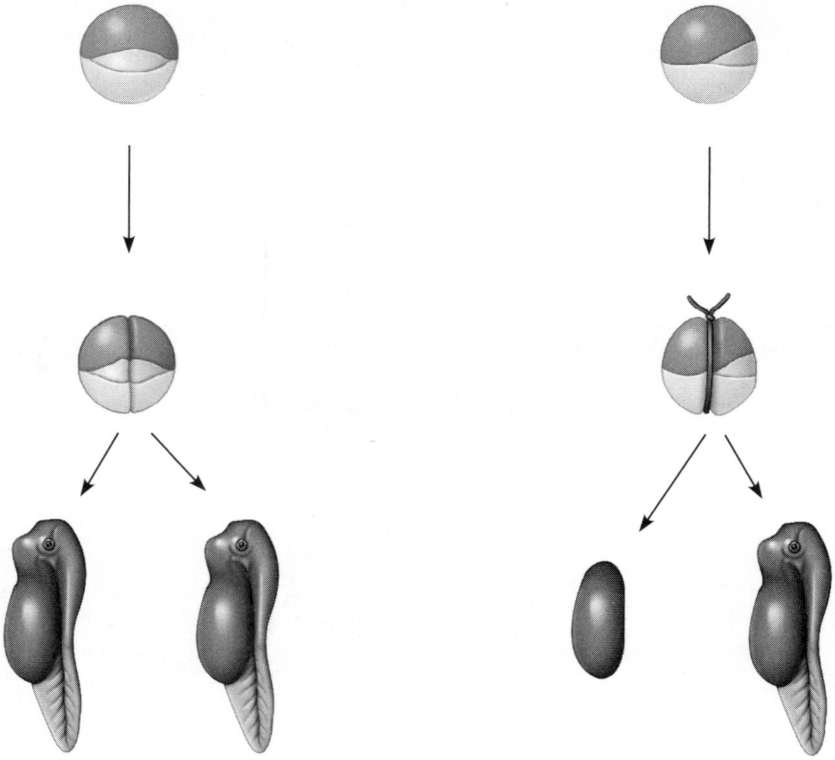

Figure 47.24 How does distribution of the gray crescent at the first cleavage affect the potency of the two daughter cells? page 1005

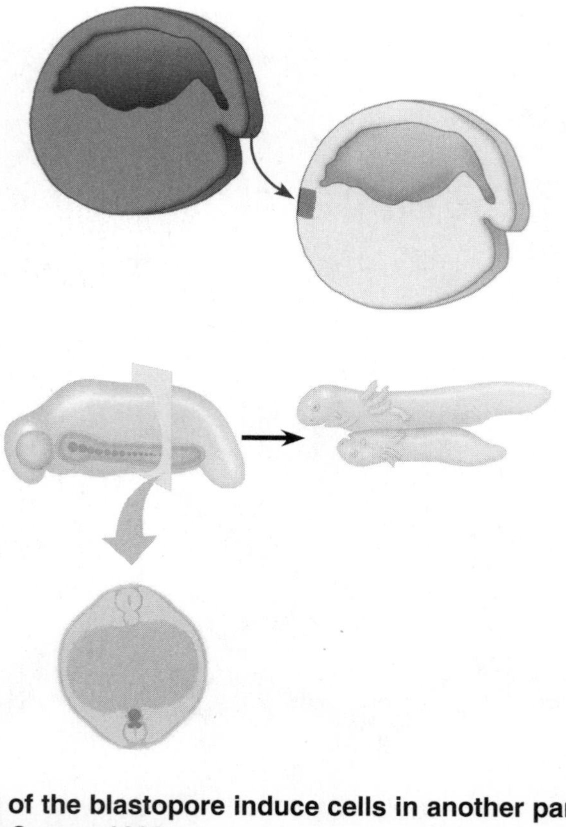

Figure 47.25 Can the dorsal lip of the blastopore induce cells in another part of the amphibian embryo to change their developmental fate? page 1006

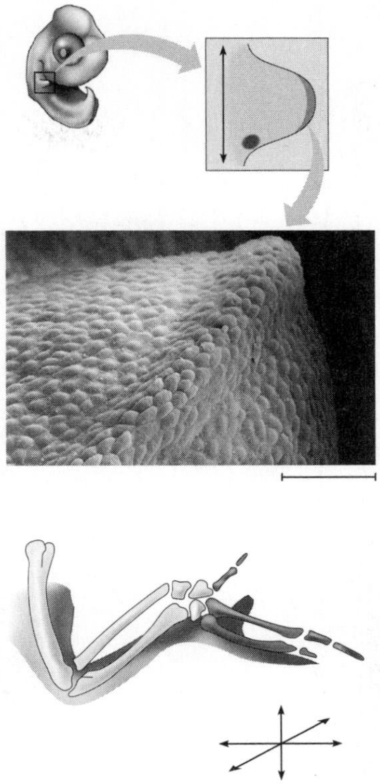

Figure 47.26 Vertebrate limb development, page 1007

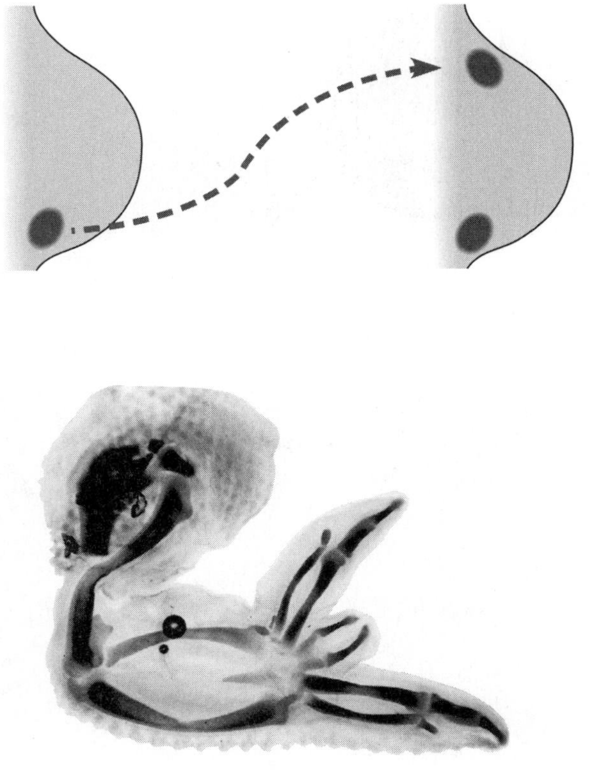

Figure 47.27 What role does the zone of polarizing activity (ZPA) play in limb pattern formation in vertebrates? page 1008

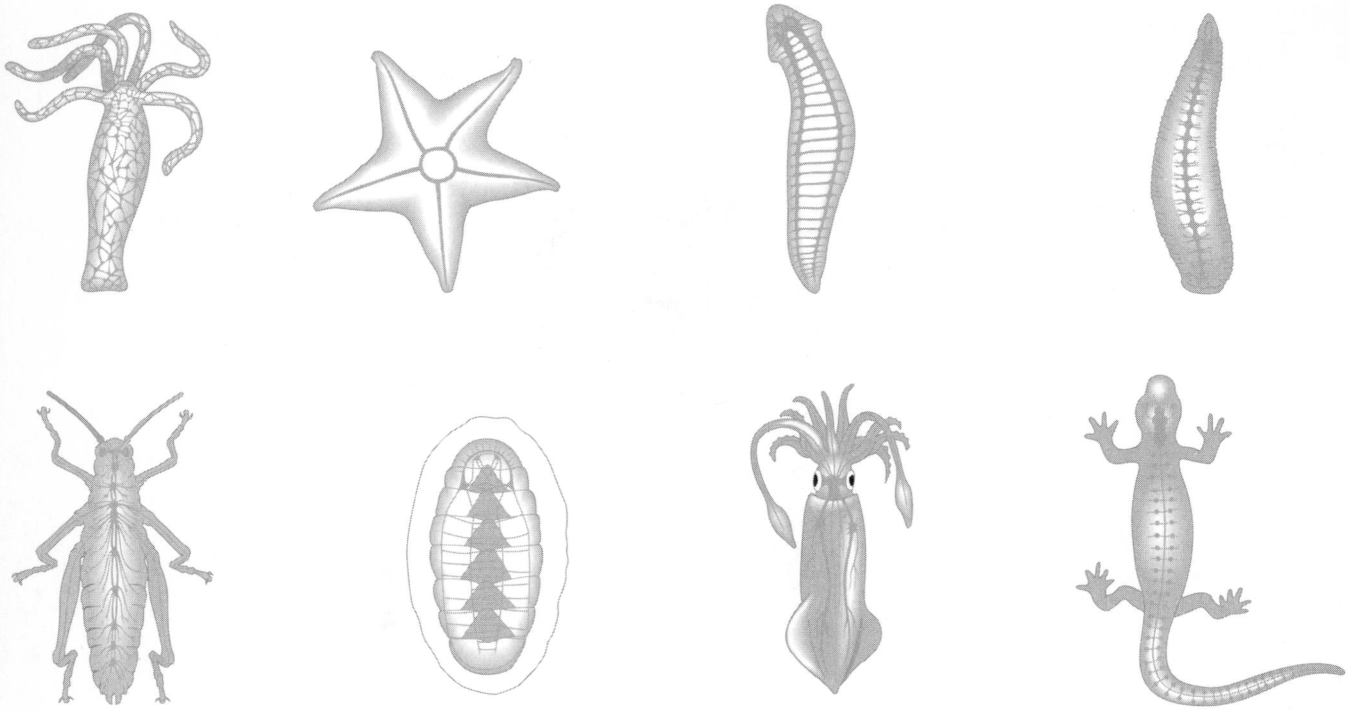

Figure 48.2 Organization of some nervous systems, page 1012

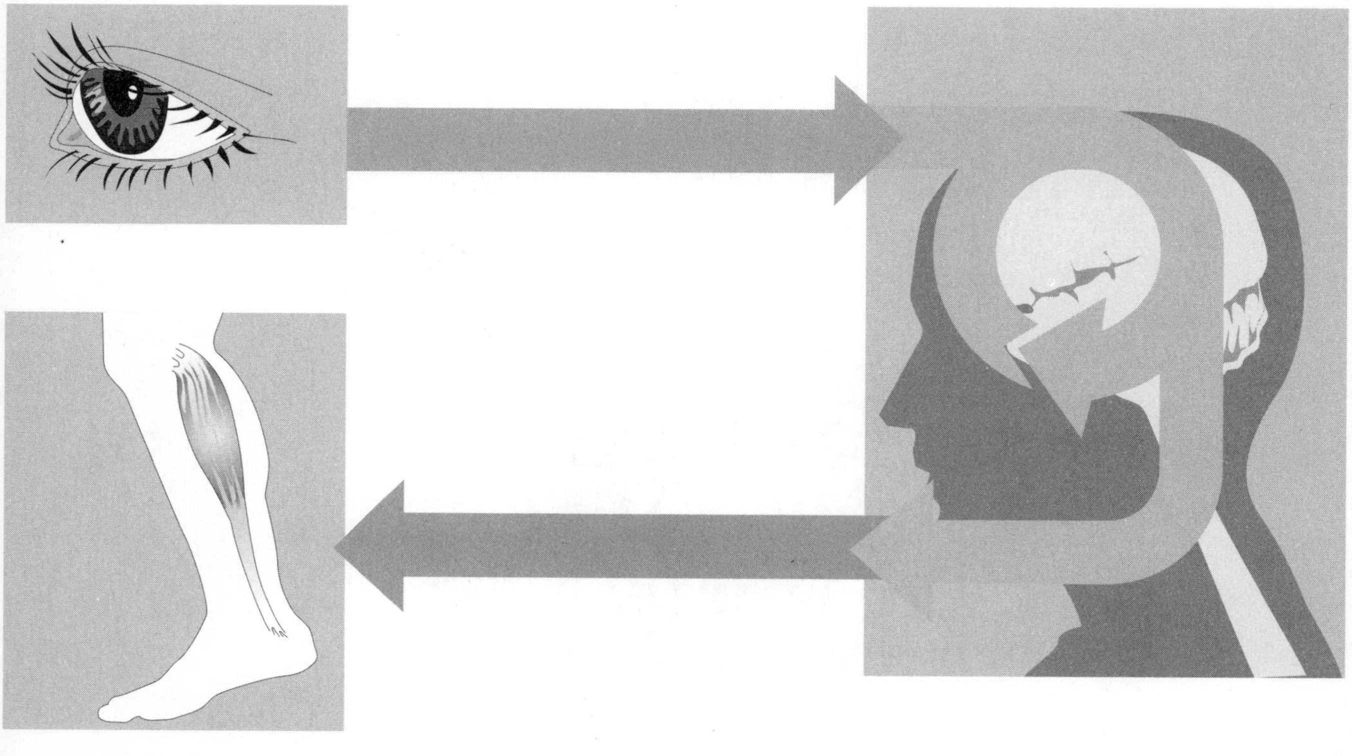

Figure 48.3 Overview of information processing by nervous systems, page 1013

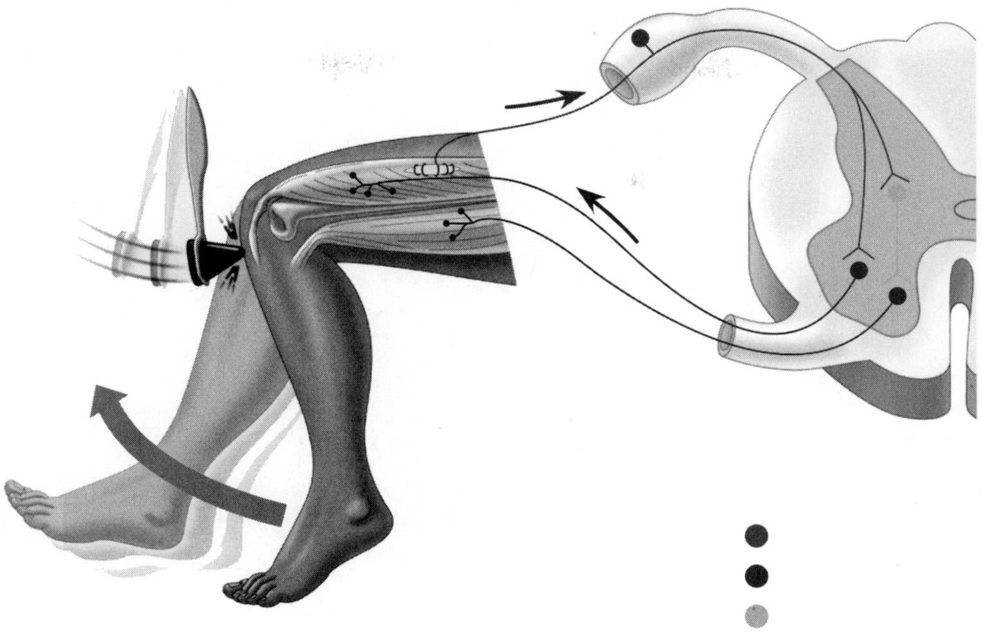

Figure 48.4 The knee-jerk reflex, page 1013

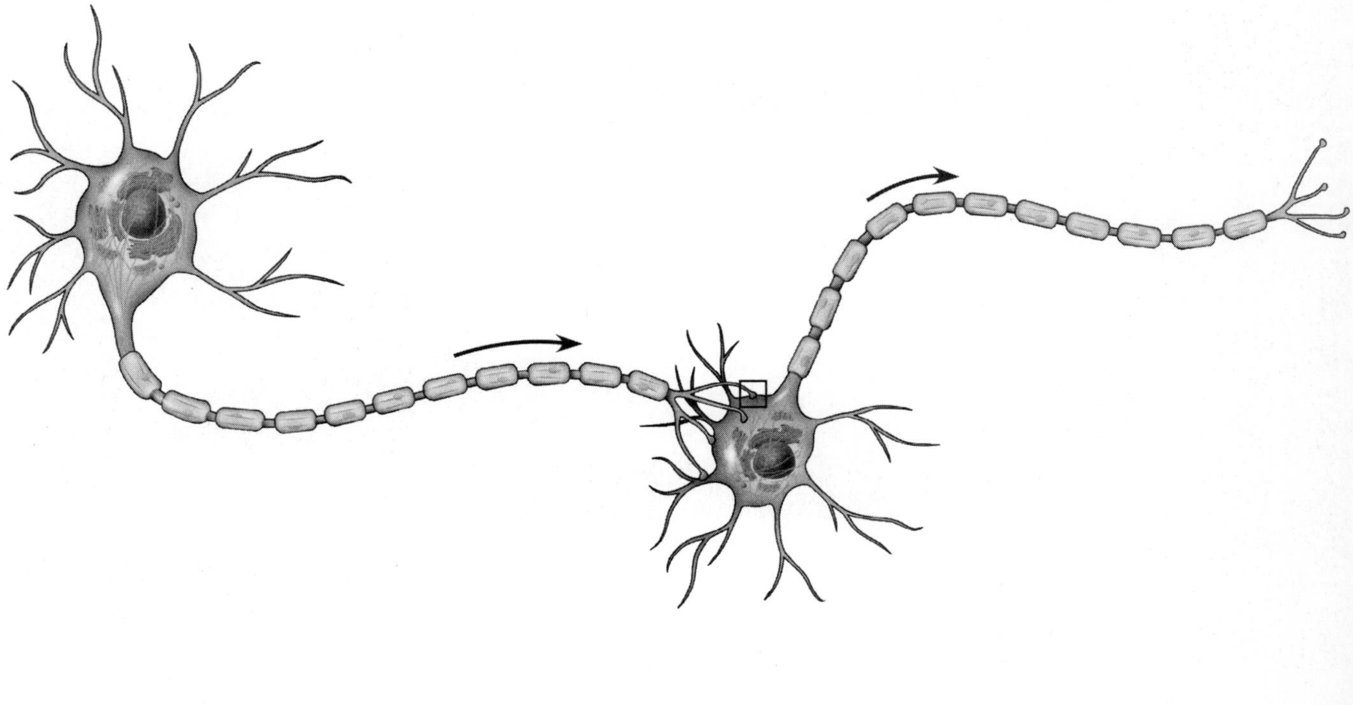

Figure 48.5 Structure of a vertebrate neuron, page 1014

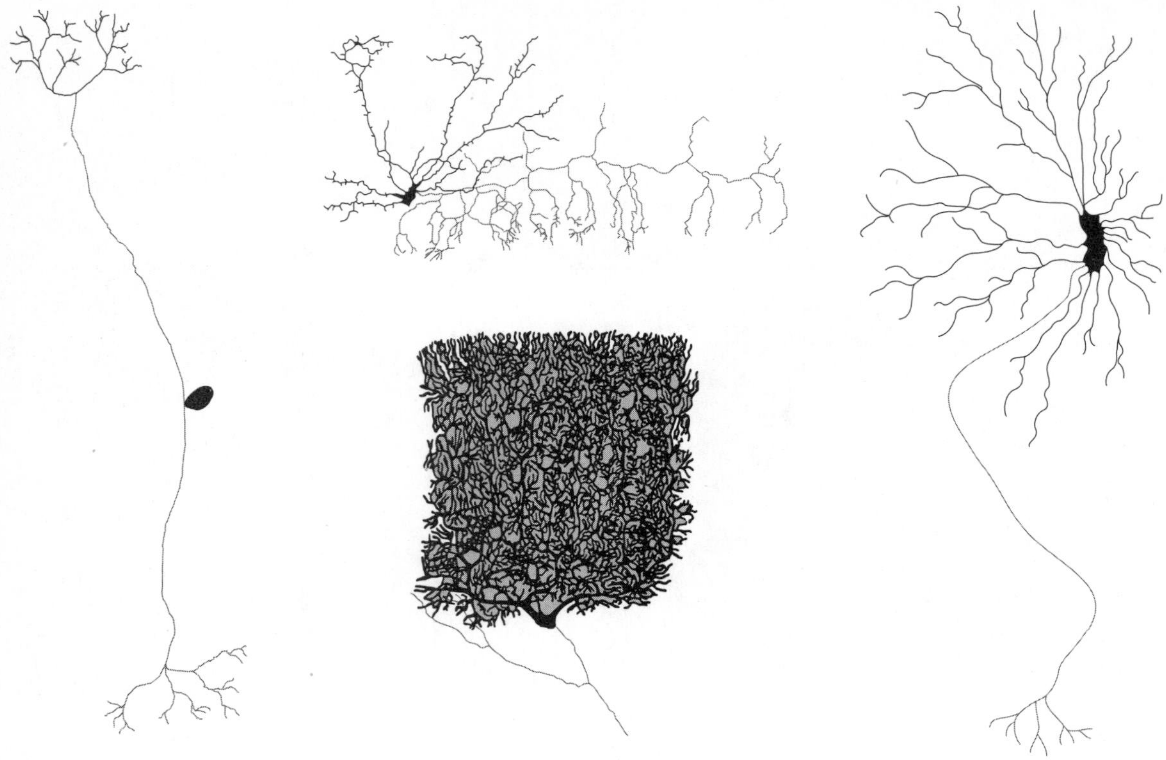

Figure 48.6 Structural diversity of vertebrate neurons, page 1014

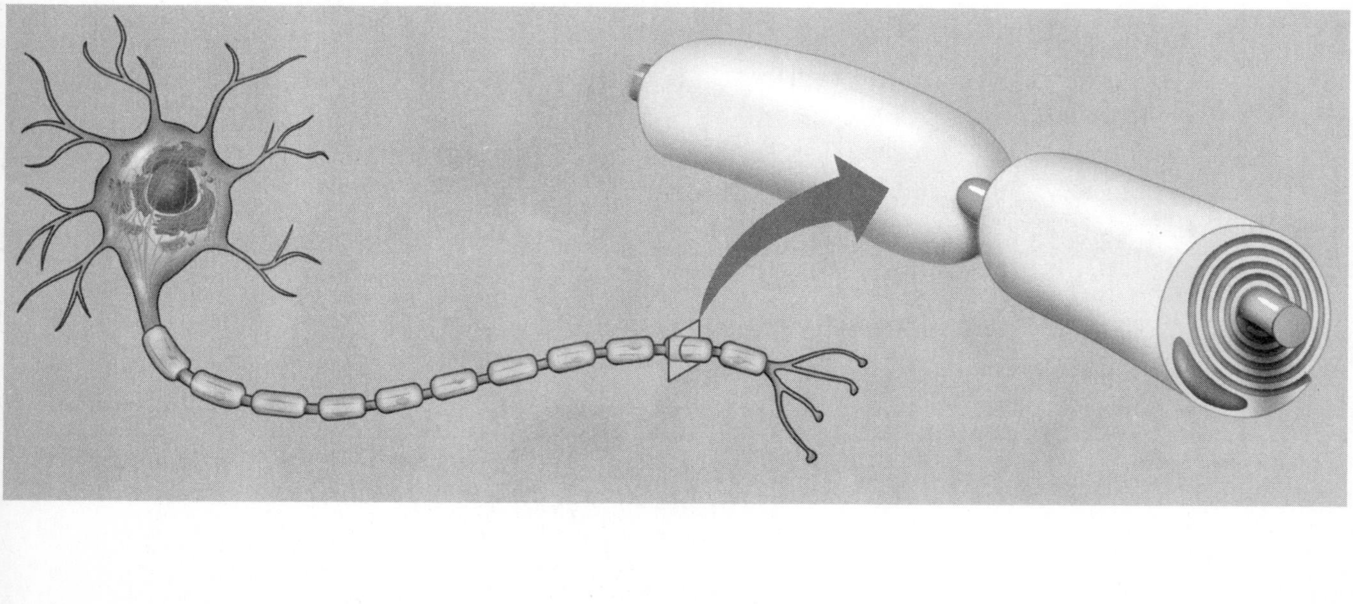

Figure 48.8 Schwann cells and the myelin sheath, page 1015

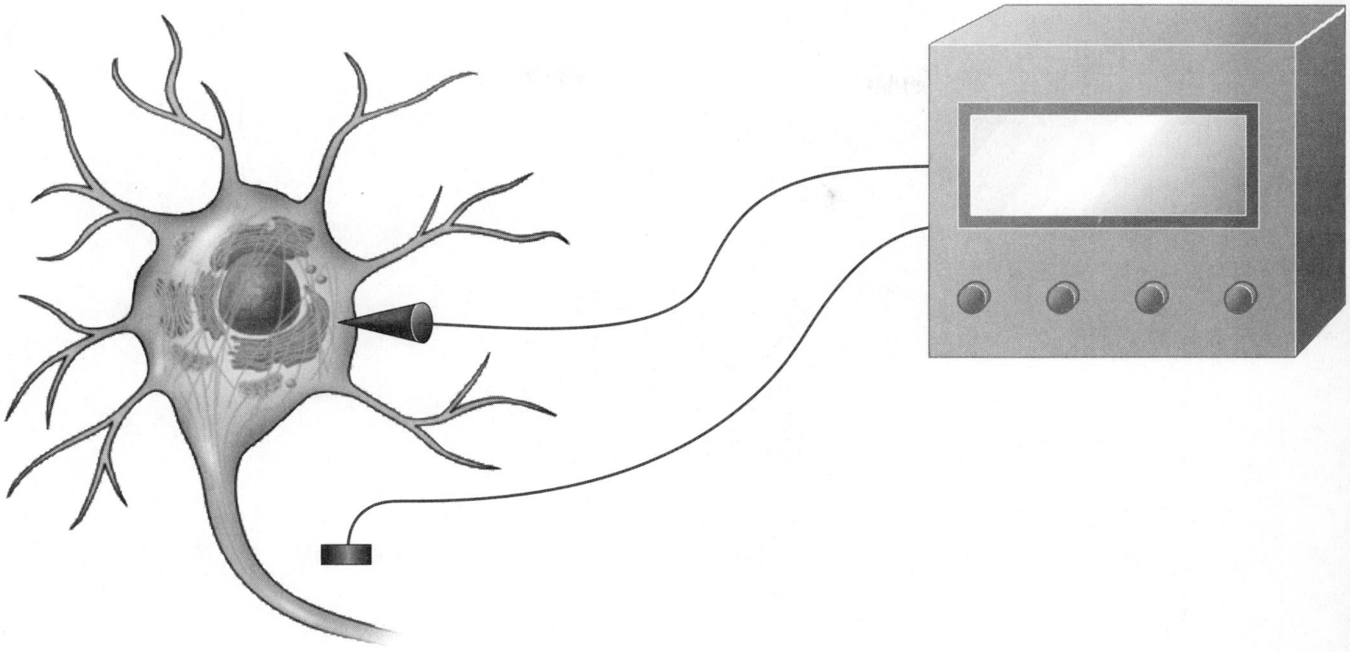

Figure 48.9 Intracellular recording, page 1016

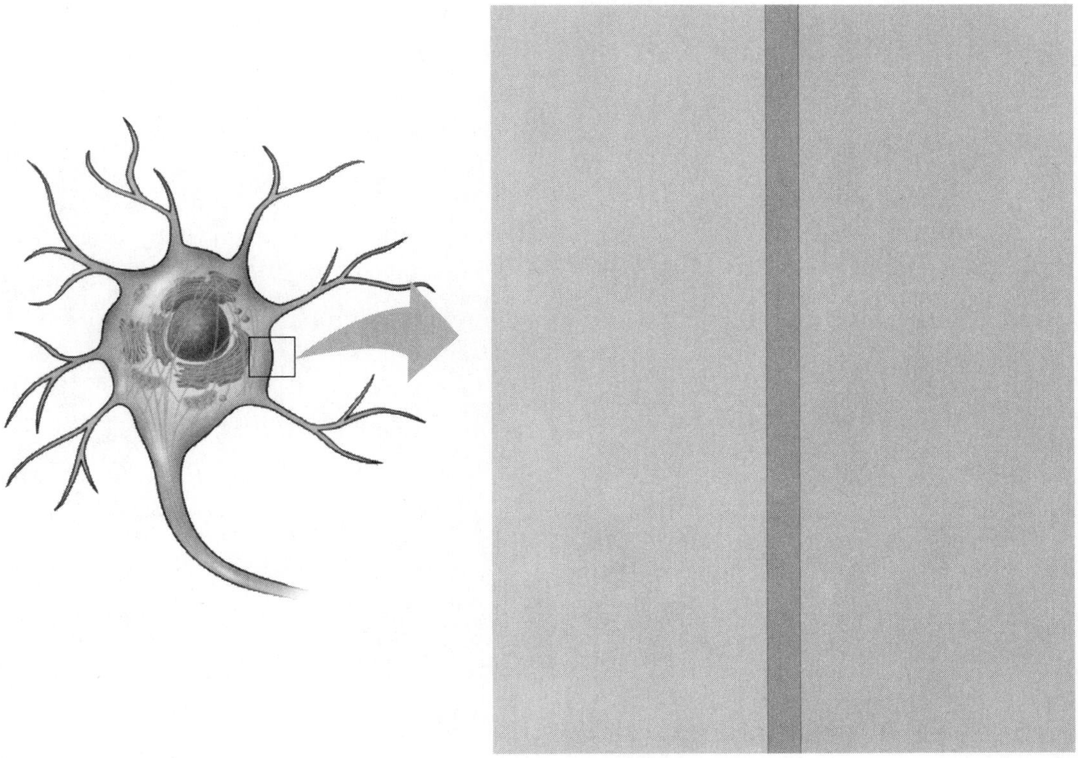

Figure 48.10 Ionic gradients across the plasma membrane of a mammalian neuron, page 1016

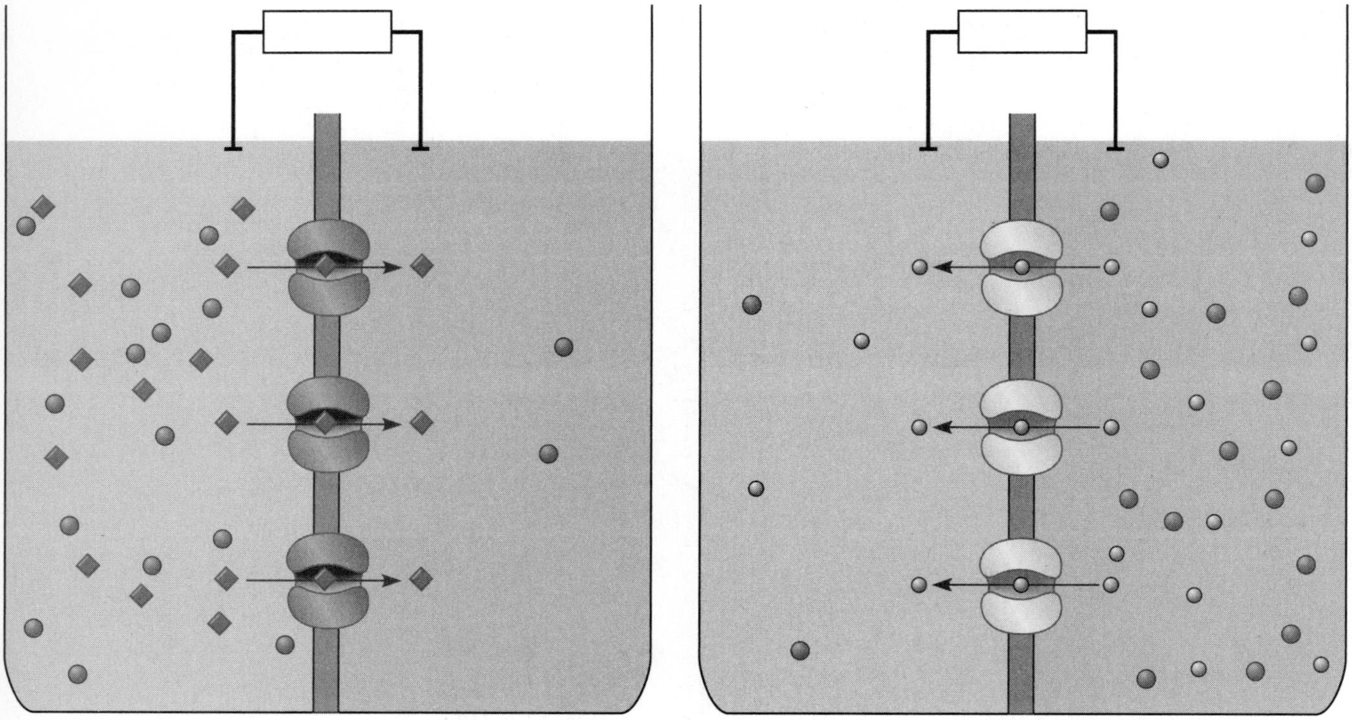

Figure 48.11 Modeling a mammalian neuron, page 1017

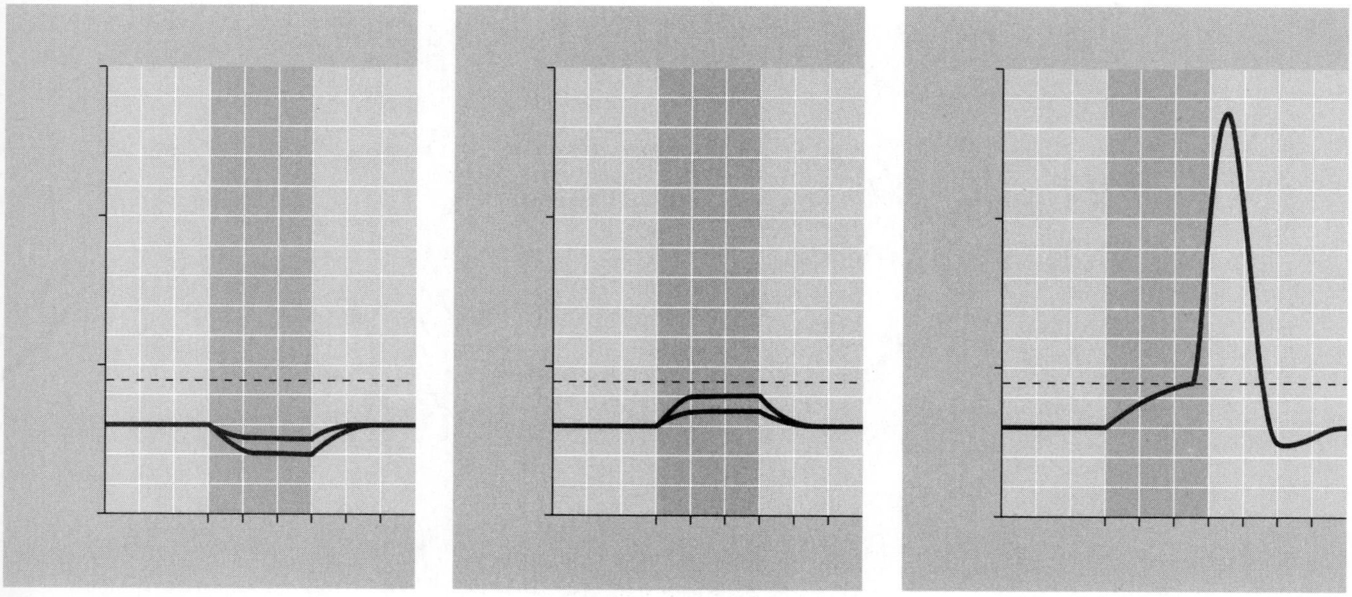

Figure 48.12 Graded potentials and an action potential in a neuron, page 1018

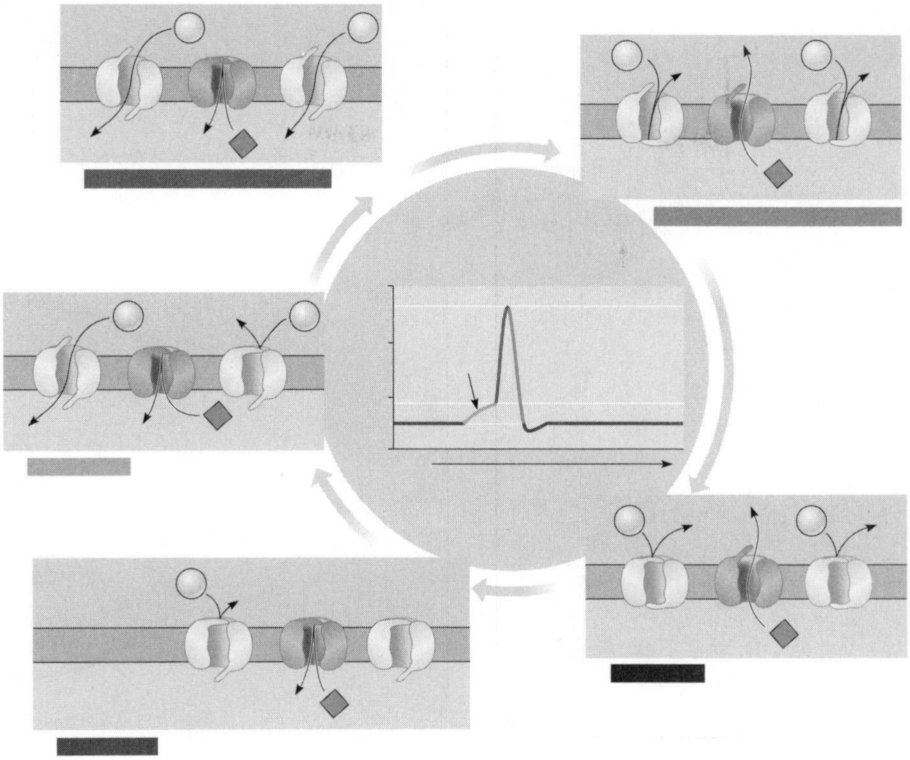

Figure 48.13 The role of voltage-gated ion channels in the generation of an action potential, page 1019

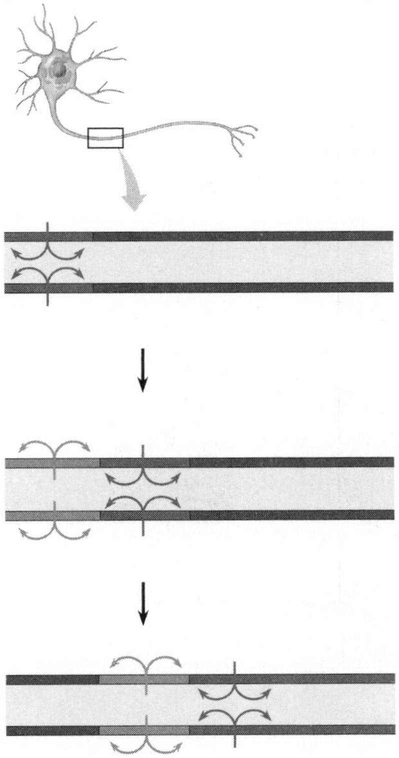

Figure 48.14 Conduction of an action potential, page 1020

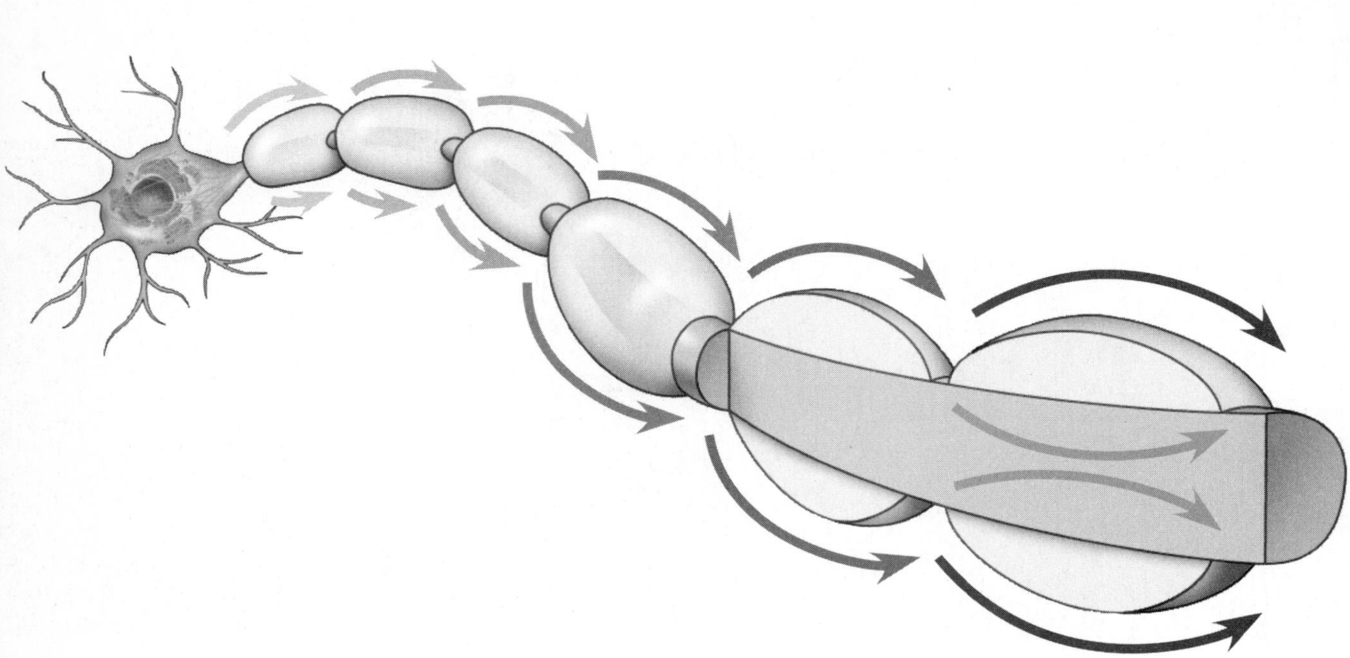

Figure 48.15 Saltatory conduction, page 1021

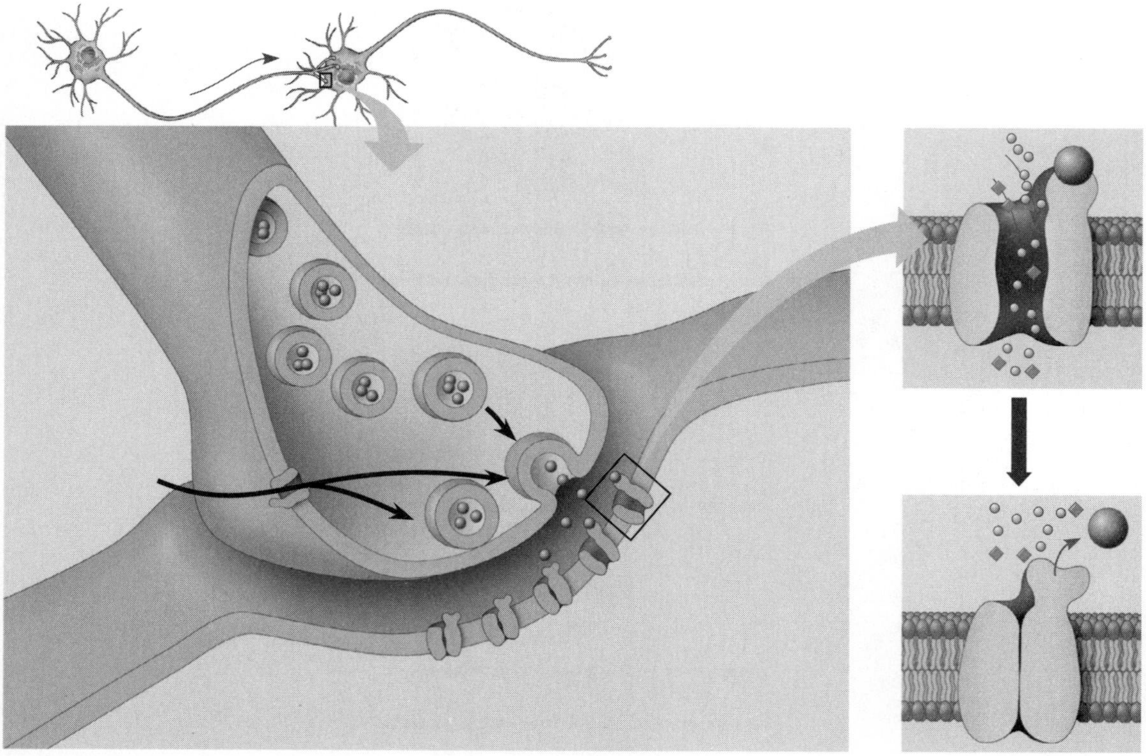

Figure 40.17 A chemical synapse, page 1022

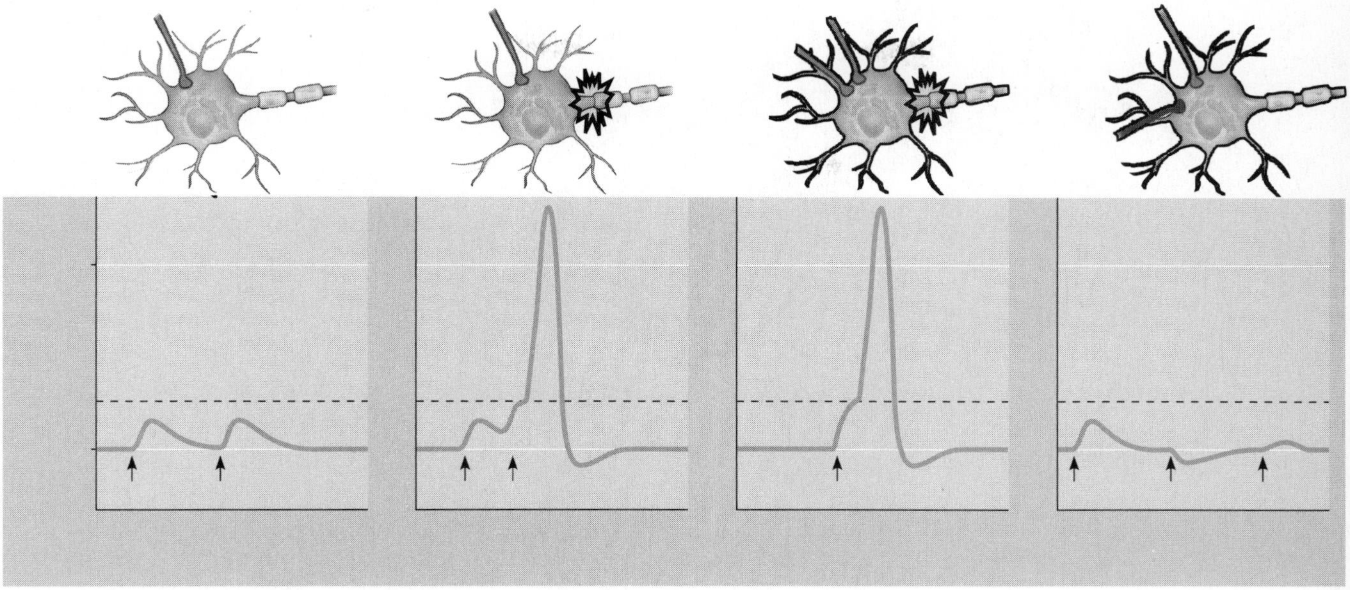

Figure 48.18 Summation of postsynaptic potentials, page 1023

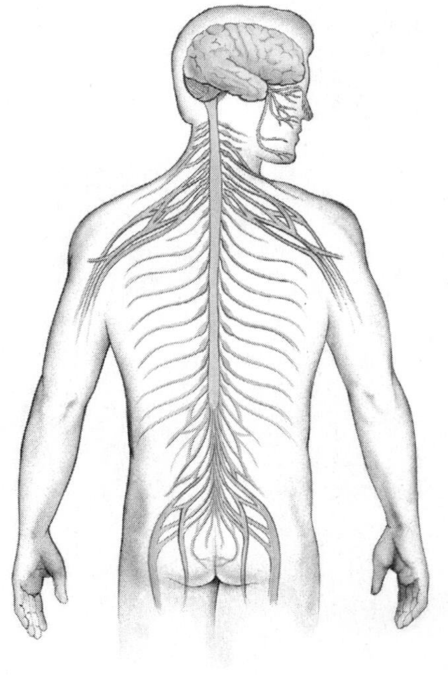

Figure 48.19 The vertebrate nervous system, page 1026

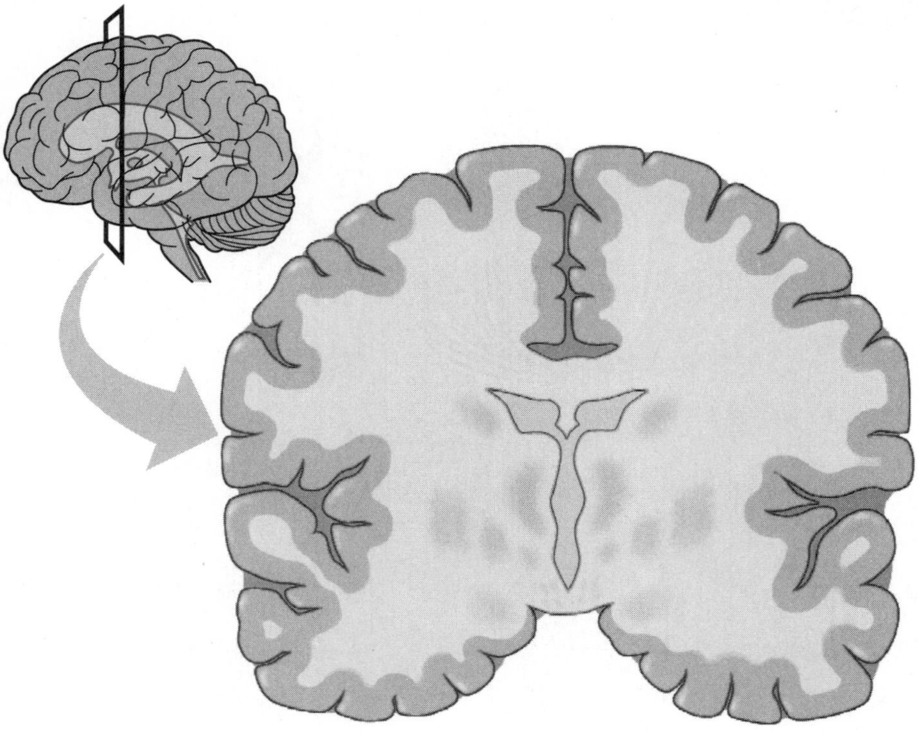

Figure 48.20 Ventricles, gray matter, and white matter, page 1026

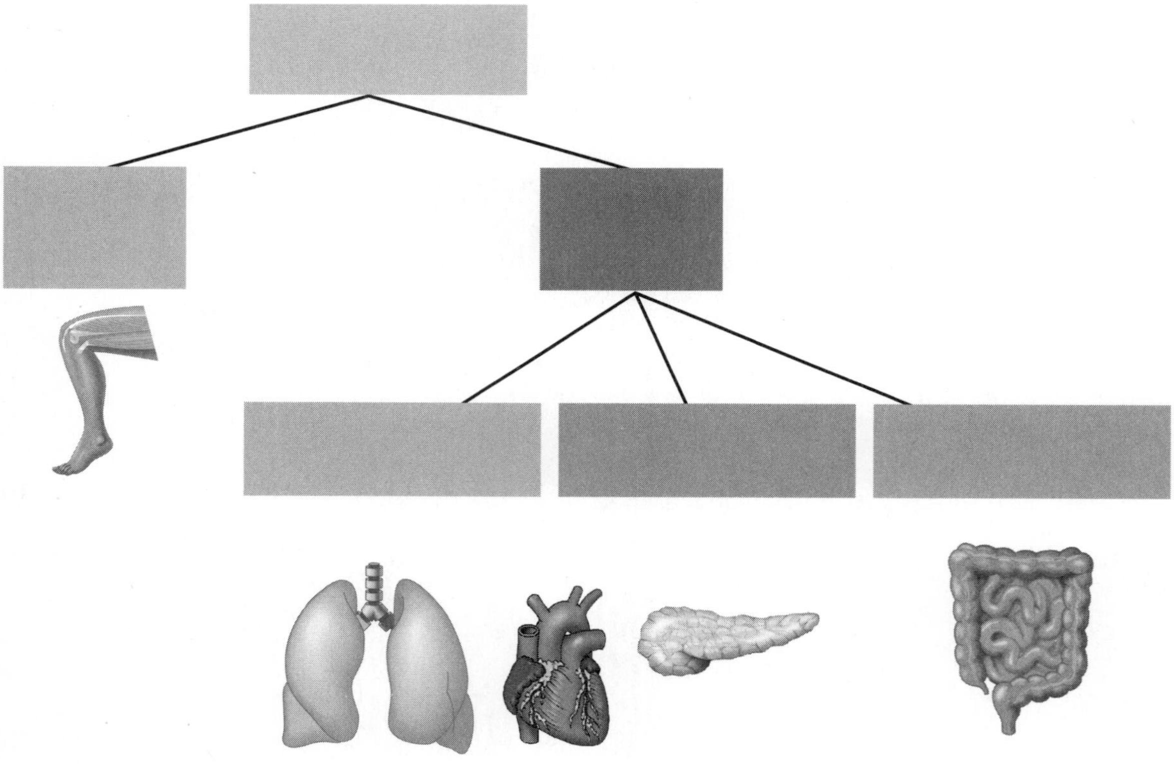

Figure 48.21 Functional hierarchy of the vertebrate peripheral nervous system, page 1027

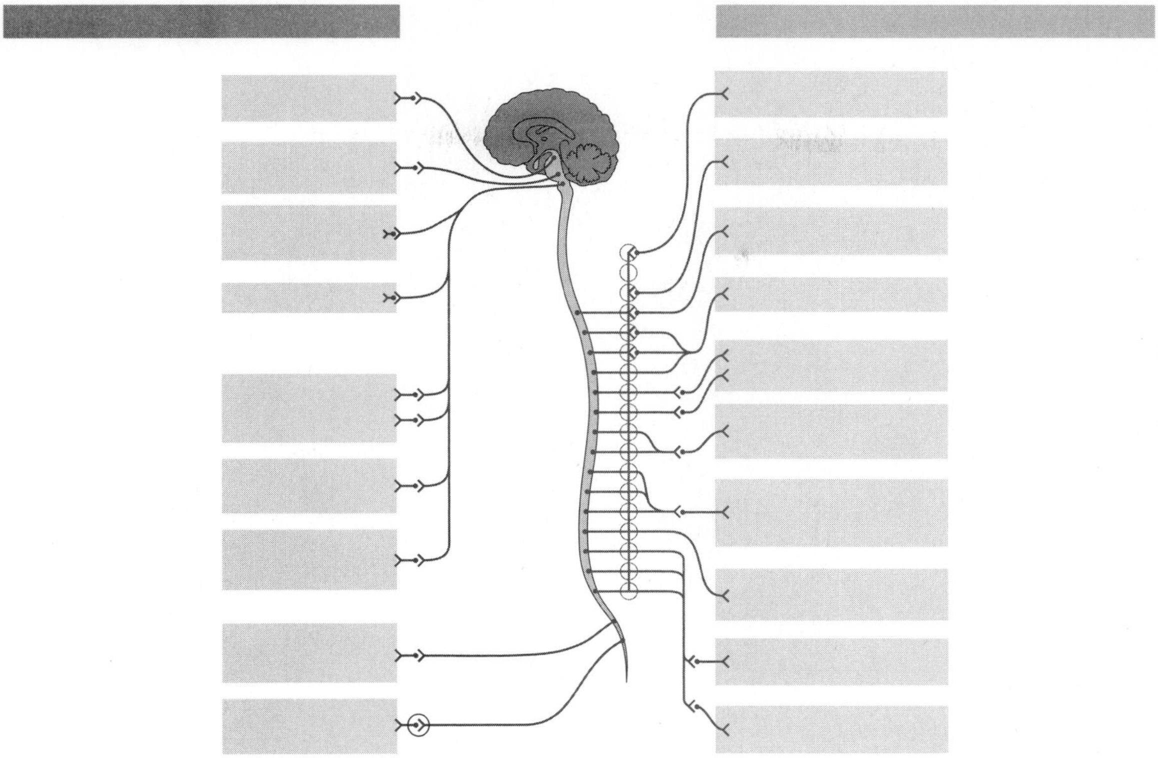

Figure 48.22 The parasympathetic and sympathetic divisions of the autonomic nervous system, page 1027

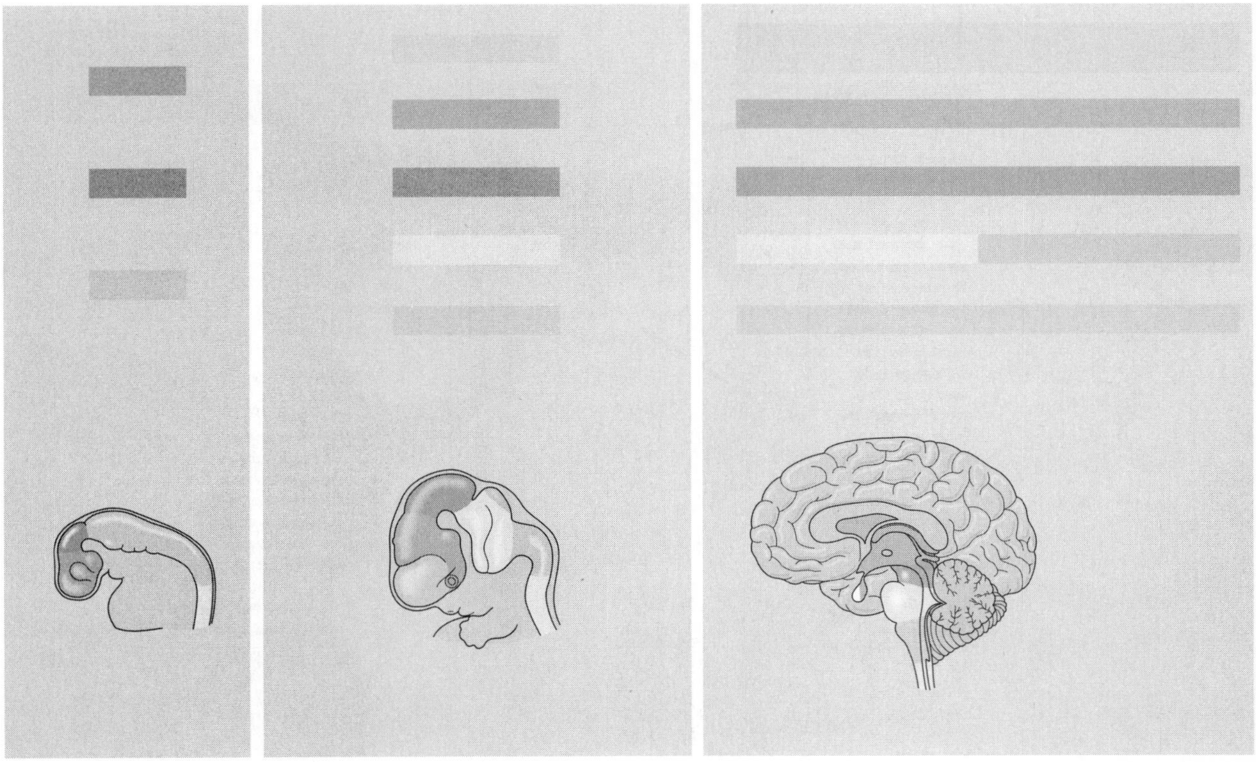

Figure 48.23 Development of the human brain, page 1028

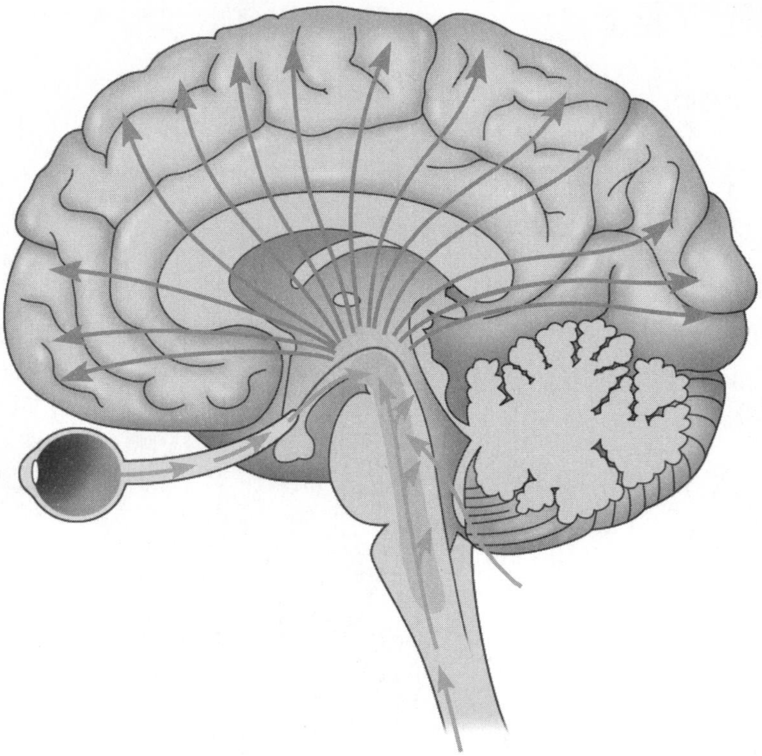

Figure 48.24 The reticular formation, page 1029

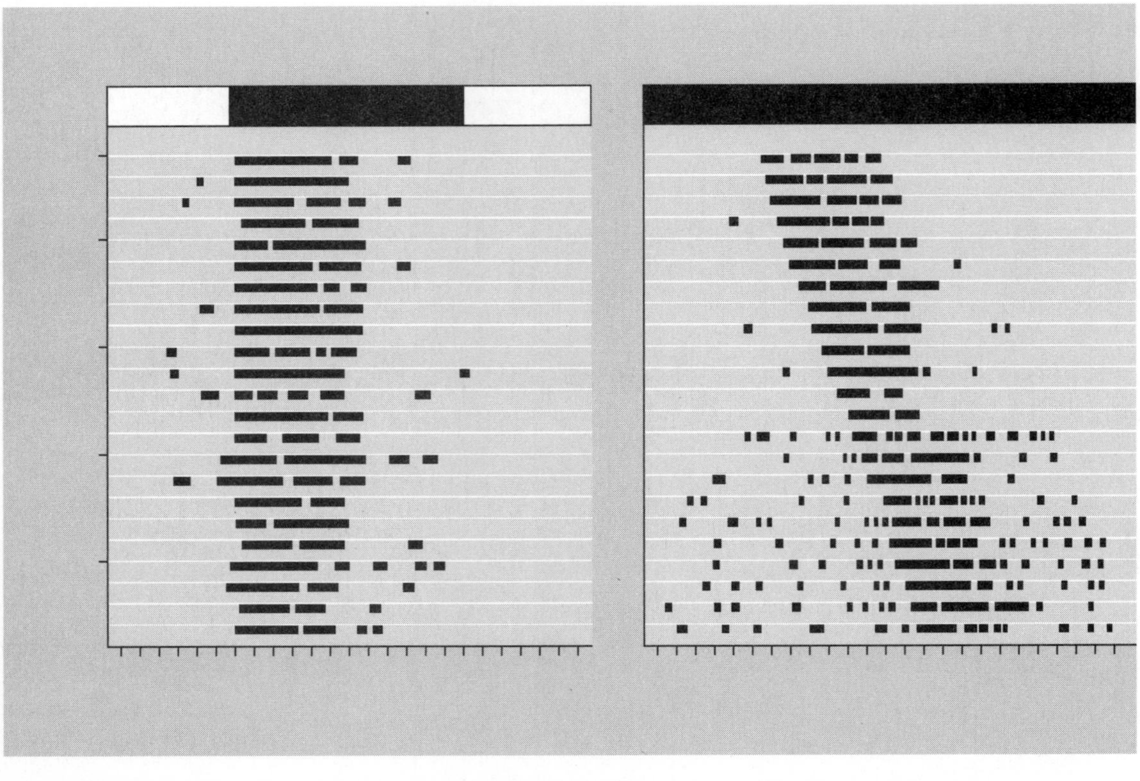

Figure 48.25 Are mammalian biological clocks influenced by external cues? page 1031

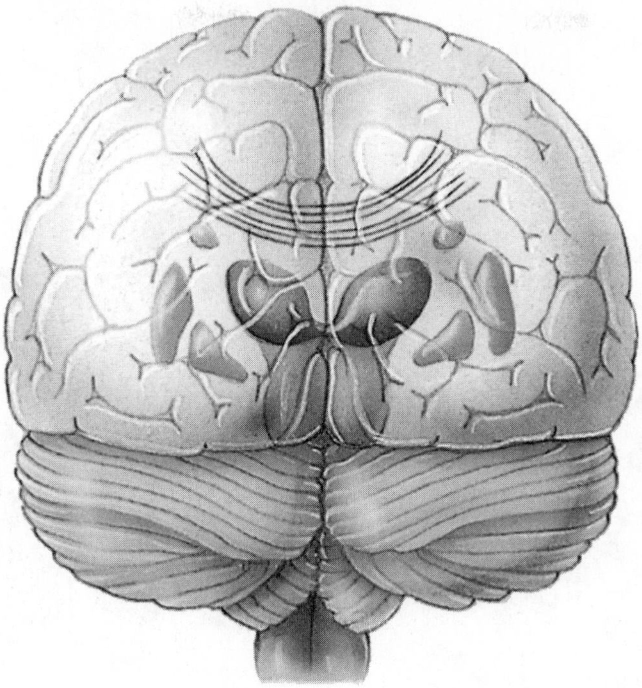

Figure 48.26 The human cerebrum viewed from the rear, page 1031

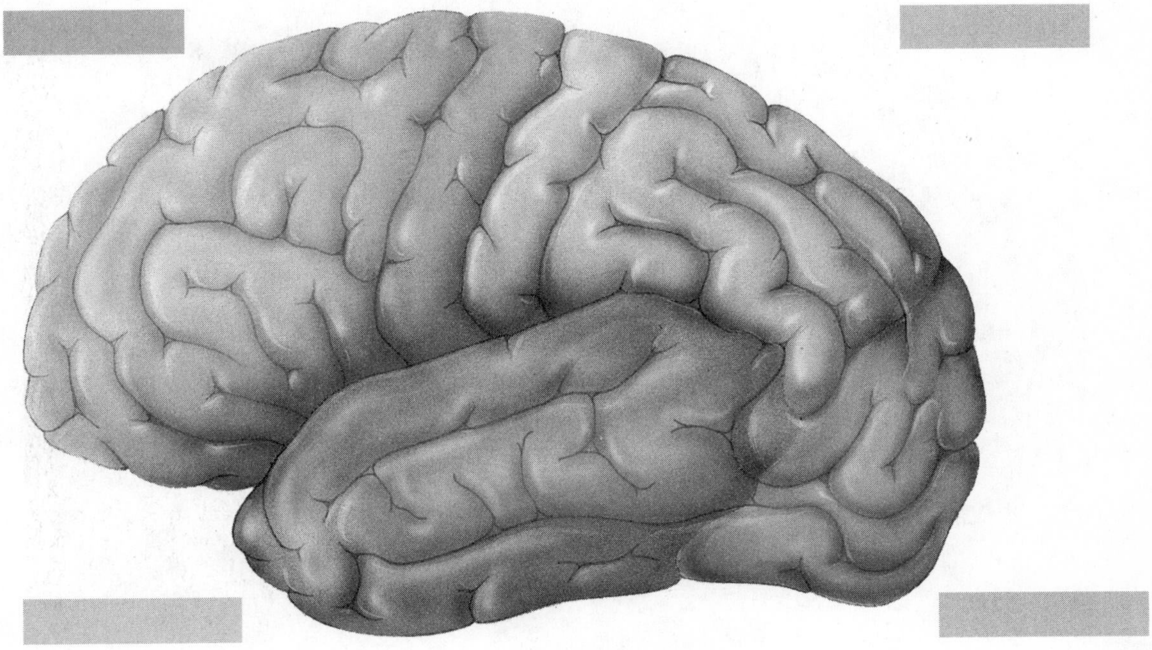

Figure 48.27 The human cerebral cortex, page 1032

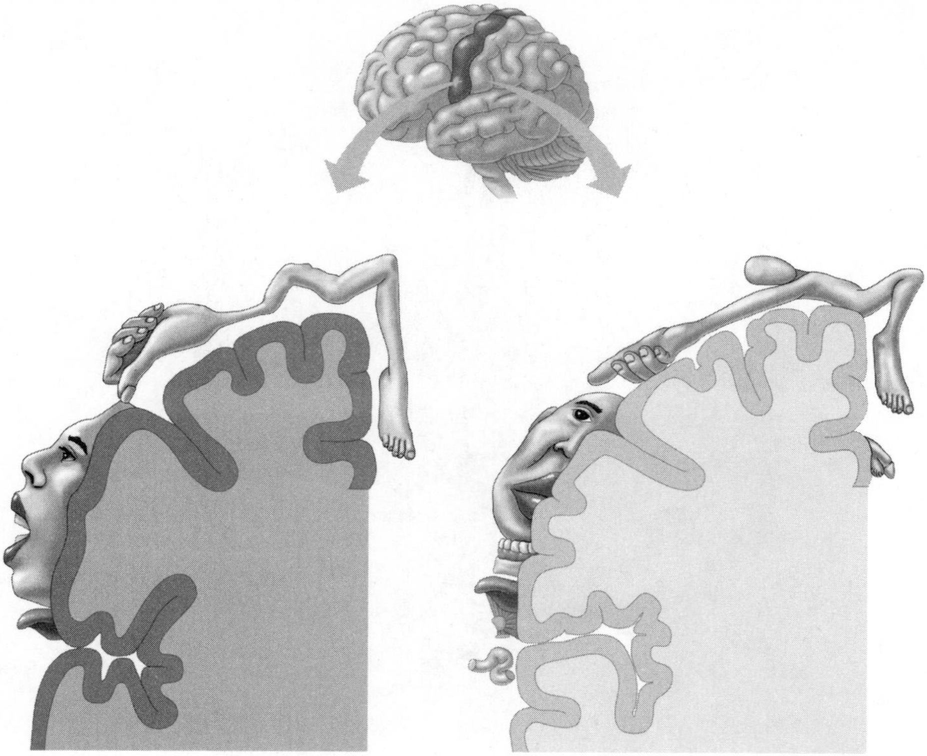

Figure 48.28 Body representations in the primary motor and primary somatosensory cortices, page 1033

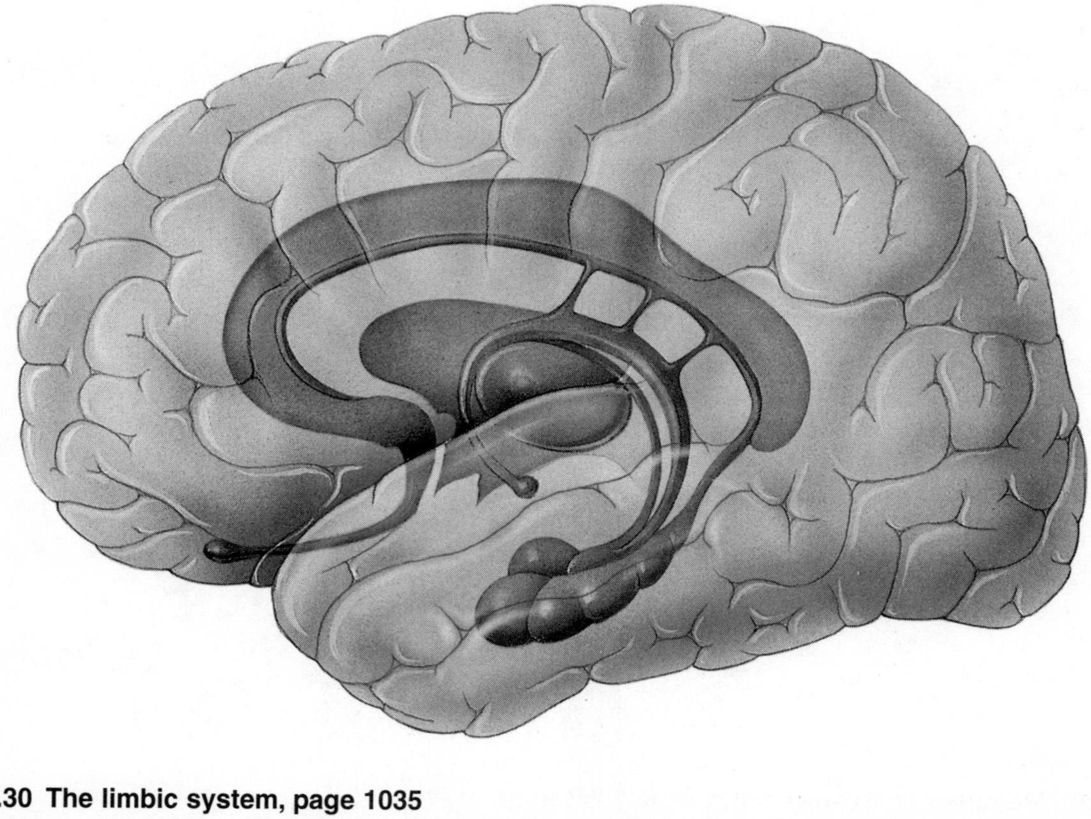

Figure 48.30 The limbic system, page 1035

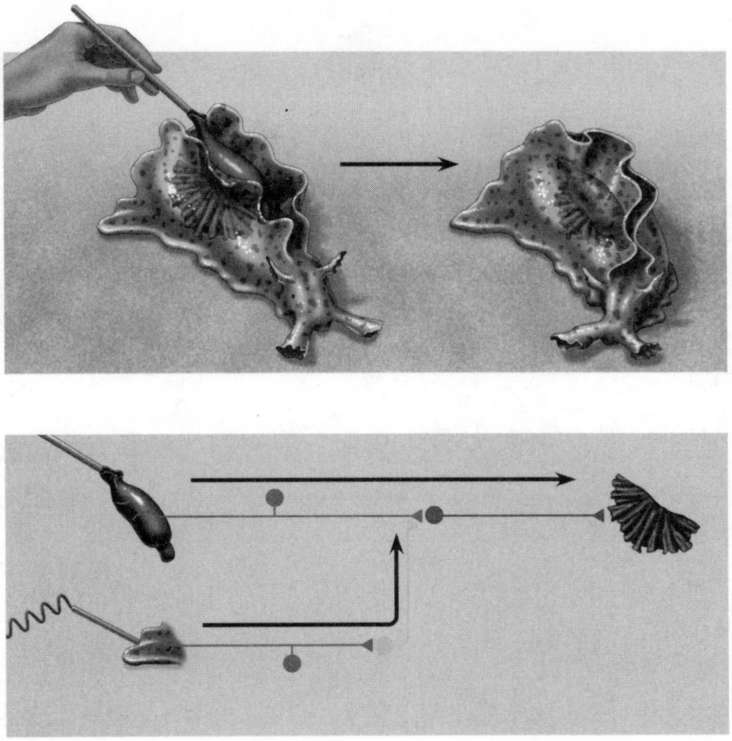

Figure 48.31 Sensitization in the sea hare (*Aplysia californica*), page 1036

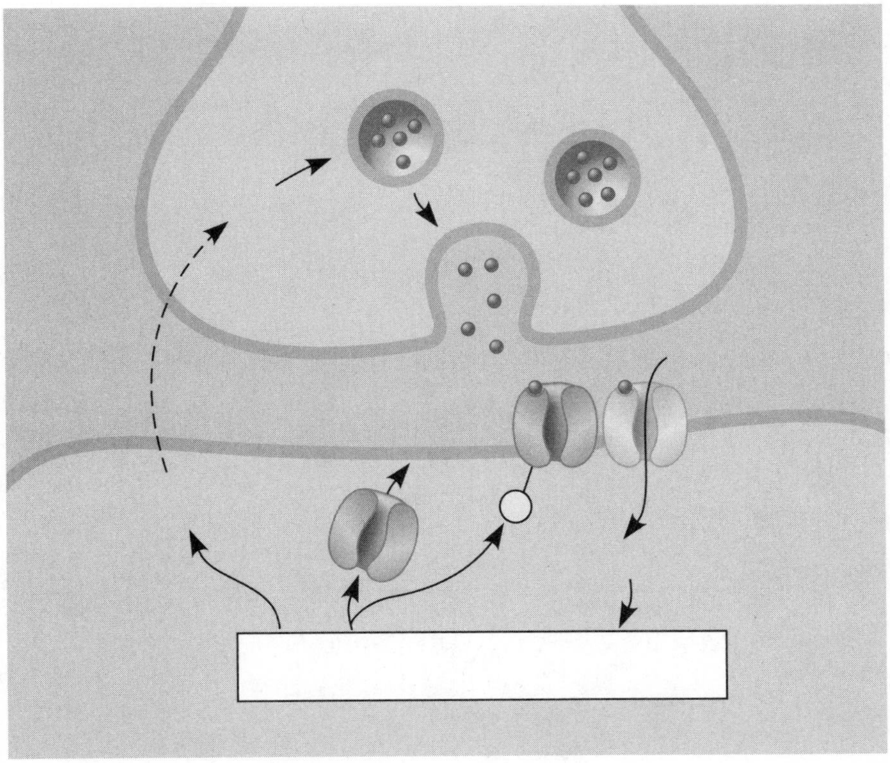

Figure 48.32 Mechanisms of long-term potentiation in the vertebrate brain, page 1037

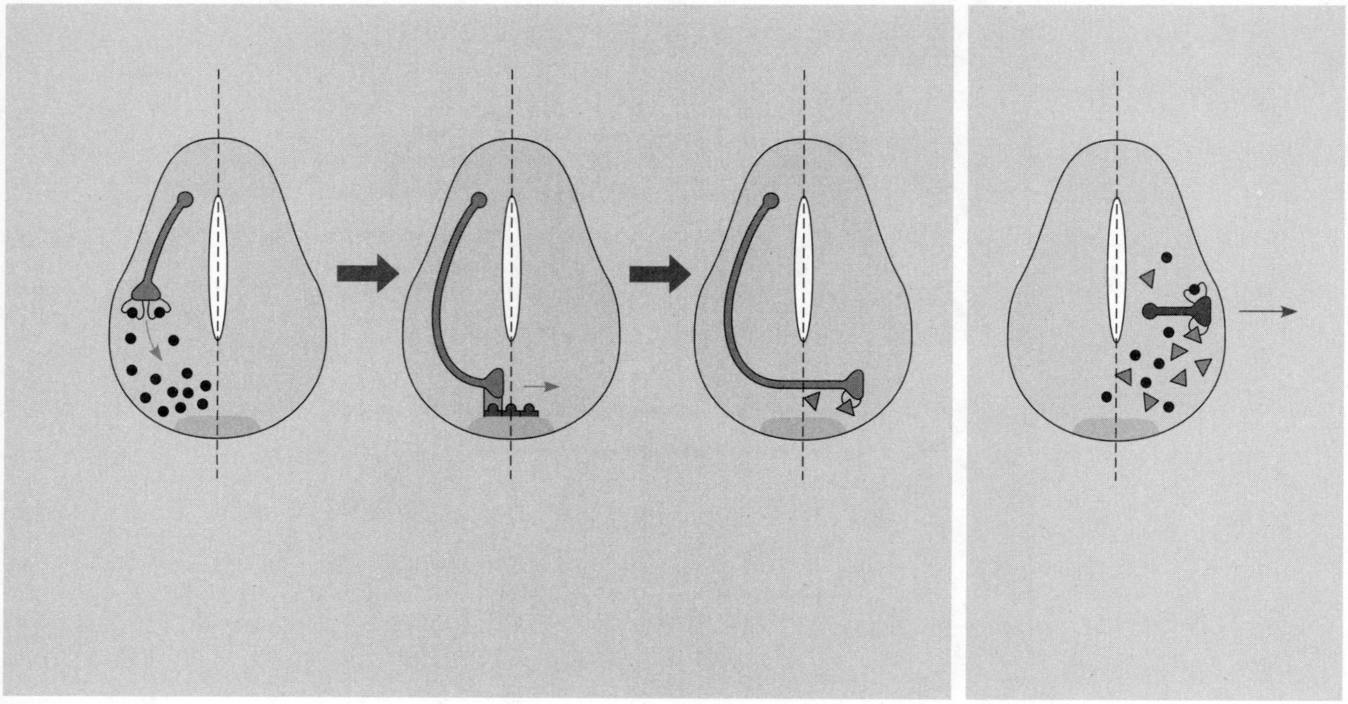

Figure 48.33 Molecular signals direct the growth of developing axons, page 1038

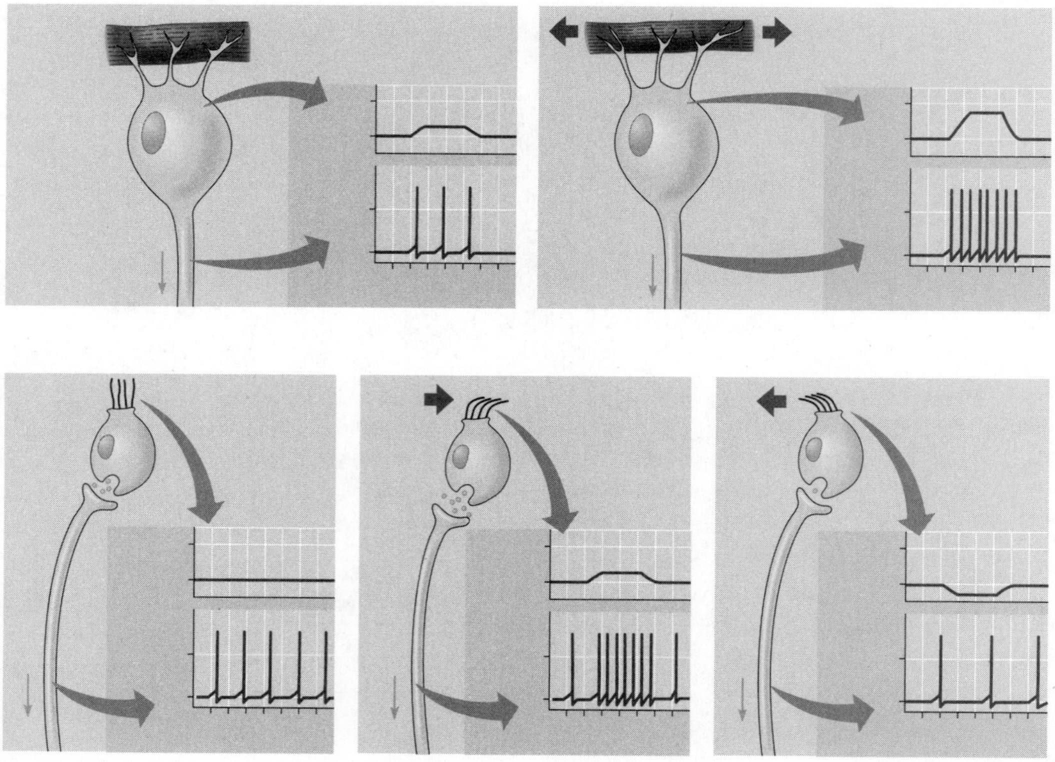

Figure 49.2 Sensory reception: two mechanisms, page 1047

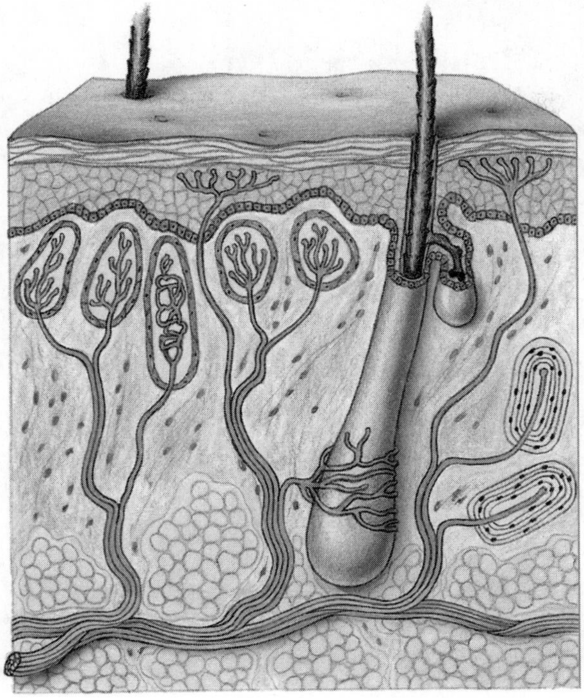

Figure 49.3 Sensory receptors in human skin, page 1048

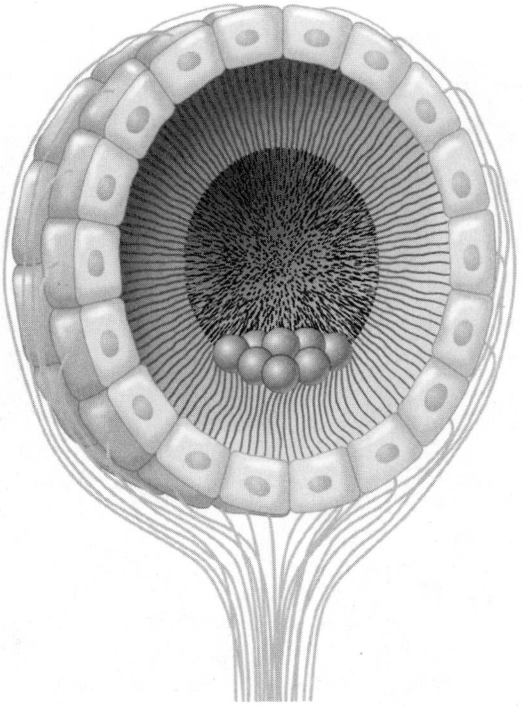

Figure 49.6 The statocyst of an invertebrate, page 1050

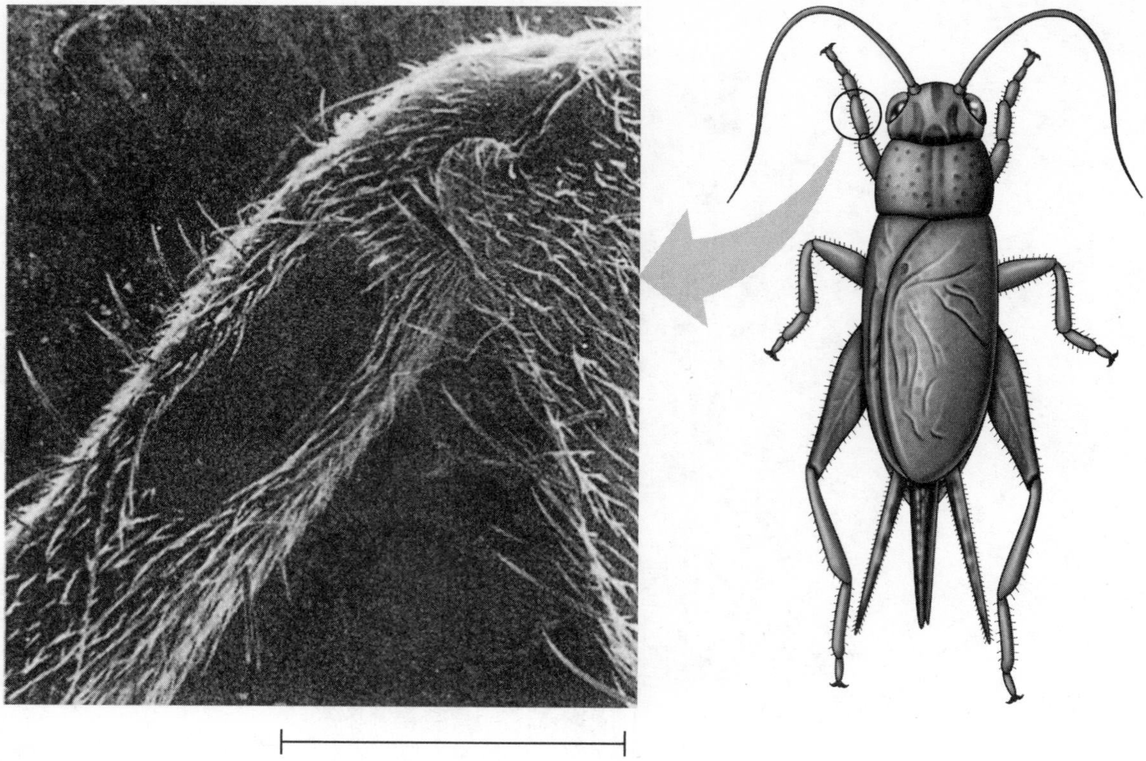

Figure 49.7 An insect ear, page 1050

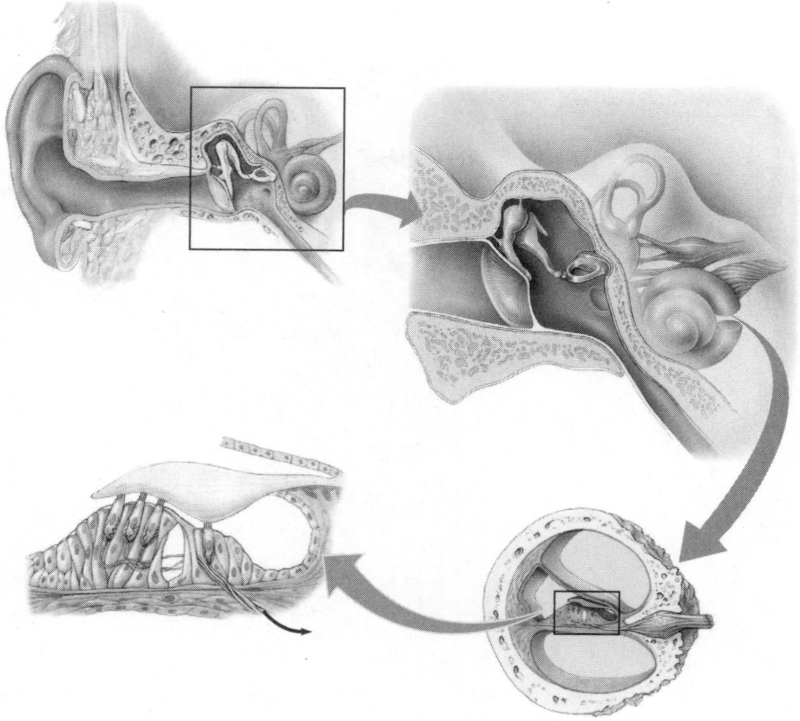

Figure 49.8 The structure of the human ear, page 1051

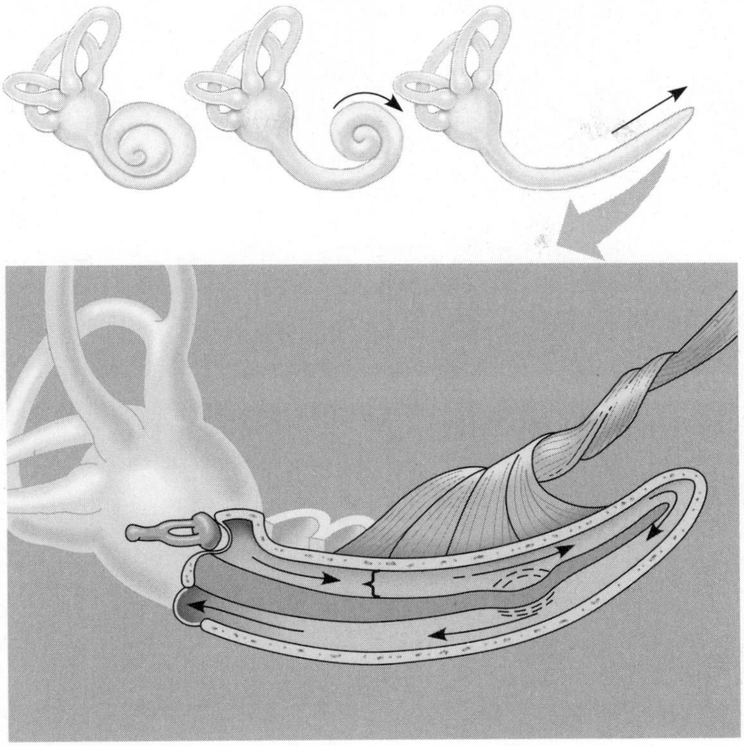

Figure 49.9 Transduction in the cochlea, page 1052

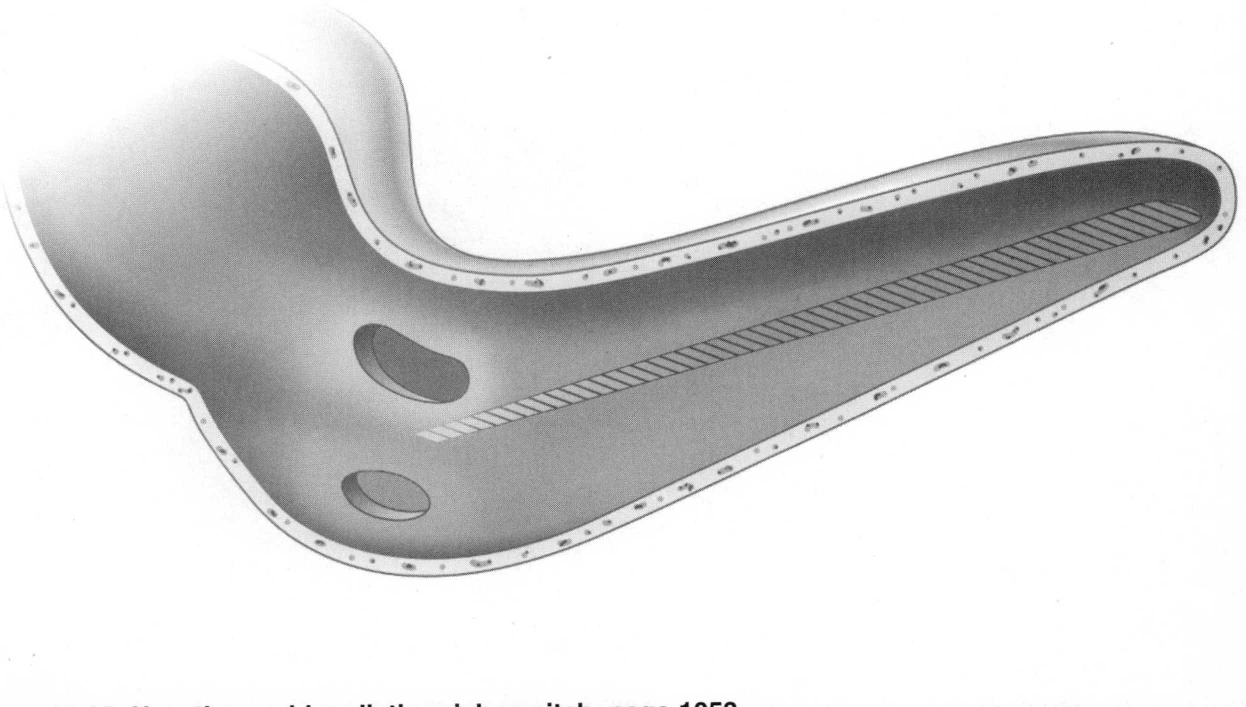

Figure 49.10 How the cochlea distinguishes pitch, page 1053

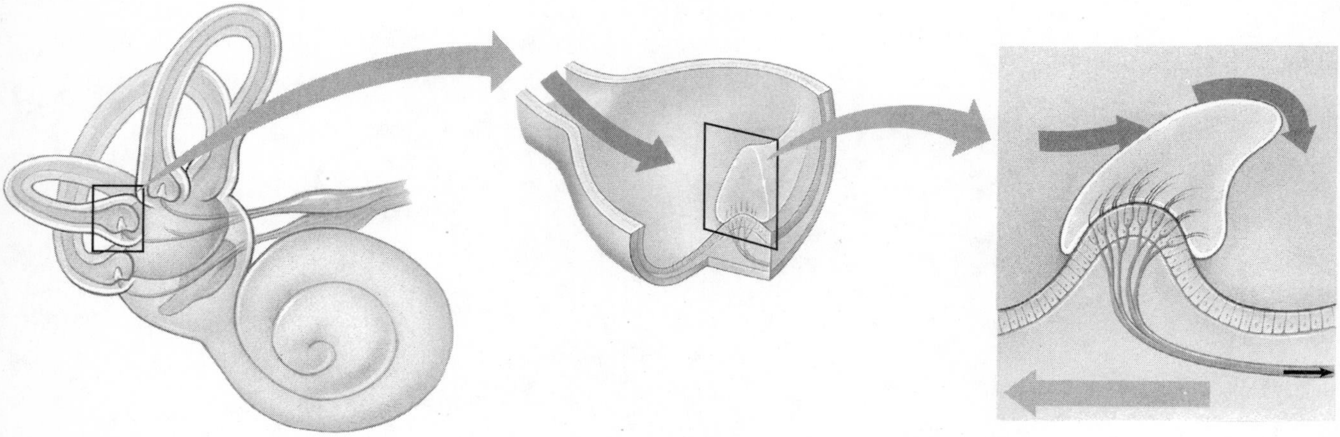

Figure 49.11 Organs of equilibrium in the inner ear, page 1053

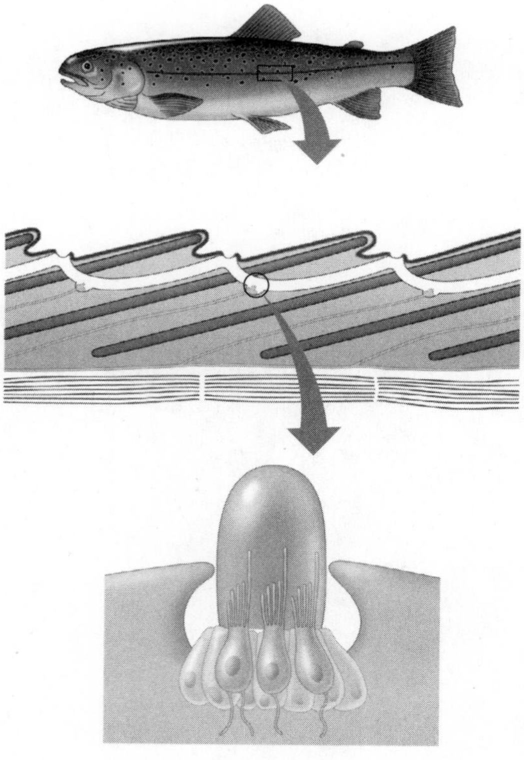

Figure 49.12 The lateral line system in a fish, page 1054

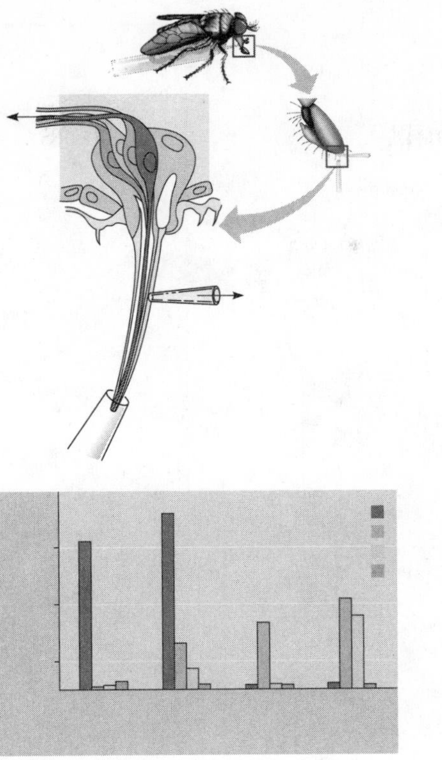

Figure 49.13 How do insects detect different tastes? page 1055

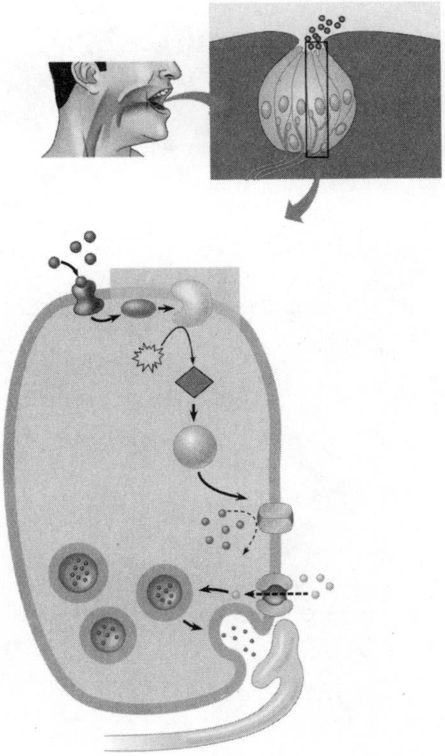

Figure 49.14 Sensory transduction by a sweetness receptor, page 1056

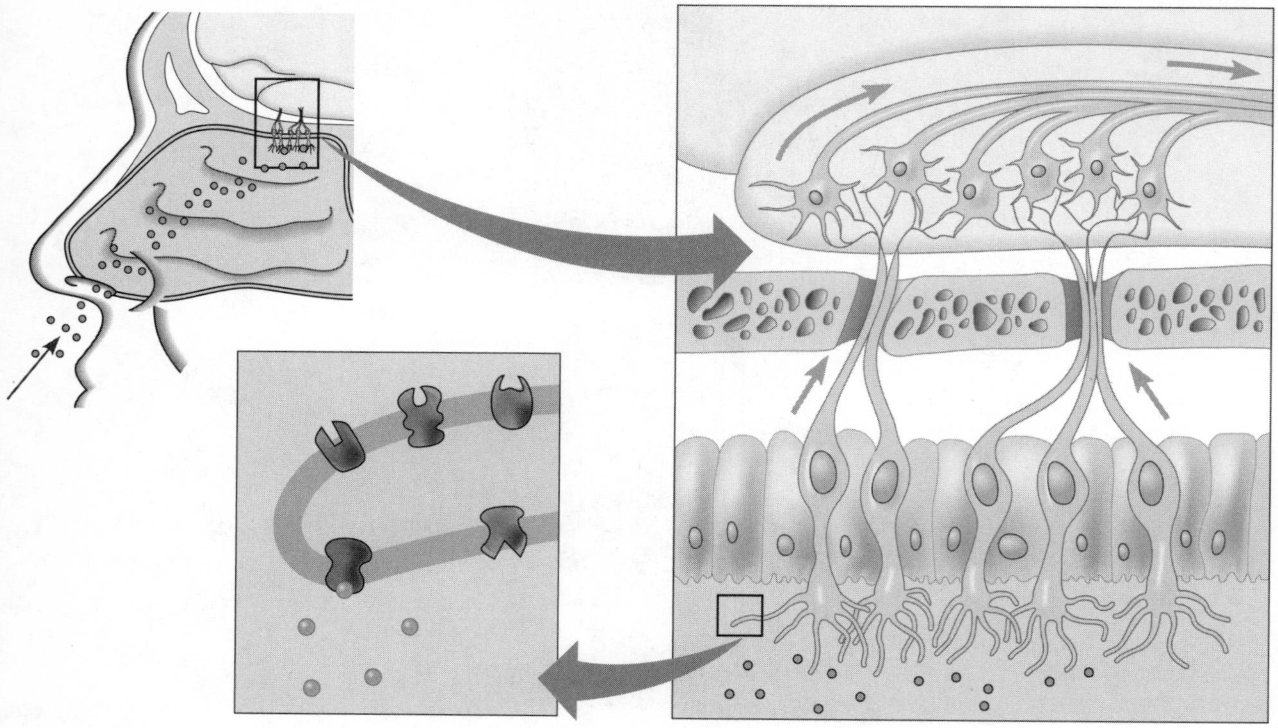

Figure 49.15 Smell in humans, page 1057

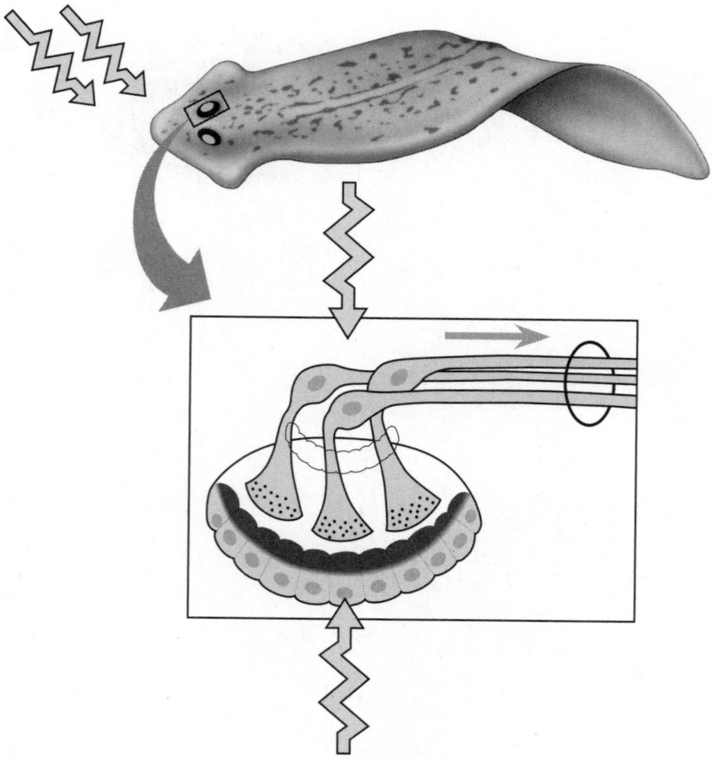

Figure 49.16 Ocelli and orientation behavior of a planarian, page 1057

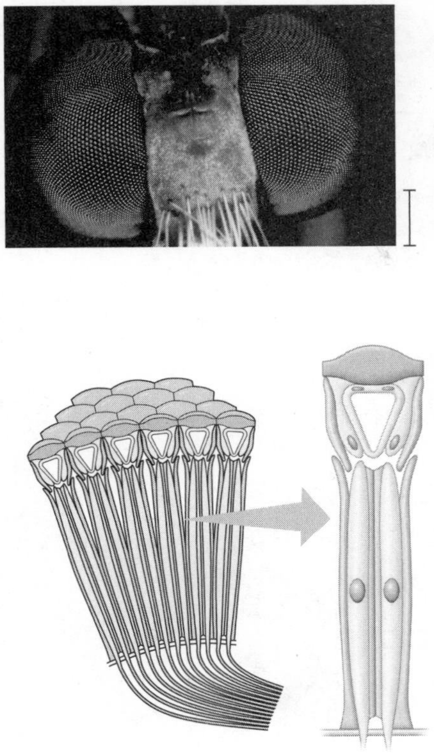

Figure 49.17 Compound eyes, page 1058

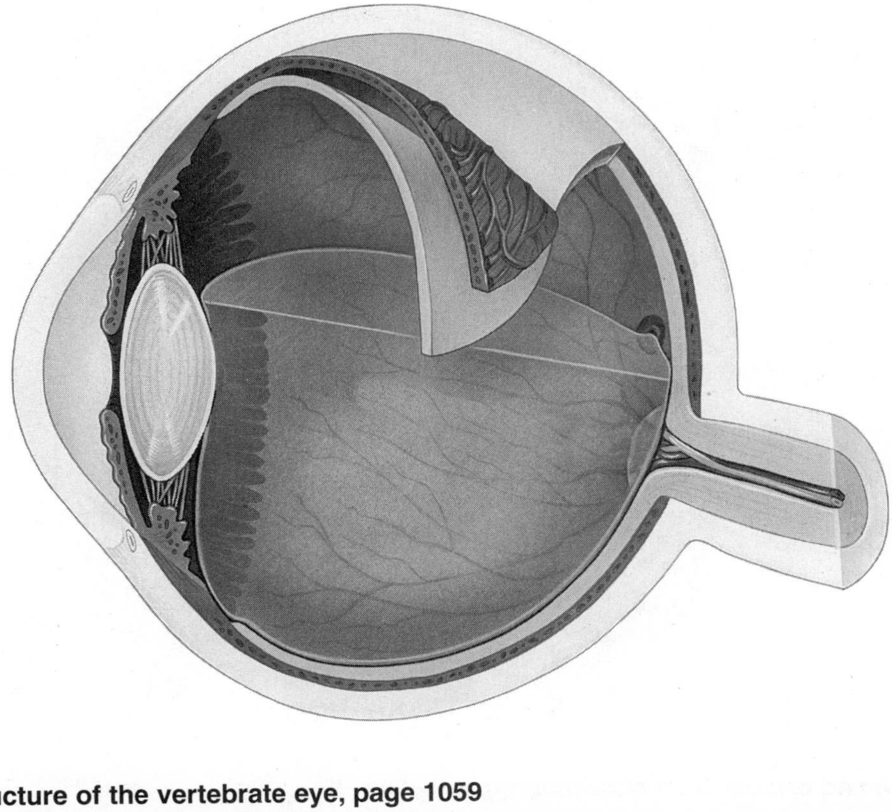

Figure 49.18 Structure of the vertebrate eye, page 1059

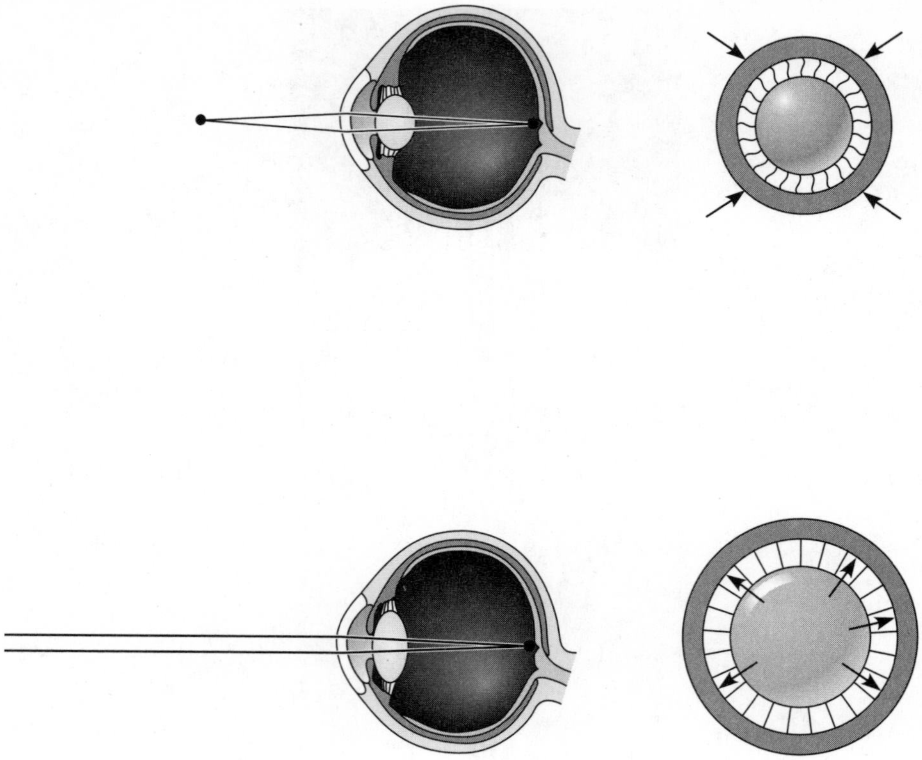

Figure 49.19 Focusing in the mammalian eye, page 1059

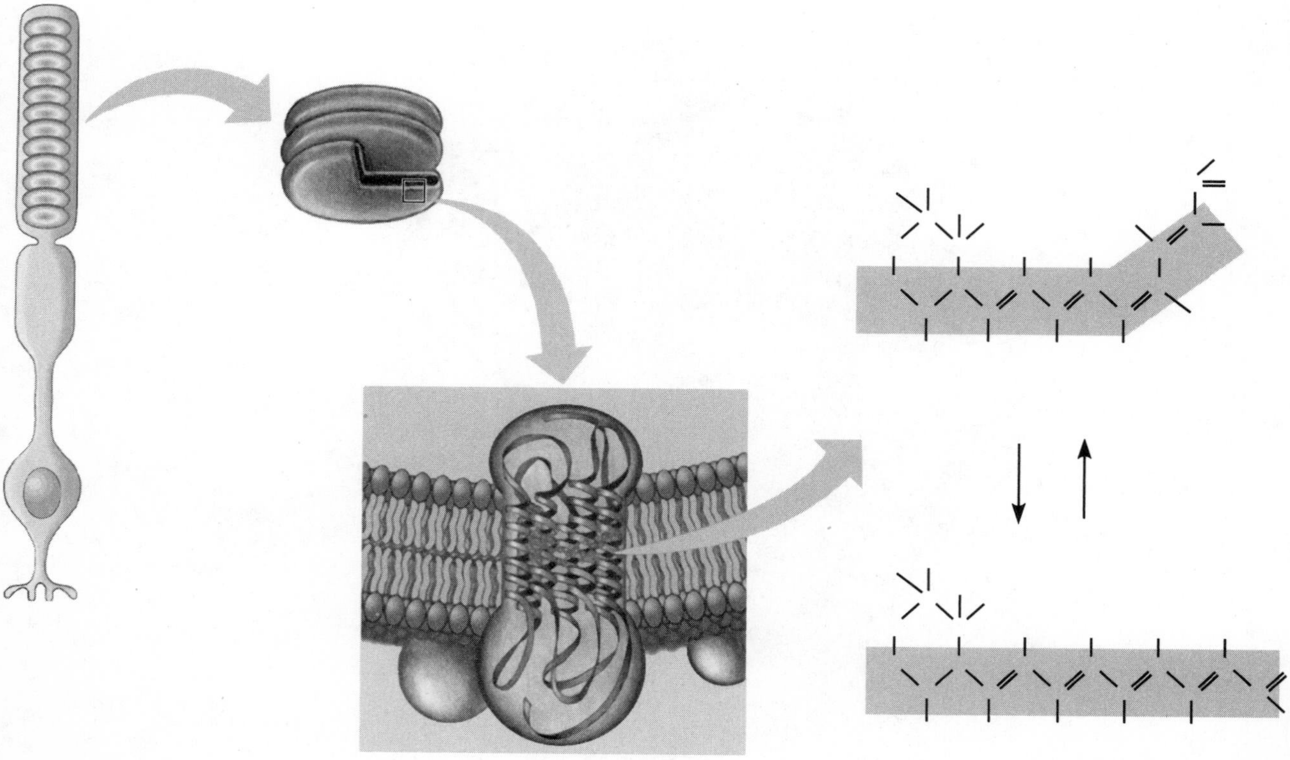

Figure 49.20 Rod structure and light absorption, page 1060

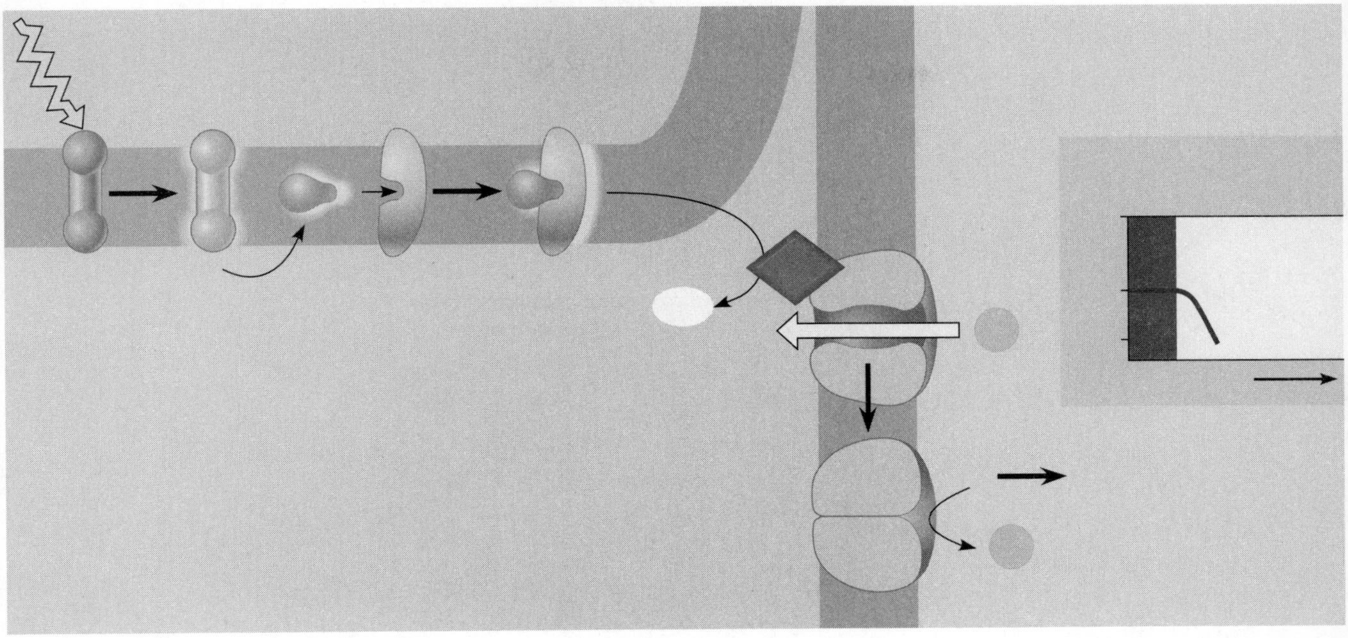

Figure 49.21 Production of a receptor potential in a rod, page 1061

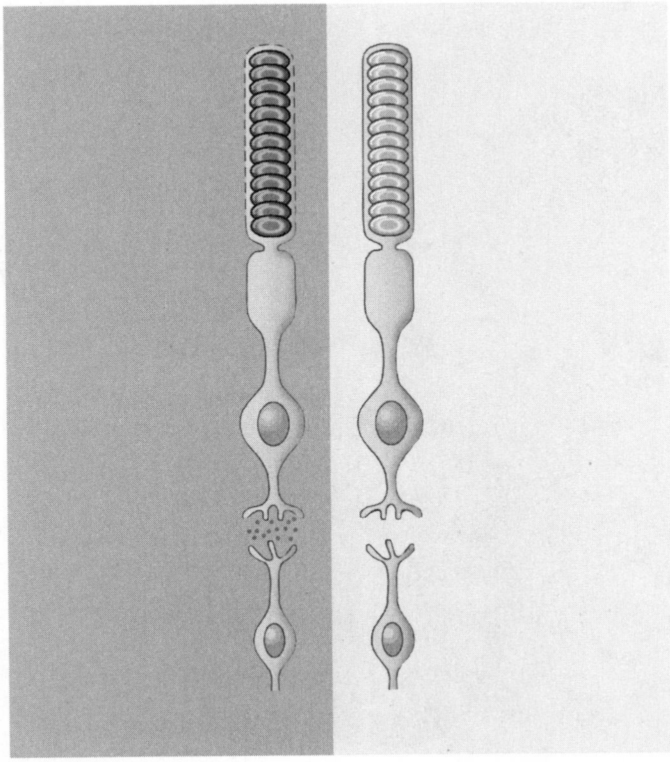

Figure 49.22 The effect of light on synapses between rod cells and bipolar cells, page 1061

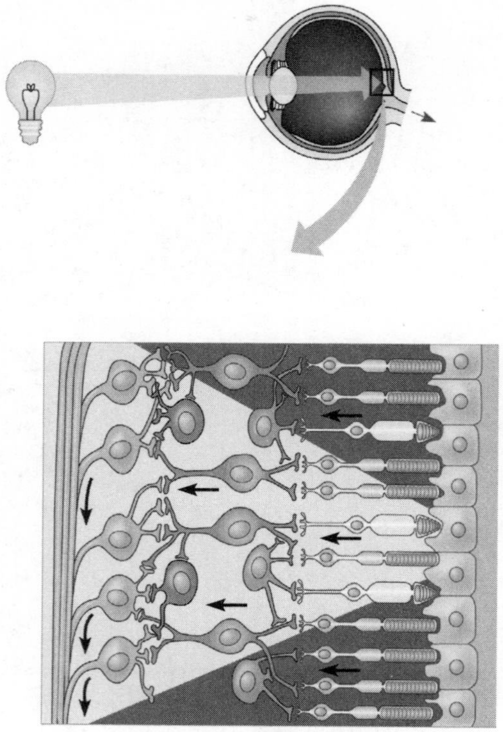

Figure 49.23 Cellular organization of the vertebrate retina, page 1062

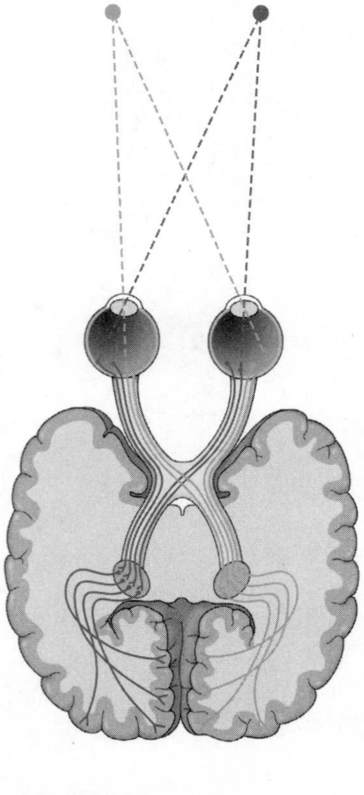

Figure 49.24 Neural pathways for vision, page 1062

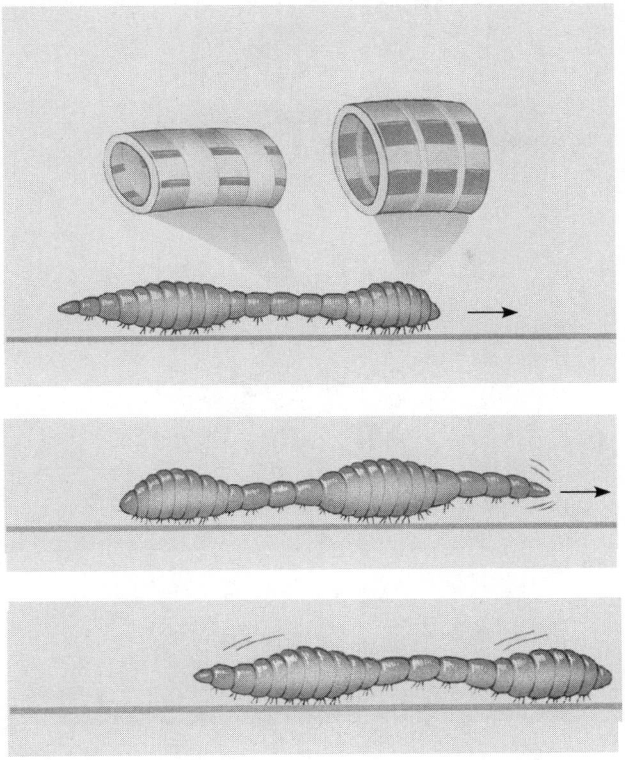

Figure 49.25 Peristaltic locomotion in an earthworm, page 1064

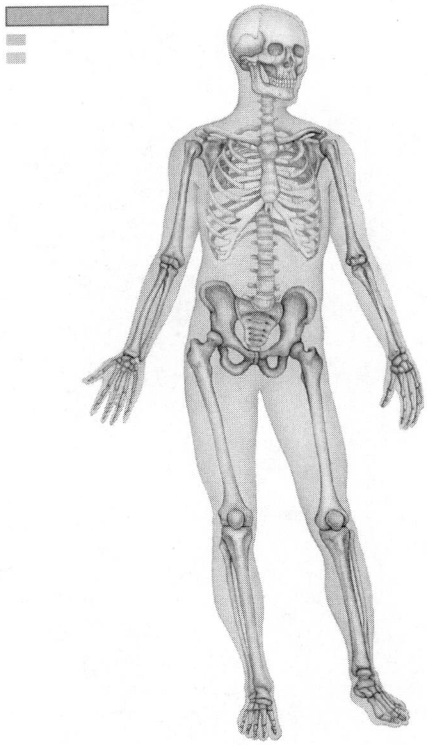

Figure 49.26 The bones and joints of the human skeleton (part 1), page 1065

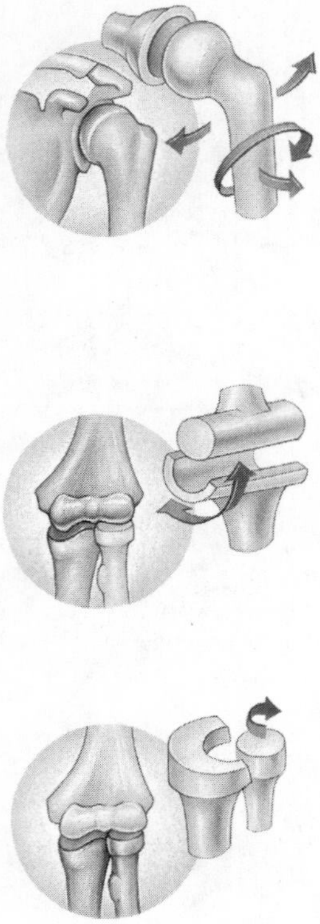

Figure 49.26 The bones and joints of the human skeleton (part 2), page 1065

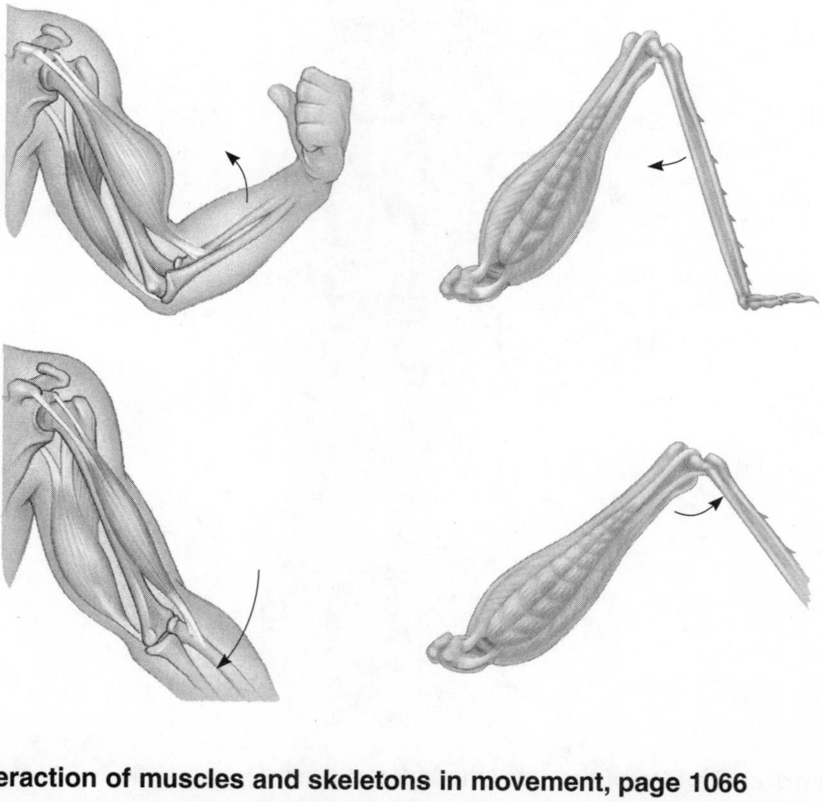

Figure 49.27 The interaction of muscles and skeletons in movement, page 1066

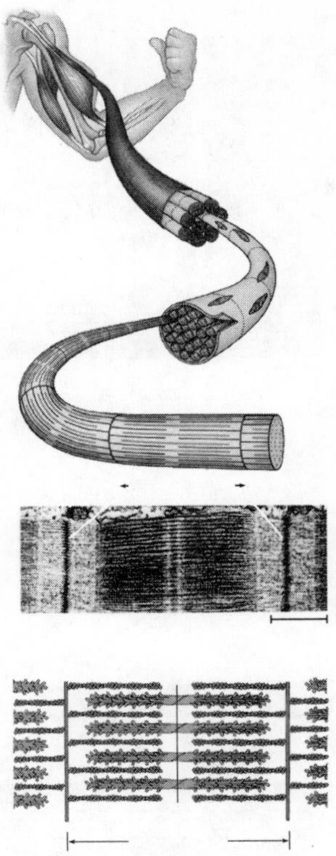

Figure 49.28 The structure of skeletal muscle, page 1067

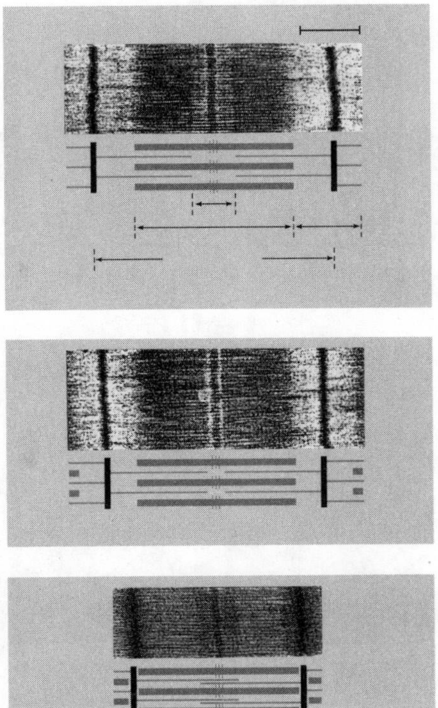

Figure 49.29 The sliding-filament model of muscle contraction, page 1067

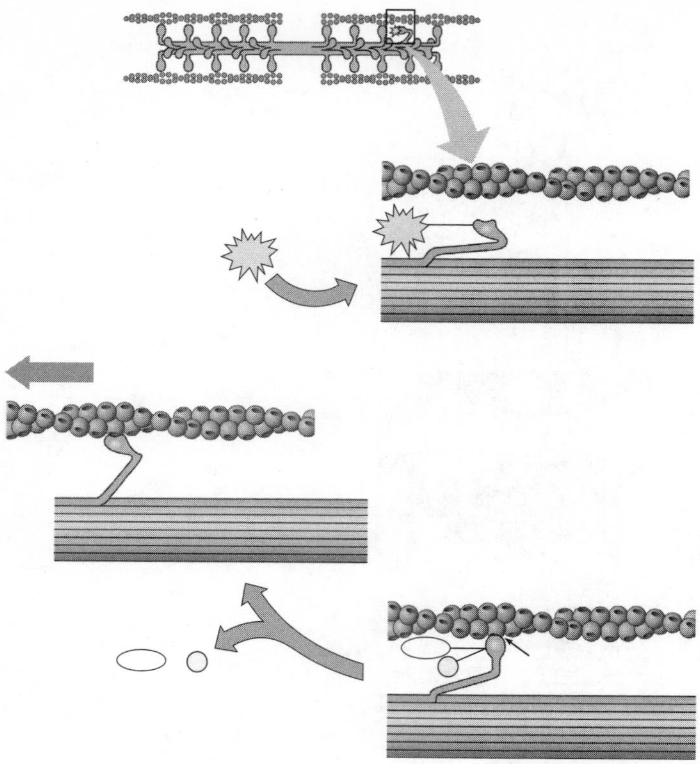

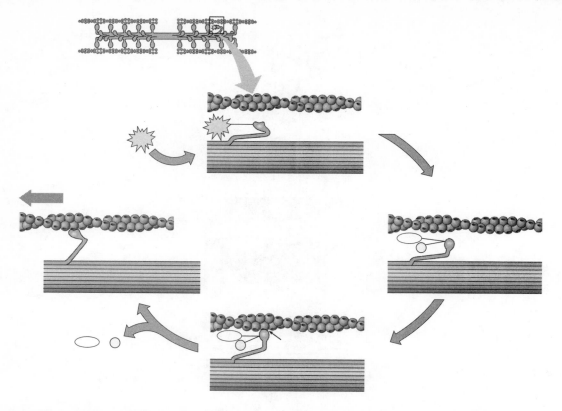

Figure 49.30 Myosin-actin interactions underlying muscle fiber contraction, page 1068

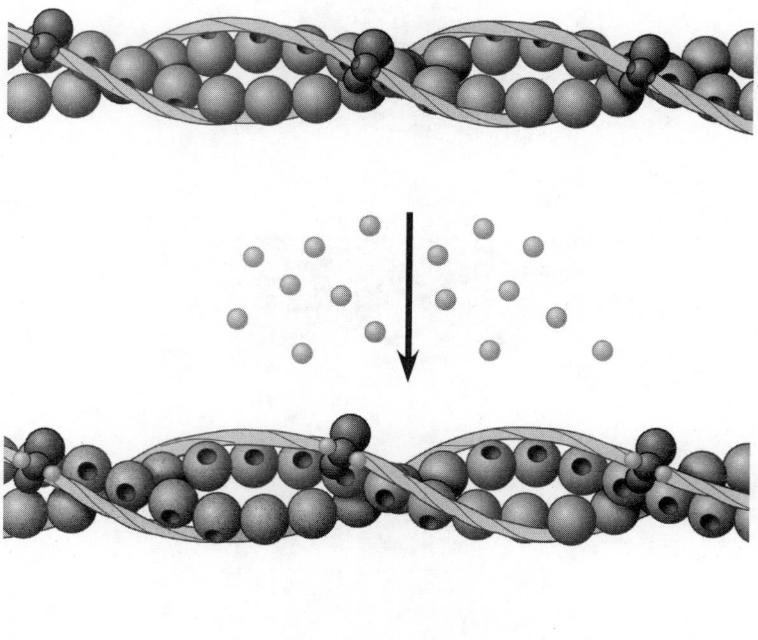

Figure 49.31 The role of regulatory proteins and calcium in muscle fiber contraction, page 1069

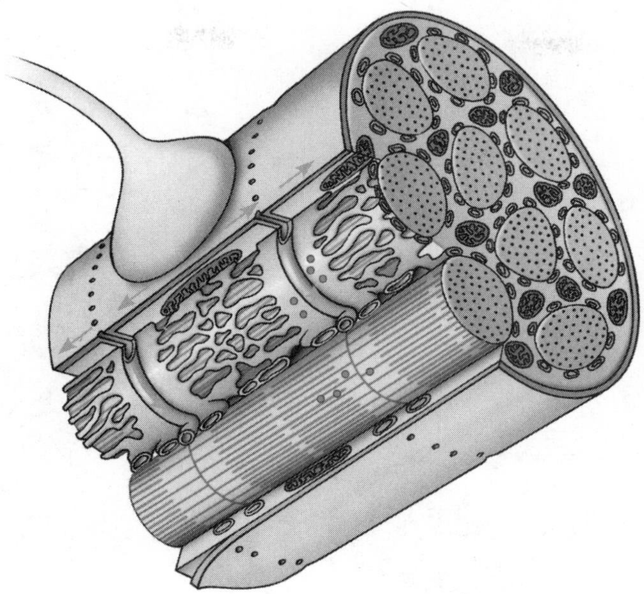

Figure 49.32 The roles of the sarcoplasmic reticulum and T tubules in muscle fiber contraction, page 1069

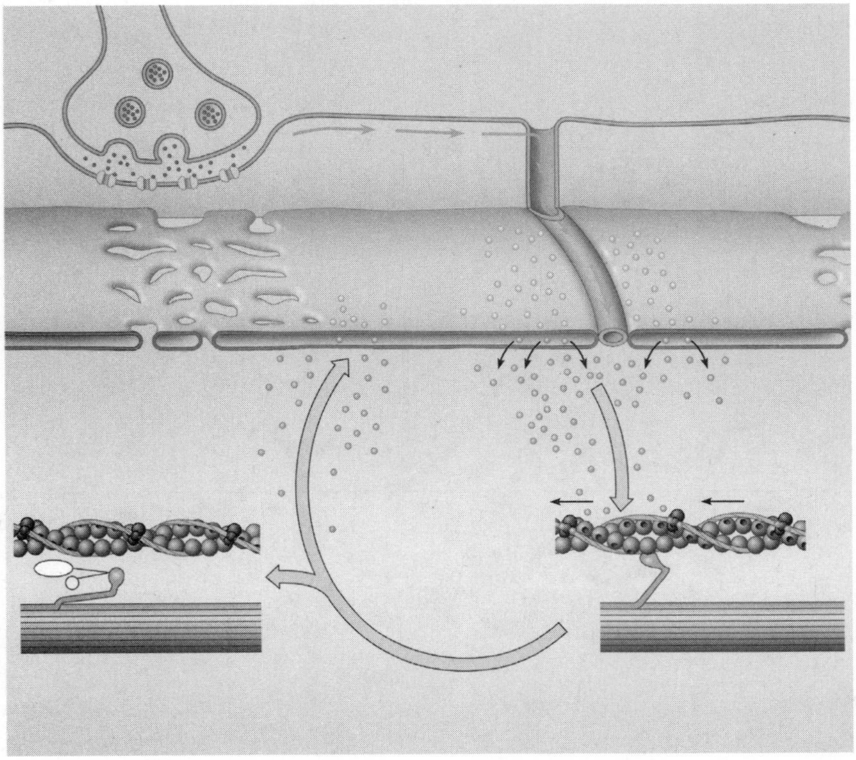

Figure 49.33 Review of contraction in a skeletal muscle fiber, page 1070

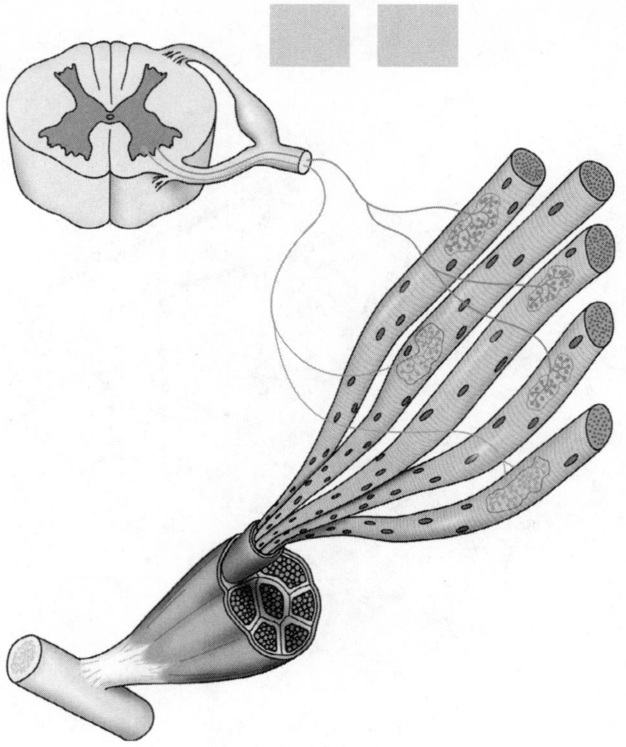

Figure 49.34 Motor units in a vertebrate skeletal muscle, page 1071

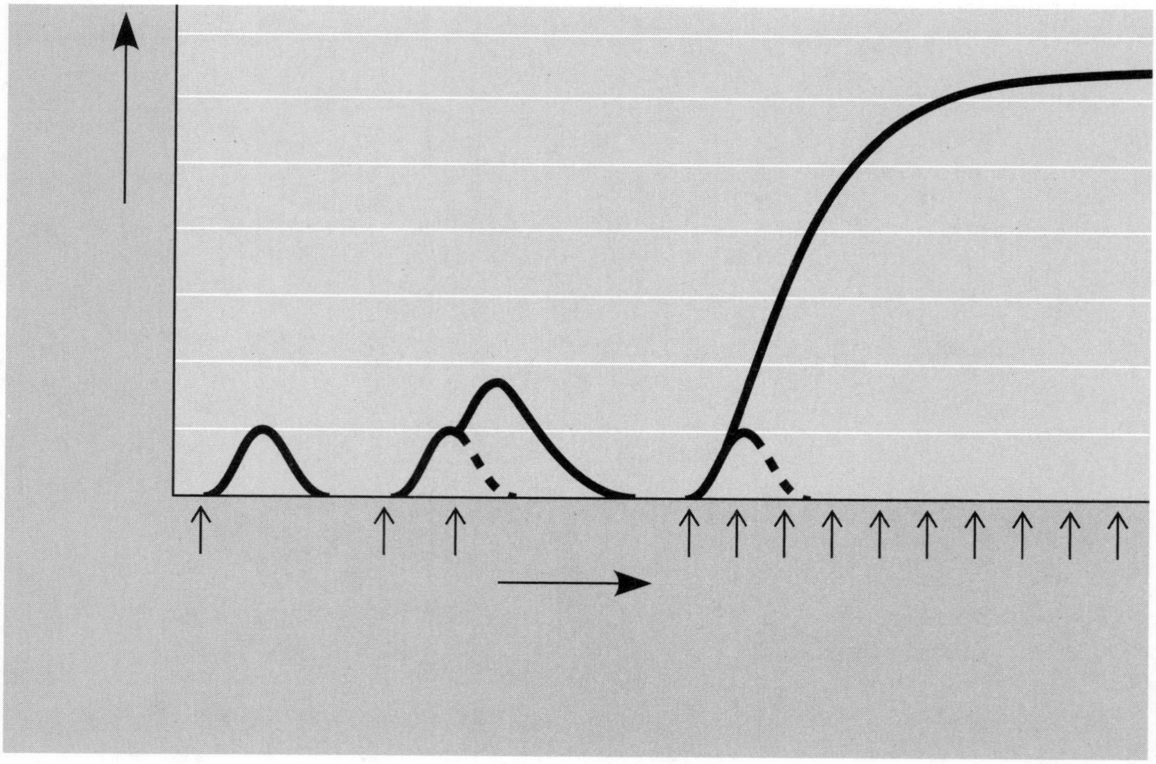

Figure 49.35 Summation of twitches, page 1071

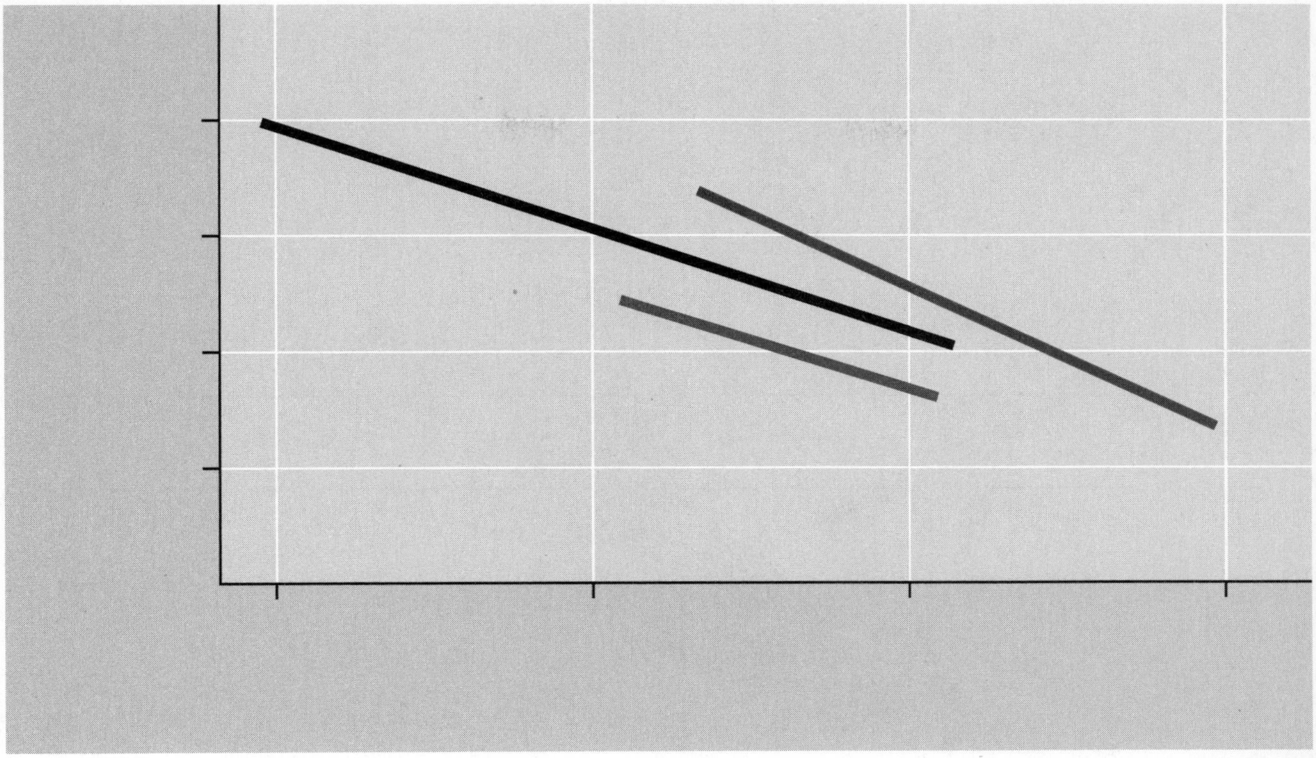

Figure 49.37 What are the energy costs of locomotion? page 1074

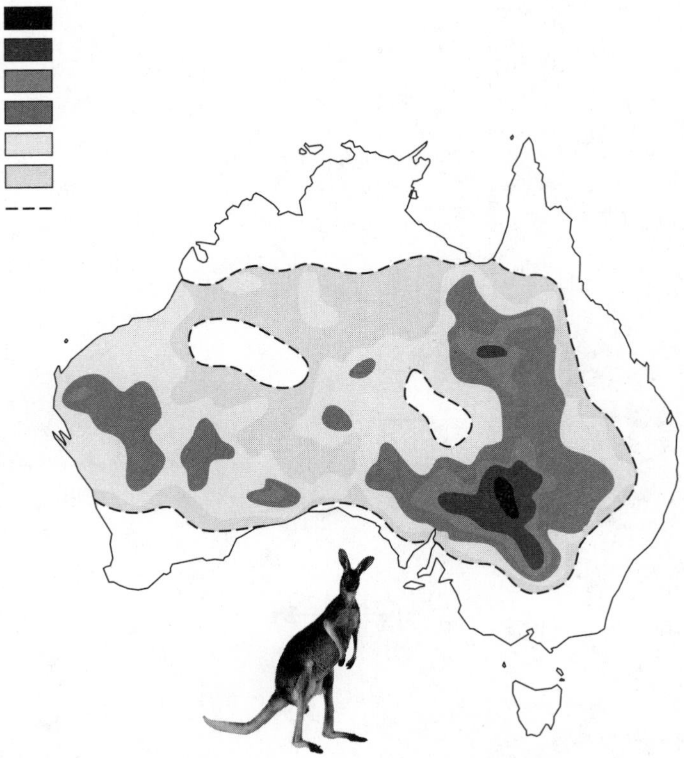

Figure 50.2 Distribution and abundance of the red kangaroo in Australia, based on aerial surveys, page 1081

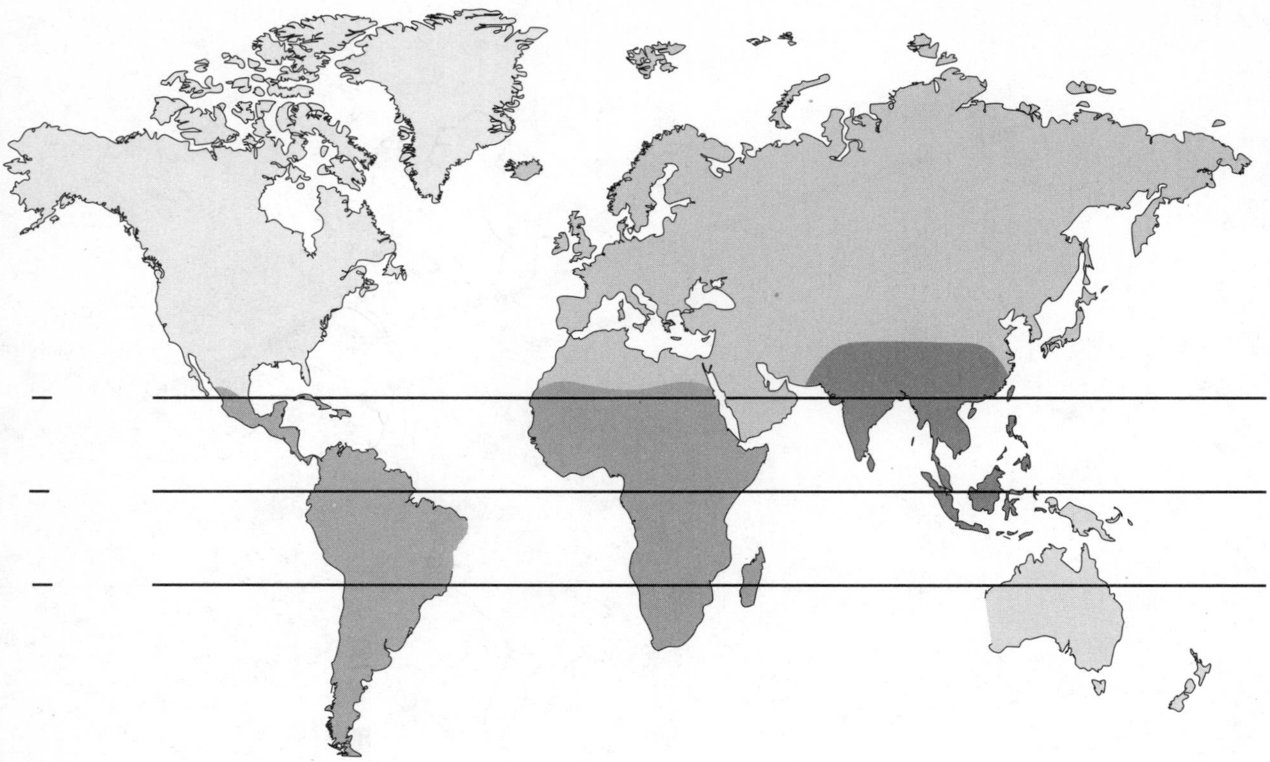

Figure 50.5 Biogeographic realms, page 1084

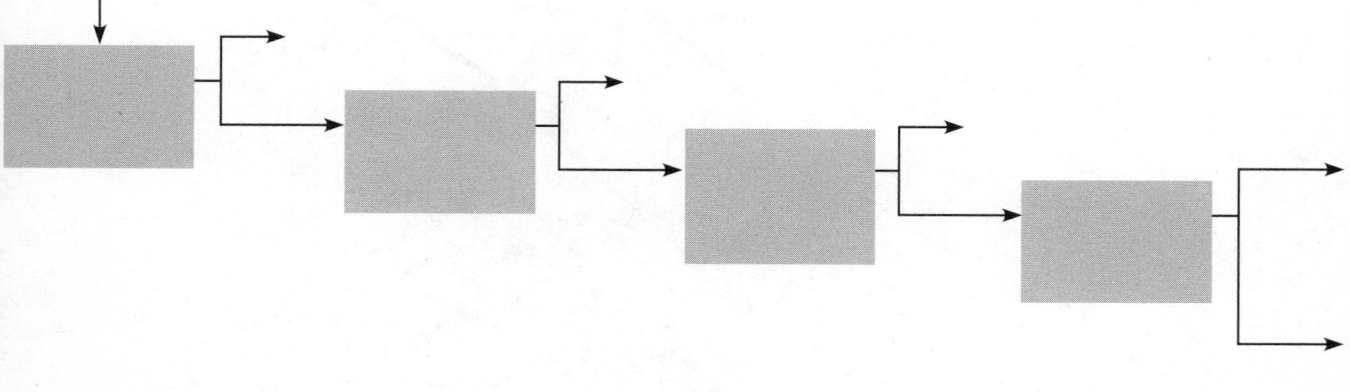

Figure 50.6 Flowchart of factors limiting geographic distribution, page 1084

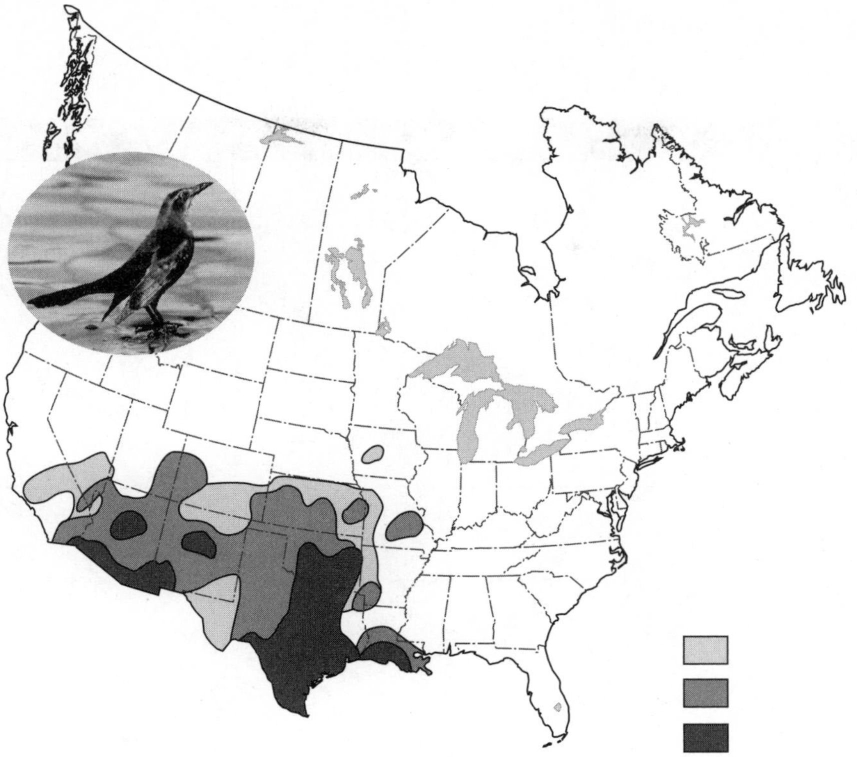

Figure 50.7 Spread of breeding populations of the great-tailed grackle in the United States from 1974 to 1996, page 1085

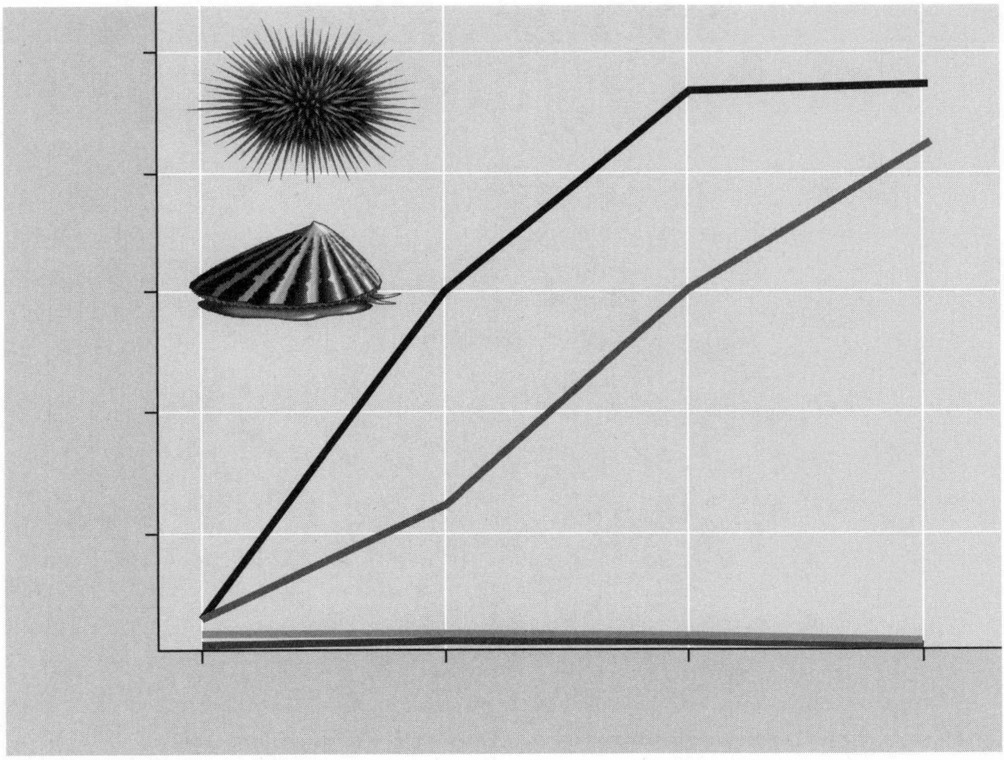

Figure 50.8 Does feeding by sea urchins and limpets affect seaweed distribution? page 1086

Figure 50.10 Global climate patterns: Latitudinal variation in sunlight density, page 1088

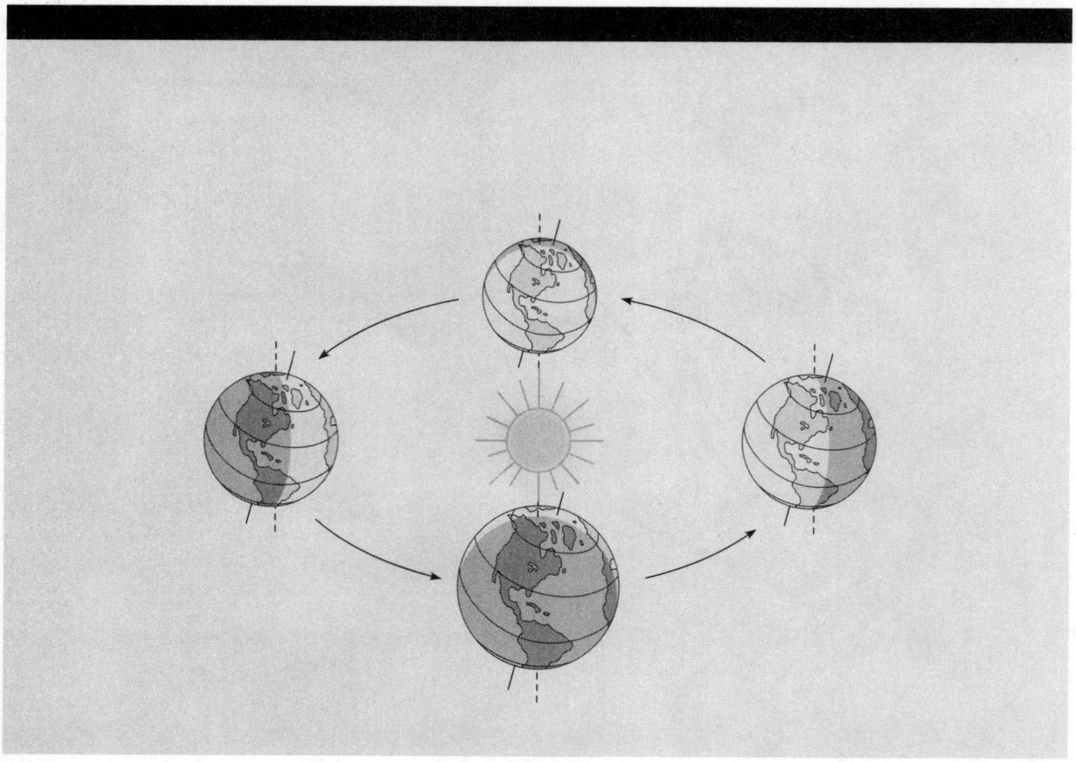

Figure 50.10 Global climate patterns: Seasonal variation in sunlight intensity, page 1088

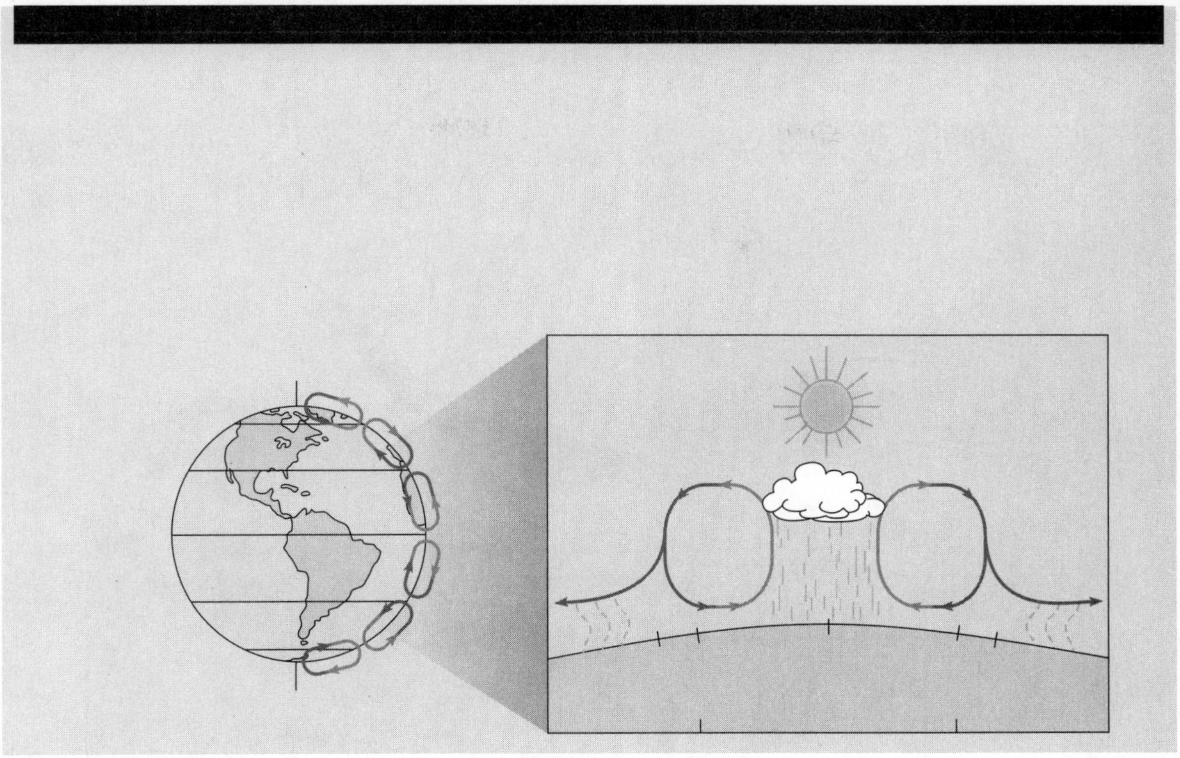

Figure 50.10 Global climate patterns: Global air circulation and precipitation patterns, page 1089

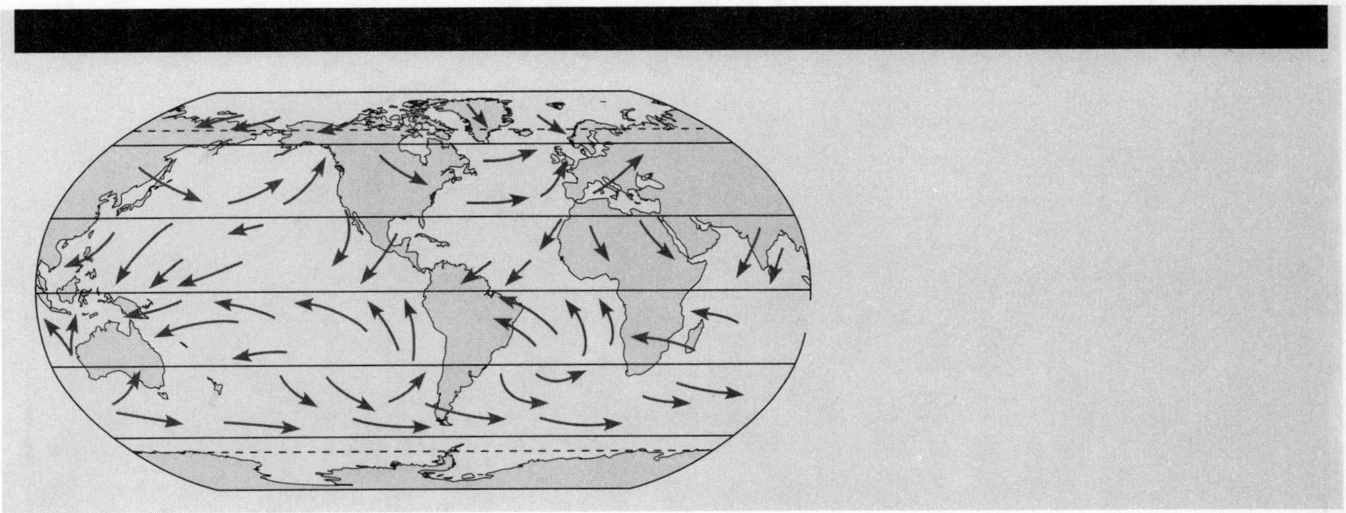

Figure 50.10 Global climate patterns: Global wind patterns, page 1089

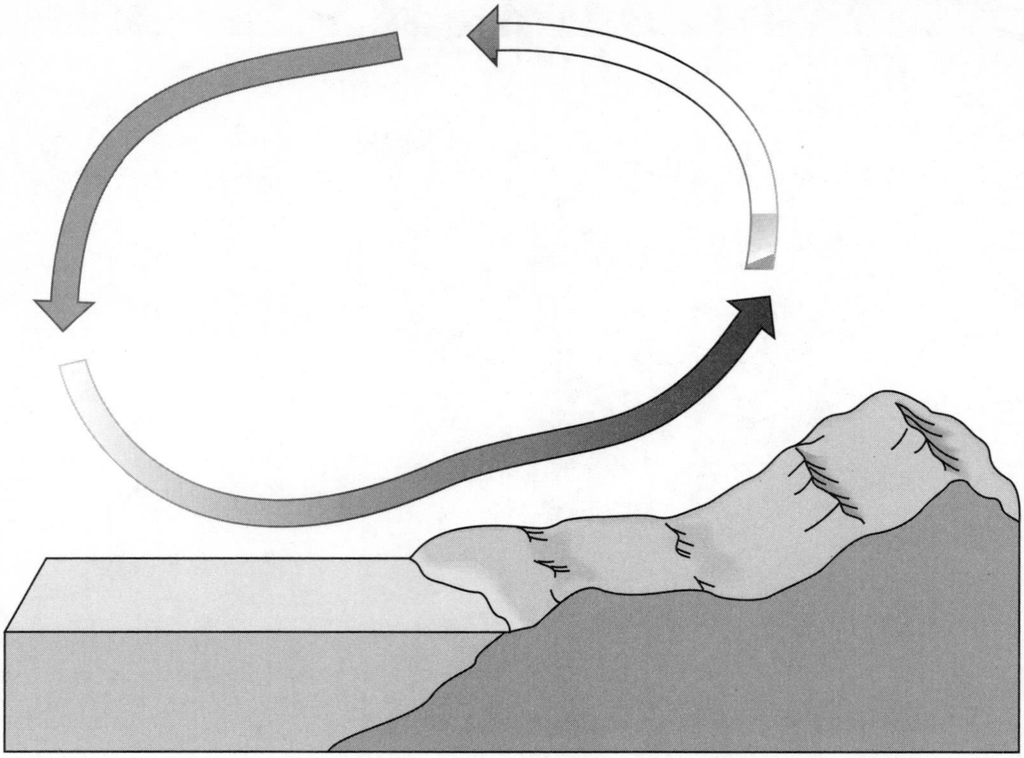

Figure 50.11 Moderating effects of large bodies of water on climate, page 1090

Figure 50.12 How mountains affect rainfall, page 1090

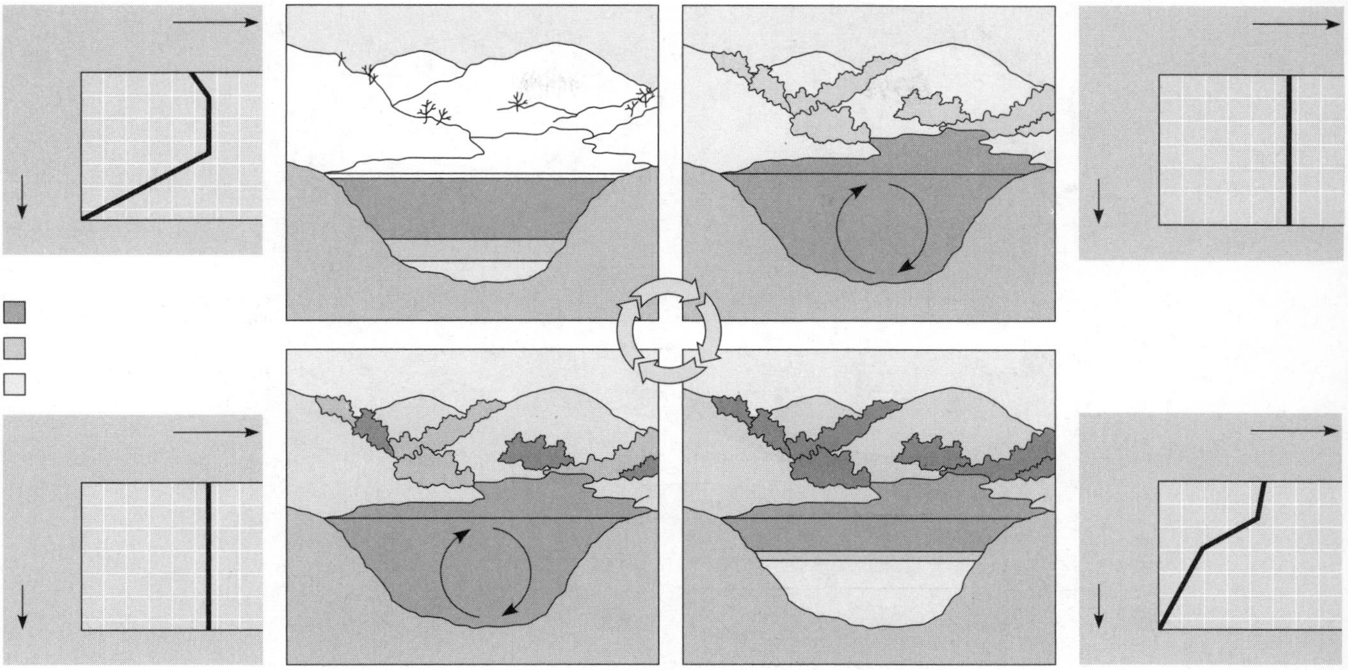

Figure 50.13 Seasonal turnover in lakes with winter ice cover, page 1091

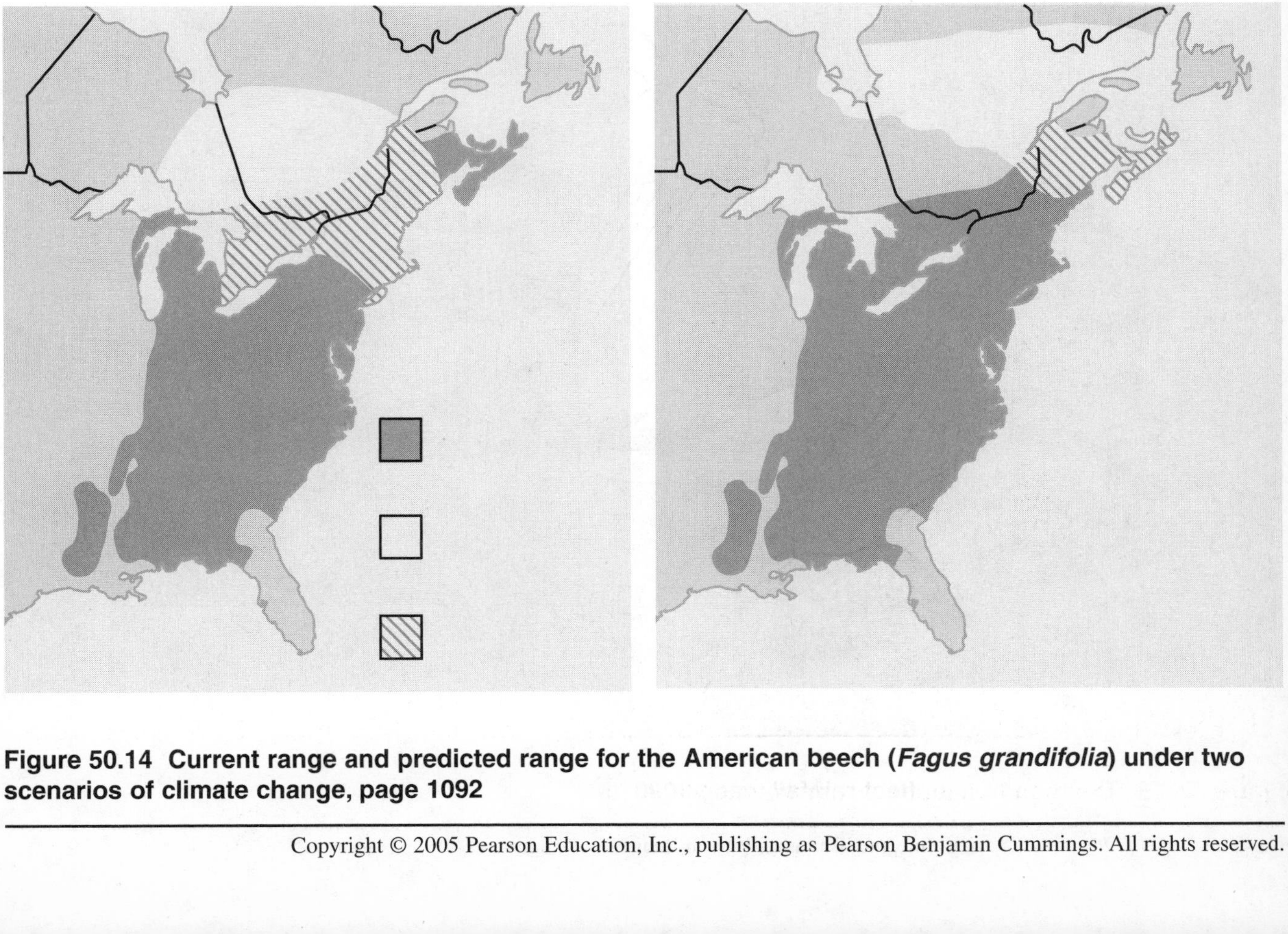

Figure 50.14 Current range and predicted range for the American beech (*Fagus grandifolia*) under two scenarios of climate change, page 1092

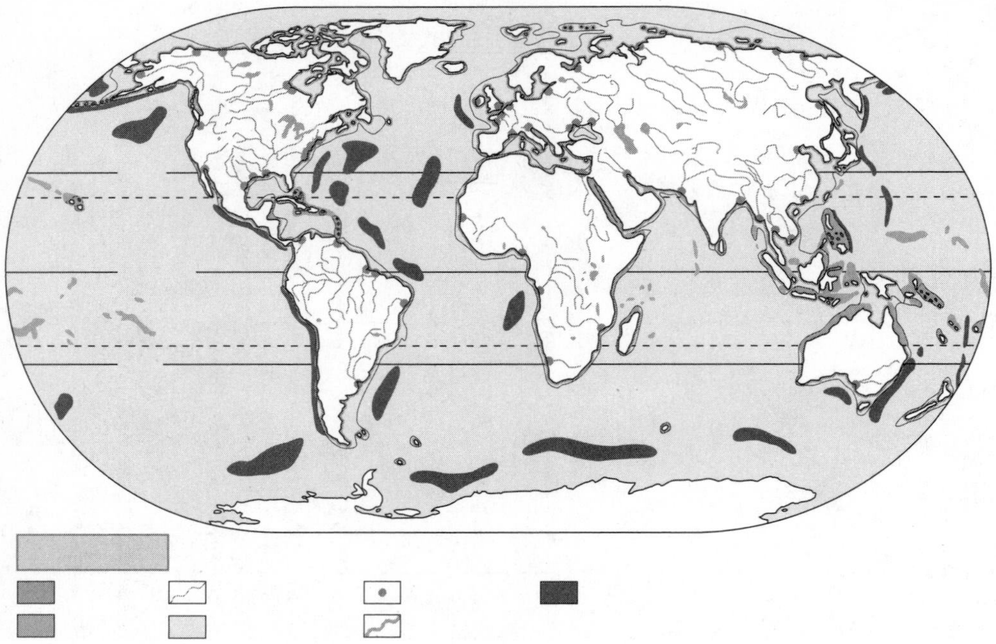

Figure 50.15 The distribution of major aquatic biomes, page 1092

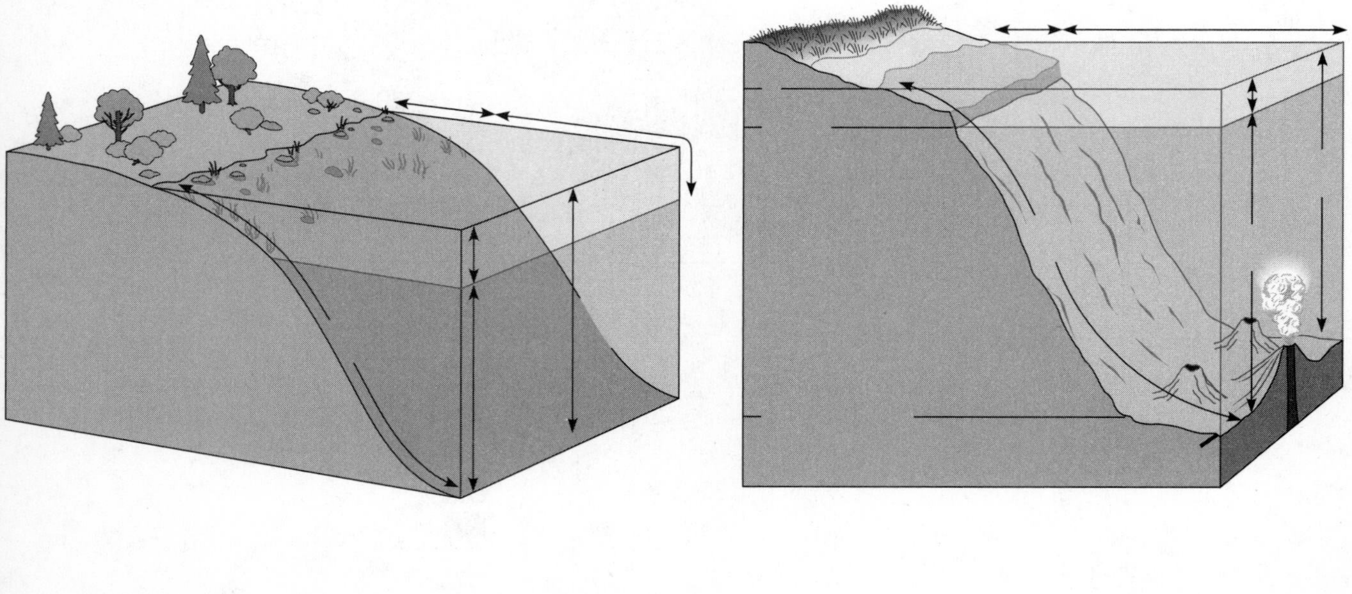

Figure 50.16 Zonation in aquatic environments, page 1093

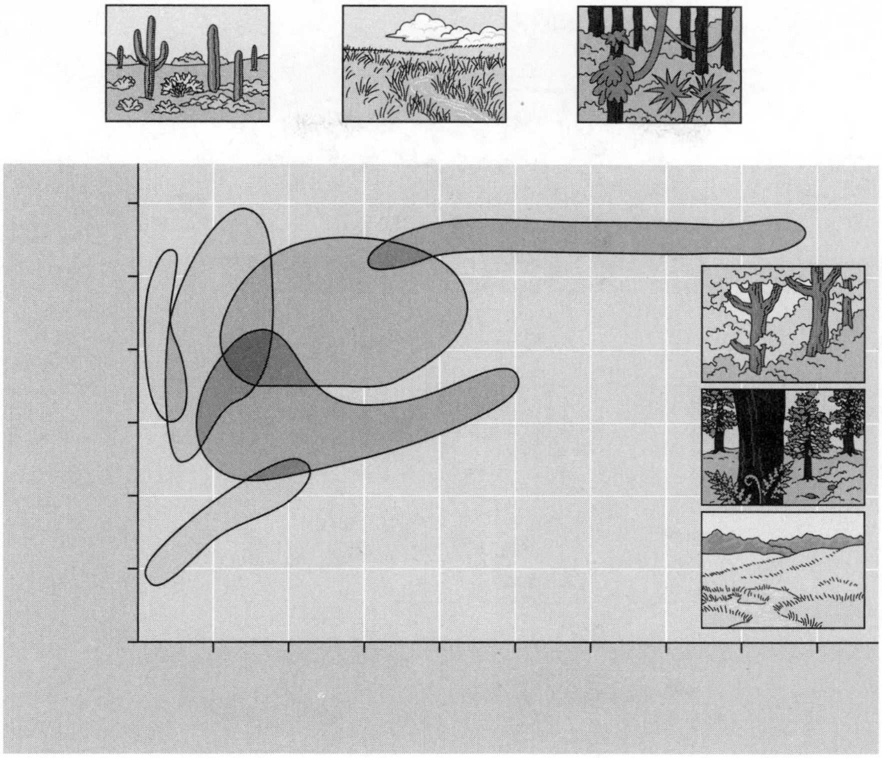

Figure 50.18 A climograph for some major types of biomes in North America, page 1098

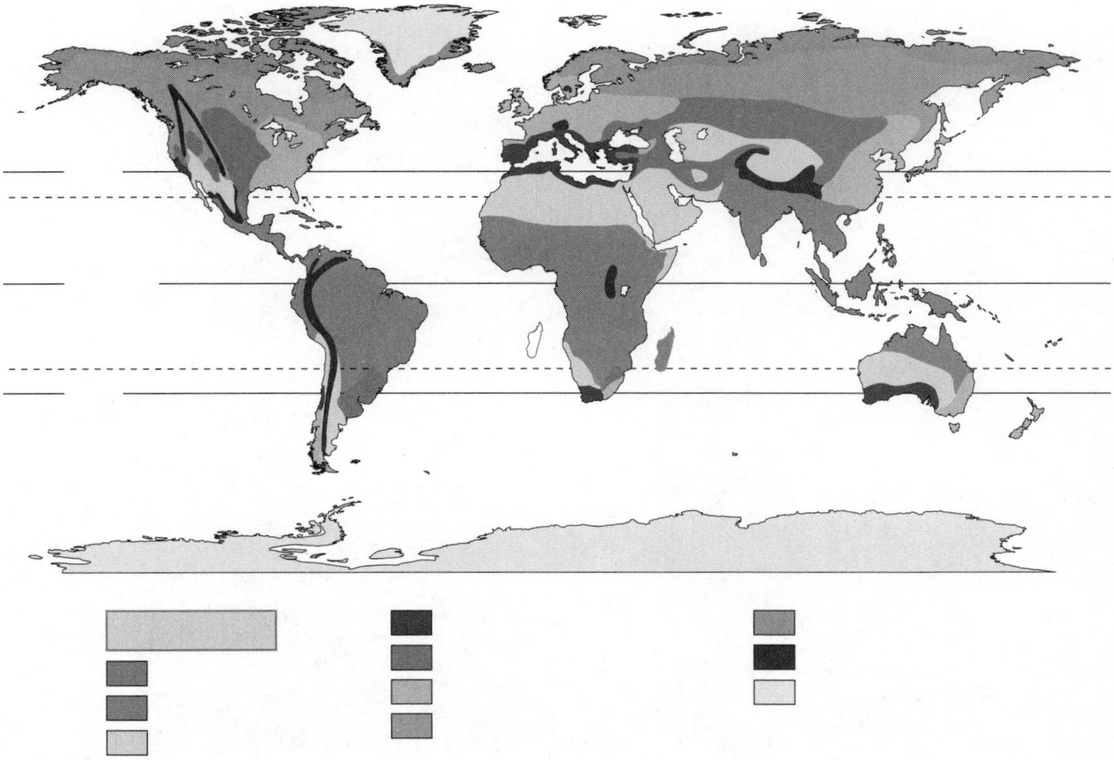

Figure 50.19 The distribution of major terrestrial biomes, page 1099

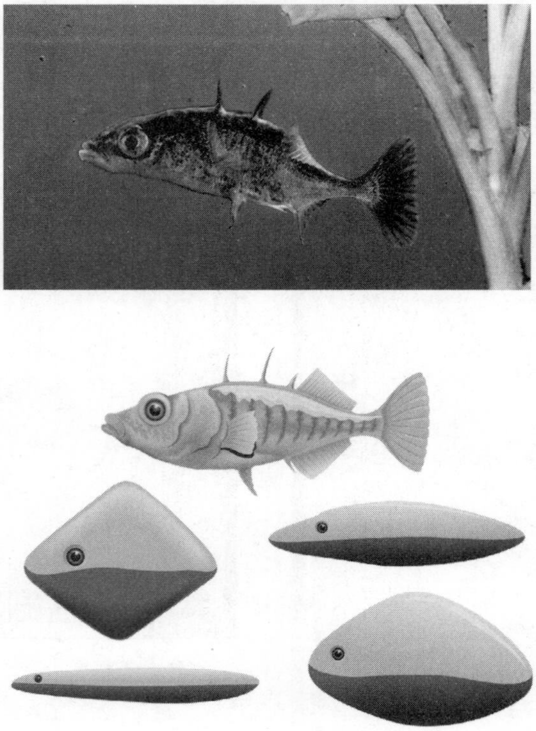

Figure 51.3 Sign stimuli in a classic fixed action pattern, page 1108

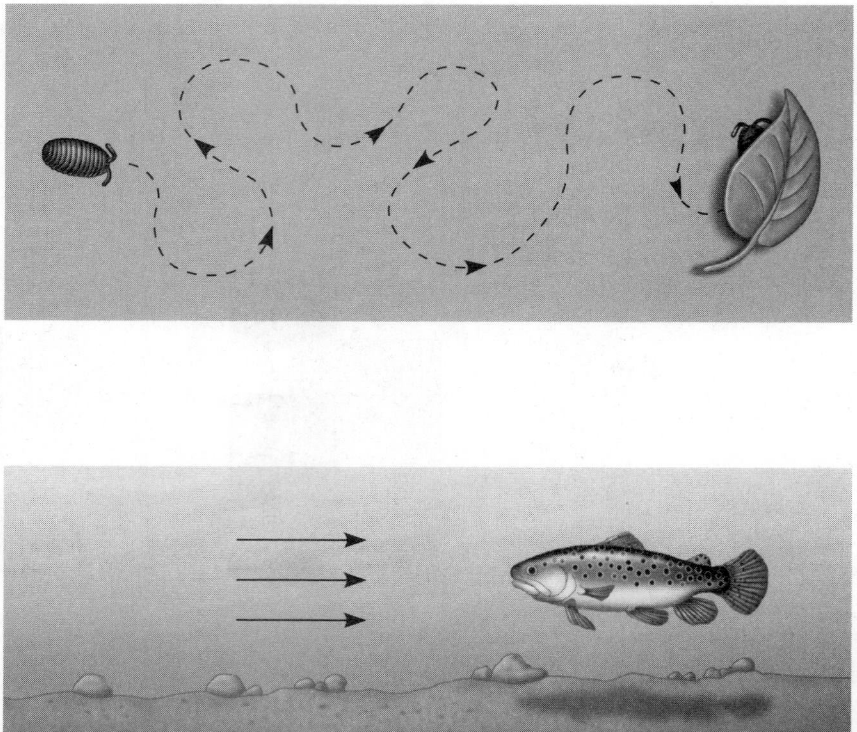

Figure 51.7 A kinesis and a taxis, page 1110

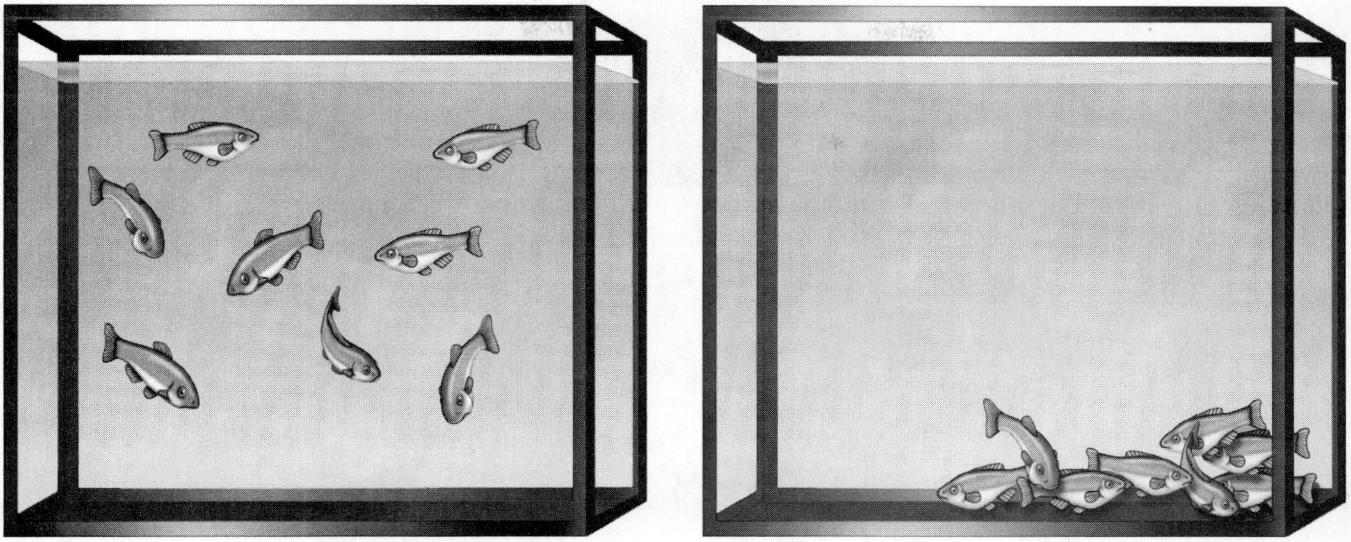

Figure 51.9 Are the different songs of closely related green lacewing species under genetic control? page 1112

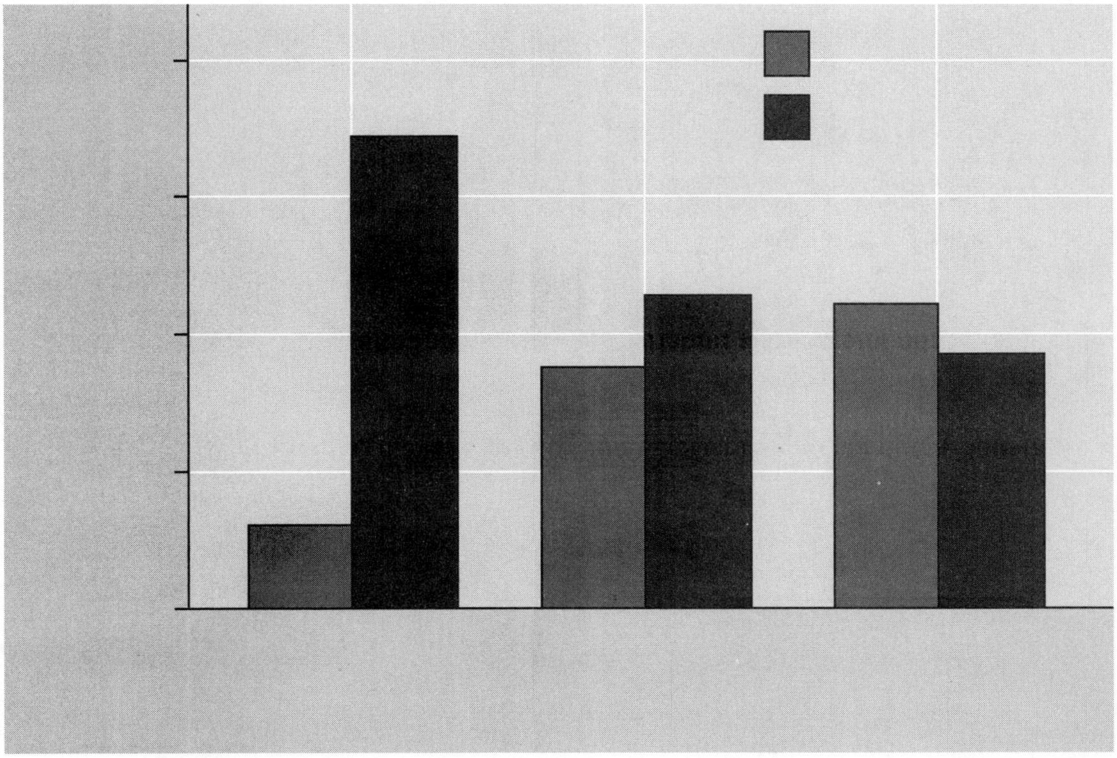

Figure 51.12 How does dietary environment affect mate choice by female *Drosophila mojavensis*? page 1113

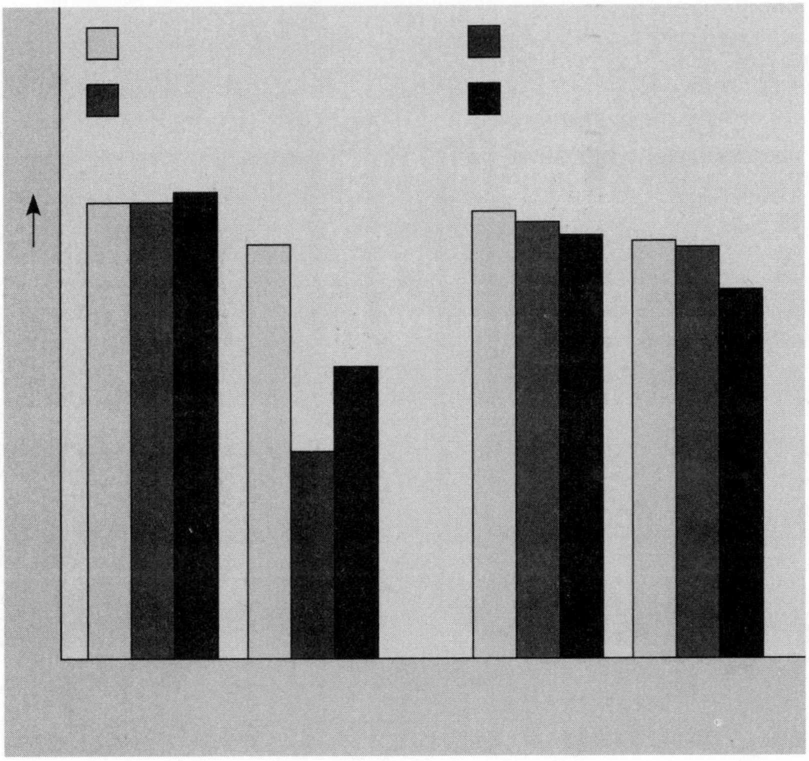

Figure 51.15 Associative learning in zebrafish, page 1117

Figure 51.19 Aggressiveness of the funnel web spider (*Agelenopsis aperta*), living in two environments, page 1119

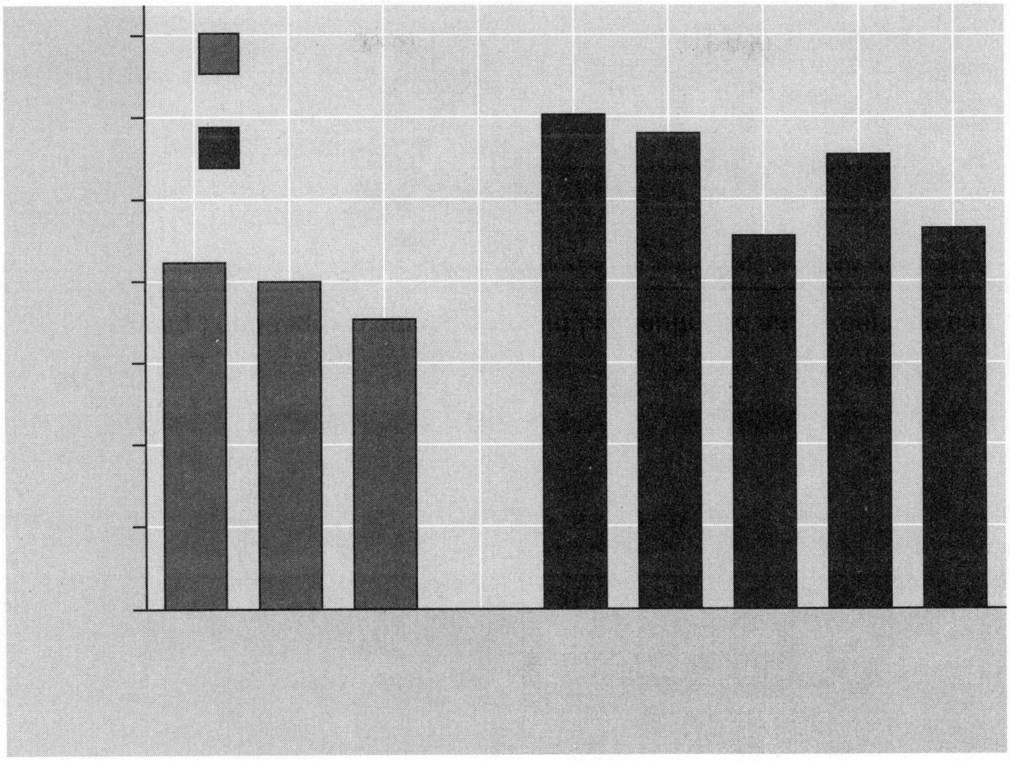

Figure 51.20 Evolution of foraging behavior by laboratory populations of *Drosophila melanogaster*, page 1120

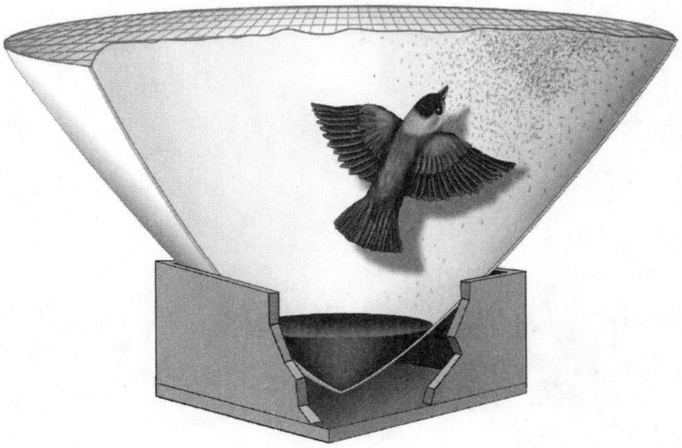

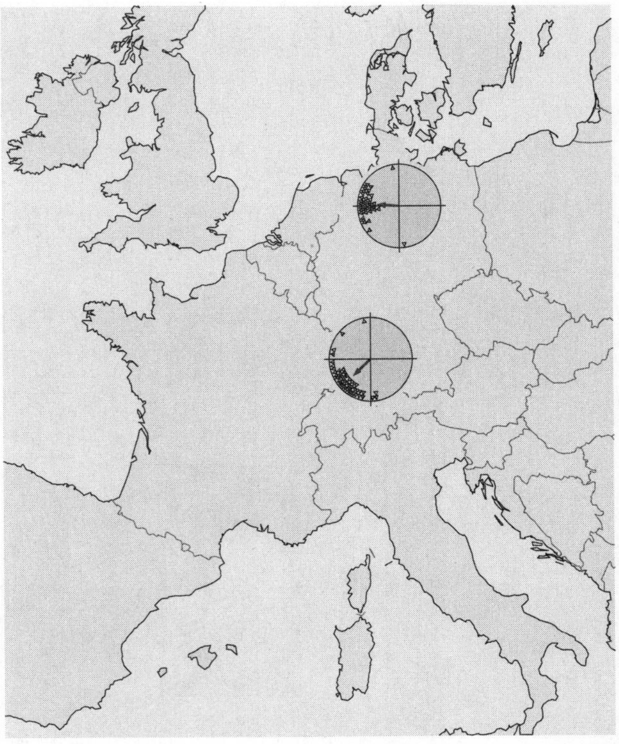

Figure 51.21 Evidence of a genetic basis for migratory orientation, page 1121

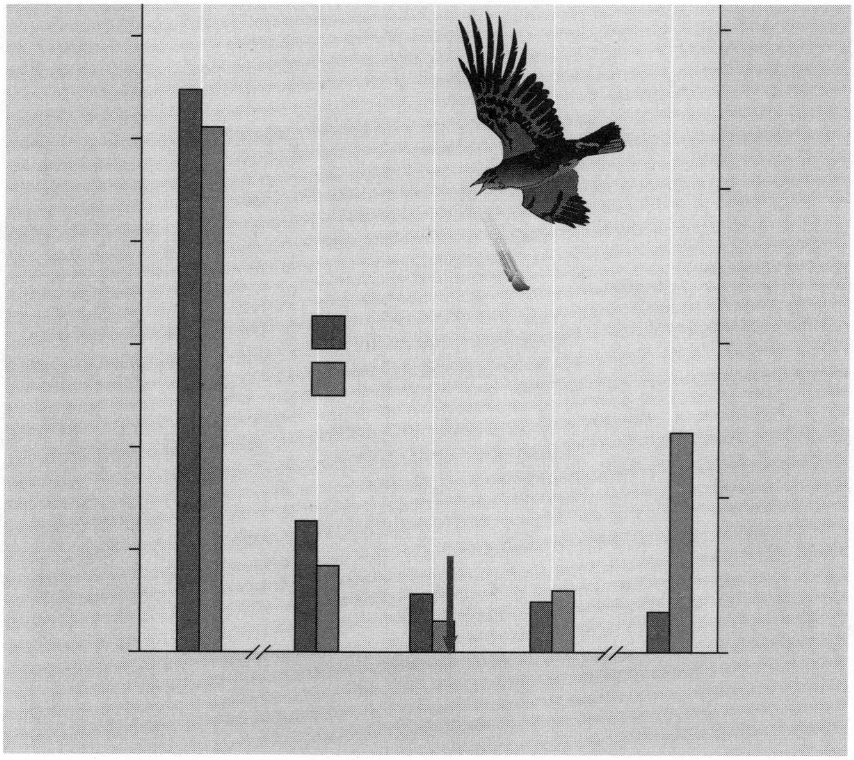

Figure 51.22 Energy costs and benefits in foraging behavior, page 1122

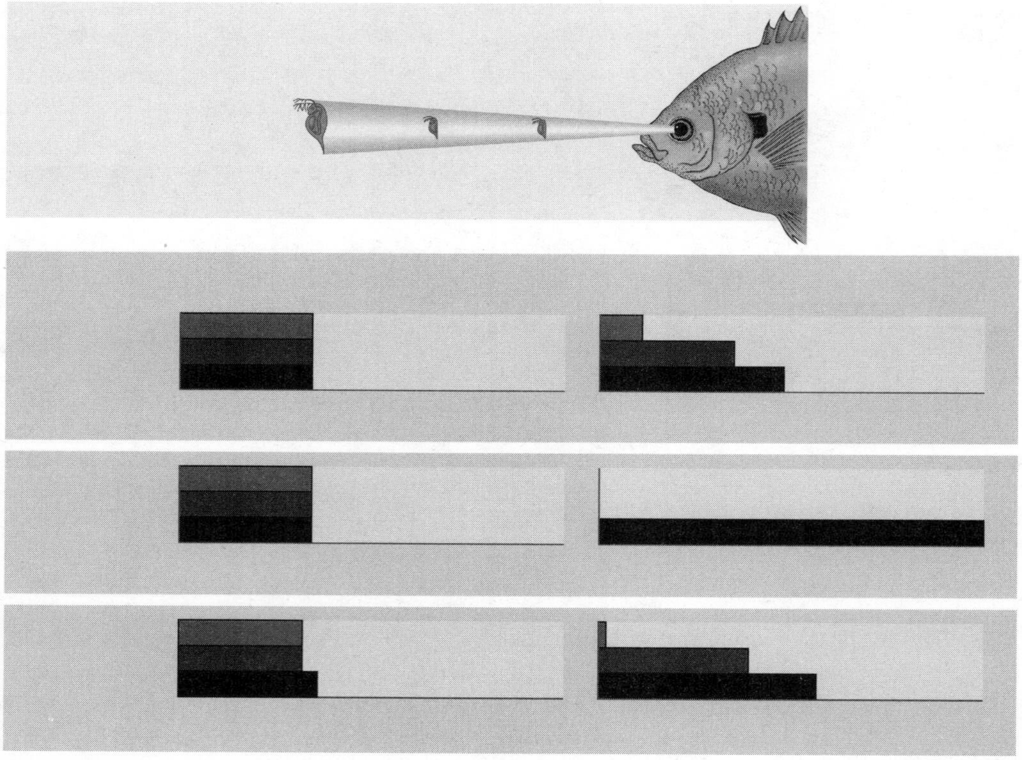

Figure 51.23 Feeding by bluegill sunfish, page 1123

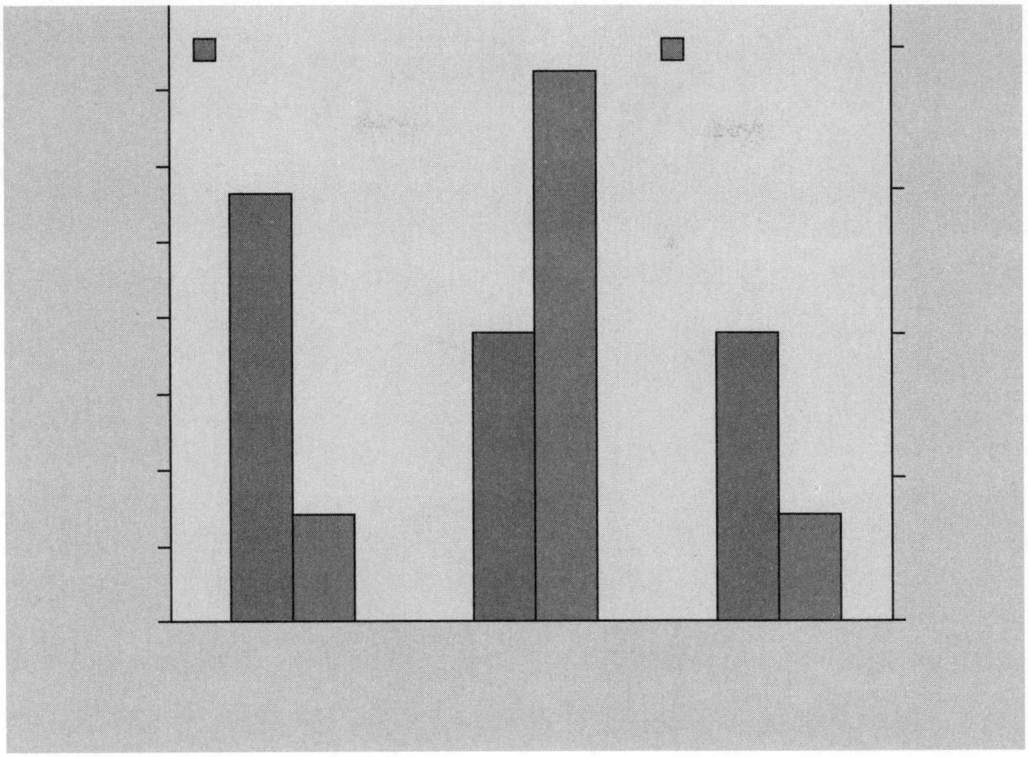

Figure 51.24 Risk of predation and use of foraging areas by mule deer, page 1123

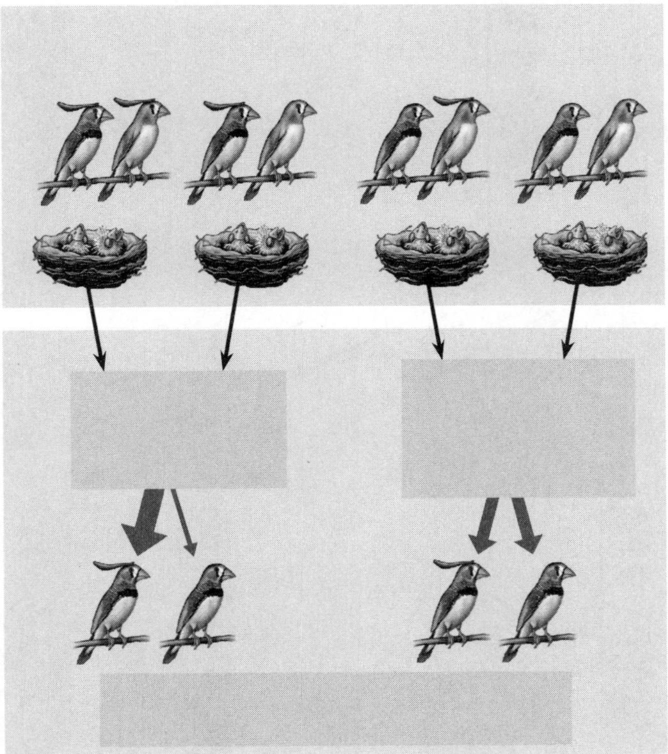

Figure 51.28 Sexual selection influenced by imprinting, page 1126

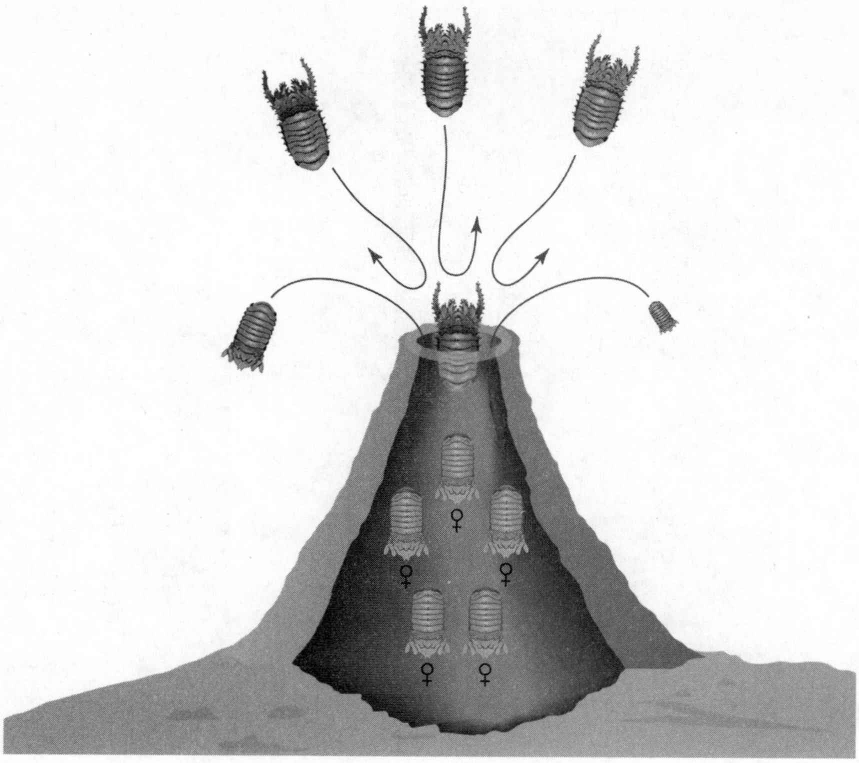

Figure 51.31 Male polymorphism in the marine intertidal isopod *Paracerceis sculpta*, page 1127

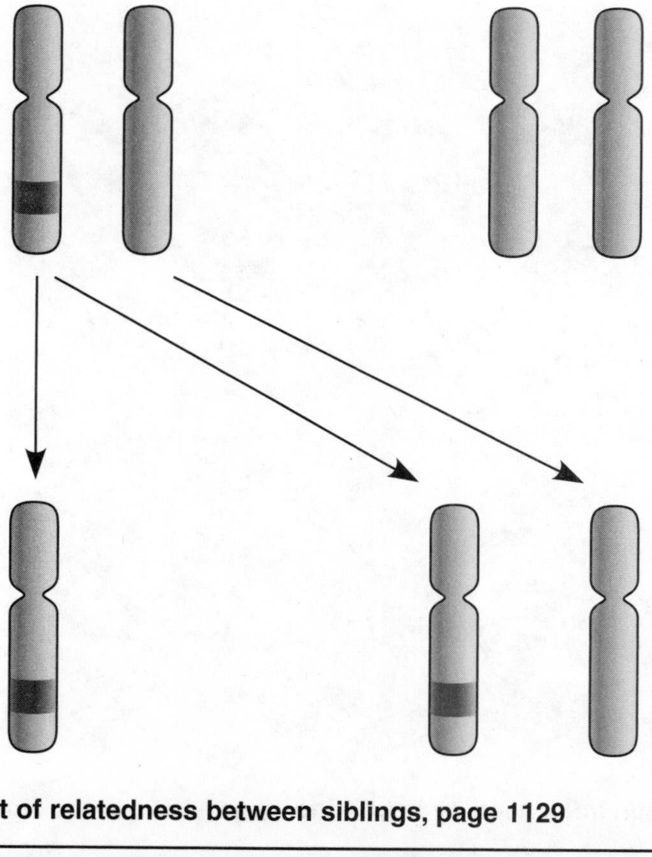

Figure 51.34 The coefficient of relatedness between siblings, page 1129

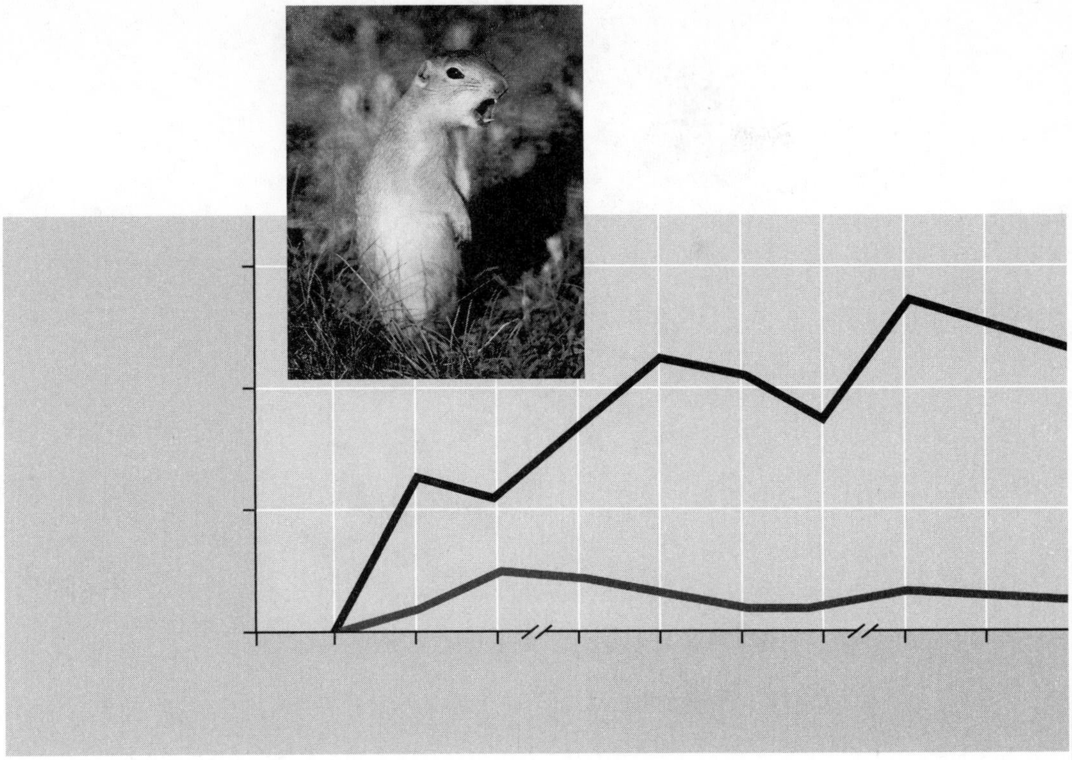

Figure 51.35 Kin selection and altruism in Belding's ground squirrel, page 1130

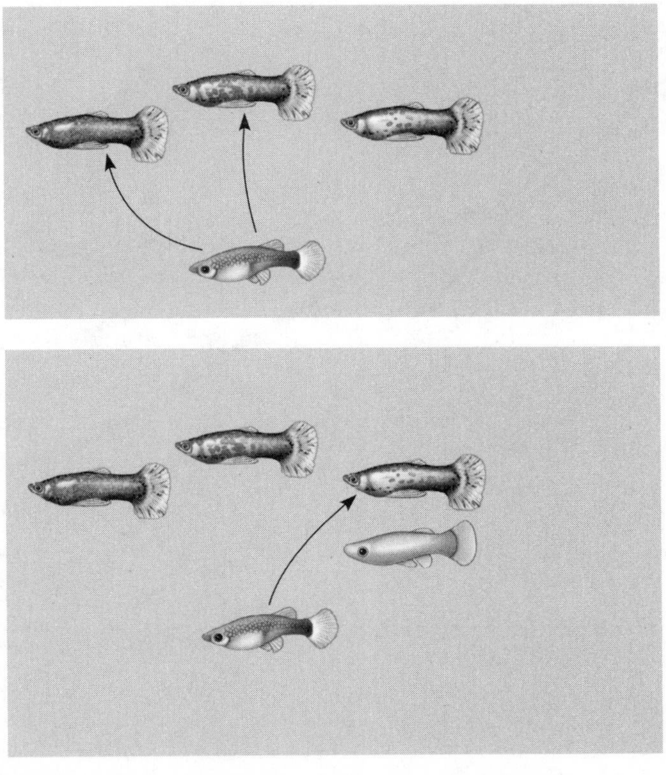

Figure 51.36 Mate choice copying by female guppies (*Poecilia reticulata*), page 1131

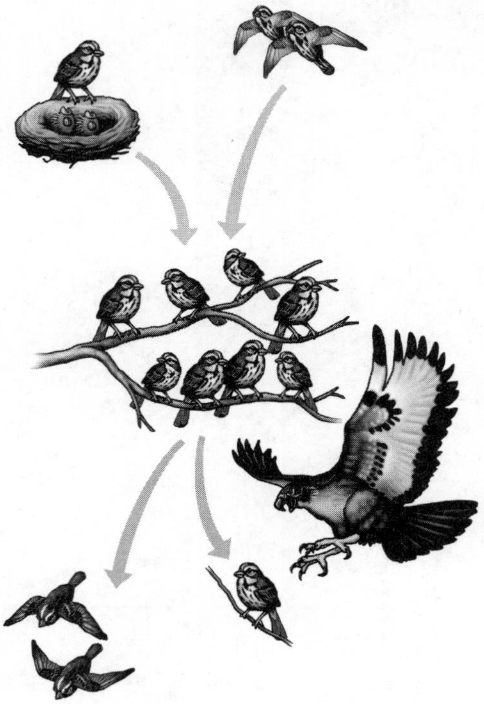

Figure 52.2 Population dynamics, page 1137

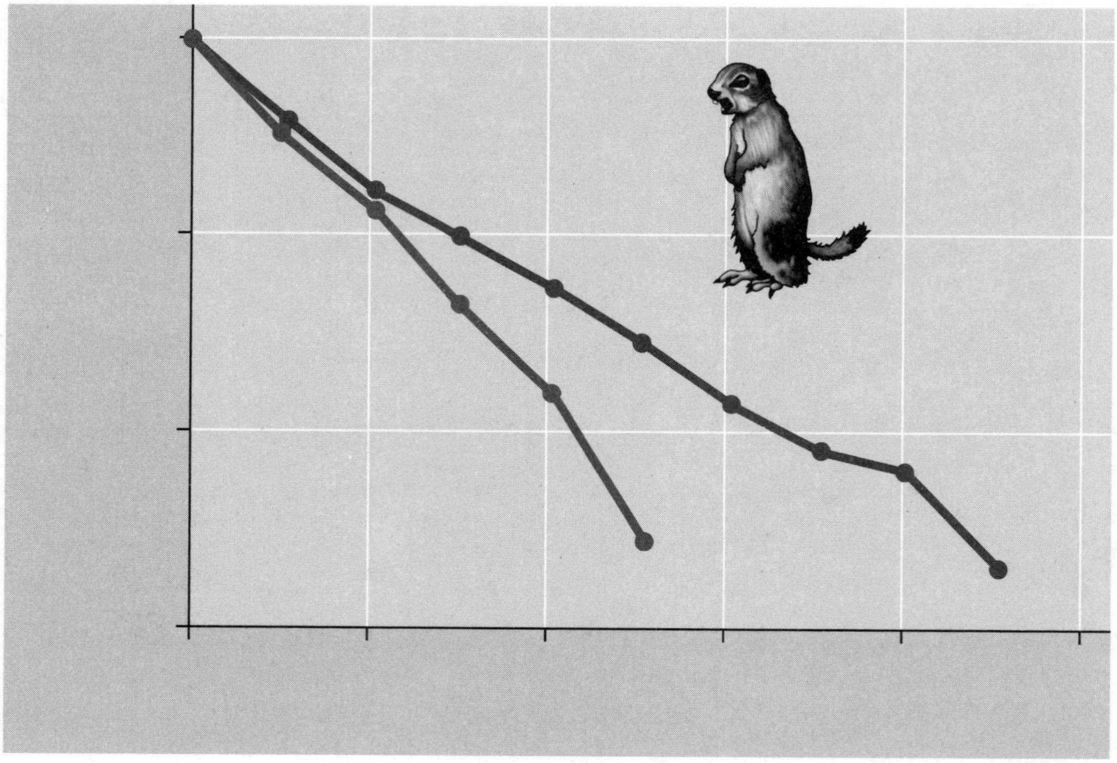

Figure 52.4 Survivorship curves for male and female Belding's ground squirrels, page 1140

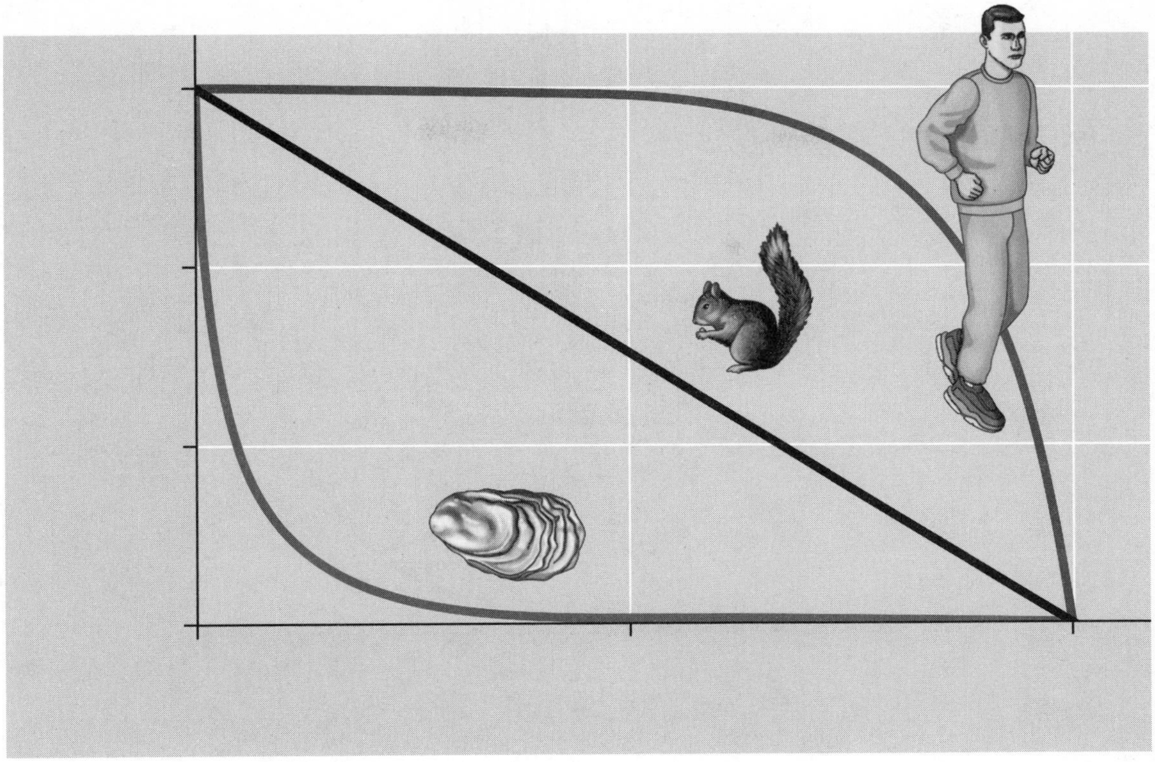

Figure 52.5 Idealized survivorship curves: Types I, II, and III, page 1140

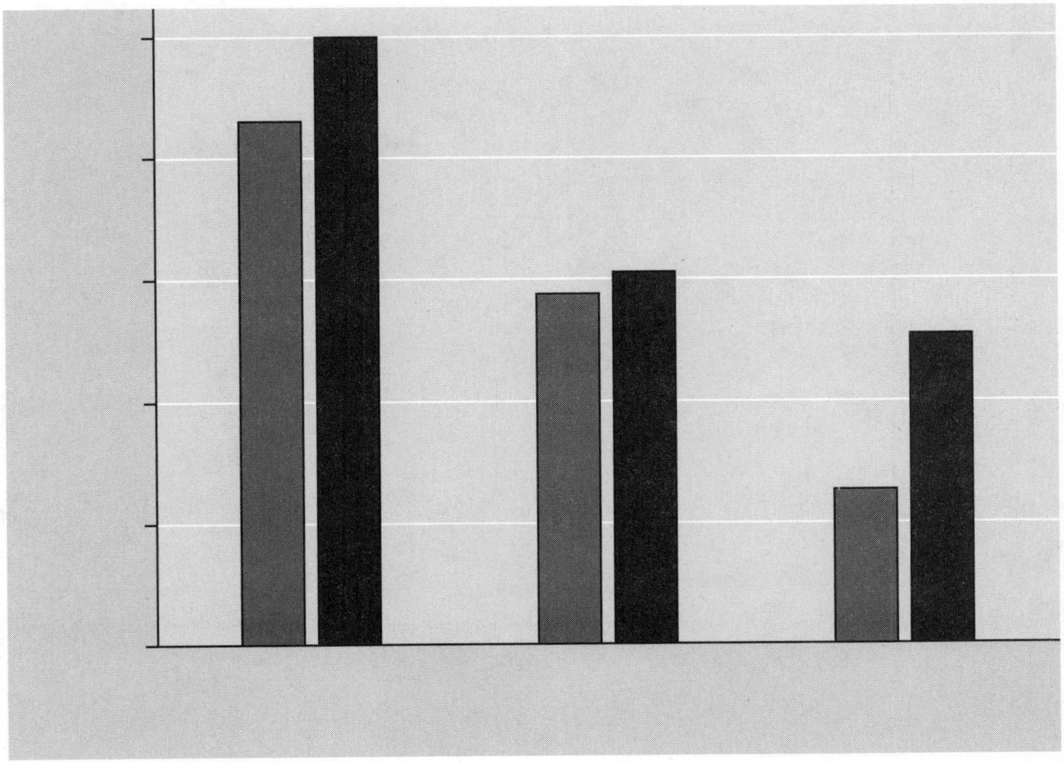

Figure 52.7 How does caring for offspring affect parental survival in kestrels? page 1142

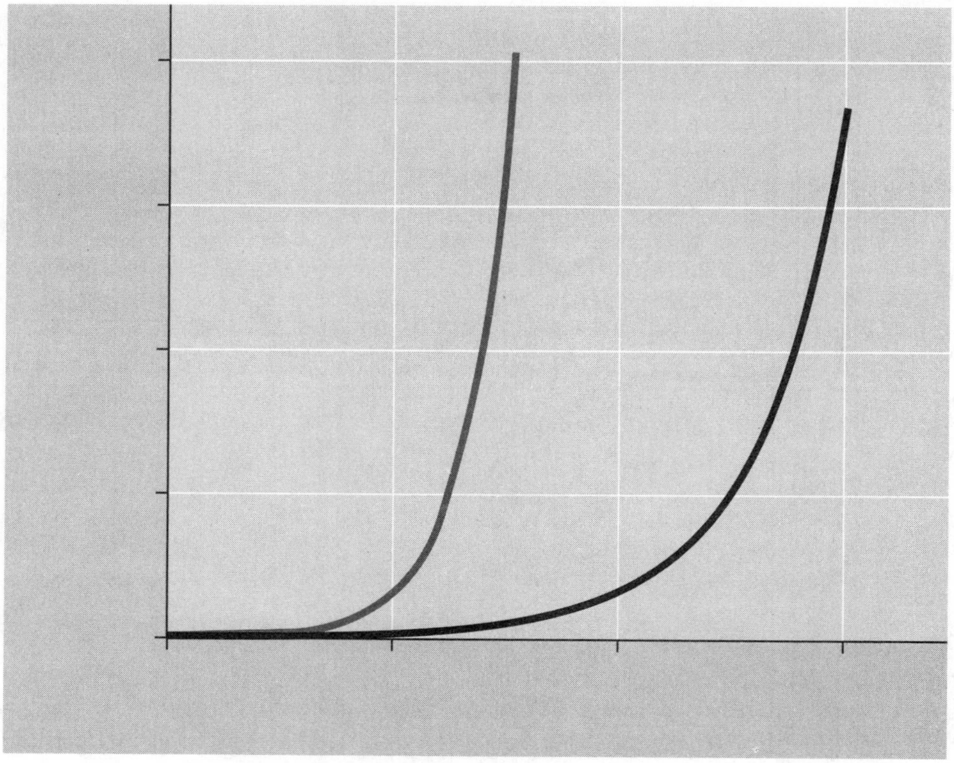

Figure 52.9 Population growth predicted by the exponential model, page 1144

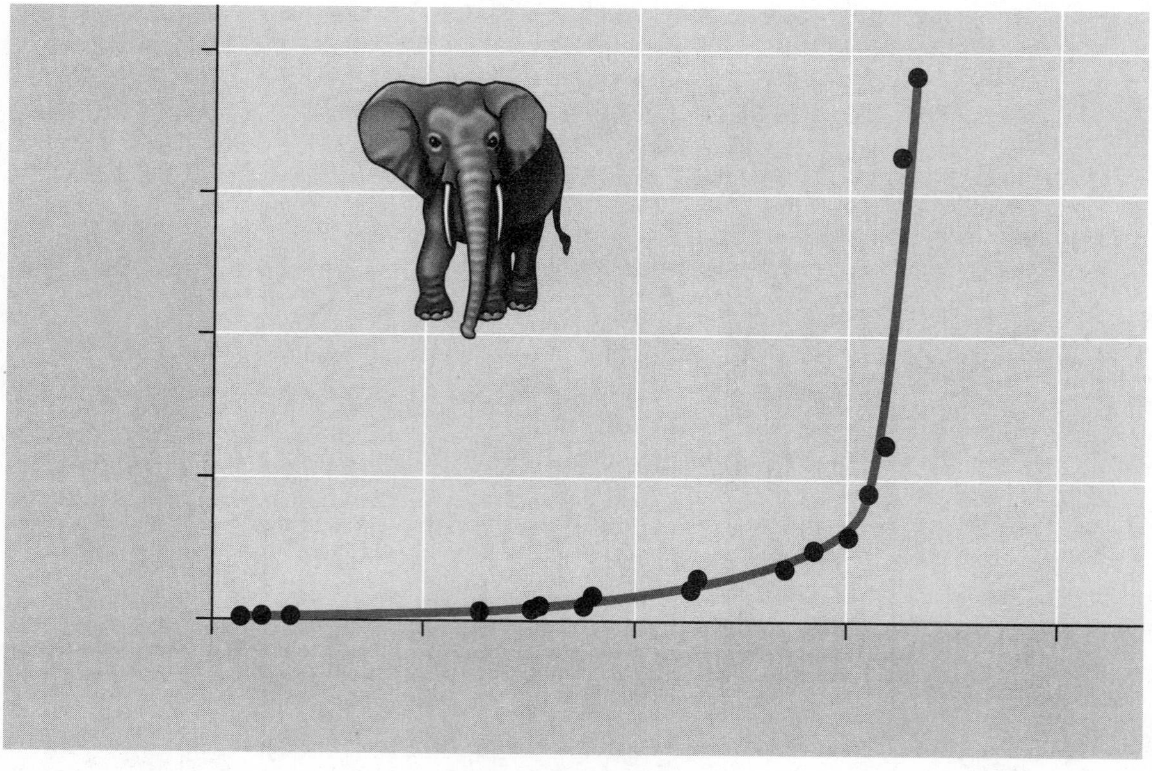

Figure 52.10 Exponential growth in the African elephant population of Kruger National Park, South Africa, page 1144

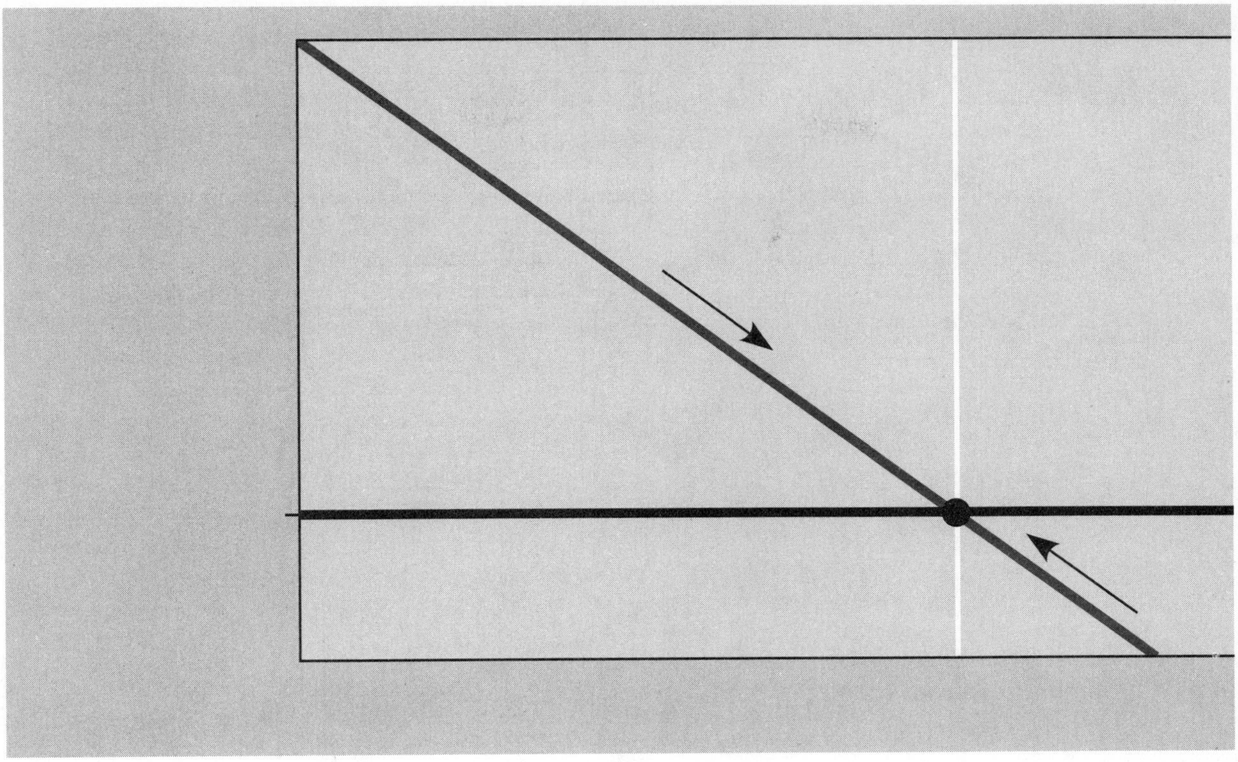

Figure 52.11 Influence of population size (*N*) on per capita rate of increase (*r*), page 1145

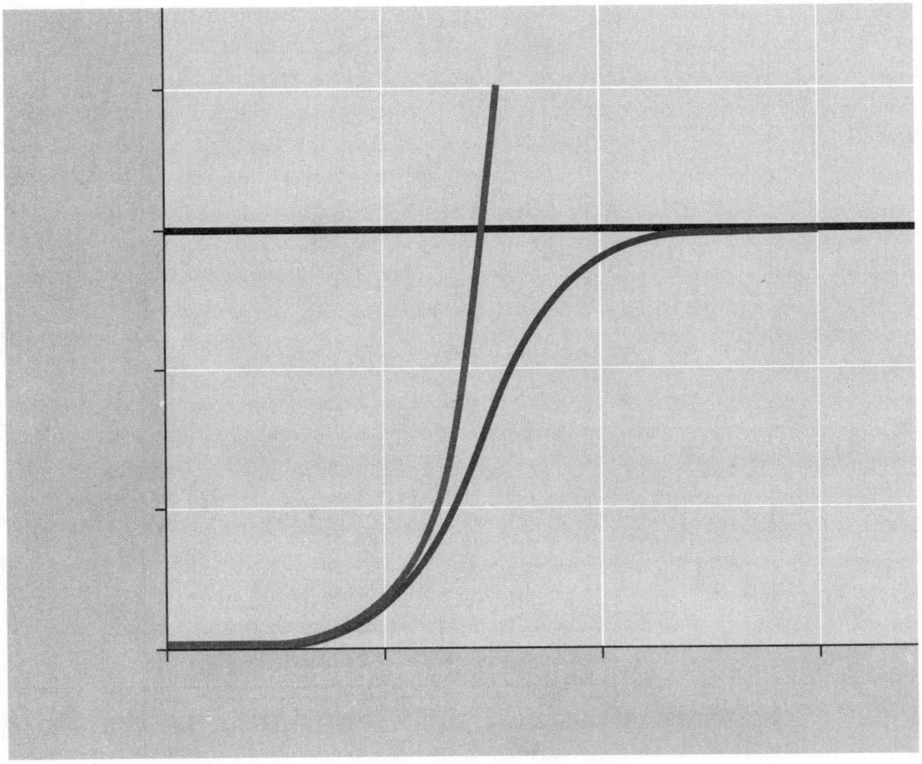

Figure 52.12 Population growth predicted by the logistic model, page 1146

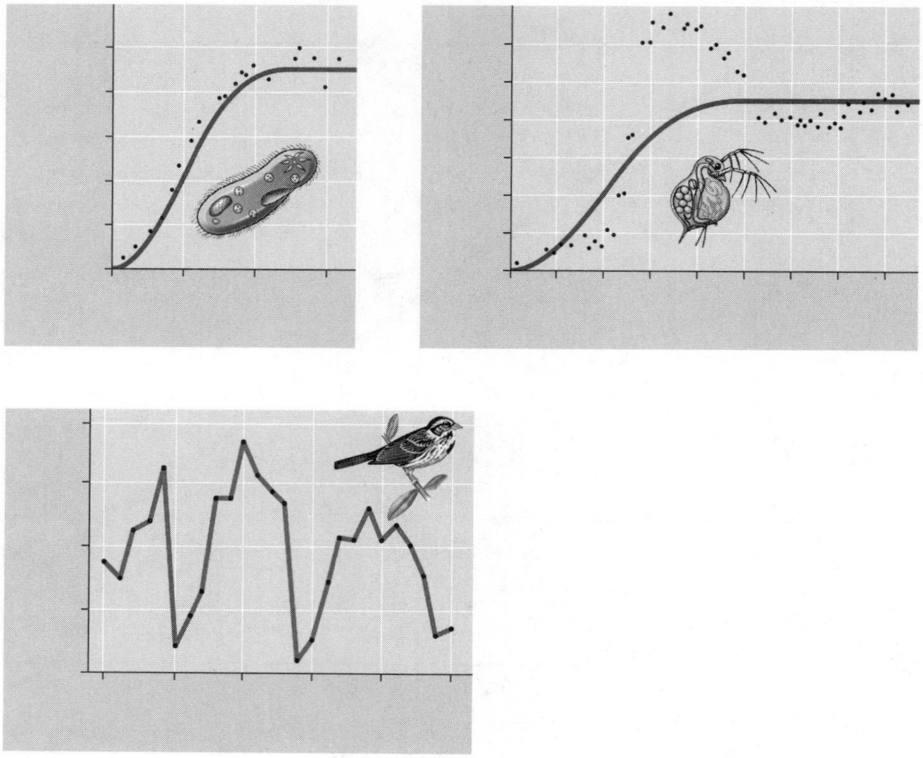

Figure 52.13 How well do these populations fit the logistic growth model? page 1147

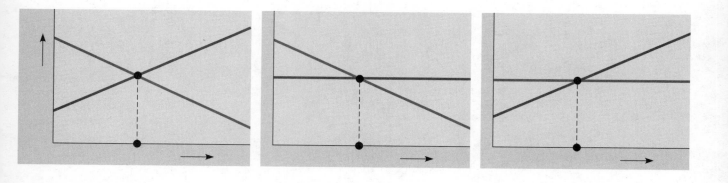

Figure 52.14 Determining equilibrium for population density, page 1148

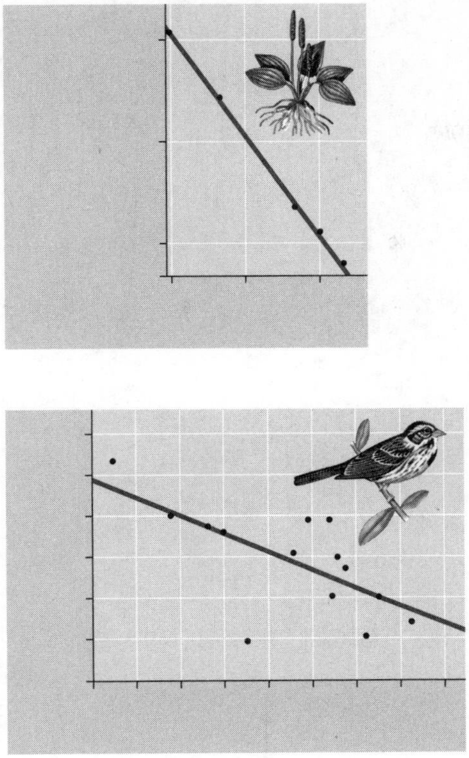

Figure 52.15 Decreased reproduction at high population densities, page 1149

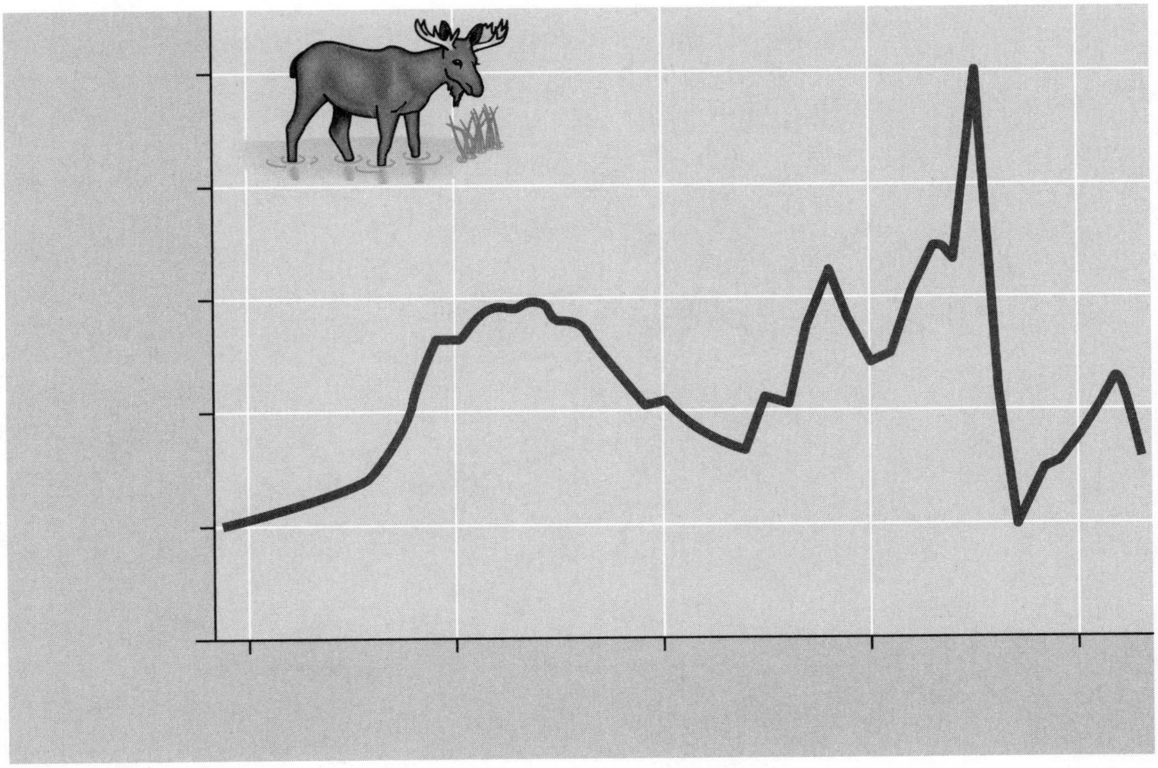

Figure 52.18 How stable is the Isle Royale moose population? page 1150

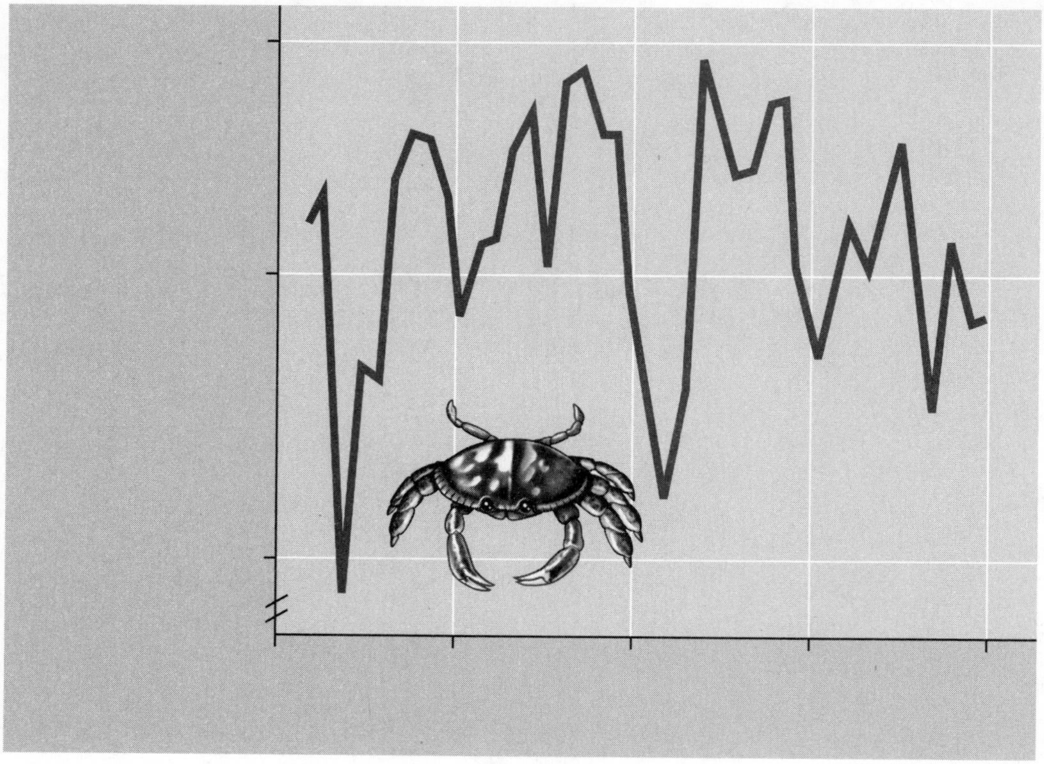

Figure 52.19 Extreme population fluctuations, page 1151

Figure 52.20 Song sparrow populations and immigration, page 1151

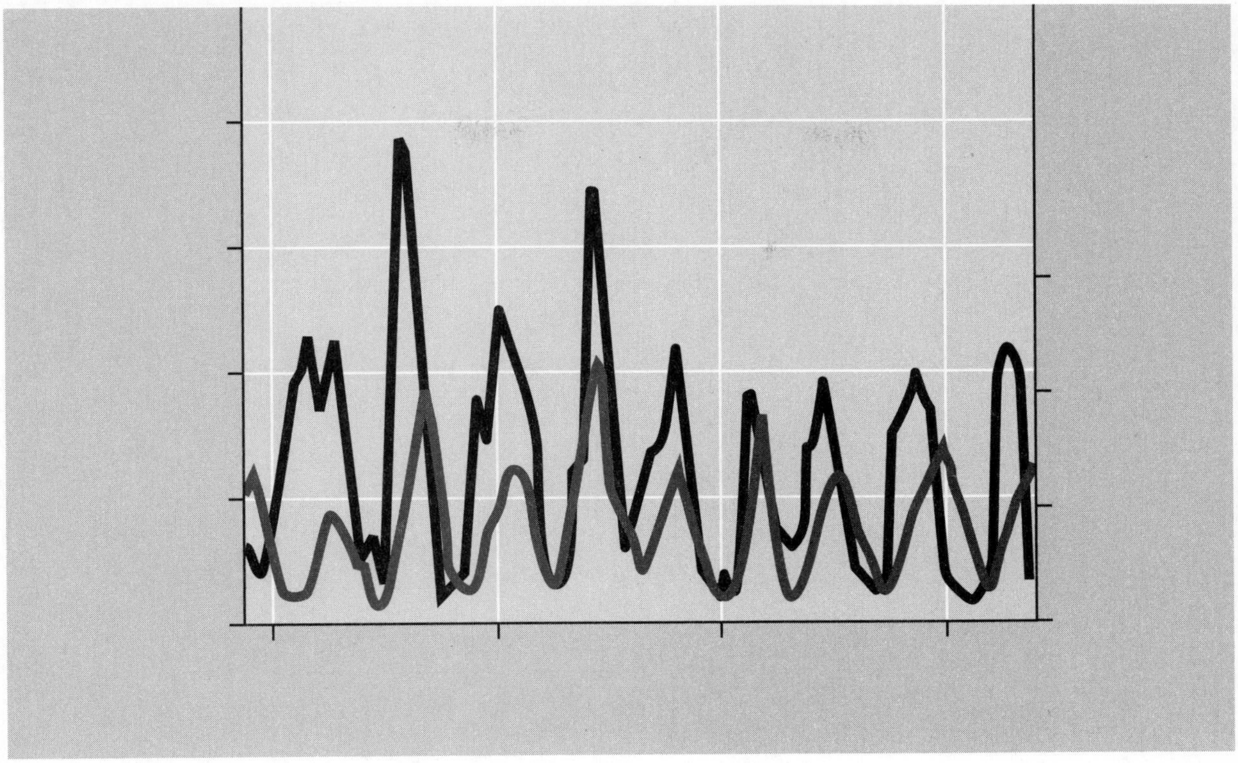

Figure 52.21 Population cycles in the snowshoe hare and lynx, page 1152

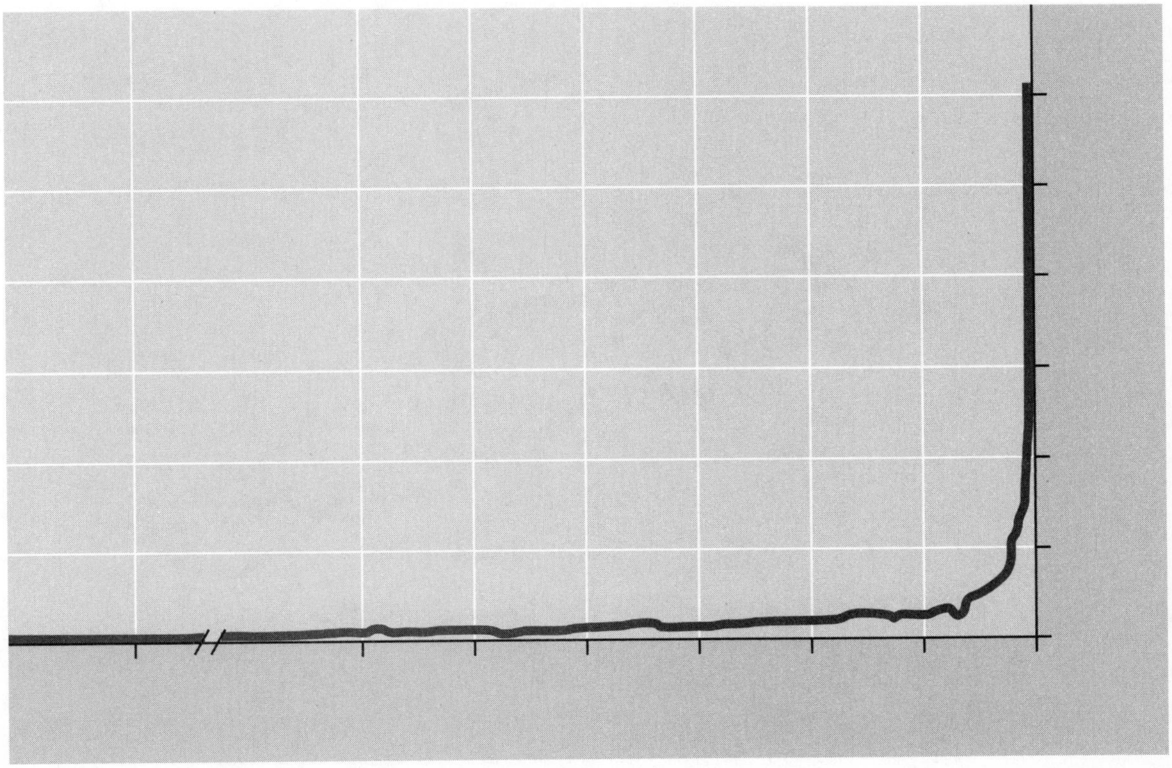

Figure 52.22 Human population growth, page 1153

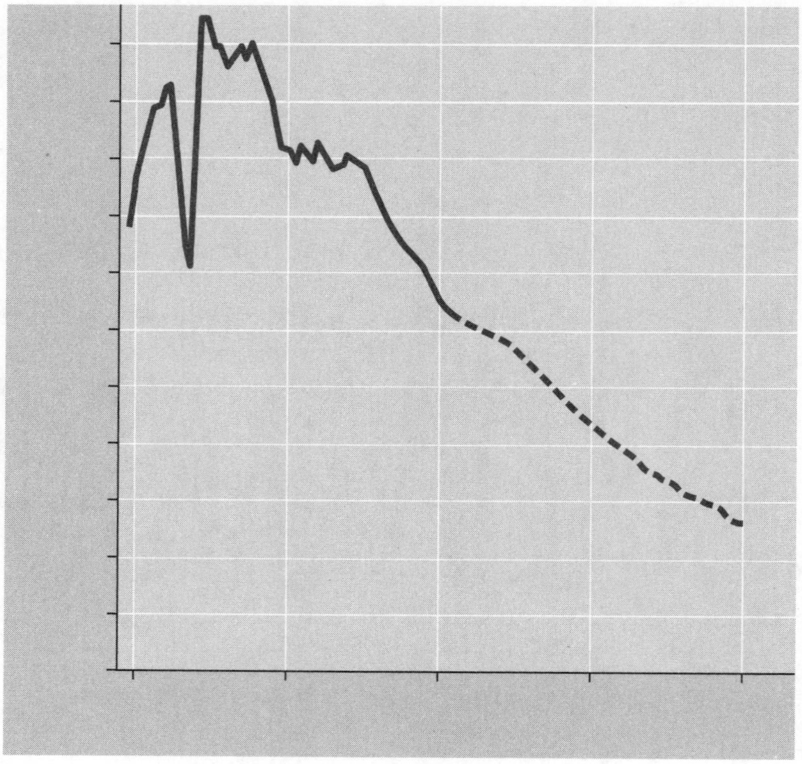

Figure 52.23 Percent increase in the global human population, page 1153

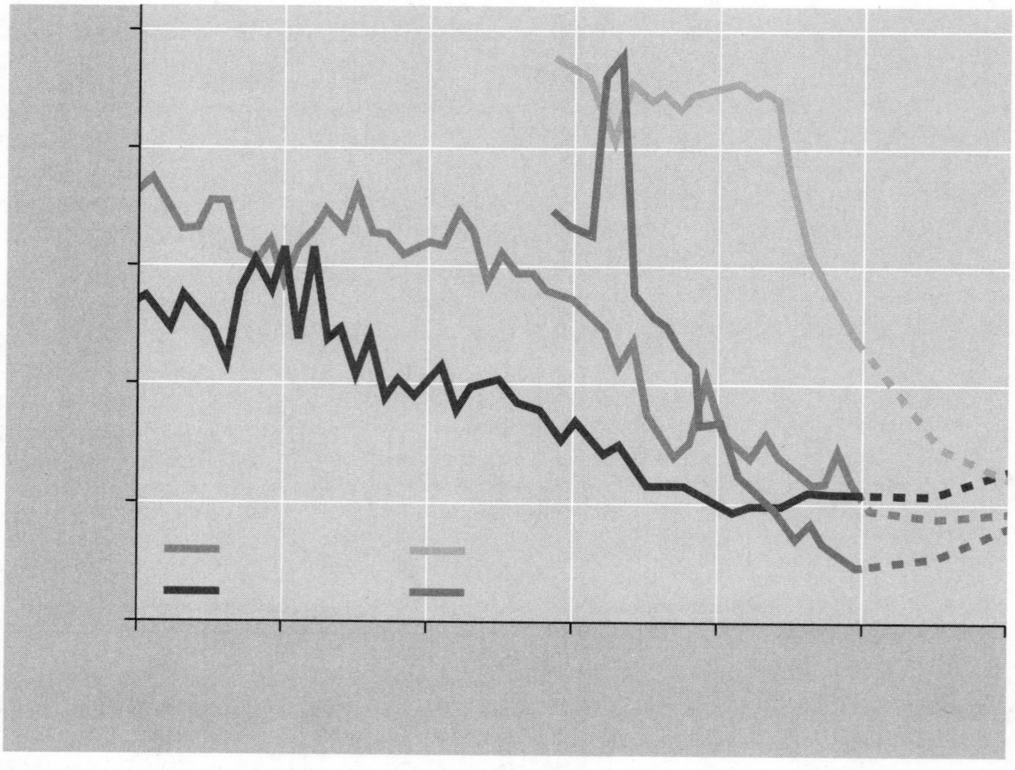

Figure 52.24 Demographic transition in Sweden and Mexico, 1750–2050 (data as of 2003), page 1153

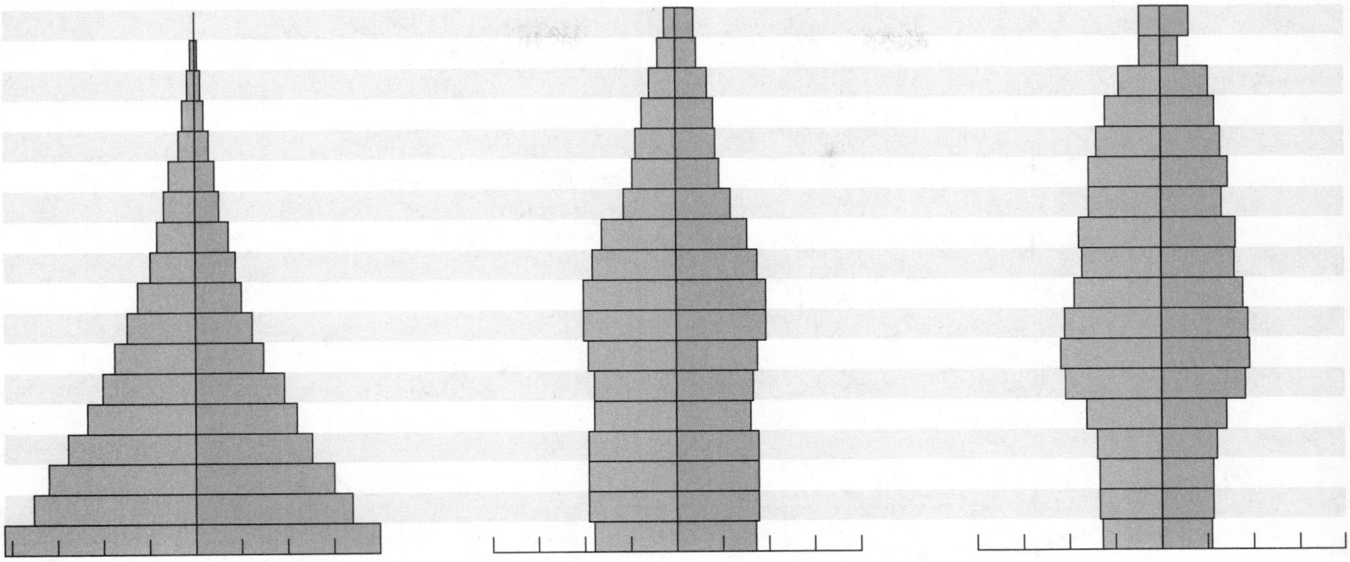

Figure 52.25 Age-structure pyramids for the human population of three countries, page 1154

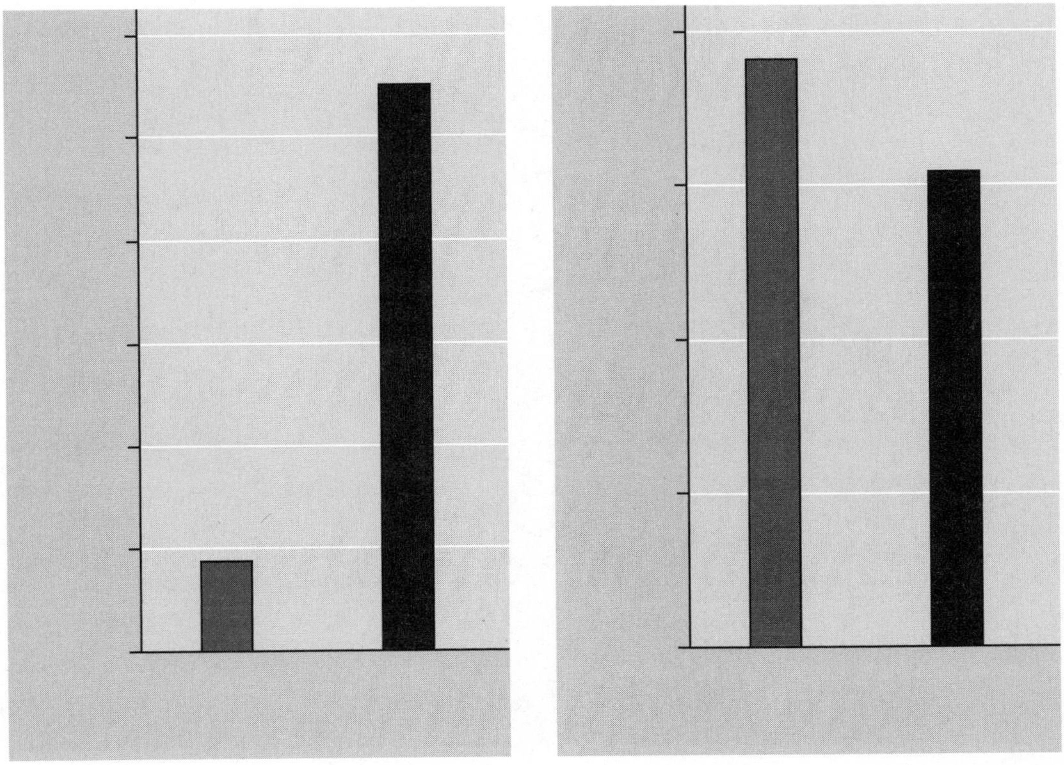

Figure 52.26 Infant mortality and life expectancy at birth in developed and developing countries, page 1155

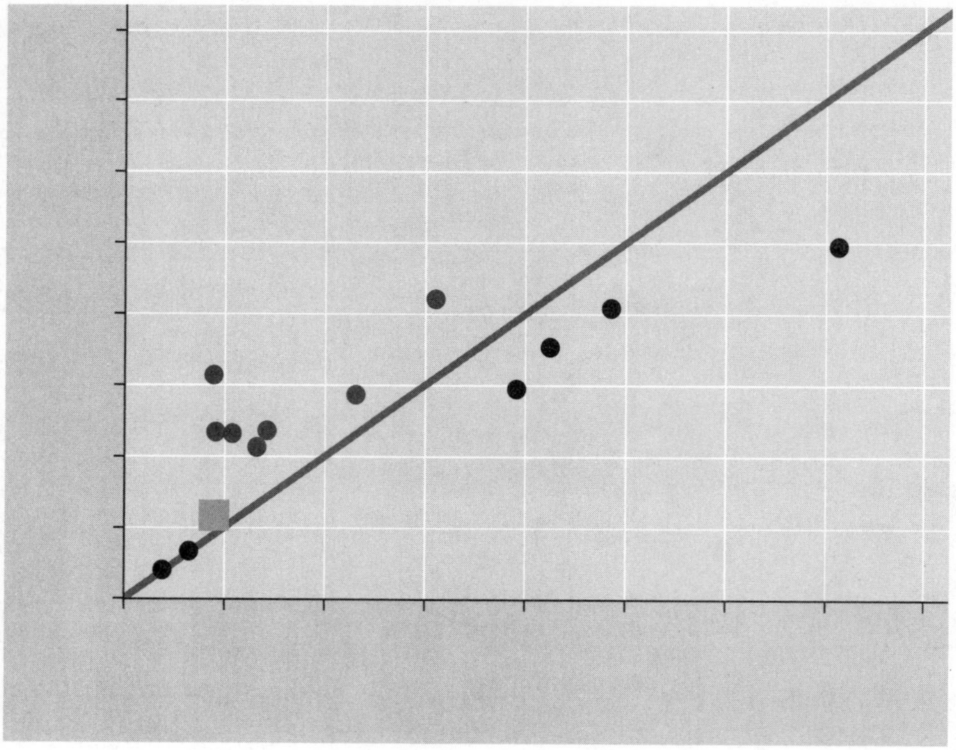

Figure 52.27 Ecological footprint in relation to available ecological capacity, page 1156

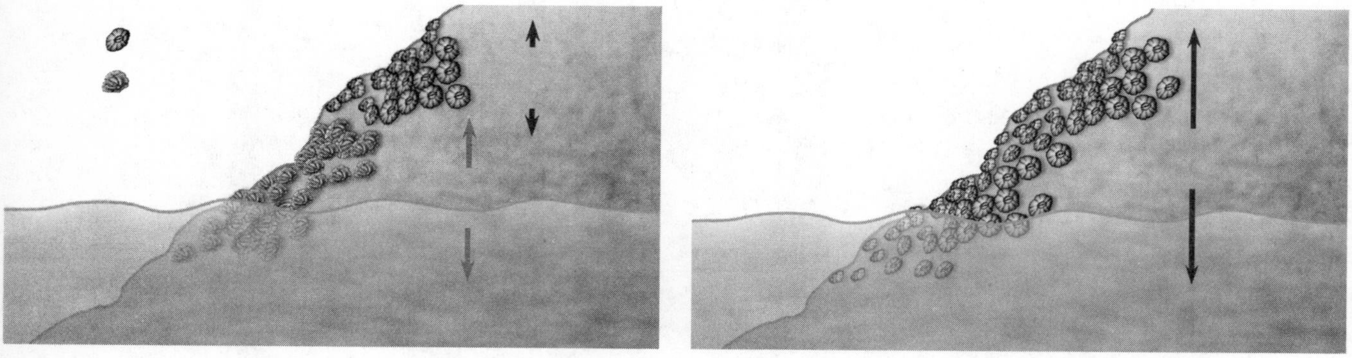

Figure 53.2 Can a species' niche be influenced by interspecific competition? page 1160

Figure 53.3 Resource partitioning among lizards in the Dominican Republic, page 1161

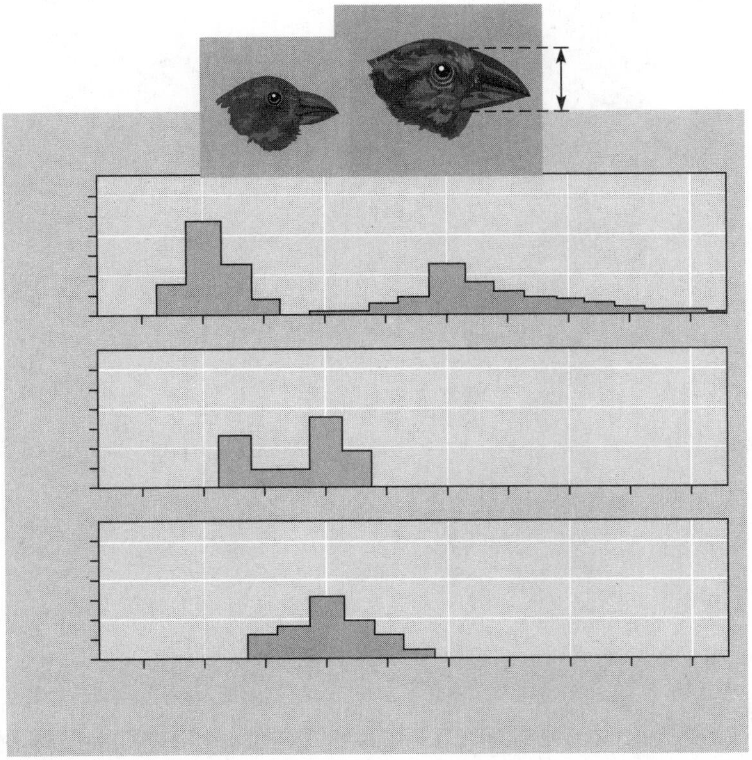

Figure 53.4 Character displacement: indirect evidence of past competition, page 1161

Figure 53.11 Which forest is more diverse? page 1165

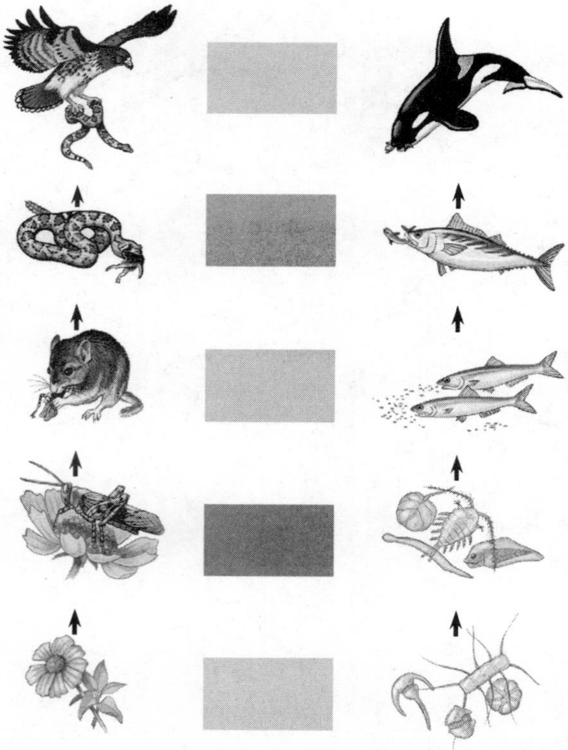

Figure 53.12 Examples of terrestrial and marine food chains, page 1166

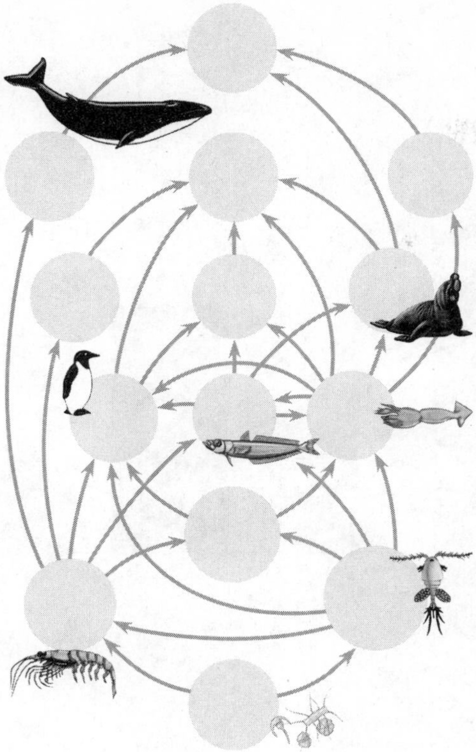

Figure 53.13 An antarctic marine food web, page 1166

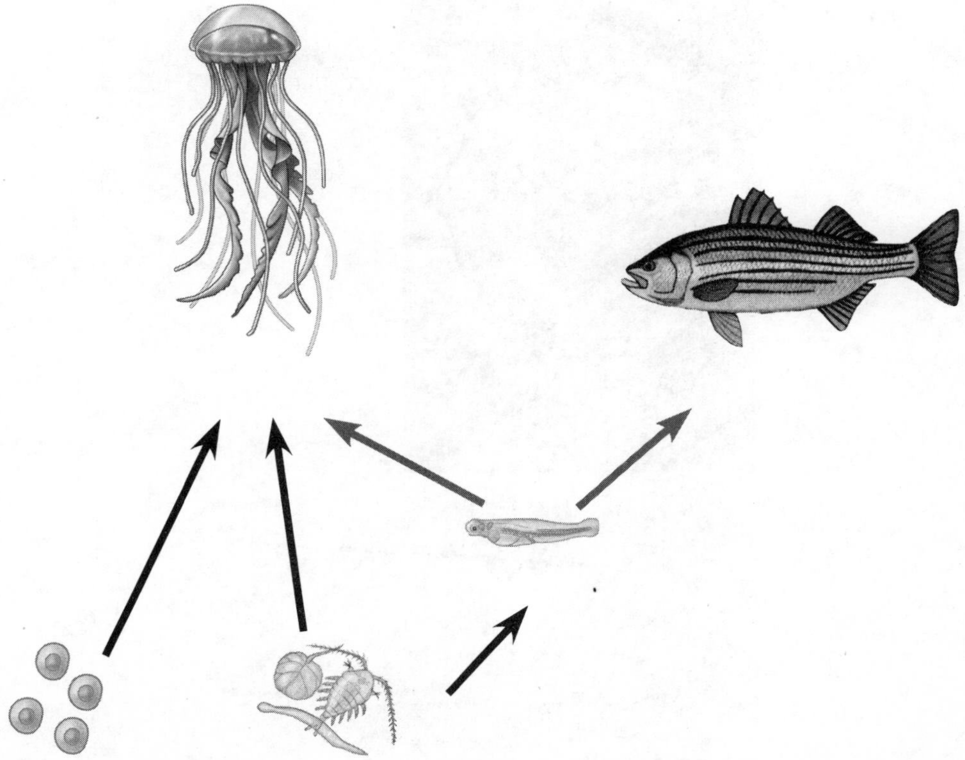

Figure 53.14 Partial food web for the Chesapeake Bay estuary on the U.S. Atlantic coast, page 1167

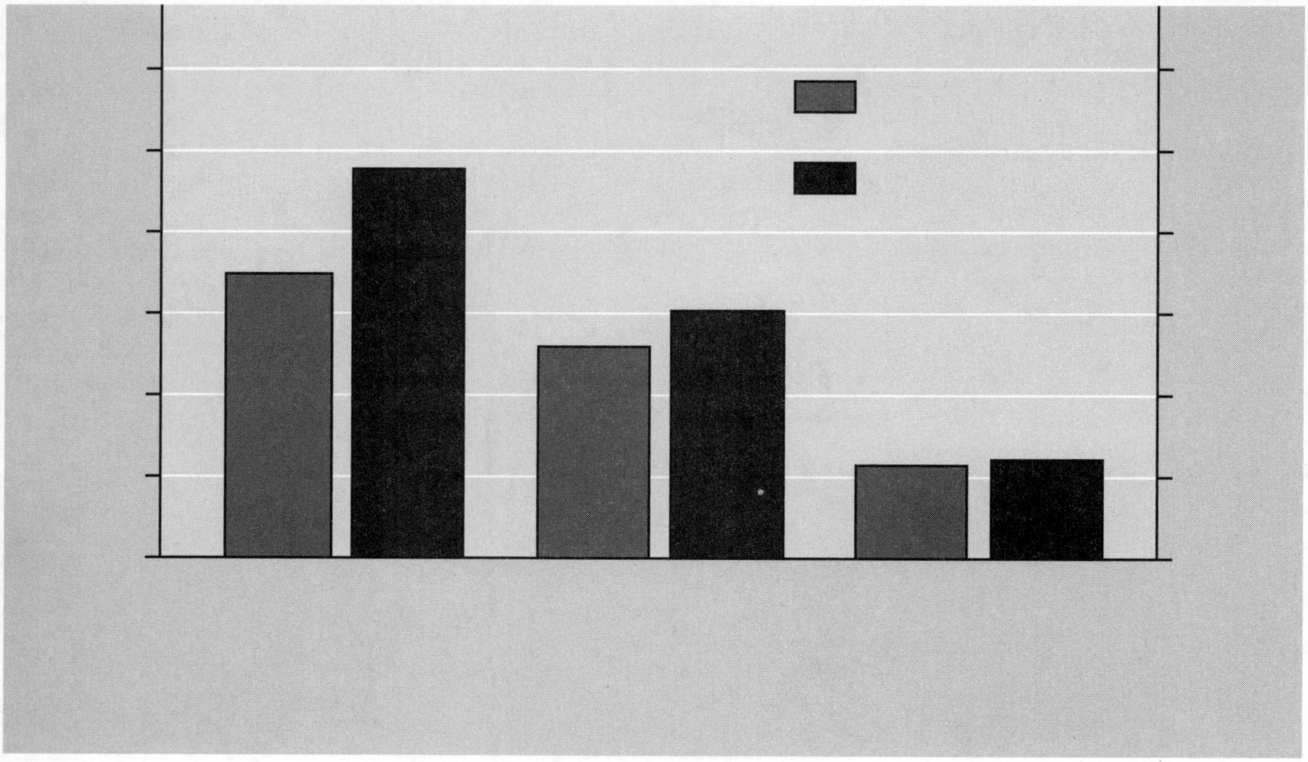

Figure 53.15 Test of the energetic hypothesis for the restriction of food chain length, page 1167

Figure 53.16 Testing a keystone predator hypothesis, page 1168

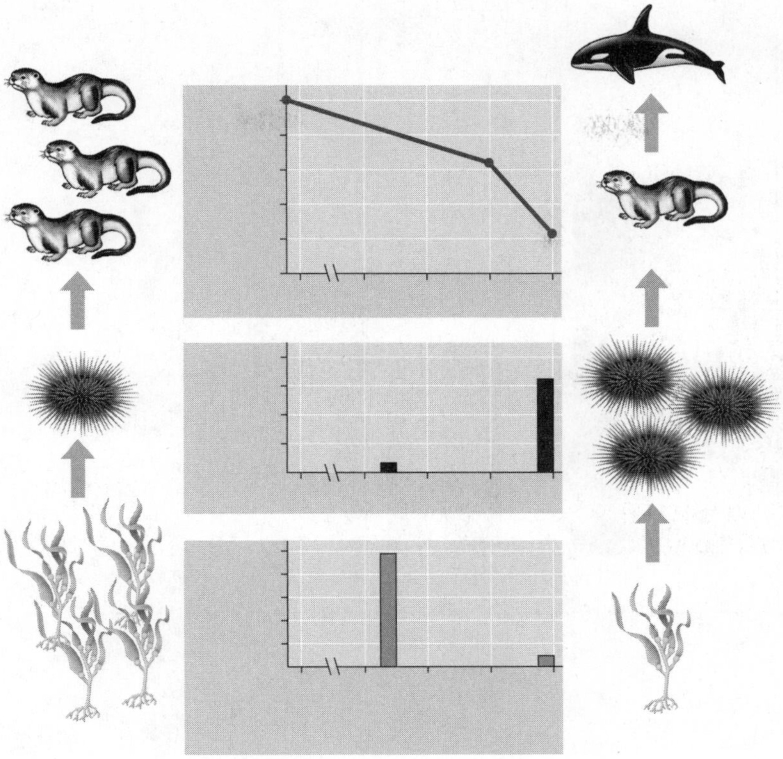

Figure 53.17 Sea otters as keystone predators in the North Pacific, page 1169

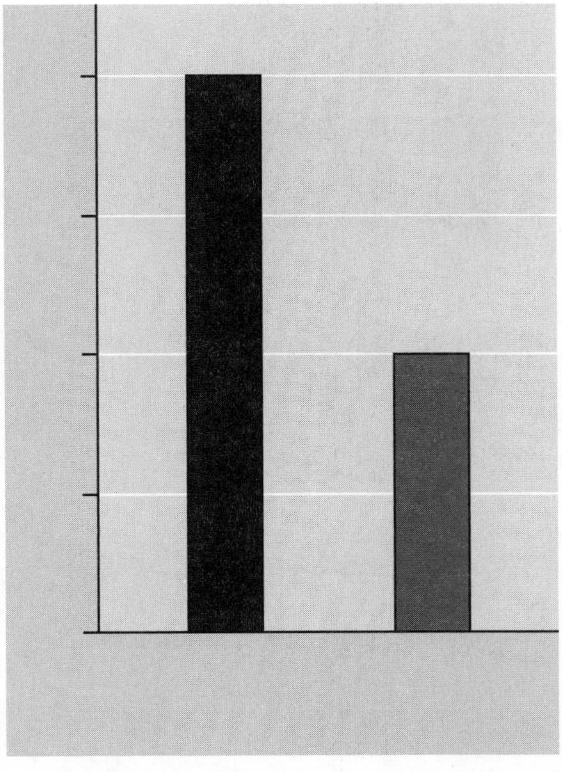

Figure 53.19 Facilitation by black rush (*Juncus gerardi*) in New England salt marshes, page 1170

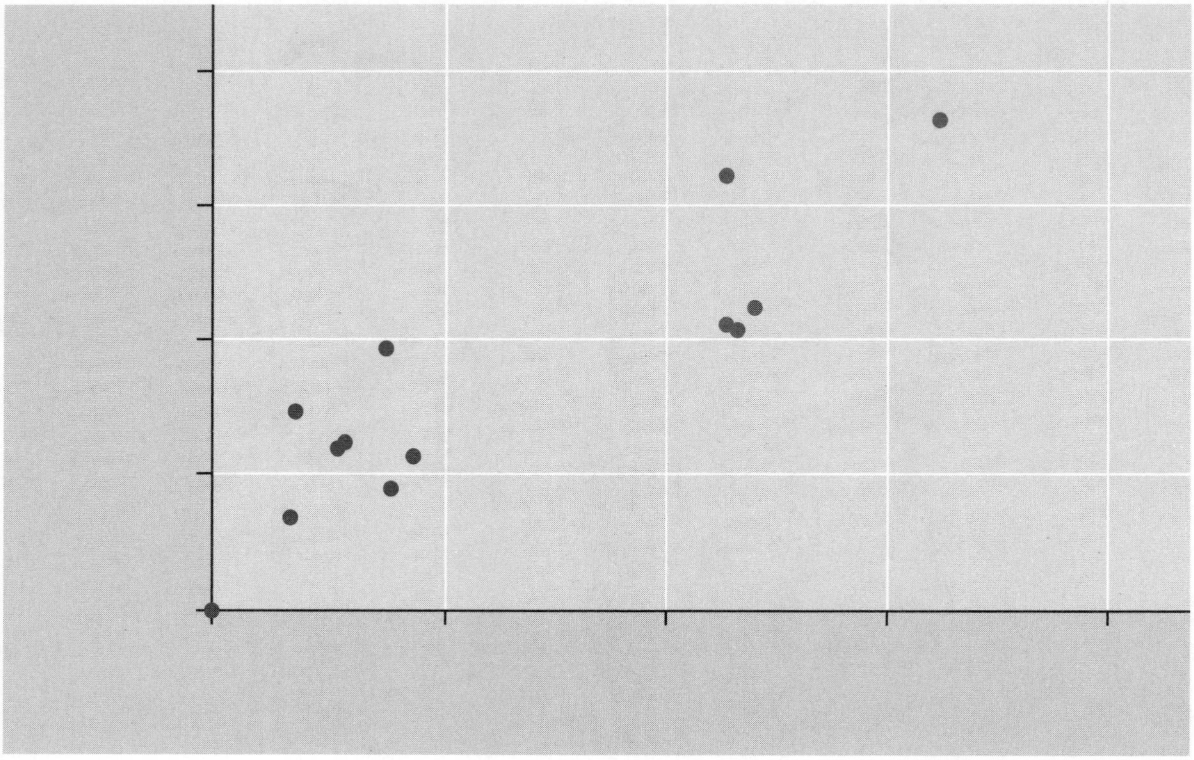

Figure 53.20 Relationship between rainfall and herbaceous plant cover in a desert shrub community in Chile, page 1171

Figure UN1 Biomanipulation diagram, page 1171

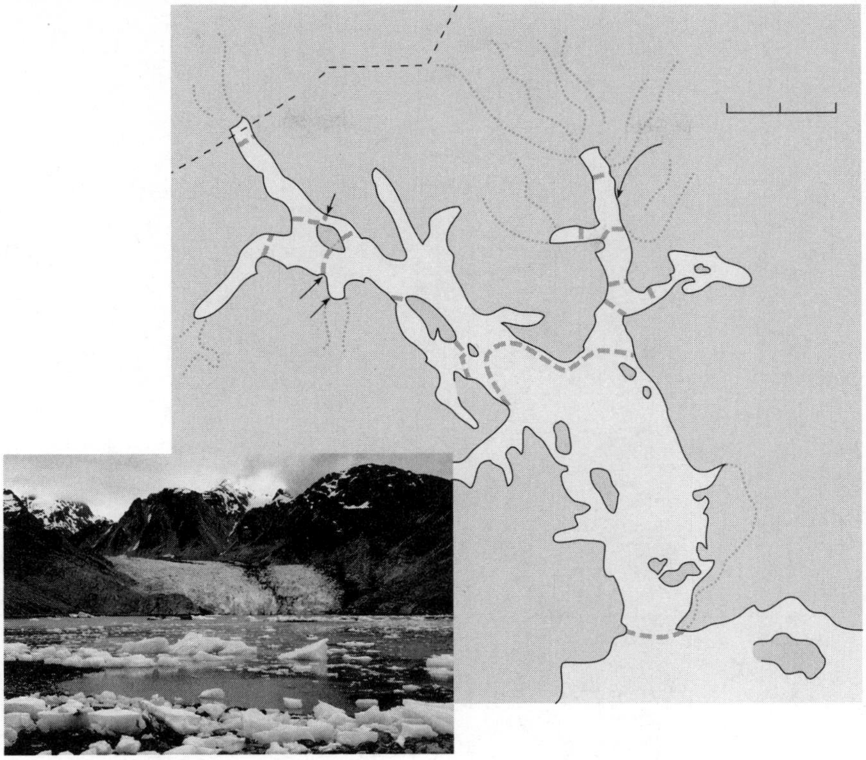

Figure 53.23 A glacial retreat in southeastern Alaska, page 1174

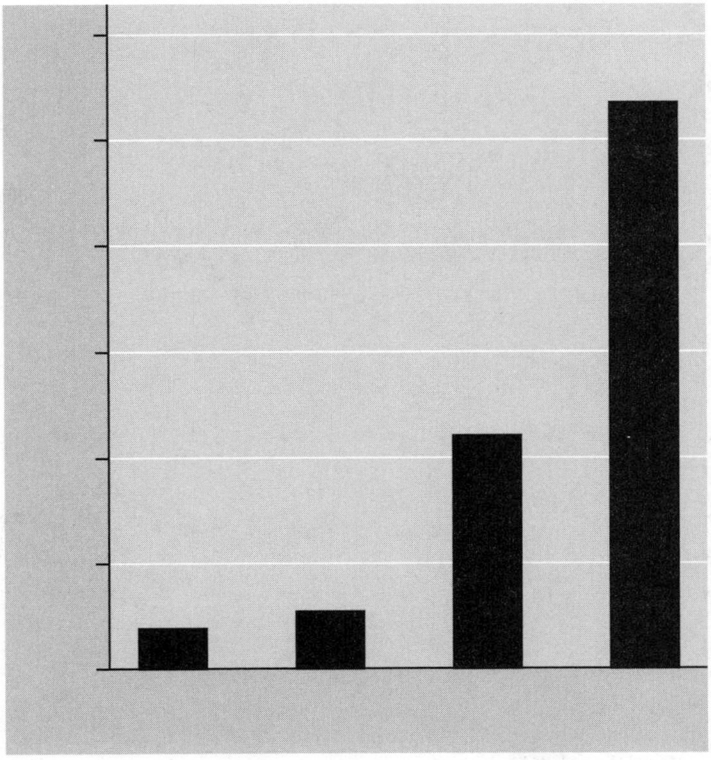

Figure 53.24 Changes in plant community structure and soil nitrogen during succession at Glacier Bay, Alaska, page 1175

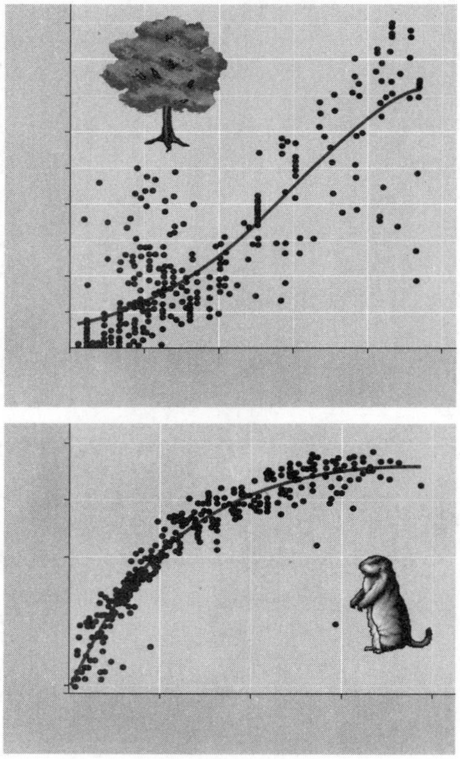

Figure 53.25 Energy, water, and species richness, page 1176

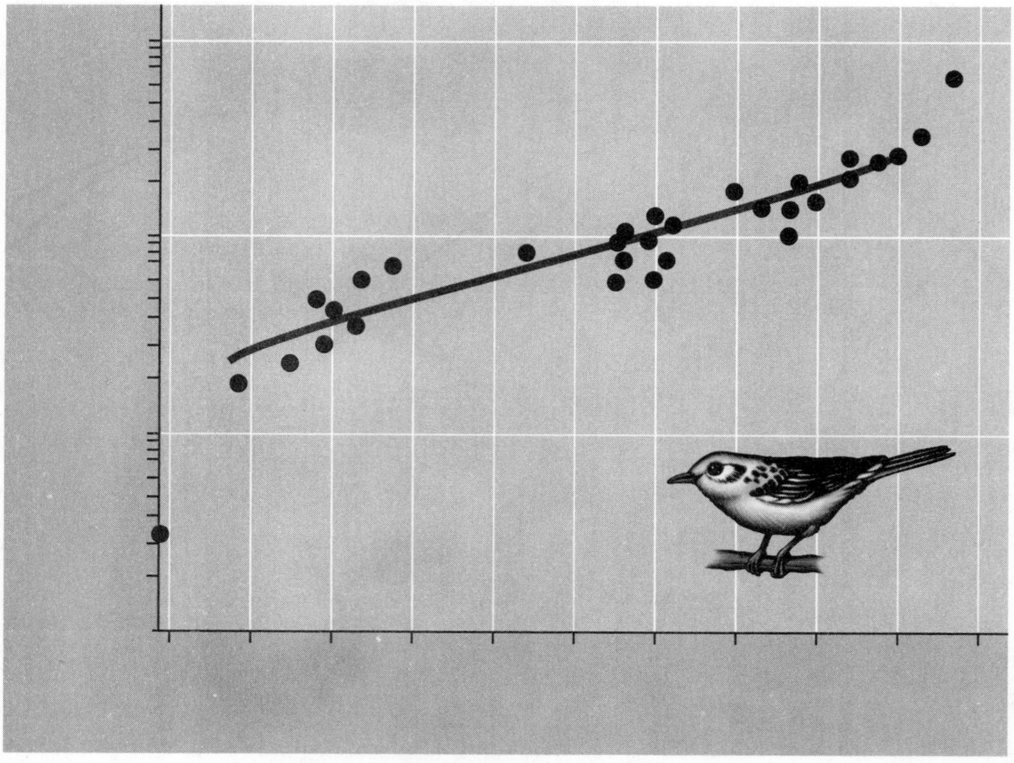

Figure 53.26 Species-area curve for North American breeding birds, page 1177

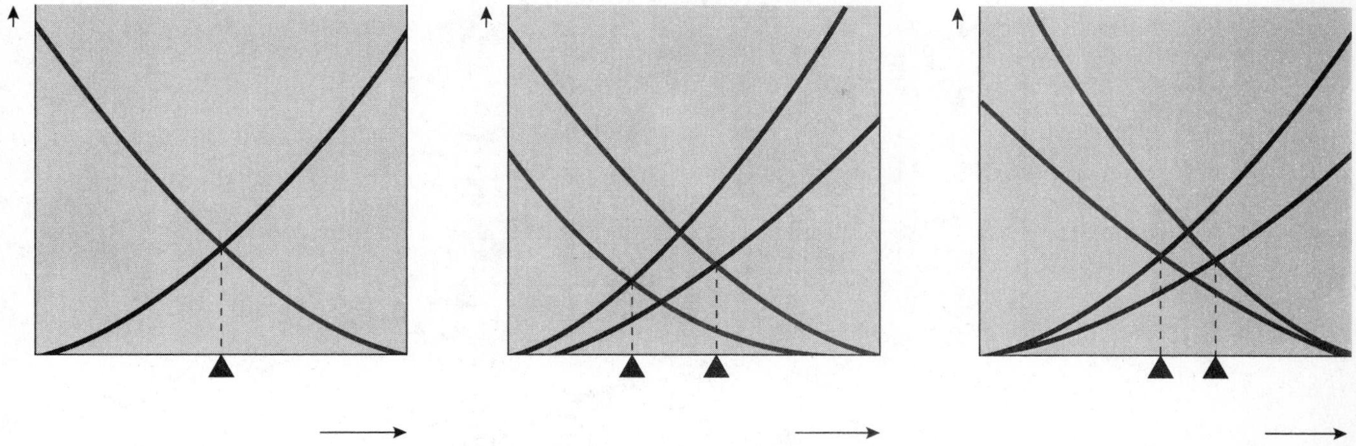

Figure 53.27 The equilibrium model of island biogeography, page 1177

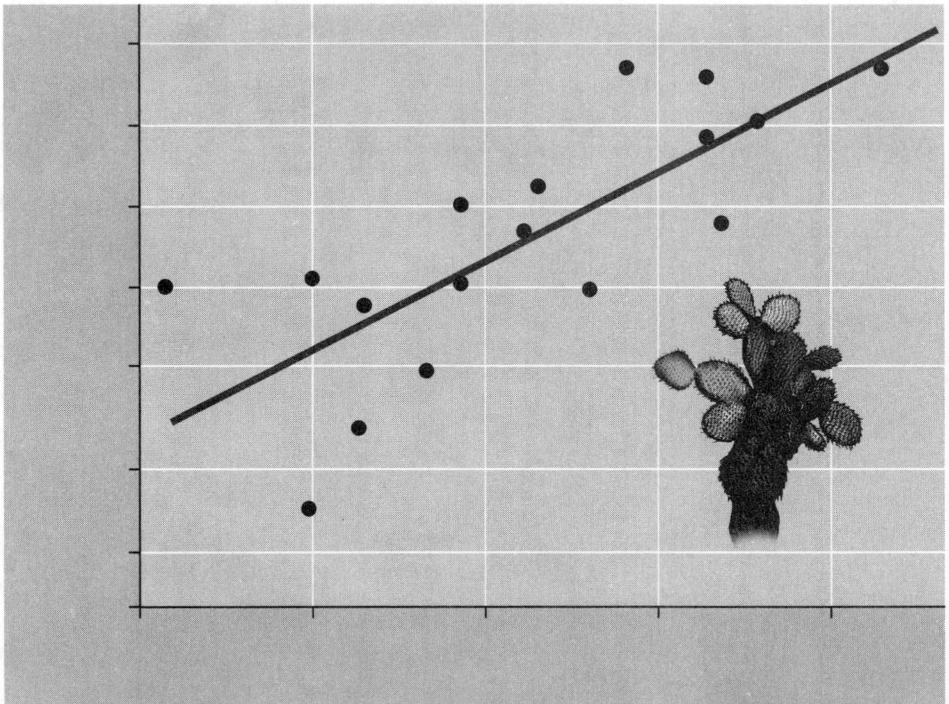

Figure 53.28 How does species richness relate to area? page 1178

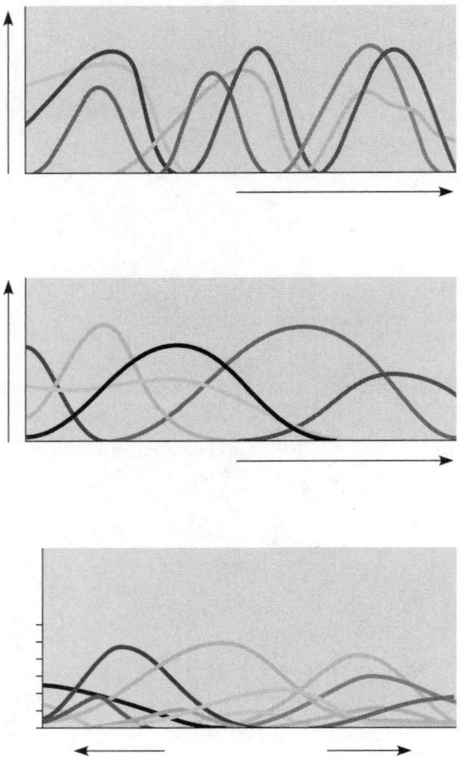

Figure 53.29 Testing the integrated and individualistic hypotheses of communities, page 1179

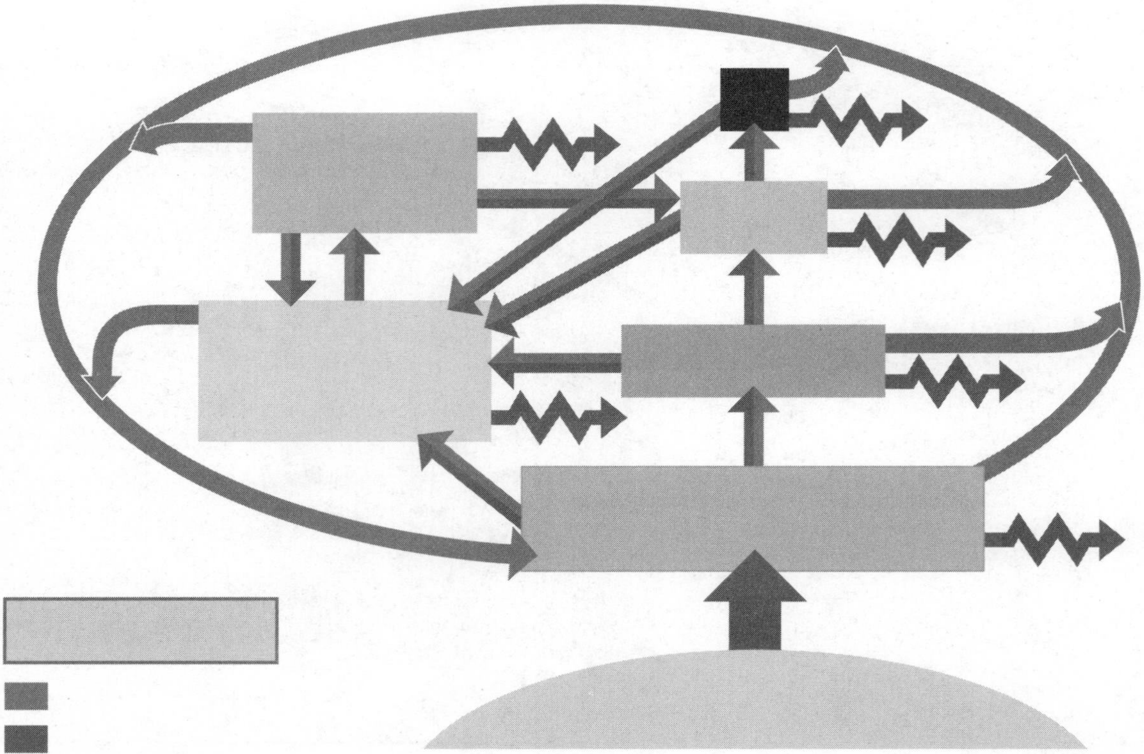

Figure 54.2 An overview of energy and nutrient dynamics in an ecosystem, page 1185

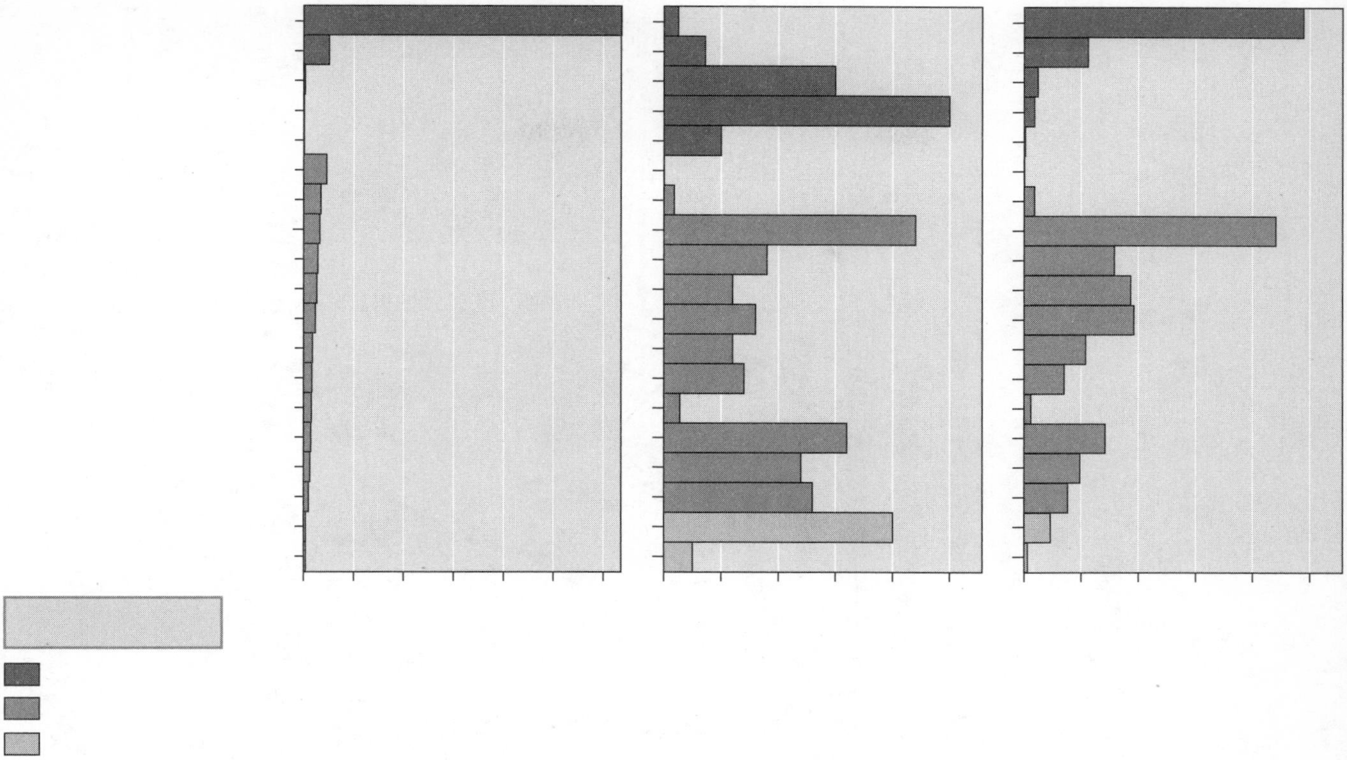

Figure 54.4 Net primary production of different ecosystems, page 1187

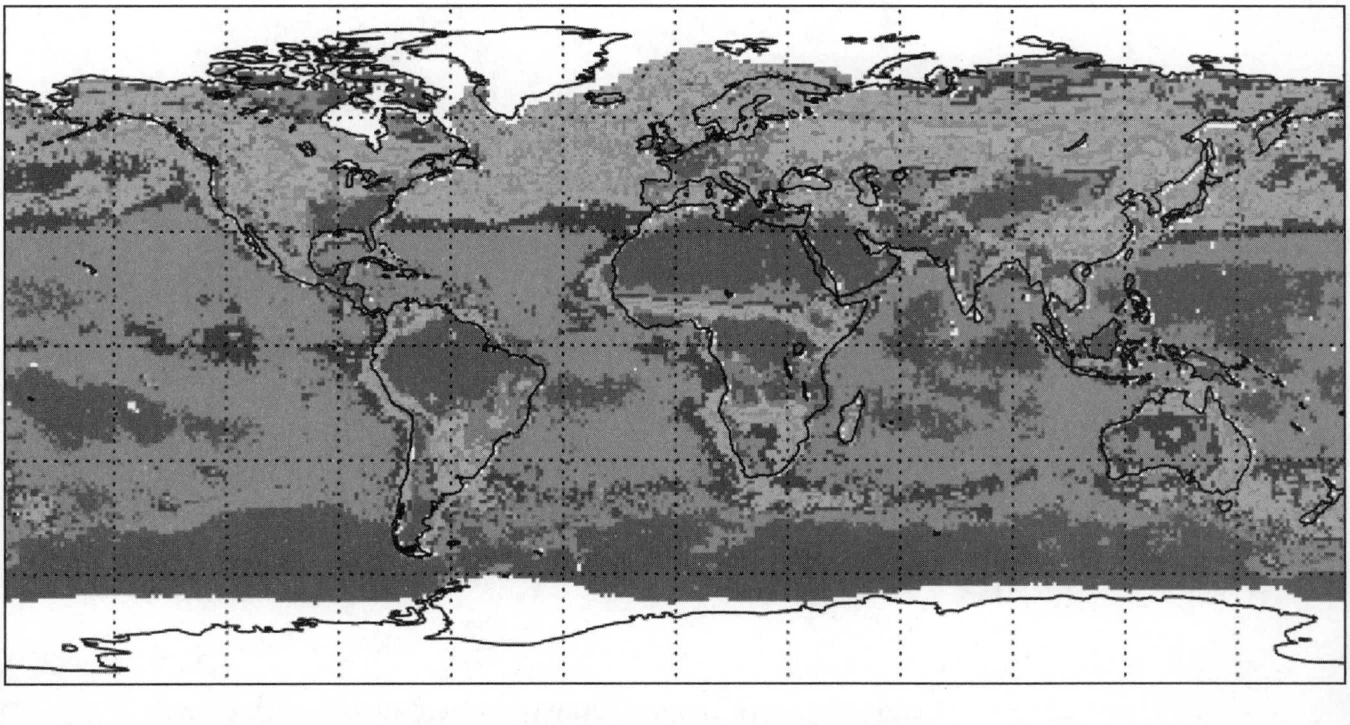

Figure 54.5 Regional annual net primary production for Earth, page 1188

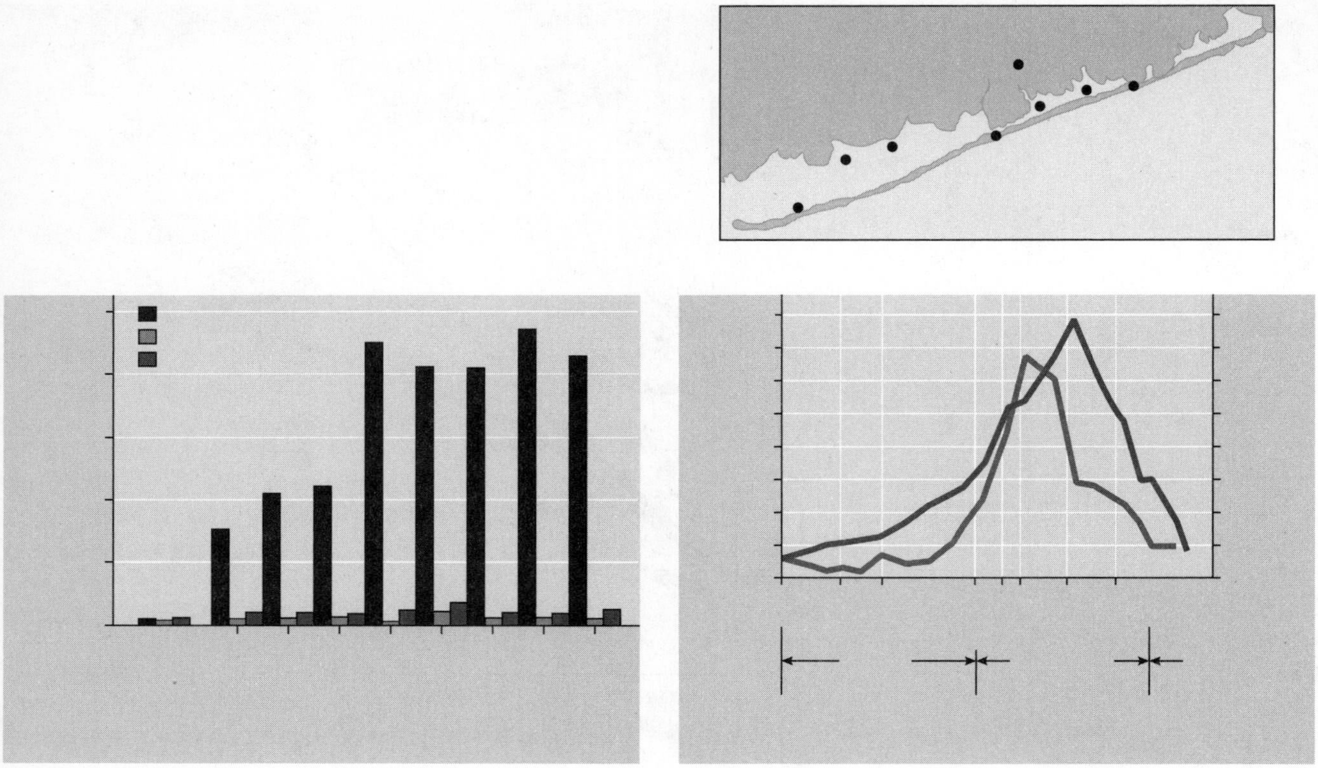

Figure 54.6 Which nutrient limits phytoplankton production along the coast of Long Island? page 1189

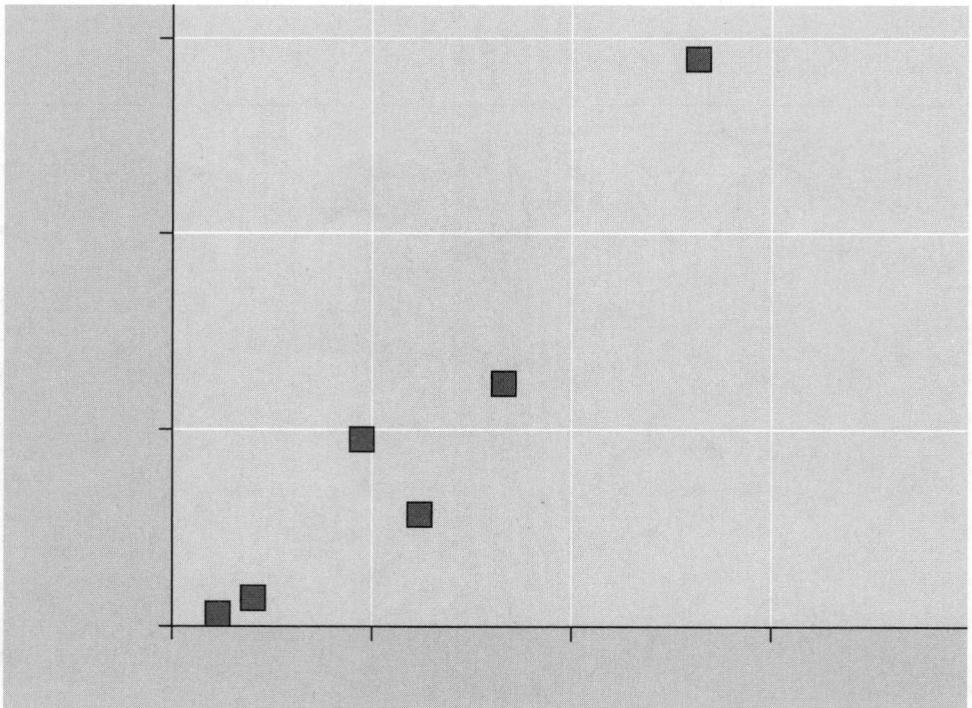

Figure 54.8 Actual evapotranspiration (moisture and temperature) related to terrestrial primary production in selected ecosystems, page 1190

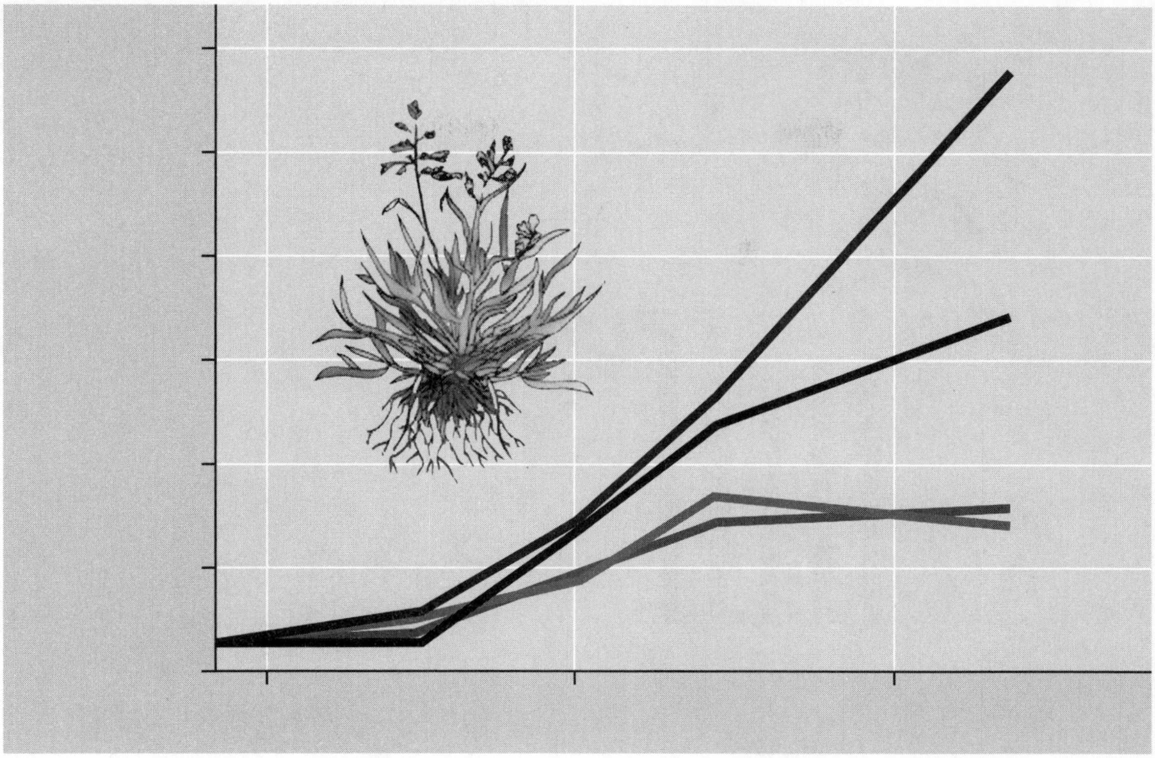

Figure 54.9 Is phosphorus or nitrogen the limiting nutrient in a Hudson Bay salt marsh? page 1191

Figure 54.10 Energy partitioning within a link of the food chain, page 1191

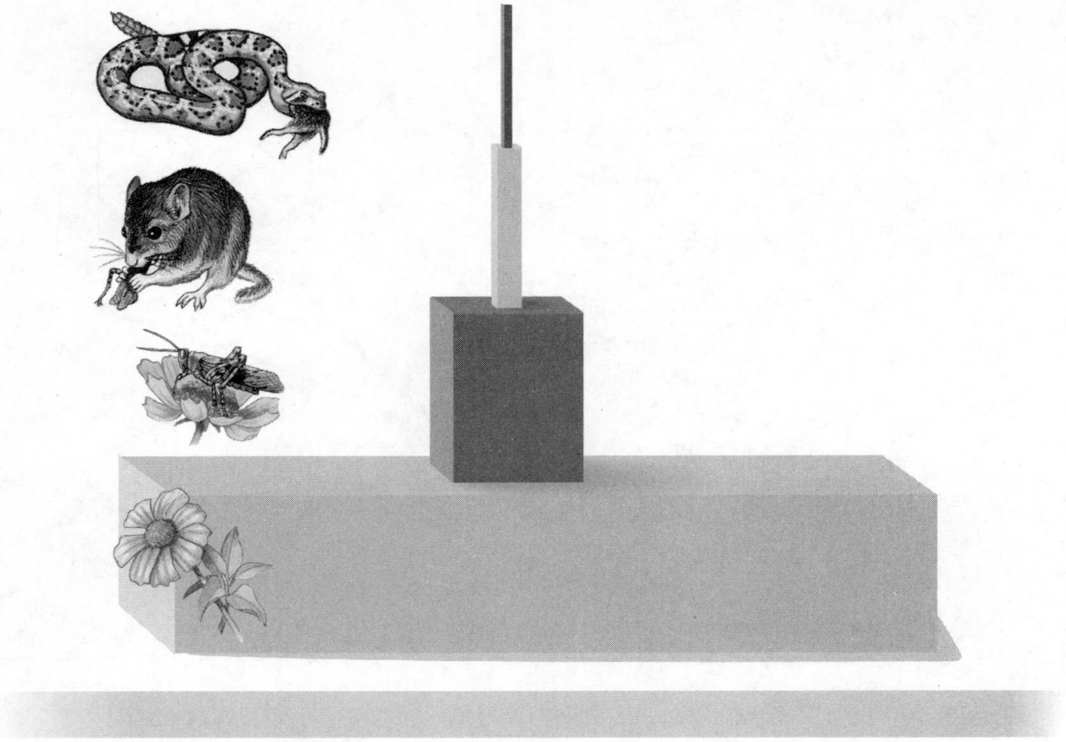

Figure 54.11 An idealized pyramid of net production, page 1192

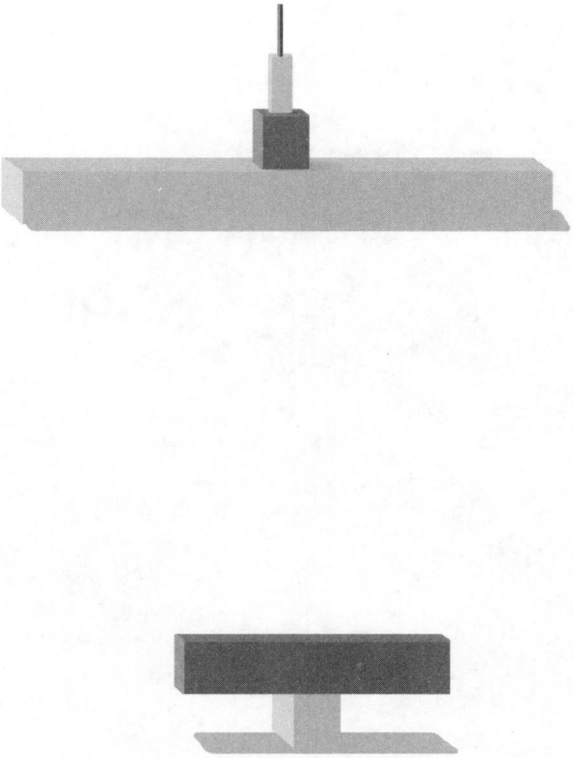

Figure 54.12 Pyramids of biomass (standing crop), page 1193

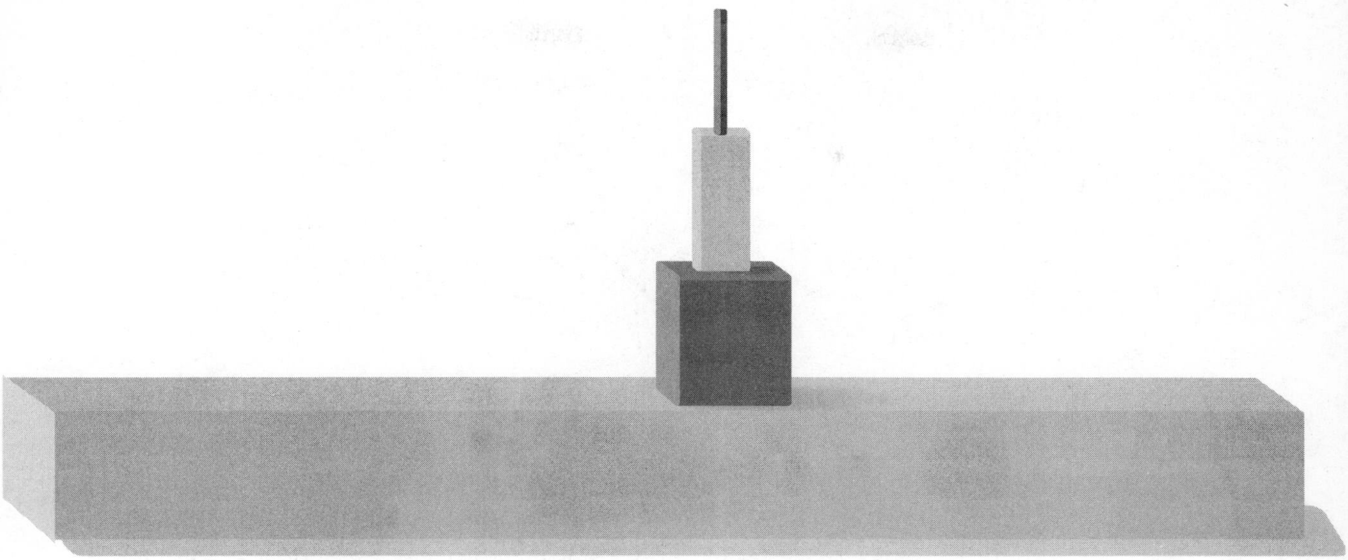

Figure 54.13 A pyramid of numbers, page 1193

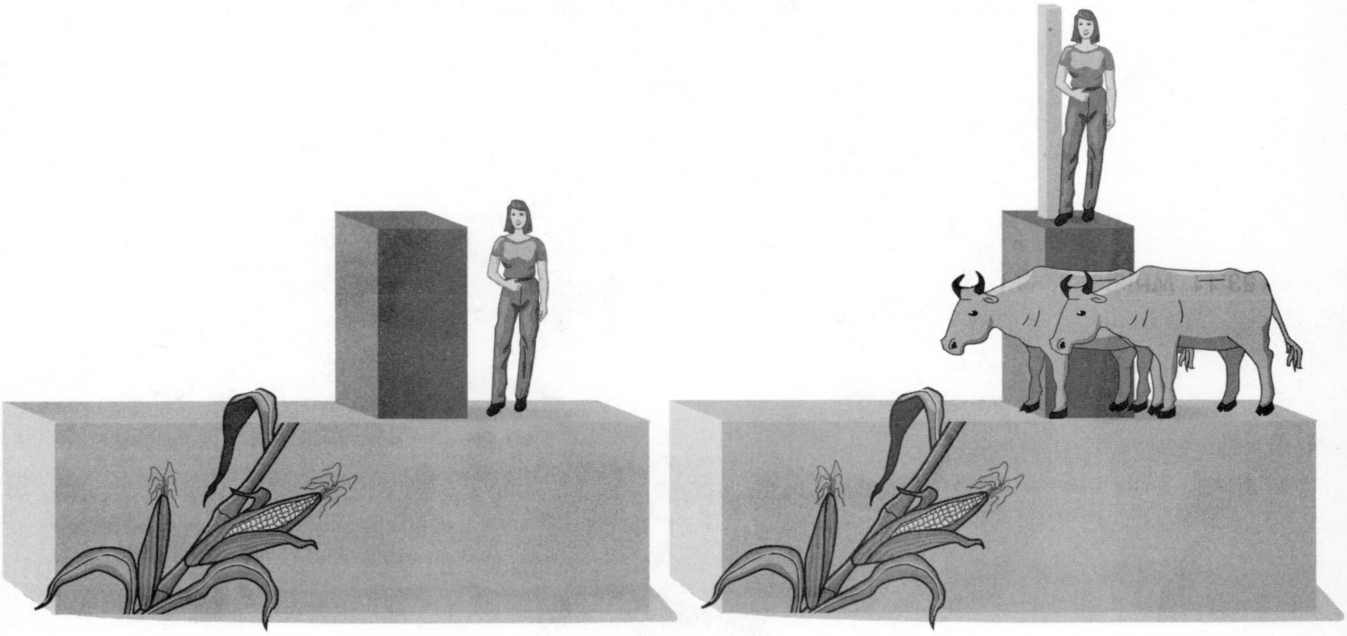

Figure 54.14 Relative food energy available to the human population at different trophic levels, page 1193

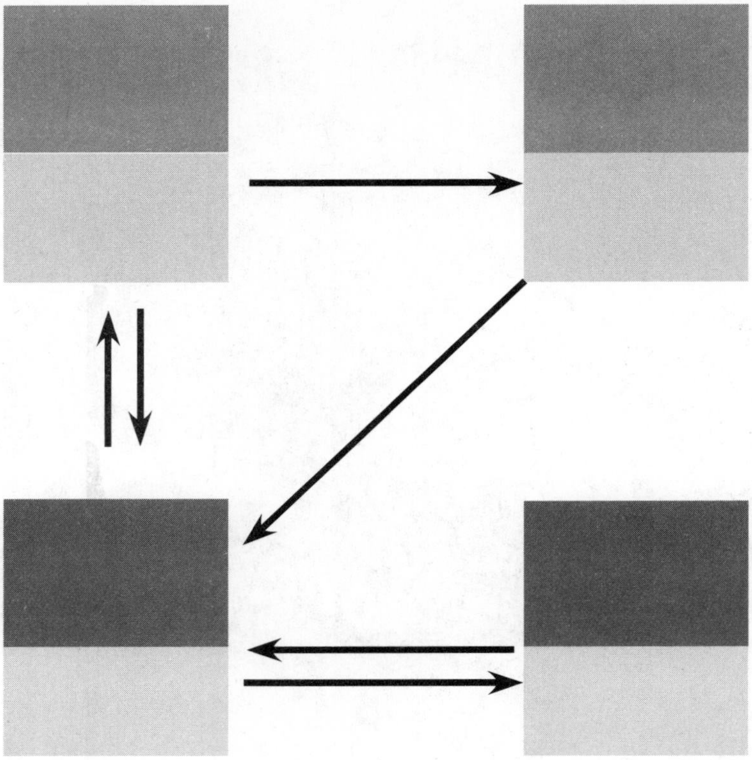

Figure 54.16 A general model of nutrient cycling, page 1195

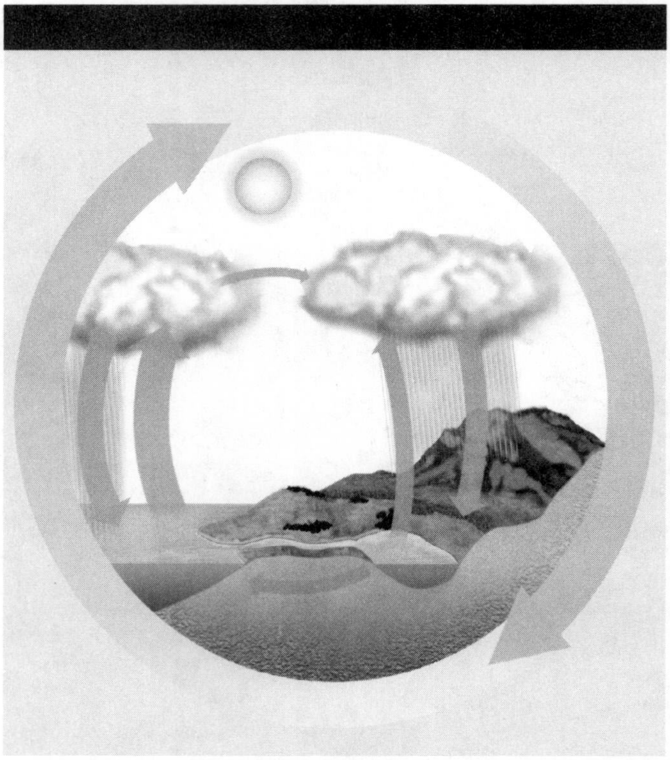

Figure 54.17 Nutrient cycles: the water cycle, page 1196

Figure 54.17 Nutrient cycles: the carbon cycle, page 1196

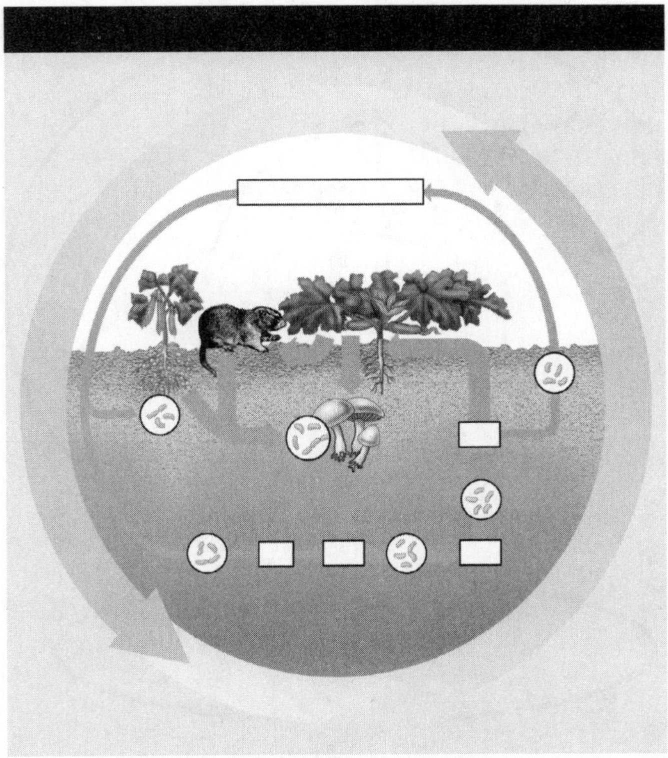

Figure 54.17 Nutrient cycles: the nitrogen cycle, page 1197

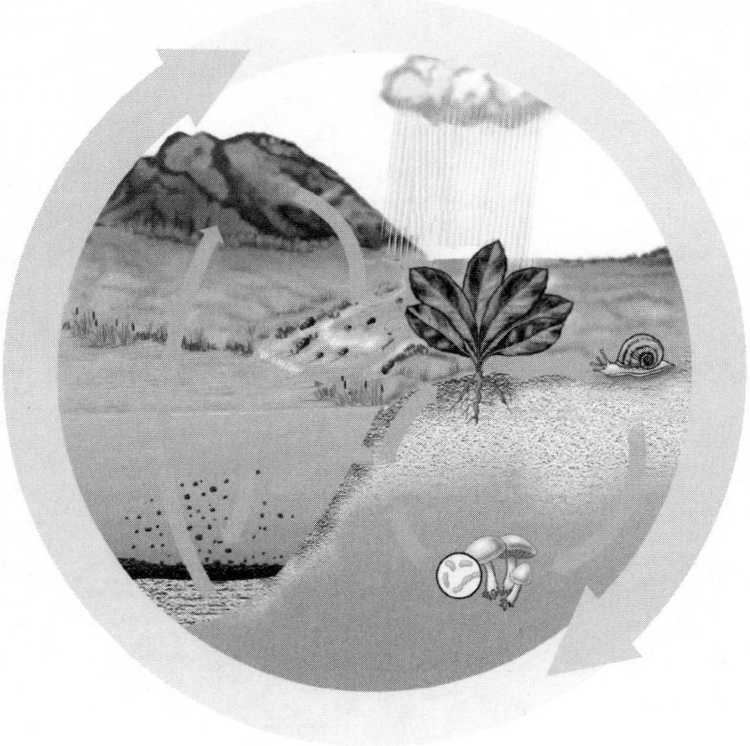

Figure 54.17 Nutrient cycles: the phosphorus cycle, page 1197

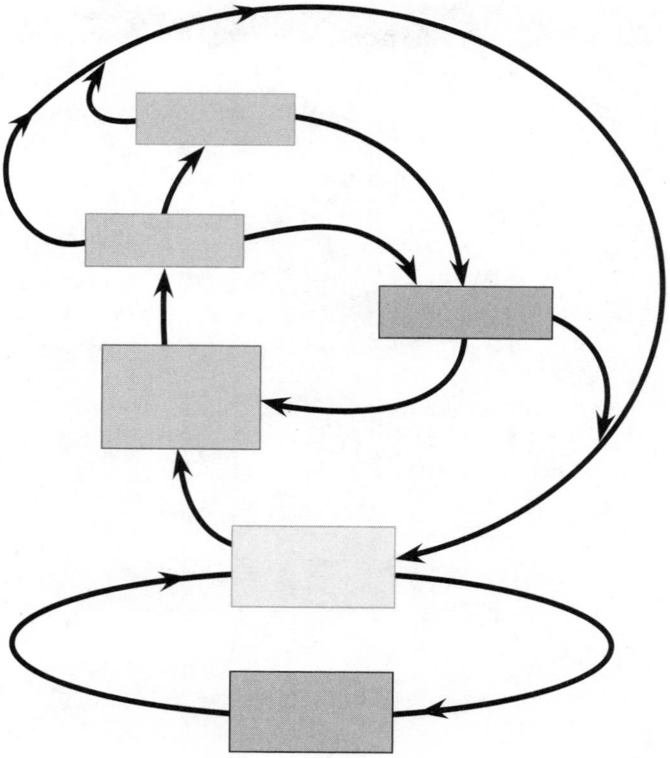

Figure 54.18 Review: Generalized scheme for biogeochemical cycles, page 1198

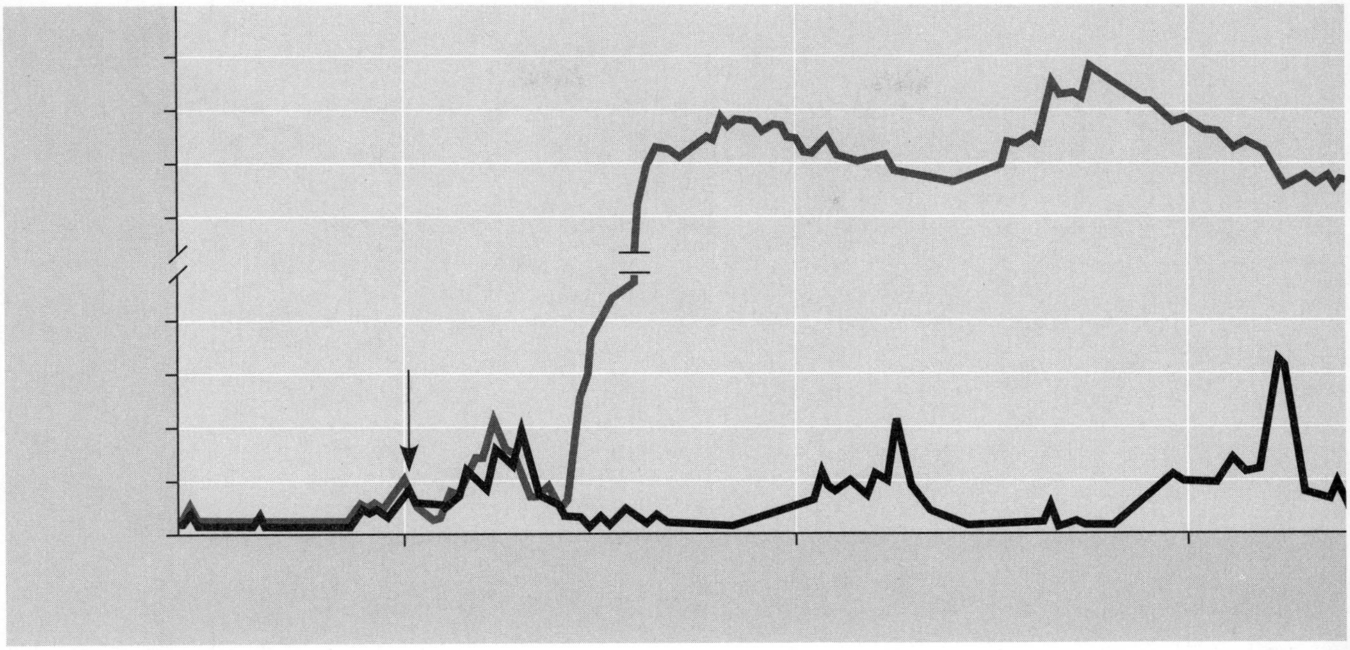

Figure 54.19 Nutrient cycling in the Hubbard Brook Experimental Forest: an example of long-term ecological research, page 1199

Figure 54.21 Distribution of acid precipitation in North America and Europe, 1980, page 1201

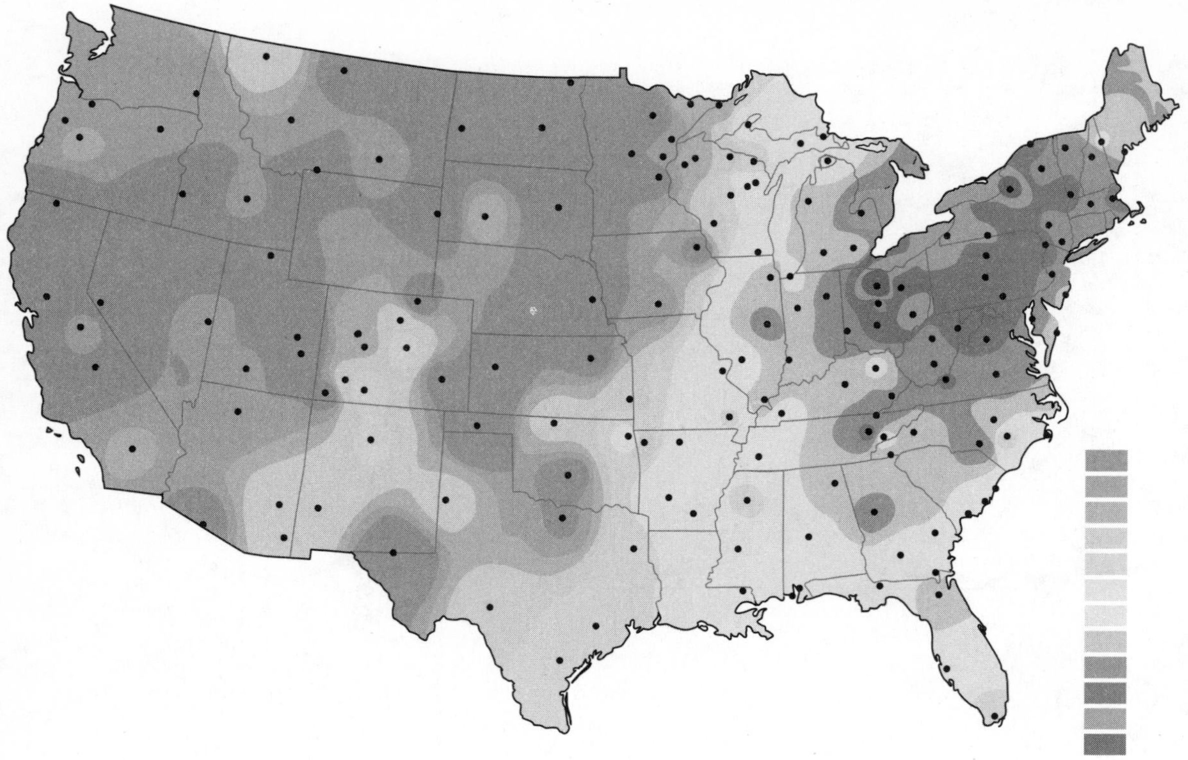

Figure 54.22 Average pH for precipitation in the contiguous United States in 2002, page 1202

Figure 54.23 Biological magnification of PCBs in a Great Lakes food web, page 1202

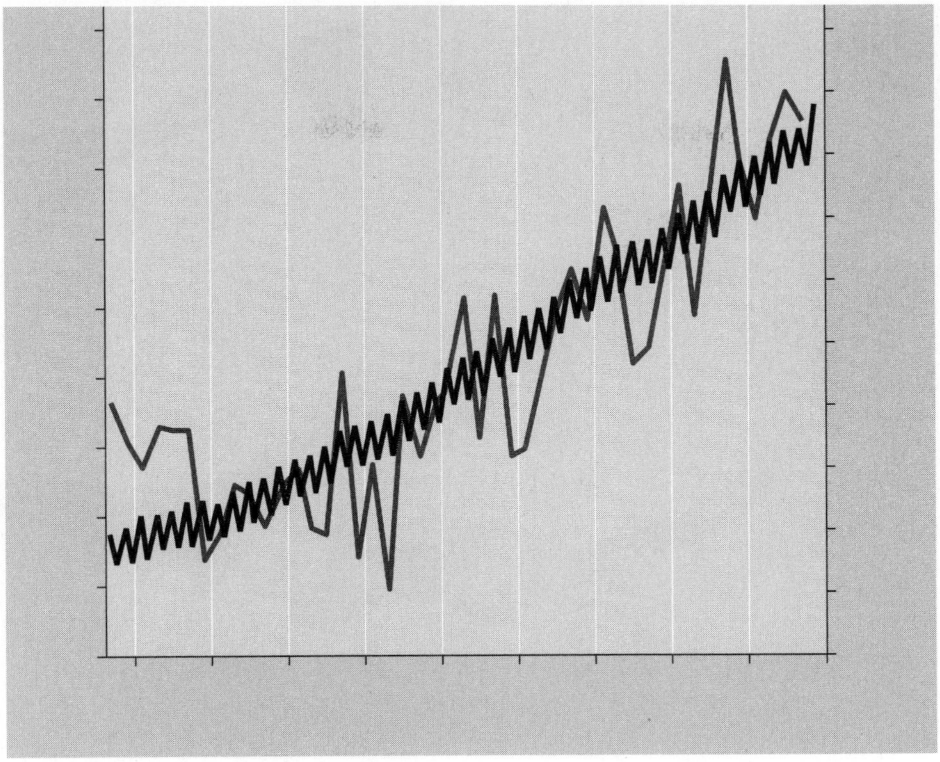

Figure 54.24 The increase in atmospheric carbon dioxide at Mauna Loa, Hawaii, and average global temperatures over land from 1958 to 2004, page 1203

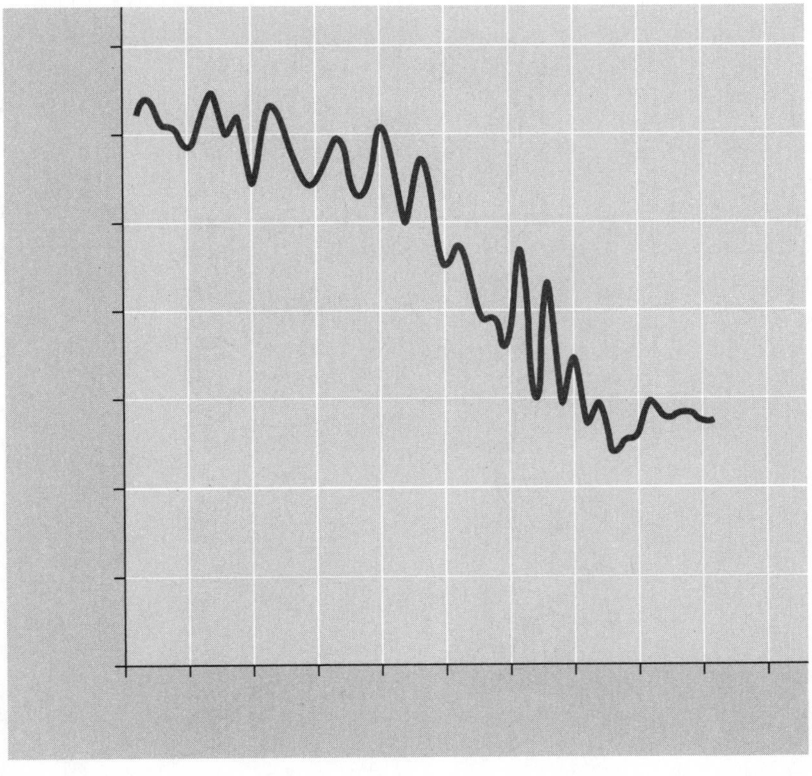

Figure 54.26 Thickness of the ozone layer over Antarctica in units called Dobsons, page 1205

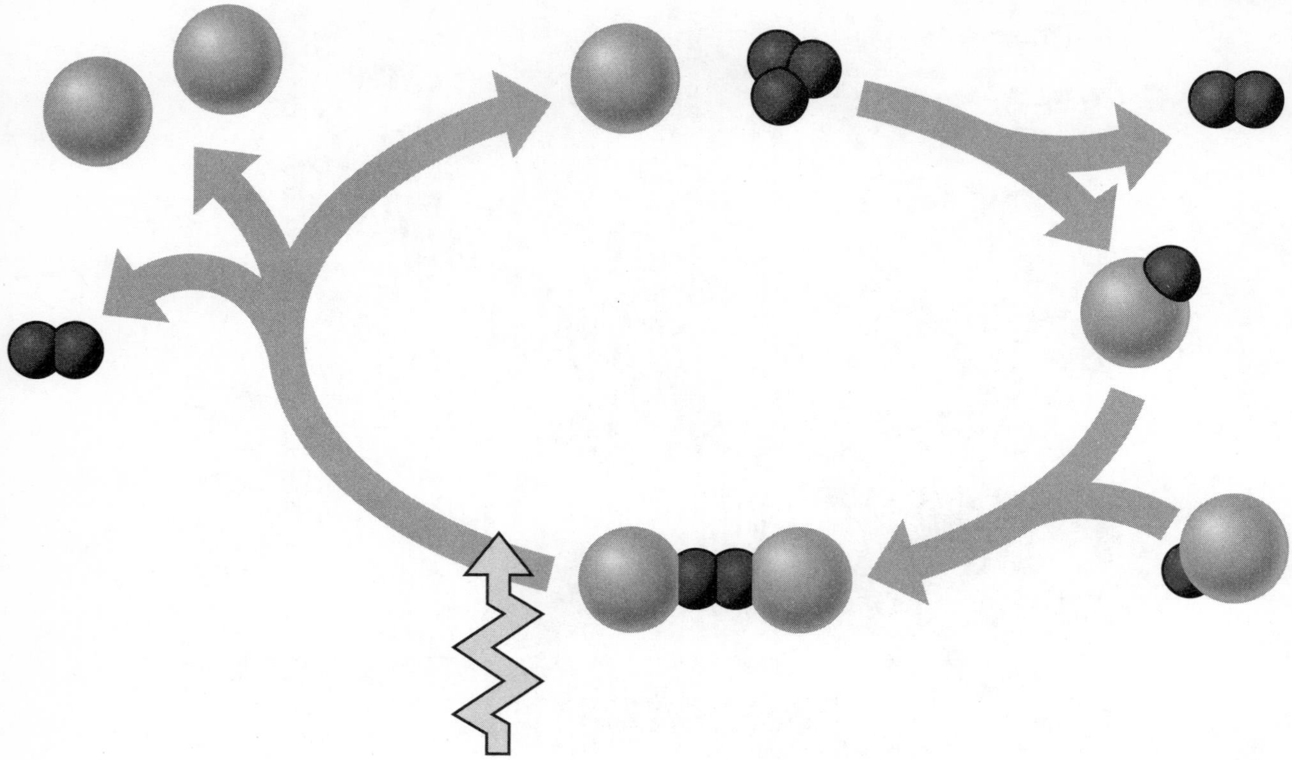

Figure 54.27 How free chlorine in the atmosphere destroys ozone, page 1205

Figure 55.2 Three levels of biodiversity, page 1210

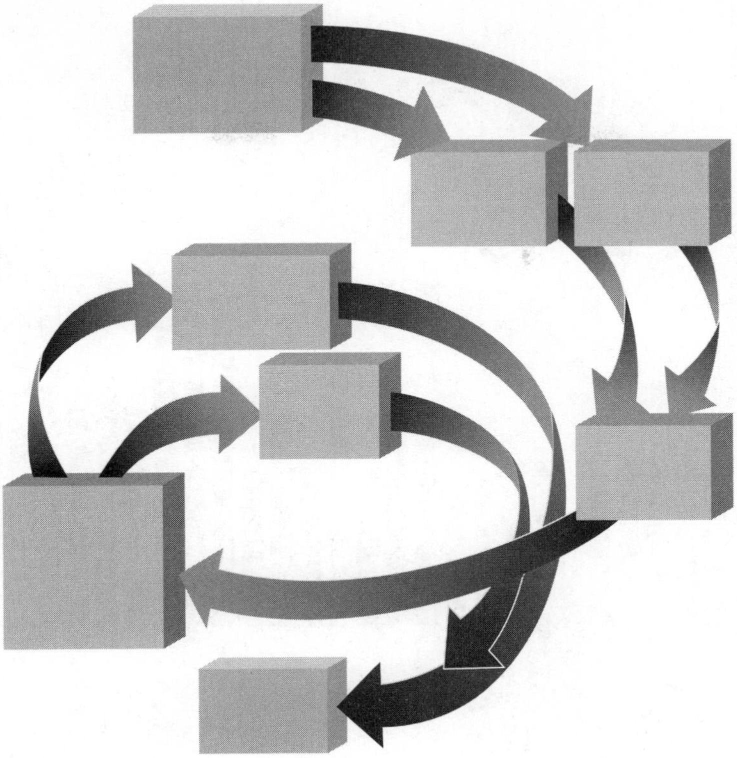

Figure 55.9 Processes culminating in an extinction vortex, page 1215

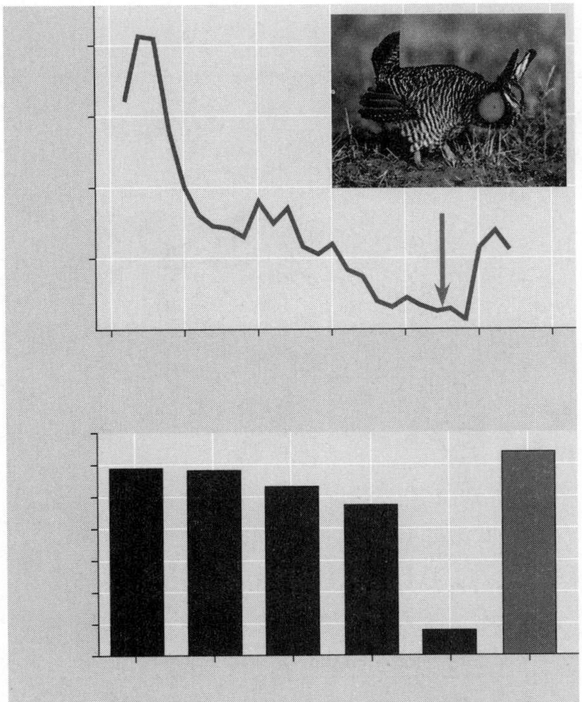

Figure 55.10 What caused the drastic decline of the Illinois greater prairie chicken population? page 1216

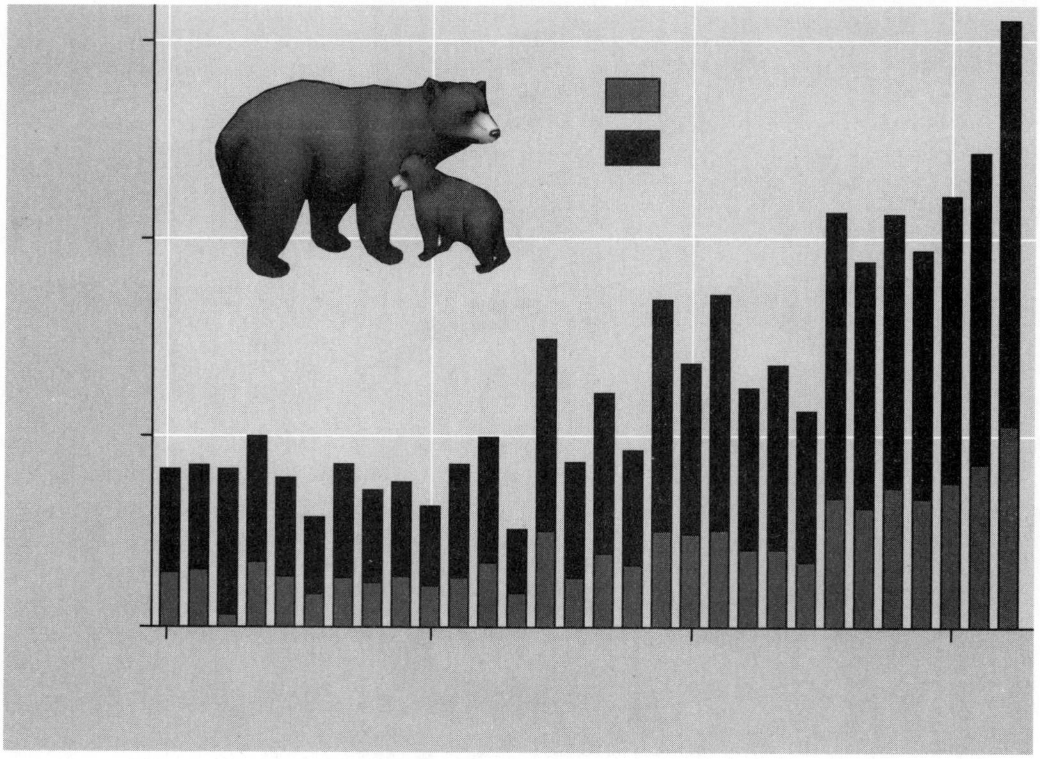

Figure 55.12 Growth of the Yellowstone grizzly bear population, as indicated by the number of females observed with cubs and number of cubs, page 1217

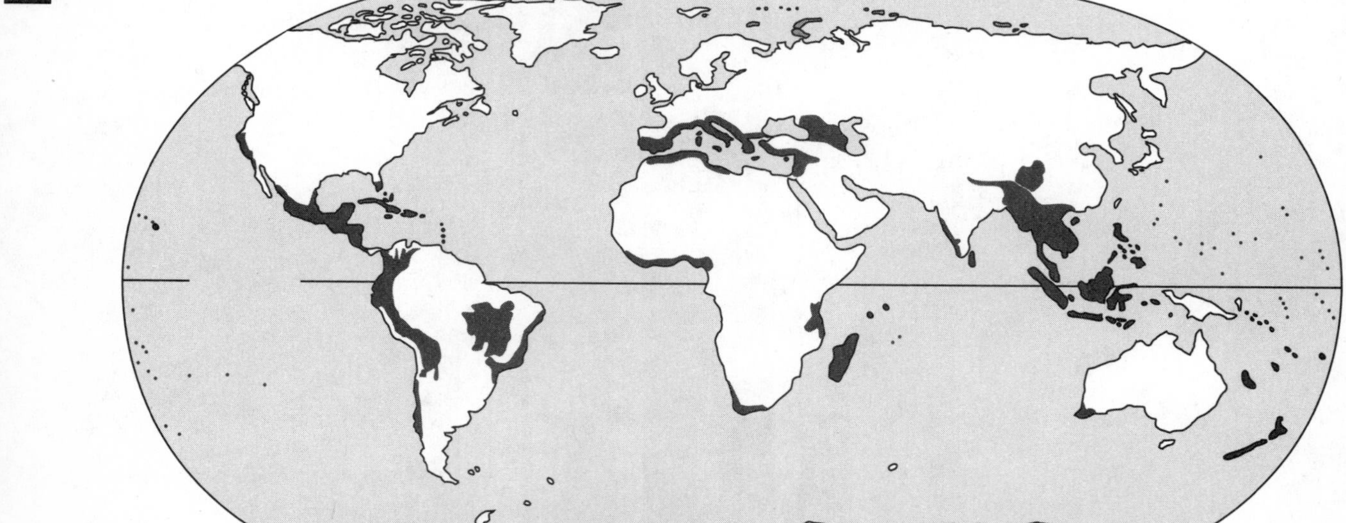

Figure 55.17 Earth's terrestrial biodiversity hot spots, page 1222

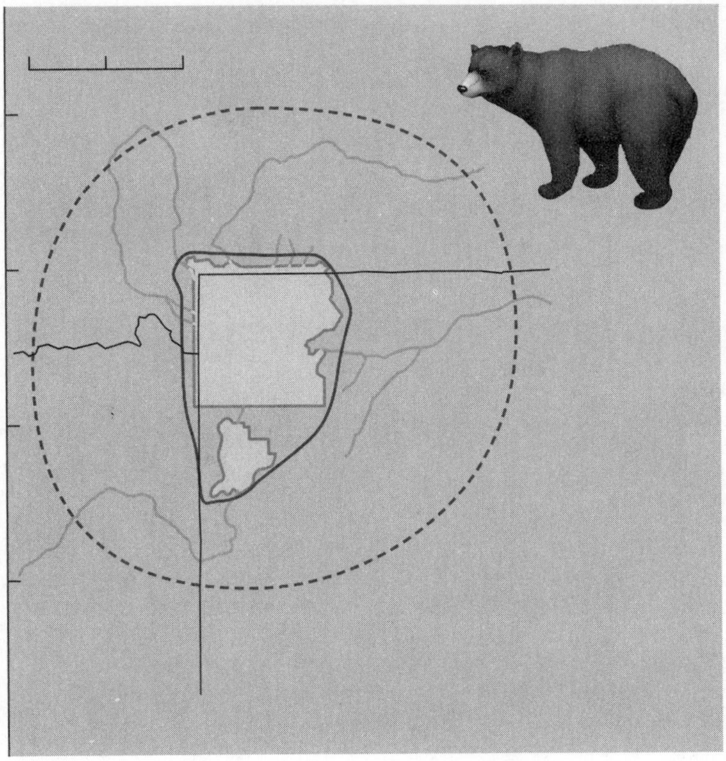

Figure 55.18 The legal (green border) and biotic (red border) boundaries for grizzly bears in Yellowstone and Grand Teton National Parks, page 1223

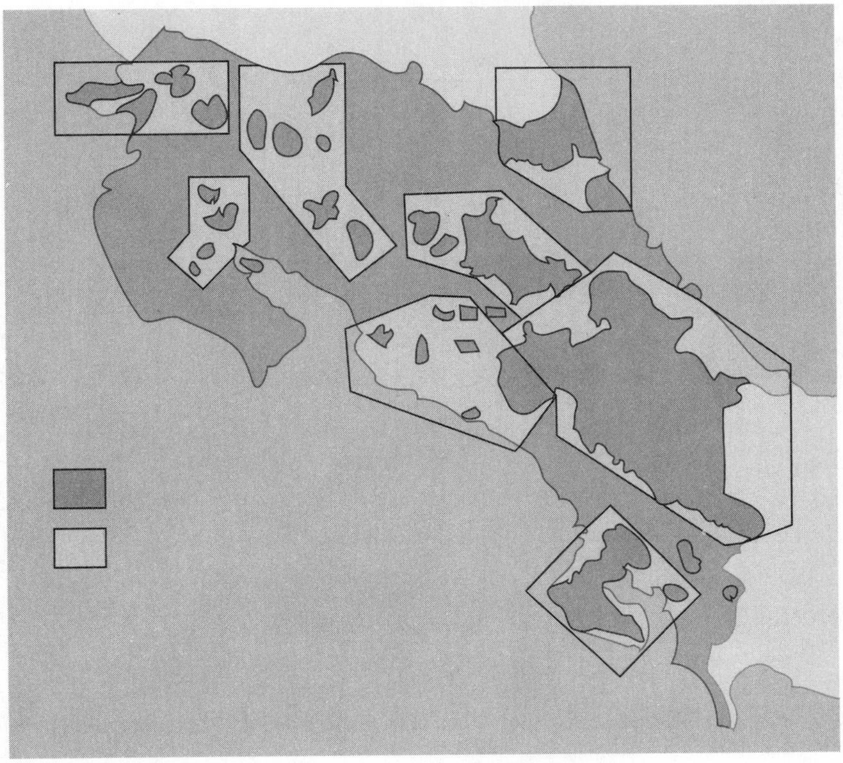

Figure 55.19 Zoned reserves in Costa Rica, page 1223

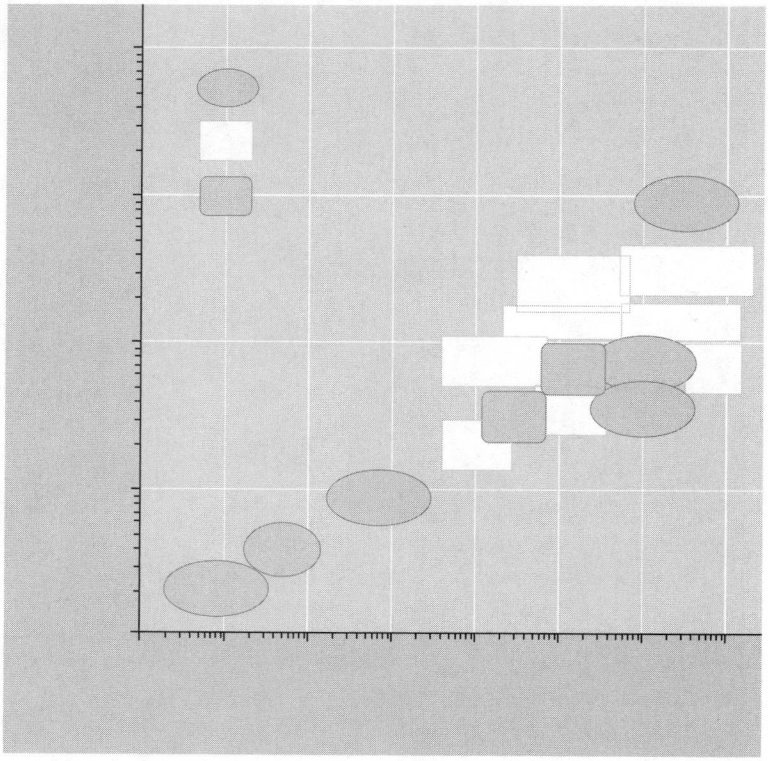

Figure 55.21 The size-time relationship for community recovery from natural and human-caused disasters, page 1225

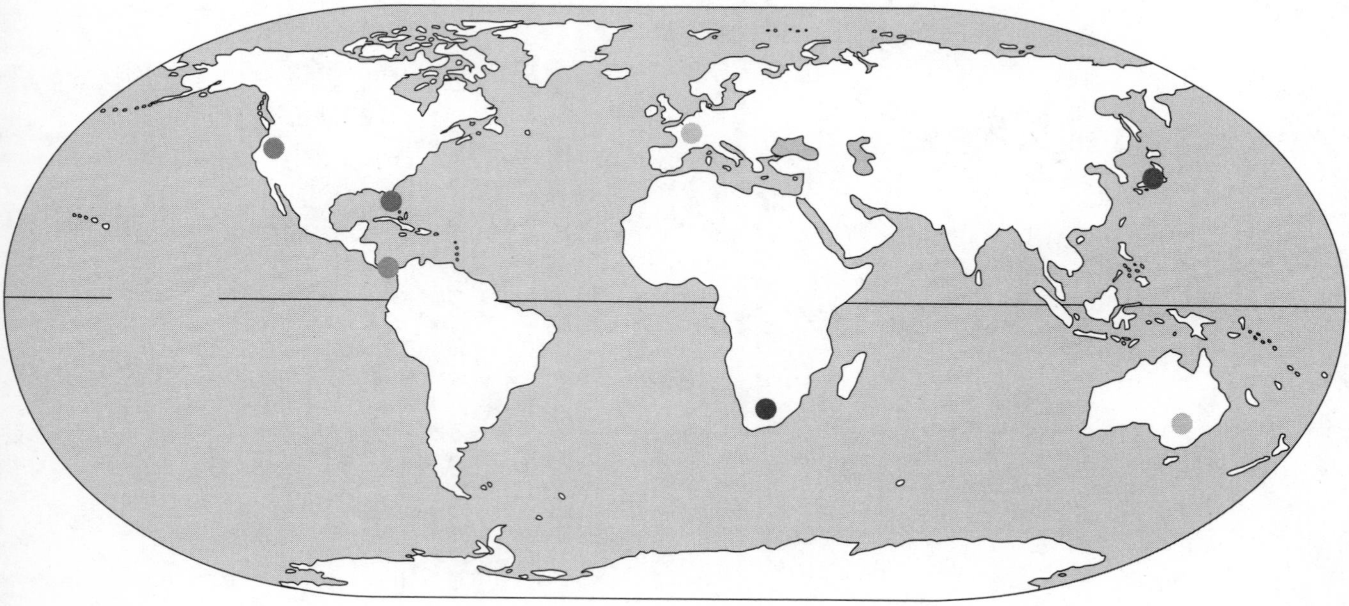

Figure 55.22 Restoration worldwide, page 1226

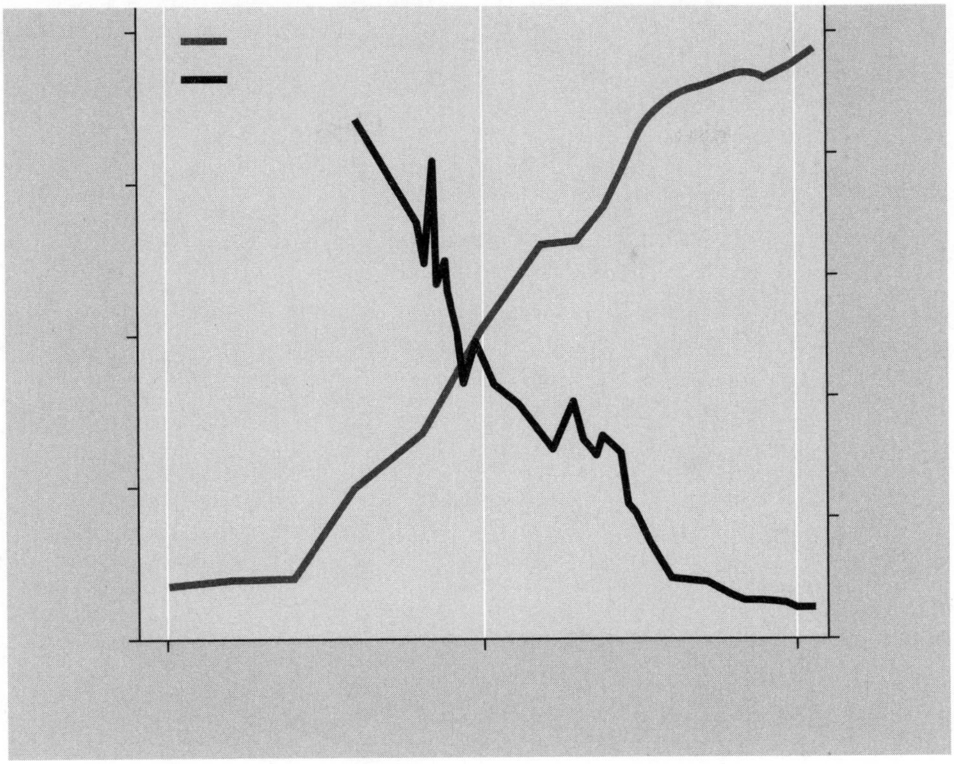

Figure 55.23 Infant mortality and life expectancy at birth in Costa Rica, page 1228

Credits

Illustration Credits

The following figures are adapted from Christopher K. Matthews and K. E. van Holde, *Biochemistry*, 2nd ed., Menlo Park, CA: Benjamin Cummings. © 1996 The Benjamin Cummings Publishing Company, Inc.: **4.6, 9.9,** and **17.16b** and **c**.

The following figures are adapted from Wayne M. Becker, Jane B. Reece, and Martin F. Poenie, *The World of the Cell*, 3rd ed., Menlo Park, CA: Benjamin Cummings. © 1996 The Benjamin Cummings Publishing Company, Inc.: **4.7, 6.7, 7.8, 11.7a, 11.10, 17.10, 19.13, 19.16,** and **20.7.**

Figures **6.9** and **6.23** and cell organelle drawings in Figures **6.12, 6.13, 6.14,** and **6.20** are adapted from illustrations by Tomo Narashima in Elaine N. Marieb, *Human Anatomy and Physiology*, 5th ed., San Francisco, CA: Benjamin Cummings. © 2001 Benjamin Cummings, an imprint of Addison Wesley Longman, Inc. Figures **6.12, 49.10,** and **49.11** are also from Human Anatomy and Physiology, 5th ed.

Figures **46.16, 48.22, 48.24, 49.26, 49.29,** and **49.33** are adapted from Elaine N. Marieb, *Human Anatomy and Physiology*, 4th ed., Menlo Park, CA: Benjamin Cummings. © 1998 Benjamin Cummings, an imprint of Addison Wesley Longman, Inc.

The following figures are adapted from Murray W. Nabors, *Introduction to Botany*, San Francisco, CA: Benjamin Cummings. © 2004 Pearson Education, Inc., Upper Saddle River, New Jersey: **30.12j, 38.3c, 39.13,** and **41.10.**

Some illustrations used in *BIOLOGY,* Seventh Edition, are adapted from Neil Campbell, Brad Williamson, and Robin Heyden, Biology: *Exploring Life,* Needham, MA, Prentice Hall School Division. © 2004 by Pearson Education, Inc., Upper Saddle River, NJ. Artists: Jennifer Fairman; Mark Foerster; Carlyn Iverson; Phillip Guzy; Steve McEntee; Stephen McMath; Karen Minot; Quade and Emi Paul, Fivth Media; and Nadine Sokol.

1.10 Adapted from Figure 4 from L. Giot et al., "A Protein Interaction Map of *Drosophila melanogaster,*" *Science*, Dec. 5, 2003, p. 1733 Copyright © 2003 AAAS. Reprinted with permission from the American Association for the Advancement of Science; **1.27** Map provided courtesy of David W. Pfennig, University of North Carolina at Chapel Hill; **1.29** Map provided courtesy of David W. Pfennig, University of North Carolina at Chapel Hill. Data in pie charts based on D. W. Pfennig et al. 2001. Frequency-dependent Batesian mimicry. *Nature* 410: 323.

3.7a Adapted from *Scientific American,* Nov. 1998, p. 102.

4.8 Adapted from an illustration by Clark Still, Columbia University.

5.13 From *Biology: The Science of Life,* 3/e by Robert Wallace et al. Copyright © 1991. Reprinted by permission of Pearson Education, Inc.; **5.20a** and **b** Adapted from D. W. Heinz, W. A. Baase, F. W. Dahlquist, B. W. Matthews. 1993. How amino-acid insertions are allowed in an alpha-helix of T4 lysozyme. *Nature* 361:561; **5.20e** and **f** © Illustration, Irving Geis. Rights owned by Howard Hughes Medical Institute.

9.5a and **b** Copyright © 2002 from *Molecular Biology of the Cell,* 4th ed. by Bruce Alberts et al., fig. 2.69, p. 92. Garland Science/Taylor & Francis Books, Inc.

10.14 Adapted from Richard and David Walker. *Energy, Plants and Man,* fig. 4.1, p. 69. Sheffield: University of Sheffield. © Richard Walker. Used courtesy of Oxygraphics.

12.12 Copyright © 2002 from *Molecular Biology of the Cell,* 4th ed., by Bruce Alberts et al., fig. 18.41, p. 1059. Garland Science/Taylor & Francis Books, Inc.

17.12 Adapted from L. J. Kleinsmith and V. M. Kish. 1995. *Principles of Cell and Molecular Biology,* 2nd ed. New York, NY: HarperCollins. Reprinted by permission of Addison Wesley Educational Publishers.

19.17b © Illustration, Irving Geis. Rights owned by Howard Hughes Medical Institute. Not to be reproduced without permission; **Table 19.1** From A. Griffiths et al. 2000. *An Introduction to Genetic Analysis,* 7/e, Table 26-4, p. 787. New York: W. H. Freeman and Company. Copyright © 2000 W. H. Freeman and Company.

20.9 Adapted from Peter Russell, *Genetics,* 5th ed., fig. 15.24, p. 481, San Francisco, CA: Benjamin Cummings. © 1998 Pearson Education, Inc., Upper Saddle River, New Jersey; **20.11** Adapted from a figure by Chris A. Kaiser and Erica Beade.

21.15 Copyright © 2002 from *Molecular Biology of the Cell,* 4th ed., by Bruce Alberts et al., fig. 21.17, p. 1172. Garland Science/Taylor & Francis Books, Inc.; **21.23** Adapted from an illustration by William McGinnis; **21.24** Brine shrimp adapted from M. Akam. 1995. Hox genes and the evolution of diverse body plans. *Philosophical Transactions B*. 349:313-319. © 1995 Royal Society of London. In Wolpert et al. 1998. *Principles of Development,* fig. 15.10, p. 452. Oxford: Oxford University Press.

22.13 Adapted from R. Shurman et al. 1995. *Journal of Infectious Diseases* 171:1411.

23.13 Adapted from A. C. Allison. 1961. Abnormal hemoglobin and erythrocyte enzyme-deficiency traits. In *Genetic Variation in Human Populations*, ed. G. A. Harrison. Oxford: Elsevier Science.

24.7 Adapted from D. M. B. Dodd, *Evolution* 11: 1308-1311; 24.14 Adapted from M. Strickberger. 1990. *Evolution.* Boston: Jones & Bartlett; **24.16** Adapted from L. Wolpert. 1998. *Principles of Development.* Oxford University Press; 24.18 Adapted from M. I. Coates. 1995. *Current Biology* 5:844-848.

25.18 Adapted from S. Blair Hedges. The origin and evolution of model organisms. fig. 1, p. 840. Nature Reviews Genetics 3: 838-849.

26.5 Figure 4c from "The Antiquity of RNA-based Evolution" by G.F. Joyce et al., *Nature*, Vol. 418, p. 217. Copyright © 2002 Nature Publishing Co.; **26.7** Adapted from D. Futuyma. 1998. *Evolutionary Biology,* 3rd ed., p. 128. Sunderland, MA: Sinauer Associates; **26.8** Data from M. J. Benton. 1995. Diversification and extinction in the history of life. *Science* 268:55; **26.10** Adapted from David J. Des Marais. September 8, 2000. When did photosynthesis emerge on Earth? *Science* 289:1703-1705; **26.17** Data from A. H. Knoll and S. B. Carroll, June 25, 199. *Science* 284:2129-2137; **26.18** Map adapted from http://geology.er.usgs.gov/eastern/plates.html.

27.6 Adapted from Gerard J. Tortora, Berdell R. Funke, and Christine L. Case. 1998. *Microbiology: An Introduction*, 6th ed. Menlo Park, CA: Benjamin Cummings. © 1998 Benjamin Cummings, an imprint of Addison Wesley Longman, Inc.

28.3 Figure 3 from Archibald and Keeling, "Recycle Plastics," *Trends in Genetics,* Vol. 18, No. 1, 2, 2002, p. 352. Copyright © 2002, with permission from Elsevier; **28.12** Adapted from R. W. Bauman. 2004. *Microbiology,* fig. 12.7, p. 350. San Francisco, CA: Benjamin Cummings. © 2004 Pearson Education, Inc., Upper Saddle River, New Jersey.

29.13 Adapted from Raven et al. *Biology of Plants,* 6th ed., fig. 19.7.

34.19 Adapted from C. Zimmer. 1999. *At the Water's Edge.* Free Press, Simon & Schuster p. 90; **34.20** Adapted from C. Zimmer. 1999. *At the Water's Edge.* Free Press, Simon & Schuster p. 99; **34.32** Adapted from Stephen J. Gould et al. 1993. *The Book of Life.* London: Ebury Press, p. 96. Reprinted by permission of Random House UK Ltd; **34.41** Drawn from photos of fossils: *O. tugenensis* photo in Michael Balter, Early hominid sows division, *ScienceNow,* Feb. 22, 2001, © 2001 American Association for the Advancement of Science. *A. ramidus kadabba* photo by Timothy White, 1999/Brill Atlanta. *A. anamensis, A. garhi,* and *H. neanderthalensis* adapted from *The Human Evolution Coloring Book. K platyops* drawn from photo in Meave Leakey et al., New hominid genus from eastern Africa shows diverse middle Pliocene lineages, *Nature*, March 22, 2001, 410:433. *P. boisei* drawn from a photo by David Bill. *H. ergaster* drawn from a photo at www.inhandmuseum.com. *S. tchadensis* drawn from a photo in Michel Brunet et al., A new hominid from the Upper Miocene of Chad, Central Africa, *Nature*, July 11, 2002, 418:147, fig. 1b.

35.21 Pie chart adapted from *Nature,* Dec. 14, 2000, 408:799.

39.17(graph) Adapted from M. Wilkins. 1988. *Plant Watching.* Facts of File Publ.; **39.29** Reprinted with permission from Edward Framer, 1997, Science 276:912. Copyright © 1997 American Association for the Advancement of Science.

40.17 Adapted from an illustration by Enid Kotschnig in B. Heinrich, 1987. Thermoregulation in a winter moth. *Scientific American* 105; 40.20 Adapted with permission from B. Heinrich, 1974, Science 185:747-756. © 1974 American Association for the Advancement of Science.

41.5 Adapted from J. Marx, "Cellular Warriors at the Battle of the Bulge," *Science,* Vol. 299, p. 846. Copyright © 2003 American Association for the Advancement of Science. Illustration: Katharine Sutliff; **41.13** Adapted from Lawrence G. Mitchell, John A. Mutchmor, and Warren D. Dolphin. 1988. *Zoology.* Menlo Park, CA: Benjamin Cummings. © 1988 The Benjamin Cummings Publishing Company; **41.15** Adapted from R. A. Rhoades and R. G. Pflanzer. 1996. *Human Physiology,* 3/e., fig. 22-1, p. 666. Copyright © 1996 Saunders.

43.7 Adapted from Gerard J. Tortora, Berdell R. Funke, and Christine L. Case. 1998. *Microbiology: An Introduction,* 6th ed. Menlo Park, CA: Benjamin/Cummings. © 1998 Benjamin Cummings, an imprint of Addison Wesley Longman, Inc.

44.5 Kangaroo rat data adapted from Schmidt-Nielsen. 1990. *Animal Physiology: Adaptation and Environment*, 4th ed., p. 339. Cambridge: Cambridge University Press; 44.6 Adapted from K.B. Schmidt-Nielsen et al., "Body temperature of the camel and its relation to water economy," *American Journal of Physiology,* Vol. 10, No. 188, (Dec.), 1956, figure 7. Copyright © 1956 American Physiological Society. Used with permission; **44.8** Adapted from Lawrence G. Mitchell, John A. Mutchmor, and Warren D. Dolphin. 1988. *Zoology.* Menlo Park, CA: Benjamin Cummings. © 1988 The Benjamin Cummings Publishing Company.

47.20 From Wolpert et al. 1998. *Principles of Development,* fig. 8.25, p. 251 (right). Oxford: Oxford University Press. By permission of Oxford University Press; 47.23b From Hiroki Nishida, *Developmental Biology* Vol. 121, p. 526, 1987. Copyright © 1987, with permission from Elsevier; **47.25 Experiment** and **left side of "Results":** From Wolpert et al. 1998. *Principles of Development,* fig. 1.10, Oxford: Oxford University Press. By permission of Oxford University Press; **Right side of "Results":** Figure 15.12, p. 604 from *Developmental Biology*, 5th ed. by Gilbert et al. Copyright © 1997 Sinauer Associates. Used with permission.

48.13 From G. Matthews, *Cellular Physiology of Nerve and Muscle.* Copyright © 1986 Blackwell Science. Used with permission; **48.33** Adapted from John G. Nicholls et al. 2001. *From Neuron to Brain,* 4th ed., fig. 23.24. Sunderland, MA: Sinauer Associates Inc. © 2001 Sinauer Associates.

49.19 Adapted from Bear et al. 2001. *Neuroscience: Exploring the Brain,* 2nd ed., figs. 11.8 and 11.9, pp. 281 and 283. Hagerstown, MD: Lippincott Williams & Wilkins © 2001 Lippincott Williams & Wilkins; **49.22** Adapted from Shepherd. 1988. *Neurobiology,* 2nd ed., fig. 11.4, p. 227. Oxford University Press. (From V. G. Dethier. 1976. *The Hungry Fly.* Cambridge, MA: Harvard University Press.); **49.23 (Lower)** Adapted from Bear et al. 2001. *Neuroscience: Exploring the Brain,* 2nd ed., fig. 8.7, p. 196. Hagerstown, MD: Lippincott Williams & Wilkins. © 2001 Lippincott Williams & Wilkins; **49.27b** Grasshopper adapted from Hickman et al. 1993. *Integrated Principles of Zoology,* 9th ed., fig. 22.6, p. 518. New York: McGraw-Hill Higher Education. © 1995 The McGraw-Hill Companies.

50.2 Adapted from G. Caughly, N. Shepherd, and J. Short. 1987. *Kangaroos: Their Ecology and Management in the Sheep Rangelands of Australia,* fig. 1.2, p. 12, Cambridge: Cambridge University Press. Copyright © 1987 Cambridge University Press; **50.7a** Data from U. S. Geological Survey; **50.8** Data from W. J. Fletcher. 1987. Interactions among subtidal Australian sea urchins, gastropods and algae: effects of experimental removals. *Ecological Monographs* 57:89-109; **50.14** Adapted from L. Roberts. 1989. How fast can trees migrate? *Science* 243:736, fig. 2. © 1989 by the American Association for the Advancement of Science; **50.19** Adapted from Heinrich Walter and Siegmar-Walter Breckle. 2003. *Walter's Vegetation of the Earth,* fig. 16, p. 36. Springer-Verlag, © 2003.

51.3b Adapted from N. Tinbergen. 1951. *The Study of Instinct.* Oxford: Oxford University Press. By permission of Oxford University Press; **51.10** Adapted from C. S. Henry et al. 2002. The inheritance of mating songs in two cryptic, sibling lacewings species (Neuroptera: Chrysopidae: *Chrysoperla*). *Genetica* 116: 269-289, fig. 2; **51.14** Adapted from Lawrence G. Mitchell, John A. Mutchmor, and Warren D. Dolphin. 1988. *Zoology.* Menlo Park, CA:

Benjamin/Cummings. © 1988 The Benjamin/Cummings Publishing Company; **51.15** Adapted from N. L. Korpi and B. D. Wisenden. 2001. Learned recognition of novel predator odour by zebra danios, *Danio rerio,* following time-shifted presentation of alarm cue and predator odour. *Environmental Biology of Fishes* 61: 205-211, fig. 1; **51.19** Adapted from M. B. Sokolowski et al. 1997. Evolution of foraging behavior in *Drosophila* by density-dependent selection. *Proceedings of the National Academy of Sciences of the United States of America.* 94: 7373-7377, fig. 2b; **51.21a** Adapted from a photograph by Jonathan Blair in Alcock. 2002. *Animal Behavior,* 7th ed. Sinauer Associates, Inc., Publishers; **51.21b** From P. Berthold et al., "Rapid microevolution of migratory behaviour in a wild bird species," *Nature,* Vol. 360, 12/17/92, p. 668. Copyright © 1992 Nature Publishing, Inc. Used with permission; **51.28** K. Witte and N. Sawka. 2003. Sexual imprinting on a novel trait in the dimorphic zebra finch: sexes differ. *Animal Behaviour* 65: 195-203. Art adapted from http://www.uni-bielefeld.de/biologie/vhf/KW/Forschungsprojekte2.html.

52.4 Adapted from P. W. Sherman and M. L. Morton, "Demography of Belding's ground squirrels," *Ecology,* Vol. 65, No. 5, p. 1622, 1984. Copyright © 1984 Ecological Society of America. Used by permission; **52.13c** Data courtesy of P. Arcese and J. N. M. Smith, 2001; **52.14** Adapted from J. T. Enright. 1976. Climate and population regulation: the biogeographer's dilemma. *Oecologia* 24:295-310; **52.15b** Data from J. N. M. Smith and P. Arcese; **52.18** Data courtesy of Rolf O. Peterson, Michigan Technological University, 2004; **52.19** Data from Higgins et al. May 30, 1997. Stochastic dynamics and deterministic skeletons: population behavior of Dungeness crab. *Science;* **52.20** Adapted from J.N.M. Smith et al., 1996, "A metapopulation approach to the population biology of the song sparrow *Melospiza melodia,*" *IBIS,* Vol. 138, fig. 3, pp. 120-128; **52.23** Data from U. S. Census Bureau International Data Base; **52.24** Data from Population Reference Bureau 2000 and U. S. Census Bureau International Data Base, 2003; **52.25** Data from U. S. Census Bureau International Data Base; **52.26** Data from U. S. Census Bureau International Data Base 2003; **52.27** Data from J. Wackernagel et al. 1999. National natural capital accounting with the ecological footprint concept. *Ecological Economics* 29: 375-390.

53.3 A. S. Rand and E. E. Williams. 1969. The anoles of La Palma: aspects of their ecological relationships. *Breviora* 327. Museum of Comparative Zoology, Harvard University. © Presidents and Fellows of Harvard College; **53.13** Adapted from E. A. Knox. 1970. *Antarctic marine ecosystems. In Antarctic Ecology,* ed. M. W. Holdgate, 69-96. London: Academic Press; **53.14** Adapted from D. L. Breitburg et al. 1997. Varying effects of low dissolved oxygen on trophic interactions in an estuarine food web. *Ecological Monographs* 67: 490. Copyright © 1997 Ecological Society of America; **53.15** Adapted from B. Jenkins. 1992. Productivity, disturbance and food web structure at a local spatial scale in experimental container habitats. *Oikos* 65: 252. Copyright © 1992 Oikos, Sweden; **53.17** Adapted from J. A. Estes et al. 1998. Killer whale predation on sea otters linking oceanic and nearshore ecosystems. *Science* 282:474. Copyright © 1998 by the American Association for the Advancement of Science; **53.19** Data from S. D. Hacker and M. D. Bertness. 1999. Experimental evidence for factors maintaining plant species diversity in a New England salt marsh. *Ecology* 80: 2064-2073; **53.23** Adapted from R. L. Crocker and J. Major. 1955. Soil Development in relation to vegetation and surface age at Glacier Bay, Alaska. *Journal of Ecology* 43: 427-448; **53.24d** Data from F. S. Chapin, III, et al. 1994. Mechanisms of primary succession following deglaciation at Glacier Bay, Alaska. *Ecological Monographs* 64: 149-175. **53.25** Adapted from D. J. Currie. 1991. Energy and large-scale patterns of animal- and plant-species richness. *American Naturalist* 137: 27-49; **53.26** Adapted from F. W. Preston. 1960. Time and space and the variation of species. *Ecology* 41: 611-627; **53.28** Adapted from F. W. Preston. 1962. The canonical distribution of commonness and rarity. *Ecology* 43: 185-215, 410-432.

54.2 Adapted from D. L. DeAngelis. 1992. *Dynamics of Nutrient Cycling and Food Webs.* New York: Chapman & Hall; **54.6** Adapted from J. H. Ryther and W. M. Dunstan. 1971. Nitrogen, phosphorus, and eutrophication in the coastal marine environment. *Science* 171:1008-1013; **54.8** Data from M. L. Rosenzweig. 1968. New primary productivity of terrestrial environments: Predictions from climatologic data, *American Naturalist* 102:67-74. **54.9** Adapted from S. M. Cargill and R. L. Jefferies. 1984. Nutrient limitation of primary production in a sub-arctic salt marsh. *Journal of Applied Ecology* 21:657-668; **54.17a** Adapted from R. E. Ricklefs. 1997. *The Economy of Nature,* 4th ed. © 1997 by W. H. Freeman and Company. Used with permission; **54.21** Adapted from G. E. Likens et al. 1981. Interactions between major biogeochemical cycles in terrestrial ecosystems. In *Some Perspectives of the Major Biogeochemical Cycles*, ed. G. E. Likens, 93-123. New York: Wiley; **54.22** Adapted from National Atmospheric Deposition Program (NRSP-3) National Trends Network. (2004). NADP Program Office, Illinois State Water Survey, 2204 Griffith Dr., Champaign, IL 61820. http://nadp.sws.uiuc.edu; **54.24** Temperature data from U. S. National Climate Data Center, NOAA. CO2 data from C. D. Keeling and T. P. Whorf, Scripps Institution of Oceanography; **54.26** Data from British Antarctic Survey.

55.9 Adapted from Charles J. Krebs. 2001. *Ecology,* 5th ed., fig. 19.1. San Francisco, CA: Benjamin Cummings. © 2001 Benjamin Cummings, an imprint of Addison Wesley Longman, Inc.; **55.10** Adapted from R. L. Westemeier et al. 1998. Tracking the long-term decline and recovery of an isolated population. *Science* 282:1696. © 1998 by the American Association for the Advancement of Science; **55.12** Data from K. A. Keating et al. 2003. Estimating numbers of females with cubs-of-the-year in the Yellowstone grizzly bear population. *Ursus* 13:161-174 and from M. A. Haroldson. 2003. Unduplicated females. Pages 11-17 in C.C. Schwartz and M. A. Haroldson, eds. Yellowstone grizzly bear investigations. Annual Report of the Interagency Grizzly Bear Study Team, 2002. U.S. Geological Survey, Bozeman, Montana; **55.17** From N. Myers et al., "Biodiversity hotspots for conservation priorities," *Nature,* Vol. 403, p. 853, 2/24/2000. Copyright © 2000 Nature Publishing, Inc. Used with permission; **55.18** Adapted from W. D. Newmark. 1985. Legal and biotic boundaries of western North American national parks: a problem of congruence. *Biological Conservation* 33:199. © 1985 Elsevier Applied Science Publishers Ltd., Barking, England; **55.21** Adapted from A. P. Dobson et al. 1997. Hopes for the future: restoration ecology and conservation biology. Science 277:515. © 1997 by the American Association for the Advancement of Science; **55.23** Data from Instituto Nacional de Estadistica y Censos de Costa Rica and Centro Centroamericano de Poblacion, Universidad de Costa Rica.

Photo Credits

Cover Image Linda Broadfoot

Chapter 1
1.4 Photodisc; **1.6** Camille Tokerud/Stone; **1.7** Photodisc; **1.16 top** VVG/SPL/Photo Researchers; **1.16 middle** W. L. Dentler, University of Kansas/Biological Photo Sevices; **1.16 bottom** OMIKRON/Photo Researchers; **1.27 top and bottom** Breck P. Kent; **1.27 middle** E. R. Degginger/Photo Resarchers.

Chapter 2
2.1 Thomas Eisner; **2.5** Terraphotographics/Biological Photo Service; **2.14** Benjamin Cummings

Chapter 3
3.5 Flip Nicklin/Minden Pictures

Chapter 4
4.6 Manfred Kage/Peter Arnold; **4.9 both** Digital Vision

Chapter 5
5.6a John N. A. Lott, McMaster University/Biological Photo Service; **5.6b** H. Shio and P.B. Lazarow; **5.8** J. Litray/Visuals Unlimited; **5.10b** F. Collet/Photo Researchers; **5.10c** CORBIS; **5.12a** Dorling Kindersley; **5.12b** Photodisc Green; **5.20** Wolfgang Kaehler/Liaison; **5.21 both** Eye of Science/Photo Researchers; **5.23** P. B. Sigler from Z. Xu, A. L. Horwich, and P. B. Sigler, Nature (1997) 388:741-750. ©1997 Macmillan Magazines, Ltd.; **5.24a** Marie Green, University of Califonia, Riverside

Chapter 6
6.3a Biophoto Associates/Photo Researchers; **6.3b** Ed Reschke; **6.3c-d** David M. Phillips/Visuals Unlimited; **6.3e** Molecular Probes; **6.3f both** Karl Garsha; **6.4a-b** William L. Dentler, University of Kansas/Biological Photo Service; **6.6** S. C. Holt, University of Texas Health Center/Biological Photo Service; **6.8** Daniel Friend; **6.10 top left** From L. Orci and A. Perelet, Freeze-Etch Histology. (Heidelberg: Springer-Verlag, 1975) © 1975 Springer-Verlag; **6.10 bottom left** From A. C. Faberge, Cell Tiss. Res. 151 (1974):403. © 1974 Springer-Verlag; **6.10 right** U. Aebi et al. Nature 323 (1996):560-564, figure 1a. Used with permission; **6.11** D. W. Fawcett/Photo Researchers; **6.12** R. Bolender, D. Fawcett/Photo Researchers; **6.13** Don Fawcett/Visuals Unlimited; **6.14a** R. Rodewald, University of Virginia/Biological Photo Service; **6.14b** Daniel S. Friend, Harvard Medical School; **6.15** E. H. Newcomb; **6.17** Daniel S. Friend, Harvard Medical School; **6.18** W. P. Wergin and E. H. Newcomb, University of Wisconsin, Madison/Biological Photo Service; **6.19** From S. E. Fredrick and E. H. Newcomb, The Journal of Cell Biology 43 (1969):343. Provided by E. H. Newcomb; **6.20** John E. Heuser, Washington University School of Medicine, St. Louis, MO; **6.21** B. J. Schapp et al., 1985, Cell 40:455; **Table 6.1 left** Mary Osborn, Max Planck Institute; **middle** Frank Solomon and J. Dinsmore, Massachusetts Institute of Technology; **right** Mark S. Ladinsky and J. Richard McIntosh, University of Colorado; **6.22** Kent McDonald; **6.23a** Biophoto Associates/Photo Researchers; **6.23b** Oliver Meckes &

Nicole Ottawa/Photo Researchers; **6.24a** OMIKRON/Science Source/Photo Researchers; **6.24b-c** W. L. Dentler, University of Kansas/Biological Photo Service; **6.26** From Hirokawa Nobutaka, The Journal of Cell Biology 94 (1982):425 by copyright permission of The Rockefeller University Press; **6.28** G. F. Leedale/Photo Researchers; **6.30** W. P. Wergin, provided by E. H. Newcomb; **6.31 top** From Douglas J. Kelly, The Journal of Cell Biology 28 (1966):51 by copyright permission of The Rockefeller University Press; **6.31 middle** From L. Orci and A. Perrelet, Freeze-Etch Histology. (Heidelberg: Springer-Verlag, 1975) copyright 1975 Springer-Verlag; **6.31 bottom** From C. Peracchia and A. F. Dulhunty, The Journal of Cell Biology 70 (1976):419 by copyright permission of The Rockefeller University Press

Chapter 7
7.4 both D. W. Fawcett/Photo Researchers; **7.14a-b** Cabisco/Visuals Unlimited; **7.20 top** R. N. Band and H. S. Pankratz, Michigan State University/Biological Photo Service; **7.20 middle** D. W. Fawcett/Photo Researchers; **7.20 bottom, both** M. M. Perry and A. B. Gilbert, J. Cell Science 39 (1979) 257. Copyright 1979 by The Company of Biologists Ltd.

Chapter 8
8.3a Joe McDonald/CORBIS; **8.3b** Manoj Shah/Stone; **8.16a-b** Thomas Steitz, Yale University; **8.22** R. Rodewald, University of Virginia/Biological Photo Service

Chapter 10
10.3 middle M. Eichelberger/Visuals Unlimited; **10.3 bottom** W. P. Wergin and E. H. Newcomb, University of Wisconsin/Biological Photo Service; **10.11b** Christine L. Case, Skyline College; **10.20a** David Muench/CORBIS; **10.20b** Dave Bartruff/CORBIS

Chapter 12
12.4 Biophoto/Photo Researchers; **12.6 all** Conly Rieder; **12.7 top** Matthew Schibler, from Protoplasma 137 (1987):29-44; **12.7 bottom** Richard Mcintosh; **12.9a** David M. Phillips/Visuals Unlimited; **12.9b** Micrograph by B. A. Palevitz. Courtesy of E. H. Newcomb, University of Wisconsin; **12.17** Gunter Albrecht-Buehler, Northwestern University; **12.18a-b** Lan Bo Chen

Chapter 13
13.3 top Veronique Burger/Photo Researchers; **13.3 bottom** CNRI/SPL/Photo Researchers; **13.11** Carolina Biological/Visuals Unlimited

Chapter 14
14.14a, both Photodisc; **14.14b, both** Benjamin Cummings

Chapter 15
15.11 Dorling Kindersley

Chapter 16
16.3 Oliver Meckes/Photo Researchers; **16.7c** Richard Wagner, UCSF Graphics; **16.12** From D. J. Burks and P. J. Stambrook, The Journal of Cell Biology 77 (1978): 762 by copyright permission of The Rockefeller University Press. Photo provided by P. J. Stambrook

Chapter 17
17.16a Joachim Frank, Howard Hughes Medical Institute; **17.20** B. Hamkalo and O.L. Miller, Jr.; **17.22** Reprinted with permission from O. L. Miller , Jr., B. A. Hamkalo, and C. A Thomas, Jr., Science 169 (1970):392. Copyright ©1970 American Association for the Advancement of Science.

Chapter 18
18.4a-b, d Robley C. Williams, University of California Berkeley/Biological Photo Service; **18.4c** G. Murti/Visuals Unlimited; **18.10 all** C. Dauguet/Institute Pasteur/Photo Researchers

Chapter 19
19.2a top S. C. Holt, University of Texas, Health Science Center, San Antonio/BPS; **19.2a bottom** Courtesy of Victoria Foe; **19.2b** Barbara Hamkalo; **19.2c** From J. R. Paulsen and U. K. Laemmli, Cell 12 (1977):817-828; **19.2d both** G. F. Bahr/AFIP; **19.17a** O. L. Miller Jr., Department of Biology, University of Virginia

Chapter 20
20.8 Repligen Corporation; **20.14** Incyte Pharmaceuticals, Inc., Palo Alto, CA, from R. F. Service, Science (1998) 282:396-399, with permission from Science; **20.17** Cellmark Diagnostics, Inc., Germantown, MD

Chapter 21
21.14a both Wolfgang Driever; **21.14b both** Dr. Ruth Lahmann, The Whitehead Institution; **21.15** J.E. Sulston and H.R. Horvitz, Dev. Biol. 56 (1977):110-156; **21.17** Dr. Gopal Murti/Visuals Unlimited; **21.20** Dwight Kuhn

Chapter 22
22.5 left ARCHIV/Photo Researchers; **22.5 right** National Maritime Museum, London; **22.10** Jack Wilburn/Earth Scenes/Animals Animals; **22.15 left** Phototake; **22.15 right** Lennart Nilsson/Albert Bonniers Forlag AB; **22.17** Tom Van Sant/Geosphere Project, Santa Monica/Science Photo Library/Photo Researchers; **22.18** Philip Gingerich, Discover Magazine

Chapter 23
23.3 top Michio Hosino/Minden Pictures; **23.3 bottom** James L. Davis/ProWildlife; **23.6** Eastcott Momatiuk/ Stone; **23.8** Kennan Ward/CORBIS; **23.14 all** Alan C. Kamil, George Holmes University; **23.15** Frans Lanting/ Minden Pictures

Chapter 24
24.4a Joe McDonald/Bruce Coleman; **24.4b** Joe McDonald/CORBIS; **24.4c** Roger Barbour; **24.4d** Stephen Krasemann/Photo Researchers; **24.4e** Barbara Gerlach/Tom Stack & Associates; **24.4f** Mike Zens/CORBIS; **24.4g** Dennis Johnson, Papilio/CORBIS; **24.4h** William E. Ferguson; **24.4i** Charles W. Brown; **24.4j** Photodisc; **24.4k** Ralph A. Reinhold/Animals Animals/Earth Scenes; **24.4l** Grant Heilman/Grant Heilman Photography; **24.4m** Kazutoshi Okuno, National Institute of Agrobiological Sciences, Tsukuba; **24.12 all** Gerald D. Carr; **24.16a** Gary Meszaros/Visuals Unlimited; **24.16b** Tom McHugh/Photo Researchers

Chapter 25
25.2 left Photodisc Green; **25.2 middle, right** Dorling Kindersley

Chapter 26
26.4 F. M. Menger and Kurt Gabrielson, Emory University

Chapter 27
27.3 Jack Bostrack/Visuals Unlimited; **27.6** J. Adler; **27.10** Sue Barns; **27.13.1** L. Evans Roth/ Biological Photo Service; **27.13.2** Yuichi Suwa; **27.13.4 left** Phototake; **27.13.4 right** Alfred Pasieka/Peter Arnold; **27.13.5** Photo Researchers; **27.13.6** Moredon Animal Health/SPL/Photo Researchers; **27.13.7** CNRI/SPL/Photo Researchers; **27.13.8 top** Frederick P. Mertz/Visuals Unlimited; **27.13.8 bottom** David M. Phillips, Visuals Unlimited; **27.13.9** T. E. Adams/Visuals Unlimited

Chapter 28
28.8 Michael Abbey/Visuals Unlimited; **28.11** Masamichi Aikawa; **28.12** M. Abbey/Visuals Unlimited; **28.14** Fred Rhoades/Mycena Consulting; **28.21** J. R. Waaland, University of Washington/Biological Photo Service; **28.26** R. Calentine/Visuals Unlimited; **28.27 both** Robert Kay, MRC Cambridge; **28.28a** D. P. Wilson, Eric & David Hosking/Photo Researchers; **28.31** W. L. Dentler, University of Kansas

Chaper 29
29.5 p 576 left Ed Reschke; **right** F. A. L. Clowes; **29.5 page 577 top left** Alan S. Heilman; **top right** Michael Clayton; **middle** Barry Runk/Stan/Grant Heilman Photography; **bottom left** Karen Renzaglia, Southern Illinois University; **29.8** Richard Kessel & Gene Shih/Visuals Unlimited

Chapter 30
30.11a David Dilcher and Ge Sun; **30.11b** K. Simons and David Dilcher; **30.12 page 602 top** Stephen McCabe; **bottom left** Dorling Kindersley; **bottom middle** Bob & Ann Simpson/Visuals Unlimited; **bottom right** Andrew Butler/Dorling Kindersley; **30.12 p 603 left top to bottom** Eric Crichton/Dorling Kindersley; J.Dransfield; Dorling Kindersley; Terry W. Eggers/CORBIS; **right top to bottom** Ed Reschke/PeterArnold; Dorling Kindersley; Tony Wharton; Frank Lane Picture Agency/CORBIS; Dorling Kindersley, Gerald D. Carr.

Chapter 31
31.2 top Hans Reinhard/Taxi; **31.2 bottom** Fred Rhoades/Mycena Consulting **31.2 inset** Elmer Koneman/Visuals Unlimited; **31.4** George L. Barron; **31.12 top left both** Barry Runk/Stan/Grant Heilman Photography; **31.12 bottom left** George Barron; **31.12 right** Ed Reschke/Peter Arnold; **31.20** Biophoto Associates/Photo Researchers; **31.24** V. Ahmadijian/Visuals Unlimited

Chapter 32
32.6 J. Sibbick/The Natural History Museum, London; **32.13a** Carolina Biological/Visuals Unlimited

Chapter 33
33.4 Andrew J. Martinez/Photo Researchers; **33.8** Robert Brons/Biological Photo Services; **33.11** Center for Disease Control; **33.12** Stanley Fleger/Visuals Unlimited; **33.23** A.N.T./NHPA

Chapter 34
34.4a Robert Brons/Biological Photo Service; **34.5** Runk Schoenenberg/Grant Heilman Photography; **34.28a** Stephen J. Kraseman/DRK Photo; **34.28b** Janice Sheldon;

Chapter 35
35.9 p 718 top Brian Capon; **middle** Graham Kent/Benjamin Cummings; **bottom both** Graham Kent/Benjamin Cummings; **35.9 p 719 top** Richard Kessel and Gene Shih/Visuals Unlimited; **bottom both** Graham Kent/Benjamin Cummings; **35.12** Carolina Biological Supply/Phototake; **35.17b-c** Ed Reschke; **35.18 left** Michael Clayton; **35.18 right** Alison W. Roberts; **35.21** Janet Braam, from Cell 60 (9 February 1990) copyright 1990 by Cell Press; **35.23 all** Susan Wick, University of Minnesota; **35.24** B. Wells and Kay Roberts

Chapter 37
37.12a both Gerald Van Dyke/Visuals Unlimited; **37.12b** Carolina Biological Supply/Phototake NYC

Chapter 38
38.3 top left to right Karen Tweedy-Holmes/CORBIS; Dorling Kindersley; Craig Lovell/CORBIS; John Cancalosi/Nature Photo Library; **bottom left to right** D. Cavagnaro/Visuals Unlimited; David Sieren/Visuals Unlimited; Marcel E. Dorken (2 photos); **38.4a left** Ed Reschke; **38.4a right** David Scharf/Peter Arnold; **38.4b** Ed Reschke

Chapter 39
39.17 all Malcolm Wilkins, University of Glasgow

Chapter 40
40.5 p 824 CNRI/SPL/Photo Researchers; **40.5 p 825 top left** Nina Zanetti; **top right** Chuck Brown/Photo Researchers; **middle left** Science VU/Visuals Unlimited; **middle right** Nina Zanetti; **bottom left** Nina Zanetti; **bottom right** Dr. Gopal Murti/SPL/Photo Researchers; **40.5 p 826 top to bottom** Nina Zanetti; Manfred Kage/Peter Arnold; Gladden Willis, M. D./Visuals Unlimited; Ed Reschke; **40.6** Dr. Richard Kessel & Dr. Randy Kardon/Tissues & Organs/Visuals Unlimited; **40.22** John Gerlach/Visuals Unlimited

Chapter 41
41.10 both Digital Vision/CORBIS; **41.17** Fred E. Hossler/Visuals Unlimited; **41.27 both** PhotoDisc

Chapter 42
42.2 Norbert Wu/Mo Young Productions; **42.9** Dr. Richard Kessel & Dr. Randy Kardon/Tissues & Organs/Visuals Unlimited; **42.13** CNRI/Phototake; **42.14** Lennart Nilsson, The Body Victorious, Dell Publishing Company; **42.17** Science Source/Photo Researchers; **42.20a** Frans Lanting/Minden Pictures; **42.20b** Peter Batson/Image Quest; **42.20c**

H. W. Pratt/Biological Photo Service; **42.20d** Dave Haas/The Image Finders; **42.22b** Thomas Eisner; **42.23 left** Dr. Richard Kessel & Dr. Randy Kardon/Tissues & Organs/Visuals Unlimited; **42.23 right** CNRI/ Photo Researchers; **42.25** Hans Rainer Dunker, Justus Leibig University, Giessen

Chapter 43
43.15 David Scharf/Peter Arnold; **43.16** Lennart Nilsson/Albert Bonniers Forlag AB; **43.17** Gopal Murti/Phototake; **43.20** Lennart Nilsson/Albert Bonniers Forlag AB

Chapter 44
44.13b Lise Bankir. From Urinary concentrating ability: insights from comparative anatomy, Bankir and de Rouffignac, Am J. Physiol Regul Integr Comp Physiol, 1985; 249: 643-666; **44.13d** Dr. Richard Kessel & Dr. Randy Kardon/Tissues & Organs/Visuals Unlimited

Chapter 46
46.3 David Crews, photo by P. de Vries

Chapter 47
47.4 all Jerry Schatten et al; **47.9 both** Dr. Richard Kessel & Dr. Gene Shih/Visuals Unlimited; **47.11** Charles A Ettensohn, Carnegie Mellon University; **47.14a** CABISCO/Visuals Unlimited; **47.14c** Thomas Poole, SUNY Health Science Center; **47.15** Carolina Biological Supply/Phototake; **47.23 both** Hiroki Nishida, Developmental Biology 121 (1987): 526. Reprinted by permission of Academic Press; **47.26a** Kathryn Tosney, University of Michigan; **47.27** Courtesy of Dennis Summerbell

Chapter 48
48.8 Alan Peters, from Bear, Connors, and Paradiso, Neuroscience: Exploring the Brain © 1996, p. 43

Chapter 49
49.7 John L. Pontier/Animals Animals; **49.16** From Richard Elzinga, Fundamentals of Entomology 3 ed. ©1987, p. 185. Reprinted by permission of Prentice-Hall, Upper Saddle River, NJ; **49.29 all** Dr. H. E Huxley; **49.37** Vance A. Tucker

Chapter 50
50.2 PhotoDisc; **50.7** Richard Ditch

Chapter 51
51.3 Kim Taylor/Bruce Coleman; **51.10** Robert Pickett/CORBIS;

Chapter 52
52.7 H. Willcox/Wildlife Pictures/Peter Arnold; **52.21** Joe McDonald/CORBIS

Chapter 53
53.3 left Joseph T. Collins/Photo Researchers; **right** Kevin de Queiroz, National Museum of Natural History; **53.16a** Bill Curtsinger/National Geographic Image Collection; **53.19** Sally D. Hacker, Oregon State University; **53.23** Tom Bean/CORBIS; **53.24a** Charles Mauzy/CORBIS; **53.24b** Tom Bean/DRK; **53.24c** Glacier Bay National Park and Preserve

Chapter 54
54.5 C. B. Field, M. J. Behrenfeld, J. T. Randerson, and P. Falkowski. 1998. "Primary production of the bio-sphere: Integrating terrestrial and oceanic components." Science 281:237–240; **54.19a-b** Hubbard Brook Research Foundation

Chapter 55
55.10 William Ervin/SPL/Photo Researchers; **55.19** Frans Lanting/Minden Pictures; **55.22 p 1126 left** Stewart Rood, University of Lethbridge; **right** South Florida Water Management District (WPB); **55.22 p 1227 top left** Daniel H. Janzen; **bottom left** Bert Boekhoven; **top right** Jean Hall/Holt Studios/Photo Researchers; **bottom right** Kenji Morita/Environment Division, Tokyo Kyuei Co., Ltd.